The
Planet
We
Live
On

Illustrated
Encyclopedia
of the
Earth Sciences

The Planet We Live On

Cornelius S. Hurlbut, Jr., Editor
Professor of Mineralogy, Emeritus, Harvard University

Harry N. Abrams, Inc., Publishers, New York

**Planned, prepared, and produced by
Chanticleer Press, Inc.**

Library of Congress Cataloguing in Publication Data
Main entry under title:

The Planet We Live On: Illustrated Encyclopedia
of the Earth Sciences

1. Earth sciences — Dictionaries.
QE5.P55 1976   550'.3   75-29977
ISBN 0-8109-0415-2

Library of Congress Catalogue Card Number: 75-29977
Published in 1976 by Harry N. Abrams, Incorporated, New York
All rights reserved. No part of the contents of this
book may be reproduced without the written permission of
the publishers

Printed in Israel

Contents

Subjects Included in the Encyclopedia

Crystallography: The study of the growth, structure, physical properties, and classification of crystals.

Economic Geology: The study of geologic materials profitably utilized by man, including fuels, minerals, and water.

Environmental Geology: The application of geological principles to the solution of problems created in the physical environment by human occupancy.

Geochemistry: The study of the distribution of chemical elements and changes in minerals, ores, rocks, soils, water, and the atmosphere.

Geomorphology: The study of the nature, origin, and development of landforms and their classification.

Geophysics: The application of the principles of physics to the study of the earth as a planet.

Glacial Geology: The study of glaciers and ice sheets and the effects and features created by glacial erosion and deposition.

Historical Geology: The study of the evolution of earth and its environment from the time of its origin to its present state.

Hydrology: The study of the properties, circulation, and distribution of continental water from the time of precipitation until it returns to the atmosphere. The two main areas of study are running water and ground water.

Marine Geology: A branch of oceanography concerned with the features of the ocean floor and the ocean-continent border.

Meteorology: The study of the earth's atmosphere.

Mineralogy: The study of the formation, occurrence, properties, and classification of minerals.

Oceanography: The study of the chemical, biological, physical, and geologic features of the oceans and seas.

Paleontology: The study, based on fossils, of life in past geologic periods.

Petrology: The study of the origin, history, occurrence, and structure of rocks.

Sedimentology: The study of sedimentary rocks and the processes by which they were formed.

Stratigraphy: The study of the arrangement, correlation, and distribution of rock strata and the interpretation of their significance in geologic history.

Space Geology: The study of condensed matter and gases in the solar system, including the study of surface features of other planets and of impact phenomena and objects on the earth's surface believed to be of extraterrestrial origin.

Structural Geology: The study of the relationships of rock units to one another and their deformational history.

Tectonics: The study of the origin, history, and classification of structural features of the earth's crust. Tectonics is similar to structural geology but broader in scope.

Volcanology: The study of volcanic phenomena and their causes.

Contributors and Editors

Editor: **Cornelius S. Hurlbut, Jr.**
Editor-in-Chief: **Milton Rugoff**
Project Editor: **Lynne Williams**
Advisory Editor: **Georg Zappler**

Contributing Editors

Cornelius S. Hurlbut, Jr., Professor of Mineralogy, Emeritus, Harvard University; author of *Minerals and Man* and *Dana's Manual of Mineralogy.*

Subjects: Mineralogy, crystallography, economic geology

William H. Matthews III, Regent's Professor of Geology, Lamar University, Texas; author of *Invitation to Geology* and *Fossils: An Introduction to Prehistoric Life.*

Subjects: Historical geology, paleontology, stratigraphy

Robert L. Nichols, Professor of Geology, Emeritus, Tufts University and Eastern Kentucky University.

Subjects: Geomorphology, glacial geology, volcanology

James W. Skehan, Director, Weston Observatory, and Profes of Geology, Boston College; co-editor of *Science and the Future of Man.*

Subjects: Structural geology, geophysics, tectonics, petrology

Robert E. Boyer, Chairman, Department of Geological Sciences, University of Texas, Austin; author of *Earth Science and Energy.*

Subjects: Environmental geology and energy resources

Contributing Specialists

Richard A. Davis, Jr., Professor and Chairman, Department of Geology, University of South Florida; author of *Principles of Oceanography.*

Subjects: Oceanography and marine geology

Miles F. Harris, American Meteorological Society; author of *Getting to Know the World Meteorological Organization.*

Subject: Meteorology

Marie Morisawa, Professor of Geology, State University of New York, Binghamton.

Subject: Hydrology

Jeffrey L. Warner, NASA, Johnson Space Center, Houston, Texas.

Subject: Space geology

Charles H. Simonds, The Lunar Science Institute, Houston, Texas.

Consultant

Vincent Manson, Director, Department of Mineralogy, American Museum of Natural History.

Text for illustrations by Susan Rayfield.

The
Planet
We
Live
On

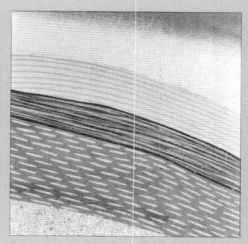

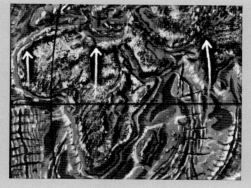
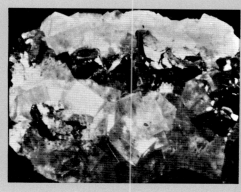

Introduction

Man and the Earth

During our long evolutionary development we have become increasingly dependent on earth materials. Now, more than ever, our destiny is inextricably linked to these materials. Despite this dependence, most of us have only a limited knowledge of our planetary habitat and even less appreciation of it.

Meanwhile, we enjoy the products of soils that have been formed from weathered rock and we delight in the beauty of precious stones and rare minerals. We get most of our energy supply from coal and petroleum derived from the remains of prehistoric plants and animals. And our industries rely heavily on mineral resources such as iron, lead, limestone, clay. These vital products have been made more readily available through a better understanding of the earth and the application of the principles of earth science and geologic engineering.

The earth has also provided us with areas of extraordinary beauty. Who does not marvel at the vastness of the Grand Canyon, the silent mystery of a great cavern, the grandeur of an Everest, the sweep of a mighty river? Each of these has been produced by basic geologic processes that began to shape the earth shortly after its birth — about 4.5 billion years ago — and are still at work today. Anyone will be richer for an understanding of this ancient but relatively fragile planet, this intricate, multifaceted environment that is our home.

The Earth Sciences

This encyclopedia covers a group of disciplines collectively called the earth sciences, each of which contributes to our understanding of the planet earth. In the broadest sense they deal with the composition, process, and history of the earth and its environment in space — in short with its air, land, and water, and its plant and animal life.

As late as the early eighteenth century the study of the earth was regarded as the domain of the natural scientist. It was a time when a single individual could still achieve some mastery of man's accumulated knowledge. But as more observations were made and the science of geology assumed an identity apart from other natural sciences, it became impossible for any one individual to be conversant with all that was known in this field. Soon students of the earth began to specialize and the major subdivisions of earth science gradually evolved. Yet none of these subdivisions is completely divorced from the mother subject; each interacts with and overlaps the others.

Today there is an increasing effort to cut across the rather arbitrary boundaries of the earth sciences in order to present a picture of the earth as a whole. This is not a return to the so-called natural science of 200 years ago. Rather, it is a more natural and meaningful integration of the most significant aspects of each subdivision. The need for such an approach can be illustrated, for example, by considering the composition of the earth. If asked what our planet is made of, many would probably answer "rocks and soil." This is not surprising, for most of us think in terms of the solid earth. But solids are not the only types of matter present. Gases and liquid matter—the air and water so essential to life and geologic processes — are as much a part of the earth as the solid surface.

These three types of matter — gases, solids, and liquids — are commonly described in terms of three rather distinct zones or spheres. The atmosphere is a thick, life-giving envelope of gas that completely surrounds our globe: it is the bailiwick of the meteorologist. He is interested in the atmosphere and in all phenomena—especially climate and weather—related to it. Another earth scientist, the geologist, is primarily interested in the lithosphere, the solid, "rocky" part of our planet. The rocks of the lithosphere provide the foundation upon which we live. They also supply man with much-needed natural resources. The hydrosphere is a widespread sheet of water that, in the form of oceans, covers almost 71 percent of the earth's surface and is so basic that life could not exist without it. This universal sea is the special concern of the oceanographer. In a larger sense the hydrosphere extends onto the continents and includes the water held in the ground and in swamps, lakes, and rivers. It is the hydrologist who studies the distribution, circulation, and composition of the precious freshwater reserve.

Although these three zones of matter are clearly defined in composition, the boundaries between them are not so distinct. Rather, there is continual interaction between them as rock comes in contact with air, water with rock, air with water. This constant intermingling is a main concern of the earth scientist, for important changes commonly occur at the interface or boundary between these zones.

Because of the life-giving qualities of air, water, and soil, the earth is inhabited by countless plants and animals. This mass of organisms makes up the biosphere, the organic realm that is just as significant as are the physical and inorganic zones of the earth. This vibrant sphere of life is intricately interrelated with the air, land, and water, for most organisms live in the narrow zone where these gases, solids, and liquids meet.

Although the biosphere would appear to fall more logically within the scope of the biologist or life scientist, the long-standing and ongoing interaction between the earth's organic and inorganic spheres makes life no less important to the earth scientist. Throughout the more than 3 billion years that life has been present on the earth, it has spread to all parts of the globe. And during this part of geologic time the organisms of the biosphere have interacted with the atmosphere, hydrosphere, and lithosphere and have become involved in a number of earth processes. For example, the biosphere is responsible for most of the oxygen we breathe, the formation of petroleum and coal, and many rocks of organic origin. The record of past life on earth, as revealed by fossils, has also provided valuable clues to the history of the planet and the evolution of life.

The Geologic Sciences

Meteorologists and oceanographers, as well as astronomers, have long had their special interests in our planet, but it is the geologist for whom the earth holds the greatest fascination. Significantly, the public is increasingly sharing this interest. Concerned citizens are becoming more and more aware of our dwindling natural resources and the role of the geologist is to locate and conserve them. In addition, many people of all ages are learning about the earth and its materials through the increasingly popular hobby of rock and mineral collecting. Teachers see the earth sciences in general, and geology in particular, as ideal for illustrating a host of scientific principles. They have also found that subjects such as earthquakes, volcanoes, dinosaurs, glaciers, and the oceans are of particular interest to curious young minds. It is not surprising, therefore, that most of the material in this encyclopedia deals with subjects closely related to geology.

In one sense, of course, geology *is* earth science, for the word is derived from the Greek *geo,* meaning earth, and *logos,* meaning study. Geology, then, is the study of the earth; its origin and development through time; its composition, shape and size; the processes that now operate or formerly operated within it and on its surface; and the origin and evolution of its inhabitants. As students of the earth, geologists make systematic observations and measurements to compile an organized body of knowledge about our globe. But their goal is not merely to assemble a mass of unrelated facts. Rather, it is to interpret these data and to develop principles and hypotheses that will explain their findings in relation to the earth.

Geology embraces such a broad spectrum of knowledge that it overlaps other sciences; thus astronomy, chemistry, physics, and biology play significant roles. Astronomy tells us where our planet fits into the universe and how it relates to other celestial objects. Astronomers have also formulated hypotheses to explain the origin of earth and our solar system. Now, with the coming of the Space Age, astronomy has assumed an even more exciting role in geology. A new type of earth science is combining elements of astronomy and geology in what is called astrogeology or planetary geology. Astrogeology applies tech - niques of classical geology to space-related problems. In its broad sense it includes the study of the form, structure, composition, history, and the processes at work on the surfaces of other solid objects in our solar system. But it differs from classical geology in a number of ways; for example, the earth's thick atmospheric envelope gives rise to several geologic processes not found on the moon. The processes that created the lunar landscape have been equally active on the earth, but weathering and erosion have subdued or obliterated the resulting features. For this reason, surface features on the moon cannot always be interpreted in the light of analogous surface features on our planet.

Chemistry is invaluable in studying earth materials, for the earth consists of rocks and minerals composed of chemical elements and compounds. The earth scientist who specializes in the chemistry of rocks, their composition, and the chemical changes they undergo is the geochemist. His work straddles the boundaries of chemistry and geology. Geochemists are interested not only in the chemicals that make up the earth's crust but also in the occurrence and distribution of chemical elements in minerals and rocks. These may serve as clues in locating valuable mineral deposits. Geochemistry also plays an important role in geochronology — the study of time in relation to earth history.

Physics is utilized in many phases of the geologic sciences. Geophysics, like geochemistry, is an in-between field that relies on the methods and principles of both physics and geology. Geophysical investigations deal largely with the earth's magnetism, gravity, electrical properties, and similar physical characteristics. Seismologists, specialists in the study of earthquakes, also depend on geophysical principles and methods. Exploration geophysics, geological prospecting or exploration using the techniques of physics and engineering is an important application of geophysics. Sophisticated geophysical devices can be used to locate the measure variations in magnetism, gravity, radioactivity, and seismic properties of rock formations. These often provide clues to the location of valuable deposits of mineral resources.

To learn more about the nature of prehistoric plants and animals, the earth scientist turns to biology. By comparing the remains of fossil species with similar organisms now living, it is often possible for a scientist to reconstruct ancient plant and animal assemblages and to make informed guesses as to the environment in which they lived. In dealing with geologic processes that erode the earth's surface and create landforms, the concepts of geography may be used. Although geology has drawn heavily from the other sciences and overlaps them in many areas, it has certain characteristics that distinguish it from the other basic sciences. Perhaps the most striking of these differences is the concept of geologic time, for vast stretches of time are necessary to explain past geologic events in terms of geologic phenomena of today.

The scope of geology is so wide that it has been separated into two major divisions. Physical geology deals with the composition of the earth, its structure, movements on and within its crust, and the geologic processes by which the surface changes. This is a very broad branch of geology, and as geologists have learned more about the earth's physical characteristics, specialized fields have developed within physical geology. The mineralogist, for example, concentrates on minerals, those chemical elements and compounds that are the building units of the earth's crust. The rocks, composed of large aggregates or masses of minerals, are studied by the petrologist. The geomorphologist investigates surface features in order to determine the origin and development of landforms. The structural geologist analyzes the various types of rock deformation that have taken place within the solid earth. Seismologists are interested in earthquakes and, through an interpretation of earthquake vibrations, draw conclusions regarding the nature of the earth's interior. Volcanologists monitor the activities of volcanoes, making temperature measurements of erupted lava and analyzing the composition of the liberated gases. Glaciologists are concerned with glaciers — their formation, advance, accompanying erosion, retreat, and the deposits left behind. Still more specialized are the geochemists and geophysicists, who apply the principles of chemistry and physics in their analysis of the earth.

The dangerous combination of abuse of our environment and a rapidly expanding world population has given rise specifically to environmental geology, but it is of concern also to all earth scientists. Environmental geology is dedicated to the conservation of earth resources and the application of geology to human needs. A major goal is to make others aware of the need to consider geologic problems in planning for the future.

There are also specialized fields of historical geology. In order to interpret earth history, the geologist must study the rocks and gather evidence of the great changes in geography, climate, and life that took place in the geologic past. By means of fossils buried in the earth, the paleontologist describes and classifies prehistoric plants and animals. He also traces their rise throughout geologic history to determine the evolutionary relationship between seemingly different organisms. The stratigrapher is concerned with the origin, composition, and proper sequence of layered rocks. Using the process of correlation, the stratigrapher matches the rock strata of one area with those of another and uses the evidence to place geologic events in their proper chronological order. The paleogeographer uses data compiled by the paleontologist and the stratigrapher to prepare paleogeographic maps that reveal the position and extent of ancient seas and landmasses. The cosmogonist is interested in the most ancient of all history. This highly specialized astronomer speculates on the origin of the universe and birth of our planet. Each of these historians plays an important part in piecing together the geologic history of the earth.

Like the major subdivisions of earth science, the main branches of geology overlap and are interdependent. The physical geologist uses mineralogy and petrology to determine what types of rocks are present and how they were formed. The historical geologist may study the same rocks to determine what kinds of organisms were living at the time the rocks were formed, the environment they inhabited, and the type of climate that prevailed.

Because the processes by which minor elements are concentrated into economically valuable deposits are complex, the economic geologist, in his search for mineral resources, draws heavily on most of the more specialized branches of earth science. Likewise, the petroleum geologist, in his efforts to locate hidden reserves, relies on the work of both the physical and historical geologist. Thus, the unification and interaction of physical and historical geology leads not only to the location of valuable earth materials but ultimately to a better understanding of the composition and history of the earth. This is, of course, the major objective and ultimate goal of all earth science.

Why an Encyclopedia of Earth Sciences? A hallmark of modern science is the amount of new information constantly generated by research. The earth sciences — like all sciences — are undergoing many important and revolutionary changes. Fresh ideas are being introduced to explain old problems. Outmoded ideas are being updated and modified — or completely abandoned — in the light of new discoveries. Equally important, new applications are being found for older methods of investigating the earth. Modern techniques are being developed to probe the earth's interior, its air, land, and water, and to understand more clearly its basic structure. Lunar landings have resulted in a better understanding of the earth-moon relationship. Oceanographers probe the sea to add to our knowledge of the great world oceans. Meteorologists strive to learn more about the earth's threatened atmosphere and to work with other earth scientists in an attempt to help preserve our endangered environment. Continental drifting, sea floor spreading, and plate tectonics have revolutionized geologic thought with respect to mountain building, ancient climates and the origin of continents and ocean basins. These are but a few of the developments that have precipitated a great upsurge of interest in the earth sciences.

Events follow each other rapidly in a revolution, and scientific revolutions are no exceptions. Such developments have rendered many otherwise authoritative books obsolete and have led to the constant revision of others. The pages that follow are thus rich in entries that record the remarkable recent advances in the fields of geology, meteorology, oceanography and lunar science. These dynamic events are having a profound effect not only on our conceptions of our planetary home but on all modern scientific thought. Perhaps more important, they offer many clues concerning the future of our fragile planet and its precious load of plants and animals.

This encyclopedia will, of course, provide the reader with detailed information on earth science subjects. But beyond that, we hope it will serve as an introduction to the earth, its place in time and space, and the role that the earth sciences play in our daily lives.

Cornelius S. Hurlbut, Jr.
William H. Matthews III

Notes on Use of the Encyclopedia This encyclopedia includes entries on all the earth sciences in one alphabetical sequence. In addition, the Guide to Entries by Subject in the Appendix gives an alphabetical list of the entries for each subject.

A cross reference is indicated by an asterisk preceding a term.

The initials at the end of longer entries are those of contributors listed on the contributors page at the front of the book.

The metric system of weights and measures has been used throughout, but conversions to English/American equivalents have been provided where it was thought that they would be especially helpful to the reader. In addition, conversion tables are given in the Appendix.

A Color Portfolio of the Earth Sciences

Spouting lava and steam, a volcanic island rose from the ocean off the coast of Iceland in 1963. One of the few visible portions of the Mid-Atlantic Ridge, Surtsey provided dramatic new evidence of the dynamic nature of the earth's crust.

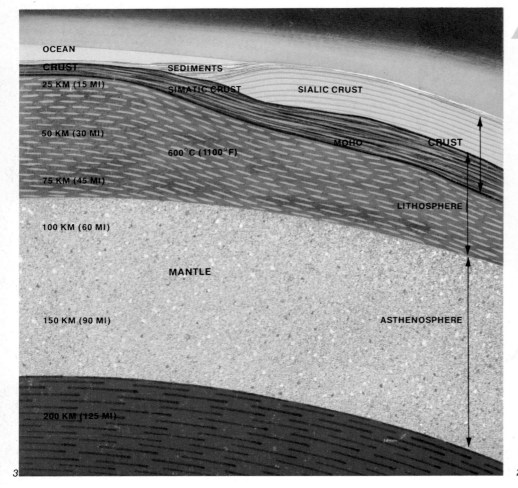

1

CRUST
16- 40 KM   10-25 MI

MANTLE
2895 KM   1800 MI

LIQUID OUTER CORE
2220 KM   1330 MI

SOLID INNER CORE
1255 KM   750 MI

WAVE FRONTS

FOCUS

TRAVEL PATHS

OCEAN
CRUST
25 KM (15 MI)
SEDIMENTS
SIMATIC CRUST   SIALIC CRUST

50 KM (30 MI)
600°C (1100°F)
MOHO   CRUST

75 KM (45 MI)
LITHOSPHERE

100 KM (60 MI)

MANTLE

150 KM (90 MI)
ASTHENOSPHERE

200 KM (125 MI)

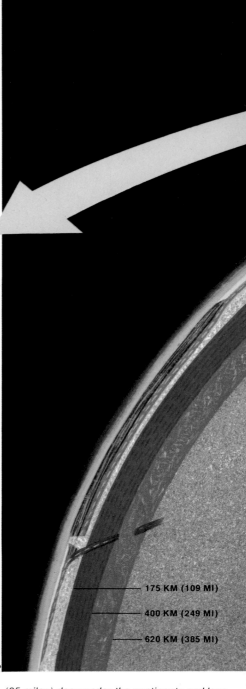

175 KM (109 MI)

400 KM (249 MI)

620 KM (385 MI)

In ancient times, men imagined the earth's interior as a realm of darkness, or as filled with water or fire. Then, in a series of experiments in 1798, an English physicist, Henry Cavendish, "weighed" the earth for the first time and realized that the core must be made of material far heavier than the surface matter.

1
Since then, much has been learned about the interior from the way earthquake waves bend as they pass through the earth. Some waves accelerate when moving through denser material; others disappear when they encounter liquids or gases. By studying these waves, scientists have concluded that the earth is composed of concentric layers: crust, mantle, liquid outer core, and solid inner core.
2
The upper layer of the earth, the crust, is a relatively thin shell of solid rock, up to 40 km

(25 miles) deep under the continents and less than 16 km (10 miles) thick beneath the ocean floor. It is composed of two parts: huge blocks of granitic rock, forming the continents, and an underlying layer of heavy basalt.

The mantle, which extends downward to a depth of about 2900 km (1800 miles), makes up over 80 percent of the earth's volume. It is believed to be composed of igneous rock.

The outer part of the core is molten nickel-iron, surrounding an inner core made up of the same material and less than 1255 km (750 miles) in radius. The exact center of the earth is

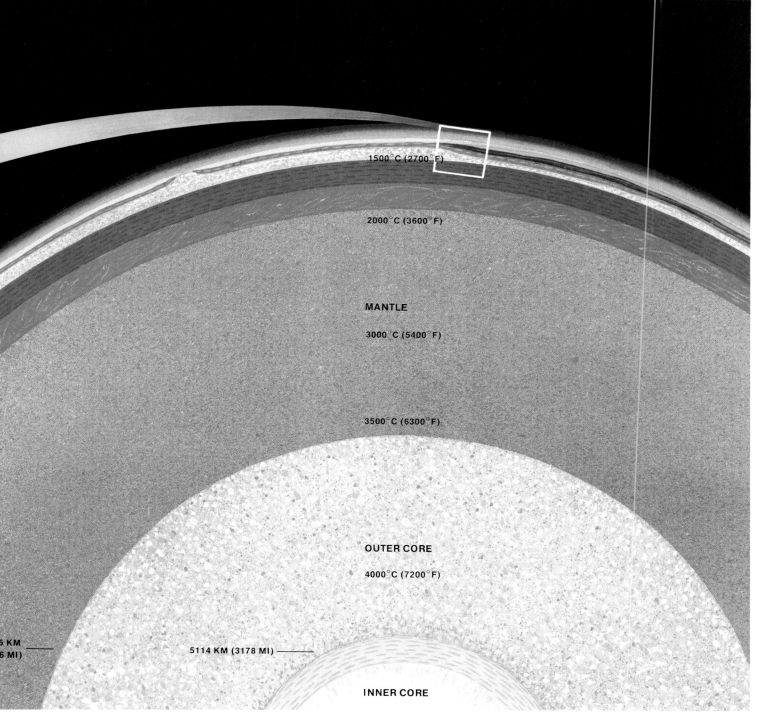

1500°C (2700°F)

2000°C (3600°F)

MANTLE

3000°C (5400°F)

3500°C (6300°F)

OUTER CORE

4000°C (7200°F)

5 KM
6 MI)

5114 KM (3178 MI)

INNER CORE

so dense that each side of a cubic inch supports 20,000 tons of pressure, and the temperature is almost as hot as the surface of the sun.

3
A magnified cross section of the earth's top layers. The outer crust behaves as though it were floating on a yielding plastic material of greater density. For this reason, it is believed that the crust and uppermost mantle comprise a strong shell, called the lithosphere, which overlies a softer shell, the asthenosphere, where the temperature and pressure are much higher and rocks flow like molasses. The boundary between crust and mantle is called the Mohorovicic Discontinuity, or Moho. The crust is composed of two layers. The uppermost, or continental crust, consists of granitic rocks rich in silicon and aluminum and is called the sial layer. Below it is the oceanic crust, composed primarily of basalt, which is rich in silicon and magnesium. This layer is therefore called the sima.

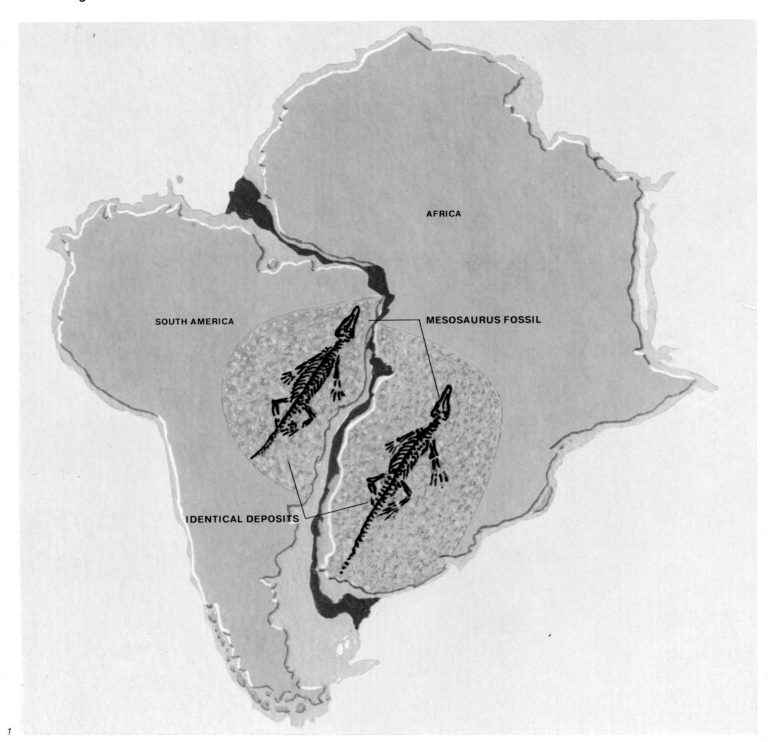

AFRICA

SOUTH AMERICA

MESOSAURUS FOSSIL

IDENTICAL DEPOSITS

1

Geographers have long been aware that if the continents could be moved they would fit together like a jigsaw puzzle. In 1911, a German scientist, Alfred Wegener, proposed a theory of "wandering continents," claiming they were once part of a supercontinent that broke into pieces which drifted to their present positions over millions of years. Wegener's theory evoked much controversy but was largely disregarded until after World War II.

1
It has since been demonstrated that Africa and South America fit together when the match is made along the continental slope (shown in light blue) at a depth of about 1000 m (3000 feet). In addition, various radioactive isotope tests indicate that distinctive rock formations on the western shoulder of Africa match rock layers along the east coast of South America. Finally, fossils of a small reptile called Mesosaurus that lived during the Permian Period more than 250 million years ago have been found only in the areas shown in orange.

It is thought that these small creatures could not have migrated across thousands of miles of ocean, and so must have once shared a common freshwater environment.
2
Viewed from 650 km (400 miles) in space, Arabia, left, is slowly tearing away from Africa. As the two landmasses separate, the Red Sea, center, its floor split by an active rift, is widening and may eventually form a new ocean.

2

Overleaf:

Maps show the presumed position of the continents from the end of the Permian to the present. Arrows indicate the direction of movement of the landmasses.

1

According to the drift theory, only 225 million years ago there was just one supercontinent— Pangaea, meaning "all lands"—surrounded by a universal sea, Panthalassa.

2

By the end of the Jurassic Period, approximately 130 million years ago, Pangaea is thought to have split into a northern landmass, called Laurasia, and a southern mass, known as Gondwanaland.

A rift had appeared between what is now Africa and South America—the beginning of the Atlantic Ocean. The fragment which would eventually become India was already separating from the huge southern continent.

3

In the Paleocene, 65 million years ago, Australia was still part of Antarctica but most of the other masses had drifted apart and were acquiring their present shape.

4

Later, Australia separated from Antarctica and Laurasia broke apart into North America and Eurasia. India drifted north, colliding with Eurasia and thrusting up the Himalayas. It is said that a map of the future, some 50 million years from now, will probably show Australia pushing north, the Atlantic and Indian oceans continuing to widen, and California west of the San Andreas Fault detached from the mainland and sliding toward the abyss of the Aleutian Trench.

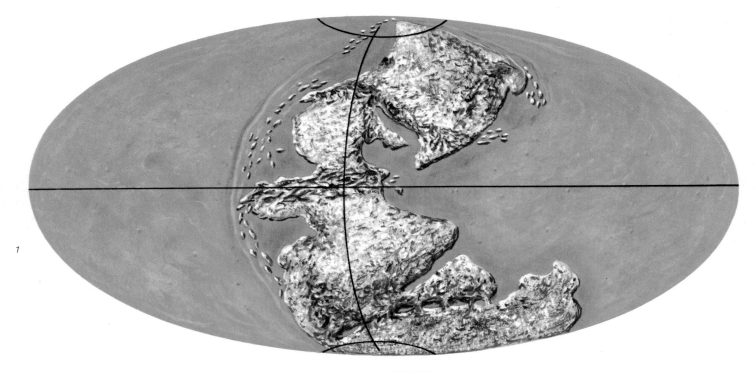

END OF PERMIAN: 225 MILLION YEARS AGO

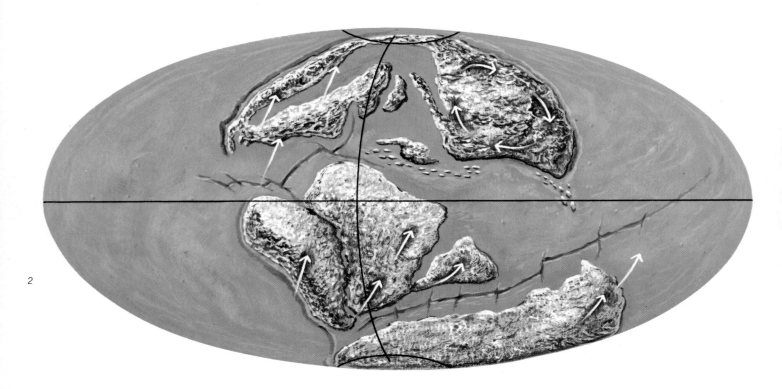

END OF JURASSIC: 130 MILLION YEARS AGO

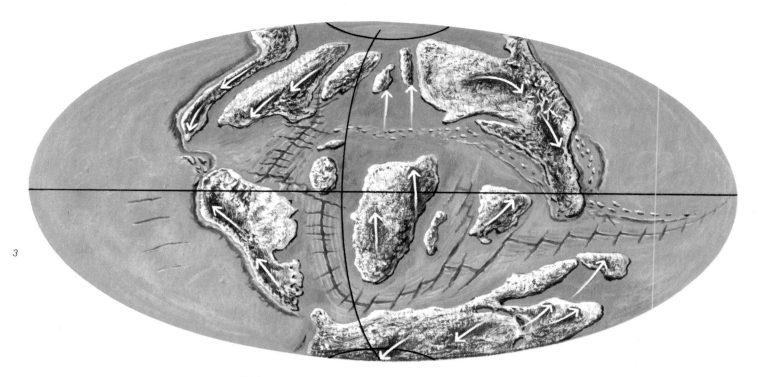

BEGINNING OF PALEOCENE: 65 MILLION YEARS AGO

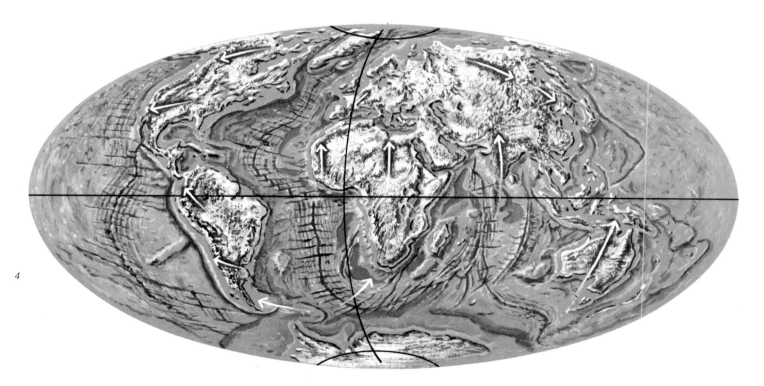

RECENT

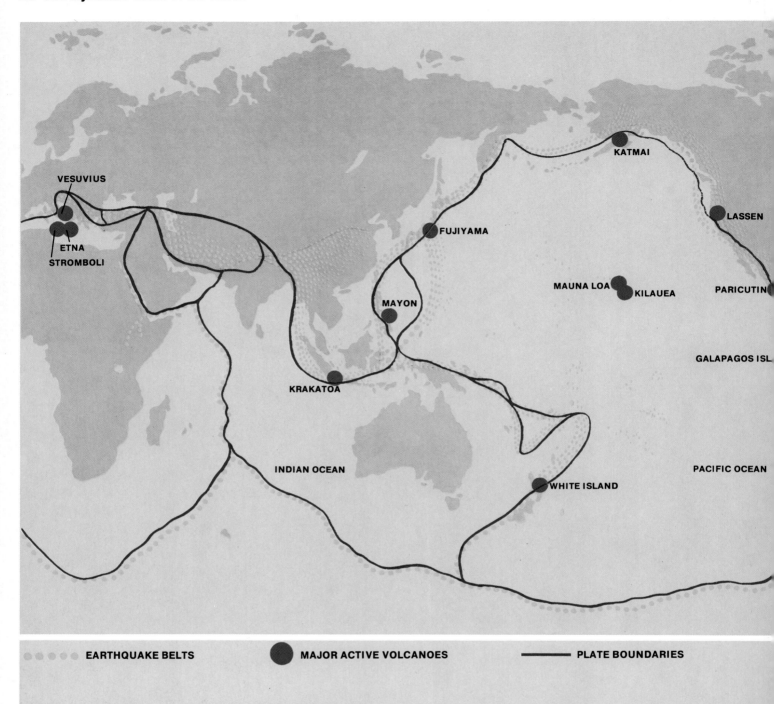

VESUVIUS

ETNA
STROMBOLI

KATMAI

LASSEN

FUJIYAMA

MAUNA LOA KILAUEA

PARICUTIN

MAYON

GALAPAGOS ISL.

KRAKATOA

INDIAN OCEAN

PACIFIC OCEAN

WHITE ISLAND

○○○○○ EARTHQUAKE BELTS   ● MAJOR ACTIVE VOLCANOES   ▬▬▬ PLATE BOUNDARIES

1

1

For centuries the earth's crust was thought to be rigid, holding the continents like rocks in cement. Then, in 1969, core samples taken from the ocean bottom revealed that new sea floor material wells up from great rifts, spreads outward, and hardens to form new crust, a process called sea floor spreading. Crustal plates (outlined in red) drift about like huge rafts, carrying the continents and ocean basins with them. It is thought that there are seven large plates and at least twenty smaller ones from 50 to 160 km (30 to 100 miles) thick.

Not all plate boundaries coincide with the edges of continents. Some plates consist in part of granite crust that forms most of the landmasses, and in part of oceanic crust that forms the ocean floor. Scientists have learned that almost all earthquake and volcanic activity, which mark the "active zones" of the earth's crust, occur in narrow belts at the boundaries of the plates. As the plates grind against each other, they may create shock waves and thrust up new mountain chains. One of the more unstable regions is located around the edge of the Pacific Ocean, where the margins of the

plates plunge into deep trenches that extend beneath the crust. In this area, called "the ring of fire," friction and intense heat produce changes that may generate earthquakes and volcanic explosions.

2

A submarine mountain range 64,000 km (40,000 miles) long winds through all the ocean basins around the globe like a seam on a baseball. Along the crest of this ridge is a rift through which molten material wells up from the earth's interior, and pushes outward, giving rise to the term sea floor spreading. A series of

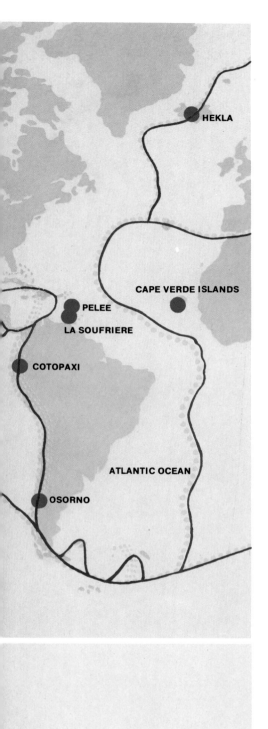

HEKLA

CAPE VERDE ISLANDS

PELEE

LA SOUFRIERE

COTOPAXI

ATLANTIC OCEAN

OSORNO

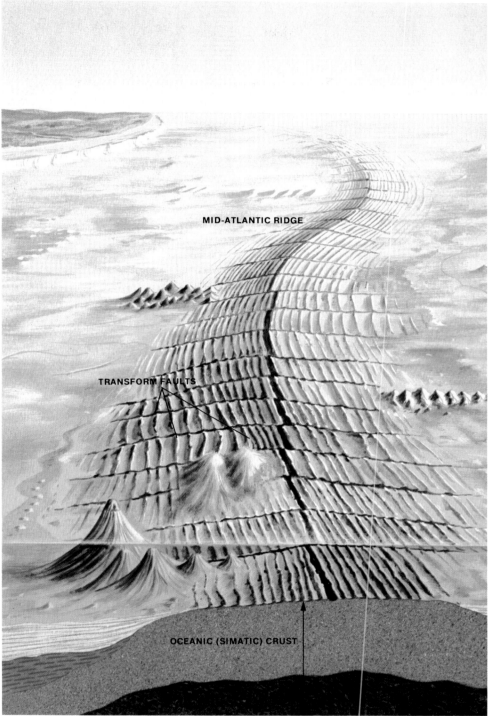

MID-ATLANTIC RIDGE

TRANSFORM FAULTS

OCEANIC (SIMATIC) CRUST

2

*deep fractures, or transform faults, cuts across
the ridge, dividing it into short, straight
sections.*

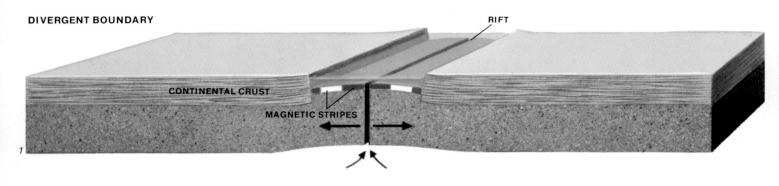

DIVERGENT BOUNDARY

RIFT

CONTINENTAL CRUST

MAGNETIC STRIPES

1

DIVERGENT BOUNDARY

CONTINENTAL SHELF

OCEANIC CRUST

CONTINENTAL RISE

2

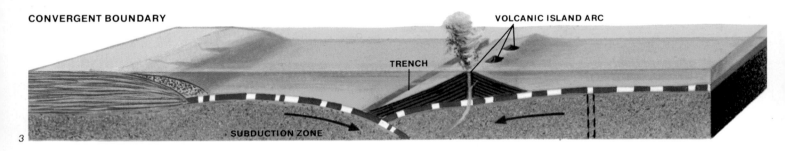

CONVERGENT BOUNDARY

VOLCANIC ISLAND ARC

TRENCH

SUBDUCTION ZONE

3

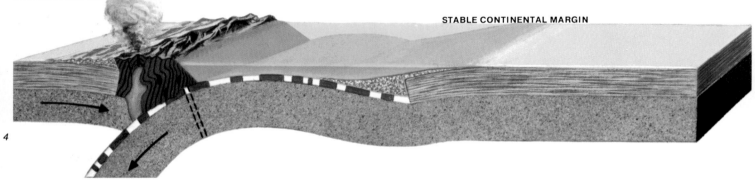

CONVERGENT BOUNDARY

STABLE CONTINENTAL MARGIN

4

As they move across the earth's surface the crustal plates jostle one another, collide and spread apart, slowly but dramatically altering the face of the globe. In the profiles above, arrows indicate the direction of plate movement.

1

Vast ridges mark the site of spreading, or divergent plate boundaries. Here, fiery molten material wells up from the interior through a narrow central rift, moves outward and cools, pushing apart the continental plates and creating new ocean floor. Stripes indicate reversals in the direction of magnetic fields over the ages.

2

Another type of divergent boundary occurs in an older ocean basin, created by the breakup of a continent.

3

Convergent boundaries are found at trenches, where a crustal plate plunges downward and is destroyed. Here a plate has slipped into a trench alongside an island arc which is separated from the continent by a sea basin, as are Japan and Hawaii. The intense heat and friction created as the plates grind past each other result in violent earthquakes and volcanic eruptions.

4

Another type of convergent boundary occurs when an oceanic plate sinks beneath a

COLLISION BOUNDARY   COLLIDING CONTINENTAL BLOCKS

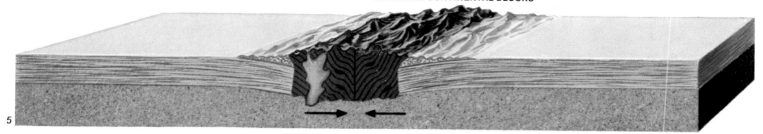

5

COLLISION BOUNDARY   NEW CONTINENTAL CRUST

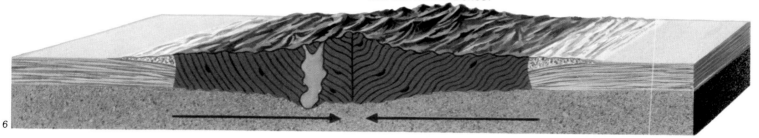

6

NEW DIVERGENT BOUNDARY   RE-OPENING OCEAN

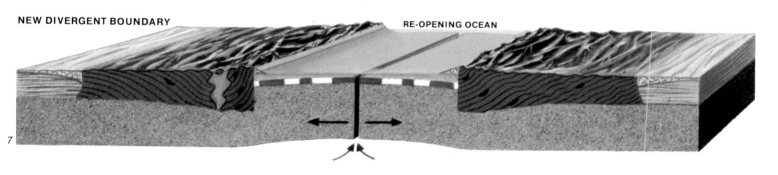

7

GREATEST DIVERGENT BOUNDARY   WIDE OCEAN

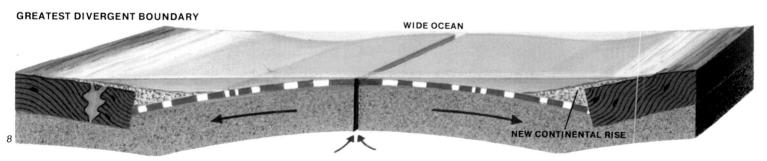

NEW CONTINENTAL RISE

8

continental plate. Here the impact has
crumpled the leading edge of the landmass,
creating a massive uplift of mountains. As the
crust of the descending plate melts in the heat
of the earth's mantle, tremendous pressure
forces it back to the surface through volcanic
vents. Such a collision along the western edge
of South America pushed up the Andes
Mountains and still triggers earthquakes in
Chile and Peru.
5
Two continental plates sometimes thrust into
each other head on, causing enormous

buckling of land along their edges. Thus
millions of years ago India, traveling northward,
thrust into the bottom of Asia in a tremendous
collision which thrust up the Himalayan
Mountain system.
6
When two continental plates collide, they may
sometimes create a new continental crust from
uplifted sea floor deposits.
7
After a collision between two plates, a new rift
may open—as in the Red Sea—eventually
widening to form an ocean.

8
Periods of spreading and converging may allow
oceans to open and close repeatedly, as may
have happened in the Atlantic, shown here in its
present stage.

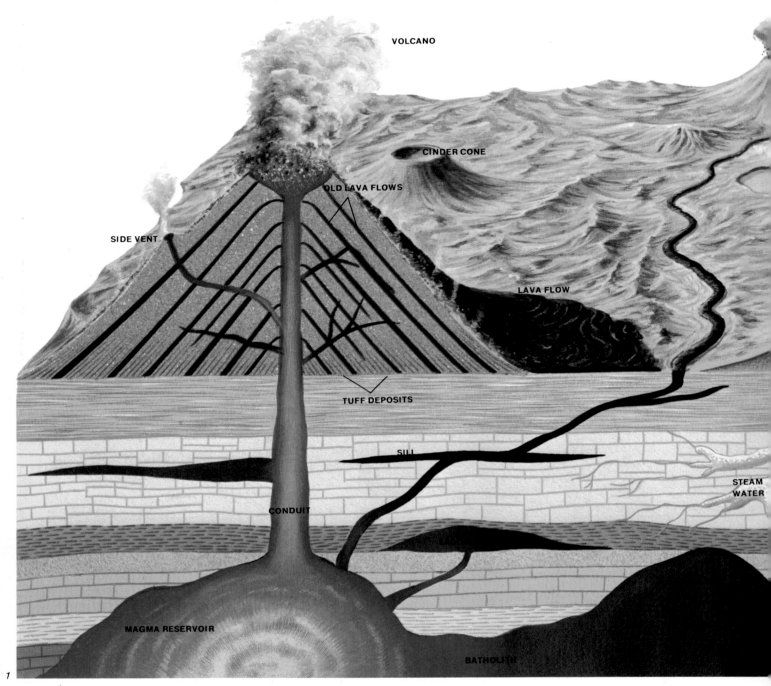

VOLCANO

CINDER CONE

OLD LAVA FLOWS

SIDE VENT

LAVA FLOW

TUFF DEPOSITS

SILL

STEAM WATER

CONDUIT

MAGMA RESERVOIR

BATHOLITH

1

1

Volcanoes and lava flows are powerful events that are constantly changing the surface of the earth. They are formed when magma from the earth's interior rises through the crust and solidifies on the surface. Many volcanic peaks are composite cones, built up of lava flows interbedded with deposits of tuff and breccia. If the central chimney or conduit is blocked, the magma and gases may force their way through a vent on the side of the volcano, forming a parasitic cone. Some volcanoes, such as Paricutin, in Mexico, are cinder cones created by the eruption of blocks, cinders and ash. In some volcanic regions the rock at shallow depth may be hot enough to boil groundwater, which rises to the surface as hot springs. In other areas boiling water and steam are emitted intermittently, forming geysers, or if only gases are emitted, fumaroles.

Most magma never reaches the surface but cools and crystallizes within the earth's crust as intrusive bodies. When magma is forced horizontally between rock layers it forms tabular sills which may be many kilometers long and hundreds of meters thick.

Mushroom-shaped structures called laccoliths result when viscous magma is intruded between strata and pushes up the overlying rocks into domes. When magma is forced into existing fractures in the rock, it may create dikes. After softer surrounding rocks have been eroded away, dikes often appear on the surface as prominent ridges. When the magma chamber itself solidifies, a batholith is formed. Such features are often thousands of kilometers long. A smaller version of a batholith is called a stock.

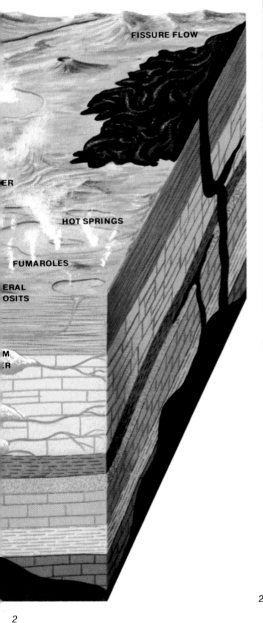

OLCANO

FISSURE FLOW

ER

HOT SPRINGS

FUMAROLES

ERAL
OSITS

M
ER

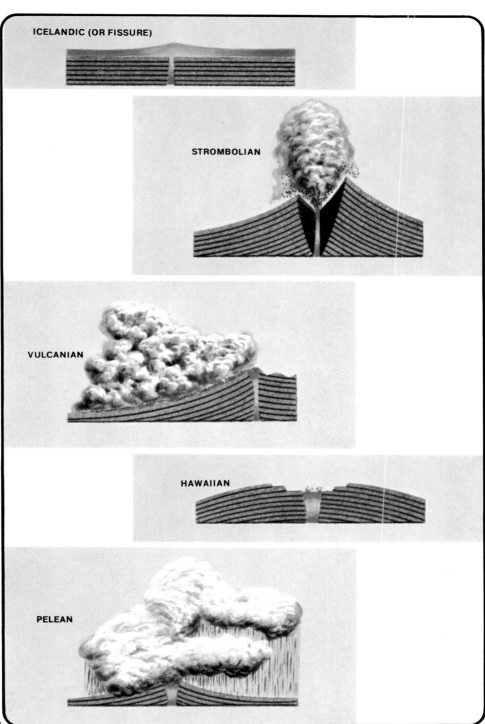

ICELANDIC (OR FISSURE)

STROMBOLIAN

VULCANIAN

HAWAIIAN

PELEAN

2

2

There are five basic kinds of volcanic eruptions. In the Icelandic, or fissure, type, lava forms extensive sheets on both sides of a fracture in the earth's crust.

Strombolian eruptions, named for Stromboli Island northwest of Sicily, are characterized by fire fountains of lava hurled from a crater. Hawaiian volcanoes, such as Mauna Loa, are known for relatively quiet eruptions of lava issuing from wide craters called calderas or from narrow vents along their sides. Calderas often fill with lava and form seething lakes.

Vulcanian eruptions are far more violent than the preceding types. In such spectacular outbursts, which characterize the activity of Mt. Vesuvius and similar volcanoes, the tephra collects around the vent, forming a cone. Pelean eruptions, named after Mt. Pelée in the West Indies, are also violent. Great, fiery clouds of superheated gases and fine tephra accompany avalanches of coarser tephra, which rush down the slopes at great speeds. In 1902, an eruption of this type from Mt. Pelée destroyed a city of about 30,000 inhabitants.

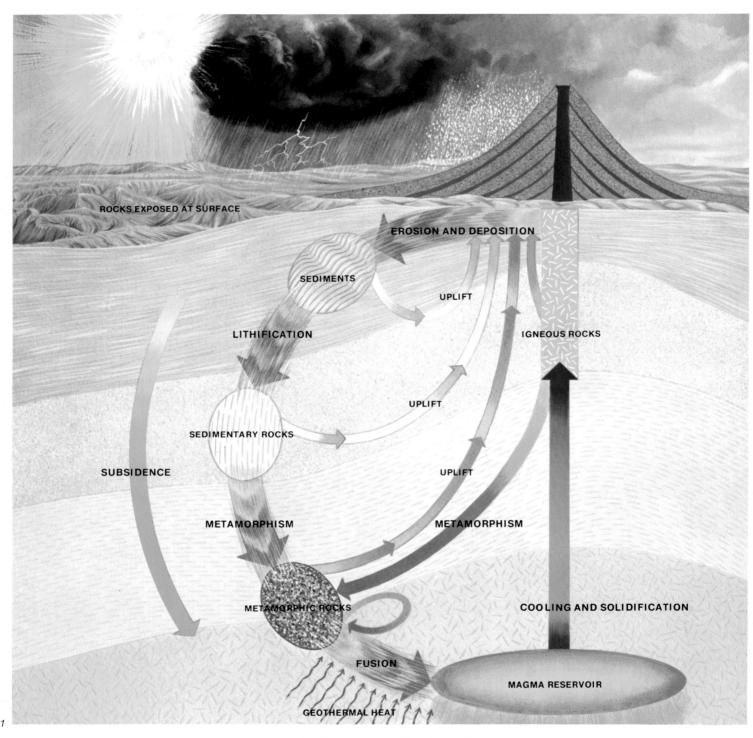

ROCKS EXPOSED AT SURFACE

EROSION AND DEPOSITION

SEDIMENTS

UPLIFT

LITHIFICATION

IGNEOUS ROCKS

UPLIFT

SEDIMENTARY ROCKS

SUBSIDENCE

UPLIFT

METAMORPHISM

METAMORPHISM

METAMORPHIC ROCKS

COOLING AND SOLIDIFICATION

FUSION

MAGMA RESERVOIR

GEOTHERMAL HEAT

1

1
The rocks found in the earth's crust are of three types—igneous, sedimentary and metamorphic—based on the way they were formed. The textures and characteristics of each kind are distinct and often easily recognizable.

All rocks are subjected to a perpetual cycle of change. As hot magma wells up from the earth's interior, it cools and hardens into igneous rock. Exposed at the surface to wind, water and ice, the rock is broken up and eroded, eventually washing into the ocean as sediments, where it may be consolidated into layers of sedimentary rock. When it subsides into regions of much greater pressure and temperature, sedimentary rock recrystallizes and is transformed into metamorphic rock. At still greater depth, higher temperatures fuse the metamorphic rock into magma. When thrust upward this molten material cools to become an igneous rock, called intrusive if it is still buried, or extrusive if it reaches the surface.

As the smaller arrows indicate, the complete cycle may be interrupted at any point. Some igneous rock is never exposed to surface weathering, but is directly converted to metamorphic rock. Other short circuits in the cycle may occur when sedimentary or metamorphic rocks are uplifted and subjected to erosion again.

2
Sedimentary rocks, such as sandstone and limestone, top, cover nearly three fourths of the visible surface of the earth. They are formed from sediments that are the products of erosion. Deposited in layers and later covered by new sediments, the material becomes compacted and is cemented with silica, iron oxide and other minerals in the circulating

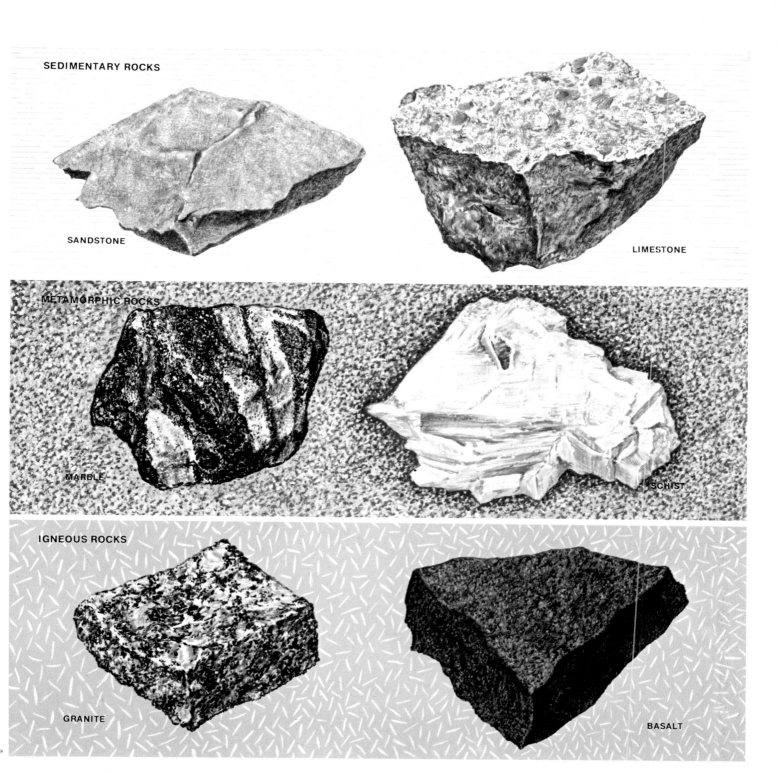

SEDIMENTARY ROCKS

SANDSTONE

LIMESTONE

METAMORPHIC ROCKS

MARBLE

SCHIST

IGNEOUS ROCKS

GRANITE

BASALT

groundwater. Sandstone is composed primarily of quartz grains. Most limestones consist of grains of calcite precipitated from seawater.

Metamorphic rocks, center, such as marble and schist, are formed at great depths within the earth, where great heat and pressure combines with chemically active fluids and vapor to change the composition of the original rock. Marble usually consists of metamorphosed limestone, and most schists were originally shale. Further metamorphism may change the schist to gneiss.

Igneous rocks, such as granite and basalt, bottom, are formed of interlocking grains crystallized from a liquid magma. Igneous rocks are classified according to their mineral composition and for each coarse-grained type there is a corresponding fine-grained equivalent. Granite, a coarse-grained rock, and rhyolite (fine-grained) consist essentially of orthoclase feldspar and quartz. They crystallize from magma rich in silica. Basalt and its coarse-grained equivalent, gabbro, form from magma low in silica. These rocks consist of plagioclase feldspar and ferromagnesian minerals.

There are over 2000 kinds of minerals. Most of them occur as small nondescript grains but some have striking colors, shapes and textures. Minerals are usually composed of two or more chemical elements, but a few of the elements, such as gold, silver and copper, are themselves minerals. All have an ordered internal atomic structure that often shows up as a geometrical arrangement of crystal faces. Crystals develop from melts, solutions and vapors as atoms arrange themselves layer by layer around a nucleus. Many are microscopic but some are tens of meters in size.

1
A fracture surface of malachite reveals its radiating silky-fibrous texture. It is found in the USSR, Zaire, South-West Africa, and the southwestern United States. The green patina that frequently covers copper statues is malachite, formed by the action of carbon dioxide and moisture.

2
Microcline is a member of the feldspar group of minerals. A green variety, amazonite, is sometimes cut as gems. More common white to gray microcline is used in ceramic glazes.

3
Apatite, from the Greek word meaning "to deceive," comes in such an array of forms and colors that early mineralogists confused it with other minerals. It is found with quartz, feldspar and iron ores.

4
Chalcopyrite, an abundant primary copper ore, occurs in brass-yellow crystals. It is often mistaken for gold.

5
Galena, commonly associated with zinc, copper, and silver minerals, has been the

source of lead for centuries. Here cubic
crystals of galena occur with amber-colored
fluorite.

6
Pink clusters of rhodochrosite encrust a
specimen of quartz from a Colorado silver
mine.

7
Orange-red crystals of crocoite crisscross a
cavity in a specimen of limonite from Tasmania.
Crystals of this rare and striking mineral form
when chrome-bearing solutions react with lead
minerals.

8
Calcite, the most abundant of the carbonate
minerals, comes in a wide variety of colors and
crystal forms. Here, amber calcite crystals
occur with hematite, an iron ore.

9
Amethyst is a variety of quartz whose violet
color is due to traces of iron. Its name, from the
Greek for "not drunken," derives from an
ancient belief that it prevented intoxication.

10
Native copper, found in excavations dating
back as far as 6000 B.C., may have been the first

metal used by man. Essential as a conductor of
electricity, it is now extracted from other
copper minerals such as chalcopyrite (4).

11
Selenite, a form of gypsum, occurs in
transparent crystals. Most gypsum forms as a
fine-grained sedimentary rock, precipitated
from evaporating seawater.

12
Prehnite commonly forms in radiating compact
masses of light green crystals. It occurs most
commonly in cavities in basaltic rocks
associated with zeolites.

1

The great features of the land—mountains, canyons, plateaus—seem permanent and immutable. In fact, water, ice, wind, and weathering are constantly breaking down and wearing away existing features and creating new forms on the earth's surface.

Running water is the most effective erosional agent, smoothing jagged contours, carving valleys and deep canyons and washing countless tons of sediment into the seas each year. The rocks of the earth's crust are also under continual attack by weathering, a process of physical and chemical breakup

resulting from exposure to frost, rain and temperature changes.

Although wind is a minor geologic agent, it can transport fine-grained particles long distances. When winds carrying large amounts of sand lose energy and drop their load, dunes are formed. They may appear to be stationary, but many of them are actually in motion, with sand constantly being carried from their windward to their leeward sides. These remarkable features can even migrate up the sides of hills and in places bury forests and at times villages.

1

Eroded by water and mass wastage along vertical joints in the rock, the sculptured columns of Bryce Canyon are all that remain of a limestone plateau. The bright colors are caused by iron oxides in the rocks.

2

Delicate Arch, in Arches National Monument, Utah, has had a complicated history involving many geologic agents. Differential erosion was the principal process; the opening was cut through where the rock was less resistant to weathering and erosion.

2

3

3
*Dunes owe their shape mainly to the strength
and direction of the wind. Here, a Sahara salt
caravan passes a sea of dunes in the Tenere
Desert in the Niger Republic of Africa.*

Running water does more to change the face of the land than all other erosive agents combined. Most of the 420,000 cubic kilometers (100,000 cubic miles) of moisture absorbed by the atmosphere each year falls back into the oceans, but about a third reaches land as rain, snow or hail. Much of this water is evaporated or drains underground, but some 38 million billion liters (ten million billion gallons) run off the surface, cutting canyons and valleys and creating broad flood plains and deltas and carrying away millions of tons of sediment as it flows back to the sea.

Underground, water percolates through soil, joints and cavities in bedrock. Groundwater near the water table sometimes forms passages and chambers. These are drained when the water table falls, creating caverns.

Along coasts, storm waves may batter the shore with a force of 29,000 kilograms per square meter (three tons per square foot) and in places the average rate of coastal retreat is more than a meter a year. The action of running water, if not offset by periodic continental uplifts, would reduce the continents to low-lying plains in some 20 million years.

1
A river meanders through a valley in Alaska. Its shifting curves were developed by continual erosion on the outside of the river's bends and by deposition on the inside.
2
Carlsbad Cavern, in New Mexico, was formed during the Tertiary. Gradual solution of the limestone at the water table was followed by regional uplifts and downcutting, which caused the water table to fall. This exposed vast chambers over 200 m (650 feet) below the surface. Stalactites hanging from the ceiling,

and stalagmites rising from the floor of the
Chinese Theater, above, were created by
deposition from dripping water containing
calcium bicarbonate.

3
A sea arch on the island of Malta was formed
when waves and currents cut through both
sides of a narrow headland. When such arches
collapse, they form columns called sea stacks,
which stand isolated from the shore.

About a million years ago, much of the Northern Hemisphere was covered by vast ice sheets, in places more than a mile thick, burying all but the tops of the highest mountains. When the world's climate slowly warmed some 11,000 years ago, the ice caps retreated, revealing a landscape scraped bare of vegetation and reshaped by glaciers.

Today, glaciers cover less than a tenth of the land surface of the earth but they still serve as powerful geologic agents. They carve mountain peaks and bulldoze valleys, gouge out lake basins and spread rock waste over the surface

of the land. Glaciers also contain 80 per cent of the earth's freshwater which, if melted, would raise the level of the oceans by 60 m (200 feet). One type, continental glaciers, covers Greenland and most of Antarctica. Another type is the valley glacier—a river of ice that flows from snowfields high in the mountains. These are from 1500 m to 120 km (1 to 75 miles) long and reach their greatest size in southern Alaska.

A glacier's movement depends mainly on the steepness of the slope and the thickness of the ice. Some glaciers creep a few centimeters a

day; others move as much as 10 m per day. Like rivers, they flow fastest in the center and slower along their margins; and more slowly near the bottom than near the surface. These uneven movements create large crevasses, some measuring 15 m (50 feet) across and more than 40 m (130 feet) deep.

In temperate zones, a glacier flows downward until it reaches an elevation where melting and evaporation destroy it. Here the glacier terminates in a slope or wall from which meltwater pours, often colored milky blue or green by finely ground rock called glacial flour.

3

If the ice front is stationary, a moraine of rock debris accumulates. In polar regions, glaciers flow into the sea where large masses of ice break away from them and form icebergs.

1
The high sheen on granitic rocks in Yosemite National Park, California, testifies to the abrasive power of a moving glacier. Elsewhere, rock fragments embedded in ice have scraped across rocks leaving long, deep grooves.

2
Steele Glacier grinds its way down a mountainside in Kluane National Park, Yukon Territory, Canada. The dark stripes on its surface are moraines, rock fragments that fell onto the glacier as it scraped against the valley walls.

3
A field of boulders deposited by a glacier is strewn across a slope in Yosemite National Park. Called erratics, some of these rocks were picked up by the glacier scores of kilometers away, then dropped as the ice receded.

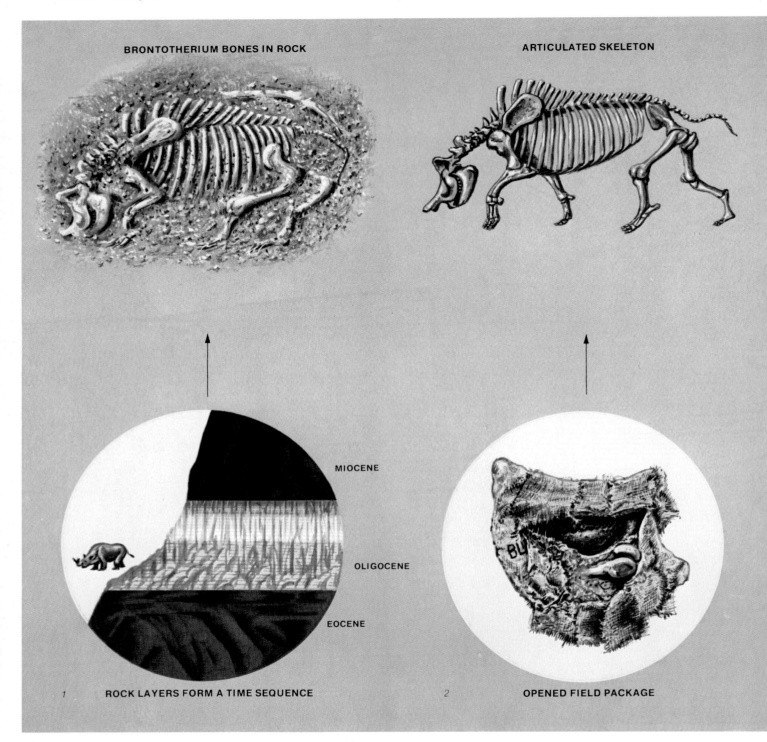

BRONTOTHERIUM BONES IN ROCK

ARTICULATED SKELETON

MIOCENE

OLIGOCENE

EOCENE

1   ROCK LAYERS FORM A TIME SEQUENCE

2   OPENED FIELD PACKAGE

An important function of the paleontologist is to reconstruct the life of the past from fossil fragments.

1

Each layer of rock, laid down as sediment over many millions of years, may contain a record of past plant and animal life as well as climatic conditions that prevailed long ago. The newest layers are at the top; deeper down, the strata hold more primitive forms of life. Using dating techniques based on radioactivity, scientists have determined the age of rock layers billions of years old.

As shown here, the remains of a titanothere, Brontotherium, *the largest land mammal ever found in North America, were uncovered in the badlands of South Dakota, one of the world's richest fossil-hunting grounds. The skeleton rested in a layer of soft clay and sandstone dating from the Oligocene Epoch, some 35 million years ago.*

2

After the bones have been excavated, cleaned, coated with a toughening agent, numbered and packed in field bags, the skeleton is assembled. The shape and size of the bones tell a great deal about the animal's way of life. Teeth, the hardest part of an animal's body, reveal what it ate and how it caught its prey.

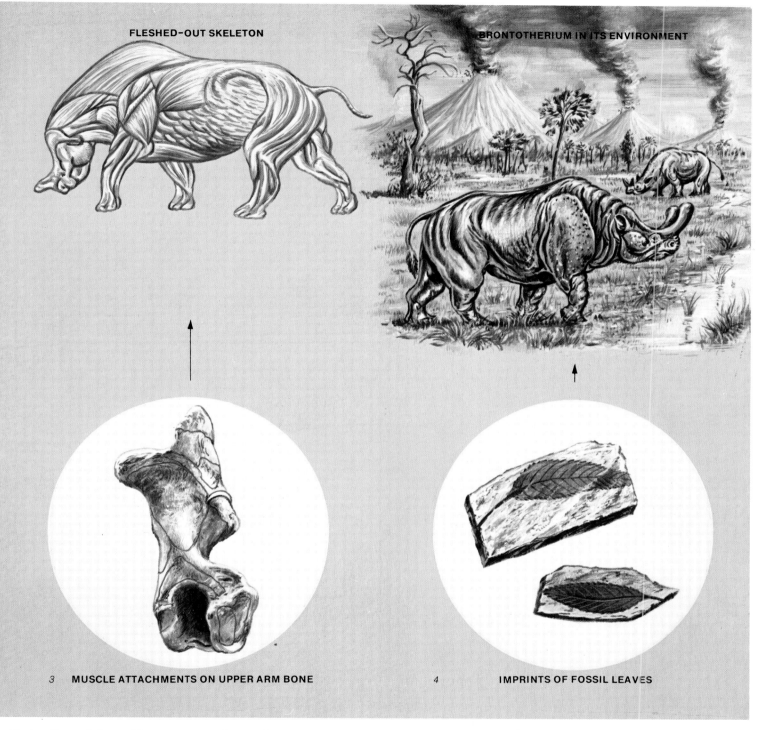

FLESHED-OUT SKELETON

BRONTOTHERIUM IN ITS ENVIRONMENT

3 **MUSCLE ATTACHMENTS ON UPPER ARM BONE**

4 **IMPRINTS OF FOSSIL LEAVES**

The location and shape of the eyes, ears and nose can be determined by sockets in the skull. The foot structure indicates how the animal moved. Joints reveal its size while the backbone and ribs tell something of its weight.

From the bones it is known that Brontotherium was a huge, cumbersome and heavy creature much like a modern rhinoceros. A large pronged horn projected from the front of its skull. It stood more than 2 m (7 feet) tall, was almost 5 m (15 feet) long and roamed across open plains browsing on leaves and grass.

3
The skeleton alone does not convey an accurate picture of an animal's appearance. Layers of muscle or fat and projections such as wattles or humps may considerably alter its body shape. But the skeleton provides clues to these features. In the drawing above, fine ridges and rough spots (indicated by red lines) show points of muscle attachments on a bone. Long spines on the shoulder vertebrae provided attachments for powerful muscles that held up the titanothere's head. A lack of other marks suggests that the animal was virtually hairless.

4
The final step is reconstructing the habitat in which Brontotherium lived. Fossil leaves near the site, carbonized in volcanic ash, indicate that the climate was semitropical 35 million years ago. Palm trees flourished and swampy plains covered the area, dotted here and there by active volcanoes. Why Brontotherium became extinct during this relatively mild period is still a mystery, but paleontologists speculate that its teeth remained unspecialized and could not withstand the wear of chewing the tough grasses of a changing habitat.

SYCAMORE LEAF; EOCENE EPOCH

MESOSAURUS; PERMIAN PERIOD

ALGAE; CRETACEOUS PERIOD

The word fossil (from the Latin meaning "dug up") originally referred to rocks and minerals as well as prehistoric plant and animal remains. The process of fossilization usually takes place so slowly that only the hard parts of an animal remain, such as bones, teeth and shells. Occasionally a rock layer consolidates more rapidly, permitting the preservation of the more delicate parts of an organism. A famous example of this is *Archaeopteryx,* the earliest known bird, identified by feathers perfectly preserved in the Jurassic limestone of a quarry in Bavaria.

The remains of animals have also been found in amber (fossil tree resin), ice and in frozen soil, but most fossils result from other types of preservation. The most common process, mineralization, occurs when minerals in water around a dead organism permeate its remains, eventually replacing its body structures. Plants are commonly fossilized by the process of carbonization, in which bacteria break down the organic compounds, leaving the plant remains as a film of carbon. Worm borings, and animal tracks, for example, are also part of the fossil record.

1

This Plantanus *leaf drifted to earth some 50 million years ago. Layers of silt compressed it into a thin film of carbon, preserving its outline and fragile veins.*

2

250 million years ago the small reptile Mesosaurus *abounded in streams and ponds, capturing fish with its needle-toothed jaws.*

3

A reef of algae, calcified in limestone during the Cretaceous Period some 100 million years ago, testifies to the warm climate of that time.

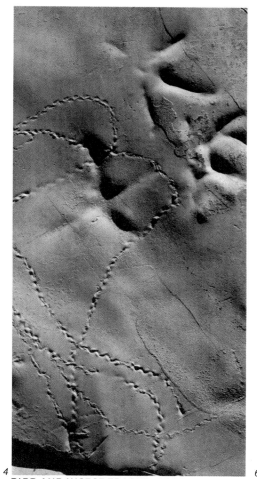

4
BIRD AND INSECT TRACKS; EOCENE EPOCH

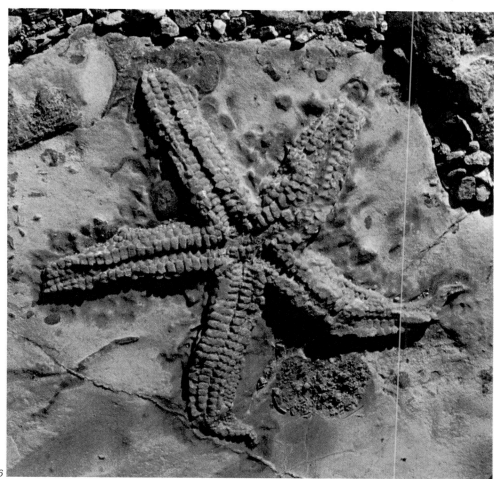

6
STARFISH; ORDOVICIAN PERIOD

5
SPIDER; MIOCENE EPOCH

7
BRACHIOPOD SHELLS; PENNSYLVANIAN PERIOD

4

An insect and a bird left delicate impressions of
their journey across a riverbank during the
Eocene, some 50 million years ago. As the mud
hardened, a mold of the tracks was formed.

5

Over 30,000 insects and their relatives, such as
the spider shown here, have been uncovered in
the shale at Florissant, Colorado, world famous
as the valley of fossil insects. The rocks were
formed from compressed volcanic ash that
settled at the bottom of a shallow lake 30
million years ago.

6

This cast of a rare fossil starfish was preserved
in the hills of southwest Utah. The rows of tube
feet along the grooves of each arm propelled
this echinoderm through tropical seas in the
Ordovician Period, over 450 million years ago.

7

During the Pennsylvanian Period, 300 million
years ago, the seas continued to support rich
invertebrate life, including the brachiopods
shown here. Much of the land was covered by
great swamps and deltas, surrounded and often
flooded by shallow water.

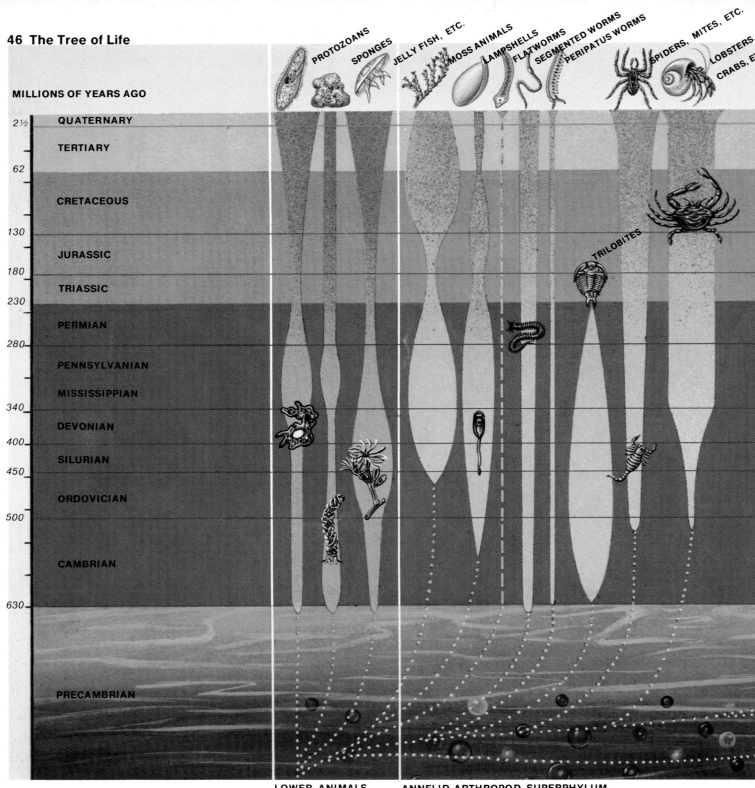

MILLIONS OF YEARS AGO

PROTOZOANS · SPONGES · JELLY FISH, ETC. · MOSS ANIMALS · LAMPSHELLS · FLATWORMS · SEGMENTED WORMS · PERIPATUS WORMS · SPIDERS, MITES, ETC. · LOBSTERS, CRABS, ET · TRILOBITES

2½	QUATERNARY
	TERTIARY
62	
	CRETACEOUS
130	
	JURASSIC
180	
	TRIASSIC
230	
	PERMIAN
280	
	PENNSYLVANIAN
	MISSISSIPPIAN
340	
	DEVONIAN
400	
	SILURIAN
450	
	ORDOVICIAN
500	
	CAMBRIAN
630	
	PRECAMBRIAN

LOWER ANIMALS   **ANNELID-ARTHROPOD SUPERPHYLUM**

Both the plant and animal kingdoms are divided into major groups called phyla based on their physical structure. The fossil record is incomplete, but enough is known to suggest the general evolution of the most important groups. In this chart, showing the relationships among major animal phyla, dotted lines mark breaks in the fossil record. The thickness of each column indicates numbers of species. Phyla in the same color range probably evolved from common ancestors.

The time-units at the left of the chart are based on the formation of certain rocks; these form a geologic calendar of the earth's history. During Precambrian time, the first life, probably algae, fungi and soft-bodied animals, developed. The Cambrian saw the appearance of brachiopods, mollusks, and crustaceans. Trilobites were common. During the Ordovician, corals and trilobites became abundant and the first fishes appeared. In Silurian time, plants and scorpionlike arthropods invaded the land.

The Devonian is known as the "Age of Fishes," but it also saw the appearance of the first insects and amphibians. During the Mississippian and Pennsylvanian, amphibians and flying insects multiplied, and the first reptiles appeared. In the Permian, reptiles and plants diversified. By the Triassic, dinosaurs and early mammals emerged.

During Jurassic time, aquatic reptiles dominated the seas, flying reptiles were common and the first birds appeared. Modern mammals became dominant and bony fishes were abundant. In the Quaternary, which continues into the present, four glacial advances spread ice over much of the Northern Hemisphere and man made his first appearance.

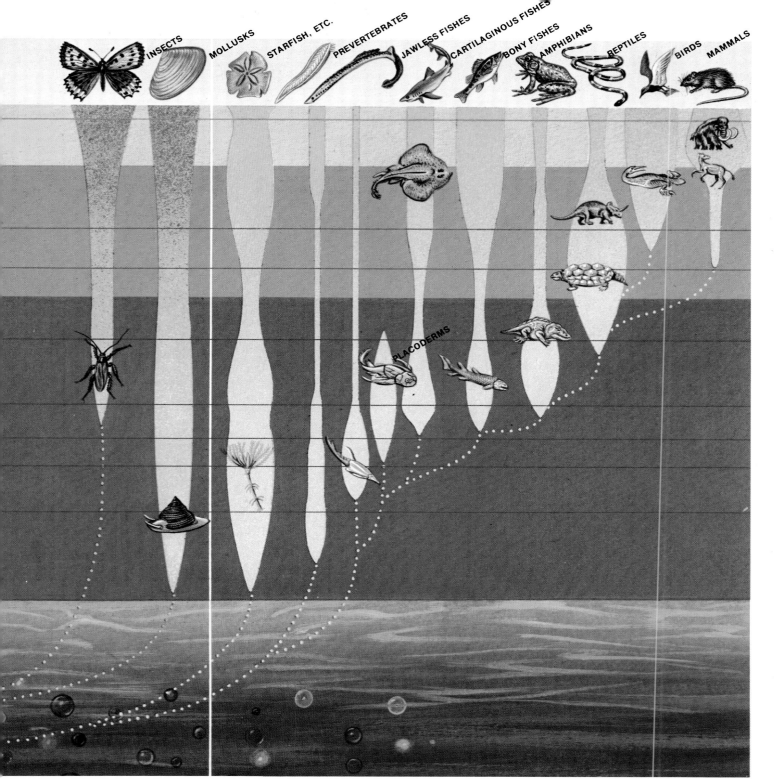

INSECTS MOLLUSKS STARFISH, ETC. PREVERTEBRATES JAWLESS FISHES CARTILAGINOUS FISHES BONY FISHES AMPHIBIANS REPTILES BIRDS MAMMALS

PLACODERMS

ECHINODERM-VERTEBRATE SUPERPHYLUM

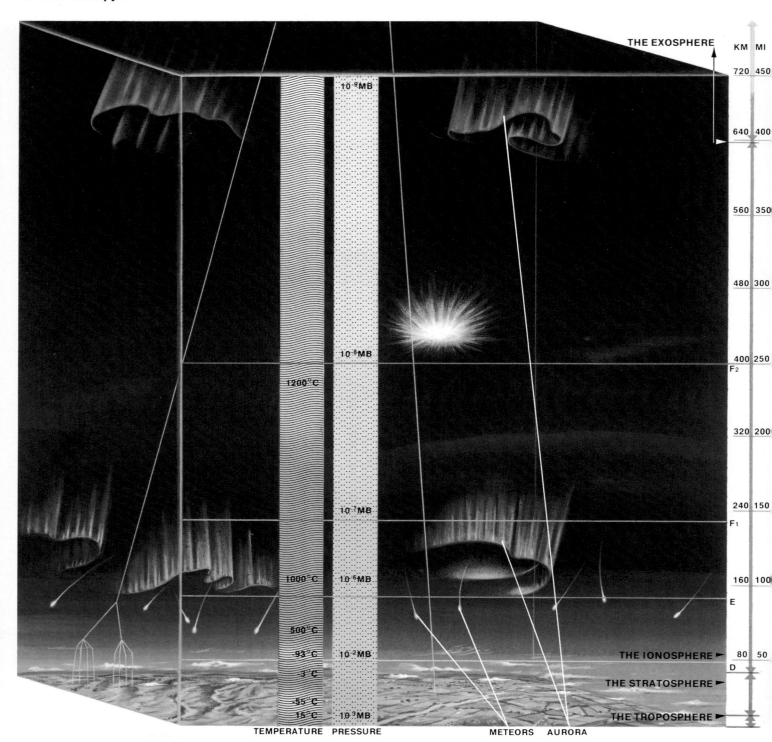

THE EXOSPHERE

	KM	MI
	720	450
	640	400
	560	350
	480	300
F2	400	250
	320	200
F1	240	150
E	160	100
THE IONOSPHERE	80	50
D		
THE STRATOSPHERE		
THE TROPOSPHERE		

10⁻⁹MB

10⁻⁸MB

1200°C

10⁻⁷MB

1000°C   10⁻⁶MB

500°C

93°C   10⁻²MB

3°C

-55°C

15°C   10⁻³MB

TEMPERATURE   PRESSURE   METEORS   AURORA

The earth's atmosphere, extending upward more than 1600 km (1000 miles), consists of four layers, each different in chemical composition, temperature and density. Just above the surface lies the troposphere, a shallow layer of air composed primarily of oxygen and nitrogen, where weather and clouds develop. Above it is the stratosphere, reaching a height of 50 km (30 miles), a stable, dry layer. The stratosphere contains a band of ozone, which acts as a shield against most short-wave radiation that bombards the earth from space. Farther up, the ionosphere contains electrically conductive layers (D, E, F, F₁ and F₂) capable of reflecting radio waves. A phenomenon of the ionosphere is the aurora borealis, which appears as flickering sheets of color at polar latitudes. The outermost layer of atmosphere, the exosphere, consists primarily of the lighter gases, helium and hydrogen. Little is known about this layer, which merges with the interplanetary medium.

The atmosphere protects the surface of the earth from many harmful elements. Meteors from outer space, hurtling at incredible speeds, are heated to incandescence by friction with air of the upper atmosphere and usually vaporize into gases before they come within 50 km (30 miles) of earth. Many forms of radiation cannot penetrate the atmosphere, but visible light, infrared radiation (or heat), some ultraviolet light and radio waves from space do reach the earth's surface.

A Z

A

Ablation. Loss of snow, *névé or ice from snowfields, glaciers, sea ice and similar features. It results from (1) the melting caused by rain and warm air, (2) *evaporation, (3) sublimation, or evaporation without a liquid phase, which is especially important in Antarctica where the air temperature is too cold for melting and the humidity is low, (4) deflation, which is the removal of snow by wind, (5) calving, which is the breaking off of icebergs and smaller masses of ice from glaciers that terminate in lakes or oceans.

Abrasion. Mechanical wearing away of *bedrock by particles transported by wind, running water, waves or glacial ice. The *erosion or *corrasion is accomplished by scratching or scouring the rock, leaving smoothed, polished surfaces in a process similar to sandpapering. Abrasion refers also to the wearing away of particles in transport so that they diminish in size as they move.

Abstraction. See Piracy.

Abyssal Hills. A low-lying, dome-shaped feature of deep ocean basins. It may rise as much as one kilometer above the surrounding *abyssal plain and may be up to several kilometers wide at the base. Such hills occur in all *oceanic basins but are most prevalent in the Pacific, where they occupy nearly half of the sea floor. They are also common along the outer margins of the *Mid-Atlantic Ridge. The origin of abyssal hills is in doubt. Most marine geologists and geophysicists favor the theory of volcanic origin, but it is also possible that the hills are compacted and lithified sedimentary material. Because the hills are covered with a veneer of sediment, it is difficult to sample the *bedrock that comprises them. (R.A.D.)

Abyssal Plain. The flat floor of the deep-sea environment; it is defined by some authorities as having a slope ratio of less than 1:1000. These plains are widespread in all *ocean basins as well as in many of the *mediterraneans such as the Gulf of Mexico, the Arctic Basin and the Mediterranean Sea. In the early days of ocean-floor exploration, it was thought that the entire oceanic basin beyond the continental margin was a flat, featureless plain. It is now known that abyssal plains comprise less than half of all ocean basins.

Originally the topography of the ocean floor probably consisted of extensive areas of low-lying *abyssal hills. The spreading out of sediment-laden *turbidity currents from the outer continental margin caused an accumulation of fine-grained sediment in the low areas between crests of abyssal hills. Eventually the sediment completely buried the abyssal hills over broad areas of the ocean floor, thus forming the abyssal plains. (R.A.D.)

Abyssal Zone. The ocean floor between depths of about 2000 to 6000 meters, including all areas below the base of the continental slope with the exception of oceanic trenches. It may also be defined as the zone, excluding trenches, where water temperature never exceeds 4°C (39°F).

The abyssal zone, with a total area of over 500 million square kilometers, covers more of the earth's surface than any other environment. Because of its great depth, conditions there are rather harsh, and organisms living there generally exhibit specific adaptations to it. Overall it is quite a uniform environment. The water is cold $(0° - 4°C; 32° - 39°F)$ with no seasonal variations, and *salinity is constant at normal marine concentrations, that is, 35 parts per thousand. Depending on the depth, the pressure is several hundred times that at sea level. Food and oxygen to support life are derived from the shallow zones of the ocean. Descending masses of cold water are the sole source of oxygen, and organic debris drifting down from the shallow depths is the prime source of food.

It was once believed that the floor was devoid of life. Although life at such depths is meager compared to that of continental-shelf areas, it is abundant even in the deepest abysses. While a great variety of organisms, including sponges, coelenterates, various types of worms, crustaceans, echinoderms, mollusks and vertebrates are found there, all are rather small and have a drab coloration. Nearly all modes of life are represented, but deposit feeders predominate.

Bottom sediments in the abyssal zone are uniformly very fine-grained. They may be terrigenous (land-derived) or *oozes of calcareous or siliceous tests (external coverings) of planktonic organisms. The rate of accumulation is extremely slow—about 1 cm per 1000 years. (R.A.D.)

Accessory Mineral. A mineral present in a rock in such a small amount that it may be disregarded in classifying the rock.

ACF Diagram. A triangular diagram used to represent mineral assemblages (metamorphic facies) produced by metamorphism of chemically diverse rocks. Several diagrams have been developed, each one indicating the minerals that will form as the result of subjecting a rock to a limited range of temperature and pressure. An equilateral triangle with the corners designated A, C and F represents most chemical components found in the majority of rocks. Corner A represents the percentage of Al_2O_3 plus Fe_2O_3 contained in the rock; C the percentage of CaO; and F, the percentage of FeO, MgO and MnO. Because other chemical constituents found in rocks are not shown on the diagram, certain allowances must be made before the rock composition can be plotted. One such requirement is that ACF diagrams may be used only for those rocks in which silica is sufficiently abundant to form not only all the silicate minerals present but also to form quartz. (J.W.S.)

Achondrite. A stony meteorite lacking chondrules, which are the small rounded bodies that characterize *chondrites. Achondrites are less common than chondrites and are composed of *hypersthene, *plagioclase, *diopside, *olivine and a small amount of nickel-iron.

Acoustics, Underwater. The transmission of sound in water. Sound travels much faster in fresh water than in the atmosphere (1445 m per second as compared to 333 m per second in air) but in salt water it increases up to 1550 m per second because the velocity is affected by salinity, temperature and pressure. Increase in these three variables causes a corresponding increase in sound velocity, which reaches its maximum in the *thermocline.

Behavior of sound waves in water is important in the use of SONAR (Sound Navigation and Ranging). If water is of uniform temperature, the velocity increases with depth, resulting in refraction or bending of sound waves toward the surface. Because of the thermocline, however, there is a zone where sound waves are absent due to the decrease in velocity in both directions away from the sound source. In naval warfare this feature enables submarines, which commonly move along the thermocline, to escape detection.

Perhaps the most widely used aspect of underwater acoustics is in the precise measurement of depth. Most precision depth recorders are based on the reflection of sound waves from the ocean floor. The type of bottom is therefore important in obtaining good reflection: a smooth, firm ocean floor is best; a soft bottom or an irregular surface tends to absorb or scatter sound waves. (R.A.D.)

Actinolite. A mineral, hydrous calcium, magnesium and iron silicate. It is a green, fibrous *amphibole commonly found in crystalline *schists in association with epidote and chlorite. In chemical composition actinolite is similar to tremolite,

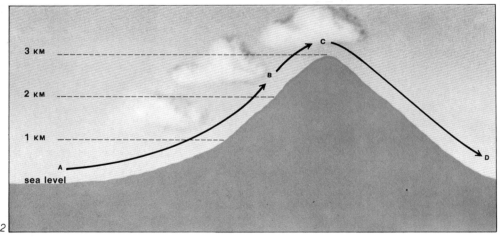

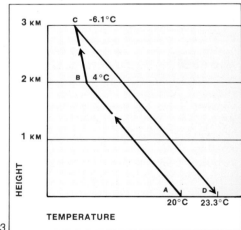

Adiabatic Process

1 As expanding moist air rises, it cools and condenses to form clouds over the Alps at La Fouly, Switzerland.

2 Points of adiabatic temperature change as air rises over and descends down a mountain.

3 As an air mass containing water vapor ascends one side of a mountain, the air cools at the dry adiabatic rate until condensation of water vapor occurs, as shown in the temperature change between points A and B. The rate of drop is 9.8°C per km, so over a rise of 2 km the air mass drops from 20°C to 4°C. When condensation occurs (at point B), the air, as it continues to ascend, cools at the slower saturated adiabatic rate of 6.5°C per km. Thus between points B and C, a distance of 1 km, the air mass drops another 6.5°C, from 4°C to -6.1°C. At point C the air begins to descend the lee side of the mountain, and its temperature rises.

Precipitation often occurs at the top of the mountain, causing the air temperature to rise (again at the dry adiabatic rate of 9.8°C per km) as the air mass sinks. Thus from point C to point D, a distance of 3 km, the temperature rises 29.4°C; the total drop was 26.1°C. The total temperature change has been a rise of 3.3°C.

the calcium magnesium silicate, but contains at least 2 percent iron substituting for magnesium. Tremolite is white; actinolite is green because of the iron. The nephrite type of *jade is a compact variety of tremolite-actinolite. See Mineral Properties; Appendix 3.

Acute Bisectrix. In *crystallography, the optical direction in a crystal that bisects the acute angle between the optic axes. This angle is referred to as 2V, the axial angle or the optic angle.

Adamantine Luster. Having the luster of a diamond.

Adiabatic Process. A change of state in which there is no transfer of heat or mass between a system and its environment. Adiabatic changes of state account for a number of important atmospheric phenomena. For example, in the lower atmosphere the typical decrease of temperature with increasing elevation is largely the result of adiabatic mixing of the air. Most cooling that leads to condensation, clouds, and precipitation is caused by the cooling of rising air; clear, dry weather is usually associated with the warming and drying out of sinking air. This cooling of rising air and the warming of sinking air are largely due to adiabatic expansion and adiabatic compression of the air, respectively. Heating and cooling that involve heat transfer, by processes such as radiation and conduction, are termed diabatic processes.

When a confined gas is compressed and there is no loss or gain of heat across the boundary of the container, the gas undergoes an adiabatic process and is warmed. The additional heat energy is equivalent to the energy expended in compressing the gas. A familiar example of heating by compression is the rise in air temperature in the chamber of a bicycle pump when the pump handle is worked energetically. When a gas is allowed to expand without gain or loss of heat in the system, it is cooled adiabatically; the heat energy transformed is equivalent to the work done by the gas in expanding against the pressure of the environment. The cool air rushing from the valve of a bicycle tire is an indication of cooling due to expansion.

Pressure in the atmosphere is proportional to the total mass of air above a given level. Like the density, therefore, the pressure decreases with increasing elevation. Consequently, when a mass of air rises, it expands under the lower pressure and cools. Conversely, when air sinks it is compressed by the additional weight of the atmosphere through which it descends, and

is adiabatically warmed. Air that contains no liquid water, rising through the atmosphere, cools at the dry-adiabatic lapse rate of 9.8°C per kilogram; sinking, it warms at the same rate. However, if rising air remains saturated, and condensation of water vapor continues within the rising mass, it cools at a rate less than the dry-adiabatic rate, for the latent heat of the condensed water vapor is added to the system. When sinking air is kept at saturation by the evaporation of liquid water it warms at the lesser rate.

When a deep layer of air, containing water vapor, is forced rapidly over a mountain range by large-scale circulation, the air cools at the dry-adiabatic lapse rate until condensation occurs, and afterward at the saturated-adiabatic lapse rate. If precipitation occurs, the air descends the lee side of the range with less water substance than it had during its ascent. Warming at the dry-adiabatic lapse rate as it descends, it often reaches lower elevations as a warm, dry wind, characteristic of many mountain areas. In the Alps, and to most meteorologists, it is known as the *foehn wind; in the Rocky Mountains it is called the chinook. Since the air can lose heat through radiation, foehn winds are usually associated with cyclones that force a deep layer of air over a mountain range in a relatively short time. In such a large mass of air, radiational cooling is small in comparison with the heating of the air during its rapid descent. (M.F.H.)

Adularia. A colorless, transparent to translucent variety of orthoclase *feldspar. It is usually found in veins in crystals which, although they belong to the *monoclinic system, appear to have orthorhombic symmetry.

Aerolite. A stony *meteorite composed principally of silicate minerals.

Aftershock. Earth tremors that follow the main shock of an earthquake. Aftershocks generally originate from the same location within the earth as that of the major earthquake. There are usually a series of aftershocks following large earthquakes, each being less intense than the former. The duration of aftershocks may be days or even months.

Agate. A fine-grained variety of *quartz composed of alternating layers of *chalcedony of different colors and transparency. Agate usually occurs as cavity fillings, with the individual bands arranged concentrically to conform to the walls of the cavity. Cross sections of agate nodules deposited in irregular cavities may show

unusual features, and they have been named accordingly: fortification agate, landscape agate, star agate, eye agate and brecciated agate. The natural color of agate ranges from white through various shades of gray to black, but may less commonly be pale shades of red, brown, blue, green or lavender. Moss agate is a gray to white translucent chalcedony that has incorporated black manganese oxide or red-to-brown iron oxides in forms resembling moss.

Agate is widely used in jewelry and in ornamental objects; much of it when so used is artificially colored. Although the crystal structure is identical to that of coarsely crystalline quartz, the fibrous nature of the chalcedony permits it to absorb dyes. The agate is first cut to the desired shape and then immersed in various chemical dyes or pigmenting solutions. The solutions permeate the agate throughout, giving it a vivid and permanent color, although the absorption may not be uniform, since some bands are more porous than others. (C.S.H.)

Age. A geologic *time unit that is a subdivision of an *epoch and the smallest unit of time normally employed. The sequence of rocks formed during an age is called a *stage.

Agglomerate. A *pyroclastic rock formed by the consolidation of volcanic detritus that contains many rounded fragments greater than 32 mm (1¼ inches) in diameter, together with a finer-grained matrix of *tuff. Agglomerates are not well stratified and are common in *volcanic necks. The well-rounded fragments are volcanic *bombs; less well-developed rounding is probably due to attrition of angular fragments during eruption. Since large fragments cannot be thrown very far from volcanic vents, agglomerates have a limited distribution. All gradations exist between agglomerates with many rounded fragments and *volcanic breccias with many angular fragments. (R.L.N.)

Agglutinate. A *pyroclastic rock composed essentially of fragments formed from clots of liquid lava ejected from a volcanic vent. The clots are sufficiently plastic on landing so that later clots adhere to them as they solidify. An agglutinate is commonly formed by *lava fountains and curtains. See Spatter Cone.

Aggradation (or Alluviation). Deposition of material carried in transport. When the transporting medium is no longer able to move its debris load, it deposits the particles in a channel, at the foot of a slope, on a *floodplain or in a lake or an ocean. Causes include anything that lowers the stream gradient, increases the supply or caliber of load, or decreases the discharge. Rates of aggradation vary. The Nile River aggrades its valley 9 mm (0.35 inches) annually whereas the Tigris-Euphrates Valley receives annual deposits averaging 18 mm (0.7 inches). In the semiarid western United States, rates of deposition may reach 40 mm (1½ inches) a year. Man may influence the rate of aggradation: for example, hydraulic mining in California has doubled the rate of aggradation of local stream channels. (M. M.)

A-horizon. See Horizons, A-, B-, and C-.

Air. The mixture of gases that makes up the earth's atmosphere. Pure air is assumed to be free of gaseous and other contaminants and also free of suspended dust or water particles, either liquid or solid. Dry air is air with all water vapor removed. Certain gases are observed in relatively constant amounts in clean, dry air near sea level. In addition to these constant or "permanent" gases of the atmosphere, there are variable ones. Some of the variable gases occur naturally, but these may also be regarded as contaminants when they are introduced into the atmosphere by industrial processes or other human activities. The most important variable gases, meteorologically, are water vapor, carbon dioxide and ozone.

The abundances of elements in our solar system indicate that the earth's primitive atmosphere must have been quite different from that of the present, consisting largely of hydrogen, helium and water vapor. The lighter gases, helium and hydrogen, were lost—as they are still being lost from the earth's atmosphere—by diffusing upward to regions where the air density is so low that a certain percentage of the molecules fail to collide with others and can escape the earth's gravitational field. Large amounts of water vapor and carbon dioxide are believed to have entered the atmosphere by degassing through magmatic and metamorphic action on rocks in the earth's crust and interior, that is, as "excess volatiles." Most of the water vapor condensed to form the oceans, while the carbon dioxide was used in life processes and deposited as limestone. The mass of carbonates in limestone is so great that it suggests that the earth's primitive atmosphere must have been quite dense. The large amount of nitrogen in the present atmosphere has not been completely accounted for, but presumably nitrogen also entered the atmosphere through degassing of the rocks. Being inert, it has

remained in elemental form; and being heavy, it has not been lost by diffusion. The large amount of oxygen in the earth's present atmosphere has been explained by some scientists as resulting from photolysis—the photochemical decomposition of water vapor; others believe that this process accounted only for the first traces of oxygen, and that the present abundance is the result of plant photosynthesis.

Although the gases of the atmosphere have thus changed quite radically during geologic time, some may be regarded, for all practical purposes, as "permanent." These include nitrogen, oxygen, argon, neon, helium, trypton, xenon, hydrogen and nitrous oxide. They are observed in almost exactly the same proportions all over the earth and up to an altitude of about 100 km. This constant proportion of the fixed gases indicates that mechanical mixing of the air by turbulent motion occurs up to that level. In the absence of mixing, the heavier gases would decrease more rapidly with height than the lighter ones, resulting in a layering effect. Satellite observations suggest that stratification of the gases begins above 100 km. If it is assumed that all of the atmosphere's oxygen was produced over a period of 100 million years, the change during 100 years would be only one millionth of the present value; recent measurements have shown that the oxygen content of the atmosphere has indeed changed by less than 0.010 percent during the past 60 years. On the other hand, the gases that create air pollution problems are extremely variable; usually expressed in parts per million, they occur in quite small quantities compared with the permanent, nonharmful gases. Those of greatest concern—carbon monoxide, sulfur dioxide, nitric oxide, nitrogen dioxide and ozone—fluctuate rapidly in time and space, varying by more than 10 percent in an hour and changing by orders of magnitude over a few city blocks. (M. F. H.)

Air Pollution. The presence in the atmosphere of substances harmful to living things or damaging to property. Although the term often refers only to substances in the air in concentrations larger than occur naturally, it is estimated that more than one-fourth of the suspended particles in the atmosphere are produced naturally. Natural contaminants include pollens, spores, terpenes (hydrocarbons produced and emitted by trees and plants), salt from sea spray, dust from blowing soil, smoke from forest fires ignited by lightning or other natural means and gases and dust from volcanoes. However, pollutants from natural sources in quantities large enough to cause harmful consequences tend to oc-

cur only locally or temporarily. By far the primary cause of air pollution is the incomplete combustion of fuels of all kinds and the consequent introduction of gaseous, solid or liquid impurities into the earth's atmosphere.

Prior to the recent widespread replacement of coal by petroleum derivatives for heating, transportation and power generation, the main combustion products causing serious air pollution problems were soot, sulfur dioxide and fly ash. Because water vapor condenses readily on coal-smoke particles at humidities of appreciably less than 100 percent, a combination of smoke and fog, termed smog, was common in many industrialized areas until the 1960s, when control measures reduced the emission of the unburned products of coal combustion. The substitution of gas and oil as fuel led temporarily to cleaner air—notably in Pittsburgh, Pa., and in London, England—but their wider use, combined with the proliferation of automobiles, produced a different kind of air pollution—photochemical smog. This was first observed in the 1950s and studied intensively in the Los Angeles area, where it was found that the nitric oxide and hydrocarbons in automobile exhaust were important contributors. In sunlight, nitric oxide unites with atmospheric oxygen to produce nitrogen dioxide, a noxious gas that gives this pollution its typical brownish color. In addition to nitrogen dioxide, the photochemically produced combinations of nitric oxide with gaseous hydrocarbons form other secondary pollutants, including ozone. These photochemical oxidants cause eye and lung irritation; damage to buildings, vegetation and fabrics; offensive odors; and thick haze.

The percentages of pollutants, by weight, introduced into the atmosphere by various sources in the United States have been estimated as follows: transportation, 42 percent; fuel combustion in stationary sources, 21 percent; industrial processes, 14 percent; forest fires, 8 percent; solid waste disposal, 5 percent; and miscellaneous sources, 10 percent. The yearly emissions in the United States are estimated to total over 200 million tons; these include carbon monoxide, 47 percent; sulfur oxides, 15 percent; hydrocarbons, 15 percent; particulates, 13 percent; and nitrogen oxides, 10 percent.

The length of time that pollutants of this sort remain in the atmosphere is at most a few months. If they were immediately mixed uniformly throughout the world's *troposphere, air pollution would be a matter of small concern. However, the time required for dilution of pollutants is not only considerable but quite variable; it varies markedly, depending on meteorological conditions and the topography in the vicinity of the source. Meteorologically, diffusion in the atmosphere, and the consequent dilution of contaminants, is related to the speed and steadiness of the wind and to the hydrostatic stability—that is, to the susceptibility of the atmosphere to vertical mixing. High, gusty winds that fluctuate rapidly in speed and direction efficiently mix the air laterally, while calm conditions or light, steady winds allow contaminants to accumulate. An unstable air mass—one in which the temperature decreases rapidly with height above the ground—permits upward and downward motion of the air and mixing of contaminants through a deep layer of the atmosphere. A stable air mass—one in which the temperature decreases less rapidly, or even increases with height—suppresses vertical transport and causes pollutants to be confined to a shallow layer near the ground. Normally, the atmosphere is most stable near sunrise, when the ground is coldest, and least stable in the afternoon, when it is warmest; wind speed and turbulence follow the same diurnal pattern—lightest at sunrise and at a maximum after noon. Day-to-day conditions vary widely, however, depending on the large-scale weather situation. *Anticyclonic conditions—marked by relatively high pressure, subsiding air, fair weather and the presence of a temperature inversion (an increase in temperature with height) some distance above the ground—combined with a cold underlying surface such as water are favorable to air pollution. At middle latitudes, these weather conditions are most frequent in the fall season. The opposite set of conditions, favoring strong mixing of the atmosphere and precipitation "scavenging" or the washout of pollutants, are most typical of spring. Topographical features that obstruct the mixing of air, such as a deep valley ringed by mountains in contrast to an exposed location, also exert strong control over air pollution. (M.F.H.)

Alabaster. A compact, fine-grained variety of *gypsum used for carving and sculpture.

Albedo. The fraction (or percentage) of the incident or incoming solar radiation reflected by a body. The fraction reflected by the earth has not been firmly established, but recent calculations based on satellite observations indicate a value of about 0.30.

Reflection from clouds accounts for more than half the earth's albedo, the remainder being reflected by the cloudless atmosphere and the earth's surface. The

Air Pollution
Smog fills California's Fontana-San Bernardino basin when automobile exhaust and other pollutants are trapped close to the earth's surface by a temperature inversion.

Allochthon

1 Round Hill, in southeast New York, is the eroded remnant of an overthrust block pushed from its original location during the formation of the Taconic Mountains some 480 million years ago.

Albedo

2 On an average day only some 30 percent of the solar radiation striking the earth is reflected back into space by clouds, atmospheric gases and particles, and the earth's surface. The remaining energy is absorbed by the earth's surface and converted into heat.

Alluvial Fan

3 This fan-shaped apron of silt, sand and gravel was deposited when a stream spilled out from the canyon and spread down into Death Valley, Calif. The dark line crossing the fan is a highway.

10% REFLECTED BACK BY IMPURITIES IN THE AIR

INCOMING SOLAR ENERGY 100%

20% REFLECTED BACK BY CLOUDS

30% REFLECTED BACK IN TOTAL (ALBEDO)

20% ABSORBED BY AIR AND DUST

50% ABSORBED BY THE GROUND

albedo varies in different portions of the solar spectrum, that is, in the visible, the infrared and the ultraviolet. Representative visible-wavelength values of the albedo for surface features are: bare ground, 10–20 percent; green forest, 3–10 percent; dry sand, 18 percent; fresh snow, 80–85 percent; fields, 3–25 percent. The reflection from old snow may be much less than for fresh snow, a factor that must be considered in accounting for the melting of glaciers and snow packs. The albedo of clouds varies greatly with the thickness and type of cloud, ranging from negligible reflection to about 80 percent.

The albedo of the lunar surface varies with its topography. The lunar seas (maria) have a low albedo, which means their surfaces have a low reflectivity and thus appear dark. Lunar highlands have a high albedo or reflectivity and seem light. (M.F.H.)

Albite (from the Latin *albus*, meaning "white"). A mineral, sodium aluminum silicate, a common rock-forming *feldspar that is usually white. It is the sodium end-member of the *plagioclase feldspar series. See Mineral Properties.

Albite Twinning. A type of twinning characteristic of all *plagioclase feldspars in which the *b* crystallographic axis is the twin axis. It is evidenced by the presence of striations on one crystal face or corresponding cleavage surfaces. See Twin Crystal.

Albitization. A process by which hydrothermal solution transforms all or part of a rock's minerals to *albite feldspar.

Alexandrite. A gem variety of *chrysoberyl that has the unusual property of appearing emerald green by daylight and red by artificial light. Discovered in the Ural Mountains in 1833, it was named after the czarevitch who later became Alexander II of Russia.

Alkali Feldspar. Feldspar rich in either potassium or sodium, or a member of a series in which both of these elements are present. Alkali feldspars include *microcline, *orthoclase, *sanidine, *anorthoclase and *albite.

Alkalic Igneous Rock. A rock in which the ratio of alkalies ($K_2O + Na_2O$) to silica is greater than 1 : 6. It usually contains sodium-rich *pyroxenes, *amphiboles and/or *feldspathoids.

Allochthon. A mass of rock that has been moved a large distance from its original site of formation due to forces acting on the earth's *crust; used in comparison with

*autochthon. An allochthon is generally formed when rock is thrust over adjacent sections of the crust or is moved by folding. The Taconic Mountains in eastern New York State are allochthonous, having been thrust from western New England in a southeast direction to their present position. (J.W.S.)

Alluvial Fan. Material deposited by flowing water where it debouches from a steep canyon onto a level area. The sudden change in gradient and the spreading of water over the flat results in a loss of the water's capacity to transport its debris load; the dropped load builds up and outward from the mouth in a low fan or cone shape. As the stream moves from the canyon over the fan, it may change to form *distributaries or *braided streams, or the water may sink into the fan to emerge downslope as a spring. Often the main channel is entrenched into the fan surface. Generally the stream flows only during floods, and it is then so heavily loaded with debris that it becomes a *mudflow or a debris flow. Thus the fan consists of a variety of deposits: braided channel gravels and sands, mudflows and debris flow gravels.

The size and slope of alluvial fans are controlled by the size of the *watershed, the character of the rocks in the basin, the climate (and thus the vegetation and hydrology) of the region, the slope of the valley sides and the gradient of the stream, the tectonics of the region and the geometry of adjacent fans and basin of deposition. Fans are characteristic of arid or semiarid regions but are sometimes found in humid regions. (M.M.)

Alluvium. Material laid down by rivers. The deposits consist of clay, silt, sand or gravel and are generally but not always stratified. They have been classified as channel fill deposits, vertical accretion deposits, floodplain splays, lateral accretion deposits, lag deposits and colluvial deposits. They form features such as point bars, channel bars, *floodplains, *levees, cones, fans, *bahadas, *deltas and terraces. Alluvium from glacial meltwater streams forms *outwash plains, *valley trains, *kames and *eskers. (M.M.)

Almandite. A mineral, an iron aluminum *garnet, characteristically found in mica schists. When transparent, its deep red to violet-red color makes it an attractive gemstone, called carbuncle since Biblical times. Almandite of gem quality comes from many parts of the world, but the principal source is the stream gravels of Ceylon and India. See Mineral Properties.

Alunite (or Alumstone). A mineral, hydrous potassium aluminum sulfate, used in producing alum. It is usually formed through the reaction of sulfuric acid solutions on rock that is rich in potassium *feldspar. In the common massive form alunite is difficult to distinguish from *limestone, *dolomite and other massive minerals such as *anhydrite and *magnesite.

Amazonite (or Amazonstone). A green variety of microcline *feldspar used as an ornamental stone and gem material because of its pleasing color. It is found widely, but of particular note in the United States at Pikes Peak, Colo., and Amelia Court House, Va., as well as in Madagascar, Norway and the Ural Mountains.

Amber. A type of fossil resin consisting largely of carbon, hydrogen and oxygen. A hard, somewhat brittle compound, amber has a resinous luster, a hardness of 2.25, and is yellow to red in color. Amber acquires an electric charge when rubbed, and its Greek name is *elektron*—the origin of the word "electricity."

The early Greeks and Romans used amber for jewelry and to ward off disease and evil spirits. It is still made into jewelry, ornaments, cigarette holders and pipe mouthpieces. Much amber has come from *Tertiary rocks along the coast of the Baltic Sea; other deposits occur in Denmark, Sweden, Britain and Italy. Some deposits of amber yield specimens that contain fossilized remains of extinct insects and spiders. These ancient creatures were trapped in the sticky antiseptic resin that oozed from cone-bearing trees in the geologic past. The resin hardened with the passing of time, leaving the animals undamaged and in a perfect state of preservation. Even fossil spider silk has been preserved in this manner. (W.H.M.)

Amblygonite. A mineral, lithium aluminum fluophosphate, found in granite pegmatites associated with *spodumene, *tourmaline, *lepidolite and *apatite. It has been used as a source of lithium.

Amethyst. A purple gem variety of *quartz. The color, which apparently results from small amounts of ferric iron, is rarely homogenous and is usually distributed in sheets parallel to the crystal faces of the rhombohedron. When heated, amethyst may be converted to citrine, a yellow-brown variety of quartz.

Ammonoid. A *cephalopod similar to the modern *Nautilus but now extinct and only known from the fossil record. Ammonoids branched off from the nautiloids in early

1

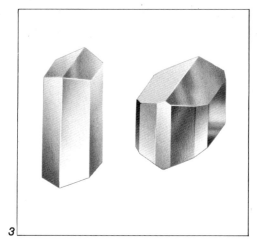

3

Ammonoid

1 Ceratites *was a 5-cm-wide (2 inch)*
cephalopod that lived during the Middle
Triassic Period, some 200 million years
ago.

2 *These specimens of* Aegoceras
Capricornus, *from the Lower Jurassic,*
were found in Yorkshire, England.

Amphibole

3 *The most common member of this*
group of complex hydrous silicates is
hornblende, a dark green to glassy-
black mineral found in igneous rocks.

Amphidromic Point

4 *Tides and the earth's rotation tilt the*
surface of a water basin, such as an
ocean or bay. The motionless center
around which the water circulates is the
amphidromic point.

2

Devonian time, becoming extremely varied and abundant during the Mesozoic Era. However, like the dinosaurs that were so numerous at the same time, the ammonoids were extinct by the end of the Cretaceous Period. Both ammonoids and nautiloids have shells that may be conical, or partially or tightly coiled. But these two groups of "headfooted" animals differed considerably in the internal structure of their shells. Some Cretaceous ammonoids coiled upwards into a spiral, and their shells closely resembled those of certain Cretaceous *gastropods. (W.H.M.)

Amphibole. A group of minerals found as common ferromagnesian constituents in both igneous and metamorphic rocks. They are similar in chemical composition and appearance to the *pyroxenes, another rock-forming mineral group. The two groups can be distinguished by their prismatic *cleavage: the amphibole cleavages intersect at angles of 56° and 124°; the pyroxenes, at 87° and 93°.

The amphiboles are essentially calcium, iron and magnesium aluminum silicates, with some included water. There is considerable chemical diversity among members of the group. It is difficult to distinguish between them without X-ray or optical tests. Of the several amphiboles, only anthophyllite is *orthorhombic. The remaining are *monoclinic and include: the cummingtonite-grunerite series, the tremolite-actinolite series, the glaucophane-riebeckite series, and hornblende.

Amphibolite. A metamorphic rock consisting mainly of the minerals hornblende and plagioclase, with quartz sometimes present in small amounts. The minerals may be segregated into bands, producing a laminated structure; the elongated hornblende grains often form parallel planes that break off easily. Amphibolites are formed from igneous rocks rich in magnesium and iron, such as *basalt (traprock) or *gabbro. When buried deep within the earth's crust and subjected to temperatures from 450° to 700°C (842° to 1292°F) the mineralogic components of the original basaltic rocks react to form hornblende; a more sodic plagioclase; epidote; and chlorite minerals that have greater stability under the conditions of relatively lower temperature and pressure in the presence of water. Amphibolites may also form from limestones containing small amounts of mud and silt. (J.W.S.)

Amphidromic Point. The focus around which tidal crests within an *ocean basin rotate and from which several co-tidal lines or lines connecting points of equal tidal

level radiate. The tidal range at each of the amphidromes is zero. Tidal circulation of this type is a result of the earth's rotation, which causes the lines to rotate counterclockwise in the Northern Hemisphere and clockwise in the Southern Hemisphere. Excellent examples of this type of tidal circulation take place in *estuaries, where the *Coriolis effect and tidal currents combine to produce a tilting of the water surface in the estuary. This tilt moves counterclockwise in the Northern Hemisphere around an amphidromic point during a tidal cycle and in the opposite direction in the Southern Hemisphere. (R.A.D.)

Amygdule (or Amygdale). A gas cavity or steam hole, called a vesicle, in glassy or fine-grained igneous rock, that is completely or partly filled with one or more secondary minerals, and found in *extrusive or shallow *intrusive rocks. It is spherical, ellipsoidal or tubular and generally less than an inch in diameter. An amygdule is formed by precipitation from solutions derived from cooling lava, and by precipitation from groundwater long after solidification of the vesicular rock. The deposition starts at the walls of the vesicle and proceeds inward. Some are banded because of changes in the minerals that were deposited. Simple amygdules are formed by one mineral, composite amygdules usually by two. Common amygdule-forming minerals are agate, chalcedony, quartz, calcite, epidote and the zeolites. The basaltic lava flows on Keweenaw Point, Mich., contain copper amygdules and are copper ores. Volcanic rocks (generally basalts or andesites) containing numerous amygdules are called amygdaloids. (R.L.N.)

Analcime. A mineral, hydrous sodium aluminum silicate, a *zeolite. It is commonly found in white to colorless trapezohedral crystals associated with other zeolites in cavities in basaltic rocks. Analcime resembles *leucite in color and crystal form but can usually be distinguished by its free-growing crystals. Leucite as a rock-forming mineral is embedded in a fine-grained matrix. In places analcime is an alteration of volcanic glass, and it has also been found as a primary constituent of igneous rocks. See Mineral Properties.

Andalusite. A mineral, an aluminum silicate with the same chemical composition as *kyanite and *sillimanite. It is a metamorphic mineral formed by both contact and regional *metamorphism. Andalusite is most commonly found in rough, elongated crystals that in many places have altered to *mica. Unaltered andalusite was once mined in large quan-

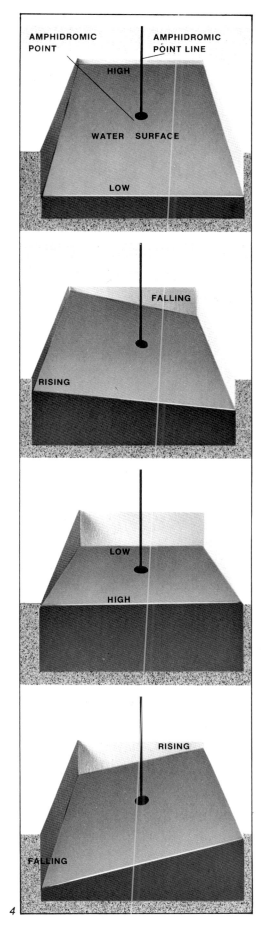

Andalusite

1 *Cross sections of chiastolite, a variety of andalusite, reveal the dark cross formed by carbon impurities trapped as the crystal grew. Some peoples believe these crystals have magic powers and wear them as amulets.*

tities, principally in California, for the manufacture of spark plugs and other high-quality porcelains. In Minas Gerais, Brazil, reddish-brown pebbles of andalusite found in stream gravels have been cut as gemstones. They are of interest because of their strong dichroism, whereby a given stone appears green or red depending on the direction of the transmitted light. Chiastolite is a variety of andalusite with symmetrically arranged carbonaceous inclusions that appear as a black cross in a cross section of an elongated crystal. Sections cut from such crystals have been worn as amulets because of their presumed magical powers. See Mineral Properties; Appendix 3. (C.S.H.)

Andesine. A mineral, one of the *plagioclase feldspars, the major constituent of the rocks *diorite and *andesite. Its composition in terms of percentages of the sodium and calcium end-members of the plagioclase series, albite (ab) and anorthite (an), varies from 70ab, 30an to 50ab, 50an.

Andesite. A *volcanic rock composed essentially of *plagioclase feldspar (oligoclase-andesine) and one or more dark minerals, frequently present as *phenocrysts, which may be hornblende, augite or biotite. Andesite is named for the Andes Mountains of South America where it is abundant.

Andesite Line. A line separating the occurrence of igneous rocks of basaltic composition from those of andesitic composition. When traced on the surface of a globe the line extends from offshore Alaska southwest to Japan and then south on the oceanward side of the Marianas and Palau islands, the Bismarck Archipelago, and the Fiji and Tonga islands. The line is less clearly traced on the eastern margin of the Pacific Ocean but probably runs along the coast of North and South America.

Andradite. A mineral, a calcium iron *garnet characteristically occurring in metamorphosed, impure, calcareous sedimentary rocks. It may be various shades of yellow, green or brown. Demantoid is a transparent green variety with a brilliant luster, highly prized as a gem. See Mineral Properties.

Angiosperm. A flowering plant, typically with broad leaves and rather complex seeds. Angiosperms comprise such familiar forms as grasses, hardwoods, vegetables, flowers and fruit; they are the most common and widely distributed of all land plants. There are also a large number of fossil forms. They first appeared in the Triassic Period, expanded during the Cretaceous Period, and have dominated the world's vegetation since that time. Paleontologists cannot explain their origin and rapid evolutionary expansion.

Angle of Repose. The steepest slope, approximately 30°, at which unconsolidated material will stand permanently. Talus slopes, the foreset beds of deltas, the slopes of cinder cones and the lee slopes of dunes are commonly at the angle of repose.

Anglesite. A mineral, lead sulfate, a minor ore of lead found as a secondary mineral in the oxidized zone of lead sulfide deposits. It is usually colorless, white or pale yellow, and may be recognized by its *adamantine luster and high *specific gravity. See Mineral Properties.

Anhedral. A mineral grain not bounded by crystal faces. By contrast, *euhedral indicates a mineral fragment which is bounded by crystal faces.

Anhydrite. A mineral, anhydrous (waterless) calcium sulfate. It is most often formed as an *evaporite when it occurs in bedded deposits associated with gypsum, halite and limestone. Crystals of anhydrite are rare and it is usually massive, making it difficult to distinguish at sight from similar aggregates of other minerals. Where other sources of sulfur are scarce or are lacking, as in Great Britain and Germany, the sulfur has been extracted from anhydrite to produce sulfuric acid. It is also used as a soil conditioner and to a lesser extent as a setting retarder in Portland cement. See Mineral Properties; Appendix 3.

Anisotropic. Having unequal physical properties in different crystallographic directions, as shown by all crystals except those of the *isometric system. Used particularly in crystal optics to designate crystals with more than one refractive index.

Ankerite. An iron-bearing member of the *dolomite group of minerals. See Mineral Properties.

Annular Drainage. See Drainage Patterns.

Anorthite. A mineral, the calcium end-member of the *plagioclase feldspar series and comparatively rare. At the other end of the series is the sodium-rich member *albite. With increasing amounts of calcium, the composition varies continuously from albite to anorthite. See Mineral Properties.

Anorthoclase. A mineral, a sodium potassium aluminum silicate, one of the *feldspar group in which sodium exceeds potassium. It is a major constituent in the rock larvikite from Larvic, Norway. Because of its attractive internal bluish iridescence it is used extensively for facing public buildings.

Anorthosite. A coarse-grained igneous rock composed essentially of *plagioclase, principally labradorite. It would thus be classified as a type of *gabbro but, unlike most gabbros, it is frequently light colored because of the paucity of dark minerals.

Antecedent Stream. A waterway able to maintain its course as the land was uplifted beneath it, having cut its channel before the formation of the geologic structure through which it flows. Slow uplift is required so that the river can keep pace by downcutting. Antecedence is occurring in several active tectonic regions of the world, such as Taiwan and Hungary. It is hard to determine whether ancient river gorges that cut through mountains are antecedent or *superimposed. The Columbia River channel cut through the Cascade Mountains is considered antecedent.

Anthracite. A hard, black, brittle *coal that normally contains more than 92 percent fixed carbon and less than 8 percent volatile constituents. It has a bright luster; will not soil the fingers; breaks with a curved or irregular fracture (rather than into blocks like *bituminous coal); and burns with a short, blue, smokeless flame capable of producing great heat. Known also as "hard coal," anthracite occurs principally in strongly folded strata where *metamorphism has changed bituminous coal (that is, "soft coal") into anthracite. During this change both moisture and volatile matter are lost while the relative amount of fixed carbon increases. In the United States it is found in eastern Pennsylvania, Utah, Colorado, Massachusetts, Rhode Island and New Mexico; other important producing areas include Wales, China, France, Spain and Korea. (W.H.M.)

Anticline. A convex *fold in rock. In the simplest anticline, the sides of the fold dip away from its axis. In others the sides, or limbs, may dip in the same direction as the axis, may be horizontal, or may take a more complicated form. The rocks are progressively older toward the interior; the internal structure can be seen where roadcuts and rivers have cut through the anticline. Anticlines result from forces within the earth's *crust that laterally compress the crustal layering and cause buckling. Anticlines are

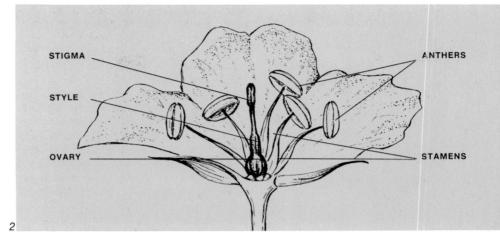

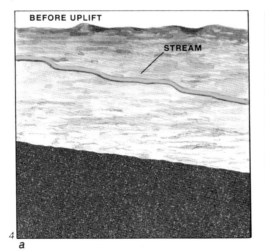

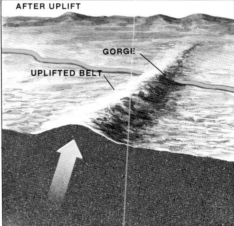

Angiosperm
2 In this cutaway view of a flower, the female reproductive parts at center consist of a style with a seed-containing ovary at its base and a stigma at top to catch pollen. The male stamens around the style are slender filaments topped by pollen-producing anthers.

3 The elmlike tree that shed this leaf millions of years ago is one of the vast group of flowering plants called angiosperms.

Antecedent Stream
4 This type of stream maintains its original course (a) despite local uplifting of the land (b). Here the stream has cut a deep gorge across the uplifted belt.

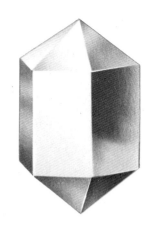

1

2

Apatite

1 *Six-sided prismatic crystals with pyramidal terminations are the most common form of apatite.*

2 *A multi-faced crystal which exhibits six other forms also occurs.*

Apollo Program

3 *Orbital experiments.*

4 *Surface experiments.*

displayed as long, low ridges in the Pennsylvania region of the Appalachian Mountain chain.

Anticlinorium. A large *arch or *anticline that is of regional extent, that is, at least several miles across. The sides or limbs of an anticlinorium differ from an anticline in that they are made up of several smaller folds. Like anticlines, an anticlinorium results from buckling of the earth's crust as a result of compressional forces acting on it. The rocks forming the Green Mountains of Vermont make up an anticlinorium.

Anticyclone. A circulation in which the air flows in approximately closed paths around a center of high pressure; it is therefore often referred to as a high. Anticyclones usually form in association with *cyclones. With a circulation opposite to that of cyclones, anticyclones are characterized by the clockwise motion of air (counterclockwise in the Southern Hemisphere) around the high-pressure center and by a component of motion outward from the center. Sinking air, *adiabatic warming and low relative humidity, as well as generally clear skies are typical of anticyclones.

Antigorite. A platy variety of the mineral *serpentine. It is the dominant constituent of the light to dark green serpentine rock used as a facing stone on buildings.

Apatite. A mineral, a fluophosphate or chlorphosphate of calcium, the most abundant and widely disseminated mineral of the *phosphate group. It is found in small accessory crystals in all classes of rock: igneous, sedimentary and metamorphic. Fine hexagonal crystals occur in a variety of colors in *pegmatite dikes and in some veins where they appear to have been deposited from *hydrothermal solutions. Apatite is distributed through the great magnetite deposit at Kiruna, Sweden, and the similar but smaller ore body at Tahawas, N.Y.

Phosphorus is one of the elements most important for plant life, and it is apatite that has contributed it to the soil as well as to the commercial phosphate deposits. When apatite is weathered at the earth's surface, it breaks down chemically. Some of the phosphorus remains in the soil and is available for plants, but much of it goes off in solution and eventually reaches the sea. Here part of the phosphorus reacts with other elements and is deposited on the sea floor. Much of the remainder is taken up by plankton to be eventually incorporated into the bony structure of fish, both the smaller ones that feed on the plankton and the

larger fish that eat the smaller. When these fish die, their hard parts settle to the bottom and in places form thick accumulations. This is the formation of rock phosphate, whose principal constituent is a cryptocrystalline apatite called *collophane. These rocks, millions of tons of which are mined annually, are the chief commercial source of phosphate for fertilizer. The United States is the leading world producer, followed by the Soviet Union, Morocco and Tunisia. About 45 percent of the Soviet production comes from rock phosphate and the balance from a remarkable deposit of crystalline apatite on the Kola peninsula. Here granular apatite associated with *nepheline syenite has been mined since 1932 to supply much of the Soviet need. In Quebec and Ontario, Canada, large, well-formed crystals in a crystalline limestone were sufficiently abundant to permit their being mined for phosphate.

Apatite is usually greenish-brown, but yellow, blue, violet and colorless varieties are found in some places. When these crystals are transparent they may be cut into gemstones; the yellow-green apatite from Durango, Mexico, and the deep purple crystals from Auburn, Maine, are the most notable examples. However, the hardness of apatite is too low to permit extensive use as a gem. See Mineral Properties; Appendix 3. (C.S.H.)

Aphanitic. A type of rock texture in which the individual mineral constituents are too small to be distinguished by the unaided eye.

Aphotic Zone. That portion of the *pelagic (open-ocean) environment that does not receive enough sunlight to allow the process of photosynthesis to take place. This zone includes most of the ocean environment because most of the light penetration is confined to only a thin upper layer. The upper limit of the aphotic zone varies, depending on latitude, seasonal or short-term conditions and location with respect to the *continental margin. For example, in clear tropical waters it may begin at a depth of 100 m, whereas in the middle latitudes it commonly starts only at about 50 m.

Due to suspended sediment in the *continental shelf area, the aphotic zone begins at a relatively shallow depth over the shelf as compared to the deep ocean. Turbulence due to storms, seasonal runoff from the land or great quantities of plankton may limit light penetration. The orientation of the earth with respect to the sun during different seasons also affects the limits of this zone. (R.A.D.)

Aplite. A light-colored dike rock composed essentially of *feldspar and *quartz and usually containing a small amount of *muscovite mica. It has a fine-grained, sugar-like texture. Although the word may be applied to any rock with this kind of texture—from *granite to *gabbro in composition—when used without modifiers it refers to a rock of granite composition.

Apollo Program. A series of American manned spacecraft flights culminating in missions that landed on the moon and returned to earth. In addition to obtaining samples from various areas of the moon, the program produced a vast amount of data, using a variety of instruments implaced on the lunar surface or in orbit.

The first exploration occurred on July 19, 1969, when Apollo 11 landed at latitude 0°41′N, longitude 23°26′E near the southwestern edge of Mare Tranquillitatus, about 80 km north of the nearest highland area. Apollo 11 collected 21.5 kilograms of samples. Half of the rocks were titanium and iron-rich basalts with clinopyroxene, feldspar, ilmenite and small amounts of silica and metallic iron. Chemical analyses of the rocks indicated that two groups were present—a high-alkali group and a low-alkali group. All of these rocks crystallized from a partial melt from the lunar interior. The age of the basalts is 3.7 billion years. The remaining Apollo 11 rocks were breccias, consisting of soil welded together by heat produced during formation of impact craters. The soil at Mare Tranquillitatus was a mixture of fragments of the two types of basalt sampled as rocks, plus a feldspathic rock ejected from the lunar highlands some 80 km to the south by meteorite impacts.

The second manned mission, Apollo 12, landed in the southeastern portion of Mare Procellarum, enabling it to examine pieces of the Surveyor III spacecraft that had landed on the moon 3½ years before. The mission returned 34.2 kilograms of samples. The rocks consisted almost entirely of iron-rich basalts consisting mostly of clinopyroxene and plagioclase, but lacking the extremely high titanium content of the Apollo 11 samples. The soils of Apollo 12 were made mostly from the local rocks but also contained an exotic component that became known as KREEP, for its high content of Potassium (K), Rare Earth Elements (REE) and Phosphorus (P). A discrete white layer was found in the soil. This layer is enriched in light-colored rope-shaped bits of glass with KREEP composition. A light-colored streak or ray of material ejected from the crater Copernicus, 300 km to the north, cuts across the Apollo 12 site; the white layer is thought to be related

ORBITAL EXPERIMENTS

Experiment	Purpose
Multispectral Photography	Photograph the moon through filters
CM Window Meteoroid	Determine micrometeorite flux by counting craters on spacecraft windows
UV Photography	Photograph the earth and moon in near ultraviolet light
Gegenschein from Lunar Orbit	Examine Moulton region of electromagnetic spectrum in the earth-sun system
Gamma-Ray Spectrometer	Measure concentrations of aluminum, silicon uranium and thorium on the lunar surface
X-Ray Fluorescence	Measure concentrations of aluminum, silican and magnesium on the lunar surface
Alpha Particle Spectrometer	Detect radon production on the lunar surface
S-Band Transponder	Determine lunar gravity field
Mass Spectrometer	Determine lunar atmosphere composition
Far UV Spectrometer	Determine atomic composition and density of lunar atmosphere
Bistatic Radar	Study electrical properties of lunar surface
IR Scanning Radiometer	Study thermal effect on lunar surface
Particle Shadows / Boundary Layer	Study earth's magnetosphere
Magnetometer	Detect magnetic anomalies on the lunar surface; determine electrical properties of the lunar interior
Lunar Sounder	Detect variations in the electrical properties of the lunar crust

SURFACE EXPERIMENTS

Experiment	Purpose
Passive Seismic	Determine structure of the lunar interior
Active Seismic	Determine near-surface lunar structure at landing site
Lunar Surface Magnetometer	Measure local magnetic fields; determine electrical properties of the lunar interior
Solar Wind Spectrometer	Determine composition of the solar wind
Suprathermal Ion Detector	Detect high-energy ions in lunar atmosphere
Heat Flow	Determine heat production of lunar interior
Charged Particle Detector	Detect ions in the lunar atmosphere
Cold Cathode Ion Gauge	Measure pressure of the lunar atmosphere
Laser Ranging Retro-Reflector	Determine with high precision the orbital motions of the moon
Solar Wind Composition	Measure rare gas portion of the solar wind
Cosmic-Ray Detector	Measure cosmic-ray tracks and particles in a variety of target materials
Portable Magnetometer	Measure local magnetic anomalies at landing site
Lunar Gravity Traverse	Determine density and structure of landing site subsurface
Soil Mechanics	Measure bearing strength and other mechanical properties of the lunar regolith
Far UV Camera / Spectrometer	Determine the distribution of hydrogen around the earth and Milky Way
Lunar Ejecta and Meteorites	Determine cosmic dust influx and production of meteoritic debris at the lunar surface
Lunar Seismic Profiling	Determine subsurface structure of landing site
Surface Electrical Properties	Determine subsurface structure in terms of electrical properties of the materials
Lunar Atmosphere Composition	Measure composition of lunar atmosphere
Lunar Surface Gravimeter	Determine magnitude of solid moon tides and detect gravity waves
Lunar Dust Detector	Measure amount of solid particles in the lunar atmosphere
Lunar Neutron Probe	Study stratification and mixing of the lunar soil and measure lunar neutron energy spectrum

1

2

to this ray. The Apollo 12 basalts date at about 3.3 billion years.

In February, 1971, Apollo 14, the fourth manned exploration, landed on the Fra Mauro formation, an area of hummocky terrain with a system of low ridges radial to Mare Imbrium. It is therefore inferred that the rocks sampled by Apollo 14 were deposited as dunes resulting from the clouds of debris thrown outward by the Imbrian impact. The expedition collected 42.7 kilograms of samples and set up a geophysical experiment package. The astronauts also climbed a hill 100 m in height formed by a meteorite impact about 29 million years ago. Virtually all of the rocks collected by Apollo 14 were complex breccias made by the welding or sintering together of fragments ranging in size from one micrometer to several centimeters. Some of the fragments were themselves breccias. Almost all of the Apollo 14 rocks and soils have the same composition as the KREEP found at Apollo 12. The Apollo 14 rocks are dated at 3.95 billion years, and this date has been accepted for the Imbrian impact.

The Apollo 15 lunar module landed in July, 1971, on a mare area known as Palas Putredinis or Marsh of Decay. A valley 1 km wide, known as Hadley Rille, cut the mare, and the study of this rille was a major objective. However, the prime objective was to sample the Apennine Mountains (or Front), which are a portion of the third ring of the Imbrian basin. Most of the rocks found along the Apennine Front were complex breccias somewhat similar to the breccias of Apollo 11, but with a different composition. The breccias contained some mare material; thus, with few exceptions, they must have been formed well after the Imbrian basin became flooded with basalt. However, some fragments found in the breccias, especially at Spur Crater, contained uncontaminated highland material. One fragment in a breccia was a 150-gram piece of anorthosite that was named Genesis Rock because of its assumed old age. It did in fact reveal an extraordinarily old age—4.5 billion years. Other highland materials found along the Front included a spectrum of breccias similar to many Apollo 14 rocks as well as the intensely recrystallized breccias and impact melt rocks found later at the Descartes (Apollo 16) landing site in the central highlands.

Like the Apollo 12 site, the Apollo 15 site lies along a ray of a major crater that penetrated mare material. The mare basalts at Apollo 15 are iron-rich and similar in age and composition to the Apollo 12 basalts.

The fifth manned lunar landing mission,

Apollo 16, landed in a hummocky floor of an old crater at 9°00'S, 15°31'E in the Descartes highlands in March, 1972—the only mission to central portions of a highland area. The sampling objectives were to return samples from the hummocky plains material, known as the Cayley Formation, and from the hills surrounding the valley; these were believed to be constructional features, possibly highly viscous and hence silicic volcanic features. The astronauts realized immediately that the rocks did not look like terrestrial volcanic rocks.

The rocks returned by Apollo 16 all come from the Cayley plains unit. Although the astronauts in the lunar rover went some 300 m up the side of Stone Mountain, all of the area sampled was covered by ray material from a very young (2-million-year-old) crater, South Ray, which penetrated only Cayley material. All of the samples returned by Apollo 16 were of impact origin, many looking similar to fall-back material in some of the larger impact craters on earth. The Apollo 16 rocks mostly date at about 3.95-4.1 billion years.

The last flight of the series, Apollo 17, landed on the moon at 20°9'N, 30°45'E in the Taurus-Littrow Valley extending off the southeast edge of Mare Serenitatis. The sampling objectives were, first, to sample the dark soil that appeared to cover the valley floor. The floor seemed to have craters covered with a young mantling deposit and thus had potentially the youngest rocks sampled by the Apollo program. The second sampling objective was to sample the highland areas in the north and south massifs that surround the valley.

The examination of these samples has gone far enough to yield a basic understanding of the geology of the landing site. The dark mantling material at the floor of the valley proved to be largely regolith made from the iron- and titanium-rich mare basalts of the valley floor, which are very similar to the alkali-poor basalts sampled by Apollo 11. At Station 4 an orange material proved to be made of transparent orange "glass" with an even higher titanium content than the typical Apollo 17 mare basalt. The orange glass occurs in all soils collected by Apollo 17 on the valley floor and along the lower edges of the north and south massifs. The orange soil formed at 3.75 billion years, approximately the same as the age of the underlying mare basalts. Thus the hope of finding the youngest lunar materials was not realized. The materials of the north and south massifs were complex impact melt rocks and breccias very similar to those sampled in the central highlands by Apollo 16. The few Apollo 17 massif rocks that have been

3

Apollo Program

1 View of the Apollo 11 command module as seen from the lunar module, which is preparing to land on the moon's surface. The lunar terrain in the background is the far side of the moon.

2 Astronaut Eugene Cernan prepares to pull a drive tube containing a core of lunar soil out of the moon's crust.

3 Astronaut Edwin Aldrin walks on the surface of the moon near a leg of the lunar module during the Apollo 11 mission. The astronaut's footprints are clearly visible in the foreground.

4 Astronaut James Irwin works at the lunar roving vehicle during the first Apollo 15 lunar surface extravehicular activity at the Hadley-Apennine landing site. Part of the lunar module appears at the left.

4

a

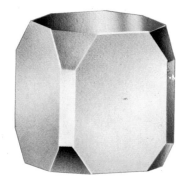

1
b

2

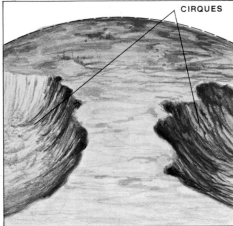

a

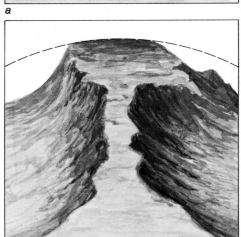

b

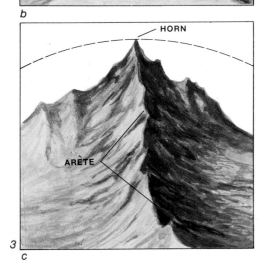

3
c

Apophylite
1 *Perfect tetragonal crystals (a and b) of this common mineral sometimes measure up to 5 cm (2 inches) in size.*

Aragonite
2 *This coarsely crystalline pseudohexagonal twin is a calcium carbonate mineral that is the principal material in coral and pearls.*

Arête
3 *When two adjacent mountain basins,*

or cirques, erode, a ridge (a) is formed between them. Continual erosion narrows the ridge (b). Finally, (c) a knife-sharp ridge, an arête develops.

Archaeopteryx
4 *A reconstruction shows the first known bird, which retained a reptilian beak, wing claws and a long, bony tail.*

5 *Fossil remains of Archaeopteryx were discovered at a slate quarry in Bavaria in 1861.*

dated yield ages of about 3.95 billion years, in the same range as those of Apollo 14 and 16 rocks. (C. H. S. and J. L. W.)

Apophyllite. A mineral, hydrous calcium potassium fluorosilicate, found as free-growing crystals in cavities in basaltic rocks, commonly associated with zeolites, datolite, calcite and pectolite.

Apophyllite is four-sided, and the equally developed prism and base frequently give crystals a cubic aspect. They can be distinguished from cubes by the presence of the pearly luster on the base and good basal cleavage. See Mineral Properties; Appendix 3.

Apparent Dip. The angle of depression of any planar feature exposed on the face of a rock *outcrop. Apparent dip refers to an angular measurement made on a dipping surface other than in a plane perpendicular to the *strike of the feature being measured. This yields results somewhat less than the actual dip of the planar feature. See Dip.

Aquamarine. The greenish-blue, transparent variety of *beryl; used as a gemstone.

Aquifer (from the Latin for "water carrier"). A geologic unit, either *regolith or *bedrock, that is sufficiently permeable to conduct and yield groundwater for springs or pumping wells. Sand and gravel are usually good aquifers; so are sandstones if not too well cemented.

Aragonite. A mineral, calcium carbonate in its orthorhombic form. Both calcite and aragonite are dimorphs, that is they have the same composition, differing only in crystallography and physical properties. Because aragonite forms under a much narrower range of pressures and temperatures than calcite, it is less common and is usually confined to near-surface deposits. Aragonite may form *stalactites in caves and occurs as deposits from hot springs and geysers. The oyster's mother-of-pearl and the pearl itself are aragonite. It occurs in crystals of two distinct types: (1) slender, tapering, single crystals called church steeple crystals by English miners, and (2) pseudohexagonal twins in which three crystals grow together. The mineral is named for Aragon, Spain, where pseudohexagonal twins are found. See Mineral Properties. (C.S.H.)

Arch. A very broad *anticline with sides that dip gently away from its axis. It is of regional extent. The Cincinnati Arch, which was a land area above water throughout most of the Paleozoic Era, extended from Tennessee to Ontario.

Archaeopteryx. The earliest known bird. The discovery of its fossil remains was extremely significant, for this creature still retains certain characteristics of its reptilian precursors. *Archaeopteryx* is known from two specimens embedded in Upper *Jurassic lithographic limestone in the quarries of Solenhofen, Bavaria. These show that the bird was about the size of a crow, with a rather long neck and an elongated, lizard-like tail. The beak was studded with sharp teeth set in individual sockets, and each of the three fingers, which terminated at the front edge of the wings, had birdlike claws. The skull, teeth, hind limbs, hip girdle and bone construction are clearly reptilian. However, the impression of feathers in the extremely fine-grained limestone is considered proof that *Archaeopteryx* was a bird and was warm-blooded.

Because birds have fragile bones and delicate feathers, they are relatively rare as fossils. Thus the discovery of *Archaeopteryx* in strata more than 140 million years old is significant in itself. More important, however, is its evolutionary significance. *Archaeopteryx* bridges an important gap in vertebrate evolution, for its reptile-like skeleton is the ideal "connecting link" in the transition of reptiles to birds. (W. H. M.)

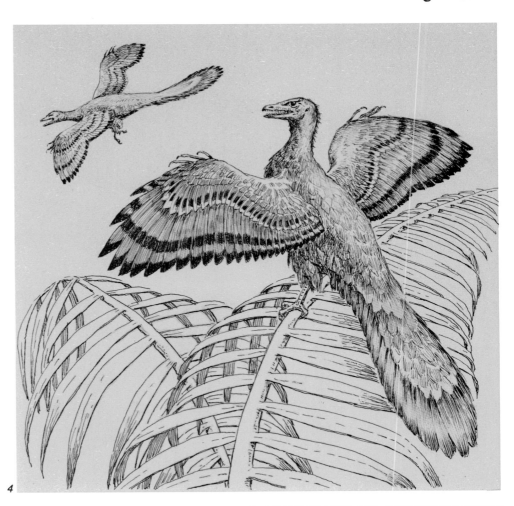

4

Archeozoic Era. The oldest known geologic era, corresponding to the earliest of *Precambrian time. It represents a vast interval of geologic history during which there is little evidence of life on earth. Most Archeozoic rocks are extremely altered volcanic and sedimentary types intruded by granite. Archeozoic rocks at least 3.6 billion years old have been found in southwestern Minnesota. These generally yield little direct evidence of prehistoric life, but some contain abundant complex carbon compounds of possible organic origin. However,*metamorphism has been so severe in the older formations that all traces of the original organisms have been destroyed.

Arête (from the French for "sharp edge"). A steep, serrated divide caused by mountain glaciation. It is formed where *cirques on opposite sides of a divide intersect or where the divide between adjacent parallel *glaciated valleys is reduced by glacial erosion to a sharp ridge. *Cols are low places along an arête; *horns are high places.

Argentite. A mineral, silver sulfide, an ore of silver. It is rarely in crystals and is usually massive or in thin coatings associated with other silver minerals. Argentite is a major

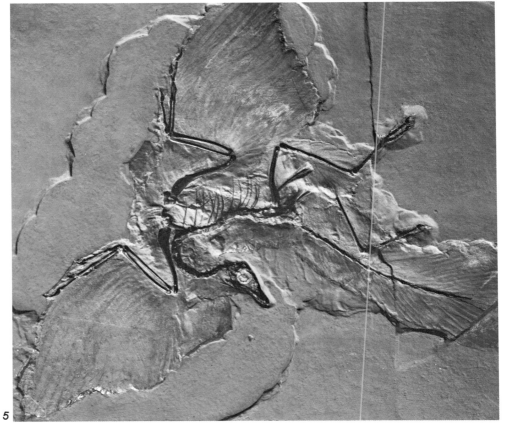

5

1

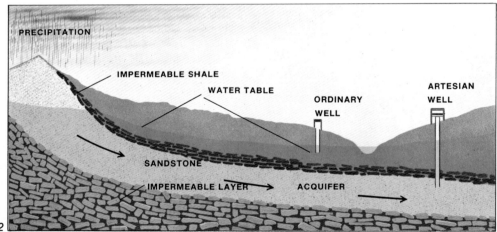

2

3

silver ore notably in Mexico, Peru, Chile and Bolivia. It is the isometric form of silver sulfide, Ag_2S, and it is stable only at temperatures above 179°C (325°F); at lower temperatures it inverts to orthorhombic acanthite. Thus the few isometric crystals that are found are actually paramorphs, possessing the external form of argentite but the internal structure of acanthite. The mineral is characterized by its high specific gravity (7.3) and high sectility (capacity to be cut easily by a knife). See Mineral Properties. (C.S.H.)

Arkose. A coarse-grained *sandstone containing at least 25 percent feldspar. It is derived from the partial disintegration of feldspar-rich rocks such as granite and gneiss. Because arkose signifies rapid erosion and deposition of sediment before feldspar had time to weather completely into clay, it is useful in recognizing certain *unconformities and in reconstructing ancient climates.

Arroyo (or Wash, Wadi). A stream channel in an arid or semiarid region. The channel is generally dry since the eroding stream is ephemeral or intermittent. The walls are steep but the channel may be wide or narrow, shallow or deep. In southwestern United States there has been since 1880 a period of rapid arroyo formation from overgrazing and/or climatic changes. In desert regions of Africa and Arabia such arroyos are called wadis. A similar dry channel called a *gully may occur in humid as well as arid regions. (M.M.)

Arsenate. A group of minerals containing arsenic and oxygen, in which each arsenic atom is linked to four oxygen atoms. This grouping, known as the arsenate ion, AsO_4^{-3}, is the fundamental building unit of the arsenates. *Erythrite (cobalt arsenate) and annabergite (nickel arsenate) are the commonest minerals in the group.

Arsenic. A chemical element that occurs as a mineral in its native form. It typically is found massive or in concentric layers with a tin-white to gray color, and is most commonly found in hydrothermal veins associated with silver, cobalt or nickel ores. As is widely known, many natural and manufactured arsenic compounds are poisonous. See Element, Native; Mineral Properties; Appendix 3.

Arsenide. A chemical compound in which *arsenic is combined with a metal. There are several mineral representatives, including niccolite and sperrylite. In the classification of minerals, arsenides are usually grouped with the *sulfides.

PRECIPITATION IMPERMEABLE SHALE WATER TABLE ORDINARY WELL ARTESIAN WELL SANDSTONE IMPERMEABLE LAYER ACQUIFER

Arsenopyrite. An iron arsenic sulfide, the most common mineral containing arsenic and the principal source of that element. It is widespread and found in abundance at many localities associated with native gold and with ores of such metals as tin, tungsten, silver, copper, lead and zinc. Arsenopyrite is *monoclinic, but the crystal in which it is found appears *orthorhombic due to twinning. See Mineral Properties.

Artesian Well. A well in which hydraulic pressure causes water to rise above the top of the confined *aquifer from which it was derived. The name comes from Artois in France, where artesian conditions were early encountered. The water may or may not reach the surface, but in places the pressure has been so great that fountains over 60 m high have been formed. An artesian well near Calais, France, has been flowing since 1126 a.d.

Conditions necessary for artesian flow are (1) an inclined aquifer that crops out on the surface at its upper end and is buried at its lower end, (2) precipitation that infiltrates the aquifer where it crops out at the surface (area of recharge), (3) impervious beds around the aquifer that prevent leakage and enable hydraulic pressure to develop, (4) a well below the area of recharge that allows water to escape from the confined aquifer (area of discharge). Artesian water is of great importance in the northern Sahara in Africa; the desert of Queensland, Australia; the Great Plains of the United States; and in such areas as Atlantic City, N.J., which is located on a *barrier beach.

An artesian spring occurs where artesian water reaches the surface from a confined aquifer along joints or faults. (R.L.N.)

Arthropod. A member of phylum Arthropoda and numerically the largest subdivision of the animal kingdom. These highly specialized invertebrates vary greatly in size and shape and include insects, scorpions, shrimp, crayfishes, crabs, centipedes and the extinct *trilobites. The typical arthropod body is segmented, bilaterally symmetrical and covered by a chitinous exoskeleton which, in some forms, contains calcium carbonate. Locomotion is by means of jointed appendages that may include legs with pincers. As arthropods grow they keep shedding their exoskeletons until maturity. Thus an arthropod might shed several external skeletons capable of fossilization. This process, called ecdysis, explains why many fossil arthropods consist of parts of the body rather than a complete specimen.

Arthropods have adapted to a wide variety of environments, living on land, in

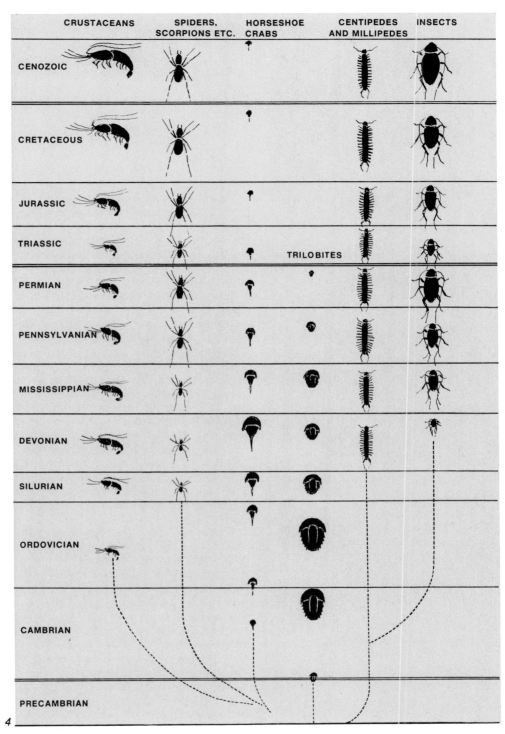

	CRUSTACEANS	SPIDERS, SCORPIONS ETC.	HORSESHOE CRABS	CENTIPEDES AND MILLIPEDES	INSECTS
CENOZOIC					
CRETACEOUS					
JURASSIC					
TRIASSIC			TRILOBITES		
PERMIAN					
PENNSYLVANIAN					
MISSISSIPPIAN					
DEVONIAN					
SILURIAN					
ORDOVICIAN					
CAMBRIAN					
PRECAMBRIAN					

Arroyo

1 Dry water-carved gullies such as this one near Chimayo, N.M., can become dangerous when runoff from torrential rain sends flash floods sweeping down from the highlands.

Artesian Well

2 Rainwater percolates down through the permeable sandstone aquifer, sandwiched between two impermeable layers of shale, and is forced upward through a well that pierces the impermeable shale caprock.

Arsenopyrite

3 This silvery metallic mineral is the most abundant ore of the deadly poison arsenic.

Arthropod

4 The family tree of the joint-legged, segmented animals known as arthropods. The size of each animal indicates the relative abundance of the group it typifies at a given time period.

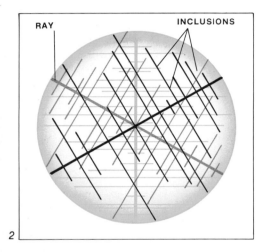

Asbestos

1 *Chrysotile, the best known variety of asbestos, is a form of serpentine that separates into heat-resistant fibers used for insulation and fireproofing.*

Asterism

2 *Three sets of inclusions intersect at 120° to produce a six-rayed star of light. Each ray forms at a right angle to a parallel set of inclusions.*

water and in the air. They are of great importance in nature today, but in spite of their long geologic history (with fossils from Cambrian to Holocene), only a few forms have been preserved and are important as fossils. These include the classes Trilobita (extinct trilobites), Merostomata (extinct *eurypterids) and Ostracoda (ostracodes). (W.H.M.)

Asbestos. A general term applied to varieties of several minerals that can be separated into slender fibers. Although different in chemical composition, they are all flexible, permitting them to be manufactured into noncombustible fabrics and other materials. *Chrysotile, known as fibrous serpentine or serpentine asbestos, constitutes about 95 percent of commercial asbestos. The other types are all *amphiboles. All are found in metamorphic rocks.

Asbestos was early manufactured into fabrics to be made into fireproof garments and later into theater curtains. These uses continue, but today far more asbestos is used in the building industry, largely as molded asbestos-cement products such as tiles, shingles, wallboard and corrugated siding. Asbestos with a bonding liquid is sprayed on walls and ceilings, giving heat insulation and acoustical properties and acting as a fire protection. The automotive industry is one of the largest consumers because asbestos has proved uniquely useful in the brake linings of automobiles.

Chrysotile occurs principally in *peridotite rocks that have altered to massive *serpentinite. The asbestos is present in cross-fiber veins with fibers essentially at right angles to the vein, and as slip-fibers that parallel the vein. In most deposits fibers vary in length from a small fraction of a millimeter to 4 cm, although they have been found as long as 30 cm. Veins of chrysotile are usually green to yellow-green but occasionally may be a golden yellow. Mining of chrysotile asbestos began in Canada in 1870 and that country has been the leading producer. With the discovery and exploitation of large deposits in Siberia, the Soviet Union has also become a major world supplier. Similar deposits but with smaller production are found in the Republic of South Africa, China and Rhodesia.

Anthophyllite is an orthorhombic amphibole which, in the asbestos variety, occurs in long, gray to brown cross fibers. Because the fibers are brittle and are of low tensile strength, anthophyllite has little commercial value.

Amosite asbestos is a variety of the amphibole cummingtonite, found only in the Republic of South Africa. It occurs in flexible fibers, 5–30 cm in length.

Crocidolite, the fibrous variety of the amphibole riebeckite, occurs in cross-fiber veins in metamorphosed ironstone at several localities in South Africa. Because the silky, flexible fibers are blue, they are commonly called blue asbestos. Some South African crocidolite has been replaced by quartz while preserving the fibrous nature of the asbestos. When cut and polished it is an attractive ornamental material used extensively in jewelry as the material called *tiger's eye.

Tremolite is white monoclinic calcium-magnesium amphibole. Iron may substitute for some of the magnesium, giving a light green color, but with increasing amounts of iron the mineral becomes dark green and is called actinolite. Thus asbestos varieties may be white or various shades of green. They are characteristically found as fine, silky fibers in slip-fiber veins. It is believed that the material to which the name asbestos was originally given was a fibrous tremolite. The mineralogical curiosities *"mountain leather" and "mountain cork" are felted intergrowths of tremolite fibers resembling leather or cork. The "leather" is found in thick, tough, flexible sheets; the cork is in thicker masses with a spongy corklike quality. (C.S.H.)

Assay. To test an ore to determine what metals it contains and the amount of each. Assay is also the word for the material on which the test or analysis is being performed.

Asterism. A property of some minerals of producing a starlike figure in either reflected or transmitted light. It results from the orientation by the crystal lattice of inclusions or cavities. The phenomenon is most strikingly shown by star rubies and star sapphires in which tiny needles of the mineral rutile are present in three crystallographic directions at 120° to each other. A domed cabochon stone cut from such a mineral shows in reflected light three intersecting beams forming a six-pointed star; each beam is at right angles to a direction of inclusions. Garnet, quartz and several other minerals may show asterism. If the inclusions are oriented in two directions at right angles or at nearly right angles to each other, a four-pointed rather than a six-pointed star results.

Some phlogopite mica contains rutile needles oriented by the pseudohexagonal lattice at nearly 120° to each other. If a point source of light is viewed through this mica, a beautiful six-pointed star results. (C.S.H.)

Asthenosphere. A shell or layer within the earth directly below the *lithosphere. The upper boundary of the asthenosphere begins approximately 100 km below the earth's surface and extends to a depth of 250 km. The upper limit is characterized by an abrupt decrease in the velocity of earthquake waves as they enter this region of the earth's interior. The asthenosphere, also referred to as the upper mantle, is believed to be weaker than the overlying lithosphere and to have a viscosity that increases with depth. According to one theory, *convection currents in the asthenosphere allow the weak material to flow plastically in the form of closed loops known as cells. The movement of the cells creates a drag on the adjacent lithosphere, causing it and the outer *crust to move over the earth's surface. (J. W. S.)

Atmsophere. The outer, gaseous envelope of the planets or the sun. Since an unconfined gas has no fixed limits, the earth's atmosphere extends upward indefinitely, mingling with the gases of interplanetary space; the latter in turn are an extension of the outer atmosphere of the sun. The gases of the earth's upper atmosphere are extremely rare, however, in comparison with those at its lower boundary, for the lower layers are compressed by the weight of all the gases above. The outermost shell or layer of the atmosphere is called the exosphere, or sometimes "the region of escape," since in this layer the air density is so low that atoms or molecules can leave the earth in appreciable numbers.

The height of the lower boundary of the exosphere, called the critical layer of escape, is variously estimated at 500 to 1000 km; above this level the mean free path of air particles is greatest for those moving upward, since in this direction there are fewer collisions with other particles. The extreme upper limit of the earth's atmosphere has sometimes been taken as 30,000 km; at about this height a molecule moving as if in rigid rotation with the earth could not be held to such an orbit by the earth's gravitational attraction. For most meteorological purposes, however, the atmosphere may be considered to have a much lower upper boundary or "top," for the bulk of its mass is confined to a layer that is quite thin in comparison with its horizontal extent, that is, with the circumference of the earth. Indeed, more than two-thirds of the mass of the earth's atmosphere lies below the elevation of its highest surface features (8.8 km) and less than one percent lies above the 33-km level.

Various schemes are used for subdividing the atmosphere vertically into concentric shells. These depend on the property under consideration, such as temperature, ionization or gaseous composition. In the homosphere, extending from the surface to approximately 100 km, the fixed or permanent gases that make up the air are so uniformly mixed that they always occur in constant proportions (see Air). In this layer, nitrogen (N_2) and oxygen (O_2) account for about 99 percent of the composition of pure, dry air. However, some of the trace gases or minor constituents, as well as the variable gas water vapor present in moist air, are highly important in determining conditions that make the earth suitable for life as we know it. Water vapor and carbon dioxide are largely responsible for the atmosphere's greenhouse effect, which keeps variations of the earth's surface temperature within a relatively narrow range. These gases permit most of the radiation from the sun to pass through the atmosphere to the earth's surface, but are also strong absorbers of the longer-wave infrared radiant energy from the earth. Absorbing energy at these wavelengths, they then reradiate a portion of it back to the surface, maintaining the surface temperature higher than it would otherwise be. Ozone (O_3), which is produced in the atmosphere as a result of the decomposition of molecular oxygen (O_2) by ultraviolet light, absorbs solar ultraviolet energy and shields plant and animal life from this potentially harmful radiation. Water vapor in the atmosphere is responsible for most phenomena that we associate with weather, and also makes the *hydrologic cycle possible.

In the heterosphere, which extends upward indefinitely above 100 km, the atmospheric gases no longer occur in constant proportions but are distributed according to their molecular weight. The heaviest gas (nitrogen) is most abundant at the bottom and the lightest (hydrogen) is most abundant at the top of the layer. The prevalence of molecular oxygen (O_2), which is broken up into atomic oxygen (O) by the sun's ultraviolet radiation, gives way to the greater abundance of the latter gas at elevations between 250 and 1000 km. Since the production of atomic oxygen depends on the absorption of solar ultraviolet radiation, which varies with the 11-year cycle of solar activity known as the sunspot cycle, the relative abundance of nitrogen, atomic oxygen, helium and hydrogen between 200 and 5000 km varies from day to night and from the maximum to the minimum of the sunspot cycle.

Electrical state and magnetic forces are the bases of other classification schemes for subdividing the atmosphere vertically. The ionosphere, a layer characterized by a relatively high number of ions and free electrons, extends upward from the top of the neutrosphere—the relatively un-ionized layer of the atmosphere below 50 km. Ultraviolet radiation from the sun is the principal source of ionization in the several regions of the ionosphere, whose variations strongly influence radio communication. The magnetosphere, the outer region of the earth's ionized atmosphere in which magnetic forces predominate over gravity in the motions of the electrically charged gases, extends upward from about 250 km. Since particles as high as 60,000 km are acted on not only by the earth's gravitational field but also by magnetic forces, and move predominantly with the earth, this region may also be considered part of the earth's atmosphere.

The best-known scheme for describing the vertical structure of the atmosphere, based on the change of temperature with height, is of considerable interest to meteorologists because this quantity controls atmospheric stability and has a profound influence on weather. The lowest layer thus defined is the troposphere, in which the temperature decreases from about 15°C at the earth's surface to about -57°C at 11 km, the average height of the tropopause. In the stratosphere above, the temperature is first constant and then increases upward to about -2°C near 50 km, the average height of the stratopause. This typical structure of the atmosphere is determined essentially by the composition of the air, which affects the transfer of solar and terrestrial radiation, and by the mixing or overturning of the atmosphere. The greenhouse effect creates a heat source at the earth's surface; the presence of the ozonosphere (10 to 50 km) results in the temperature maximum at 50 km, where ozone absorbs the incoming solar radiation most strongly. In the mesosphere, the temperature again decreases upward, reaching a minimum of about -92°C at 79 km. At an altitude of about 89 km, the base of the thermosphere, the temperature increases very rapidly upward to a maximum of about 1200°C during the middle of the sunspot cycle.

Since water vapor is concentrated mainly in the lower troposphere, and since vertical motions are favored by the relatively unstable temperature distribution in the troposphere, cloud and precipitation are confined mainly to this lower layer of the atmosphere. Evaporation of water from the earth's surface is made possible by the relatively high surface temperature; condensation results when the moist air transported upward cools by expansion under the lower pressure at higher elevations. The low temperature at the tropopause acts in a sense as a condenser; water vapor condenses and precipitates out, so that—according to most studies—the stratosphere contains relatively little water vapor. Moreover, because the temperature increases with height, the stratosphere is quite stable, and vertical motions are suppressed. As a consequence of this vertical stability and the absence of precipitation, dust and other materials introduced into the stratosphere remain there much longer than in the troposphere.

The lower atmosphere responds strongly to the regular diurnal and seasonal changes of solar radiation associated with the earth's rotation and its revolution about the sun. The spherical shape of the earth results in an excess of solar radiation at low latitudes and a deficiency in the polar regions. The thermal structure of the atmosphere, including the depth of the troposphere and the temperature of the stratosphere, varies with latitude and seasonal changes in radiation.

The temperature difference between the earth's tropics and its polar regions creates pressure differences that maintain a general or global circulation of the atmosphere. This convectively induced circulation is predominantly horizontal in character, however. This is because the depth of the atmosphere is slight in comparison with its horizontal extent, and the rotation of the earth causes the large-scale motion to be predominantly geostrophic, that is, the motions represent a balance between *Coriolis and pressure-gradient forces. The earth's climate can be explained in terms of the global circulation of the atmosphere, the distribution of solar radiation and the distribution of surface features such as mountains.

The high atmosphere reacts not only to the periodic changes in solar radiation but also to more subtle, irregular changes in radiation from the sun, which includes streams of high energy particles that make up the *solar wind. In the thermosphere, the temperature varies markedly not only between day and night but also with the sunspot cycle. The magnetosphere reacts to variations in the solar wind, and flares on the sun are followed by magnetic storms, which disrupt radio communications; the various forms of the aurora, at times of intense solar activity, provide beautiful displays that represent energy emissions from the rarefied gases of the high atmosphere. See Color Plate "The Canopy of Air." (M.F.H.)

Atmosphere, Planetary Circulation of. In the broadest sense, the motions of the earth's atmosphere. The planetary circulation includes not only the average of mean motions—often described as the general circulation—but also those statistics on the circulation required for a dynamically consistent picture of the motions. The average observed motions of the atmosphere do not in themselves offer such a consistent description, primarily because they cannot account for the exchange of heat and angular momentum (the "quantity of rotation") between the lower, or tropical, and higher, or polar, latitudes. Angular momentum is proportional to the angular velocity of the particle and the square of the distance from its center of rotation. Particles near the Equator, because of their great distance from the earth's axis of rotation, therefore have much greater angular momentum than those near the poles. Because of friction between the earth's surface and the moving atmosphere in contact with it, easterly winds at the earth's surface extract westerly angular momentum from the earth, tending to slow down the earth's rate of rotation; westerly surface winds lose momentum to the earth, tending to speed up its rotation. Since the earth and its atmosphere maintain an approximately constant rate of rotation, the angular momentum extracted by surface easterly winds must be balanced by that delivered to the earth by surface westerlies, and transported from the easterly belt to the westerly.

The necessity for a latitudinal exchange of momentum and heat is to be found in the observed worldwide distributions of *winds and *temperature. We know from existing temperature distribution that the tropical and polar regions radiate almost the same amount of energy into space. However, the tropics receive much more energy from the sun than they radiate, and the polar regions radiate more solar energy than they receive. A consistent picture of the general circulation must thus account for a transfer of the excess heat poleward from the tropics in each hemisphere. Similarly, the *wind distribution at the earth's surface shows that the atmosphere extracts westerly angular momentum from the earth at low latitudes and loses westerly momentum to the earth at higher latitudes.

Since the observed average circulation fails to account satisfactorily for the latitudinal exchanges of heat and momentum required by the observations, meteorologists have concluded, and have verified, that the migratory eddies of the atmosphere—the traveling lows, highs, and waves—perform part of that exchange and must therefore be considered an integral part of the general circulation.

The driving force of the global wind system is the latitudinal temperature contrast resulting from the excess energy received at the Equator and the deficiency at the poles. On a nonrotating earth with a uniform surface, heating at the Equator and cooling at the poles would presumably result in one large convective circulation covering each hemisphere. Air warmed at the Equator would rise, flow poleward, sink as it cooled, and return near the earth's surface to the Equator. Thus the entire circulation would take place in north-south planes cutting the earth's axis and its meridians. Averaged out over all meridians, the resulting circulation is described by meteorologists as a meridional cell.

The earth's rotation complicates this simple picture, and the nonuniform surface of land, water and mountains complicates it still further. There is a general upward flow of air near the Equator, but the air sinks to the earth's surface in the subtropical high-pressure belts near 30° latitude. As it moves poleward aloft, the air develops a component of motion from the west, relative to the earth beneath, since it tends to maintain the excess angular momentum of air at the Equator. The transfer of westerly momentum accounts in part for the strong westerly current, or jet stream, near the base of the stratosphere at subtropical latitudes.

Some of the air that sinks toward the earth's surface in the subtropical high-pressure belt returns to the Equator, developing a component of motion from the east because of the earth's rotation, and forming the northeast and southeast trade winds. Another branch turns poleward, gaining a westerly component, to meet the polar easterlies at the polar front. The polar easterlies represent the outflow from the *anticyclones that have been caused by cooling, subsiding air in the polar regions.

The meridional cell that moves equatorial air poleward at high levels is called a Hadley cell, after the English scientist George Hadley who, in 1735, first explained the easterly component of the trade winds as an effect of the earth's rotation. The Hadley cells are semipermanent features of the planetary circulation, being very persistent from day to day at any given location. The vertical circulation (upward at low latitudes and poleward aloft) of the Hadley cells accounts in large part for the transport of heat and momentum in the tropics. The return flow of these cells—the trade wind systems near the earth's surface—transports water vapor into the intertropical convergence zone.

At middle latitudes the situation is quite different. The vertical cell here is not an observed feature but rather the statistical result of transports computed on a daily

Atmosphere, Planetary Circulation of

1 Viewed from a weather satellite over the Arctic Circle, the intertropical convergence zone, where air is rising, shows as a band of clouds near the Equator. Clear sections mark sinking air within subtropical high-pressure cells. Spiraling cloud formations reveal polar-front cyclones, and vortices of tropical cyclones can be seen at lower altitudes.

2 The principal features of global circulation are convection cells which result from the variations in solar heat received on the earth from the Equator to the poles. If the earth had a smooth surface and did not rotate, there would be only one such cell; air warmed at the Equator would rise, flow poleward, sink as it cooled, and return to the earth's surface near the Equator. The earth's rotation and a surface that is not uniform complicate this picture. Friction between the surface and the moving atmosphere in contact with it produces circulation cells north and south of the Equator in the tropical zones and at mid-latitudes as well as at the poles. The earth's rotation deflects the winds produced in these cells, creating world-wide circulation patterns such as the trade winds.

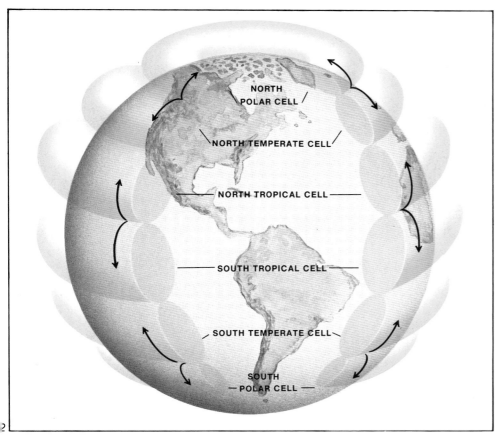

Atmospheric Electricity
Multiple forks of lightning—electric discharges as high as 100 million volts—illuminate the night sky near San Francisco, Calif.

basis. This indirect cell is driven by the Hadley circulation on one side and the cyclonic disturbances in the polar-front zone on the other. The actual transport of heat and momentum at these latitudes appears to be accomplished not by a vertical circulation *per se* but rather, in horizontal planes, by the traveling eddies of the atmosphere: the waves in the westerlies with their accompanying cyclones and anticyclones. The northerly and southerly winds on either side of the waves and cyclones transport vast quantities of cold air into the tropics and warm air toward the poles. The exchange of momentum by the horizontal eddies is accounted for by an asymmetry in the motion, indicated by a "tilt" in the axis of the planetary waves. See also Atmospheric Pressure; Atmospheric Waves; Vorticity; Winds. (M.F.H.)

Atmospheric Density. The ratio of the mass or quantity of air to the volume it occupies. The density of dry air for the standard atmosphere as defined by the temperature and pressure at sea level is 1.225 kg per cubic meter. However, since the lower atmosphere is greatly compressed by the weight of the air above, atmospheric density decreases markedly with elevation. Thus the mass of air that weighs about 2.2 pounds and normally fills one cubic meter of space at sea level would occupy 100 cubic meters at a height of 32 km, 1,000,000 cubic meters at 95 km and 1,000,000,000 cubic meters at 158 km above sea level.

Ordinary meteorologic instruments are not designed to measure the air density directly at the earth's surface or in the free atmosphere up to heights of about 30 km. Instead, the density is computed from observations of pressure and temperature by means of the equation of state for an ideal gas, which relates the density to the pressure, temperature and molecular weight of the gas. However, in the upper atmosphere the density becomes a primary measurement, since it is more readily observed than the other variables. Below 200 km the density has been measured directly by observing the motion of freely falling spheres ejected from rockets. The deceleration of such a sphere is proportional to the frictional drag of the air, which in turn is proportional to the air density. Above 200 km the atmospheric density has been determined from calculations of the observed effect of air drag on the motions of satellites. (M.F.H.)

Atmospheric Electricity. Electrical phenomena occurring in the earth's atmosphere, including the earth's fair-weather electric field, the air-earth current and luminous discharges such as lightning and St. Elmo's fire. A permanent fair-weather electric field exists in the atmosphere. The surface of the earth carries a negative charge and the upper atmosphere a positive charge. Near the surface of the earth the atmospheric field strength is about 100 volts per meter; in fair weather the field decreases to about 5 volts per meter at a height of 10 km. However, near thunderstorms and under convective clouds, the surface electric field varies widely, not only in magnitude but also in direction. Calculations have shown that if the atmospheric electric field were not maintained in some way, it would be neutralized within a few minutes by the air-earth current—a flow from the positively charged atmosphere to the negatively charged earth. Early in the 20th century the British scientist C. T. R. Wilson suggested that the earth's negative charge is maintained by thunderstorms, which can generate the necessary supply current. More than 2000 thunderstorms are in progress over the earth's surface at any given time. Most are negatively charged at their bases, and it is believed that these storms maintain the earth's negative charge by drawing a large positive charge from the surface. Support for this view is found in the fact that the total number of thunderstorms over the entire earth shows the same variation during the Greenwich day (the 24-hour period beginning at midnight, Greenwich Meridian Time) as does the earth's fair-weather electric current. The latter reaches a maximum at 1900 GMT.

As more and more electric charge builds up in the thundercloud, the voltage difference between the cloud and ground increases until the air between them can no longer prevent a discharge. Lightning is believed to be one of the means whereby a net negative charge is transferred from the atmosphere to the earth. Lightning is a large-scale example of a spark, a gaseous electrical discharge that occurs along a relatively narrow path of high ion density. It is highly luminous and of short duration. An arc of lightning passes back and forth from cloud to ground with many subsequent alterations of flow until the accumulated electric potential is relieved.

However, there is evidence that what is called a point discharge is of greater importance in balancing the air-earth current. A point discharge is a silent, nonluminous, gaseous electrical discharge that comes from a pointed conductor. During unsettled weather, especially thunderstorms, trees and other grounded objects with points or protuberances may become sources of point discharges into the atmosphere. Intermediate between the spark and point discharges is the corona discharge, often called St. Elmo's fire. This discharge is luminous and often audible; like the point discharge, it comes from pointed objects that are exposed to the atmosphere. (M.F.H.)

Atmospheric Humidity. In general, a measure of the water-vapor content of the air. The two best-known measures of humidity are relative humidity, popularly referred to simply as humidity, and dew point.

Relative humidity is, approximately, the ratio of the water vapor actually present in the atmosphere to the maximum water vapor possible at the prevailing temperature, expressed as a percentage. It is defined more rigorously as the ratio, expressed in percent, of the observed vapor pressure to the saturation vapor pressure; the latter quantity is the pressure of the vapor of a substance when the vapor is in equilibrium with a plane liquid or solid surface of the pure substance. At saturation there is no net transfer of molecules from the liquid or solid to the vapor phase, or vice versa; the saturation vapor pressure of a substance varies with the temperature alone.

Relative humidity is the most common measure used to indicate the effect of moisture on human comfort. At higher temperatures, high relative humidity limits drying and evaporative cooling of the skin. At low temperatures, high humidity increases conductive heat loss from the skin. Although relative humidity has many other practical applications, it is less useful in meteorologic work than certain other measures of humidity, since its changes may—and normally do—reflect variations of temperature as well as of water vapor.

The dew point is the temperature at which the saturation vapor pressure of the air, dependent only on its temperature, is equal to the pressure of the water vapor actually present. Ordinarily, condensation is assumed to begin when the temperature of the air is lowered to its dew point. Actually, condensation in the atmosphere may occur at temperatures above the dew point when water-soluble particles, called hygroscopic condensation nuclei, are present; or it may be inhibited when suitable condensation nuclei are not present in the air. (M.F.H.)

Atmospheric Motions. The typical motion patterns of the atmosphere, ranging in size from minute turbulent eddies from 1 to 10 cm in diameter, to the large-scale wind systems—the trade winds, for example—that form the atmosphere's general circulation. Between these two extremes are numerous motion systems of characteristic size and duration. In general, the lifetime of an identifiable pattern not fixed by the earth's topographic features is roughly proportional to its size.

Included in the range from 1 m to 1 km are the larger turbulent eddies, gusts, tornadoes and convective clouds. Fronts, squall lines and tropical cyclones are of the order of 10 to 100 km in size. Extratropical cyclones and anticyclones are roughly 1000 km in horizontal extent, while the length of a planetary wave is about 10,000 kilometers.

Although the atmosphere extends upward indefinitely, most of its mass is concentrated in a thin layer surrounding the earth, so thin that its depth with relation to the diameter of the earth has been compared to the skin of an apple. Thus the vertical extent of even the largest motion systems of meteorologic interest is little more than 20 to 30 km. (The atmosphere's tidal motions, and similar rapidly moving waves, constitute an exception.)

It is of some interest that typical wind velocities in the atmosphere are independent of scale; that is, winds of 10 to 100 m per second may be found not only in the planetary wind systems but also in small-scale phenomena such as thunderstorms, squall lines and tornadoes. The differences in wind velocity from place to place also are independent of scale. The divergence of the wind velocity and the lateral spreading out, as well as the lateral convergence of the air, are roughly proportional to velocity differences divided by a typical length. Thus it follows that the horizontal divergence is inversely related to the scale of the disturbance. As a practical consequence, the typical vertical velocities in the atmosphere are also inversely related to the scale of the phenomena, for the vertical motions of the air are controlled essentially by the divergence of the wind velocity. The vertical velocities of the larger-scale motion systems of the atmosphere are therefore small, ranging from 1 cm to 1 m per second, while the updrafts and downdrafts in smaller systems, such as severe thunderstorms, may be 50 times greater.

In addition to the atmosphere's dimensions, *solar radiation, gravity, the earth's rotation and friction are important factors in explaining the air's patterns of motion.

Unbalanced pressure forces are the immediate cause of motion in the atmosphere. Since the atmosphere is a fluid in approximate hydrostatic equilibrium—that is, the vertical acceleration is quite small compared with the acceleration of gravity—pressure forces reflect the weight of the air above. It follows that gravity is one of the fundamental forces involved in the air's motion. Ultimately, however, atmospheric motions derive their energy from solar radiation—more specifically, from the unequal distribution and absorption of solar energy at the earth's surface and within the atmosphere. This unequal absorption of energy provides the temperature differences that result in air circulation.

Most incoming solar radiation not immediately reflected to space is absorbed at the earth's surface; thus the surface, in combination with the cold atmosphere above, is one of the two primary energy sources for atmospheric motion. The towering cumulus clouds that develop over heated land on a summer afternoon offer dramatic example of this heat source at work on a small scale. The surface air, becoming heated and buoyant, is forced upward through the denser air of the environment by vertically directed pressure forces. On a much larger scale, the differential heating at low and high latitudes results in unbalanced pressure forces in the horizontal plane. It is essentially the equatorial heat source, combined with the polar energy sink, that maintains the general circulation of the atmosphere.

The earth's rotation acts to deflect the flow of air, which otherwise would stream from high to low pressure, so that the larger-scale air motions are nearly parallel to the isobars, the lines connecting points of equal pressure. The rotating earth also provides a ready source of spin for the atmosphere above; when this spin is concentrated by converging air motions, vortex patterns such as cyclones, hurricanes and tornadoes develop. The *Coriolis force and its variation with latitude also partially explain why wave patterns develop in the zonal winds of the atmospheric circulation. Atmospheric motions are continually dissipated by friction, but this loss is continually being made up by the conversion of solar energy into the kinetic energy of the winds.

The variety of motion patterns in the atmosphere is most dramatically revealed by cloud photographs from satellites. These, together with photographs from aircraft and ground observation points, show the intricate and sometimes surprising patterns of atmospheric motions. (M. F. H.)

Atmospheric Pressure. The pressure, or force per unit area, of the atmosphere. This pressure results from the gravitational attraction exerted upon a column of air above a given point. Thus it is essentially a measurement of the "weight" of the atmosphere. The pressure at any point acts equally in all directions.

The most common unit of pressure used in meteorology is the millibar, equal to the force exerted by 1000 dynes acting on an area of one square centimeter. (A force of 1 dyne is that required to give a mass of 1 gram an acceleration of 1 centimeter per second each second.) Another unit still in common use is inches of mercury, that is, the height of a column of mercury that exactly balances the weight of the column of atmosphere above the measurement point. The mercury barometer (one of many types of barometer in use) is a highly accurate instrument for measuring atmospheric pressure.

Pressure differences within the atmosphere are the immediate cause of air motion and are often referred to collectively as the pressure-gradient force. Normally the vertically directed pressure force is balanced by the force of gravity; when the temperature is known, this balance makes it possible to determine the variation of pressure with height. The horizontally directed pressure force is the immediate cause of *wind. In the free atmosphere, the large-scale motions are approximately represented by the geostrophic wind, which is the motion arising from a balance between the horizontal pressure-gradient force and the *Coriolis force, an apparent force due to the earth's rotation. Under this condition, the wind blows parallel to the isobars (lines connecting points of equal pressure). However, at the earth's surface the additional force of friction causes the wind to blow at an angle across the isobars, the flow being partly from high to low pressure but having a component parallel to the isobars.

Because of the relation between the field of pressure and the air's motion, the average distribution of pressure at various elevations is useful in studying the general circulation of the atmosphere. At the earth's surface the isobars delineate the semipermanent belts of high and low pressure. If the earth's surface were completely level and homogenous, the pressure distribution would presumably be completely zonal. This idealized zonal distribution of pressure is most nearly approximated in the Southern Hemisphere, where the earth's surface is 81 percent water. In each hemisphere, a low-pressure belt is evident at high latitudes; another, at the Equator, separates belts of high pressure girdling the earth at subtropical latitudes. High-pressure areas, or *anticyclones, overlie the polar regions.

In the Northern Hemisphere, however, the large landmasses greatly alter the underlying zonal distribution. The effects of seasonal temperature changes on the average pressure distribution are especially marked. In winter, the cold land surface cools the overlying atmosphere and the dense air settles close to the surface, creating high-pressure centers. At the earth's surface there is a net outflow of air from the continental anticyclones, particularly over the eastern and southern borders of Asia, producing the winter monsoons of India- and Southeast Asia. The winter monsoon is characteristically cold and dry. In summer, the warm continents have an upward flow of air, which moves out over the adjoining oceans, resulting in low-pressure areas over the land. The return flow of air at the surface, from the oceans into the continental low-pressure areas, produces the summer monsoons. These are characterized by cloudiness and often by copious rains.

The major wind belts at the earth's surface might be inferred in a general way from the known relation between wind and pressure field. The easterly trade winds blow obliquely from the subtropical high-pressure belts toward the equatorial low pressure. The surface westerlies blow obliquely out of the subtropical highs into the low-pressure belts at high latitudes. The polar easterlies, barely evident, blow from anticyclones over the polar regions, with a component toward the west. As the seasons change, the major wind systems of the earth reflect the seasonal migration of the high- and low-pressure belts and the contrasting pressure distributions over the continents and the ocean. (M.F.H.)

Atmospheric Radiation. Infrared radiation emitted by or propagated through the atmosphere. It is of major importance in determining the vertical temperature distribution and the heat balance of the earth-atmosphere system. Water vapor and, to a lesser extent, carbon dioxide in the air absorb infrared radiation from the earth's surface (*terrestrial radiation) at certain wavelengths. This selective absorption allows other portions of the terrestrial radiation to pass directly from the earth's surface to space. The energy absorbed is re-emitted, partly upward and partly downward; it is repeatedly absorbed and re-emitted as it passes through the atmosphere. The downward flow of energy keeps the earth's surface much warmer than it would be without it (this is known as the *greenhouse effect); the upward flux represents energy loss in the planet's heat-balance equation. (M.F.H.)

Atmospheric Tide. An atmospheric oscillation of the scale of the earth, so termed because of resemblances to the oceanic tide. Atmospheric tidal motions involve the entire atmosphere and are caused by both gravitational and thermal action. Both the

sun and the moon produce gravitational atmospheric tides, but the amplitude of these is comparatively small. The lunar tide, for example, produces a pressure wave at the earth's surface whose maximum amplitude is about .08 millibar; the solar gravitational tide is smaller. The atmosphere's thermal tides are produced by the earth's diurnal motion, which alternately exposes half the earth's surface and atmosphere to solar heating and to darkness.

The atmosphere's tidal motions are most pronounced in the upper air, where the air density is low and periodic heating and cooling is more extreme. In the mesosphere the tidal winds apparently constitute a dominant feature of the organized motion, reaching speeds of tens of meters per second. By contrast, in the lower atmosphere the thermally produced tidal winds are of the order of tens of centimeters per second.

The dominant thermally produced tide is semidiurnal, resulting in two high points and two low points of barometric pressure during each 24-hour period. The pressure wave is largest in the tropics, where its amplitude may reach 1 or 2 millibars and mask other pressure variations. At higher latitudes, the large interdiurnal variations of pressure associated with the passage of *cyclones and *anticyclones normally conceal the tidal pressure variation. (M.F.H.)

Atmospheric Waves. Generally, any pattern in the atmosphere having an identifiable periodicity in time and/or space. Sound or acoustic waves are caused by a mechanical disturbance and propagate through the atmosphere as a result of the alternating compression and rarefaction of the air. The air particles oscillate in a direction parallel to the direction in which the sound wave moves. Gravity waves, in which buoyancy or reduced gravity acts as the restoring force on particles displaced from a position of equilibrium, are familiar to most of us in the form of ripples on a pond or waves seen from the beach. However, gravity waves also occur in the atmosphere and are being increasingly studied.

In numerical predictions of *atmospheric motions, both sound and gravity waves are eliminated from the equations of motion in order to forecast the waves or other disturbances significant for weather forecasting. The planetary waves of the atmosphere are of this type. Planetary waves, also called long waves or Rossby waves, are horizontal wavelike patterns or meanders in the major belt of westerly winds. The wavelength of a planetary wave is of the order of 10,000 km, longer than that of the *cyclonic and *anticyclonic dis-

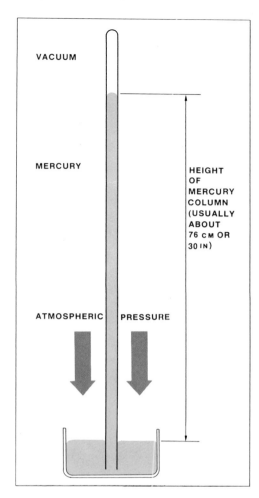

Atmospheric Pressure
In a mercury barometer, the weight of the atmosphere pressing down on the mercury in the dish drives mercury up the glass tube. Lower atmospheric pressure lets it drop.

Atmospheric Waves

1 The platelike or laminated form of this mountain wave cloud over Mount Shasta, Calif., is believed to result from variations in humidity. As the air moved over the mountain, condensation was apparently more extensive in the upper layers, perhaps because of the pattern of the streamlines.

Atoll

2 A fringing reef grows upward as a volcanic cone subsides and forms a barrier reef. Eventually the cone disappears, leaving a lagoon surrounded by an atoll.

3 Atolls vary from less than 1.5 km (1 mile) to 30 km (20 miles) wide.

turbances of the lower troposphere. The pure wave type that most closely resembles the planetary wave is the Rossby wave, which depends for its existence on the variation of the *Coriolis force with latitude.

When air is forced over a mountain range, oscillatory motions in the vertical plane are often produced downstream. In the crest of such mountain waves, *clouds may form as a result of cooling and condensation of water vapor in the lifted air. See Adiabatic Process; Geostrophic Wind. (M.F.H.)

Atoll. A circular or oval-shaped *coral reef that forms an island enclosing a body of water called a lagoon. Atolls may rise thousands of meters above the sea floor and are most abundant in the Pacific Ocean. Most atolls are only a few kilometers in diameter. The highest portion of the atoll surrounds the lagoon. Coral reefs in general, and atolls in particular, show distinct zones of environments and organisms. On the windward side, the base of the seaward slope is commonly covered with debris from the upper part of the reef. At a depth of 60 m or so, the reef fauna begins, particularly the coral and coralline or red algae that build the framework of the reef. The crest of the windward reef contains a narrow ridge of coralline algae and a broad reef flat. Reef islands may be present, composed of debris brought in by waves. Behind the island—that is, on the protected side—a steep slope leads into the lagoon; its zonation is different from that of the windward slope because there is less wave energy at the leeward side of an island and consequently fewer high-energy forms. The leeward reef may contain many of the same organisms and zonations as the windward reefs but they will be less well developed. Small patch reefs or pinnacle reefs may be present in the lagoon.

A number of theories have been suggested to explain the origin and development of atolls. The first comprehensive theory, proposed by Charles Darwin in his *Coral Reefs* (1842), has stood the test of time and is compatible with data collected in recent years. Darwin's theory is based on sinking or subsidence of volcanic islands in relation to sea level, and the simultaneous growth of a coral reef. Initially a fringing reef develops around a volcanic island. Subsidence of the volcano and upward growth of the reef produces what is known as a *barrier reef with a lagoon between it and the island itself. Continued subsidence will cause the island to submerge, leaving the atoll in the form of a reef with a central lagoon. (R.A.D.)

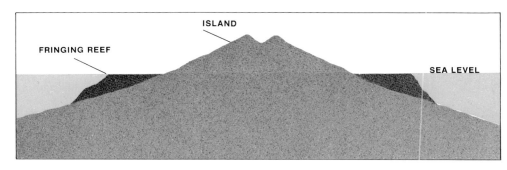

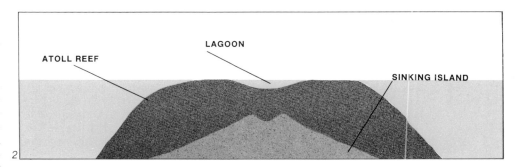

Attitude. The spatial relationship of both planar and linear features of rock to an imaginary horizontal plane. Attitudes of planar features are expressed by their *strike and *dip, while linear features are described by azimuth and plunge.

Attrition. *Abrasion or wearing away of particles carried by a stream as they rub or strike against each other and the channel bed, so that grains become smaller downstream. The rate of attrition depends upon the original size of grains, mixture of sizes in transport and the mineral character of the grains. Quartz resists attrition more than feldspars, hornblendes or pyroxenes. Gravels are more quickly reduced in size than sands. In natural streams, *weathering and additions from tributaries may also influence attrition downstream.

Augen. A large lenticular-shaped grain, or aggregates of minerals forming eye-shaped structures, as viewed in cross section of a rock. Augen are usually associated with *metamorphic rocks, especially with coarse-grained banded varieties such as *schist and *gneiss. Augen usually consist of the minerals quartz, feldspar or garnet. The term is derived from the German word for eye.

Augite. A mineral, calcium, magnesium and iron silicate, the most common of the *pyroxene group. It is an important rock-forming mineral found chiefly in dark *igneous rocks. See Minerai Properties.

Aureole. A zone of *metamorphic rock surrounding a once-molten body of rock (*magma). Aureoles are products of chemical and mineralogic changes brought about by heat and the migration of chemically active fluids from the magma into the earlier-formed solid rock. Alteration is greatest along the solid rock-magma contact (*contact metamorphism), decreasing with distance from the contact until a point is reached where chemical and mineralogic changes have not occurred in the surrounding rock. Within this zone or aureole, rings of variable mineral composition develop in response to the variable temperature, pressure, and bulk chemical composition of the rocks. (J.W.S.)

Aurora. Emissions of radiant energy from the upper atmosphere over the middle and high latitudes of the earth. Often described as one of nature's most awesome and beautiful displays, the aurora assumes a variety of rapidly moving, changing and sometimes pulsating shapes and forms, predominantly pale green with occasional patches of red and pink. The forms are usually described as arcs, rays, draperies, crowns and bands. Diffuse luminous areas often appear after intense displays of rays and curtains.

The aurora is related to streams of charged particles from the sun caught in the earth's magnetic field. Intense displays occur after the appearance of flares on the sun during periods of high solar activity. At these times the earth's magnetic field is apparently disrupted by streams of high-speed protons and electrons from the sun. The disturbances in the field cause particles trapped in the earth's *magnetosphere to escape in large numbers from the radiation belts at the points where the horns of the crescent-shaped belts are directed toward the earth. The high-speed particles, together with those from the sun, energize the gases of the upper atmosphere just as an electric current energizes the gas in a neon-filled tube and causes it to glow. The green hues of the aurora come from oxygen and the red from both oxygen and nitrogen.

In northern latitudes the auroral displays are called the aurora borealis, aurora polaris or northern lights; in the southern hemisphere they are known as the aurora australis. The aurora, most intense during great *magnetic storms, also reaches its lower latitude and greatest height at these times. Normally, in the polar regions, the displays extend from about 80 to 250 km above the earth, but during intense storms the streamers may extend upward to heights of 800 to 1000 km. (M.F.H.)

Authigenic. Minerals that originated at the time of or after the formation of the rock in which they are found. The word is applied particularly to minerals such as *tourmaline crystals that have formed in sedimentary rock at the site where they are found.

Authigenic Sediment. Minerals that are direct inorganic precipitates from seawater. A great variety of marine sediment is formed in this manner, including manganese and phosphorite nodules, zeolites, glauconite and other clay minerals, as well as some minerals that are relatively rare. Such precipitation depends on many factors, particularly ionic concentrations, circulation and the influx of sediment from land.

Autochthon. A massive body of rock essentially unmoved from its site of origin; used in comparison with *allochthon, which has been moved from its original site. Autochthonous masses may have undergone extensive deformation.

Autunite. A mineral, the yellow-green, platy, secondary hydrated phosphate of uranium. It is found in the oxidized zone overlying deposits of *uraninite. See Mineral Properties.

Available Relief. Vertical elevation from upland divide to the mouth of the main valley floor. It determines the amount of headwater erosion and the depth to which valleys will be cut, and is thus an important limiting factor in drainage development.

Avalanche. A very rapid movement of large masses of snow, ice, debris or rock down steep mountain slopes. The most spectacular and dangerous of all mass movements, an avalanche may fall, slide, roll or flow.

Snow avalanches may be composed of powdery, wet or slab snow. Some consist of clean snow; others contain rock debris. Large, dangerous avalanches originate on slopes of 30° - 45°. On steeper slopes there is little snow; they rarely occur on slopes much less than 30°. Tens of thousands take place in Switzerland every year. Their velocity may be as much as 320 km per hour, and they may create dangerous air blasts. They form minor geomorphic features including avalanche boulder tongues, chutes and striations.

Ice avalanches are common on mountain glaciers where ice falls occur or where hanging glaciers terminate on high cliffs. Here the blocks of ice that fall into the valley below may reunite and form reconstructed glaciers.

Debris avalanches are usually formed on steep slopes from *regolith containing fine-grained material. Saturation by torrential rains gives the material mobility. Earthquakes may trigger them. The light-colored scars on the forested slopes of the White Mountains of New Hampshire, the Green Mountains of Vermont and elsewhere result from debris avalanches.

Rock avalanches derive essentially from *bedrock. They originate as rock slides and falls, often triggered by earthquakes. They are fostered by *joints and *bedding planes parallel to steep slopes, by the presence of clays and shales and by heavy rains. During descent the rock is shattered and its deposition results in a chaotic topography.

Rock avalanches are accompanied by a roar. They can flow laterally for several kilometers and have been reported to climb up slopes more than 450 m. Their great mobility is thought to be due to interstitial air that reduces internal friction and to a cushion of compressed air that they ride like a hovercraft. They generate winds strong enough to pick up men and even

Avalanche

1 The Blackhawk Slide in the Mojave Desert, Calif., is a prehistoric rockfall consisting of small rock fragments not accompanied by clay or silt. It originated in the mountains in the background and its mobility may have been due to air trapped beneath it.

2 A massive cascade of wet or powdery snow, often accompanied by tornado-like winds, can reach a speed of up to 300 km (about 180 miles) per hour. The wet type shown here is known in the Alps as a Grundlawine.

automobiles, and may form giant waves if they land in a body of water. The highest landslide wave ever recorded occurred in 1958 in Lituya Bay, Alaska. A rock avalanche (set off by an earthquake) with a volume estimated at 30.5 million cubic meters fell from a height of 914 m (3000 feet) into the bay; a wave 536 m (1720 feet) high rolled up on the opposite side of the bay and destroyed a forest. A rock avalanche killed 2600 people in 1963 in the Vaiont Valley in the Italian Alps; more than 225 million cubic meters of rock debris fell into a reservoir and a series of waves rolled over the dam, flooding and destroying everything downstream for more than 20 km. A rock avalanche at Elm, Switzerland, in 1881 killed 115 people; another at Frank in Alberta, Canada, in 1903 killed 70 people.

Avalanche chutes are trenches or channels cut in the bedrock of valley walls by the action of repeated avalanches over a long period of time. (R.L.N.)

Aventurine. Varieties of both *quartz and *feldspar colored by inclusions of other minerals. Aventurine quartz may have iron oxide, chromium mica or a number of other minerals disseminated within it. It may be yellow, brown, green or red depending on the included mineral. More commonly seen as an ornamental material than natural aventurine is a glass imitation known as goldstone or aventurine glass. Light reflected from flakes of incorporated copper give it a yellow or red color. See Sunstone.

Axial Angle (or Optic Angle). In crystallography, the acute angle between the optic axes of a biaxial crystal, designated as 2V. Light rays traveling along the optic axes are refracted on leaving the crystal. The apparent axial angle, 2E, is the angle measured in air between these rays.

Axial Ratio. A ratio that expresses the relative lengths of crystallographic axes. In the *tetragonal and *hexagonal crystal systems, the length of the c axis is expressed in terms of an a axis taken as unity. For example, in quartz, $a : c = 1 : 1.102$. In the *orthorhombic, *monoclinic and *triclinic systems, the lengths of the a and c axes are expressed in terms of the b axis taken as unity. There are thus two ratios, a:b and b:c. For barite it is given as $a:b:c = 1.627 : 1 : 1.310$. See Crystal Axis.

Axis of Symmetry (or Symmetry Axis). An imaginary line through a crystal about which the crystal may be rotated and brought into identical position during a complete rotation. See Crystal Symmetry.

Azurite. A mineral, the deep blue hydrous copper carbonate. Unlike many minerals that may have a variety of colors, azurite has a constant blue color. It occurs in the oxidized zone overlying copper deposits, where it is usually found with the green copper carbonate *malachite. It is also commonly associated with iron oxide, *cuprite and native copper. Fine crystals have come from Chessy, France; Tsumeb, South-West Africa; and Bisbee, Ariz. See Mineral Properties.

B

Bahada (or Bajada). A depositional feature formed in a basin under arid or semiarid conditions. It may consist of coalescing *alluvial fans or may be built up of material washed down from the fans. It is an undulatory form composed of alluvial sands and gravels, debris flows and *mudflows. Bahadas are common in the Basin and Range province of southwestern United States and in drier mountainous areas of Asia and Australia. See Slump.

Barchan. See Dune.

Barite. A widely distributed mineral, barium sulfate. Its principal occurrence is in association with ores of lead, silver, copper, cobalt and antimony in metalliferous veins. It is also found in limestone in cavities, veins or replacement deposits as a major constituent and, due to its insoluble nature, in residual masses in clay overlying limestone. Barite is the chief source of barium for the chemical industry. Because of its high *specific gravity (4.5), its major use is as a weighting agent in oil-well drilling. The drill hole is charged with mud containing finely ground barite to prevent oil or gas from being blown from the hole. Well-crystallized specimens of barite, examples of almost model-perfect orthorhombic symmetry, have been found at many localities, but the finest have come from Cumberland and Westmoreland, England. See Mineral Properties; Appendix 3. (C.S.H.)

Barrier Beach (or Barrier Island, Offshore Bar). An elongated sand and gravel ridge roughly parallel to a coastline and separated from it by a lagoon and/or a tidal marsh. The crest of the barrier rises above high tide. It is generally composed of dunes of sand blown from the beach face. A constructional landform, it may be as much as 15 km (10 miles) long but is usually less than 1 km wide. A tidal inlet frequently separates one barrier from another. The barrier at Cape Hatteras, N.C., is more than 30 km from the mainland, but in most places barrier beaches are much closer to the coast.

They move shoreward as *wind and *waves transport material from the oceanward side to the landward side, and they may eventually merge with the mainland. They are found along coasts with gently sloping ocean bottoms. Some may have been *spits that were separated from the mainland by inlets cut through them during storms. Others were formed offshore by material thrown up from the ocean bottom by wave action. The barrier beach is in a dynamic environment. It is continually being modified by shore cur-

Barite
1 This soft but heavy sulfate mineral, shown here in a typical tabular-shaped crystal, is used in the paint and oil industries and is the principal source of the element barium.

Barrier Beach
2 A long, broken ridge of sand and gravel, separated from the coast by a lagoon and estuary such as this one near Mazatlán, Mexico, is formed by wave deposits and modified by longshore currents and wave action.

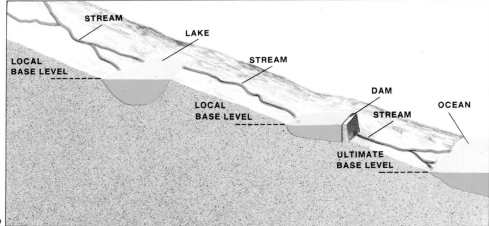

rents and wave action. Good examples are found along the Atlantic and Gulf coasts of the United States. The communities of Atlantic City, N.J., Daytona and Miami Beach, Fla., and Galveston, Texas, are all built on barrier beaches. A barrier chain is a series of barrier beaches separated by tidal inlets. (R.L.N.)

Basal Conglomerate. A coarse, *clastic sedimentary rock composed of well-sorted, homogeneous rock fragments at the base of an overlying sedimentary sequence resting on a surface of erosion. It typically forms the initial *stratigraphic unit in a marine series and marks or immediately rests on an *unconformity. It originates as advancing seas erode the land to produce a nearly flat rock surface covered by a relatively thin layer of gravel and coarse sand.

Basalt. A fine-grained, dark-colored volcanic rock, the extrusive equivalent of a *gabbro. The principal constituent of the groundmass is labradorite *feldspar, but more calcic *plagioclase may be present as *phenocrysts. The common dark minerals are augite and olivine. Augite may be found in both groundmass and phenocrysts, whereas olivine is usually found in phenocrysts. There are small amounts of hornblende and biotite in some basalts.

Basalt is the most abundant volcanic rock covering or underlying the sediments of the ocean floors. In such places as the Columbia River Plateau it forms *lava flows covering tens of thousands of square kilometers. Elsewhere, as in the Hawaiian Islands, basalt forms gigantic volcanoes. (C.S.H.)

Basalt, Plateau. See Plateau Basalt.

Base Level. The limiting level to which continents may be eroded. The grand base of erosion is sea level, but there may be temporary, local base levels at any elevation. A resistant rock, for example, may provide the temporary level to which streams locally degrade their channels and below which the *watershed cannot be eroded. A basin with interior drainage has a local base level that is the lowest elevation of the basin. Dams form the local base level for streams entering a reservoir.

Basement Complex. Any group of undifferentiated rocks that lie below the oldest identifiable rock sequence in an area. Generally, basement complexes consist of volcanic and metamorphic rocks of *Precambrian age. However, some complexes may be of Paleozoic, Mesozoic or Cenozoic age. See Craton.

Basalt Conglomerate
1 Pebbles and boulders of a basal conglomerate rest on an eroded surface of Cretaceous siltstone overlain by Tertiary sandstone in California.

Base Level
2 The ocean is the lowest, or ultimate, base level to which a stream can cut its channel. The surface of a lake or a dam that interrupts the downhill course of a stream becomes the local base level, and the stream becomes graded to it.

Basalt
3 The cliffs of the Faeroe Islands north of Scotland have eroded to reveal alternating layers of tuff and basalt, evidence of the islands' volcanic origin. Each basaltic layer represents a volcanic eruption. Tuff is consolidated volcanic ash.

Basement Rock. That portion of the *continental crust composed of *Precambrian rocks. See Craton.

Basin. A depression within the earth's *crust that has been filled with sediment. Basins are further subdivided into classifications based on (1) their location and time of formation relative to larger crustal features and (2) their geometry. See Geosyncline.

Batholith. A large body of medium- to coarse-grained rock of *igneous origin, with a surface exposure that exceeds approximately 100 square kilometers. Batholiths are thought to originate deep in the earth's crust; gravity data suggest that some have floors as much as 10 km beneath the earth's present surface. The origin and mechanics of implacement of batholiths is still not agreed upon; however, they are known to be related to areas of the earth's crust where significant downwarping and later uplift have formed mountain belts. The core of the Sierra Nevada in California and Pikes Peak and adjacent mountains in central Colorado are of batholithic origin. In southeastern Australia numerous batholiths of *Ordovician to *Permian age are found in the Tasman Geosynclinal Zone. (J.W.S.)

Bathyal Zone. The region between the edge of the *continental shelf and the ocean floor. It extends from about 200 to 2000 m in depth, or to a depth where temperatures do not go above 4°C (39°F). The upper boundary is actually better defined as the maximum depth of light penetration. The general conditions of the bathyal zone include absence of light, temperatures ranging from 4° to about 15°C (39° to 59°F) and a normal marine salinity of 34 – 36 parts per thousand. Although this zone has fairly uniform conditions, a variety of sediment accumulates. Terrigenous (land-derived) sediments including *clays, *turbidites and *glacial marine sediment, organic *pelagic oozes made up of foraminifera, pteropods (planktonic snails) and coccoliths, as well as *authigenic sediments are relatively common. (R.A.D.)

Bathymetry. The measurement and mapping of the ocean floor. Such information is essential for navigation as well as for scientific investigation of the *oceanic basin. Until about World War II, only the *continental shelf areas had been relatively well mapped. At present there is a concentrated effort by many countries to assimilate depth data and compile accurate charts of the entire ocean floor. Charting such areas

3

1

Bathythermograph

1 *This instrument is towed by an oceanographic research vessel in the upper 100 – 300 m of the sea to detect temperature variation. Here it is being hoisted aboard for a reading.*

Beach

2 *A cross-profile of a typical temperate-zone sandy beach shows the zone of breaking waves on the sea side and the zone of wind action and dune development on the landward side. In summer, waves of low height and energy build up a shelf called a berm. The stronger storm waves of winter build up a winter profile and berm farther inland.*

Baymouth Beach

3 *The baymouth beaches on the south shore of Martha's Vineyard, Mass., were formed by spits extending from the headlands.*

depends on the records of the precision depth recorders carried by many commercial ships and all scientific vessels. The excellent physiographic charts of the ocean floor prepared by Bruce C. Heezen and Marie Tharp of the Lamont Geological Observatory result from such records.

Bathyscaph. A free-moving vessel used to explore the ocean depths. The first such vessel, the *Trieste,* was designed and constructed in 1954 by the Swiss physicist Auguste Piccard and his son Jacques. The *Trieste* consisted of a 2-meter-wide observation sphere attached to the base of a submarine-like hull that housed the engine and propellers and served as a ballast tank. The *Trieste* and her successor, *Trieste II,* have descended to depths of more than 9100 m. See Deep-Sea Research Vehicle (DRV).

Bathythermograph. An instrument that measures and records temperature to depths of 300 m. A torpedo-shaped apparatus just under 1 m in length, it is towed behind a moving ship. It is thus restricted to use in the upper waters, and is commonly employed in studying the *thermocline (a thin zone of temperature and density change 100 – 200 m below the surface). Because the bathythermograph can be used while a ship is under way, whereas other measuring instruments require the ship to be stationary, it saves oceanographers much time and money. (R.A.D.)

Bauxite. The ore of aluminum. Bauxite has been the only commercial source of aluminum since the beginning of the 20th century. At that time it was believed to be a mineral and was assigned a chemical composition and described in terms of its physical properties. Much later, mineralogic study showed it to be a fine-grained mixture of several minerals, and thus more properly designated a rock. Its principal mineral constituents are the hydrous aluminum oxides, gibbsite, boehmite and diaspore, any of which may be dominant.

Bauxite forms residual deposits in tropical or subtropical regions by the weathering of rocks high in alumina. In temperate zones clay, essentially aluminum silicate, is the weathering product of such rocks, but in the Tropics the silica also is removed, leaving only aluminum and iron oxides. These residual concentrations, known as *laterites, form a surface blanket over vast areas in regions where the *bedrock is rich in alumina and iron or iron alone. If the underlying rock contains only small amounts of iron, bauxite results; if it has high iron and low alumina, an iron ore is formed. (C.S.H.)

Baymouth Beach (or Baymouth Bar, Bay Barrier). A ridge of sand, gravel or both built by waves and shore currents completely or almost completely across the mouth of a bay from one headland to another. It is usually formed by a *spit growing from one or both of the headlands. Such beaches are common along irregular coastlines resulting from submergence. With the growth of the beach the bay will be transformed into a lagoon. A complete closure is made only if shore currents are stronger than currents running in and out of the lagoon. The lagoon may slowly fill with material swept by waves and wind from the beach, with *alluvium carried into it by rivers and with the accumulation of plants growing in it. *Deltas, mudflats and marshes may eventually fill it. The headlands will be eroded and the beach will retreat across the lagoon or over mudflats or marshes until it joins the mainland. Good examples are found on the south side of Martha's Vineyard, Mass.; along the Rhode Island coast; and particularly on the Baltic coast of Poland and Lithuania, where one baymouth beach is approximately 100 km long. (R.L.N.)

Bayou. A lake or sluggish watercourse in a half-closed, abandoned channel. It may be on a *floodplain or a *delta. It results when a river overflows and cuts a new channel, abandoning the old one, which gradually silts up.

Beach. A deposit of sand and/or gravel along the coast of a lake, sea or ocean. It extends from the outermost breaking *waves to the upper limit of wave action, is devoid of vegetation and slopes downward toward the water. It is made of rock fragments, shells, other hard parts of marine organisms, quartz and other minerals.

The shore is a dynamic environment, with beach sediments washed landward by wave action, moved seaward by the backwash and drifted along the shore by the currents. This continuous movement results in attrition which tends to round the fragments and reduce their size. When more material is brought in than is removed, the beach builds forward (prograding beach). When less material is brought in than is removed, the front of the beach is eroded and retreats landward (retrograding beach). Ripple, *rill and swash marks, cusps, *cross-bedding and other features of interest to sedimentologists are formed on beaches. Sand *dunes are found along the sandy beaches of both lakes and oceans. Beaches have been mined for gold, diamonds and other substances. See Baymouth Beach; Spit; Tombolo. (R.L.N.)

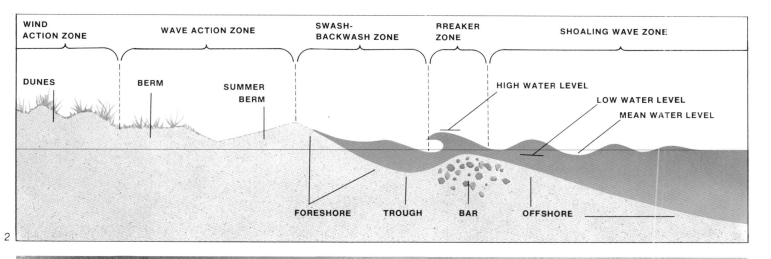

WIND ACTION ZONE | WAVE ACTION ZONE | SWASH-BACKWASH ZONE | BREAKER ZONE | SHOALING WAVE ZONE

DUNES | BERM | SUMMER BERM | HIGH WATER LEVEL | LOW WATER LEVEL | MEAN WATER LEVEL

FORESHORE | TROUGH | BAR | OFFSHORE

2

3

1

Bead Tests. Qualitative tests for certain metals made by dissolving the unknown material in a glassy drop of flux. The commonest flux in use is borax. With a Bunsen burner the borax is fused to a glassy bead in a loop of platinum wire. A tiny bit of dissolved test material may impart to the borax a characteristic color. For example, cobalt yields a blue bead, chromium a green bead.

Becke Test. A test used in optical mineralogy to compare one *index of refraction with another. In 1893 Friedrich J. K. Becke observed that if two transparent minerals in contact with each other are examined under a microscope, it is possible to determine which has the higher refractive index. The test is to focus sharply on the contact and then to raise the microscope tube to throw the image slightly out of focus. A bright line—the Becke line—will move into the adjacent mineral or other surrounding material, such as a liquid, of higher refractive index. (C.S.H.)

Bedding. A layered or planar feature associated with *sedimentary rocks but also found in *igneous and *metamorphic rocks. In sedimentary rock, bedding is the result of the changes in the grain size and mineral composition of materials as they are being deposited. Metamorphic rocks may display bedding that was not completely destroyed during the metamorphism of a sedimentary rock. Some igneous rocks also show bedding due to repeated volcanic *lava flows. Bedding can also be produced by crystals settling to the floor of a molten body of rock and forming layers of a particular mineral. A well-known example is the Palisades Sill along the Hudson River in New Jersey. About 6 m from the base of the cliff is a distinctive layer rich in olivine in an otherwise homogeneous body of rock. (J.W.S.)

Bed Load. The solid debris transported by a stream on or near its bed. Because this material is too heavy to be carried in suspension, it is moved by rolling, sliding or saltation (sudden jumps) along the bottom. The intermittent movement causes the particles to move more slowly than the water. Movement results from the instantaneous drag caused by differences in velocity between water at the top and at the bottom of the grains, or from lift caused by vortices. The movement of the bed load depends on the friction of the grains, the gradient of the stream and the velocity of the water.

It is very difficult to measure a stream's bed load because any measuring trap will itself interfere with the transport. However, it is possible to study bed load movement in an artificial channel or flume, where underwater dunes form. Particles move up from the back of the dune and slide or roll down the dune front into the trough, with material and dunes thus moving downstream. At higher velocities, where *Froude number equals 1, the dunes disappear and the bed is flat; then, as Froude number becomes greater than 1, *sand waves (antidunes) form on the bed. Particles are then scoured from the front of one antidune and moved forward to rest on the back slope of the next antidune. Although the grains are moving downstream, antidune configuration is displaced upstream. The water waves break when they have twice the amplitude of the antidunes, causing turbulence that results in much lifting and transporting of grains. (M.M.)

Bedrock. The continuous shell of solid rock that may crop out at the earth's surface or be buried by a *regolith of variable thickness. When visible at the earth's surface it is called an *outcrop. It usually appears in mountain districts and on steep slopes in arid regions where erosion is prevalent. It may also be buried by hundreds of meters of unconsolidated sediments, as in Death Valley and other undrained basins in the American Southwest. Bedrock consists of *igneous, *sedimentary and *metamorphic rocks; it may contain coal, oil, minerals, ores and many other economically important substances. Most knowledge of the planet earth has been obtained from studying bedrock.

Beheaded Stream. A watercourse that has lost its tributary or some of its drainage basin to another river by stream capture. See Piracy.

Belemnoid (or Belemnite). An extinct *cephalopod, probably similar to the present-day cuttlefish, abundant in the seas of *Jurassic and *Cretaceous time. Belemnoids were among the first fossils noted by early man. Their sharp, dartlike appearance led to the superstition that they represented the points of thunderstorms. Because of their shape they have also been referred to as fossil cigars or finger stones. The oldest known belemnoid was found in rocks of Late Mississippian age, and the group became extinct at the end of Cretaceous time. They are especially useful as *guide fossils for Jurassic and Cretaceous strata. (W.H.M.)

Benioff Zone (or Subduction Zone). A plane that dips beneath a continent or continental margin at an angle of about 45°. According to the theory of *plate tectonics, plates of the *lithosphere sink into the upper *mantle through this zone. The location of a Benioff Zone is indicated on the surface by a deep-sea trench. Earthquake foci tend to occur along such zones. The feature was named for Hugo Benioff, an American seismologist who located it.

Benthos. The organisms that live at the bottom of the sea. The benthos represent a tremendous diversity of life and may be either plants (phytobenthos) or animals (zoobenthos). A variety of modes of life are present, including sessile forms, that is, attached and immobile, and vagrant forms that move about on the sediment surface or burrow within it. Benthonic organisms may be slow-moving, such as sea urchins, snails and crabs, or they may be sedentary, such as sponges, corals and anemones; as a result, they generally have planktonic (floating) larval stages, which ensure widespread distribution. (R.A.D.)

Bentonite. A clay formed by the alteration of volcanic ash or *tuff, composed chiefly of two clay minerals, montmorillonite and beidellite. Bentonite has the unique property of absorbing water, causing it to swell to several times its original volume. The ensuing instability can cause complications when a building foundation rests on this kind of clay.

Bergschrund. A deep, semicircular *crevasse, or a group of closely spaced, almost parallel crevasses, beween the head of a mountain glacier and the *bedrock and snowfield above it. A bergschrund is usually filled with snow and bridged over during the winter, but is open during the summer. Its downvalley wall is a glacier, and its upper wall is ice and/or bedrock. It develops when a glacier moves downvalley away from either bedrock or a stationary snowfield. *Frost action is particularly vigorous in the bergschrund in the summer because the *meltwater that accumulates in it during the day may freeze during the night. Where freezing and thawing take place in cracks in the bedrock, fragments are detached and carried away by the glacier. The bedrock is thus eroded and the glacier bites its way back into the mountain, forming a *cirque. (R.L.N.)

Berm. A small ephemeral bench or terrace found on the backshore of a *beach. It is characterized by a flat or a gently landward-sloping surface cut off on the seaward side by a steep slope. It is built by material thrown up by constructive waves and is cut back by destructive waves. It is common to find a higher winter berm with a lower summer berm seaward of it.

2

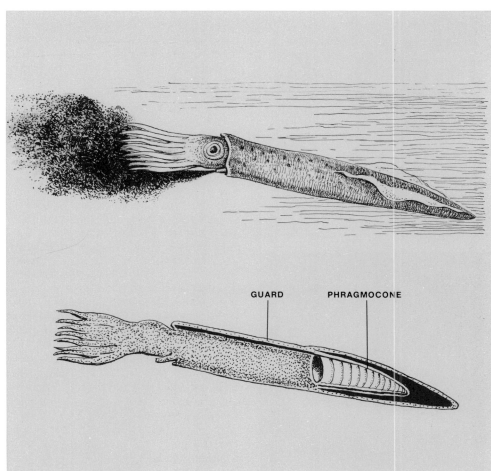

3

Bedding

1 *Distinctive weathering characteristics expose the different layers, or beds, in an outcrop of shale and limestone.*

Belemnoid

2 *The oval depression at the center of this fossil belemnite is an imprint of its ink sac.*

3 *These squidlike cephalopods from the Mesozoic Era released a cloud of ink when disturbed. A cross section shows the long, cigar-shaped internal skeleton, known as a guard, and a chambered section toward the rear of the animal, called a phragmocone.*

GUARD   PHRAGMOCONE

1

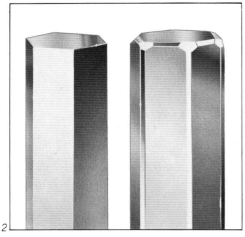

2

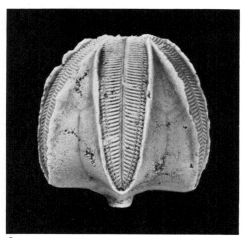

a

3
b

Beryl. A mineral, an aluminum silicate of the rare element beryllium and its principal ore. Beryl has been known and fashioned into highly prized gems since ancient times. The advancing technology of the 20th century made it possible to extract beryllum from beryl for use as a structural and alloying metal. Today beryl is sought as an ore of that element as well.

The principal occurrence of beryl is in granite *pegmatites, where it is usually found in pale green hexagonal crystals embedded in coarse *feldspar and *quartz. Although the average crystal measures only a few centimeters in maximum dimension, in places giant crystals 8 m long and 2 m across have been found. Such material is nontransparent and flawed and has value only as an ore of beryllium. Far less common are the transparent gem varieties, "pocket beryl," found as free-growing crystals in cavities in the pegmatites. Although frequently colorless, beryl is also found in several colors to which variety names are given. Most common and best known of these are the blue to sea-green aquamarine. Crystals with a rich golden yellow color are golden beryl; morganite is pink to rose.

The deep green beryl, emerald, is so highly prized that it is grouped with diamond, ruby and sapphire as a precious stone. Its occurrence differs from the other gem varieties. In places it is associated with pegmatites, but it is found in the adjacent rock and not in the pegmatite itself. Elsewhere, as at Muzo, Colombia, it occurs in veins in a black carbonaceous limestone associated with quartz and pyrite crystals. In the Ural Mountains, South Africa, Rhodesia and Egypt, emeralds have been found in mica schist.

Beryl is a *cyclosilicate composed of beryllium, aluminum, silicon and oxygen. Small amounts of other elements may be present, and to these is attributed the color of some of the gem varieties. Emerald contains minute amounts of chromium or vanadium, and the pink of morganite is believed to result from the presence of cesium. Beryl is almost invariably in hexagonal crystals whether they be large or small, and it can be recognized by the presence of the hexagonal prism and base. Small gem crystals may be highly modified by other forms. See Mineral Properties; Appendix 3. (C.S.H.)

B-Horizon. See Horizons, A-, B- and C-.

Biaxial Crystal. A crystal having two optic axes, that is, directions along which light will move with no double refraction. Only crystals of the orthorhombic, monoclinic and triclinic systems can be biaxial.

Bifurcation. The separation or branching of a *stream. Bifurcation ratio is the ratio of the number of streams of a given order to the number of streams of the next higher order. The term "order" designates the relative position of streams in a drainage network, with the smallest unbranched tributaries as the first order and the main trunk stream as the highest order.

Biogenic Sediment. Any sediment composed of skeletal material derived from either plants or animals. Many tropical beaches are biogenic, and debris associated with reefs is wholly biogenic. Deep-sea sediments are also largely biogenic and are frequently made up of plant or animal skeletons derived from small floating organisms. After the individual organism dies, its test (skeleton or shell) settles to the ocean floor. Skeletal material of silica or calcium carbonate may be present. (R.A.D.)

Biosphere. The zone at and near the earth's surface in which all living things are found; the "sphere of life." Ranging from submicroscopic viruses to giant sequoia trees, this horde of organisms has adapted to almost every environment, from hot springs to glacial ice. Members of the biosphere have been photographed on the ocean floor and more than 15 km above the earth's surface. Simple life forms have even been found in water from oil wells more than a mile inside the earth.

The biosphere is as important as the earth's physical zones, the *lithosphere, *hydrosphere and *atmosphere. Plants and animals continually interact with other parts of the earth and are involved in many important earth processes. Coal and petroleum are formed from the remains of prehistoric organisms. Many rocks, especially some limestones, are of organic origin. Bacteria play a key role in the development of certain types of iron ore. Finally, the study of *fossils has provided much information about earth history and the development of life. (W.H.M.)

Biostratigraphic Unit. A *stratum or group of rock strata that is unified and identified by its fossil content or paleontologic character. It may be based on a single fossil species or a distinctive assemblage of fossils without regard to the physical or structural properties of the enclosing rocks. They include biozones (or range zones), which are composed of rocks formed during the lifespan of a particular kind of fossil, and assemblage zones, defined by a group of associated fossils rather than by the range of a single *guide fossil.

Biotite. A common rock-forming mineral of the *mica group. It is green, black or dark brown. See Mineral Properties.

Biozone. A depth division within the ocean characterized by particular conditions and organisms. See Marine Environment.

Bituminous Coal. A dark brown to black *coal that contains approximately 70 to 80 percent fixed carbon and 15 to 20 percent volatile matter. It is soft enough to soil the fingers, typically breaks into irregular rectangular blocks, and burns with a smoky flame. Known also as "soft coal" to distinguish it from *anthracite ("hard coal"), it is the most abundant kind of coal and is widely used in industry. Major producing areas in the United States include West Virginia, Pennsylvania, Kentucky, Illinois, Ohio, Virginia, Indiana, Alabama, Utah and Tennessee; it also occurs in Europe, Africa, Asia and Australia.

Blastoid (or Sea Bud). A member of class Blastoidea of phylum Echinodermata. These extinct, short-stemmed *echinoderms had a small, symmetrical, globe-shaped to budlike calyx or body typically formed of thirteen prominent plates in a characteristic five-sided pattern. The mouth was in the center of the calyx, surrounded by five openings called spiracles. Blastoids first appeared in the *Ordovician Period and were extinct by the end of *Permian time. One species, Pentremites, is especially common in rocks of *Mississippian age and is a good *guide fossil to geologic formations of this age. (W.H.M.)

Blowpipe. A tube through which air is directed into a luminous flame to accelerate combustion and produce a hot flame. By placing a substance in the flame, information may be obtained about its general nature. Since the middle of the 19th century the blowpipe has been used to make qualitative tests of the common elements found in minerals. In the classic tapered blowpipe, air from the lungs is directed through a small orifice into either a candle or an alcohol flame. In modern blowpipes, a combustible gas and air are mechanically directed together to produce a sharp hot flame. (C.S.H.)

Body Wave. Any *seismic wave that travels through the crust of the earth. Body waves may be compressional waves (P Waves) or shear waves (S Waves).

Bog Iron Ore. The name given to an impure mixture of hydrous iron oxides, consisting principally of the mineral *goethite. It is found in layers or nodular masses at the bottom of swamps or bogs, and in the 18th century was used extensively as an ore of iron.

Bolson. A large, undrained basin in an arid or semiarid region, commonly surrounded by fault-block mountains. In the lowest part of large bolsons there are usually *playas or *playa lakes surrounded by alluvial aprons formed by converging *alluvial fans. Vegetation is sparse in the lower areas but increases on the mountains. Although features formed by wind are common, most of the landforms result from water. Streams are intermittent, however. Many bolsons were occupied by *pluvial lakes during the Pleistocene. They are found in the Basin and Range Province of southwestern United States and northern Mexico, and the term is chiefly used in reference to these areas. (R.L.N.)

Bomb. A volcanic feature formed when a clot of lava ejected from a volcanic vent solidifies. Bombs range in size from a few millimeters to several meters. Many contain *vesicles (gas cavities) and tend to be rounded due to gyrating and spinning of the clot during passage through the air and from flattening on impact with the ground. A cored bomb is formed when liquid lava solidifies around a solid core of either volcanic or nonvolcanic rock plucked from the walls of a *volcanic chimney by the magma as it moves upward.

Bonanza. A miner's word for a rich mass of ore. It is also used to signify good luck.

Bond, Chemical. A force that holds together atomic particles. These forces are usually described as belonging to one of four types of bonds: ionic, covalent, metallic and van der Waals'. A consideration of bonds is important in mineralogy because the physical properties of minerals are directly related to the type and intensity of the binding forces. For example, although diamond and graphite are both composed of atoms of carbon, diamond is hard because its atoms are held together by the very strong covalent bond, whereas graphite is soft because adjacent sheets of the carbon atoms that make up its structure are held to one another by the very weak van der Waals' bond.

The ionic, or electrostatic, bond joins ions together by the attraction of their unlike electrostatic charges. The inert gases—helium, neon, argon, krypton and xenon—contain the maximum number of electrons possible in their outer shells. When atoms join together they tend to adjust their electrons to obtain the configuration of one of the inert gases. One

4

Beryl

1 Such well-formed hexagonal beryl crystals as this are frequently of gem quality and are cut as aquamarines and emeralds.

2 A crystal of this hard mineral, the ore of the rare element beryllium, is typically six-sided.

Blastoid

3 A side view (a) of a 330-million-year-old Mississippian fossil, Petrimites godoni, shows one of five lancet (spear-shaped) plates with a longitudinal groove and a series of transverse grooves used for food-gathering. The mouth and surrounding spiracles (the largest of which is also the anal opening) are visible when the fossil is viewed from above (b).

Bomb

4 This type of viscous lava clot, which erupts from a volcanic vent and forms a cracked surface as it cools, is known as a "bread crust" bomb.

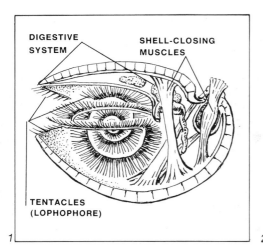

DIGESTIVE SYSTEM

SHELL-CLOSING MUSCLES

TENTACLES (LOPHOPHORE)

1

2

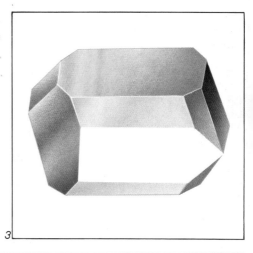

3

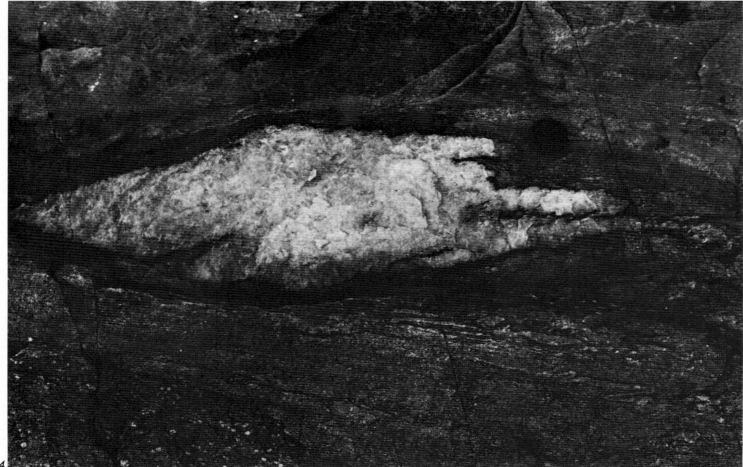

4

Brachiopod

1 The basic anatomy of this bivalved mollusk includes lophophore (tentacles), the digestive system, and shell-closing muscles.

2 Among the earliest forms of life, these ancient small marine animals of shallow seas are enclosed in a pair of unequal-sized shells, a fossil of which is shown here, at the top of an anchoring stalk.

Borax

3 This crystal is typical of the mineral borax, which is concentrated as crusts on alkali lakes and is used in the glass, enamel and chemical industries.

Boudinage

4 This quartz and feldspar boudin in gneiss at Pelham Bay Park, N.Y., is part of a "sausage-string" structure that resulted when tension within the earth's crust stretched the rock.

way they accomplish this is by losing or gaining electrons. For example, the sodium atom has one electron in its outer shell that is easily lost, making the atom a positively charged ion. On the other hand, a chlorine atom, by gaining a single electron, achieves a stable electron configuration and becomes a negatively charged ion. Therefore, sodium and chlorine join together in sodium chloride by the attraction of their opposite electrostatic charges. In ionic bonded compounds a positive ion is not paired with a given negative ion, but ions of opposite charge surround each other and distribute their charge equally among their nearest neighbors. In sodium chloride the positive charge of each sodium ion is shared by six surrounding chlorine ions; likewise, the negative charge of chlorine is shared by six sodium ions so that the crystal as a whole is electrically neutral. The ionic bond is the principal binding mechanism in many minerals.

In covalent, or homopolar, bonding, atoms achieve the electron configuration of a noble or inert gas by sharing electrons in their outer shells. For example, a chlorine atom lacking one electron in its outer shell may unite with another chlorine atom so that one electron from each does double duty in the outer electron shells of both. Thus the inert gas configuration is approximated in both atoms. The covalent bond is the strongest chemical bond. It is most common in organic compounds. Its best representative in minerals is diamond. The four vacancies found in the outer electron shell of carbon are filled by sharing electrons with four other carbon atoms. Every atom is thus linked to four others, forming a continuous network.

Van der Waals' bond is a weak bond that holds uncharged structural units and neutral molecules. In minerals, it binds these particles in a crystal lattice. It arises from the small residual charges on the surfaces of the essentially neutral units. This is the weakest chemical bond common only in organic compounds and solidified gases, where it accounts for their low hardness and low melting points. When present in minerals it is combined with other types of bonding and is a direction of easy cleavage and low hardness. *Graphite is composed of sheets of carbon atoms joined by a covalent bond, but the sheets are held together by van der Waals' bond.

A metallic bond is the type to which metals owe their coherence. The atoms of the true metals easily lose their outer electrons to other atoms. The structural units of metals are positively charged atoms held together by the aggregate charge of electrons that surround the nuclei. Because the electrons are free to move through the structure, they give metals their high electrical and thermal conductivity. In minerals, pure metallic bonding is present only in native metals. (C.S.H.)

Boracite. A mineral, essentially magnesium borate, found in crystals in *evaporite deposits. Boracite is *isometric above 265°C (509°F) but *orthorhombic at lower temperatures. The crystals show isometric forms indicating formation as the high-temperature variety.

Borate. A type of mineral containing the chemical group BO_3^{-3}. Over 100 borate minerals are known, but the boron and boron compounds of industry are recovered from only a few. The most abundant borates are *borax, *colemanite, *kernite and *ulexite.

Borax. A mineral, a hydrated sodium borate, mined as an ore of boron. It forms as an *evaporite and is thus characteristically found in arid regions in *playa deposits and as a surface efflorescence. Borax is frequently found in clear, transparent crystals. However, in a dry environment these crystals quickly lose water and turn white with the formation of tincalconite. See Mineral Properties.

Bornite. A mineral, a copper iron sulfide and an ore mineral of copper. The fresh surface has a bronzelike color that on exposure quickly tarnishes to a variegated purple and blue, giving rise to the name peacock ore. Although in places bornite is found as a secondary replacement of earlier-formed copper minerals, it is most commonly primary and may occur in intimate mixtures with chalcopyrite and chalcocite. It is not as important a copper ore as these other two minerals. See Mineral Properties.

Bort. The black variety of diamond known as carbonado. This black diamond is cryptocrystalline and thus less brittle than crystals. The name bort is also given to other diamonds either so badly flawed or of such poor color as to render them worthless as gems.

Botryoidal. A mineral aggregate resembling a bunch of grapes (as the Greek root of the word signifies).

Boudinage. A structure in *sedimentary and *metamorphic rocks resembling a string of sausages. It results from the stretching of layered rock by tensional forces within the earth's crust. Individual boudins form from a layer of relatively brittle material that thins out and breaks into segments in a manner analogous to the necking that develops in a steel rod as it is pulled apart. The layers of rock sandwiching the brittle layer are more ductile and thus stretch and flow plastically around the individual boudins.

Boulder Train (or Indicator Fan). A fan-shaped area containing distinctive, easily recognized glacial *erratics derived from an *outcrop at the apex of the fan. The fragments, broken and abraded during transit, get smaller with increasing distance from the source. A boulder train is the best indication of the direction of glacial motion. If the area of the source outcrop is known, and if the volume of the erratics in the train can be estimated, the amount of glacial erosion of the source outcrop can be calculated. The angle at which the margins of the train diverge is a measure of the maximum change in glacial motion. Since not all the erratics in a train are of boulder size, some authorities prefer the term indicator fan for this feature. (R.L.N.)

Brachiopod. A member of the phylum Brachiopoda, a group of bivalved, exclusively marine animals with shells composed of two pieces called valves. The valves, which may be of chitinous, calcareous or phosphatic material, enclose and protect the animal's soft parts. Adult brachiopods are typically attached to the ocean floor by means of a fleshy stalk called the pedicle. The pedicle normally projects through an opening in the pedicle, or ventral, valve. The opposing valve, which is usually smaller, is called the brachial, or dorsal, valve. Phylum Brachiopoda has been divided into two classes. Class inarticulata includes the rather primitive "unhinged" brachiopods. Their valves do not possess hinge teeth, but are held together by muscles. Most are oval or tongue-shaped and lack a pedicle opening. Members of class Articulata have a well-developed hinge structure for articulation of the valves. There are teeth on one valve that are engaged in sockets on the opposing valve. Articulate brachiopods are characterized by sturdier shells and more complex structure than the inarticulates. Brachiopods range from Early *Cambrian to *Holocene (Recent) in age, but were especially abundant during the *Paleozoic Era. (W.H.M.)

Bragg's Law. A formula used in crystallography to determine *crystal lattice spacings. The angle at which X rays of a given wavelength are "reflected" from a family of atomic planes in a crystal reveals the spacing of the planes. In the Bragg equation, $n\lambda = 2d\sin\theta$, λ is the wavelength, d is the spacing of atomic planes and θ is the angle at which the X rays are reflected from the planes. The n is a whole number, and the angle of reflected X rays varies with its value. When $n = 1$, it is called the first-order reflection; when $n = 2$, the second-order reflection, etc.

Braided Stream. A stream with numerous channelways that divide and rejoin. Braided streams occur under conditions of highly variable discharge and an abundant *bed load. These conditions are provided where a river flowing through a semiarid region receives most of its discharge from high mountainous headwaters. Precipitation or snow melt causes high runoff which acquires a large load as it flows through the semiarid region. As the water recedes seasonally or downstream, the river becomes braided, since the reduced discharge is no longer able to carry all the bed load. The Platte River in Nebraska is an example of this type of braided river.

Another type of braiding occurs where glaciers are melting and the supply of debris is consistently large in amount and in caliber whereas the discharge fluctuates as the ice melts and refreezes. This causes channelways to shift and divide around material that is alternately moved and deposited as the relation of load to discharge varies. Examples of this kind of braided river can be found in Iceland, Alaska, the Alps or wherever melting glaciers appear.

A third kind of environment in which braided streams occur is found in Japan, where rivers have their sources in steep mountains, flow out onto level plains, and are completely contained in a humid climate. The explanation here seems to be in the caliber of the bed load, which is very coarse, and in the change in stream gradient. The high discharge in the mountains combined with a steep slope results in erosion of coarse material, which is carried onto the flat areas. There the load is dropped because of the reduced gradient, the loss of energy and braiding develops.

Recognition of braided stream deposits in older sediments yields information concerning former environments. Thick alluvial deposits in the Susquehanna River in New York and Pennsylvania and other streams in northern United States and Europe are relics of the melting ice of the *Pleistocene Epoch. Rivers such as the Murrumbidgee in Australia illustrate from their ancient braided pattern that their older channels were formed in a climate much different from today's. (M.M.)

Breccia. A *clastic rock composed of broken, angular rock fragments larger than 2 mm in diameter and enclosed in a fine-grained matrix. Unlike *conglomerate, in which the fragments are rounded, breccias consist of particles that became embedded in a matrix and were not transported and worn by *abrasion. Sedimentary breccias are relatively rare but may represent evidence of terrestrial mudflows or submarine landslides. Tectonic or fault breccias consist of fragments produced by rock fracturing during faulting and other crustal deformation. Igneous breccias can be formed by the cementation of sharp-cornered particles of volcanic rock shattered by volcanic explosions (volcanic breccia). Rock shattering may also occur during the intrusion of plutonic rocks. If the broken fragments of the intruded rock become cemented together they form an intrusive breccia. (W.H.M.)

Breccia, Flow. Found in *lava flows and characterized by a glassy or fine-grained matrix and by angular fragments of rock 3 cm or more in diameter. The fragments may be torn from the walls of the chimney up which the *lava moved or picked up from the terrain over which the lava flowed, but they much more commonly result from the breakup of the solidified crust of the flow. These fragments are then engulfed in the still liquid part of the flow. A zone of flow breccia is usually present at both the top and bottom of aa and block lava flows. They are significant because they furnish information on rocks older than the flow and on the way the flow moved. (R.L.N.)

Breccia, Impact. See Lunar Geology.

Breccia, Lunar. See Lunar Geology.

Breccia Pipe. An elongated pipelike structure consisting of broken pieces of rock known as *breccia. In cross section the pipes are usually circular or polygonal, but may be tabular or of very irregular shape. The pipes are generally several times longer than they are wide, and the fragments formed in pipes may range from very small pieces to masses weighing several tons. It is believed that pipes form by the removal of *magma that forced its way into the overlying crust, allowing slumping of the overlying crust into the void. Two methods to explain the removal of the magma have been suggested: (1) the nearby eruption of a volcano or lava flow, causing the magma to drain from the pipe, and (2) a pulsation of the magma, causing a cycle of recession and advance. In some pipes there is evidence that the injection of magma and its removal have occurred on a

Braided Stream
Located on a glacial outwash plain, this waterway has become a complex tangle of channels, sandbars and islands.

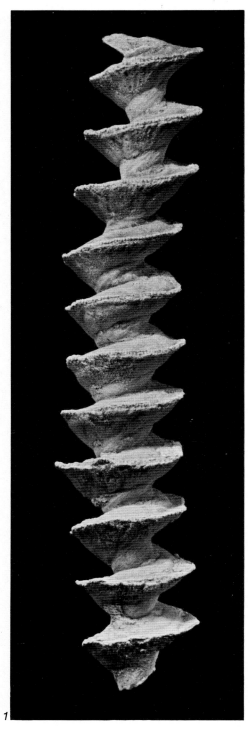

cyclic basis. Each withdrawal of magma created more fracturing of the wall rocks to form breccia.

In many pipes the spaces between rock fragments have been filled with ore minerals, and the interest in breccia pipes lies in the localization of these important minerals. Some porphyry copper deposits in the Basin and Range Province of the southwestern United States are located in breccia pipes, and diamonds are found in the pipe of the famous Kimberly, South Africa, deposits. (J.W.S.)

Breccia, Pyroclastic. A pyroclastic rock containing abundant angular fragments more than 64 mm (2½ inches) in diameter imbedded in a finer-grained matrix. It consists mainly of consolidated volcanic detritus formed by the volcanic shattering and disintegration of older volcanic and nonvolcanic rocks and by the solidification of clots of lava thrown out during eruption. Pyroclastic breccia has a limited distribution since large fragments cannot be thrown very far from volcanic vents; it is therefore common in *volcanic necks and is generally not as well stratified as *tuff.

Care must be used in identifying these breccias since they bear a resemblance to other kinds of breccias. Tuff breccia contains fragments larger than 64 mm in diameter, but much of the matrix material is of smaller diameter. (R.L.N.)

Brucite. A mineral, magnesium hydroxide, usually occurring in platy crystals. It is commonly associated with serpentine, dolomite and magnesite and may form by secondary alteration, as through the action of groundwater on magnesium silicates, especially serpentine. See Mineral Properties; Appendix 3.

Bryozoan. A member of phylum Bryozoa (also known, as phylum Polyzoa). Bryozoans form lacelike branching or spiral colonies as much as 60 cm (2 feet) across and are commonly found matted on shells, rocks, fossils and other objects. They abound in modern seas, and a few species live in fresh water. Each secretes its own cuplike shell or exoskeleton, which may be preserved as a fossil. The individuals live in small boxlike chambers that resemble pits on the surface of the colony. Bryozoans range from *Ordovician to *Holocene (Recent) in age, although possible *Cambrian forms have been noted.

Butte. A flat-topped hill bounded by cliffs and steep slopes and composed of horizontal sedimentary rocks. A butte closely resembles a *mesa but has a smaller summit area.

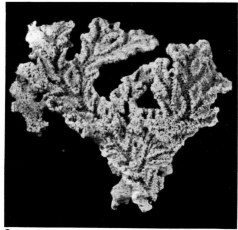

a

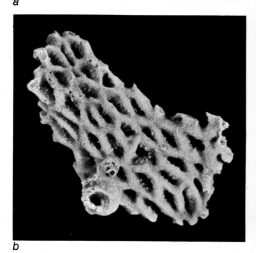

b

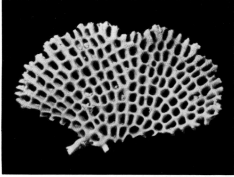

c

Bryozoan

1 This screwlike calcium structure was secreted by a colony of Archimedes some 300 million years ago.

2 The basic interior anatomical structure of a bryozoan consists of a simple U-shaped digestive tube and muscle system.

3 Bryozoan specimens (a, b, and c), which silicified in limestone during the Permian Period 250 million years ago, are from the Glass Mountains, Texas.

C

Calaverite. A mineral, the gold *telluride. Gold occurs most commonly as the native metal, and at most localities where the telluride is found it is a rarity. But at Cripple Creek, Colo., and Kalgoorlie, Australia, in the late 19th and early 20th centuries rich deposits were mined with calaverite a major ore mineral. See Mineral Properties.

Calcareous. Composed of or pertaining to calcium carbonate and used to describe the hard parts of invertebrates that consist of such material. Reef-building corals secrete exoskeletons of calcium carbonate, which comprise the bulk of most coral reefs. Oysters, clams and snails also secrete calcium carbonate shells, which may then be preserved as fossils. See Limestone.

Calcite. A mineral, the common form of calcium carbonate, dimorphous with *aragonite. The common carbonate minerals are frequently divided into two groups, the *rhombohedral typified by calcite and the *orthorhombic by aragonite. Calcite is not only the most abundant of the carbonate minerals but is unique among all minerals in its great diversity of crystal forms. Various combinations of forms give single crystals widely differing shapes, from flat tabular to slender pointed.

All calcite, whether in single or twinned crystals, shows a perfect rhombohedral cleavage, the mineral's most characteristic feature. This physical property is shared by the other rhombohedral carbonates, but the angles differ at which the cleavage surfaces intersect. In calcite the angle is 74°55'. Chemically pure calcite that is clear and colorless is called *Iceland spar. Most commonly it is milky white, but it may be yellow, blue, green or red. Calcite shows better than any other mineral the property of *double refraction. This breaking of light into two rays is well illustrated in the two images observed when an object is viewed through a cleavage rhombohedron of Iceland spar.

Well-crystallized calcite is deposited from solution under a wide range of geologic conditions. It is a common *gangue mineral associated with ore minerals in hydrothermal veins. Fine crystals are found in limestone caverns and in cavities and veinlets in basic *igneous rock. However, calcite is far more abundant as a rock-forming mineral, for as a fine-grained material it is the principal mineral of great thicknesses of limestone and chalk. In marbles (metamorphosed limestones), the calcite has been recrystallized into coarser, cleavable grains. Calcite is also found in caves as stalactites, stalagmites and incrustation. The *traver-

1

2

Calcareous

1 *Exposed by glacial erosion, these lime-secreting algae once flourished in seas that covered Montana more than 600 million years ago.*

Calcite

2 *This group of crystals illustrates the diversity of forms that can be assumed by calcite.*

tine or tufa deposited by both hot and cold calcareous spring waters is also calcite.

Calcite, mostly as limestone, is an important industrial raw material. It is used to manufacture cements, lime and mortars, with the greatest amount used in the production of Portland cement. Other uses are as a dimension stone for building; a flux for melting metallic ores; an aggregate in concrete; and, finely ground, a fertilizer and for whiting and whitewash. Because of its strong double refraction, calcite in the form of Iceland spar has long been used in specialized optical equipment, such as the Nicol prism to produce polarized light. See Mineral Properties; Appendix 3. (C.S.H.)

Caldera. A large, nearly circular, steep-walled depression at the summit of a *volcano. Much larger than a *crater, it may be more than several kilometers in diameter and hundreds of meters deep. Renewed activity may build a new volcano in a caldera. Large lakes, such as Crater Lake, Oreg., which occupies a caldera about 9 km in diameter and 900 – 1200 m deep, are commonly found in calderas. Calderas have been attributed to explosion, subsidence or explosion followed by collapse. Formerly it was thought that most were formed by violent eruptions that destroyed the top of the volcano, but only smaller calderas such as Bandaisan, Japan, are formed by explosion alone.

The large depressions at the summits of Mauna Loa and Kilauea in Hawaii, and Masaya in Nicaragua, are calderas formed by collapse unrelated to explosive activity. Supporting this is the fact that they are bordered by *fault scarps and very little *pyroclastic debris. The collapse is due to the extrusion of flows low on the flanks of these volcanoes and to the intrusion of dikes, both of which lower the magmatic chambers and leave the cones above the magmatic chambers unsupported. It is believed that most of the larger calderas are formed by cataclysmic eruptions followed by collapse. Krakatoa, East Indies; Mount Katmai, Alaska; Batur, Bali; and Crater Lake, Oreg., are among the well-known calderas. Buldir in the Aleutian chain is a submarine caldera, and Santorin in the Grecian Archipelago is a partially drowned caldera. (R.L.N.)

Caliche. A dull, earthy *calcite deposit that occurs in certain areas of scant rainfall. Caliche is commonly found mixed with other materials, such as clay and/or gravel. It may be firm and compact or loose and powdery. Caliche forms when ground moisture, containing dissolved calcium bicarbonate, moves upward. In dry areas this moisture evaporates, leaving a crust of calcium carbonate in the form of caliche on or near the surface of the ground. Caliche is quarried in certain regions and is used chiefly as road material and as an aggregate. It is referred to as hardpan and tepetate in parts of the United States; in Great Britain it is known as calcrete. The term has also been applied to mineral nitrates that occur in a few desert areas of northern Chile. (W.H.M.)

Cambrian Period. The oldest period of the *Paleozoic Era; it began about 600 million years ago and lasted for perhaps 100 million years. Its name derives from *Cambria*, the Latin for Wales, where its rocks were first studied and named by Adam Sedgwick in 1835. Rocks of the Cambrian Period are exposed in Africa, Asia, Australia, Antarctica, North America, South America, Greenland, and Europe. Mostly of marine origin, they are the oldest known rocks containing an abundance of well-preserved fossils.

Cambrian time was marked by a slow encroachment of the seas on the lands of the Northern Hemisphere. Although beginning as shallow embayments, the seas gradually moved inland; during Late Cambrian time as much as 30 percent of the North American continent was submerged. Most Cambrian sediments were deposited in two elongated seaways. One filled the Appalachian *geosyncline, a great trough that followed the northeast-southwest trend now occupied by the Appalachian Mountains. A western trough, the Cordilleran geosyncline, paralleled the site of the present-day Cordilleran Ranges of western North America. Similar geosynclines apparently occurred on other continents, but they are not as well known as those of North America. These ancient seaways were the site of much sedimentation, and strata with an aggregate thickness of 9000 – 15,000 m occur in the axes of certain geosynclines. Lower Cambrian rocks consist largely of sandstone, shale and some limestone. During Late Cambrian time the sea overflowed the geosynclines, depositing carbonate sediments that produced limestone and dolomite. The end of the Cambrian was a time of continental uplift and retreat of the seas, and there were localized mountain-building movements in New England and on the east coast of Canada. In Europe, the close of the period was marked by considerable volcanic activity.

There is little direct evidence of the Cambrian climate, but the presence of extensive carbonate deposits and the nature of the fossil record suggests a rather warm, mild climate. This is further supported by the hordes of invertebrates that filled the Cambrian seas. Most major groups of marine invertebrates are represented as Cambrian fossils, although there is no record of backboned creatures. The land must have been devoid of life, for there is no evidence of terrestrial or freshwater plants or animals.

Although Cambrian life was unlike that of today, it was not totally primitive, ranging from simple spongelike animals to complex *arthropods. The latter include the *trilobites, extinct relatives of modern crabs, shrimp and crayfish. They possessed jointed legs and segmented bodies, resembling the present-day sow bug. These bottom-dwelling scavengers generally ranged from less than 2.5 cm (1 inch) to 10 cm (4 inches) in length, but one giant, *Paradoxides harlani*, was about 45 cm (18 inches) long. Trilobites make up as much as 60 percent of the entire Cambrian fauna, and certain species are especially useful as *guide fossils in identifying subdivisions of the Cambrian. *Brachiopods were second in abundance. These small, bivalved animals were attached to the ocean floor by a fleshy stalk, and their shells constitute from 10 to 20 percent of known Cambrian fossils. Present also were archaeocyathids, a strange group of animals apparently related to the sponges. They are especially abundant in Lower Cambrian rocks, where they occur in reeflike masses. The remainder include sponges, worms, *gastropods, *cephalopods, *protozoans, and primitive *echinoderms such as the *cystoids. Plants were represented by blue-green algae that formed calcareous reefs on the Cambrian sea floor.

The best-known and most important Cambrian fossil discovery was made in the Burgess Shale of western Canada in 1910 by Charles D. Walcott, an American paleontologist. The fossils occur in a layer of black shale of Middle Cambrian age on Mount Wapta in British Columbia. Even the most delicate specimens are preserved in minute detail, and include sponges, jellyfish, arthropods and annelid worms. Animals of this type are seldom found as fossils because they lack preservable hard parts, and they are especially rare in rocks as old as Cambrian. This important fossil discovery furnished paleontologists with much valuable information about early soft-bodied forms and provided important clues as to what types of life may have been present in the Precambrian.

Cambrian rocks are relatively poor in economic resources. However, building stone (primarily marble and slate) has been quarried in New England and the British Isles; copper in Tennessee and Zaire; iron in Pennsylvania, France, Great

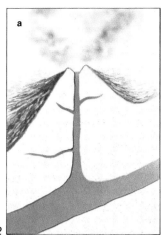

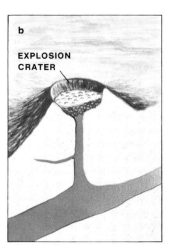

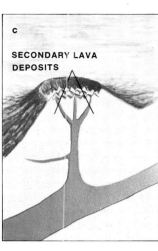

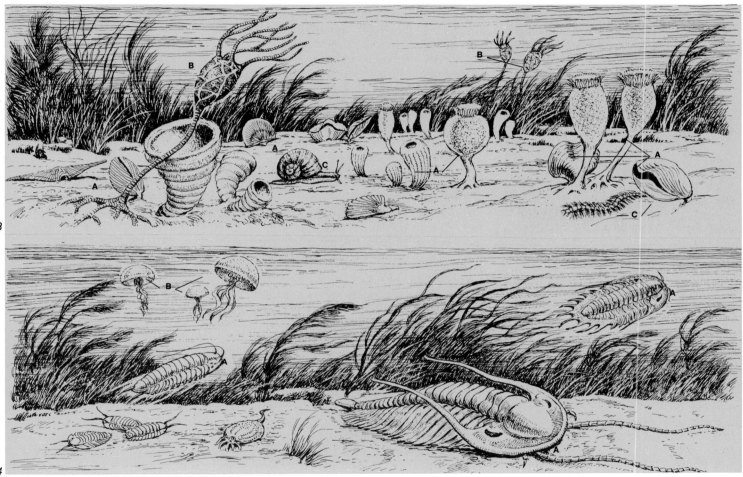

Calcite

1 The most abundant carbonate, this mineral occurs in a wide variety of crystal forms, one of which is shown here.

Caldera

2 The stages in the formation of a caldera. (a) The top of a volcano explodes during a violent eruption. (b) The sides collapse into the central reservoir. (c) Hot lava later fills the cavity.

Cambrian Period

3 Over 600 million years ago the seas teemed with invertebrate life. Primitive spongelike organisms, including the fragile glass sponges and clam-shaped brachiopods (A) were widely distributed. Stalked echinoderms called eocrinoids (B) captured food with their long tentacles. Cambrian snails and multilegged worms (C) left tracks in the mud.

4 The trilobites, early arthropods resembling horseshoe crabs, were dominant forms of life during this period. These sea-floor scavengers (A) ranged in size from 2.5 cm (1 inch) to some that were 45 cm (18 inches). Jellyfish much like modern ones drifted over dense seaweed forests on the ocean floor.

Cap Rock

1 *The soft underlying rock of this structure weathered away faster than the more durable material above it, leaving this precariously balanced feature in Arches National Monument, Utah.*

Casts and Molds

2 *A mold is formed when (a) a fossil is buried in sediment; (b) the fossil dissolves, leaving a record of its outside shape, called a mold; (c) sediment eventually covers the mold.*

3 *A cast is formed when (a) a fossil dissolves between two layers of sediment, leaving a mold of its form; (b) the mold is filled by a third material; (c) the overlying layer of sediment weathers away, exposing a cast of the fossil.*

Cassiterite

4 *The typical tetragonal crystal form of cassiterite is shown here, along with a characteristic twin crystal.*

Britain, Spain and West Germany; lead in Germany; and barite in southeastern Missouri and Virginia. (W. H. M.)

Camptonite. A variety of the *igneous rock type lamprophyre, which is composed essentially of *labradorite with sodic *hornblende, *pyroxene and *olivine.

Canadian Shield. See Craton.

Cannel Coal. A very fine-grained type of *bituminous (soft) coal composed largely of finely divided plant material, pollens, algae and spores. Because it ignites easily and burns with a high, yellow flame, it was originally called candle coal. It commonly occurs wherever coal is found.

Canyon. A spectacular, extremely deep and narrow, steep-walled river channel or valley, generally cut into bedrock. The valley walls may tower vertically for hundreds to thousands of feet above the river bottom. The western United States has many beautiful canyons, cut by rivers (especially the Colorado River and its tributaries) during the *Tertiary and the *Quaternary times, as a result of tectonic uplift and climatic change.

Capacity (of Streams). The total amount of solid debris of a given size that a river can transport. Originally defined in terms of *bed load, it now also refers to *suspended load discharge. Capacity for moving a given amount of debris increases with increased slope, increased discharge and decreased width. These factors affect the velocity and bottom roughness, which influence the turbulence and hence the stream's ability to transport. Another factor in determining capacity is homogeneity in size of material to be transported. If fine grains are added to a coarse bed load, not only is more total material moved but also more coarse grains. However, when grains coarser than those already in the bed load are added, capacity is reduced.

Arrangement of grains also has an important influence on turbulence and hence on transport capacity. Numerous large particles may create eddies that interfere with each other. However, if large grains are spaced far apart, each creates turbulence that will entrain smaller particles. The determination of capacity is thus a complex reaction of interrelated variables. However, it is certain that maximum stream capacity is seldom, if ever, reached. (M. M.)

Cap Rock. The hard, impervious covering of the salt column in a *salt dome. It consists largely of gypsum and anhydrite, with some calcite and varying amounts of sul-

fur. The cap rock of certain salt domes in the Gulf Coast area of the United States contains large deposits of sulfur formed as the halite and anhydrite of the salt dome were pushed toward the earth's surface. Some of the upper part of the halite dissolved but not the anhydrite; it was left as a protective cap on the salt column. Part of the anhydrite was later altered into gypsum, limestone and the sulfur that occurs in certain cap rocks.

Laboratory experiments suggest that the sulfur in cap rock was formed through the action of sulfate-reducing bacteria. In the presence of petroleum, these bacteria converted the sulfate in some of the anhydrite into hydrogen sulfide. Later, hydrogen sulfide was oxidized—perhaps by reaction with more of the anhydrite—to form the sulfur.

Most of the large cap rock sulfur deposits are about 450 – 750 m underground and are extracted by the *Frasch process. Petroleum may also occur in the more porous and permeable cap rocks of certain Gulf Coast salt domes. (W. H. M.)

Carbonado. See Bort.

Carbonate. A compound containing the carbonate ion, CO_3^{-2}. There are three important isostructural groups of carbonate minerals: the *calcite, the *aragonite and the *dolomite groups. In addition to minerals in these groups, the only important carbonates are the basic carbonates of copper, azurite and malachite.

Carbon Dating. See Radiometric Dating.

Carbon Dioxide. A colorless gas made up of one carbon atom and two oxygen atoms, of molecular weight 44.00995, heaviest of the four most abundant gases of dry air. A continuous exchange of carbon dioxide occurs between the oceans, which contain over 99 percent of the terrestrial CO_2, and the atmosphere. It is estimated that slightly more than half of the CO_2 in the atmosphere is dissolved into the oceans each year. The oceans return somewhat less than this amount to the atmosphere, the remainder being supplied to the air by natural and artificial combustion. The concentration of CO_2 in the air increased from about 290 parts per million (ppm) in 1860, when the industrial revolution got well under way, to about 320 ppm in 1970. About half of the carbon dioxide produced from fossil fuels apparently remains in the atmosphere. In 1970 the concentration was rising at the rate of one percent a year.

Carbon dioxide contributes to the atmosphere's *greenhouse effect, and it has been suggested that a predicted increase in world fuel consumption will result in a warming of the earth's climate by about 0.5°C (0.9°F) by the year 2000. Currently, such predictions consider only first-order effects, based on the earth-atmosphere radiation balance. Eventual adjustments of the atmosphere to an increase in CO_2—and the effect of an increase on the exchange with the ocean—are as yet unpredictable. Thus the effect on world temperature of increased CO_2 production is necessarily speculative.

At locations remote from sources of air pollution, CO_2 shows a seasonal variation caused by increased uptake by plants during the growing season, when CO_2 is used in photosynthesis to form carbohydrates in the biosphere. (M.F.H.)

Carboniferous Period. A European time unit that includes the *Mississippian (Early Carboniferous) and *Pennsylvanian (Late Carboniferous) periods and the rocks formed during this part of earth history. The term was coined by W. D. Conybeare and William Phillips in 1822 to describe the coal-bearing rocks in the British Isles.

Carbonization. A type of fossilization known also as distillation. As organic matter is buried and slowly decomposes under water or sediment, oxygen, nitrogen and hydrogen are lost by distillation. When the liquids and gases have been completely removed, only a thin film of carbon remains, somewhat like a "carbon copy" of the original plant or animal. This type of preservation is so delicate that the wings of insects, the smallest veins on leaves and even individual cell walls may be thus preserved. Plants, *graptolites, insects and fish are commonly fossilized in this way. Coal is also formed by the process of carbonization. When large concentrations of vegetation are subjected to the effects of heat, pressure and the absorbing power of enclosing material, organic volatile materials are driven off. The remaining mass of carbonized plant remains is coal.

Carnotite. A mineral, the hydrous vanadate of uranium and potassium. This ore of uranium is particularly prevalent in the Colorado Plateau region of the United States. It is a secondary mineral whose origin is usually attributed to the action of surface waters on pre-existing minerals of uranium and vanadium. It is bright yellow and is such a strong pigmenting agent that a small amount will color the rock yellow.

Cassiterite. A mineral, tin oxide, the only important ore of tin. In small amounts cassiterite is widely distributed as a constituent of granitic rocks and pegmatites.

But in only a few localities is it found in sufficient quantity to be mined as an ore. In a few places—notably Bolivia and Cornwall, England—high-temperature hydrothermal veins contain commercial cassiterite. The cassiterite from which most of the world's tin is extracted today is mined as a placer mineral, that is, as sand grains and pebbles that through countless years of *weathering have been liberated from the host rock and concentrated in stream gravels. The major producers of this stream tin are Malaysia, the Soviet Union, Thailand and China. At present Bolivia is the only country with significant production from vein deposits. See Mineral Properties. (C.S.H.)

Casts and Molds. Although casts and molds do not represent the actual remains of organisms, those of prehistoric plants and animals are considered *fossils. In this type of preservation none of the original organic material remains, nor any replacement of it. However, a trace or impression of the organisms provides evidence as to the nature of the plant or animal responsible for it. A mold is the impression of an organism or part of an organism (bone, shell, teeth, etc.) in the surrounding material. If the shell of an animal was pressed down into sediment before it hardened into rock, it may have left an impression of the outside of the shell, which would be known as an external mold. An internal mold represents the morphology and characteristics of the inner surface of the original organic remains. Internal molds, sometimes referred to as steinkerns (stone-kernel) or cores, are produced in nature when the shell of an animal becomes filled with sediment and the shell material is later removed by solution or erosion. This leaves behind the hardened sediment to reflect the internal structures of the animal.

A cast is a positive reproduction of organic remains from a mold. It may form when any substance, mineral or artificial, fills the mold formed by solution of the original hard parts. A cast occurs when a natural mold is filled while still embedded in the surrounding rocks. The formation of molds and casts has resulted in the preservation of numerous fossils and is particularly characteristic of fossil clams and snails because their shells are easily dissolved. (W.H.M.)

Cataclasite. A *metamorphic rock produced by deformation of a pre-existing solid rock without the formation of new minerals. Deformation involves crushing or shattering of brittle rock to the extent that mineral composition and texture of the original rock can still be recognized.

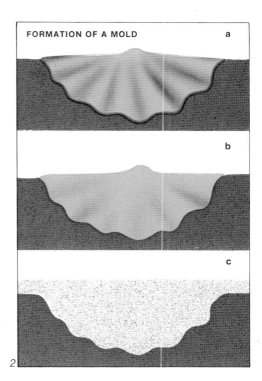

FORMATION OF A MOLD a

b

c

2

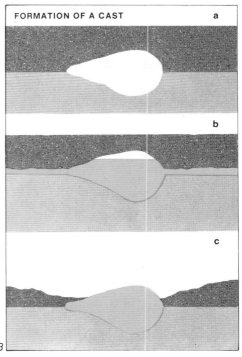

FORMATION OF A CAST a

b

c

3

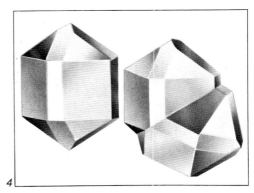

4

Cataclasites form in areas where movement between two bodies of solid rock occurs (*faults). They form at shallow depths within the earth where rock is brittle enough to be shattered and where temperatures are too low to allow chemical reactions to form new minerals. With increased deformation, cataclasites grade into *mylonites. Cataclasites also form as a result of meteorite impact. See also Lunar Geology.

Cataclastic Metamorphism. The process by which brittle rock is shattered and then reduced by crushing or grinding. See Kinetic Metamorphism.

Cataclysm. The rapid decrease in the intensity of meteorite bombardment of the moon about 4 billion years ago.

Catastrophism. A hypothesis that assumed that differences between fossils occurring in successive rock layers were the result of extinctions of old organisms by a natural catastrophe, followed by the creation of the new organisms that occur in the younger rocks above them. This hypothesis supports the idea of a young earth and a short period of creation by explaining geologic and evolutionary change in terms of violent and supernatural cataclysms such as floods. One of its main supporters was Baron Georges Cuvier (1769-1832), a noted French biologist and paleontologist who argued against organic evolution. He explained the extinction of most animals and the shaping of the landscape as the work of the Biblical Noah's Flood. The Flood and other catastrophic events were followed by special creations to replace the plants and animals thus destroyed. These earth-shaking catastrophes were worldwide in scope and not to be confused with local natural disasters such as earthquakes, volcanic eruptions, hurricanes, floods and tsunamis.

Cuvier's prominence in the 18th-century scientific community did much to prolong the acceptance of this outmoded doctrine. It was later replaced by the principle of *uniformitarianism, the assumption that the natural processes now modifying the earth and its inhabitants have operated rather uniformly and continuously through all geologic time. (W. H. M.)

Cat's Eye. A green, *chatoyant variety of *chrysoberyl highly prized as a gemstone.

Cave and Cavern. A cave is a roofed-over rock cavity large enough for a man to crawl through. Some authorities also insist that the recesses of the cavity must be so far from the entrance that no light reaches them, and that interior cross sections must be larger than the entrance. A large cave is considered a cavern. An opening whose entrance is its largest cross section and whose roof projects over the floor is called a rock shelter

Caves have interested man from remote antiquity. Primitive peoples used them for shelter and made remarkable paintings of prehistoric animals on the walls. Later, caves were exploited for the guano and chemicals they contained. The exploration and study of caves is called speleology.

Caves are formed in many ways. Waves and shore currents cut caves in cliffs along a shoreline. Large tunnel-like caves are cut in the ice at the bottom of glaciers by meltwater streams. In lava, after a crust is formed on the top, bottom and sides of a flow, the liquid in the center may drain away, leaving a tunnel several kilometers long and many meters wide. The spaces between the large blocks in coarse *talus and between large fragments deposited by glaciers form "bear-den" caves. Rock shelters are common in stratified sedimentary rocks where weathering and erosion remove less resistant layers and leave more resistant rock to form roofs and floors.

The largest and most famous caves and caverns, however, are found mainly in limestone, marble and dolomite. Although limestone is not very soluble in pure water, it is relatively soluble in water containing dissolved carbon dioxide. Water obtains some carbon dioxide from the atmosphere but much more from soil that contains decaying humus. Even though massive limestone is impervious, most limestone has vertical cracks and bedding-plane joints. The subsurface water containing dissolved carbon dioxide moves through these openings, and in time the cracks enlarge, channels form and chambers develop. It was formerly thought that they were formed above the water table by streams similar to those found at the surface. Most authorities now believe that caves are generally formed at or somewhat below the water table by slow-moving groundwater containing carbon dioxide. The progressive lowering of the water table by the downcutting of surface streams or the lowering of sea level then brings the caves above the water table. Once the caves are filled with air, *stalactites and other secondary features form. But even after a cave has emerged, streams may further enlarge it by both solution and *abrasion.

Many caves contain secondary deposits. Stalactites hang from the ceiling, stalagmites grow upward from the floor, columns are formed where stalactites and stalagmites join and "curtains" and "draperies" adorn walls and ceilings. The colors may be ghostly white or delicate hues of gray, yellow and brown. These features, together with rivers, lakes and waterfalls as well as bats, blind fishes and other organisms, make many of the great caves unique places of wonder and beauty.

Very little is known about the age of caves. They are of course younger than the youngest rocks in which they are cut and older than the oldest deposits found in them. The rate of solution of limestone, as shown by the quantity of calcium bicarbonate carried in rivers, indicates that cave formation must be a rapid process. The growth of stalactites and stalagmites varies so that they are of little use in dating; at best they give only a minimum age for a cave. Some caves can be dated if their channelways have not been tilted during dated *orogenies that affected the area. Such caves have proved to be less than 5 million years old. The bones of *Pleistocene mammals and a complex cave stratigraphy indicate antiquity for some caves. Light will be cast on this problem by correlating the sequence of events inside caves with the geologic history of surrounding areas and by finding material datable by carbon-14 measurement.

The air temperature in deep caves is more or less constant: about the mean annual temperature of the area. In view of changes in climate in recent geologic time, it is not surprising that many caves give evidence of climatic change. Calcium carbonate is precipitated from solution as *calcite, the low-temperature form, and also as *aragonite, the high-temperature form. In western United States certain caves contain stalactites and stalagmites composed of aragonite coated with calcite which is still forming. It is thought that the temperature when the aragonite was deposited was higher than at present.

The National Speleological Society of America has records of more than 11,000 caves in the United States. Carlsbad Caverns, N.M., and Mammoth Cave, Ky., both in National Parks, are perhaps the best known. The largest subterranean chambers known are in Carlsbad Caverns, where the Big Room is about 400 m (1312 feet) long, 200 m (656 feet) wide and 90 m (295 feet) high; it contains the Great Dome, a stalagmite 19 m (62 feet) tall. The explored part of the main cavern is 37 km (23 miles) long; only 5 km (3 miles) are open to the public. The total length is still unknown. A colony of several million bats inhabits part of the caverns during the summer months. The main cave and accessible passages of Kentucky's Mammoth Cave

Cave and Cavern
Smoo Cave at Sutherland, on the north coast of Scotland, was formed when water cut a deep channel in the soluble rock.

have a length of more than 240 km (150 miles) on five levels. Caves are found in the Austrian and French Alps, the Pyrenees, the Caucasus, in Wales, England, Hungary, Yugoslavia and Spain, in China—indeed, generally wherever limestones crop out at the surface. The deepest cave yet descended by man, the Gouffre Berger Cave near Grenoble, France, is at least 1.1 km (3,680 feet) deep.(R.L.N.)

Cavitation. A type of stream erosion that occurs at very high velocities. Velocity increases when a stream channel narrows, which means the pressure must decrease to maintain a constant total energy. If the pressure decreases to the pressure of water as a vapor, bubbles form. Velocity decreases when the stream widens again, so the pressure of the water increases and the bubbles collapse. The collapse causes shock waves in the water, which moves out to the channel walls, causing erosion or pitting. Cavitation rarely occurs in streams but more often in pipes and canals. (M. M.)

Celestite. A mineral, strontium sulfate, the principal source of the element strontium. It is most commonly found in small grains distributed through limestone or sandstone, or in crystals lining cavities in these rocks. Celestite frequently occurs in well-formed orthorhombic crystals. These bear such a close physical resemblance to *barite (barium sulfate) that it is almost impossible by inspection to distinguish crystals of the two minerals, except that a flame is colored red by celestite and green by barite. See Physical Properties.

Cenozoic Era. The most recent geologic era; it began with the end of Cretaceous time, about 70 million years ago. The word "Cenozoic" translates as "recent life," referring to the many modern plant and animal species that developed during this era. It is divided into two very unequal segments of geologic time: the relatively long *Tertiary Period and the more recent *Quaternary Period, which began about 2 million years ago. The Cenozoic is called the "Age of Mammals" because of the unprecedented evolution and the abundance of these warm-blooded animals during Tertiary and Quaternary time.

Cenozoic marine invertebrates resembled their Cretaceous forebears but were decidedly more modern in appearance. Foraminifera were present in great numbers and are valuable *guide fossils of the Tertiary—especially for the petroleum geologist. Corals, bryozoans, echinoderms (particularly echinoids) and arthropods were also abundant. Brachiopods, which had so dominated Early Paleozoic

seas, were now greatly diminished in number and variety. Mollusks were the dominant Cenozoic marine invertebrates, but ammonoids, which were so common throughout Mesozoic time, were replaced by an unprecedented number and species of pelecypods and gastropods, many of which resembled the clams, oysters and snails of today. Tertiary vertebrates are especially well known through the fossilized remains of fishes, amphibians, reptiles, birds and, to a greater extent, mammals. Fishes were plentiful and consisted of many bony fishes and large numbers of sharks. Some of the latter were 18−25 m long and had 15 cm teeth. The amphibians were represented by salamanders, toads and frogs. The reptilian hordes of the Mesozoic had dwindled to representative snakes, lizards, crocodiles and turtles, which were present in about the same numbers as today. The birds resembled those of the present, but because of the fragility of their bodies they are rarely found as fossils. Of particular interest are the so-called "giant" birds of the Tertiary, big ostrich-like, flightless creatures, such as *Dinornis* and *Diatryma* that were as much as 3 m tall and laid eggs over 30 cm long.

The greatest development was among the mammals. Those of the Paleocene Epoch were primitive and small, and rather unlike modern species. *Hyracotherium* (also called *Eohippus*), the earliest horse, appeared in the Eocene. Eocene forms also included the earliest rodents, camels, rhinoceroses and creodonts, forerunners of the meat-eating mammals. During Oligocene time mammals took on a more modern appearance and included dogs, cats, camels, horses, rhinoceroses, pigs, rabbits, squirrels and (in Africa) small elephants. Also present during the Eocene and Oligocene were mammals quite unlike any living today. These included the uintatheres, great rhinoceros-like beasts, some of which stood 2 m tall at the shoulder. The titanotheres, another group of gigantic Early Cenozoic mammals, appeared first in the Eocene, at which time they were about the size of sheep. They had increased to titanic proportions by Middle Oligocene time, and *Brontotherium*, the largest land mammal whose fossil remains have been collected in North America, had an elephant-like body and was as much as 2.4 m (8 feet) tall at the shoulders. A large hornlike growth protruded from the massive skull and was divided at the top. Mammals became even more varied and abundant in the Miocene, probably aided by the expansion of the grasses that blanketed Miocene plains and prairies. Horses, camels, deer, pigs, rhinoceroses and other

now-familiar mammals were present in North America. The Pleistocene also had its share of strange mammals. One of these, the giant hornless rhinoceros called *Baluchitherium*, the largest land-dwelling mammal of all time, was as much as 9 m (30 feet) long and stood 5.5 m (18 feet) high at the shoulders. The baluchitheres first appeared in the Oligocene and became extinct during Miocene time. These great beasts must have been restricted to Asia, for their remains have not been found elsewhere. Another interesting Oligocene-Miocene form were the entelodonts or giant swine, some of which were 1.8 m (6 feet) tall at the shoulders. Pliocene mammals were still more advanced than those of the early Tertiary epochs. Giant ground sloths, such as *Mylodon*, were common in the southern United States during both Pliocene and Pleistocene time; some were as much as 4.5 m (15 feet) tall and weighed a ton or more. Present also were the glyptodonts, distant relatives of the armadillos.

Most Cenozoic formations have not been deeply buried or deformed, and consist largely of thick deposits of marine and terrestrial *sedimentary rocks. In North America, marine formations occur largely in a narrow belt along the Atlantic and Gulf coastal regions and along the Pacific coast. Nonmarine sedimentary rocks occur in the Great Plains region and the western United States, and volcanic rocks are exposed in western and northwestern North America. Both nonmarine and marine Cenozoic rocks occur in Europe, South America, Africa, Asia and Australia. (W.H.M.)

Centrosphere (or Barysphere). The central portion of the earth's interior. It lies beneath the lithosphere, or crust, and consists of an upper zone called the *mantle and a deeper zone called the *core. There is disagreement regarding its exact characteristics, but it is believed to have a specific gravity of about 13 and a temperature of around 4000°C (7200°F). It is also thought that part of the core is liquid, that iron is the major constituent of the core, and that silicates rich in iron and magnesium constitute the bulk of the mantle.

Cephalopod. A member of class Cephalopoda of phylum Mollusca. The cephalopods include the squids, octopuses and the pearly *nautilus. The name, derived from two Greek words meaning "head" and "foot" respectively, refers to the many parts of the animal's foot that emerge from the head. Both the huge eight-armed octopus and the gigantic deep-sea squid have given rise to many weird legends. These squid are the largest known animals

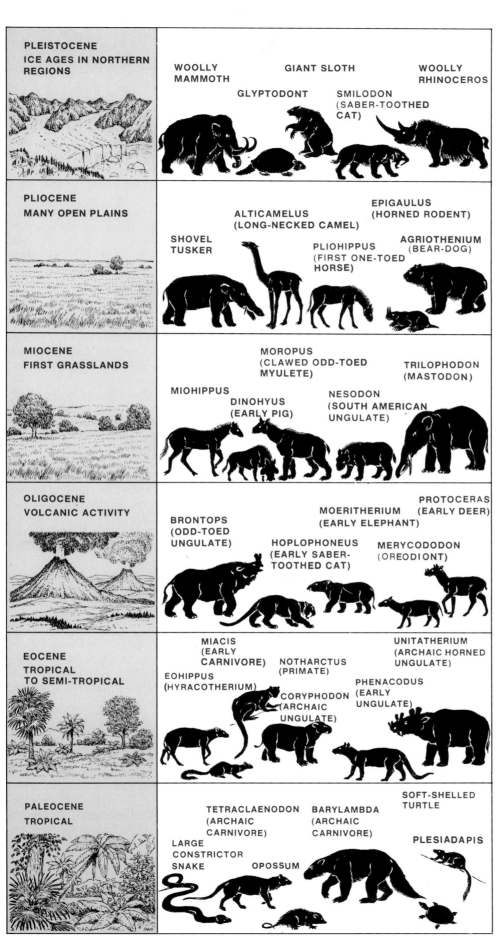

PLEISTOCENE
ICE AGES IN NORTHERN REGIONS

WOOLLY MAMMOTH
GIANT SLOTH
WOOLLY RHINOCEROS
GLYPTODONT
SMILODON (SABER-TOOTHED CAT)

PLIOCENE
MANY OPEN PLAINS

ALTICAMELUS (LONG-NECKED CAMEL)
EPIGAULUS (HORNED RODENT)
SHOVEL TUSKER
PLIOHIPPUS (FIRST ONE-TOED HORSE)
AGRIOTHENIUM (BEAR-DOG)

MIOCENE
FIRST GRASSLANDS

MOROPUS (CLAWED ODD-TOED MYULETE)
TRILOPHODON (MASTODON)
MIOHIPPUS
DINOHYUS (EARLY PIG)
NESODON (SOUTH AMERICAN UNGULATE)

OLIGOCENE
VOLCANIC ACTIVITY

PROTOCERAS (EARLY DEER)
MOERITHERIUM (EARLY ELEPHANT)
BRONTOPS (ODD-TOED UNGULATE)
HOPLOPHONEUS (EARLY SABER-TOOTHED CAT)
MERYCODODON (OREODIONT)

EOCENE
TROPICAL TO SEMI-TROPICAL

MIACIS (EARLY CARNIVORE)
NOTHARCTUS (PRIMATE)
UNITATHERIUM (ARCHAIC HORNED UNGULATE)
EOHIPPUS (HYRACOTHERIUM)
CORYPHODON (ARCHAIC UNGULATE)
PHENACODUS (EARLY UNGULATE)

PALEOCENE
TROPICAL

TETRACLAENODON (ARCHAIC CARNIVORE)
BARYLAMBDA (ARCHAIC CARNIVORE)
SOFT-SHELLED TURTLE
LARGE CONSTRICTOR SNAKE
OPOSSUM
PLESIADAPIS

Cenozoic Era

This part of earth history, consisting of the Tertiary and Quaternary Periods, and known as the Age of Mammals, covers the last 70 million years. The chart illustrates each epoch within these periods chronologically, reading from the earliest at the bottom to the most recent at the top. A typical landscape indicating the climate of each epoch is shown at left, and representative animals appear at the right. During the Paleocene Epoch, much of the world was covered with tropical jungles of tree ferns, palms and the first flowering plants. The great dinosaurs had become extinct, and only a few reptiles survived. Among the mammals were marsupials and some early carnivores called creodonts.

The Eocene Epoch brought some 20 million years of mild, damp climate, with much of the earth covered by semi-tropical jungle. Hoofed animals like Coryphodon were common. The "odd-toed" perissodactyls were represented by the first horse, Eohippus. During the Oligocene Epoch, which began some 40 million years ago and lasted 15 million years, mountains were thrust upward, the climate became cooler and the great forest regions began to give way to open grassland. Ungulates became dominant, such as the huge Brontops.

By the Miocene Epoch, some 25 million years ago, grassland covered much of the earth. Hoofed animals continued to be widespread, including an early elephant called Trilophodon.

The final epoch of the Tertiary Period, the Pliocene, lasted about 11 million years, during which the trend toward cooling approached its climax. Toward its end, summers grew short and chill winds swept over vast plains. Pliohippus (the direct ancestor of the modern horse) and horned rodents such as Epigaulus flourished.

The Pleistocene Epoch, the Great Ice Age, began some 2 million years ago. Four times during this epoch ice sheets and glaciers covered the landmasses. A trend toward gigantism developed among mammals. Large armadillo-like glyptodonts and giant ground sloths spread northward from South America. The huge woolly rhinoceros and mammoth had dense, shaggy coats enabling them to survive arctic conditions. Their contemporary, early man, painted their likenesses on the walls of his caves. The Pleistocene ended about 10 thousand years ago, giving way to the Holocene, or Recent epoch.

1

2

without backbones. Living in the Atlantic Ocean off the coast of Newfoundland, some attain an overall length of more than 15 m (50 feet) and may weigh almost 30 tons; their serpent-like arms are covered with sucker discs. Both squid and octopus swim by "jet propulsion": Water is taken into the mantle cavity and ejected through a structure known as the funnel, pushing the animal's body in the opposite direction.

The two types of cephalopods most likely to be found as fossils, the *ammonoids and the *belemnoids, died out at the end of Cretaceous time. But during the Mesozoic Era the ammonoids were very abundant in the shallow seas that covered much of the world. Because their calcareous shells were well adapted to preservation and were so numerous, ammonoids are valuable *guide fossils for Triassic, Jurassic and Cretaceous rocks. Most living cephalopods, such as the squids and the cuttlefish, have a shell inside their body; however, the more important fossil forms, the ammonoids and nautiloids, secreted an external calcareous shell. Among the externally shelled cephalopods found as fossils are forms that had straight, tapering, cone-shaped shells. Others had shells that were loosely coiled or were coiled in a spiral like a ram's horn. (W. H. M.)

Cerargyrite. A mineral, silver chloride, and a secondary alteration product of other silver minerals. Its presence near the surface indicates that primary ores lie below. "Slugs" (formless masses) of cerargyrite found in an oxidized zone led to the discovery of the great lead, zinc and silver mines at Broken Hill, New South Wales, Australia. Cerargyrite contains over 75 percent silver and in places has been a valuable ore of that metal. It is a soft mineral resembling horn and has thus been called "horn silver," a name also applied to bromyrite, the silver bromide that occurs with and has properties similar to cerargyrite. See Mineral Properties; Appendix 3. (C.S.H.)

Cerussite. A widely distributed mineral, the carbonate of lead. It is found in oxidized zones overlying primary lead deposits and has formed by the alteration of *galena. It is commonly associated with other secondary minerals such as anglesite, pyromorphite, smithsonite and limonite. Cerussite is orthorhombic and is frequently found as single crystals, pseudohexagonal twins and reticulated groups. Well-crystallized specimens have been found at many localities, but the finest are from Tsumeb, South-West Africa, and Broken Hill, New South Wales, Australia. See Mineral Properties; Appendix 3. (C.S.H.)

C-Horizon. See Horizons, A-, B- and C-.

Chabazite. A mineral of the *zeolite group found with other zeolites lining cavities in basaltic rocks. It can usually be distinguished from associated minerals by the rhombohedral form of its crystals.

Chalcanthite. A mineral, the naturally occurring hydrated copper sulfate, identical to the artificial blue vitriol. It is a rare secondary mineral found only in arid regions, where it has formed by the oxidation of primary copper sulfides. At Chuquicamata, Chile, it has been an important copper ore mineral. See Mineral Properties.

Chalcedony. Cryptocrystalline *quartz with a fibrous structure, commonly found lining or filling cavities in rocks. The name chalcedony is given to a brown to gray translucent variety with a waxy luster. Other varieties are carnelian (red), sard (brown), chrysoprase (green), bloodstone (green with red spots). *Agate and *onyx are layered chalcedony. Most colored varieties have been valued as semiprecious gems since ancient times.

Chalcocite. A mineral, a copper sulfide, and an important ore of copper. It has formed chiefly as a secondary mineral in the enriched zones overlying primary deposits of copper sulfides. Chalcocite has a lead-gray color that tarnishes to a dull black on exposure. It is distinguished from similar minerals by being imperfectly sectile. See Mineral Properties.

Chalcopyrite. A mineral, a copper iron sulfide, and one of the most important and widespread copper ore minerals. It forms under a wide range of conditions in a variety of geologic environments. It is found in *igneous rocks, *pegmatite dikes, *contact metamorphic deposits and crystalline *schists. Of economically greater importance is its occurrence in hydrothermal veins in association with other sulfide ore minerals. Equally significant is its presence in *porphyry copper deposits, where it is the chief primary mineral.

Chalcopyrite has a brass yellow color and thus is called "fool's gold," a name it shares with the more abundant *pyrite. Scratching with a knife distinguishes the three minerals: gold is sectile (that is, it can be cut by a knife); chalcopyrite is brittle; and pyrite cannot be scratched with a knife. See Mineral Properties. (C.S.H.)

Chalk (from Latin *caix*, meaning lime). A soft, fine-grained, earthy, white to greyish *limestone. it is composed largely of biochemically derived *calcite consisting

3

Chabazite
1 As is characteristic of many zeolite minerals, chabazite crystals line a cavity in a basaltic lava flow.

Chalcedony
2 Consisting of a microscopic intergrowth of quartz particles, chalcedony typically occurs in botryoidal (grape-like) pearly masses.

Chalk
3 The chalk cliffs, 106 m (350 feet) high at Etretat, on the French shore of the English Channel, are composed of the soft, compacted skeletons of microscopic oceanic plants and animals.

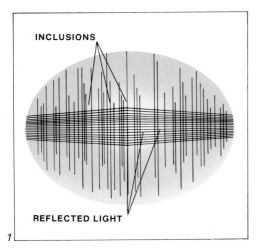

of the remains of minute plants and animals, especially *foraminifers and *coccoliths (the shells of microscopic algae) that inhabited ancient, shallow seas. Calcium carbonate (usually as calcite) may reach as much as 99 percent of the total volume of the rock. Chalk deposits are typical of rocks from the Cretaceous Period and are widespread in England, France and North America.

Challenger Expedition. The voyage made from 1872 to 1876 by H.M.S. *Challenger*, a ship outfitted by the British Admiralty to study the oceans of the world. This expedition is considered the birth of the science of oceanography in that it was the first extended voyage for the sole purpose of collecting scientific data. The chief scientist, Charles W. Thompson, headed a staff of three naturalists, a chemist and a secretary. After traveling more than 125,000 km in 3½ years, the ship returned to England with a quantity of data that resulted, two decades later, in the 50-volume Challenger Report. Among many contributions reported were the identification of nearly 5000 new species of organisms, the detailed chemical analysis of seawater and the discovery of manganese nodules. (R.A.D.)

Chandler Wobble. A wobble of the earth about its spin axis. See Nutation.

Chatoyancy. The property by which some minerals produce in reflected light a band of light resembling the eye of a cat. This property is found in minerals with a silky appearance resulting from a parallel arrangement of elongated inclusions or from closely packed parallel fibers. A cabochon gemstone cut from such a mineral scatters reflected light, producing a lustrous band across the stone at right angles to the length of fibers or inclusions. Chatoyancy is well shown in cat's eye, a gem variety of *chrysoberyl, and in tiger's eye, fibrous crocidolite replaced by quartz.

Chert. A compact, extremely hard, siliceous *sedimentary rock composed of the mineral *quartz with a crystal structure so fine that it cannot be seen under an ordinary microscope. It occurs as bands (banded chert) or as irregular masses (chert nodules). Chert has a splintery to conchoidal *fracture and is typically black, white or gray. The origin of many cherts is unknown. Some, such as *radiolarian chert, are of organic origin; others are thought to have been formed by the primary deposit of colloidal silica or as a siliceous replacement of pre-existing rocks.

Chézy Equation. A basic formula, proposed by Antoine de Chézy in 1769, for calculating the velocity of a stream; it is used widely, along with the *Manning equation. The Chézy equation relates the velocity of a stream by its slope and its depth of flow. $V = c(RS)^{1/2}$ where V is the mean velocity of the stream, R is the hydraulic radius (depth) and S is the slope. The letter c is a constant that must be determined for each stream and whose value depends on gravity, friction forces and the roughness of the stream. Once c is determined for a stream, the equation can be used to calculate the mean velocity for any depth and slope.

Chinook. A local name in the Rocky Mountain region of the United States for the foehn that occurs on the eastern slopes of the Rockies. A foehn is a dry, warm wind that results when strong winds passing over a mountain range are forced to descend on the lee side. The name was first used because the wind came from a Chinook Indian camp.

Chlorite. A representative of several minerals, collectively known as the chlorite group or the chlorites. They are *phyllosilicates with a pseudohexagonal layered structure resembling that of the *micas. The color is characteristically green, but may be orange to brown in manganese-bearing varieties and violet in chromium-bearing varieties. The chlorites are common minerals in low-grade *schists and in *igneous rocks as alterations of *ferromagnesian minerals such as pyroxene, amphiboles, biotite and garnet. Some schists are composed almost entirely of chlorite, and the green color of many igneous rocks is due to the presence of the mineral.

The chlorites are essentially magnesium silicates. Clinochlore is the most common of the members of the group, all of which arise from the substitution of varying amounts of iron for magnesium and of aluminum for silicon. In addition to the green color, chlorites are characterized by micaceous cleavage and by nonelastic cleavage flakes. See Mineral Properties. (C.S.H.)

Chondrite. A stony meteorite containing chondrules, as opposed to an achondrite, which is a stony meteorite that does not have chondrules. Chondrules are small, rounded bodies composed chiefly of *olivine and *enstatite embedded in a rock that is made up essentially of olivine, *hypersthene, *plagioclase and nickel-iron. The chondrites are the most abundant of all meteorites.

Chromate. A mineral containing the ionic group CrO_4^{-2}. The chromates are generally grouped with the sulfates in mineral classification. *Crocoite is the most common chromate.

Chromite. A mineral, an oxide of iron and chromium in the *spinel mineral group, and the only ore of chromium. Pure chromite is not found because some magnesium is always present substituting for iron, and usually some ferric iron and aluminum substitute for chromium. Chromite is a common accessory mineral in *peridotites and in *serpentines derived from them. In some of these rocks it is found in large irregular masses that serve as an ore. Extensive layered chromite deposits, such as those found in the Bushveld Igneous Complex of South Africa, are thought to have been formed by magmatic differentiation. The major chromite-producing countries are the Soviet Union, South Africa, Turkey and the Philippines. See Mineral Properties; Appendix 3. (C.S.H.)

Chrysoberyl. A rare mineral, beryllium aluminum oxide, found in granitic rocks and *pegmatites and in mica *schist. It is one of the few beryllium-bearing minerals, but is far less abundant than *beryl. Common chrysoberyl is yellow-green and occurs in both tabular crystals and in pseudohexagonal twins that are sometimes cut as gemstones. However, far more prized as gems are *cat's eye (the green chatoyant variety) and *alexandrite (green by daylight but red in artificial light). See Mineral Properties; Twin Crystal.

Chrysocolla. An amorphous copper silicate. In the strict sense it is not a mineral for it is noncrystalline and has a wide range of chemical composition. It is a minor ore of copper and is found in the oxidized zones of copper deposits, recognized by its conchoidal fracture and blue-green color. It occurs abundantly at Chuquicamata, Chile. See Mineral Properties.

Chrysotile. The fibrous variety of the mineral *serpentine. It is the most important of the minerals mined commercially as *asbestos.

Cinder Cone. A cone-shaped *volcano composed essentially of unconsolidated volcanic dust, ash, lapilli, blocks and *bombs ejected from a volcanic vent. It is commonly basaltic but may be rhyolitic. Cinder cones are usually less than 300 m high and have steep slopes largely at the *angle of repose. Found at the summit is a bowl- or funnel-shaped depression

Chatoyancy
1 Countless needle-like inclusions, or hollow tubes, lying parallel within some minerals (such as this chrysoberyl) form a band of pale opalescence that moves as the stone is turned, creating the cat's-eye, or chatoyant, effect.

Chondrite
2 Photographed in polarized light, a section of a meteorite, or chondrite, shows rounded chondrules embedded in olivine and pyroxene.

Cinder Cone
3 Amboy Crater, in California's Mohave Desert, was formed when molten rock ejected from a volcanic vent cooled and fell as black gravel around the vent.

called a *crater, which may be 100 m deep. Flows that broke out low on the slopes of the cone are often present but have less volume than the fragmental material. If the vent is elongated instead of circular, a cinder ridge is formed. Cinder cones are common in New Mexico, Arizona (especially near Flagstaff), Oregon, Idaho, California, Hawaii, Italy, and Antarctica. Paricutin, a large cinder cone, arose between 1943 and 1952 in a cornfield about 350 km from Mexico City. (R.L.N.)

Cinders. See Pyroclastic Material.

Cinnabar. A mineral, mercury sulfide, the only important ore mineral of that metal. It can usually be recognized by its vermilion-red color and high specific gravity (8.10). Because cinnabar has a strong pigmenting power, a rock with only a small percentage of the mineral may be colored red. The Chinese used it as a pigment long before it was known to contain mercury. Cinnabar is formed near surface deposits and is commonly found as rock impregnations and vein fillings near recent *volcanic rocks. In places it has been found as hot spring deposits. The most famous and productive cinnabar deposit is at Almaden, Spain, where mining has been going on since ancient times. Other important localities are Idria, Yugoslavia; Huancavelica, Peru; and Hunan, China. See Mineral Properties; Appendix 3. (C.S.H.)

Cirque. A semicircular indentation or hollow, like half of a bowl, formed on a mountain slope by processes associated with snowbanks or glaciers. Cirques occur in all glaciated mountain ranges. They are found near the summits of high mountains and below sea level in high latitudes where they have been drowned by the post-glacial rise of sea level. The headwall may be one hundred to many hundred meters high and is generally steep. A lake in a rock basin is often found on the cirque floor. Cirques tend to be larger and more numerous on the lee slopes of mountains and, in the Northern Hemisphere, on the north side where snow and ice are better protected from insolation. One or more cirques are generally found at the head of a glaciated valley, but they also occur elsewhere. They owe their origin to *frost action, *mass wasting and glacial *abrasion and plucking, processes by which snowbanks or glaciers incise themselves into a mountain slope. The floors of cirques are thought to occur close to the *snowline that existed when they were formed. Thus they have climatological significance and can be used to estimate the rise in snowline since the cirque was formed. (R.L.N.)

Clastic. A term describing rocks typically composed of broken rock fragments (clasts) or organic remains transported from their point of origin. *Conglomerate, *sandstone and *breccia are typical clastic rocks. Igneous rocks with clastic texture are called *pyroclastics. Certain biochemical sedimentary rocks (such as *coquina, which is composed of broken shells) are also described as clastic.

Clay. A soft, smooth, earthy *sedimentary rock composed of mineral fragments smaller than particles of dust. Among these particles are clay minerals consisting of silicon, aluminum and other elements. Minute particles of quartz, calcite and other minerals may also be present. Clay may be almost any color. When moist it has an earthy odor. Wet clays are usually plastic, but become firm and solid when dry. Clay is used to make bricks, tile, pottery, sewer pipes and many other products. *Kaolin (or china clay) is a special kind of white clay that can be used to make chinaware. It contains particles of the clay mineral *kaolinite plus other clay minerals. *Bentonite, another clay, is formed from weathered volcanic ash. It contains the clay mineral *montmorillonite and looks smooth and soaplike. Fresh bentonite is white, pale green or pale blue; dried-out or weathered samples are tan, brown, yellow or reddish. When wet, bentonite absorbs water, swells and then has a jelly-like appearance. It can be used to absorb unwanted coloring material in petroleum and in vegetable oils; It is then called bleaching clay. Another important use of bentonite and other clays is in the rotary method of drilling for oil and gas; mud pumped into the filled hole carries the rock cuttings up to the surface, cools the drilling tools, and coats and seals the walls of the hole. (W.H.M.)

Cleavage. The ability or tendency of a rock to split along planar surfaces that are not necessarily parallel to the original *bedding of the rock. Cleavage is a secondary *foliation produced by deformation or *metamorphism of pre-existing rock, and results in closely-spaced parallel *fractures or other parallel structures or textures. Many forms of cleavage have been recognized and classified by geologists.

Cleavage is categorized by its mode of origin, its relationship to internal textural or structural features inherent in the rock, or to features such as *folds or bedding.

Slaty cleavage, a property that allows slate to be split into large thin sheets, and *schistosity are associated with rocks whose platelike, lenslike, or rodlike grains have been recrystallized into paral-

lel, tabular arrangements. The term schistosity represents cleavage in rocks (schist) whose individual minerals may be seen without a magnifying lens. Both slaty cleavage and schistosity may result from flowage of brittle rock when subjected to pressures and temperatures that allow the rock to be deformed without fractures. During rock flowage mineral grains may rotate into planar alignment or they may recrystallize along planes parallel to the direction of flowage.

Fracture cleavage develops as closely spaced planes of fracture (measuring from millimeters to a few centimeters) and develops irrespective of the orientation of the minerals within the rock. Cleavage of this type results when the original rock is subjected to forces that cause it to *shear. that is, when segments of rock slide past one another.

In fracture cleavage displacement of the individual segments between cleavage planes does not occur.

Shear cleavage is also a shear phenomenon that results in the visible displacement of rock segments between fracture cleavage planes. It forms like fracture cleavage but with the addition of movement along the cleavage planes.

Slip cleavage, also referred to as strain-slip cleavage or crenulation cleavage, develops in schistose rocks that have been deformed into small folds with wavelengths measured in fractions of an inch. The deformation is great enough to stretch the shorter limbs of the folds and to reorient the platy minerals parallel to the plane formed by the stretched limbs. These planes are zones of weakness along which the slip cleavage develops.

Bedding cleavage is similar to slaty cleavage but is parallel to the bedding or layering of the original rock. The origin of bedding cleavage is found in the factors responsible for the various types of cleavage so far mentioned. Depending on the locality, bedding cleavage may result from deformation· recrystallization, or by compression due to the great thicknesses of overlying rock.

Geologists have intensively investigated types of cleavage and their geometric relationships to structural features. Enough has now been learned about these relationships so that cleavage characteristics can be used as a tool to work out the history of complex crustal deformation. (J.W.S.)

Cleavage, Mineral. See Mineral Cleavage.

Cleavelandite. A platy variety of *albite feldspar. It most commonly occurs in *pegmatites.

Cliff. A steep slope or face cut in *bedrock or *regolith. Cliffs are formed principally by faulting, volcanic activity, waves and shore currents, running water and glaciation. Cliffs that result from faulting are common. The eastern face of California's Sierra Nevada, which is hundreds of kilometers long and in places nearly 3 km high, is a fault scarp modified by *mass wasting, stream action and glaciation. High cliffs frequently surround *calderas. Nearly vertical cliffs a hundred meters high and more surround the calderas near the summits of Mauna Loa and Kilauea on the island of Hawaii. Those that rise above Crater Lake, Oreg., are about 600 m high and extend down into the lake a hundred meters or more.

Marine cliffs are familiar to those who live near the ocean. They are 600 m high at St. Helena in the south Atlantic, where Napoleon was imprisoned. The chalk cliffs along the English Channel, which are over 100 m high in places, have been known since Roman times. The walls of glaciated valleys, whether on land or invaded by the ocean (*fiords), can be steep and even overhanging. Those in the Yosemite Valley in California and in the Lauterbrunnen Valley in Switzerland are hundreds of meters high and world famous. So too are those in the fiords of Norway, Alaska and New Zealand. The cliffs associated with the *horns, *cirques and *arêtes of high glaciated mountain ranges are perhaps the most awe-inspiring landforms on earth.

The erosional activity of running water, together with mass wasting, is responsible for the magnificent cliffs in the Grand Canyon and in Zion Canyon. (R.L.N.)

Climate. In general, the long-term manifestation of weather; more specifically, the statistical ensemble of weather conditions over a specified area during a specific interval of time, usually several decades. Temperature, precipitation, cloudiness and the other aspects of climate show considerable variability. Some measure of this variability is usually included in the climatic description of a place. In spite of the variability, the major controls of climate are relatively stable, so large deviations of the weather elements are almost invariably followed by a return to the average or "normal" values. The earth's climates do change, as historical accounts show, over periods of tens, hundreds and thousands of years. On the scale of thousands and millions of years the earth's climate has changed radically. Perhaps the most dramatic example of such changes were the ice ages with their superimposed glacial and interglacial stages.

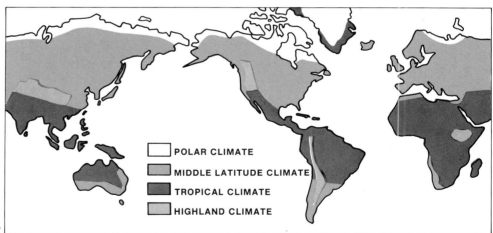

Cirque
1 These bowl-shaped depressions at the summit of the Uinta Range, Utah, developed as glaciers carried away rocks loosened from the bedrock.

Climate
2 The four major climate zones of the world are polar, middle latitude, tropical and highland.

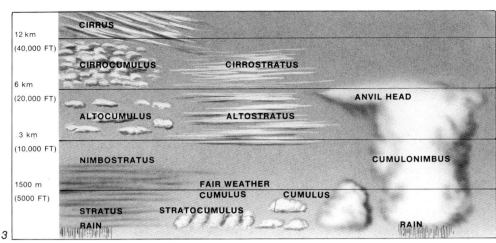

Clouds

1 Clouds called altocumulus often form a roll-like canopy, sometimes signifying rain.

2 a) Horizontal air movements produce stratiform clouds. b) Vertical air movements produce cumuliform clouds.

3 Clouds can be classified according to height and form. Cirrus clouds which develop typically at 9000 m (30,000 feet), or higher are often blown about into feathery strands called "mares' tails." Altocumulus clouds form as puffy patches or layers at about 3000 m (10,000 feet). Nimbostratus, the true rain clouds, range in height from near the earth's surface to about 2000 m (6500 feet). Towering cumulonimbus are the familiar thunderheads, whose bases may almost touch the ground while strong updrafts carry the tops to as much as 23,000 m (75,000 feet).

4 The sun shines dimly through a gray sheet of altostratus.

5 Thick nimbostratus clouds continuously disgorge rain and snow.

Overleaf

Cumulonimbus clouds (a) bring thunderstorms and sometimes hail. (b) Cirrus clouds often herald a storm. (c) Stratocumulus do not produce rain unless they fuse and become nimbostratus.

The word climate is derived from the Greek *klima*, meaning inclination, a derivation that reveals the importance the Greeks attached to latitude as a control of the earth's temperature climate. Recognizable climatic types exist because the major climatic controls (including solar radiation, the earth's motions, the composition of the atmosphere and the distribution of surface features such as land, water, mountains and land cover) are all comparatively regular and permanent. Although the addition of carbon dioxide, heat, water vapor and solid particles to the atmosphere from man-made sources has been cited as a possible cause of climatic change, artificial influences on climate are still relatively small and appear to be only local or at most regional in their effects. (M.F.H.)

Clouds. A cloud is a visible mass of either minute water droplets or ice particles—or a mixture of both—in the atmosphere above the earth's surface. (When a cloud comes in contact with the surface of the earth it is called fog.) Cloud droplets form when water vapor condenses in microscopic particles of soil, sea salts, or smoke in the atmosphere. These particles are called condensation nuclei.

At temperatures above freezing, clouds are composed entirely of water droplets. At temperatures below freezing they are a mixture of ice particles (which occur when water or water vapor condenses on minute particles called ice nuclei) and liquid water. Water in such microscopic quantities can remain liquid far below normal freezing temperatures and is then said to be supercooled. Supercooled droplets have been observed at temperatures as low as -35°C (-31°F) in the atmosphere and -40°C (-40°) in the laboratory. At extremely low temperatures, clouds consist almost entirely of ice crystals. Clouds composed of water droplets have rather sharply defined edges and a more "solid" appearance than ice-crystal clouds, which appear rather wispy.

Since condensation nuclei are ordinarily present in the air, cloud formation depends primarily on the cooling of water vapor. At any given temperature, there is a limit to the quantity of water vapor that the air can hold. When this limit is reached, the air is said to be saturated. If the amount of water vapor in the air increases through evaporation of surface water or if the air temperature falls, there is an excess of water vapor in the air, which then condenses into liquid form. Fog is usually caused by the addition of water vapor to the air through the evaporation of water; clouds usually result from a drop in temperature.

The temperature at which air is fully

4

5

a

b

c

saturated is called the dew point; below the dew point, vapor condenses and cloud particles form. For the most part, the cooling of air until the dew point is reached is caused by the upward motion of air. Rising air cools because of the decrease in air pressure at higher elevations which allows the air to expand. The individual air molecules become more diffused and do not strike one another so frequently, thus lowering the temperature. This cooling of the air as a result of volume change is called adiabatic cooling.

If the rising air diffuses horizontally, the clouds formed will be stratiform, or layerlike. Vertical movements — rapid updrafts — produce clouds with a marked vertical, or cumuliform, character, such as the great billowy mounds typical of thunderclouds. The prevailing scheme of cloud classification, first introduced by the English scientist Luke Howard in 1803, derives from these two basic cloud shapes, stratiform and cumiliform, and their altitude—high, middle or low. The highest clouds are the cirrus and cirrostratus; these are always composed of ice crystals and are feathery or tufted. A halo around the sun or moon is a pure ice-crystal, or cirrostratus, cloud. Another very high cloud, cirrocumulus, (cumulus means mound or heap) which consists of patches of small round clouds, may contain water droplets as well as ice crystals.

Of the middle altitude clouds, altostratus is nearly always mixed water and ice, but occasionally may be composed entirely of ice crystals. Altocumulus is usually a water cloud. Of the low clouds, cumulus, nimbostratus (nimbus means rain cloud) and stratocumulus are usually water clouds but may be mixed. The towering cumulonimbus, or thunderhead, which frequently is topped by a spreading "anvil" or "plume," contains both water droplets and ice crystals, as well as large water droplets, snowflakes, snow pellets and even hail.

Cloud particles are very small, having a mass about one-millionth of a typical raindrop. The proportion of a cloud droplet to a raindrop is like that of a BB shot to a basketball. It was once assumed that raindrops were simply condensed cloud droplets grown larger through continued condensation. Cloud droplets, which range in diameter from about .002 to .1 mm, grow rapidly from initial condensation to about .04 mm, but after that, growth by condensation takes place very slowly. Since only drops larger than .2 mm fall rapidly enough to escape evaporation before they reach the ground as rain, continued condensation is not an adequate explanation in the light of how easily droplets become large enough to fall as rain.

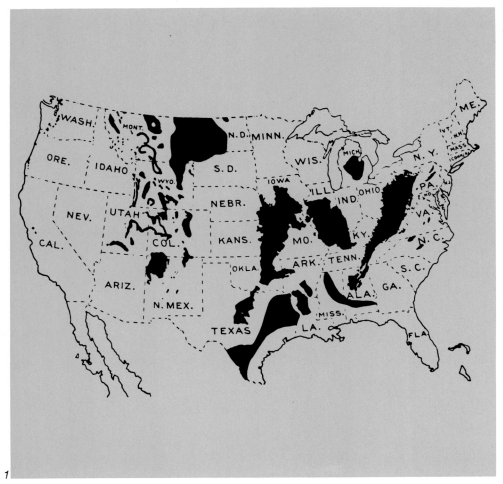

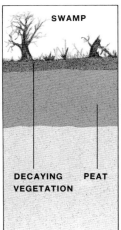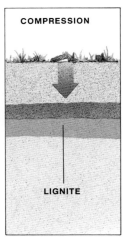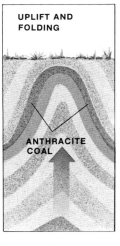

Coal

1 Over half the world's coal reserves, some four trillion tons, are found in North America. In eastern and central United States, most coal deposits were formed in the swamps of the Pennsylvanian Period.

2 The stages in the formation of coal are: (a) decaying swamp vegetation is converted by bacteria into peat; (b) peat is gradually buried and forms lignite; (c) more compression results in bituminous coal; (d) under intense heat and pressure, anthracite is formed.

3 Coal is classified by the amount of water, volatile substances and "fixed carbon" it contains. Peat (a), with 80 percent moisture, is the first step in the formation of coal. Bituminous, a soft coal (b), represents 90 percent of the coal mined. Anthracite (c) is hard coal, the highest grade. Coke (d) is the residue left when bituminous coal is heated.

It is now thought that rain is typically formed by melting of snow previously formed in a cloud. A mixture of super-cooled water droplets, water vapor and ice crystals is unstable; the ice crystals grow rapidly at the expense of the water droplets. This occurs because water droplets evaporate more rapidly than ice at the same temperature. As the droplets evaporate, their vapor condenses onto the ice crystals so that there is a continuous transfer of water in the cloud from liquid to ice. When the crystals become large enough, they fall to the ground as snow or, if they pass through warmer lower layers of air, as rain. This ice crystal theory of the formation of rain was suggested by Alfred Wegener in 1911, fully described by Tor Bergeron in 1943 and further developed by W. Findeisen. The continuous transfer of water from liquid to ice can be initiated by seeding, or artificially introducing, ice crystals into clouds. In 1946, Vincent J. Schaefer demonstrated the success of this method by dropping six pounds of crushed "dry ice" (frozen carbon dioxide) into a supercooled altocumulus cloud. This experiment quickly led to further research and to attempts to produce rain, prevent hail and forestall lightning by seeding clouds with particles of various substances, especially silver iodide, that have a physical similarity to ice.

Since rain also falls from water clouds in which the temperature is above freezing as well as from mixed water-ice clouds, the ice crystal theory does not explain all precipitation. In so-called warm clouds, condensation accounts for the initial growth of cloud droplets, but continued growth of the falling droplets is due to coalescence with smaller droplets in their path. A falling raindrop captures smaller droplets not only by collision but also be sucking other droplets into the partial vacuum that results from its rapid fall through the cloud. Raindrops grow most rapidly when there are drops of many different sizes colliding. Collisions are most likely when the clouds are tall and convective movements of the air are strong. See Condensation. (M.F.H.)

Coal. A naturally occurring, readily combustible rock derived from partially decayed plant remains. It occurs as beds with *sedimentary rocks and consists largely of carbonized vegetable matter of nonmarine origin. Coal formation proceeds in stages, beginning with the accumulation of large masses of plant remains in a humid, swamplike environment. During coalification, the percentage of carbon increases as water and volatile hydrocarbons are driven out of the deposit. The grade of coal (determined by the percentage of elemental or "fixed" carbon present) apparently depends on the depth at which the coal was buried and the amount of heat and pressure to which it was subjected. High-ranking coals are denser, contain less moisture and volatile gases, and have a higher heat value than low-rank coals.

*Peat, a brown, porous, spongy mass of partially decayed vegetation, is the first stage of coal formation. When covered with sediment, peat may gradually be converted to *lignite, a brownish-black, low-grade coal. Time and continuing pressure produced by deep burial slowly transform lignite into *bituminous (soft) coal. Bituminous coal is dark brown to black, breaks with a blocky fracture and burns with a smoky flame. It occurs in layers of varying thickness that commonly rest on a bed of underclay, a layer of altered soil in which the plants that formed the coal were apparently rooted. Some underclays are valuable as *fireclay. *Anthracite (hard coal) is very hard, black, difficult to ignite and burns with a blue, almost smokeless flame. Its fixed-carbon content is between 92 and 95 percent. It is derived from bituminous coal that has been deeply buried and subjected to *metamorphism during earth movements such as folding. In addition to its use as a fossil fuel, coal is an important source of coke, which is used in the production of steel. Coke is burned in blast furnaces, where it supplies carbon, which combines with the oxygen of iron ores to free the metallic iron.

Coal is common in rocks of *Mississippian and *Pennsylvanian age in the eastern United States, Europe, Britain, Belgium, Germany and the southern part of the Soviet Union. Coal of *Cretaceous age occurs in certain parts of the Rocky Mountains and Great Plains. (W.H.M.)

Coastal Plain. A plain of low relief sloping gently seaward. It results from the emergence of the ocean bottom and is commonly formed by a differential upward movement of the crust. Marine sediments that dip seaward usually underlie a coastal plain. It is bordered on the landward side by older and more dissected terrain. The shoreline of a coastal plain migrates seaward during the emergence so that the rocks get progressively younger with decreasing distance from the shore. If the emergence has taken place recently, elevated marine cliffs, caves and beaches may be found on the landward border. Where the emergence is ancient, erosion will have destroyed all of these features. A continuous coastal plain extends in America along the Atlantic and Gulf coasts from New Jersey to Texas and southward into Mexico. In places it is 300 km wide,

3

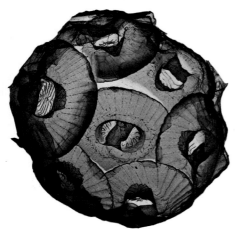

mostly less than 150 m, and occupies nearly 10 percent of the United States. (R. L. N.)

Cobaltite. A mineral, the sulfarsenide of cobalt, but usually containing some nickel. Cobaltite is frequently found in crystals resembling those of *pyrite, but is distinguished from pyrite by its silver-white color. It is found in *metamorphic rocks and in veins with other cobalt and nickel minerals. It is a major ore of cobalt, and has been mined at Cobalt, Ontario, and at Tunaberg, Sweden. See Mineral Properties; Appendix 3.

Coccolithophores (or Coccoliths). Tiny photosynthetic organisms that are part of the *plankton and are found only in warm, low-latitude waters. Each individual is only a few microns in diameter. After death it falls apart into several calcium carbonate plates which accumulate as calcareous *oozes on the ocean floor. Coccolithophores are also important constituents of ancient chalk deposits.

Coelacanth. A *crossopterygian fish abundant during *Devonian time. The coelacanth was once believed to have been extinct since the end of the Cretaceous Period, more than 70 million years ago. But in 1938 a coelacanth about 1.5 m (5 feet) in length was caught by a trawler off the southeast coast of Africa. Since then several other specimens have been caught, all belonging to the genus *Latimeria*. These have been most useful in providing additional information about a group known previously only as fossils and about the history of the vertebrates.

Coelenterate. A member of phylum Coelenterata, or "hollow-bodied" animals, including such familiar forms as the jellyfishes, sea anemones and corals. All coelenterates are water-dwelling animals, and most species live in salt water. The living animal is characterized by a saclike body cavity, a definite mouth and tentacles with stinging cells. Some forms, like the jellyfishes, have an umbrella-shaped body and are single, free-moving organisms. Others, such as the colonial corals, consist of many individuals living together in a colony.

Most zoologists and paleontologists recognize three classes of coelenterates: (1) the Hydrozoa, including small animals known as hydroids, (2) the Scyphozoa, including the jellyfishes, and (3) the Anthozoa, including the corals and sea anemones. Because of their fragility and lack of hard parts, hydrozoans and scyphozoans are not commonly found as fossils. However, rare specimens of fossil jellyfishes have been found preserved by *carbonization or as *casts and molds in very fine-grained sedimentary rocks.

The stony corals of class Anthozoa are important geologically, for the coral animal, or polyp, secretes a cup-shaped calcareous exoskeleton called a corallite. This is usually divided by radial partitions called septa (singular septum). The polyp lives in the calyx, which is the central bowl-shaped depression in the top of the corallite. Some corals, called solitary corals, build an individual corallite for each polyp. According to their shape they may be referred to as "horn corals," "cup corals" or "button corals." The colonial or compound corals have their corallites fused together to form a massive stony colony.

In tropical seas coral colonies grow and add on to each other until large bodies called reefs are formed, such as the Great Barrier Reef off the northeast coast of Australia, or the reefs at the southern tip of the Florida Keys. Because of their porous and permeable structure, certain buried fossil reefs contain much petroleum.

Fossil corals provide information about climates in the geologic past. Modern corals typically live in waters associated with tropical and subtropical climates, and seem to have always inhabited such waters. It is therefore assumed that rocks containing fossil corals were deposited in warm water. Thus the presence of fossil reef-building corals in the New Siberian Islands within the Arctic Circle suggests that the climate there was tropical to subtropical at least during Silurian time. Coral remains have been found in rocks ranging from Ordovician to Holocene (Recent) in age. They were especially abundant during the Paleozoic Era and have provided many *guide fossils for Paleozoic rocks. (W. H. M)

Coke. A combustible solid material derived from coal by *carbonization. It is used as a fuel.

Col. A low place in a divide usually originated by mountain glaciation. It is formed by headward erosion and intersection of two *cirques on opposite sides of a divide. St. Bernard, St. Gothard and other famous passes in the Alps are cols.

Colemanite. A mineral, a hydrous calcium borate, used as a source of *borax. It is usually found in Tertiary lake deposits, interstratified with *ulexite and borax and probably formed by their alteration. Colemanite is a colorless to white mineral with one highly perfect cleavage. It is found in numerous localities in California and Nevada and also in Argentina. See Mineral Properties.

Coccolithophores

1 These microscopic skeletal remains of plants compose a large portion of the deep-sea sediment called calcareous ooze.

Coelacanth

2 These lobe-finned fish were found only as fossils from Mesozoic times and were thought to be extinct until a live specimen was caught off the coast of Madagascar in 1938.

Coelenterate

3 About 250 million years ago these jellyfish, resembling certain species living today, existed in seas that covered the area that is now Cortlandt, N.Y.

Colemanite

4 One of the most important ore minerals of boron, these unusually large crystals represent the many fine specimens found in California.

Columnar Jointing
The cliffs of Staffa Island off the coast of Scotland are composed of basaltic lava that fractured into hexagonal columns as the rock cooled.

Collapse Depression. A round, trenchlike or irregular depression resulting from the partial collapse of the roof of a *lava tunnel in a pahoehoe *lava flow. Collapse depressions range from about a meter to hundreds of meters in length and are commonly 3 to 9 m deep. Some are empty and dry, others contain small ponds, and still others are floored with *alluvium and vegetation. The collapse may have taken place long after the solidification of the flow or while liquid lava was still present. Angular fragments of the broken crust usually cover the floors. Collapse depressions have been identified on lava flows on the moon. (R.L.N.)

Collophane. The massive, cryptocrystalline *apatite that makes up the bulk of rock *phosphate.

Colluvium. Unconsolidated material, derived from both *regolith and *bedrock, that accumulates on valley walls and is moved by gravity alone. Colluvium may fall, slide, roll or flow, and it can move slowly or rapidly. Its composition depends upon the source rocks. It lacks *stratification, is generally unsorted and consists of angular and subangular fragments that range greatly in size and are derived from the valley walls above. It can therefore be confused with *till. Colluvium includes all the kinds of deposits involved in *mass wasting, such as *talus, *mudflow, *solifluction and avalanche deposits. It may completely veneer and conceal the source rock from which it is derived. Colluvium may be interbedded with *alluvium on the margins of *flood plains. (R.L.N.)

Columbite. A mineral, the oxide of iron, manganese and niobium, and the principal ore mineral of the element niobium. Pure columbite is rare. Usually some tantalum partially substitutes for niobium in a series that extends to the tantalum end-member *tantalite. Thus the mineral as mined is usually a source of both niobium and tantalum; the specific gravity (5.2-7.9) increases with the tantalum content. The minerals occur in *granites and related pegmatites associated with certain relatively rare minerals: *cassiterite, *wolframite and *microlite. See Mineral Properties; Appendix 3.

Columnar Jointing (or Columnar Basalt, Polygonal Columns). Elongated, geometrically regular columns that are hexagonal or pentagonal in cross section. They have from three to eight sides, with six being the most common, and are bounded by joints. They range from several centimeters to more than one hundred meters in length

and from a few centimeters to a few meters in diameter. They are found in *lava flows and lakes as well as in dikes, *sills and other intrusive *igneous rock bodies. They occur most frequently in basalts, but also in *obsidian and other igneous rocks as well as ignimbrite. They are usually elongated perpendicular to the cooling surfaces of the rock bodies in which they occur, and are therefore usually perpendicular to the tops and bottoms of flows and sills and perpendicular to the walls of dikes. Jointing occurs when igneous rocks, solidified at several hundred degrees Centigrade, cool down to air temperature and contract. Similar columns and joints are formed when mud dries, but the contraction is then due to loss of water.

The Devils Postpile, Calif., contains spectacular basaltic columns; those at Devils Tower, Wyo., are 100 or so meters long and up to 2.5 m (8 feet) in diameter. Such columns are common in the Columbia River and Snake River basalts, and those of the Giant's Causeway in Northern Ireland and of Fingal's Cave, Staffa Island, Inner Hebrides, Scotland, are famous throughout the world. (R.L.N.)

Compaction. The reduction in thickness or volume of fine-grained sediments. Compaction results from a decrease in pore space within the sediment in response to continual burial by younger sedimentary material. When sediments of a homogeneous nature are deposited on a relatively smooth, flat-lying surface, they will be compacted to the same degree everywhere. Should such sediments be deposited on an uneven surface, the lower portions of the surface will receive more sediments than the higher areas. As other sediments are deposited over the older deposits, they will begin to take the form of the original surface. This is caused by a greater degree of compaction of sediments lying in the lower areas. The resulting structure of the sediments is the direct result of differential compaction. (J.W.S.)

Compensation Depth. The depth of the ocean at which the production of organic matter by photosynthesis is exactly balanced by the breakdown of organic matter by respiration (oxidation). Because of changes in the transparency of water, position of the sun and other factors, the depth of compensation varies, but it is generally around 100 m.

A second meaning of the term is to denote the depth above which calcium carbonate is stable and below which it goes into solution. This condition, which occurs in the vicinity of 3500 m, results from combined low temperature and high pressure.

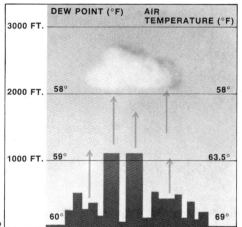

Conchoidal Fracture

1 *This specimen of obsidian, a volcanic glass, has broken in a curving, shell-like (conchoidal) fracture.*

Condensation

2 *The height at which clouds begin to form in rising air can be determined when the temperature and dew point at the ground are known. Rising air cools at the rate of 1°C for every 100 m (5.5°F for every 1000 feet); the dew point falls 0.2°C for every 1000 m (1°F for every 1000 feet).*

3 *Moisture from a jet plane's engine forms a trail of fine ice crystals over the Colorado Rockies.*

Cone of Depression

4 *When a large quantity of groundwater is pumped from a well, a conical depression is created in the water table. Excessive pumping can lower the water table in an area's wells.*

Cone-in-Cone

5 *Commonly found in shale, such a series of conical structures that fit one inside the other probably formed because of the weight of overlying sediments.*

Competency (of Streams). The largest size (diameter) of material that a stream can transport as *bed load. The competence of a river changes not only from place to place but from time to time, and depends mainly upon velocity. Competency equals average velocity to the sixth power: $c = V^{-6}$. A river can move much larger particles when flooding because the discharge—hence velocity—is high.

Compression. An inward-directed force that tends to decrease the linear dimension or volume of a body. Measured per unit area it is called compressive stress. The force producing the compressive stress acts normal to the plane along which the components are being pushed together. Many structural features found on the earth's surface, such as *folds and thrust *faults, are the direct result of compressive forces within the earth.

Most rock materials can withstand much compressive stress before failing. This property makes many kinds of rock desirable for construction purposes. A building stone such as the Venus de Milo granite from St. Ludger de Milot, Quebec, can withstand stresses as great as 2300 kilograms per square centimeter before it will be crushed. (J.W.S.)

Conchoidal Fracture. A smooth, curved type of fracture found in brittle minerals and glass. The fracture resembles the interior surface of a shell and is best illustrated by quartz.

Concordant. A term denoting the structural relationship between crustal rocks and bodies of *igneous rock that have been injected into them. Structures are said to be concordant if their contact is parallel to the planar structures of the older rocks. A *sill is a concordant body of rock because it has formed parallel to bedding planes of the rock it intruded. Other concordant structures of molten origin are *laccoliths, *lopoliths and *phacoliths.

Concretion. A hard, typically rounded or nodular rock mass differing chemically and physically from the enclosing rocks. It is typical of certain *sedimentary rocks and commonly exhibits a concentric structure indicating growth by deposition of successive layers. It consists of substances precipitated from solution, commonly around a nucleus such as a fossil or a grain of sand. Concretions, normally smooth on the exterior, vary greatly in shape and size. Hollow concretions lined with crystals pointing inward are called *geodes. These interesting mineral formations have recently become popular as decorative pieces.

Condensation. The change of a substance from a vapor to a liquid or a solid. In meteorologic usage, condensation applies only to the change from vapor to liquid, as in the case of dew. The change from water vapor directly to ice, as in the case of frost, or the reverse change, is referred to as sublimation.

Like the opposite change of phase—*evaporation—condensation is an essential process in the *hydrologic cycle. It produces not only *clouds and fog but also the phenomenon called damp haze. This occurs when small particles are present (principally salt) that accelerate condensation even in relatively dry air. Condensation does not by itself produce precipitation, with the exception of very fine drizzle; other mechanisms are needed for cloud droplets to grow into liquid drops or solid particles that reach the ground in any significant amount.

Condensation may be caused by the addition of sufficient water vapor to the air, through evaporation, to bring about saturation. However, in the atmosphere it is mainly caused by the cooling of air to the temperature at which the air, containing a given quantity of water vapor, becomes saturated. (Correctly speaking, it is not the air but the water vapor in the air that becomes saturated, but the terms "saturated" and "unsaturated" air are commonly used. Since the saturation vapor pressure of a gas is a function of its temperature alone, for every value of the vapor pressure there is a corresponding temperature, termed the dew point, at which saturation nominally occurs. When the air temperature coincides with the dew point, the relative humidity is 100 percent. However, condensation in the atmosphere does not necessarily take place precisely at a relative humidity of 100 percent, because saturation vapor pressures are determined with respect to a plane surface of pure water or ice—conditions not always present in the atmosphere. The saturation vapor pressure is lower over an ice surface than over water, a physical fact of great importance in the formation of precipitation.

In general, condensation in the atmosphere requires the presence of microscopic particles, called condensation nuclei, or other liquid with solid surfaces, such as pre-existing droplets or ice particles, for the vapor to condense on. In perfectly clean air, relative humidities of several hundred percent would be needed before condensation could occur. However, since relative humidities exceeding 102 percent are seldom observed, it may be concluded that condensation nuclei are normally present in the atmosphere. The most common natural nuclei are believed to be soil particles, smoke particles from grass and forest fires, and sea salts. Particles associated with *air pollution may also be effective as condensation nuclei, since they are often hygroscopic like sea salts; that is, they have an affinity with water molecules. Hygroscopic nuclei are very effective because when water vapor condenses upon them, they form solutions with quite low saturation vapor pressures. For example, condensation can begin on large drops of sodium chloride when the nominal relative humidity is about 78 percent. For this reason, salts carried into the air as sea spray, from which the water evaporates, are thought to explain much of the condensation in the atmosphere.

The saturation vapor pressure is related to the droplet radius, or curvature, as well as to the nature of the water solution. For this reason, very small particles are not as effective as nuclei as are the larger particles. As an extreme case, the saturation vapor pressure over a droplet with a radius of 10^{-7} cm is three times as great as over a plane water surface. A concentration of water vapor three times the nominal saturation value would be required to produce condensation on such a small water droplet. However, condensation can occur on a pure water droplet having a radius of 10^{-5} cm at a nominal relative humidity of about 102 percent. (M.F.H.)

3

Cone-in-Cone Structure. A succession of parallel nests of cones fitting one within the other, characteristic of certain thin-bedded, calcareous *sedimentary rocks. Most cones are composed of fibrous calcite, and each cone is separated by a thin film of clay. The sides of the cones are ribbed or fluted, ranging from 10 mm to 10 cm in height. Their origin is not completely understood, but it appears to be the result of pressure (from the weight of overlying sediments) aided by mineral crystallization and solution along a series of intersecting, conical shear zones.

Cone of Depression. A conical depression in a *water table surrounding a well from which water is being pumped. As the water level in the well is lowered by pumping, groundwater from the surrounding area flows into the well, lowering the water table in that area. The lowest point of the cone of depression is at the well. When pumping is stopped, the water table around the well and the water in the well rise and the cone of depression slowly fills in.

Cone Sheet. One of several sheetlike *dikes arranged in segmented concentric circles, which together form an inverted conelike structure within the earth's *crust.

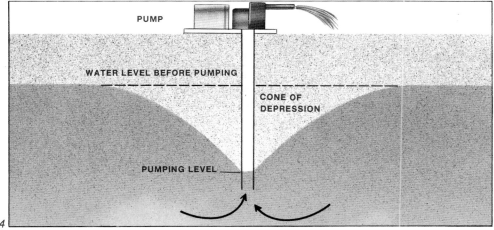

4

5

The concentric rings dip toward an axis of concentricity usually at a 30° to 45° angle from a horizontal plane. The formation of cone sheets is most likely due to the injection of a body of *magma into the brittle portion of the earth's crust. The pressure of the injected body produces tension cracks in the overlying crust into which magma from the main magmatic body is injected. Solidification of the injected material results in the formation of the cone sheets. Important examples of such structures have been found in Scotland and in the state of Wyoming. (J.W.S.)

Conglomerate. A clastic *sedimentary rock consisting of more or less rounded rock particles at least 2 mm in diameter, embedded in a fine-grained matrix of sand or silt. It resembles cemented gravel. Although the rock fragments may vary greatly in size, shape and composition, they are typically rounded and smoothed from abrasion during transportation by streams or from wave action. It is known also as puddingstone, especially in Great Britain.

Connate Water. See Subsurface Water.

Conodont. A small, amber-colored, tooth-like fossil believed to represent the hard parts of some type of extinct fish. They may be teeth or internal supports for soft tissues such as the gills. Although geologists know little about their origin, these peculiar fossils are of considerable value in *micropaleontology. Conodonts first appeared in the early Ordovician and apparently lived until Triassic time. However, most were extinct at the end of the Permian Period; some scientists believe that conodonts reported from Triassic and Cretaceous rocks are Paleozoic forms redeposited with the Mesozoic sediments. Certain species of Ordovician, Devonian and Mississippian conodonts are especially useful as *guide fossils in the work of micropaleontologists. (W.H.M.)

Conrad Discontinuity. A boundary between two zones in the outermost shell of earth known as the *crust. The boundary at present can only be recognized from the behavior of earthquake waves as they are propagated through the earth. At the discontinuity there is an increase in the velocity of the downward propagating waves from 6.1 km per second to 6.4-6.7 km per second. The depth at which the discontinuity lies is variable and is often difficult to recognize because the change in velocity of *seismic waves through the crust tends to be gradual rather than abrupt. (J.W.S.)

Consequent Stream. A stream whose course is determined by the original slope of the land surface. Such streams can be seen forming a series of parallel rivulets on hillslopes or roadcuts. On a larger scale, they form on uplifted coastal plains flowing down the gentle seaward slope of the land.

Contact Metamorphism. Chemical, mineral and physical changes in an earlier formed (*country) rock brought about by contact with *magma that has moved up from the depths of the earth's *crust. The changes result from a temperature rise caused by heat flowing from the magma and by aqueous fluids emanating from the magma. The fluids act as a catalyst for chemical reactions of great importance in the metamorphic process. The temperature range in contact metamorphism may vary from 300° to 800°C (392° to 1472°F), and the maximum confining pressure is comparable to that encountered at 10 km beneath the earth's surface. Rocks produced by contact metamorphism are generally very fine-grained and brittle (such as *hornfelds) or coarse-grained (such as *marble and *skarn).

Contact metamorphism of rock is limited to local areas. The extent to which metamorphism affects the invaded country rock can be measured from tens of meters to a few kilometers. The extent of metamorphism is controlled by the size of the magmatic body, its temperature and its composition. Larger intrusions, or those of higher temperature, allow the process to continue longer and therefore penetrate deeper into the country rock. Contact metamorphism of certain kinds of rocks and under certain conditions may form minerals suitable for mineral collections or even of gem quality. An example is *lapis lazuli, noted for its pure, intense blue color and sought after as a semiprecious gemstone. Lapis lazuli is found mainly in Afghanistan and Iran but small veins are also found on Italian Mountain, Colo., where it occurs in a contact metamorphosed limestone. (J.W.S.)

Continental Accretion. The term used for the widely—but not universally—accepted theory that the continents have increased their surface area during geologic history. The mechanism by which continental accretion occurs is related to the cycle of events that develops and destroys geosynclinal belts forming next to the stable continental crust (*craton). The cycle is as follows: A trough (or *geosyncline) develops next to a relatively stable landmass and is filled with sediments derived from the landmass, or with an accumulation of hard parts from marine organisms or

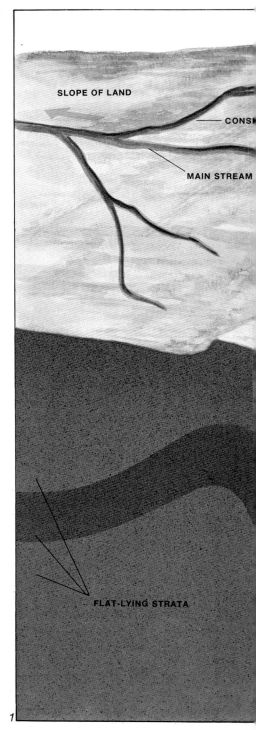

SLOPE OF LAND

CONS

MAIN STREAM

FLAT-LYING STRATA

1

2

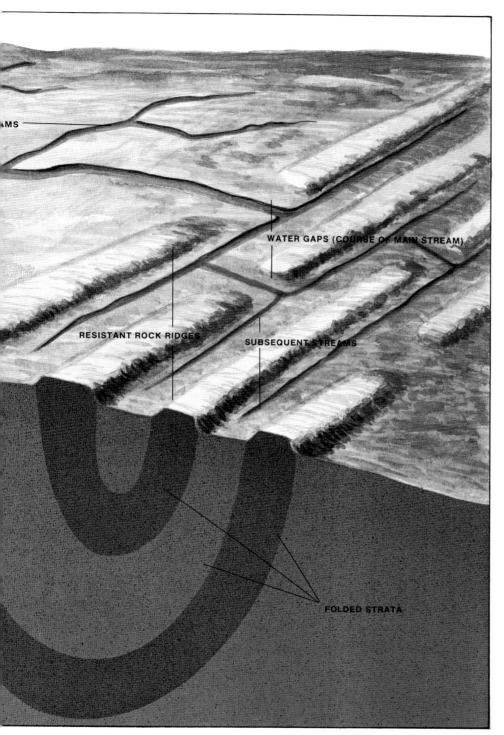

MS

WATER GAPS (COURSE OF MAIN STREAM)

RESISTANT ROCK RIDGES

SUBSEQUENT STREAMS

FOLDED STRATA

Consequent Stream

1 *The stream's pattern is determined solely by the direction of the slope of the land, and is therefore consequent. Parallel belts of weak rock in folded strata have determined the course of the tributaries, which are called subsequent streams.*

Conglomerate

2 *More than 450 million years ago, wave action laid down a layer of white quartz pebbles and fine gravel, which formed a conglomerate now exposed in the Shawangunk Mountains of New York.*

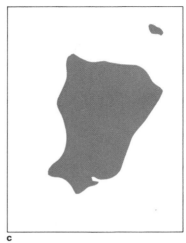
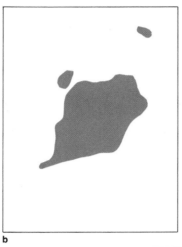

a   b   c   d

1

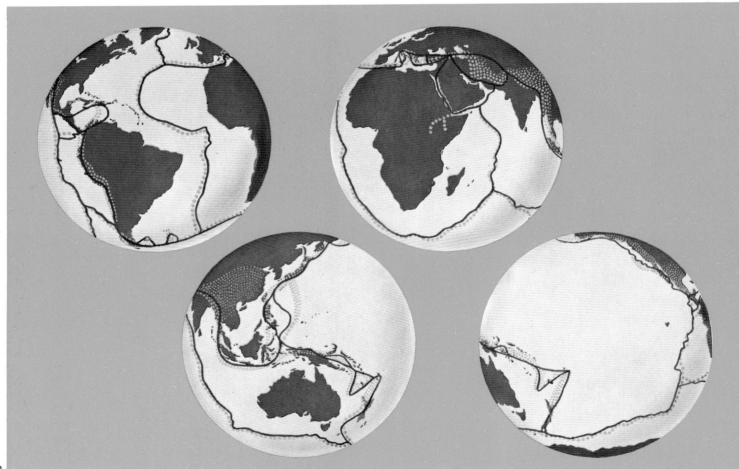

2

Continental Accretion
1 The continents are believed to have developed from several smaller land areas. Shown here are the estimated extent of North America about (a) 2.6 billion years ago, (b) 1.6 billion years ago, (c) 1.2 billion years ago, (d) and today.

Continental Drift
2 Wegener's theory of continental drift was confirmed by the evidence of sea floor spreading. Rifting in the crust and welling up of the mantle take place along the ocean ridges (solid lines). Most oceanic earthquake belts (dotted lines) occur along these ridges, with the greatest concentration at the trenches.

with debris from volcanic activity on the ocean floor. The sediment-filled trough is then subjected to crustal forces (*plate tectonics) that cause the deposits to *fold, partially melt and weld onto the primitive craton. As the cycle is repeated, a series of segments containing progressively younger rock is built around the original landmass, forming a pattern similar to the growth rings of a tree.

The evolution of continents by accretion is well displayed on North America. The original landmass of North America has been dated as older than 2.5 billion years, while the last known segments of geosynclinal deposits to be welded onto the continent are younger than 600 million years. Estimating the amount of material that has been added to the primitive crust is difficult. It is believed that much of the original crust was reworked during the first cycle of accretion. Other segments of crust added to the continent throughout geologic time have also been extensively deformed. A number of scientists contend that the reworking of the original crust of North America greatly decreased its original area. (J.W.S.)

Continental Borderland. Topographically and structurally complex areas on the margin of continents. The irregular topography is due to block *faulting and *folding. Depressions in the shallow margin trap and accumulate sediment so that little of it reaches the *continental rise. Typical of such margins are those of southern California, the Coral Sea and the South China Sea, the northwest Indian Ocean and off northern Venezuela.

Continental Crust. The thick, discontinuous portion of the outer zone of the earth; it forms the continental landmasses and has a composition similar to that of granite, being mostly quartz and feldspar. It is also known as the sialic crust because of its high silicon and aluminum content. Although the continental crust is described as having a granitic composition, it actually contains a wide range of elements and minerals, including a thick veneer of sedimentary rocks. This portion of the crust is typically 20-30 km thick. Its vertical extent can be determined by the manner and rate of transmittal of seismic waves. The continents are actually large discontinuous blocks of granitic crust resting on more dense basaltic crust. (R.A.D.)

Continental Drift. A theory proposing that the continents are not static masses but are part of the earth's *crust, which is in motion over the earth's surface. In 1628 the English scientist Francis Bacon noted the similarity between the outlines of western Africa and eastern South America, but it was not until the late 1880s that an Austrian geologist, Eduard Suess, suggested that Africa, South America, Australia and India were once part of a larger landmass, which he called Gondwanaland. This theory was based largely on the distribution of closely correlated geologic formations throughout continents of the Southern Hemisphere.

In 1912 the theory of continental drift was again proposed by the German meteorologist Alfred Wegener. Like Suess, he based his theory on the similarity of rocks and landforms now separated by expanses of ocean as well as on the distribution of fossils. He saw, as did many other geologists, that fossilized bones and plants on different continents showed a strong resemblance to each other. He believed that the number of comparable species on separated continents could not have resulted from migration across wide oceans, since it was unreasonable to expect animals to swim such distances. Most scientists were quick to disagree, relying on the widely accepted theory that animal migration did occur, not by swimming but by *land bridges that had connected the continents at various times in geologic history and had later sunk beneath the sea.

Wegener disagreed with the theory of land bridges for the following reasons: Land bridges, like the continents, would be composed of material of less density than the rocks that make up the earth's interior. If large masses of land had sunk into the ocean, the density of the crust underlying the oceans should be the same as that of the continents. But gravity measurements made at sea showed that the crust underlying the ocean was of greater density than the crust associated with the continents. If landmasses had been forced down into the oceanic crust, they should have risen again since they were lighter than the underlying crust. No indication of such re-emergence had been found. Wegener therefore argued that the North American and Eurasian continents had been merged with "Gondwanaland" in a single supercontinent, which he called *Pangea, and that this landmass began to break up at the beginning of the *Mesozoic Era, about 225 million years ago.

Much controversy arose over Wegener's drift theory, with the majority of scientists believing that large landmasses could not move laterally across the surface of the earth. The question remained unresolved until the 1950s, when scientists found that the crust of the ocean became progressively older farther away from the newly discovered ocean ridges in all ocean basins. It was believed that these ridges were zones of weakness within the crust into which molten rock from deeper portions of the earth was being injected. As the molten material was injected into this elongated fissure, it forced aside adjacent parts of the sea floor and also widened the distance between continents. This new theory, developed more or less simultaneously by Harry H. Hess and Robert S. Dietz and called *sea floor spreading, helped support the Wegener theory of continental drift. Today it is a widely accepted fact that the continents are in motion. The study of how continents spread apart and move over the surface of the earth has led to an entirely new branch of earth science known as *plate tectonics.

Now that continental drift has become a widely accepted concept, attention is being directed toward understanding the reconstruction of the continents prior to the drift. Scientists believe that the continents were joined into one major landmass, Pangea. However, some scientists believe that two major landmasses, Laurasia and Gondwanaland, existed. Laurasia, made up of North America and Eurasia, was separated by a narrow sea from Gondwanaland, which consisted of the continents of Africa, Australia, Antarctica, India and South America. What both groups agree upon is the time at which the original landmass broke up and the paths of their migration.

Those who accept the theory of a single original landmass usually estimate that the initial breakup occurred around 200 million years ago, or during the Middle Triassic Period. Prior to this, Pangea was surrounded by an ancient sea, Panthalassa, and North Africa was separated from Eurasia by a smaller body of water called the Tethys Sea. Today the Pacific Ocean and Mediterranean Sea are remnants of these ancient bodies of water. The arctic region of North America was joined to eastern and northern Europe. Northwest Africa lay adjacent to the east coast of North America, its west coast coinciding with the east coast of South America. Australia, India and Antarctica were connected to the southeast section of Africa.

The breaking up of the supercontinent began with the development of a large east-west rift system and an outpouring of basaltic *magma. The rift system ran parallel to and just north of the Equator and opened up the area of the Atlantic and Indian oceans. Remnants of this rift system remaining today are the Triassic basins of red sandstones and interlayered basalts found along the east coast of North America.

Near the close of the Triassic Period, the western Mediterranean and Caribbean

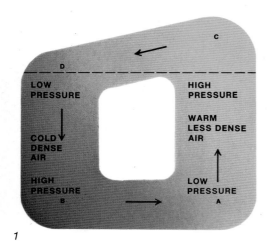

1

Convective Circulation

1 *Air warmed by contact with the earth's surface expands and rises. As it flows through the upper atmosphere its heat dissipates and eventually the air mass cools, contracts and sinks to earth again. The pressure at point A is lower than at point B because a column of expanded warm air weighs less than a column of dense cold air. However, since warm air occupies more volume (area above the dotted line) than cold air, the pressure at point C is greater than at point D.*

Copper

2 *These specimens of native copper crystals (a and b) come from the important copper mines of Michigan's Keweenaw Peninsula.*

seas had begun to form, while another large rift system began to divide Africa and South America from Antarctica, India and Australia. During the Jurassic the North Atlantic Ocean continued to open while Africa and Eurasia drifted closer together, narrowing the Tethys Sea. Near the end of the Jurassic (approximately 130 million years ago) another large rift began to divide South America from Africa, at first separating the southern regions of the two continents and later forming a narrow sea adjacent to what is now Nigeria. In the Cretaceous, the final separation of South America from Africa was completed. The North Atlantic continued to widen, and Antarctica, India and Australia drifted steadily southward from Africa. It is thought that near the end of the Cretaceous India broke away from Antarctica and began its 5000-km journey north toward the underside of Asia. At the same time the original rift that separated North America from northwest Africa continued to migrate north to form the eastern margin of Greenland.

About 45 million years ago, during the Cenozoic, India, moving northward from the Antarctic, finally collided with Asia, forming the Himalayas. At the same time Australia broke from Antarctica and began drifting northeast. During the past 45 million years the continents drifted into their present positions. Much of the data concerning the Mesozoic and later movement of plates has been derived from geophysical studies of the ocean basins. In recent years, however, data from the continental landmasses of the world indicates that pre-Mesozoic and even Precambrian movement of plates took place, but this early relative distribution of continental landmasses and their shapes is less well understood than the later reconstructions. See Convection Currents. See also Color Plate "An Ancient Supercontinent Breaks Up;" Color Plate "The Drifting Continents;" and Color Plate "The Shifting Plates of the Crust." (J.W.S.)

Continental Ice Sheet. See Glacier.

Continental Margin. A part of the submarine environment between the shoreline and the deep ocean floor. It includes the *continental shelf, *continental slope and *continental rise provinces, all of which are *sialic or continental in composition. The continental margin is part of the continent in all respects except that it is covered by seawater. The geology and physiography of the margin reflect that of the adjacent area of the continent that is above sea level. The continental margin represents one of the world's most important economic resources. Most of the life in the sea re-

sides there, and commercial fishing is largely dependent on it. In recent years much of the world's petroleum has been obtained from the continental margin, and exploration for other mineral resources is in progress. (R.A.D.)

Continental Rise. The most seaward province of the *continental margin. It is an area of fairly rapid sediment accumulation. Because of the subtle gradients between the *continental slope, rise, and *abyssal plain, the rise may be difficult to recognize topographically. The vertical to horizontal ratio averages 1:150 but ranges from 1:50 to 1:800. Although the relief (the vertical difference between high and low places) on the rise is generally low, the seaward extensions of *submarine canyons may cross the rise, and isolated *seamounts may cause high local relief. The rise is composed of a thick accumulation of fine terrigenous sediment carried to the base of the continental slope by *turbidity currents, gravity slides and bottom currents. The rise is essentially a coalescing of many submarine sediment fans. (R.A.D.)

Continental Shelf. The shallow, gently sloping zone extending from the shoreline to an abrupt change in slope at the outer limit. The gradient is only about 00°07' or 1:500. The width of the shelf ranges from only a few kilometers to hundreds. The depth at its outer limit is between 35 and 250 m, with the worldwide mean at 128 m. Relief (the vertical difference between the high and low places) on the shelf is generally 20 m or less, with reefs, glaciated channels and glacial *moraines accounting for most of it. Landward extensions of *submarine canyons may also encroach on the shelf. Because of the great utilized and potential wealth of the shelf, it is the most thoroughly studied portion of the ocean. Both its seafood and mineral resources are currently of great economic importance and will undoubtedly increase in importance in the future. (R.A.D.)

Continental Slope. A province extending from the outer edge of the *continental shelf to the upper limit of the *continental rise. It is the steepest portion of the *continental margin, having a mean inclination of 4°17' although the range is from 1° to more than 45°. The slope extends to a mean depth of 3600 m, but it may reach a depth of 7000 m where the slope extends into a deep-sea trench. Relief on the slope is in excess of 1000 m in the submarine canyons, which are common on the world's continental slopes. Because of the steep gradient on the slope little sediment accumulates there. (R.A.D.)

Contour Line (and Contour Interval). A line, commonly drawn on a map, that joins all points representing places on the earth's surface of equal elevation above a given plane, which is usually mean sea level. The line marking the intersection of the ground with a horizontal plane 3 m above sea level is the 3-m contour line, and so forth. The contour interval is the difference in elevation between two successive contour lines. If the scale is small and there are considerable differences of elevation in the area, the contour interval is large; if the scale is large and the differences of elevation are small, the contour interval is small.

A contour map shows the elevation, configuration and size of landscape features by means of contour lines drawn at regular intervals of elevation and by the scale of the map. Where the surface of the ground is steep, the contour lines are close together; where it is flat, they are far apart. Such maps are particularly valuable to engineers, city planners, geologists, foresters, the military and others concerned with the earth's surface. The U.S. Geological Survey has published thousands of topographic contour maps (with contour intervals ranging from 1 to 1000 feet) of various parts of America. See Topography. (R.L.N.)

Convection Current. A concept or mechanism that helps explain the irregular distribution of heat flow within the earth's *crust as well as the phenomenon known as *continental drift. The concept of convection currents had been suggested by some earth scientists to explain crustal deformation, but not until the early 1960s had enough information been gathered to make the concept plausible to most earth scientists. Convection currents are the result of thermal gradients within the deeper layer of the earth known as the *mantle. The change in temperature (cooler in the upper mantle and hotter at the lower depths) causes the hotter plastic material to rise toward the earth's surface. There the material is deflected and flows parallel to the surface until it has cooled sufficiently, whereupon it again begins to descend. This movement produces cells of convection within the mantle. These may be of many sizes and shapes. They are not necessarily continuous and may change speed, stop or even change their direction. Scientists generally agree that the upwelling convection currents coincide with an elongate system of suboceanic mountain ranges known as mid-ocean ridges, while the downgoing portion of convection currents coincides with deep ocean trenches. The motion of convection currents helps explain the addition of new oceanic crust along the mid-ocean ridges and the dragging down of crust into the ocean trenches. The movement causes continents to separate in some areas and to collide in others. (J.W.S.)

Convective Circulation. An organized motion characterized by the presence of convective cells, usually with upward motion (away from the heat source) in the central portion of the cell, and sinking or downward flow in the cell's outer region. Convective circulations are caused essentially by differential heating of a fluid and the unequal action of gravity on fluid masses of different density. In the atmosphere, differential heating of the earth's surface and the air above, combined with the consequent outflow of air from the heated region, results in pressure forces that act in one direction at the earth's surface and in the opposite direction aloft. For small-scale *atmospheric motions, the resulting flow is directly across the isobars (lines connecting points of equal pressure) from high to low pressure. In large-scale motions, however, the air flow, initially directed across the isobars, is deflected as a result of the earth's rotation and approaches the *geostrophic wind, which blows parallel to the isobars. Although convection is the driving force of the atmosphere's general circulation and other motions, the flow patterns in the atmosphere are much more complex than those represented by the simple convective circulation cell. (M.F.H.)

Convergence. In meteorology, the word used to describe *atmospheric motions where there is a net flow of air into a given volume of the atmosphere. See Divergence.

Copper. A chemical element that occurs in an uncombined state as a mineral. It is one of the native metals that in the past was mined extensively as an ore. Small amounts of native copper, rarely in mineable amounts, are found at many localities in the oxidized zones overlying deposits of copper sulfides, the major ore minerals of the metal. The one exception is its occurrence on the Keweenaw Peninsula, Mich., on the south shore of Lake Superior. For 160 km parallel to the axis of the peninsula, native copper has been mined as a major ore mineral. Mining began shortly before 1850, and the area continued as a major copper producer for 75 years. The only other similar type that has been mined, but of far less importance, is at Corocora, Bolivia. See Mineral Properties; Native Elements; Appendix 3. (C.S.H.)

a

2

b

Coquina

1 Composed of shell fragments cemented with calcium carbonate, this form of conglomerate is named after the Spanish word for shell.

Coral Reef

2 Throughout the world, coral reefs occur in warm, tropical and equatorial zones between 30°N and 25°S (based on mean temperature for the year). Solid lines represent extended 20°C isotherms (based on mean temperature for the coldest month); dashed lines represent restricted 20°C isotherms. Most coral reefs (shaded areas) are in the restricted zone.

3 The most developed region of a coral reef is the windward or fore reef side, which displays spur and groove formation on the upper slope.

4 Built up from the sea bottom by deposits of skeletons of coral, a small, palm-covered islet mantles part of Australia's 700-mile-long Great Barrier Reef.

Coprolite (from the Greek *kopros* meaning "dung"). The fossilized fecal pellets or droppings of animals. Typically rounded, tubular or pelletlike in shape, fossil excrement is frequently found in association with the fossilized remains of fish, reptiles and mammals. The shape and external markings may yield information about the anatomy of the animal that left them. The content may also provide clues as to the type of food eaten by the animal. Coprolites are typically high in phosphate of lime and certain organic constituents, and some are useful as fertilizer.

Coquina. A very coarse, organic limestone composed primarily of shells and shell fragments cemented together in a porous mass. Most are of relatively recent origin, such as the cemented shell deposits now forming off the coast of Florida. Older, more firmly consolidated shell masses are called fossil coquinas.

Coral Reef. A wave-resistant structure built by framework-building organisms, particularly corals and *coralline algae. Reef-building corals live in colonies and flourish only where daylight penetrates intensely enough to permit photosynthesis and consequently the existence of the algae on which the corals depend. These algae, called zooxanthallae, live symbiotically within the coral tissues and supply the corals with the calcium they require for external support.

Because coral reefs are generally restricted to depths of less than 60 m and temperatures above 20°C (72°F), they occur only in the low latitudes. Although the reefs are identified with coral, a tremendous variety of other plant and animal life lives on and around reefs. There is a distinct zonation of coral types and forms related to water depth and wave action. Because aerated waters provide abundant food and oxygen, corals thrive in areas characterized by high physical energy and show maximum growth toward the direction of most intense wave action. The massive varieties of coral grow in the zone of breaking waves whereas the more delicate and fragile forms occur in the protected areas. (R.A.D.)

Coralline Algae. The Corallinaceae comprise one of the largest families within the red algae (Rhodophyta). These marine calcareous forms are commonly associated with *coral reefs. They may be of a massive encrusting type or a branching form. Red algae can live at the greatest depths of all photosynthetic organisms because they use light rays in the blue and violet portion of the spectrum.

Cordierite. A mineral, a silicate of magnesium, aluminum and iron. It occurs as an accessory mineral in gneisses and in *schists, formed by the *metamorphism of rocks rich in alumina. Cordierite resembles quartz and is difficult to distinguish from it by physical properties. However, cordierite has a strong *pleochroism, colorless when viewed in one direction and violet in directions at right angles. For this reason, when cut as a gemstone it goes under the name dichroite. See Mineral Properties.

Core. The innermost portion of the earth. The core consists of three segments: (1) an outer liquid core approximately 1700 km thick, (2) an inner transitional zone around 500 km thick surrounding (3) a solid inner core with a radius of 1216 km (a little larger than the moon). For many years scientists suspected that a solid inner core existed, but it was not until very recently that they were able to gather strong evidence of it. An array of *seismographs located near Billings, Mont., detected *seismic waves generated in the earth that could only have been reflected back to its surface by a solid inner core.

The core is unique in yet another way. Scientists believe that fluid motion within the liquid portion of the core is responsible for generating the earth's main *magnetic field. How the core does so is still only poorly understood, but has been likened to a self-exciting dynamo. Materials present in the core are considered to be mainly iron with lesser amounts of nickel and silicon and with a density roughly 13 times greater than water. See Color Plate "The Earth in Cross Section." (J.W.S.)

Coriolis Effect. Named after an early 19th-century French mathematician, Gaspard Gustave de Coriolis, this is the deflection in the motion of an object as it moves over a rotating sphere such as the earth. From an observer's point of view, any object moving over the earth's surface is deflected to the right in the Northern Hemisphere and to the left in the Southern Hemisphere. There is no Coriolis effect for an object moving along the Equator, and the effect increases north or south from that position.

Although the Coriolis force theoretically affects any object moving over the earth, it is most obvious in the paths taken by airplanes or missiles. Also affected are fluids such as the atmosphere and the oceans, moving water, and especially the major current systems of the oceans. (R.A.D.)

Corrasion. Mechanical wearing away of a *bedrock channel by a transporting medium, accomplished by impact, scouring or scratching of particles in the water (*abra-

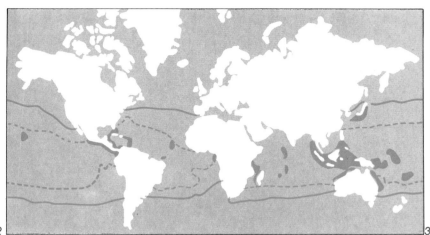

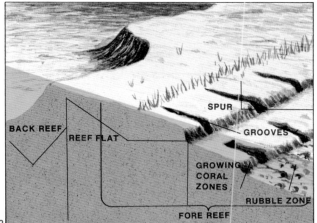

BACK REEF

REEF FLAT

SPUR

GROOVES

GROWING
CORAL
ZONES

RUBBLE ZONE

FORE REEF

SANDSTONE	
CONGLOMERATE	
LIMESTONE	
BLACK SHALE	
FOSSILIFEROUS SHALE WITH FOSSILS IN COMMON	
CONGLOMERATE	
LIMESTONE	
BLACK SHALE	
GREEN SHALE	A B C D

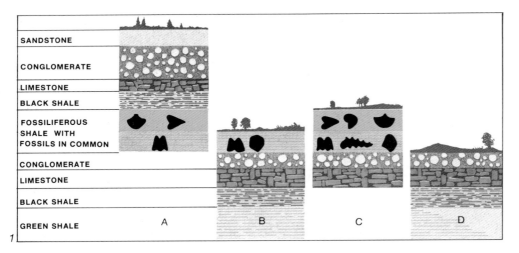

sion), by shear force of water on the bedrock (*evorsion) or by pressure effects (*cavitation). The result may be smoothing and polishing of the rock, or it may be pitting, fluting and production of *potholes.

Correlation. The process of demonstrating that certain sequences of layered rocks are closely related to each other or that they are stratigraphic equivalents. Correlating the rock records between two areas is largely a matter of establishing the existence of equivalence in geologic age, fossil content and position in the sequence of strata in the different areas. This is often necessary, for no single area can provide a rock section containing a record of all geologic time. However, since sediment is always being deposited in one place or another, it is possible to correlate widely scattered outcrops (rock exposures) to compile a composite record of geologic events. Correlation may be complicated by the presence of *faults, *unconformities and other processes that obscure the geologic record.

Rock layers that may be traced on the surface without interruption are the easiest to correlate. *Formations consisting of similar rock types and/or occupying the same position in a sedimentary sequence can also be used in correlation. Fossils, if present, are especially helpful in matching widely separated outcrops. *Guide fossils (those having a limited geologic range but a rather wide geographic distribution) are particularly useful. *Biostratigraphic units such as biozones or assemblage zones, and the stage of evolution of well-known organisms, are also employed in fossil correlation. Subsurface correlation of underground formations is necessary in mineral and petroleum prospecting. This is accomplished by comparing cores and rock samples taken from drill holes or by means of geophysical instruments that measure the electrical properties or radioactivity of the buried rocks. (W.H.M.)

Corrosion. Chemical wearing away of *bedrock by water in a stream. This is most evident where the channel flows over limestone or other easily soluble rocks. Many streams carry a large load in solution and corrosion is an important erosional agent. The total amount of material carried in solution by the rivers of the world is estimated to be 3.9 billion tons per year. Chemical erosion of Australia is slow; Europe has the fastest rate. Climate, geology, topography, vegetation and time are all important factors in determining the amount of corrosion. Other things being equal, the concentration of chemical constituents in river water is greater when the

Correlation

1 Cross sections of strata from different locations can be matched in terms of age by similarity of rock structure or fossil content. Limestone layers common to sections B, C and D are covered by conglomerate, forming a sequence of layers that can be correlated. In sections B and C the conglomerate layer is covered by a layer of fossiliferous shale that is also found in section A. Thus sections A and D can be correlated even though they have no rock layers in common.

Crater

2 The funnel-shaped depression at the top of Mount Ngauruhoe, North Island, New Zealand, was formed when the vent spewed forth volcanic gases, liquids and solids.

ratio of groundwater to surface water is large, since percolation through rock allows more contact with bedrock and more time for solution to take place. Steep slopes give less chance for solution than gentle slopes, which promote slower water movement. Vegetation, because it promotes infiltration and delays surface flow, tends to aid corrosion. The concentration of chemical constituents is less when discharge is large, as, for example, during flood stages. Chemical constituents in runoff from small watersheds reflect the composition of the bedrock and may be unique. However, because variability of bedrock increases with size of the area, rivers corroding very large drainage basins are similar in chemical content. See Karst Topography. (M.M.)

Corundum. A mineral, aluminum oxide, highly prized for its gem varieties, ruby and sapphire, which rank with diamond and emerald as precious stones. But far more abundant is common corundum, which is found in many localities and has been mined in quantity as an abrasive. Corundum is a hard mineral, 9 on *Mohs' scale, and for this reason has been the standard abrasive material for centuries. Its extreme hardness also enhances its use as a gem, permitting it to take a high polish that remains bright through years of wear.

Crystals of corundum are common. Although they vary considerably in habit, the prism is usually present, revealing the hexagonal symmetry. The crystals may be long hexagonal prisms, tabular plates outlined by the prism, or rounded barrel shapes resulting from several dipyramids of varying slope. Pure corundum is white or colorless; however, it is usually colored by small amounts of impurities. Ordinary corundum may be a shade of brown, gray, pink or blue; even a wider color range is found in the transparent gem material. In addition to the red of ruby and the blue of sapphire, it may also be green, yellow, purple or colorless. All of these color varieties of gem corundum may contain minute needles of *rutile, oriented by the hexagonal crystal lattice in three directions at 60° to each other. When such crystals are cut in smooth, rounded cabochon stones they show the property of asterism, that is, light is scattered by the oriented inclusion in the form of a six-pointed star, as in star ruby and star sapphire.

Corundum is often found as a primary constituent in igneous rocks containing *feldspathoids such as *nepheline syenite. It also results from the metamorphism of rocks rich in aluminum, such as *bauxite. The fine rubies found in Burma are formed from aluminous impurities during the recrystallization of a limestone and are found both in the limestone and in the overlying weathered zone. Because corundum is chemically inert and has a high specific gravity (4.0), it collects in *placers. From such alluvial deposits have come the sapphires and rubies for which Ceylon is famous.

Artificial corundum is produced on a large scale by fusing bauxite in an electric furnace and allowing it to crystallize. This material and manufactured silicon carbide have largely replaced natural corundum as an abrasive. Synthetic rubies and sapphires, colored by small amounts of chromium and titanium, are made by melting powdered alumina in an oxyhydrogen flame which on cooling forms single crystal "boules." It is difficult for the untrained person to distinguish these synthetic gems from the natural stones. See Mineral Properties. (C.S.H.)

Cosmic Sediment. Particles of extraterrestrial origin, commonly found in the sediment of the world's ocean basins. Most occurs in the form of minute metallic or silicate spherules measuring from about a micron to 0.5 mm in diameter. These particles probably represent metallic meteorites that have broken up as they entered the atmosphere. Small fragments of stony meteorites (*chondrites) composed of olivene and pyroxene are also found in marine sediments. Estimates of the annual influx of cosmic sediment range around a few thousand tons per year. (R.A.D.)

Cosmogony. The study of scientific hypotheses or cultural myths regarding the origin or creation of the solar system, including the planet earth, and the organization and nature of the universe. See Earth, Origin of.

Coulee. A broad, steep-sided stream channel that may or may not contain water. In the southwestern United States it is synonymous with *arroyo or dry *gully. In northwestern United States it is applied to an ancient glacial meltwater spillway such as the Grand Coulee or Dry Coulee in Washington.

Country Rock. The main mass of rock in which a mineral deposit or ore vein is enclosed; also, the wall rock surrounding an igneous intrusion.

Covalent Bond. See Bond. Chemical.

Covellite. A mineral, a copper sulfide, usually found in the zone of sulfide enrichment. Associated with other copper minerals—chiefly *chalcocite, *chalcopy-rite and *bornite—it forms as a product of their alteration. Covellite can be distinguished by its platy *cleavage and indigo-blue color. See Mineral Properties.

Crater. A circular or elliptical, funnel-, bowl- or saucer-shaped depression at the summit of a volcano. It is usually less than 1.5 km in diameter and may be a few hundred meters deep, with a vent at the bottom from which volcanic gases, liquids and solids may issue. It is surrounded by a rim that separates the outer slope of the volcano from the inner crater-wall slope. The walls are composed of *lava, *pyroclastic material or both. Volcanoes and craters are formed by volcanic material that builds a wall or ring around the vent. As the wall is enlarged, a volcanic cone and a crater are formed; they grow upward and are therefore constructional landforms. After an eruption the bottom of the crater has approximately the same diameter as the underground chimney that fed the volcano; however, subsequent landslides fill it in and widen it. A nested cone or nested crater results when a cone with a crater develops inside an older and larger crater.

Crater lakes on high mountains are common. Volcanic heat is still present if such lakes do not freeze during the cold winters. good examples of craters are found in the western United States at Sunset Crater National Monument near Flagstaff, Ariz., and Cinder Cone, Lassen Volcanic National Park, in northern California. Craters formed by impact are found in great numbers on the moon and the planet Mars as well as on the earth. (R.L.N.)

Craton. That portion of the continental *crust composed of *Precambrian rocks (600 million years or older and referred to as basement rocks) that have not been subjected to major deformation since the end of the Precambrian. The basement rocks represent areas that were at one time mountainous but were subjected to extensive erosion, gradually becoming relatively flat, low-lying areas. Cratons are subdivided into two landforms: shields and platforms. Shields are those areas of a continent in which a large, gently convex surface of basement rock is exposed. Platforms consist of basement rocks that have been covered by essentially flat-lying layers of stratified rocks, predominantly sedimentary. Platform areas generally surround shields.

The distribution of cratons is usually restricted to the interior of the continents. They are considered the nuclei from which continents grow. In the cratonal area of North America, both shield and platform

1

2

landmasses are found. Cratons also make up large parts of other continents. The north-central part of South America and the northwestern and central portions of Africa are cratonal areas. The western half of Australia and Scandinavia are covered by Precambrian rock. The Canadian Shield is a large expanse of exposed eroded Precambrian rock. The shield covers more than 5 million square kilometers of Canada and forms the Labrador Peninsula; parts of the provinces of Quebec, Ontario, Manitoba and Saskatchewan; and much of the Northwest Territories and part of the Arctic Islands. The shield has been of great importance to geologists, for erosion has exposed much of the mineral wealth contained in the older rocks. Large deposits of gold, uranium, nickel, cobalt and iron have been discovered, making mining one of Canada's largest industries.

The relationship between shield and platform areas is best displayed along the southern periphery of the Canadian Shield area. Just to the south of the Great Lakes the Precambrian basement disappears beneath the undeformed, stratified rocks of the interior lowlands of the United States. The lowlands represent the platform area of the craton and extend from the Appalachians westward to the Rocky Mountains and south to the Gulf of Mexico. Just like the Canadian Shield, the platform of the interior United States represents a wealth of natural resources. The area contains not only mineral wealth in the form of oil, coal and ore deposits, but offers vast areas of rich soil, making it excellent for agriculture. (J.W.S.)

Creep. A slow, almost imperceptible downslope movement of the *regolith. Creep is due primarily to gravity but is augmented by wetting and drying, freezing and thawing, the growth and decay of plants, the burrowing of animals and the colloid content of the regolith. Annual surface movements of 1 mm to more than 25 mm have been measured. Creep occurs widely on both steep and gentle slopes regardless of the amount of vegetation, and can damage telephone poles, fences and retaining walls. Although not spectacular, it is a major process in land denudation. (R.L.N.)

Creodont. A primitive, meat-eating mammal of the order Creodonta. Appearing in Late Cretaceous time, these ancestral carnivores reached their maximum development early in the Tertiary Period. The creodonts underwent early specialization: Some species were catlike while others resembled dogs. Although their teeth and claws were specialized for certain functions, their brain was probably about half as large as that of a modern carnivore of equal size, which possibly accounts for their extinction during Early Pliocene time. (W.H.M.)

Cretaceous Period. The last period of the *Mesozoic Era; it began approximately 135 million years ago and lasted for about 70 million years. Its name derives from the Latin *creta*, meaning "chalk," and was first used in 1822 to describe the chalk cliffs along the English channel. The Cretaceous was a time of widespread submergence. The Atlantic and Gulf coastal plains were beneath the ocean, and a great inland sea extended from the Gulf of Mexico to the Arctic Ocean, separating North America into two landmasses. During the maximum inundation, almost 50 percent of the present land surface was submerged, but by the close of the period the continent was approximately its present shape and size.

Cretaceous rocks are widely distributed and occur in all the continents. Although there are both nonmarine and marine deposits, the latter are more representative of Late Cretaceous time. European Cretaceous rocks are widespread and have been studied in great detail. Some are chalky and very fossiliferous, and in places contain masses of dark flint. Lower Cretaceous fossiliferous marine strata occur in southern Germany, France, Switzerland, and the northern part of the Soviet Union. In England, the Soviet Union and northern Germany, Lower Cretaceous formations are largely freshwater or continental in origin. The Upper Cretaceous rocks consist of marine sediments, and massive fossiliferous chalk beds occur in France and England. Cretaceous rocks are also well known in South America, Australia and Asia (northern Siberia, southern India, the East Indies and Japan). In North America, the Cretaceous constitutes the thickest and most extensive of the Mesozoic systems. These rocks are common in the Rocky Mountain and Great Plains areas (Colorado, South Dakota, Wyoming, Montana, Kansas, Iowa, Nebraska, Utah and eastern Idaho); along the Pacific coast (from Baja California to British Columbia and Alaska); on the Atlantic coast in a narrow band extending southwestward from New Jersey to Florida; and the Gulf coastal plain in a curving belt that can be followed from Georgia into southeastern Arkansas across southern Oklahoma and Texas and into Mexico. The Cretaceous Period (and the Mesozoic Era) was closed by the *Laramide orogeny. During this period of crustal deformation the sediments of the Rocky Mountain *geosyncline were folded, faulted and uplifted to form the great

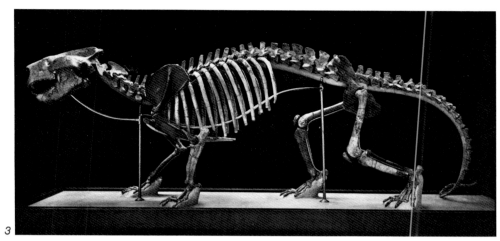

3

4

Creep

1 Evidence of this imperceptible form of mass wasting, in which the soil surface moves slowly downhill, can be seen in tilted fence posts and trees, whose root systems spread out behind the more rapidly moving trunks.

2 These trees near Lake Stahahe, N.Y., are bent because they grew upward while the slowly moving mantle rock tilted them downhill.

Creodont

3 Andrewsarchus, from Mongolia, was a giant among early carnivores, with a skull more than 1 m (3 feet) long.

4 Patriofelis, a lion-sized North American carnivore, roamed the land during the Eocene Epoch, some 50 million years ago.

Rocky Mountain System. This mountain-building revolution was accompanied by much volcanic activity in the western part of the United States, and volcanic materials such as lava and ash are common in many parts of this region.

Cretaceous climates were essentially mild and temperate, although probably cooler than those of the Jurassic. There is evidence of climatic differentiation, particularly in the early part of the period. In Late Cretaceous time, when widespread seas covered western North America, climates were probably milder and more equable than today.

Cretaceous life was marked by a number of major advances in plant and animal evolution. Plant life expanded, and the first *angiosperms (flowering plants) appeared in Early Cretaceous time. By the middle of the period such familiar trees as walnut, elm, oak, magnolia and maple were well represented. The grains and grasses also went through a period of rapid expansion; these, as well as the more advanced forms of trees, flowers and shrubs, provided food and protection for birds and mammals and thus probably aided greatly in their development. Mollusks were the predominant marine invertebrates and *gastropods and *pelecypods (both clams and oysters) were very abundant. *Cephalopods were in their last stages of development, and the *ammonoids were present in many shapes and sizes. Most of their shells were in the form of the typical flat coil, but some were straight, loosely coiled or coiled in a high spiral. Many were covered with ornate spines, ribs and nodes, while others were perfectly smooth. This extreme degree of specialization foreshadowed the extinction of the group at the end of the Cretaceous. *Echinoderms were common, especially the sea urchins.

Reptiles continued to rule land, sea and air. Many new species appeared, only to become extinct at the close of the Cretaceous. Among the *dinosaurs, duckbilled forms such as *Trachodon* (known also as *Anatosaurus*) and the armored ankylosaurs were common. The horned ceratopsian dinosaurs included *Triceratops*, a heavy-set, plant-eating creature as much as 9 m (30 feet) long. Its 2-m skull had a parrotlike beak; a frilled, bony shield that protected the neck; and three horns. Probably the most famous dinosaur of the Age of Reptiles was *Tyrannosaurus rex*, largest of the flesh-eating reptiles, who stood 6 m (20 feet) tall, was 12 to 15 m (40 to 50 feet) long and weighed many tons. Flying reptiles were highly advanced, although less numerous and varied. *Pteranodon*, a typical Cretaceous *pterosaur, had a wingspread as great as

Cretaceous Period

1 *During the Cretaceous many new species of reptiles appeared, though most became extinct by the end of the period. One of the most awesome of the horned dinosaurs, Triceratops, B, was 9 m (30 feet) in length. Tyrannosaurus, A, the largest land carnivore that ever lived, was 12 m (40 feet) long, 6 m (20 feet) high and weighed eight tons. Ankylosaurus, C, was covered with plates and spikes, and a great mass of bone at the end of its*

tail formed a powerful club. Parasaurolophus, D, a "duck billed" dinosaur, was a semiaquatic reptile with webbed toes and a strong tail for sculling through the water.

Crevasse

2 *As a glacier moves, stresses develop in the ice and create steep-walled cracks as much as 30 m (100 feet) deep and 12 m (40 feet) wide, such as Mendenhall Glacier, Alaska.*

7.6 m (25 feet) but weighed only about 6 kg (14 pounds). Marine reptiles were common, and the *ichthyosaurs and *plesiosaurs were joined by the *mosasaurs—huge lizardlike reptiles as much as 15 m long. Giant sea turtles, some 3.4 m (11 feet) long and 3.6 m (12 feet) across the flippers, were also numerous. The end of the Cretaceous Period, like the end of the Paleozoic, was marked by the extinction of many once-numerous Mesozoic species as changing climates and geographic patterns took their toll. There was a drastic reduction in the numbers and varieties of reptiles, and the dinosaurs, pterosaurs, ichthyosaurs, plesiosaurs and mosasaurs all became extinct. The ammonoids and *belemnoids also died out at the end of Cretaceous time.

Cretaceous rocks are rich in a variety of mineral resources, including coal. Oil and gas are produced from petroleum-bearing strata. Clays, shales and limestones are used to manufacture Portland cement and ceramic products, and as construction materials. (W. H. M.)

Crevasse. A wedge-shaped, nearly vertical fracture in the upper part of a glacier. A crevasse decreases in width with increasing depth; it may be hundreds of meters long, many meters wide and over 30 m deep. It is formed when brittle ice fractures because of stresses resulting from glacial motion. Although open and easy to see during the summer, a crevasse may be concealed by layers of freshly fallen snow, called a crevasse lid, during the winter. Crevasses make travel on some glaciers both difficult and dangerous. Below ice falls, where they are commonly large and numerous, glaciers may be impassable. (R. L. N.)

Crinoid. A member of class Crinoidea of phylum Echinodermata. Known as sea lilies, their name derives from the calyx which, with its numerous branching arms, resembles a long-stemmed flower. Certain living crinoids are stemless and free-living in the adult stage. The crinoid skeleton is composed of the stem, the body or calyx, and the arms. The calyx is typically cup-shaped, with five grooves radiating outward along the segmented arms as channels to convey food to the mouth. The calyx, composed of many calcareous plates, encloses the animal's soft parts. The crinoid stem serves for support and attachment, and consists of a relatively long, flexible stalk composed of numerous calcareous, disclike or buttonlike segments or columnals. Many crinoids have stalks as much as 15 m (50 feet) long, and when the animal dies the columnals are

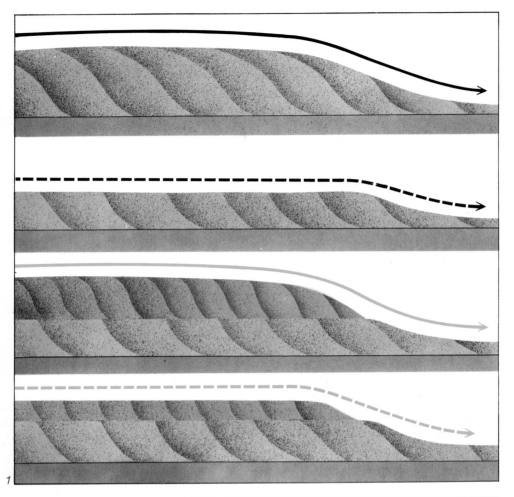

1

2

scattered about on the ocean floor. Paleozoic limestones often contain such great numbers of crinoid columnals that they are called crinoidal limestones.

The earliest known crinoids occur in rocks of Ordovician age, and their remains are particularly abundant in Paleozoic rocks. Most crinoids living today are stemless, free-moving feather stars, much less abundant than the stalked crinoids. (W.H.M.)

Cristobalite. A mineral, a high-temperature isometric form of silica. It crystallizes as the stable form above 1470°C (2678°F). On cooling it inverts to a low-temperature, metastable form between 275° and 163°C (527° and 325°F). As the low-temperature form it occurs in *lavas, both in spherical aggregates in cavities and as a constituent of the fine-grained groundmass. See Mineral Properties; Appendix 3.

Crocoite. A mineral, the rare lead chromate, with a bright hyacinth-red color. It is found in slender prismatic crystals in the oxidized zone of lead deposits in association with *pyromorphite, *cerussite and *wulfenite. The most famous localities for crocoite are Dundas, Tasmania, and Beresovsk in the Ural Mountains. See Mineral Properties; Appendix 3.

Cross-Bedding (or Cross-Strata; Inclined Bedding). Strata inclined with respect to a thicker rock layer within which they occur. Cross-beds are laid down at an angle to the horizontal and occur typically in granular sediments, especially sandstone. They are indicative of sediments deposited in deltas and sand dunes and by melting glaciers, and are useful in reconstructing ancient climates.

Cross Fibers. Closely packed fibrous minerals that have grown with the length of the fibers at right angles to the vein in which they occur. Although most commonly applied to *asbestos, the term is also used for other minerals such as the satin spar variety of *gypsum.

Crossopterygian. Primitive lobe-finned fish of the order Crossopterygii. They appeared in Middle Devonian time and are believed to be the forerunner of the modern lungfishes, probably possessing similar respiratory organs. Their four lobe-shaped fins had a fleshy upper portion that contained bones resembling those of the limbs of land vertebrates. Their teeth were conical in shape, with a complex internal structure similar to that of certain labyrinthodont amphibians.

Eusthenopteron, a Devonian crossopterygian, was an elongated, carnivorous fish whose skull pattern bears a marked resemblance to the earliest amphibians. The only living member of this order is the *coelacanth fish known as *Latimeria*. The coelacanths were thought to have become extinct about 70 million years ago, but since 1938 several living specimens have been caught off the southeast coast of Africa. (W.H.M.)

Crust. The outermost layer or shell of the earth. Its lower limit is the boundary known as the Mohorovicic Discontinuity or Moho. This is the boundary at which *seismic waves generated by earthquakes increase abruptly in velocity as they enter the *mantle, the material directly beneath the crust.

The crust of the earth is not homogeneous in composition or in distribution. Crustal material forming the continents is composed of two sublayers, which together have a thickness of 30 to 40 km or greater under regions of large mountain ranges. The upper layer (often called the sial) is exposed on the surface of the earth in *cratons (parts of the crust which have remained undeformed since the Precambrian). The sial is composed mainly of relatively less dense igneous rock such as granite and of metamorphic rock such as gneiss. The mean *density of these rocks is 2.8 and their overall chemical composition shows that they contain abundant amounts of silica (61.9 percent) and alumina (15.6 percent). Separating the upper and lower crust is another boundary defined by an increase in the velocity of earthquake waves as they pass from the upper crust into the layer below. This boundary, known as the Conrad Discontinuity, is believed to represent a change in composition of the lower crust. Since the lower layer, representing the last 15 to 20 km of crustal material, lies below the depth to which drill holes have penetrated, scientists are able only to speculate on its composition. However, their speculation is based on some firm evidence. Scientists have measured the speed with which earthquake waves propagate through earth materials of different densities. The speed of the waves traveling through the lower crustal layer indicates that it is composed of material with a density of about 3.0. A rock found on the surface of the earth having a similar density is *gabbro, a coarse-grained rock composed of lesser amounts of silica than most of the overlying rocks of the upper crust, and containing great amounts of iron and magnesia. The fine-grained equivalent of gabbro is *basalt, which is known to form the crust beneath oceans. Therefore, scientists reasonably speculate that the material forming the lower crust is basalt.

This assumption has recently been challenged on the basis of the knowledge that basalt converts to rock of another mineral composition (eclogite) as a result of the high temperature and pressures encountered in the lower crustal layer. The newly formed rock has a density of about 3.5, which is inconsistent with data on the velocity of earthquake waves in the lower layer. It has also been suggested that the lower layer is of the same composition as the upper layer but that the minerals formed at the lower depths have greater densities, which would account for the increase in velocity of earthquake waves. The crust of the earth found below the oceans is quite unlike that of the continents. Its thickness is much less than that of the continental crust, measuring about 10 km, and much more uniform. The average composition of oceanic crust (often termed the *sima) is predominantly silica (48.7 percent), alumina (16.5 percent), iron (8.5 percent), magnesia (6.8 percent) and lime (12.3 percent). Its density is about 3.0 and greater than that of the overall continental crust. Unlike the older age of the continental crust, most of the oceanic crust is very young, less than 200 million years old, and is constantly being generated along ridges dividing the ocean basins. See Sea Floor Spreading. (J.W.S.)

Crustal Plate. See Plate Tectonics.

Cryolite. A mineral, sodium aluminum fluoride. Although *monoclinic, it commonly occurs in rough pseudocubic blocks that are the result of parting. The mineral is usually snow-white, but may be reddish or brownish from included impurities. Small amounts of cryolite have been found at several localities, but the only important deposit is at Ivigtut on the west coast of Greenland. Here, surrounded by granite, is a large mass associated with *siderite, *galena, *sphalerite and *chalcopyrite. It has been actively mined largely for use as a flux in the electrolytic process to produce aluminum from bauxite. Cryolite is also important in the glass and enamel industries, as a filler in bonded abrasives, and in the making of insecticides.

Crystal. A regular geometric form bounded by smooth plane surfaces that are the external expression of an ordered internal atomic arrangement. The word is also applied to any solid having an ordered atomic structure. See Crystallography.

Crystal Axis. A line of reference used to describe atomic planes, crystal faces and vectorial properties of crystals. The axes that are usually chosen are parallel to and

Cross-Bedding

1 When periods of deposition (indicated by solid lines) alternate with brief periods of erosion (indicated by broken lines), cross-bedding occurs.

2 Sweeping layers of sandstone in Hoskininni Mesa, Ariz., are compressed sand dunes piled up by the wind and later covered by water.

Crystal

3 Calcite (a) displays more crystal forms than any other mineral. The shape of the crystals shown here is scalenohedral, one of the most common. Barite crystals (b) display nearly perfect orthorhombic symmetry. Halite, or common salt, (c) usually occurs in granular masses; perfectly formed cubic crystals with shining faces such as this are rarities.

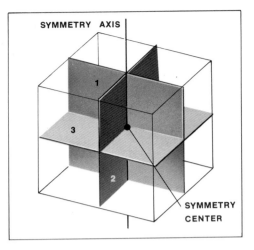

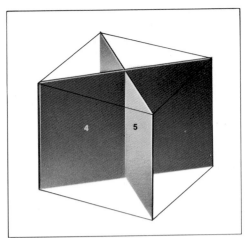

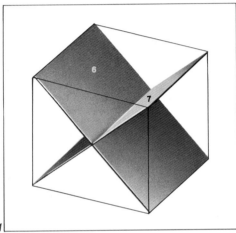

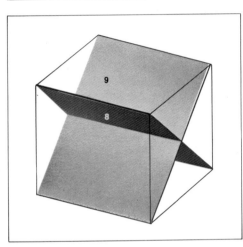

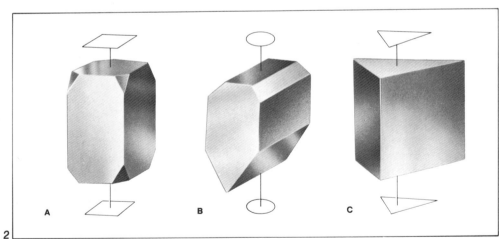

their lengths proportional to the edges of the unit cell of the crystal.

Crystal Chemistry. The science relating the internal structure of *crystalline substances with their chemical composition and physical properties.

Crystal Classes. The 32 possible combinations of symmetry in which crystals can form. These combinations are the only possible arrangements of symmetry planes and axes in which planes and axes all intersect at a common point. See Crystal Systems.

Crystal Habit. The characteristic shape or form of a mineral crystal or crystalline aggregate. Some minerals almost universally are bounded by faces of the same crystal form or forms and thus· have a constant shape or habit. Depending on the crystal forms present, a mineral at one locality may have a completely different appearance or habit from the same mineral in another locality.

Crystal Lattice. A framework on which the atoms, ions and molecules that form crystals are arranged. The French crystallographer Auguste Bravais first showed that there can be only 14 basically different kinds of three-dimensional arrays in crystals; for this reason they are commonly called the Bravais lattices.

Crystalloblastic Series (or Idioblastic Series) An orderly listing of minerals according to their tendency to develop crystal faces during *metamorphism. Mineral grains are mostly without crystal faces in metamorphic rock, and they have irregular surface textures, called crystalloblastic, regardless of their internal structures. During alteration through stresses in the earth's crust, certain minerals show a greater tendency than others to form grains with well-developed crystal faces, called idioblasts, when adjacent to certain other minerals.

The German petrologist Friedrich Becke in 1913 arranged the minerals in a series according to their tendency· to develop idioblasts when in contact with any minerals that are lower in the series. Sphene, first on the list, almost always forms well-developed crystal faces; quartz and those below it, almost never. The series is as follows: (1) sphene, rutile, magnetite, hematite, ilmenite, garnet, tourmaline, staurolite, kyanite; (2) epidote, zoisite; (3) augite, hornblende; (4) breunnerite, dolomite, albite; (5) mica, chlorite; (6) calcite; (7) quartz, plagioclase; (8) orthoclase, microcline.

Crystal Symmetry

1 The nine symmetry planes of a cube.

2 Crystals are shown with the following axes of symmetry: (A) tetragonal crystal with a fourfold symmetry axis, (B) orthorhombic crystal with a twofold symmetry axis, (C) trigonal crystal with a threefold symmetry axis.

Crystal Systems

3 (A) Isometric crystals have three mutually perpendicular axes of equal length. (B) Tetragonal crystals have three perpendicular axes, two of equal length. (C) Orthorhombic crystals have three perpendicular axes, all of different length. (D) Monoclinic crystals have three axes, all of different length, only two mutually perpendicular. (E) Triclinic crystals have three axes, all of different length and none perpendicular. (F) Hexagonal crystals have four axes, three of equal length at a 60° angle to each other, the fourth a different length and perpendicular to the others.

A complete explanation of why metamorphic minerals behave in this manner has not yet been found. What is thus far known about mineral growth in metamorphic rocks is,(1) the crystalloblastic series does not indicate the relative age of crystallization, i.e., a mineral listed early in the series does not necessarily begin to grow earlier than minerals listed below it; (2) although there are some exceptions, it appears that minerals possessing high density, hardness and surface energy will develop crystals displaying their natural outlines. Many geoscientists as well as metallurgists are now at work on the phenomenon of mineral growth. See Xenoblastic. (C.S.H.)

Crystal Optics. The branch of *crystallography that deals with the behavior of light as it passes through crystals. Since the reaction of transmitted light differs from crystal to crystal, each is characterized by its own set of optical properties that are important in crystal description and identification. The most important optical property is *index of refraction, which can be easily calculated. When light passes from air into a crystal at an oblique angle it is refracted, that is, bent from the path of the incident (incoming) beam, and its velocity is decreased. The velocity is a function of the angle of refraction, and the refractive index is the reciprocal of the velocity.

Optically nonopaque substances are divided into two groups: isotropic and anisotropic. The isotropic includes crystals of the isometric system and noncrystalline substances such as glass. Light moves through them in all directions with a uniform velocity, and they then have a single refractive index. The anisotropic group includes all crystals except isometric. In them the velocity of light varies with crystallographic direction, with a corresponding variation in refractive index. Light on entering a random section of an anisotropic crystal is doubly refracted, that is, broken into two polarized rays traveling at different velocities. For such a section there are thus two refractive indexes. In tetragonal and hexagonal crystal, light moving parallel to the c axis is not doubly refracted. Since there is only one such direction, the optic axis, these crystals are called uniaxial. Maximum and minimum refractive indices are measured for them, one with light vibrating parallel to the c axis and the other with light vibrating at right angles to c.

Orthorhombic, monoclinic and triclinic crystals are optically biaxial since there are two optic axes, directions of no double refraction. For these crystals the maximum, minimum and intermediate refractive indices are reported, measured with light vibrating in three mutually perpendicular directions. Also reported is the optic angle (2V), the angle between the two optic axes. (C.S.H.)

Crystal Symmetry. In crystallography, the outward expression of crystal symmetry is seen in the regular arrangement of crystal faces and the angles between them. The regularity may be evidenced in three ways by the arrangement of like faces: across a plane, about a line, about a point.

A symmetry plane is an imaginary plane that divides a crystal into two halves, each of which in the model-perfect crystal is the mirror image of the other. For each face, edge or point on one side of the plane there is a corresponding face, edge or point on the other side. Some crystals of the isometric system, for example the cube, have nine symmetry planes; other crystals may have one, two, three, four, five, six or seven symmetry planes, and some have none.

A symmetry axis is an imaginary line through a crystal about which the crystal may be rotated and brought into identical position during a complete rotation. If after a 180° rotation the crystal repeats itself, the axis is of twofold symmetry; if the repeat is after a 120° rotation, there is a threefold symmetry axis. In crystals no symmetry axes other than one-, two-, three-, four- and sixfold may be present.

A symmetry center is an imaginary point at the center of a crystal. If a line beginning at any point on the surface is passed through this central point, it will emerge at a similar point opposite. (C.S.H.)

Crystal Systems. Certain of the 32 crystal classes have symmetry characteristics in common which permit them to be referred to the same crystallographic axis. There are six of these larger groupings, called crystal systems: *isometric, *tetragonal, *orthorhombic, *monoclinic, *triclinic and *hexagonal. See Crystal Optics; Crystal Symmetry.

Crystal Twin. See Twin Crystal.

Crystalline. Having an ordered internal atomic arrangement. This is contrasted with amorphous, or lacking internal order. Minerals are crystalline; the rare amorphous substances that occur in nature are called *mineraloids.

Crystallography. The science that deals with the interatomic arrangement of solid matter and the laws that control the growth, external form and internal structure of such matter. This definition does not convey the

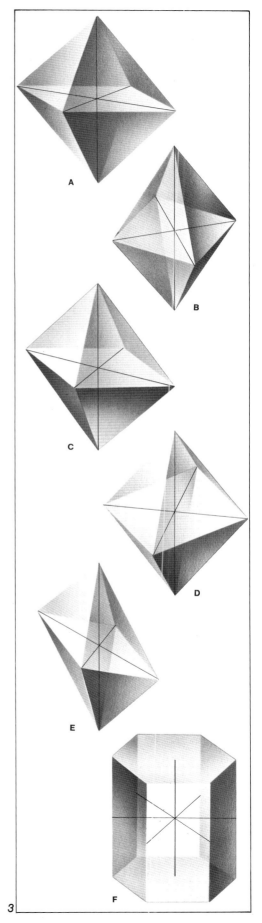

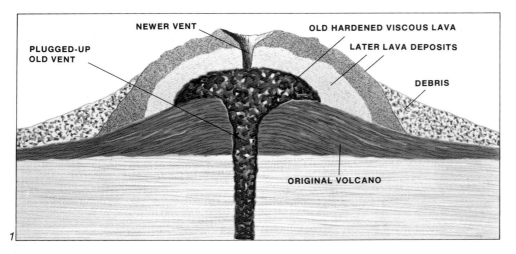

many ramifications of the science, however. Crystallography had its beginning with the study of the outward shapes of mineral crystals, and morphological crystallography remains an important branch of the science. Through a study of external geometry, crystals can be grouped into 32 symmetry classes and into larger divisions, the six *crystal systems.

For a long time crystallography has also included a study of the properties of crystals. Some of these—such as hardness and electrical and heat conductivity—are vectorial, that is, the magnitude of the property changes with crystallographic direction. The speed of light through a crystal is perhaps the most important vectorial property because its reciprocal, the refractive index, is easily measured. Optical crystallography, which deals in general with the behavior of transmitted light, is a major branch of the science.

It was long recognized that the various physical and morphological properties of crystals were related to the internal atomic arrangement. However, the relationship was not understood until after 1912, when X rays were used to study the internal structure of crystals. Structural crystallography has since become of prime importance. The size and arrangement of atoms within the crystal, the distance between atomic planes and the dimensions of the basic unit of construction or the *unit cell of the crystal can be determined. These data, when considered with the chemical composition, give a meaningful interpretation to the many other properties of crystals. See Crystal Optics; Crystal Symmetry. (C.S.H.)

Cube. A form of the *isometric crystal system composed of six mutually perpendicular and symmetrically equivalent faces. In the geometrically perfect cube, each face is a square.

Cubic System. See Isometric System.

Cuesta. A low, irregular and somewhat linear ridge with a steep cliff face on one side and a gentle back slope. This topographic feature is the result of erosion on gently sloping sedimentary beds. It is similar to a *hogback but has a more gentle dip slope. The top is formed on a resistant layer and the cliff face is eroded back as the underlying beds are eaten away beneath it. Cuestas are found primarily under two types of geologic conditions: (1) On recently uplifted coastal plains where the sediments dip gently seaward. Cuestas of this type occur on the Gulf coastal plain of the southern United States and in the Paris region of France. (2) On gently dipping

Cumulo-Dome

1 This volcanic structure results from the accumulation of lava too viscous to spread very far from the vent.

Cycad

2 Remnants of one of the most primitive groups of plants, living cycads such as the Mexican genus Dioon bear seeds in a large terminal cone.

Cyclone

3 The early stage of a middle-latitude cyclone develops when (a) a cold front begins to push under a mass of warm air. (b) In the open stages, cold air continues to push back the warm air, which bulges out at front into a low-pressure area. (c) During the occluded stage, the cold front overtakes the warm front, lifting it completely off the ground. (d) The swirling low-pressure cell eventually dissolves as air pressure equalizes.

sediments which have been uplifted over a broad continental region and erosion has stripped away the central part leaving gently dipping ridges surrounding the area. Examples are the Niagara cuesta in northern New York and the English Wolds. (M.M.)

Culmination. Structurally, the highest point on an upfolded portion of the earth's *crust. In a mountain produced by folding, the culmination would be the highest point along the axis of the *fold. Several culminations may be found along the length of a mountain range. In a fold mountain that has been subjected to erosion, the highest topographic point may or may not coincide with the culmination. In folded mountain chains with many culminations, the lower portions of the range that separate the culminations are known as *depressions.

Cumulo-Dome. A steep-sided volcanic dome, rounded or roughly level on top, usually without a *crater. It may or may not be located inside an older volcanic crater or *caldera. Although not as large as *shield or composite volcanoes, these are larger than *spatter or *cinder cones. They are formed by the protrusion of very stiff, nonbasaltic lava too viscous to spread very far from the volcanic vent. They grow by expansion from within like a bubble blown from a bubble pipe. The crust stretched during the growth of the dome is broken into blocks, which accumulate as *talus on its steep sides. External growth also takes place when viscous lava is thrust up through cracks in the stretched crust, forming spines. Cumulo-domes are not as common as cinder or composite cones. Lassen Peak, Calif., one of the largest known, is 800 m high and has a basal diameter of 2400 m. Puy-de-Dome in the Auvergne District of France, and Bogoslof in the Aleutians, are cumulo-domes. They are also found on Mount Pelée, Martinique; on Ascension Island; on the lower slope of Mount Shasta, northern California. See Plug Dome; Tholoid. (R.L.N.)

Cuprite. A secondary copper mineral, cuprous oxide. It occurs in the oxidized zone of copper deposits associated with *limonite and other secondary copper minerals. Because it commonly occurs in ruby-colored isometric crystals, it has been called ruby copper. In places it has been mined as a minor copper ore. See Mineral Properties. (C.S.H.)

Curie Point. A temperature reached in the cooling process of molten material at which any magnetic minerals present will begin to produce a magnetic field. Above the Curie temperature, thermal agitation destroys the existing oriented magnetic field.

Cutoff. A formation that occurs when a meandering river breaks through a bend. There are two kinds of cutoffs: chute and neck. A chute cutoff occurs during floods when the stream flows across the inner low part of a bar that has been deposited on the inside of a bend. Such a cutoff may be abandoned when the flood recedes and the stream returns to its old course. A neck cutoff forms when a *meander loop has developed in such a way that the meander neck is very thin and is eroded through. Such a cutoff is compatible with the main flow of the water, and since it shortens the stream course it is retained. In this situation the old channel becomes an *oxbow lake. (M.M.)

Cycad. A representative of a large group of plants comprising the order Cycadales. Cycads are palmlike plants believed to be among the most primitive of living seed plants. They have short, pithy trunks with a thin woody covering, naked seeds borne in simple cones, and leaflets arranged on either side of a common stem. Living cycads are called coco palms or sago palms and are found in limited numbers in many tropical areas today; their fossils have been recovered from rocks of Mesozoic age. Living species have cylindrical, erect trunks and large, fernlike leaves on short, thick stems. There is evidence that primitive cycads were living as far back as Late Permian time. (W.H.M.)

Cyclone. A circular system of air movement around a center of low pressure; often referred to as a "low." The direction of air motion in a cyclone is counterclockwise in the Northern Hemisphere, clockwise in the Southern, with a component of motion toward the center. Originally used in a generic sense to describe any circular or highly curved *wind system, the term cyclone now usually refers to the extratropical cyclones that move, at intervals of several days to about a week (depending on latitude and season), in a generally eastward direction across the earth's surface outside the tropics. In the Indian Ocean, tropical cyclones also are commonly referred to as cyclones, but in other parts of the world they have regional names, such as hurricane and typhoon.

Unlike the occasional cyclones of the tropics, extratropical cyclones (which in their fully developed form normally occur in combination with *anticyclones) are a regular feature of day-to-day weather. These two major types of cyclones differ not only in frequency but also in their struc-

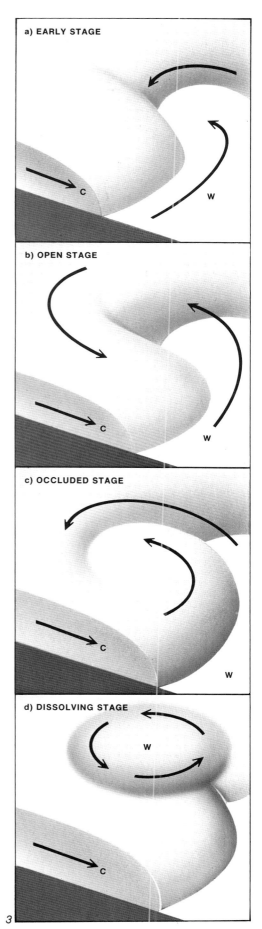

a) EARLY STAGE

b) OPEN STAGE

c) OCCLUDED STAGE

d) DISSOLVING STAGE

3

1

2

ture and formation. A tropical cyclone originates within a homogeneous mass of warm moist air. Most extratropical cyclones, on the other hand, form along a front (a zone between cold dry and warm moist air masses); for this reason, meteorologists often refer to them as frontal cyclones.

In structure, a mature tropical cyclone closely resembles a circular vortex whose axis of rotation extends vertically upward through the *troposphere. The axis of an extratropical cyclone is displaced horizontally, often a distance of several hundred miles or more, in the direction of colder air. Its circulation, especially in the developing stages, is asymmetrical; and within a few kilometers above the earth's surface its closed wind pattern usually gives way to the wave patterns of motion typical of the middle-latitude upper *westerly winds.

Early in the 20th century, a group of Norwegian meteorologists based a theory of the origin and development of cyclones on the growth of wave disturbances in the *polar front. Present-day meteorologists, while recognizing the importance of a zone of strong temperature contrast for cyclone development, look upon the behavior of the fronts within a cyclone as an incident of development rather than a cause.

The necessary conditions for the development of cyclones appear to be: (1) a source of rotation and (2) an outflow of air in the middle troposphere, or at higher elevations, accompanied by an inflow of air directly below that does not quite compensate for the outflow. In an extratropical cyclone, the outflow aloft is provided by the divergence that typically occurs in the poleward-streaming air of an upper wave in the westerly winds. Since nature abhors a vacuum, air in the lower troposphere flows inward or converges into the incipient low-pressure center created by the outflow and turns upward to feed the diverging winds aloft. The source of rotation is the rotation of the earth itself, for on the rotating globe any large-scale convergence of air (except near the Equator) results in cyclonic spin. As it converges toward a point, the air maintains the rotation, or vorticity, that characterizes the earth itself around the local vertical axis at that particular latitude. As the axis of rotation of the disc of air is shortened, the converging air begins to rotate relative to the earth beneath.

The modern theory of middle-latitude cyclone formation has the advantage that it explains not only the prevalence of cyclones at these latitudes but also the formation of the accompanying anticyclones.

Just as the poleward wave current aloft provides the divergence necessary to cyclone formation, the Equatorward flow provides a pattern of motion in which convergence takes place. This convergence causes a rise of pressure directly below, to the west of the developing cyclone, resulting in divergence of air. For the same reason that convergence under the forward part of the wave results in cyclonic rotation, the large-scale divergence in the rear leads to spin in the reverse or anticyclonic direction and the development of an anticyclone in the lower troposphere.

Upward air motion, adiabatic cooling (cooling due solely to the expansion of the ascending air), *condensation leading to cloud formation and precipitation, and, frequently, strong winds are typical of cyclones. Thus they are often referred to simply as storms. See Convergence; Cyclone, Tropical; Polar Front. (M.F.H.)

Cyclone, Tropical. A rotating storm, ranging in size from 100 to 1500 km across, that originates over tropical oceans. Tropical cyclones form in areas where the water temperature exceeds 27°C. Their typical paths take them westward and poleward, so that they often strike the Equator side or eastern coastline of continents. Tropical cyclones do not occur in the South Atlantic Ocean, because the water temperature there does not become sufficiently high and because the intertropical convergence zone, where the trade winds of the two hemispheres meet, always lies north of the Equator in the Atlantic Ocean. Nor do they form at latitudes within 5 degrees of the Equator, possibly because the *Coriolis force there is too small to produce a cyclonic circulation.

Severe tropical cyclones in the North Atlantic Ocean, the Caribbean Sea, the Gulf of Mexico and the Pacific Ocean off the west coast of Mexico are called hurricanes. 'n other parts of the world they have various regional names—for example, typhoons in the western Pacific, cyclones in the Indian Ocean and baguios in the Philippines. By international agreement, tropical cyclones are classed as hurricanes when their winds exceed 32 m per second; as tropical storms when their maximum winds are between 17 and 32 m per second; and as tropical depressions when wind speeds are not as great as 17 m per second.

A unique three-dimensional view of the cloud structure of a tropical cyclone, Hurricane Gladys, was obtained by the astronauts of the Apollo 7 spacecraft on October 17, 1968. This view, taken when the storm was about 200 km west of Florida, reveals the typical cloud bands, composed of separate cumuliform clouds for the most part, that spiral inward toward

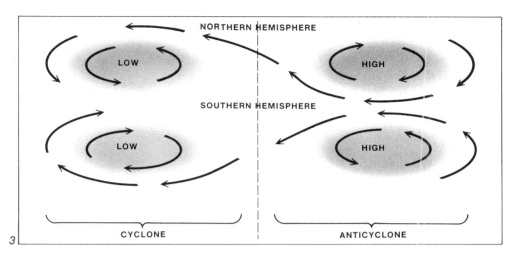

Cyclone, Tropical

1 World distribution of tropical cyclones is seasonal and is confined to six regions, all over tropical and subtropical oceans where water temperature exceeds 27°C (80°F).

2 Hurricane Ava, as photographed by weather satellite on June 7, 1973, displays the swirl of clouds around the "eye" of a low-pressure system of a tropical cyclone.

3 In the Northern Hemisphere winds spiraling inward toward the center of low pressure (counterclockwise) form a cyclone. Winds spiraling outward from the center of high pressure (clockwise) form an anticyclone. In the Southern Hemisphere, the wind directions of cyclones and anticyclones are reversed.

4 Storm-driven waves and winds up to 150 miles an hour batter the beaches of Miami, Fla., during a hurricane.

the center of a hurricane. Near the center, an inner band of extremely tall cumulonimbus clouds, which may reach heights of about 15 km, surrounds a central portion of the storm, called the eye. The eye of the storm, which is sometimes free of clouds, is in this case concealed by a circular cap of cirrostratus clouds, composed of ice crystals, that spread from the glaciated tops of the wall clouds.

The eye of a tropical storm, in contrast to the band of high winds surrounding it—which may reach velocities of 100 m per second (224 miles per hour) — is relatively calm. The eye, which ranges from about 5 to 50 km in diameter, is usually several degrees warmer than the surrounding part of the storm, since it is composed of descending air that is warmed adiabatically, that is, by compression as the air descends to levels of higher atmospheric pressure. Because of this warm core, the low pressure at the center of a tropical cyclone must give way to a high-pressure area or *anticyclone aloft. The spiraling winds converge toward the storm center but become nearly tangential at the wall cloud, where the converging air rises most rapidly. At upper levels, the air flows outward from the anticyclone that presumably overlies the cyclone below. The structure of the tropical cyclone is thus quite different from that of other cyclones, which have cold cores or centers and are not topped by high-pressure areas.

Although the precise mechanism by which tropical cyclones form is not understood, it is known that they derive their energy from the latent heat released by condensation within the rising air currents. This apparently explains why tropical cyclones originate only over areas of very warm water, which provides an ample and continuous source of energy through evaporation from the sea surface.

Although the rain and cloud patterns of hurricanes vary from storm to storm, the banded structure of the rain showers accompanying a hurricane is usually detectable in satellite photographs or on radarscopes. The pressure at the center of an intense tropical cyclone may be as much as 200 millibars or 20 percent lower than the surrounding pressure (about 1000 millibars). As a tropical cyclone approaches a coastline, the strong winds drive the water ahead, piling it up in the shallow coastal region and raising the level of the sea in what is often popularly called a tidal wave but is technically referred to as a storm surge. The low pressure at the center of an intense storm also contributes to the rise of water. Flooding the coastal lowlands, the storm surge and the battering waves frequently produce more damage and many

more deaths than do the high winds. The Bay of Bengal cyclone that flooded the lowlands at the mouth of the Ganges River on November 12, 1970, when several hundred thousand people were drowned, is a tragic example of the catastrophic potential of the storm surge.

When a hurricane moves inland it loses much force because its energy supply (the warm ocean water) is cut off; however, heavy rains often continue well inland along the path, producing major floods such as those associated with Hurricane Agnes in 1972, when it moved northward over the eastern United States. (M. F. H.)

Cyclosilicate. The group of silicates in which the silicon-oxygen tetrahedra are linked together by sharing oxygen ions to form rings. There are three-, four- and six-membered rings, and all of them have a 1:3 ratio of silicon to oxygen. Beryl, $Be_3 Al_2 (Si_6 O_{18})$, is an example. See Silicate Structure and Classification.

Cyclothem. A series of beds, deposited during a single sedimentary cycle, which typically consists of (1) an accumulation of plant material, (2) a deposition of limestone, (3) a deposition of sandstone and (4) erosion of land to sea level and weathering of rocks to soil. These beds, typical of certain *Pennsylvanian formations, probably result from unstable crustal conditions that produced periodic advances and retreats of the sea. The lower half of a cyclothem consists of continental (nonmarine) sediments and the upper half consists of marine deposits. The so-called "ideal cyclothem" has been described in Pennsylvanian rocks in western Illinois and appears to represent deposits laid down in a delta environment. It is a theoretical sedimentary series consisting of the following rock units in descending order from the most recent to the earliest (unit 1 being the earliest rock and the first stage in the cycle): (10) gray marine shale with limestone *concretions; (9) fossiliferous, marine limestone; (8) hard, black, thin-layered shale with limestone concretions; (7) fossiliferous, fine-grained, impure limestone; (6) gray marine shale; (5) coal; (4) underclay; (3) freshwater limestone; (2) shale or sandy shale; (1) fine-grained sandstone with *unconformity at the base. Distinct cycles of cyclothems or a group of related cyclothems are called megacyclothems. (W. H. M.)

Cystoid. A member of class Cystoidea of phylum Echinodermata. These primitive, attached *echinoderms were relatively common in early Paleozoic time. The typical cystoid has a somewhat globular or

saclike calyx (the main body skeleton) composed of numerous, irregularly arranged, calcareous plates. The plates composing the calyx are usually perforated by pores or slits, which were probably used in excretion or respiration. The calyx was attached to the sea bottom by a short stem. Cystoids range from Cambrian to Devonian in age and were especially abundant during Ordovician and Silurian time.

D

Dacite. A volcanic rock that is the extrusive equivalent of *quartz diorite. The principal minerals are *plagioclase feldspar (oligoclase to andesine), quartz, *augite or *hornblende. All these may occur as phenocrysts (coarse crystals) in a fine-grained or glassy groundmass. Dacite is a light-colored rock difficult to distinguish from *rhyolite without microscopic examination.

Darcy's Law. A law stating that the velocity of water through a porous material is directly proportional to the hydraulic gradient times the permeability of the material. This relationship is represented by the equation $V = K[(h_2-h_1) \div L]$ where V is the velocity of flow, K is a measure of the permeability, and $(h_2-h_1) \div L$ is the slope (hydraulic gradient) of the water table. Because of friction, groundwater moves far slower than does surface water. Rates have been measured by introducing salts or dyes in one well and measuring the time necessary for them to reach another. A velocity of 7 m (23 feet) per year was measured in an aquifer in Florida by determining the age of the groundwater (by means of carbon-14 measurements) along a series of wells. Velocities of 6 m (20 feet) per day are reached in highly permeable material; one of the highest velocities measured in the United States was 235 m (770 feet) per day. However, in many areas velocities of less than 30 m (100 feet) per year are common. (R.L.N.)

Dark Mineral. A rock-forming mineral having a dark color. The principal dark minerals are *amphibole, *pyroxene, *olivine and *biotite. Because all of these contain iron and magnesium, they are also called the *ferromagnesium minerals.

Datolite. A hydrous calcium borosilicate mineral found usually in crystals with many flashing faces. It commonly occurs in cavities in basaltic rocks associated with *zeolites, prehnite and apophyllite. Occurrences in the United States are principally in the trap rocks that extend through Massachusetts and Connecticut into New Jersey. The outstanding localities are at Westfield, Mass., and Bergen Hill, N.J. In addition to well-crystallized datolite there is a compact variety resembling unglazed porcelain. It has been found associated with the native copper of the Keweenaw Peninsula, Mich. See Mineral Properties. (C.S.H.)

Deep-scattering Layers. Certain reflecting layers in the sea that register on the echograms of depth-recording devices. The reflections are generally scattered, giving a diffuse pattern rather than the sharply defined reflection caused by the bottom. One of the peculiarities of these layers is their daily migration; they move up almost to the surface during the night and descend to depths generally of 200 to 600 m during the day. These layers, which occur in all oceans, are apparently caused by various organisms. To fit the characteristics of the layers, these organisms must swim, occur in vast numbers and reflect sound. Squid, fishes and small shrimplike crustaceans, called euphausids, all fulfill these characteristics, with the euphausids appearing to comprise most of the deep-scattering layers. (R.A.D.)

Deep-sea Drilling Project (DSDP). An investigation of the history of the world's ocean basins by means of drilling in deep water from a specially designed ship, the *Glomar Challenger*. The project is supported by the National Science Foundation and is an outgrowth of Project Mohole, which was a deep-drilling project in the early 1960s designed to penetrate to the earth's mantle. Project Mohole was abandoned because of lack of funds and the *JOIDES group took the lead in the field of deep-water drilling. A successful program in 1963 provided deep-water drilling data from the outer continental margin of the southeastern United States.

The *Glomar Challenger*, the first ship designed for deep-drilling of the ocean floor, was completed in 1968. It permitted the start of the Deep-sea Drilling Project, which was administered by the Scripps Institution of Oceanography, University of California. The drilling and core recovery capabilities of the ship exceeded original expectations and its work has continued. The basic pattern of operation consists of two-month cruises with the scientific staff changed after each cruise. The *Glomar Challenger* has covered the entire globe except for the Arctic Sea. The project has provided much important data, including confirmation of the geologic youth of the oceanic crust (less than 200 million years old) as well as establishing the concept of *sea-floor spreading. (R.A.D.)

Deep-sea Research Vehicle (DRV). A highly specialized miniature submarine designed to enable one or two persons to investigate the ocean depths. It evolved from the *Trieste,* an early deep-sea exploration vessel. During the 1960s dozens of different models such as *Deepstar, Alvin* and *Shelf Diver* were constructed. Although most DRVs are designed with depth limitations, a few can descend to the bottom of the deepest ocean trench. Some models are equipped with mechanical arms for taking samples of the bottom, but

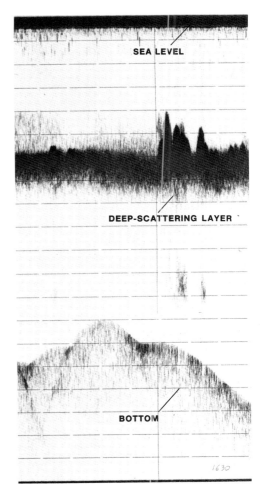

Deep-scattering Layer
In this echogram recorded off Cape Largo, Mexico, the dark band at top represents the outgoing signal, the curving pattern below is the actual profile of the bottom, and the heavy line at center is the "false bottom", or deep-scattering layer. The peaks are probably schools of fish.

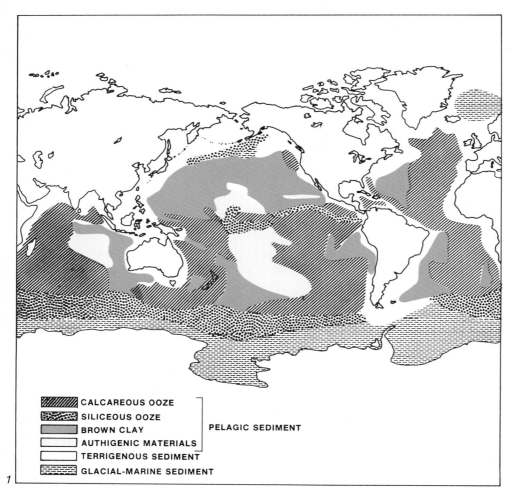

CALCAREOUS OOZE
SILICEOUS OOZE
BROWN CLAY
AUTHIGENIC MATERIALS
TERRIGENOUS SEDIMENT
GLACIAL-MARINE SEDIMENT

PELAGIC SEDIMENT

1

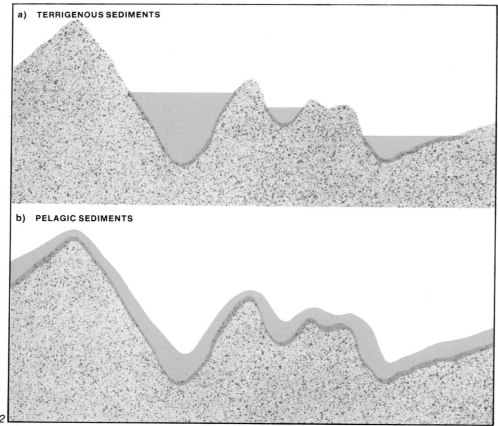

a) TERRIGENOUS SEDIMENTS

b) PELAGIC SEDIMENTS

2

their primary use is to make possible direct observation and photography of the depths. ee Deep-submergence Rescue Vehicle. (R.A.D.)

Deep-sea Sediments. Sediments accumulating on the deep ocean floor beyond the *continental margin or terrace. Although the deep-sea environment is relatively homogeneous, it contains a great variety of sediment types. The two major categories are *terrigenous and *pelagic. Terrigenous sediments are derived from the continents and contain significant amounts of silt and sand. Pelagic sediments consist of clays, volcanic and cosmic particles, *authigenic and *biogenic sediments. Deep-sea sediments are generally fine-grained and accumulate slowly in thin layers, the rate of sedimentation being on the order of 1 cm per 1000 years. (R.A.D.)

Deep-submergence Rescue Vehicle (DSRV). A special type of miniature deep-sea submarine designed for deep-water rescue operations. The primary function of the DSRV is to reduce such loss of life as resulted from the destruction of the American submarines *Thresher* and *Scorpion* in the mid-1960s. Undoubtedly future research missions will also utilize these vehicles. They are small enough to be quickly transported by large planes to any place in the world. See Deep-sea Research Vehicle.

Deflation. The removal of material by wind. This is an important geologic process on sandy beaches, deserts, dry washes in arid regions, among active dunes and in humid areas where vegetation is destroyed by plowing or natural causes. Clay, silt and sand are easily moved by wind. During the Dust Bowl days of the 1930s, dust blown from the Great Plains was carried out over the North Atlantic. Similarly, red dust from the Sahara falls in the Alps and occasionally in England.

Deflation basins (blowouts) are depressions formed by the removal of fine-grained deposits by the wind. A deflation basin covering thousands of square meters and more than 3 m deep has been found in a *playa in the Mohave Desert of California. Many *mesas about 3 m high occur where playa deposits are veneered with gypsum too resistant to be deflated. Big Hollow in the Laramie Basin of Wyoming, 14.5 km (9 miles) across at its widest point and 45 m (150 feet) deep, and the P'ang Kiang hollows in the Gobi Desert of Mongolia, 8 km (5 miles) across and 60 − 120 m (200 − 400 feet) deep, are attributed to deflation. The deflation of sand from dunes sparsely covered with vegetation results in small blowouts near the summit of the dunes. Their lack of vegetation, if formed recently, contrasts sharply with the greener areas around them.

Deflation was also an important process during the *Pleistocene glacial period, when very great quantities of *loess were formed from the fine-grained material blown from *valley trains and *outwash plains. (R.L.N.)

Degradation. In a stream, the process of downcutting, resulting in the removal of rock and loose debris. Sometimes it is synonymous with denudation, but degradation includes only active erosion processes. In streams it results in the flattening of the gradient to a slope at which no further degradation takes place. The amount of degradation depends upon the discharge and velocity of the stream and the type of rock over which it flows. It is the opposite of *aggradation. (M.M.)

Delta. A deposit formed when a river enters a body of water such as a pond, lake or ocean. The sediments pile up to sea level as streams flow over the delta plain, and they grow outward as *distributaries (bifurcating channels) deposit the load at the mouths on the delta front. The material deposited may be gravel, sand, silt or clay. The surface of a delta is a fairly flat plain with lagoons, lakes and marshes.

Distributaries flow through the delta to open water, depositing bars in channel and mouth and building up *levees along the banks. The *thalwegs of distributary channels have extremely low gradients and the channel bottom may be below base level for much of the distributary length. The distributary network on a delta surface is a dynamic, shifting one; new distributaries are always opening up and old ones filling in. Crevassing (breaching) of a distributary levee during floods is part of deltaic evolution whereby the new channel builds outward and the old one is abandoned.

The form of a delta depends upon the predelta coastline and the interrelationship of load, rate of deposition and action of the waves and currents. Most deltas having many distributaries are arcuate (fan shaped) or are shaped like the Greek letter (Δ). The Nile represents the classic delta form and the Niger is a typical arcuate delta. Deltas built up within narrow walls of a submerged mouth, such as on Canada's Mackenzie River, are estuarine. Streams carrying large amounts of fine material maintain dominant single channels with few minor distributaries and tend to build fingers of material along the major channels, which result in digitate ("bird's foot") deltas. The Mississippi, whose load con-

3

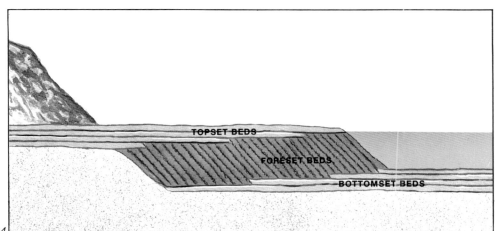

4

TOPSET BEDS

FORESET BEDS

BOTTOMSET BEDS

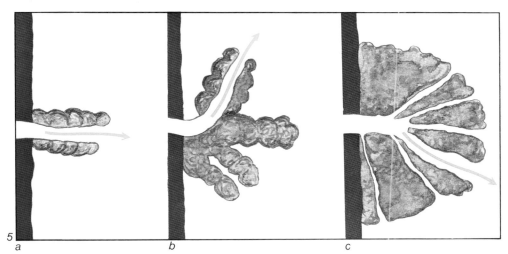

5

a   b   c

Deep-sea Sediment

1 The worldwide distribution of the main groups of deep-sea sediments.

2 Terrigenous sediments (a) derive directly from the land and fill in low areas. Pelagic sediments (b) settle evenly on the ocean floor.

Delta

3 Montana's Flathead River drops much muddy debris at its mouth, forming a delta as it enters Flathead Lake.

4 Cross section of a delta. Coarse material is deposited first, forming a series of slopes called forest beds. Finer material is swept farther along to settle as bottomset beds. Overlying deposits of sediment are topset beds.

5 Stages in the formation of a delta are: (a) A river flows into a lake, creating a channel as it deposits sediment. (b) The river shifts position, abandoning its original channels. (c) During floods the river forms radiating distributaries.

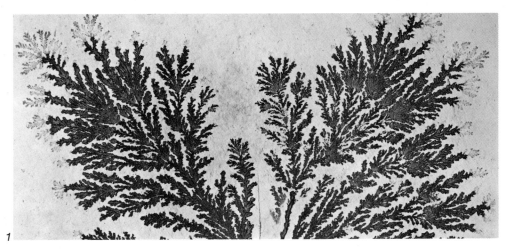

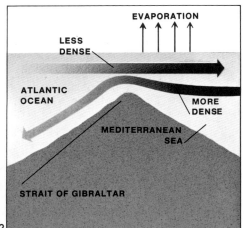

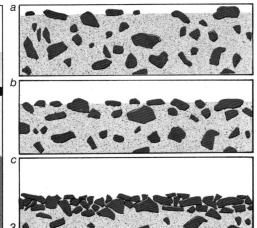

sists primarily of clay-silt, has such a delta. Most present-day deltas are fairly recent, having begun development during the Late *Pleistocene, after stabilization of sea level. (M.M.)

Dendrite. A branching, treelike pattern formed in a mineral or on a rock surface by the crystallization of a foreign mineral. They may occur in fractures in fine-grained *igneous rocks as flat, branching patterns similar to frost crystals on a window pane. They may also resemble plant remains and are commonly mistaken for fossils. Oxides of manganese commonly crystallize in dendritic patterns and are responsible for the mosslike design seen in moss agate.

Density. The degree of compactness of a substance; the mass per unit volume; usually measured in grams per cubic centimeter or in pounds per cubic foot. Sometimes the term *specific gravity is used interchangeably with density but the use of the term "density" properly requires the citation of units. The specific gravity of a substance is the ratio of the density of the substance to the density of an equal volume of water. The density of all substances is compared to the density of pure water at 4°C, which equals 1 gram per cubic centimeter. The density of a mineral is determined by its atomic structure and its chemical composition, and is affected by temperature and pressure. A rise in temperature decreases density while an increase in pressure increases density. (J.W.S.)

Density Current. In oceanography, any circulation which is caused by a density gradient or difference from one location to another. Water masses are typically characterized by a particular density which is susceptible to significant changes as a result of variations in temperature, salinity or suspended sediment. When two water masses converge, gravity induces currents because it causes the denser water to sink under the less dense mass. An excellent example of a density current is the tongue of salty, heavy Mediterranean Sea water that flows through the Straits of Gibraltar, sinks, and spreads several hundred kilometers into the less salty, lighter water of the Atlantic Ocean. A *turbidity current is a special type of density current caused by suspended sediment. Essentially all of the deep-water circulation in the oceans is due to density currents. (R.A.D.)

Deposit Feeder. A great variety of bottom-dwelling invertebrates that ingest sediment. They include burrowers or grazers such as various snails, clams, worms, sea

Dendrite
1 The fernlike appearance of this mineral material results from the precipitation of manganese-oxides from groundwater.

Density Current
2 The Mediterranean, which loses a great deal of water each year through evaporation, has saltier, heavier water than the Atlantic. At the Straits of Gibraltar the denser water of the Mediterranean sinks beneath the Atlantic water and flows into it as a subsurface density current.

Desert Pavement
3 Desert pavement, or deflation armor, develops when: (a) rock fragments are interspersed throughout the soil; (b) wind removes fine particles at the surface; (c) large pebbles and rock fragments become concentrated in the soil; eventually they form a continuous cover, preventing further deflation.

4 Dry winds have swept away all fine-grained particles, leaving this stone-littered surface at Death Valley, Calif.

cucumbers, sea urchins and burrowing shrimp. These invertebrates derive nourishment from organic matter and detritus in the sediment. Nearly all the sediment in some marine environments passes through such organisms and is excreted in pellet form.

Depression. Within a mountain range produced by *folding, a depression represents a low point toward which fold axes plunge. They commonly occur in embayments and are large-scale features encompassing entire regions. Portions of the southern Appalachian Mountains form depressions.

Desalination. The process of extracting freshwater from seawater. Freshwater that can be derived from the sea is a prime natural resource. Although in most cases desalination is still not economically feasible, it is currently in use in several hundred installations around the world. The first desalination plant in the United States, established in 1954, is still in operation at Freeport, Texas. The largest, which is at Key West, Fla., supplies the city with water. Among the various methods of desalination currently employed are freezing of the seawater, ion exchange and, most commonly, distillation. (R.A.D.)

Desert. An area where precipitation is less than the potential evaporation. Deserts can be hot or cold but are always characterized by a relative scarcity of vegetation. They cover approximately 30 percent of the land surface of the world. The occurrence in the stratigraphic record of extensive and thick wind-deposited sandstones and *evaporites such as salt and gypsum indicates that deserts have existed throughout much of geologic time.

Hot deserts owe their existence to either being (1) far from sources of moisture, (2) cut off from moisture by high mountains, (3) in coastal areas along which there are cold-water currents and onshore winds, (4) in high-pressure areas characterized by warm, dry air due to descending air masses. Cold deserts have temperatures so low that the atmosphere cannot hold much moisture. Hot deserts are characterized by sand *dunes, *loess, *ventifacts, *desert pavement (all resulting from wind) and by such features as evaporites, salt lakes, *salinas and desert varnish. Cold deserts are characterized by *solifluction deposits, patterned ground, felsenmeere (see Frost Action) and other periglacial features, and commonly by glacial features. The hot deserts are found chiefly in the southwestern United States, Saudi Arabia, northern and southwest Africa, central Asia and the west coast of South America. The cold deserts are found in Antarctica, Greenland and elsewhere. (R.L.N.)

Desert Pavement (or Deflation Armor, Desert Armor, Lag Gravel). A layer of stones concentrated at the earth's surface due to the removal of silt and sand by wind. The process ends when stones completely veneer the surface and wind no longer has access to fine-grained material.

Devonian Period. The fourth oldest period of the *Paleozoic Era; it began about 345 million years ago. The Devonian System was originally studied in Devonshire in southwestern England, where it is composed of marine sedimentary rocks. In Wales, Scotland and western England, however, the Devonian consists of a thick section of terrestrial deposits known as the Old Red Sandstone. In Early Devonian time North America was a low-lying continent and the seas were quite limited. The Appalachian *geosyncline received sediments throughout most of Devonian time, and rocks of this age occur in the eastern Mississippi Valley and Great Lakes area, northwestern Canada and the Appalachian region. Devonian strata are especially well developed in New York State, where the succession of beds is almost complete and the rocks contain abundant fossils. In western United States there are exposures of Devonian rocks in Arizona, Colorado, Utah, Wyoming, Montana, Idaho and Nevada. Most of these formations are of Middle and Late Devonian age. Devonian rocks also occur in the British Isles, Germany, France and the Soviet Union; China and elsewhere in Asia; southern Africa; Australia and New Zealand; and South America. In North America the end of the period was marked by the climax of the Acadian *orogeny, a mountain-building movement that began in Middle Devonian time. This uplift, accompanied by considerable volcanic activity, raised mountains from the Appalachian region through New England to the Maritime Provinces of Canada.

Climate appears to have been mild and temperate throughout most of the Devonian. The fossil record suggests moderate temperatures as far north as the Arctic regions. Devonian life was characterized by the expansion of the land plants, which ranged from small, leafless species to giant tree ferns as much as 12 m (40 feet) tall. Marine invertebrates were very abundant, consisting of reef-building corals, sponges, *echinoderms, *mollusks and numerous *brachiopods. *Trilobites were greatly reduced in number, but some very large species attained a length of 74 cm (29 inches). The first insects appeared on the land, and freshwater clams and snails were also present. Vertebrates underwent an almost explosive development, and both freshwater and marine fishes were numerous. These included the primitive, jawless *ostracoderms; the jawed, armored *placoderms; and the true sharks. Also present were the sharklike arthrodires, which were heavily armored, had jointed necks and attained a length of 9 m (29 feet). New forms included the lungfishes, a primitive group that possessed gills but had developed lungs as accessory breathing organs. Some representatives of this group are still living today, forming an important link between the gill-breathing fishes and the air-breathing amphibians. Not only was the swim bladder modified to serve as a primitive lung, but some of these fishes developed paired flipper-like fins, enabling them to live out of water for a short period and to have a limited degree of mobility on land. In view of the unprecedented development of these vertebrates, the Devonian has been called the Age of Fishes. Of special importance was the appearance of the first tetrapod (four-footed vertebrate), a primitive amphibian that appears to have evolved from the *crossopterygians.

Certain Devonian rocks are of considerable economic importance, yielding glass sand in the Appalachian region, building stone in Ontario and in New York, Ohio and Indiana, and oil and natural gas in many parts of North America. (W.H.M.)

Diagenesis. The physical and chemical changes that occur in a sediment during and after *lithification. The term has been used with several other meanings, and may refer to the collective processes that affect sediments either after deposition or while deposition is in progress. These changes —which include *compaction, cementation, recrystallization and replacement—operate over long periods of time to produce *sedimentary rocks.

Diamond. A mineral, the native element carbon, generally recognized as the most precious gemstone. The esteem in which a gem is held depends on its abundance and peculiar properties. A stone that is common, no matter how attractive, is not held in high regard. Diamond is far from abundant; to recover a single crystal, miners must remove 10 to 30 million times as much rock. Further, several of its properties combine to make it an attractive stone. Its hardness, the greatest of all minerals, permits it to take a high polish that is undimmed by years of wear. It is further enhanced by a brilliant luster—matched

1

a

b

c

2

only by a few other minerals—that is called *adamantine, a word that refers to diamond. It also has an unusually strong dispersion, as evidenced by flashes of the spectral colors when white light is refracted through a cut stone.

Most diamonds used as gems, which are only 20 percent of those mined, are colorless. The average stone is either off-color (usually some shade of yellow or brown) or so badly flawed that it has value only for industrial purposes. The rare gemstone with a deep color, such as the blue Hope diamond or the yellow Tiffany diamond, commands far higher prices than colorless stones of the same size.

Chemically, diamond is pure carbon, a composition it shares with *graphite, a soft, flaky, black mineral. The vast difference in the properties of these minerals arises from arrangement of their carbon atoms and the forces holding them together. In graphite, sheets of carbon atoms are held to one another by the weak van der Waals' bond, but in diamond each carbon atom is held to four others by a strong covalent bond (see Bond, Chemical). Diamond is isometric; is usually found in crystals, most frequently as an octahedron; and has a perfect octahedral cleavage. Thus, in spite of its hardness, it is brittle and cleaves easily parallel to one of the four octahedral directions. The gem cutter takes advantage of this in cutting any sizable stone.

Diamond, one of the most chemically inert substances, is insoluble in acids and alkalies. Because of this, as well as its hardness and high specific gravity (3.52), in its natural state it can be transported long distances by flowing water and can accumulate in deposits called *placers. All the early diamonds came from such alluvial (waterborne) deposits, first from India, then beginning in 1725 from Brazil, and in 1866 from South Africa. It was not until 1869 near Kimberley, South Africa, that diamonds were found in the rock in which they formed. As mining of these deposits progressed, it was found that *kimberlite, the rock containing the diamond, which had a roughly circular surface expression extended downward in pipe-shaped fashion. Many "diamond pipes," mostly in Africa but some also in Siberia, are today actively worked, including the world's largest diamond mine, the Premier in the Transvaal, South Africa. But production from diamond pipes merely supplements the production from alluvial deposits. Nearly all the diamonds produced in Zaire, the country with the largest production, are recovered from placers. However, only about 5 percent of the Zaire diamonds are of gem quality, whereas in South-West Africa more than

80 percent of the 100,000 carats of diamonds taken each month from ancient beach sands are of gem quality.

Because of their high hardness, diamonds have many important industrial uses. As a fine powder they are used to polish diamonds and other gemstones. Larger crystals are used in diamond drills, an indispensable tool in obtaining rock samples from deep within the earth's surface. Metal discs with diamond-impregnated rims (diamond saws) will cut the hardest of materials: Saws an inch or two in diameter can cut gemstones while saws several feet in diameter can slice granite blocks. Diamond dies, consisting of single crystals with holes of fixed diameter drilled through them, are used for drawing wire; the diamond is so resistant to wear that the diameter of the many miles of wire drawn through the die remains constant. See Mineral Properties; Appendix 3. (C.S.H.)

Diaspore. A mineral, hydrous aluminum oxide. In a fine-grained massive form it is a major constituent of *bauxite, the ore of aluminum. As tabular crystals it is associated with *corundum in emery rock and in chlorite schist.

Diastrophism. All major movements of the earth's *crust, including the movements responsible for the formation of mountains, ocean basins and deep ocean trenches.

Diatom. A minute, one-celled aquatic plant that secretes a delicate, siliceous bivalved exoskeleton. Known from *Jurassic to the *Holocene (Recent), the shells of diatoms accumulate in large numbers on the bottom of present-day seas to form thick deposits of diatomaceous ooze. Fossil diatom oozes may be altered to form a dense, hard, siliceous rock called diatomite. Diatoms also comprise the bulk of diatomaceous earth. This very fine-grained, white, yellow or light gray powdery rock is the unconsolidated equivalent of diatomite. It is used as an absorbent, an abrasive, a filtering material and an insulator. Major deposits occur in California, Oregon, Washington and Nevada; Brazil, Germany and Algeria are also large producers. (W.H.M.)

Diatreme. An elongated pipelike structure consisting of breccia. See Breccia Pipe.

Dichroism. In crystallography, the property possessed by some crystals of showing two different colors when polarized light is transmitted through them in certain crystallographic directions. See Pleochroism.

Differential Thermal Analysis. A means of determining the temperature and magnitude of thermal reactions in crystalline materials. The method involves the continuous heating of a sample to an elevated temperature and recording the differences in temperature between it and a sample of inert material. The record, usually made automatically on a moving chart, shows whether an endothermic reaction (absorption of heat) or an exothermic reaction (liberation of heat) has taken place. During an exothermic reaction the reacting substance is briefly raised to a temperature higher than that of the inert sample; the reverse is true during an endothermic reaction. This method is particularly useful in studying clay minerals. (C.S.H.)

Dike. A table-like body of rock formed from the solidification of molten material (*magma) injected into the overlying *crust from deep within the earth. The magma is injected into fractures that cut across any planar structure within the intruded rock. Dikes may vary in size from a few millimeters in width and a few centimeters long to hundreds of meters in width and even hundreds of kilometers in length. Dikes are divided into categories based on their spatial arrangement and number. Groups of parallel dikes are known as sets; numerous parallel dikes are termed swarms. One such area is located near the Firth of Clyde in southwest Scotland. Other dikes are associated with centers of volcanic activity and radiate from a central point. In south-central Colorado a spectacular array of radiating dikes is found in the Spanish Peaks. (J.W.S.)

Dike Ridge. A topographic feature formed when a dike is more resistant to erosion than the *country rock. If the dike is less resistant, a trench will be formed. If both have the same resistance, both will remain at the same height. A remarkable group of more than 500 dike ridges—from less than 1 m to 18 m (between 2 and 60 feet) wide, up to 40 m (125 feet) high and as much as 24 km (15 miles) long—is found around the Spanish Peaks in southern Colorado. A spectacular dike ridge several kilometers long and about 15 m high radiates from Ship Rock in New Mexico. (R.L.N.)

Dimorph. In crystallography, a chemical element or compound found in two distinct crystal structures. Examples of dimorphous minerals are diamond and graphite, both of which are carbon; and pyrite and marcasite, both of which are iron sulfide. See Polymorph.

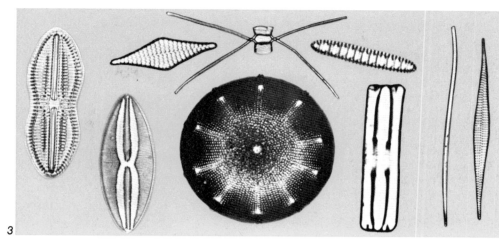

3

4

5

Diamond

1 This crystal shows an octahedral form with many triangular growth marks.

2 Three common forms of diamond crystals are (a) an octahedron modified by a hexoctahedron (a form with 48 faces), (b) an hexoctahedron with curved faces, (c) an octahedral twin crystal.

Diatom

3 These microscopic algae live in either fresh or sea water. Some of their shells are shown here. When the organisms die the shells settle and form rock.

Dike

4 Magma forced from deep within the earth's crust has enlarged and filled a crack to form this dike in a marble ledge.

Dike Ridge

5 This long wall of basalt in Colorado's Sangre de Cristo mountains became exposed when the rock on both sides of a basaltic dike was eroded away.

Dip

1 The slope of tilted beds measured at right angles to the strike is called the dip; the direction of a horizontal line on the bedding plane of the exposed outcrop is the strike. In this diagram the beds strike north and dip 40° east.

Discharge Rating Curve

2 In determining stream flow, the height of the water surface is recorded by a gauge in the stream (a). Discharge is determined from the area of the stream's cross section and the velocity. One way of measuring discharge is to suspend a current meter from a cable car at several points. From these measured discharges, a rating curve (b) is constructed that permits stream discharge to be estimated directly from gauge height.

Dodecahedron

3 The dodecahedron, a crystal form with 12 faces, is a common form for many minerals crystallizing in the isometric system.

Divergence

4 A mass of sinking air (a), spreads out or diverges horizontally (b) as it approaches the earth's surface. Because of the earth's rotation, the diverging air develops anticyclonic spin (c) (clockwise in the Northern Hemisphere; counterclockwise in the Southern Hemisphere).

5 In a column of upward air motion (a) the air converges (b) near the lower boundary of the atmosphere. As it converges it develops cyclonic spin (c) (counterclockwise in the Northern Hemisphere; clockwise in the Southern Hemisphere).

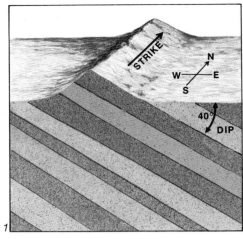

1

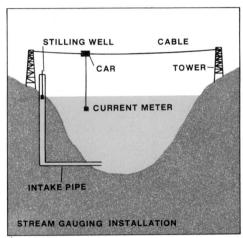

d

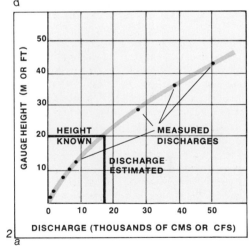

2
a

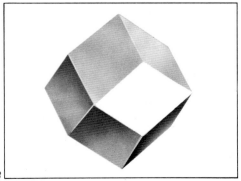

3

Dinosaur. See Fossil Reptile; Jurassic Period; Triassic Period.

Diopside. A mineral in the *pyroxene group, calcium magnesium silicate, which forms a complete solid solution series with hedenbergite, calcium iron silicate. The color, white to light green, deepens with increase of iron. Diopside is a *contact metamorphic mineral found in crystalline *limestones and *dolomites. See Mineral Properties.

Diorite. A coarse-grained *igneous rock composed essentially of *plagioclase feldspar (oligoclase-andesine), *hornblende and *biotite. Usually dark minerals are present in sufficient amount to give the rock a dark color. Diorites are found in many parts of the world and usually occur in relatively small circular bodies called *stocks or in tabular bodies called *dikes.

Dip. The spatial orientation of any layering or planar feature, measured in degrees with respect to an imaginary horizontal plane. When measured in a vertical plane perpendicular to the *strike of the planar feature, it is known as true dip. Measurements made in any other plane not perpendicular to the strike will result in values less than true dip and are referred to as *apparent dip.

Dip Joint. A *joint that strikes parallel to the dip of a planar feature.

Discharge. The amount of water flowing down a stream channel. It is measured in cubic meters per second (cms) and is defined as equal to the cross-sectional area multiplied by the mean velocity of water through the section. The formula is $Q = AV$, where Q is discharge in cms, A is channel area (width times depth) in square meters and V is mean water velocity in meters per second.

Discharge Rating Curve. A curve used in estimating streamflow. The depth, width and velocity of a stream channel are measured for a number of discharges at a given point on the riverbank. *Discharge is calculated (using $Q = AV$) and a graph is plotted relating discharge to depth of water. Using the curve, discharge can be read for any depth. The curve is only valid if the channel is stable and does not change its depth or width. See Discharge.

Discontinuity. In general terms, an abrupt change in a physical property or process. Geophysicists apply the term to zones within the earth in which the velocity of *seismic waves changes abruptly as the waves propagate down through the earth.

To geologists studying the processes by which sediments are deposited in ocean basins, a discontinuity represents a period of time in which sedimentation has ceased and/or erosion has occurred, producing a break in the sedimentary record. In structural geology discontinuities represent abrupt changes in rock types, caused for example by *faulting or mountain-building activity. See Conrad Discontinuity; Mohorovicic Discontinuity. (J.W.S.)

Discordant. A term describing the structural relationship of a body of rock originating from molten material to any planar feature inherent in the intruded rock. Discordant intrusive bodies cut across planar structures and generally take the form of *dikes, *cone sheets, *batholiths and *stocks.

Dislocation. The process by which brittle rock is shattered and then reduced by crushing or grinding.

Dispersion. In optical crystallography, the separation of white light into its respective colors. Also the variation of optical properties, particularly refractive index, with the wavelength of light.

Distributary. See Delta.

Divergence. In meteorology, the net flow of air out of a given volume. At the earth's surface, air diverging horizontally requires the downward movement of air from aloft. Conversely, *convergence of air at the earth's surface requires upward motion, since the air's compressibility is relatively small.

*Cyclones or low-pressure areas are characterized by a net inflow of air near the earth's surface; this inflow produces predominantly upward motions, while a net outflow of air aloft permits the inflow to continue. During the development of a cyclone, the divergence aloft more than compensates for the inflow at lower levels. *Anticyclones are characterized by winds that blow outward from a center of high pressure near the earth's surface. In this case, the influx of air above the anticyclone more than compensates for the divergence lower down while the high-pressure area is developing.

The development and movement of cyclones and anticyclones are closely related to divergence and to vorticity. In medium- or large-scale motion systems in the atmosphere, convergence is normally associated with an increase in cyclonic vorticity. Divergence, on the other hand, results in a decrease of cyclonic vorticity

and the development of anticyclonic spin. The relationship between divergence and vorticity or spin has an analogy in the way a spinning dancer controls his speed. When the dancer draws in his arms (convergence of mass), his speed of rotation increases; by extending his arms (divergence of mass), the dancer decreases his rate of spin. (M.F.H.)

Divide. The boundary between one drainage basin and another. Rain falling on one side of the divide flows in one direction and into one group of streams whereas rain on the other side flows in another direction and into another group of streams. The divide is generally, but not always, marked by a ridge of higher ground, the highest part of the drainage basins it separates. Stream piracy may cause drainage divides to shift position. The continental divide of the United States marks the division between streams flowing eastward into the Gulf of Mexico (Atlantic Ocean) and those flowing westward to the Pacific. (M.M.)

Dodecahedron. A form of the *isometric crystal system composed of twelve faces. In the geometrically perfect rhombic dodecahedron each face is a rhomb; in the pentagonal dodecahedron, more commonly called the *pyritohedron, each face is a pentagon.

Doldrums. An equatorial trough of low pressure lying roughly between 5°S and 5°N latitude. The doldrums were formerly called the equatorial belts of *winds and calms. Because of the absence of strong pressure gradients (variations in pressure which induce a flow of wind), this zone has no prevailing surface winds. Calms prevail as much as a third of the time and violent squalls are common.

Dolomite. A mineral, calcium magnesium carbonate. Although the amount of calcium and magnesium may vary, the composition usually corresponds closely to the formula $CaMg(CO_3)^2$. Dolomite occurs in rhombohedral crystals, commonly with curved faces that characterize the mineral. The luster of some varieties is pearly, giving rise to the name "pearl spar." The color is usually some shade of pink but may be colorless, gray, brown or black. Although dolomite crystallizes in the rhombohedral crystal class with lower symmetry than *calcite, the two minerals have similar rhombohedral cleavage and also resemble each other in massive form. However, they may be distinguished readily by the fact that dolomite does not effervesce in cold dilute hydrochloric acid.

The name dolomite, as it was first used,

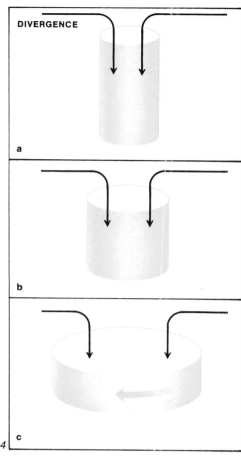

DIVERGENCE

a

b

c

4

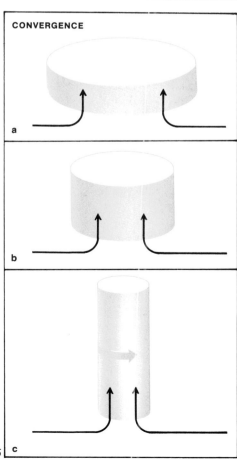

CONVERGENCE

a

b

c

5

referred to a *sedimentary rock. It was later applied to the mineral when it was discovered that both rock and mineral were the same chemically. It is as a sedimentary rock and the metamorphosed equivalent, dolomite marble, that the mineral is found most abundantly and extensively. As such it occurs in many parts of the world, including the Dolomite Alps of northern Italy. As a rock, dolomite is believed to have formed from *limestone by the replacement of half of the calcium by magnesium.

Frequently the replacement is incomplete and the dolomite rock is composed of a mixture of the minerals calcite and dolomite. Dolomite, usually in recognizable crystals, occurs as a *gangue mineral in lead and zinc veins associated with calcite, barite and fluorite. See Mineral Properties; Appendix 3. (C.S.H.)

Dome. An upfolded body of rock in which the layered elements such as *bedding dip away from the crest of the uplift like a series of inverted matching bowls. Domes may be circular or oval and may cover many square kilometers. They result from upward-thrusting forces from within the earth's *crust and are classified according to their origin. Salt domes are formed by plastically flowing *rock salt that rises from within the crust because its *specific gravity is less than that of the surrounding rock; it acts like a punch and deforms the overlying crust into a dome. Salt domes are of importance to economic geologists because they contain sulfur as well as salt and form traps in which oil and gas may accumulate. They have been found in the Gulf Coast of the United States, beneath the Gulf of Mexico, in Colorado and Utah, and in such other places as the Soviet Union, France and the Middle East.

Domes may also form from the injection of molten rock (*magma) between layers of existing crustal rock. The magma forces the weaker overlying crust to buckle upward into a dome. Such volcanic domes are formed when magma accumulates in a restricted area because it is too viscous to flow any great distance. Other rather oval-shaped domes may form when compressional forces cause the crust to buckle. Such domes are generally called *anticlines. See Cumulo-Dome; Fold. (J.W.S.)

Double Refraction. The property of transparent *crystals (other than those of the isometric system) of splitting a beam of ordinary light into two mutually perpendicular polarized rays that traverse the crystal at different velocities. A random section of such a crystal will have two refractive indices, one associated with each ray.

Dolomite
1 The classic dolomite location is the Dolomitic Alps, located on the border between northern Italy and Austria.

Double Refraction
2 This transparent calcite block splits transmitted light into two rays, as the reflected double image shows.

Drainage Basin
3 The Mississippi and its tributaries form a vast drainage basin.

4 Stream patterns are affected by the rock on which they develop. Dendritic patterns (a) occur when streams flow over flat strata. Trellis patterns (b) are common where the edge of sedimentary rocks form parallel belts. Parallel patterns (c) develop on coastal plains and tilted plateaus. Radial patterns (d) occur when streams flow from a high central zone. Centripetal patterns (e) flow in toward the center of a basin. Rectangular patterns (f) result from joints and faults in the rock.

Drainage Basin (or Watershed). The area drained by a given river and its tributaries. Precipitation falling over the region flows out through the main stream. The basin is bounded by a divide or *interfluve from adjoining watersheds. Most drainage basins debouch directly or eventually into the oceans. Those that do not are said to have interior drainage and debouch into lakes.

Drainage Density. A measure of dissection of a *drainage basin (watershed). It describes the relative spacing of a stream network, expressed as the ratio of total length of streams in a watershed to area of the basin. Drainage density depends upon the climate, the geology and the physical characteristics of the drainage basin. Climate affects the vegetation and stream discharge. Rock and soil types determine the resistance of the basin to erosion and amount of infiltration. Topography and relief influence both runoff and the ability of the stream to erode the surface. In general, a region underlain by sandstone has a lower drainage density than one underlain by shale. Low drainage density (3 − 4 km of channel per square km) denotes a region of coarse texture and is typical of watersheds underlain by resistant rock, such as in the Appalachian Plateau. Medium drainage density basins (8 − 16 km per square km) are found in the humid middle and eastern United States on many different rock types. High drainage density (30 − 100 km per square km) characterizes basins underlain by weak rock in semiarid regions. Ultrafine texture with drainage density greater than 200 km per square km is characteristic of badlands. (M.M.)

Drainage Patterns. The geometric arrangement of a channel network, including a main stream and all its tributaries; a spatial configuration of the drainage of a basin. There are a number of types of drainage patterns. A *dendritic pattern is a treelike, branching arrangement in which the network is randomly oriented with no dominant control over the direction of flow. The dendritric pattern is common in the midwestern United States, where the streams form on horizontal sediments. A parallel drainage pattern is formed by a series of streams flowing in essentially the same direction, generally down a pronounced slope. These types are found on coastal plains or tilted plateau regions. A trellised pattern is found in regions of tilted or folded sediments such as in central Pennsylvania. Here major valleys tend to be elongated and located along the strike of the nonresistant layers, with shorter tributaries flowing in at right angles. This kind of pattern may also be

2

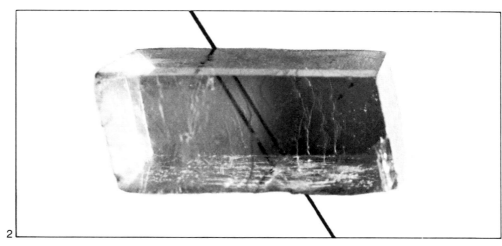

3

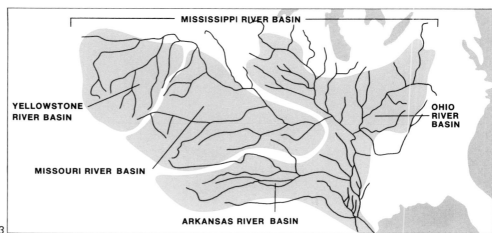

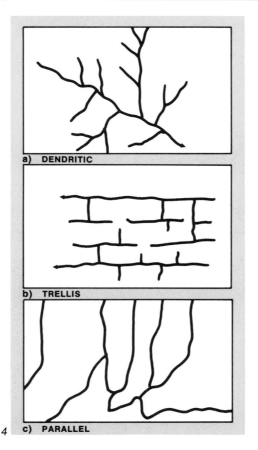

a) **DENDRITIC**

b) **TRELLIS**

c) **PARALLEL**

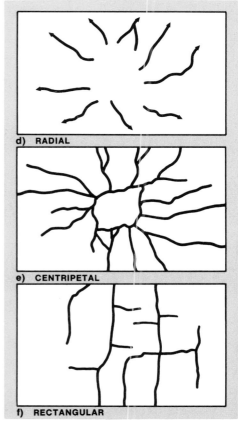

d) **RADIAL**

e) **CENTRIPETAL**

f) **RECTANGULAR**

4

1

2

3

formed by drainage in a glaciated area where the streams tend to flow between oriented hills of glacial deposits, such as in the drumlin area of New York State near Palmyra. Annular drainage is found where streams erode the weaker tilted rocks around a *dome, creating a concentric trellised pattern, as in the Black Hills of North Dakota and the Weald in England. Rectangular drainage occurs where systems of joints or faults control the direction of streamflow. Examples are found in Zion National Park. Radial drainage is a pattern where valleys drain outward from a common center. Controlled by slope, the network is found on domes and on volcanoes such as Fuji in Japan and Mauna Loa on Hawaii. Centripetal drainage occurs where streams flow in toward the center of a basin. Such a pattern is also controlled by surface slope and is found in large-size structural *basins, in *craters and in *calderas. It is also characteristic of basins in arid or semiarid regions where drainage moves toward the interior basin. (M.M.)

Dredge. In oceanography, heavy-duty apparatus designed to take samples from the ocean bottom. Usually towed behind a ship, the dredges are generally basket-shaped devices made of heavy mesh or interwoven chains. They are dragged along the ocean floor to collect large rocks, *manganese or phosphate nodules, or even to break off pieces of *bedrock. They are among the most useful research devices available to oceanographers.

Dreikanter. See Ventifact.

Drift Bottle. A container designed to determine patterns of ocean currents. The bottles, released at a predetermined location at sea, contain ballast and a card that offers a reward to anyone who returns it with the date and location of the recovery. A chart showing where and when the bottles have been found permits determination of general current directions and velocities. Plastic floating cards are now also used in mapping currents. (R.A.D.)

Drip Stone. See Stalactite.

Drogue. In oceanography, a submerged device that moves with the current in order to measure both velocity and direction of water movement. Commonly it consists of a cross, parachute or other device that presents resistance to water. The drogue is connected by a fine wire or nylon line to a surface float, which may be equipped with a flag, light, radar reflector or radio transmitter to facilitate tracking as well as observation.

Drowned Valley. The lower end of a valley flooded by an ocean, lake or reservoir. It may result from a sinking of the land, a rise of sea level or a combination of both. Examples are the Delaware and Chesapeake bays where the lower ends of the Delaware and Susquehanna river valleys were drowned. The Thames Valley and Bristol Channel in England and the Río de la Plata in South America are also drowned valleys. Estuaries are valleys drowned by the ocean and are characterized by a mixing of fresh and salt water.

Drumlin. A smooth, elongated, streamlined hill formed by glacial ice and composed essentially of *till. The long axis may be two or more times as long as the shorter axis and is parallel to the direction of glacial motion. Drumlins may be 60 m high and more than 1.5 km long. Single drumlins are the rule, but coalescing ones are not uncommon. Clusters containing scores or even hundreds of drumlins with their long axes parallel are found in upper New York State, Massachusetts, Wisconsin, Michigan, Nova Scotia, Northern Ireland and elsewhere. Two famous drumlins in the United States are Bunker Hill and Breed's Hill.

Although it is agreed that drumlins are formed beneath actively moving ice, some geologists think they result from glacial erosion of older drift, others think they have been formed by the accretion of till. (R.L.N.)

Druse. A rock cavity lined with minute crystals. The cavity is usually small and the encrusting crystals are said to be drusy, as for example, drusy quartz.

Duboys Equation. An equation used to calculate the *tractive force needed to entrain loose grains lying on a stream bed. It is valid only for small particles or when the stream velocity is low.

Dune. A mound, hill or ridge of windblown sand. Dunes are found where there is a source of sand, a wind strong enough to carry it and a land surface on which to deposit it. These conditions are commonly found inland from sandy beaches with onshore winds, close to streams with sandy bottoms exposed during the dry season, and in desert areas where the disintegration of sandstone and other rocks provides the sand. Dunes may be as little as 1 m or over 100 m high. Unless fixed by vegetation, they migrate because sand is blown from the windward to the leeward side, but their movement is slow—usually less than 30 m per year. Dunes may bury forests, houses and even whole villages,

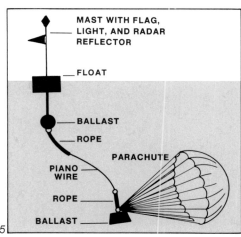
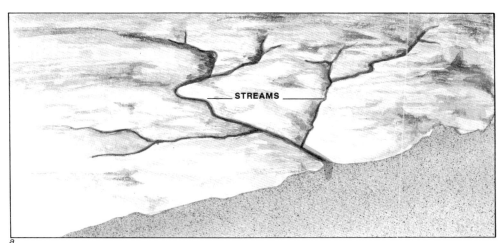

Drainage Patterns
1 Treelike dendritic streams flow across the shore of California's Salton Sea.

2 Aerial view of a stream offset where it crosses the San Andreas Fault.

Drumlin
3 An ice sheet molded deposits of hard glacial till into smooth hills, or drumlins.

Drift Bottle
4 To study currents, drift bottles are set afloat containing a card requesting the time and place of recovery.

Drogue
5 Currents near the surface can be traced by lowering a parachute to a known depth, where its resistance causes it to move at the same velocity as the water.

Drowned Valley
6 A coastal stream-valley system (a) has become inundated (b) through a rise in sea level or crustal subsidence.

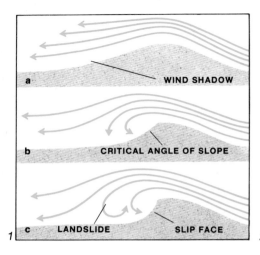

which may later be uncovered if the dune moves on.

Dunes vary in shape depending on such factors as the quantity and distribution of the sand and the strength and direction of the wind. Transverse dunes are ridges perpendicular to the wind. Barchans are crescentic dunes characterized by a crest that separates a flat windward slope from a lee slope at the *angle of repose, and by *horns that point downwind. Fore dunes are sand ridges parallel to a beach from which the sand is derived. In general, fore dunes do not move very far inland, since they are fixed by vegetation.

The dunes of the Arabian Peninsula, the Sahara and the Kalahari of southwest Africa are among the largest and most famous in the world. Coastal dunes occur along the European coasts, the Atlantic coast from Massachusetts to Florida, and on the Texas, California, Oregon and Washington coasts. There is an area of dunes in Indiana south of Lake Michigan; and dunes are common in the Southwest in association with dry washes and lake beds. The White Sands National Monument in New Mexico and the Great Sand Dunes National Monument in Colorado contain two of the most spectacular dune fields in the United States.

A sand sheet is a layer of windblown sand so thin that it reflects the topography on which it is deposited. It may cover an area of a few thousand square meters or, as in the Libyan Desert, hundreds of square kilometers.

Eolian sandstones resulting from the consolidation of dunes have been formed in many geologic periods. A well-known example is the Permian Coconino sandstone found at and around the Grand Canyon; it covers 83,000 square kilometers (32,000 square miles), is as much as 300 m thick and has a volume of approximately 12,000 cubic kilometers (3000 cubic miles). Since this area is much more extensive than any area of dunes now found in the United States, the Conconino desert of Permian time must have been similar to today's Sahara and the Arabian desert. (R. L. N.)

Dune

1 *The development of a slip face on a dune occurs when wind, blowing across the dune crest, eventually creates a wind shadow (a) in the lee of the dune. (b) Sand falls until a critical angle of slope is reached. (c) Then a landslide occurs and the slip face is formed.*

2 *Transverse dunes, such as this one at Gironde, France, are common along coasts and lake shores. They align at right angles to the wind direction.*

3 *Crescent-shape dunes called barchans, such as these at El Paso, Texas, are formed in desert regions with a small amount of sand and moderate winds blowing in one direction.*

4 *In Saudi Arabia long, narrow ridges of sand called seifs increased in height and width due to cross winds and grew in length due to winds blowing parallel to the seifs.*

Dunite. A variety of the rock type peridotite, composed almost completely of *olivine but containing small amounts of *pyroxene and *chromite as accessory minerals. Dunite masses in North Carolina, South Carolina and Georgia are of importance because of commercial deposits of *corundum associated with them. Larger bodies are present in the Ural Mountains, U.S.S.R., and in New Zealand, notably in the Dun Mountains, for which the rock is named.

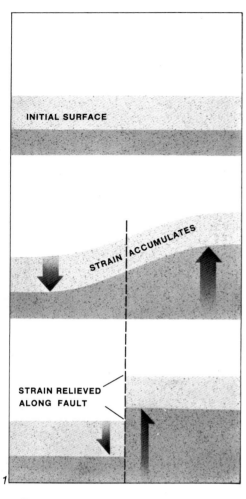

INITIAL SURFACE

STRAIN ACCUMULATES

STRAIN RELIEVED ALONG FAULT

1

Earthquake

1 When internal pressures push the crust in opposite directions it is slowly bent into an S-shape. According to the theory of elastic rebound, energy accumulates in the bent rock until it snaps, creating a vertical fault scarp and releasing energy into the earth's crust in the form of seismic waves.

2 Dust rises in Sycamore Canyon near Awin, Calif., during an aftershock of the 1952 Tehachapi quake.

that bonded the two segments together. At the point of strain release, the adjacent blocks, Reid said, would snap past each other, releasing enough energy to cause an earthquake. The elastic rebound theory is now accepted as an explanation of shallow earthquakes, and with some modifications, possibly of deeper earthquakes.

Because of the destruction and an average annual loss of 14,000 lives caused by earthquakes, a large effort is being made to predict and control them and to reduce hazards associated with them.

One method suggested for controlling or perhaps modifying large, destructive earthquakes involves drilling a series of wells along a seismically active fault and pumping water into these wells where it could migrate along the fracture zone. The water would act as a lubricant, helping the bodies of rock adjacent to the zone slip past each other in a series of small jerky motions rather than one major displacement. The smaller motions would cause earthquakes of intensities too small to cause property damage or loss of life. See Earthquake Prediction. (J.W.S.)

Earthquake, Measurement of. The determination of the relative energy released by earthquakes according to two scales in standard use. The first scale, called the Mercalli Scale (after an Italian scientist, Giuseppi Mercalli, who developed it in 1902), is designed to measure intensity as determined by the effects that earthquakes have on man and on man-made objects. Numbers on the scale range from I (for an earthquake detectable only by sensitive instruments) to XII (for an earthquake producing nearly complete destruction). An intensity number on the Mercalli Scale assigned to a place near the source of an earthquake indicates roughly how much the earth shook at that location. For each earthquake, such numbers are determined in many places and the highest value recorded is generally given as the intensity of the earthquake.

The second scale is known as the earthquake magnitude or Richter Scale. It is based on the amount of energy released at the earthquake's focus as recorded by *seismographs. The Richter Scale therefore describes an earthquake independently of its effect on locations of objects on the earth's surface, and thus there is only one number for each earthquake. Developed by the American seismologist Charles F. Richter, the scale has a range of values from −3 to 9. The smaller numbers represent earthquakes of very low energy that are barely detectable on seismograph records. The numbers at the other end of the scale represent earthquakes capable

of severe destruction such as the Alaska earthquake of 1964, the most powerful recorded in North America, which registered 8.4 on the Richter Scale. (J.W.S.)

Earthquake Prediction. The techniques of forecasting earth tremors and estimating the time of their occurrence, their magnitude and the location of their origin. This rapidly developing technology is based largely on recorded changes in rock properties preceding *earthquakes. Within the past few years, several related phenomena have been monitored as earthquake predictors. They include (1) increases in electrical conductivity of the ground; (2) changes in chemical composition of groundwater from deep wells; (3) slight vertical and horizontal movements of the earth's surface; (4) seismological indicators such as stress orientation changes, shifts in velocity and frequency composition of *seismic waves, and increased microseismic activity.

Earthquakes are tremors resulting from movements that occur as rocks fracture and shift along *fault planes. Prior to fault movement, stresses in the rocks gradually build up and the physical properties of the rocks change. These changes alter the seismic patterns that precede the earthquake and, if recognized, can serve as a signal of the imminent movement.

For two decades, Soviet and Japanese scientists have been most active in investigating the seismic patterns preceding earth tremors. At the same time United States scientists have investigated extensively the mechanical properties of rocks in laboratory studies in which rocks are subjected to deforming stresses. A slight volume increase in the rocks, called dilatancy, is the apparent cause of stress reorientation and seismic-wave changes. As stress increases and the rocks approach rupture, microscopic cracks and tiny voids develop, resulting in total volume increase. But evidence that this occurs underground is still inconclusive.

One hypothesis is that as underground rocks dilate, minute cracks form. Groundwater gradually seeps into these openings and the fluid pressure in the rocks rises. The rocks will then fracture more readily. An alternative hypothesis suggests that dilatancy is more likely to occur in fluid-filled, porous rocks. Water is drained into the new cracks that develop and overall fluid pressure is therefore reduced. This tends to delay pending fault movement until enough groundwater flows into the region to return fluid pressures to their previous levels. Whatever the mechanism, dilatancy and groundwater migration offer a possible explanation of the

features observed as earthquake precursors. Because electrical conductivity depends largely on the amount of water in the rocks, conductivity would rise as dilatancy develops. Likewise, groundwater flow increases during dilatancy, and this could promote the release of radon, a natural gas formed from the activity of short-lived isotopes. The microscopic cracks could produce the microseisms that are known to increase in intensity prior to an earthquake. As cracking takes place and fluid pressure changes, stress reorientation would also occur.

The most widely used predictor of earthquakes is the variation in seismic-wave velocities first noted by Russian scientists and then by U.S. seismologists. Even the slightest rock movements produce two basic forms of sound waves: compressional (P) waves and shear (S) waves. They differ in their movement and consequently in the speed with which they pass through rocks. Normally P waves travel about 1.75 times faster than S waves. However, in the region of an impending quake, the ratio of the velocities of P waves to S waves (Vp/Vs) has been observed to decrease some weeks, months or even years before the quakes. Gradually, the ratio Vp/Vs increases until it reaches essentially its normal level just prior to the earthquake. The explanation may be related to the microcracks not filled with water, since the velocity of P waves decreases in rocks not saturated with fluid. Some seismologists foresee monitoring of P-wave velocities as a way to predict earthquakes. Furthermore, the time interval between the drop in Vp/Vs ratio and the quake appears related to magnitude; longer time intervals signal a stronger earthquake.

Fluid pressure may also be a key not only to earthquake prediction but perhaps to earthquake control. U.S. Geological Survey scientists at the Rangely Oil Field, Colo., tested the role that fluid pressure might play in triggering earthquakes. The area was already under fluid pressure well above normal because water had been injected into the oil field as part of a secondary recovery program. Pressure was first reduced by removing fluid, resulting in a marked decrease in the frequency of small earthquakes. Then water was reinjected into the field and the pressure again raised. A substantial increase of small quakes followed. Now being explored is the feasibility of controlling fluid pressure in earthquake regions, especially heavily populated areas such as along the San Andreas Fault in California. This could help reduce the extent or magnitude of quakes.

New insight into crustal movements on a global scale has helped to explain earth-quakes. Recent theory holds that the earth's crust and uppermost mantle are divided into gigantic segments, called plates, that are slowly moving over the earth's surface. Plate boundaries mark the areas of concentrated earthquake activity produced by movements of one plate against another.

Despite the important new information about earthquakes, accurate prediction still presents problems. There is no assurance that conditions applicable to the smaller, shallow earthquakes apply to the very large events or those generated at greater depths in the crust. Moreover, an extensive network of seismic stations is necessary to monitor earthquake precursors in any area. China, which is particularly vulnerable to earthquakes, has set up an extensive system of 250 seismic stations and deep-well monitoring points; it has reported some successful earthquake prediction, including the evacuation of a town of sizable population before a major earthquake, through the monitoring of radon increase in deep wells in combination with seismic-wave monitoring. In the United States a network of 100 seismograph stations is now in operation along a 300-km segment of the San Andreas Fault in central California. Computerized systems allow accurate location of each small earthquake in a matter of minutes. Anomalies in the relative velocities of P and S waves are also being recorded. These data may enable scientists to predict large quakes in a densely populated earthquake area. Even if technology cannot prevent such events, it may afford the warning that will allow precautions to be taken. (R.E.B.)

Earth's Planetary Motions. Such motions as the earth rotating on its axis at speeds of more than 1100 km/hr at the Equator, and at the same time revolving around the sun at 106,000 km/hr. These motions produce changes experienced on earth in the form of day and night and seasons of the year.

Day and night are caused by the earth's rotation with one complete turn of 360° on its axis every 24 hours. Only half of the earth faces the sun; the other half is in darkness, and as the earth rotates from west to east, the first appearance of the sun each day is to the east, the sun appearing to rise in the east and set in the west.

The axis of the earth's rotation is inclined at 23½° from the perpendicular and therefore makes an angle of 66½° with the plane of the earth's orbit. This inclination of the axis of the earth results in changes in the length of day and night as the earth revolves around the sun. Unequal distribution of sunlight because the earth is inclined causes the hemisphere that leans toward the sun to have longer days than nights and the hemisphere leaning away, shorter days than nights. Lengths of day and night change gradually, and on the day known as summer solstice (June 21 or 22) the North Pole leans the full 23½° toward the sun, providing the longest day of the year in the Northern Hemisphere and the shortest daylight for the Southern Hemisphere. At summer solstice daylight increases from 12 hours at the Equator to periods of weeks within the Arctic Circle. At the North Pole the "day" is six months long. Winter solstice (December 21 or 22), the shortest day in the Northern Hemisphere, occurs when the North Pole leans 23½° away from the sun.

Two days each year, midway between the solstices, the earth's axis is perpendicular to the plane of its orbit and the sun's rays extend an equal distance north and south of the Equator. This happens on March 21 or 22 and September 22 or 23. Each date is known as an equinox, meaning equal night, because day and night are then equal in length over the entire earth.

The earth undergoes a wobbling motion as it turns on its axis. This motion, called precession, is an extremely slow process; one cycle is completed in 26,000 years. The precession moves clockwise and is caused mainly by the moon's gravitational effects on the earth. The change in position of the earth's axis during precession results in a gradual shift in the time of equinox and solstice. However, the rate is slow, and it takes 13,000 years for the tilt to reach the opposite position, at which time summer and winter solstices are reversed.

The combination of the earth's revolution around the sun and the earth's constant inclination of 23½° to the plane of its orbit results in our seasons. Warmer weather comes to the Northern Hemisphere each year when it receives a greater daily share of the sun's rays. The warming trend begins after the vernal equinox (March 21 or 22) because from then until the autumnal equinox (September 22 or 23) the Northern Hemisphere day is longer than its night. Summer is due to longer days and the stronger sunshine resulting from a larger share of sun's rays close to the vertical on the earth. At the same time, the winter season prevails over the Southern Hemisphere. Autumnal equinox begins a change in seasonal patterns. The Southern Hemisphere receives more than 12 hours of daylight, and the sun's rays over the Northern Hemisphere become more slanting and weaker. At winter solstice, the Southern Hemisphere encounters its longest period of daylight. The pattern then gradually shifts as the

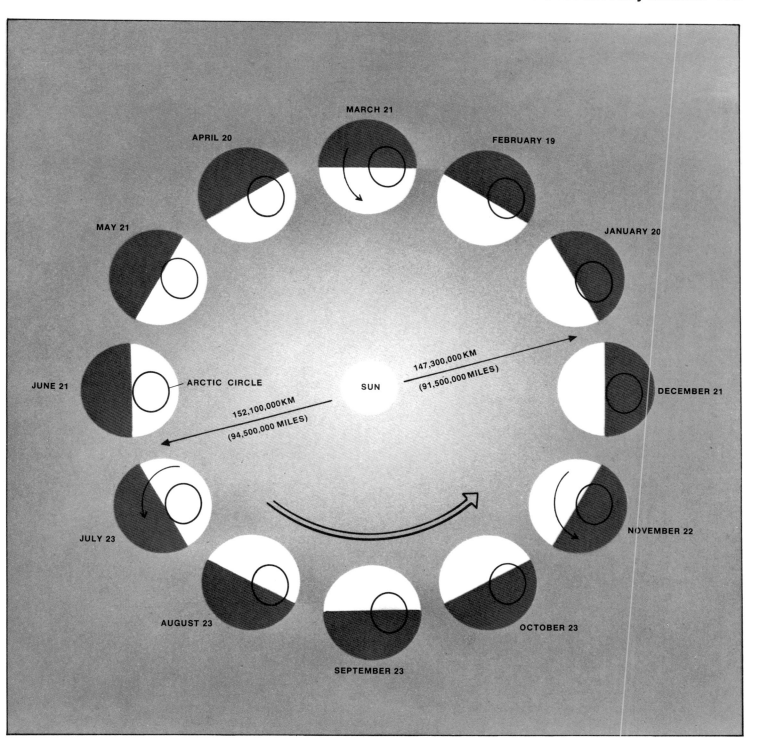

Earth's Planetary Motions
The seasons result from a fixed tilt in the earth's axis of rotation, which turns the poles alternately toward the sun. As the earth sweeps around its orbit, the Arctic Circle receives 24 hours of daylight on June 21 and almost total darkness on December 21.

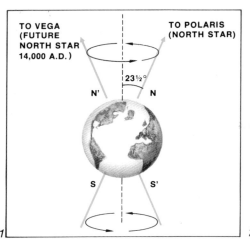

TO VEGA (FUTURE NORTH STAR 14,000 A.D.)

TO POLARIS (NORTH STAR)

23½°

N' N

S S'

1

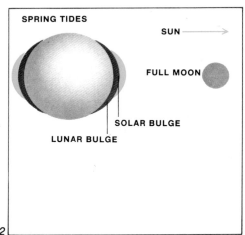

SPRING TIDES

SUN

FULL MOON

SOLAR BULGE

LUNAR BULGE

2 a

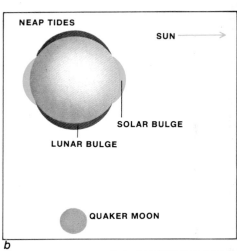

NEAP TIDES

SUN

SOLAR BULGE

LUNAR BULGE

QUAKER MOON

b

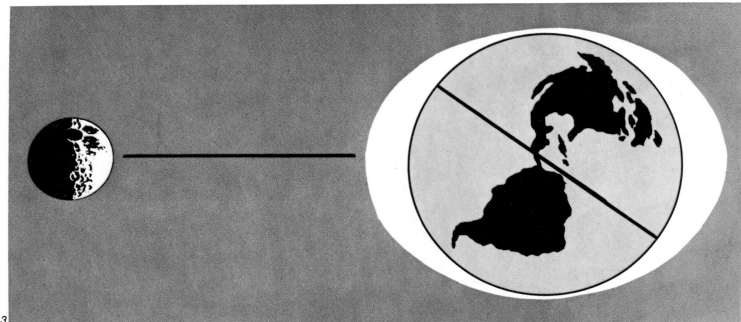

3

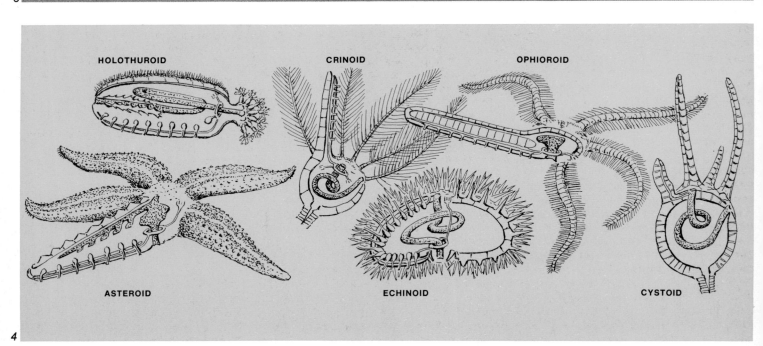

HOLOTHUROID

CRINOID

OPHIOROID

ASTEROID

ECHINOID

CYSTOID

4

seasonal cycle is completed and the Northern Hemisphere shifts toward spring.

Seasonal patterns do not reflect the proximity of the earth to the sun. Because the earth's orbit is not a circle, distance between the earth and sun changes constantly. The earth is closest to the sun about January 1 and farthest away about July 1. However, the maximum difference in the earth's distance from the sun during one complete revolution is only 3 million miles, whereas its average distance from the sun is about 92.9 million miles. For this reason directness of the sun's rays and total daylight time, rather than distance, control our seasons. (R.E.B.)

Earth Tides. Deformations produced in the earth's *crust by the gravitational pull of the sun and the moon. The crust is not completely rigid and shows tidal movements like those of the oceans. The rhythmic rise and fall of points on the earth's surface is not as large as that of the oceans, however; it amounts to a maximum displacement of less than 1 m. Displacement is measured with a *gravimeter, which records changes in the earth's *gravitational potential. As the station on which the gravimeter is set rises and falls due to crustal deformation, the instrument records the changes in gravity as the distance between it and the center of the earth changes. (J.W.S.)

Easterly Winds, Polar. A system of prevailing *winds in the two polar zones. Many scientists consider the concept misleading, since winds in the polar regions move in a great variety of directions.

Echinoderm. A member of the phylum Echinodermata, a large group of exclusively marine animals, most of which exhibit a marked fivefold radial symmetry. Living echinoderms have well-developed nervous and digestive systems and a distinct body cavity and are a relatively complex group of organisms. A typical echinoderm has a skeleton composed of numerous calcareous plates intricately fitted together and covered by a leathery outer skin called the integument. The echinoderm body commonly is star-shaped, but some types may be shaped like hearts, biscuits or cucumbers.

The phylum Echinodermata is divided into two subphyla: the Pelmatozoa, or attached forms, which includes the classes Cystoidea (*cystoids), Blastoidea (*blastoids) and Crinoidea (*crinoids); and the Eleutherozoa, or free-moving forms, which includes the classes Stelleroidea (starfishes and "serpent stars"), Echinoidea (*echinoids) and Holothuroidea (sea cucumbers). Echinoderms range from

*Cambrian to *Holocene (Recent) in age and are abundant as fossils in many marine formations all over the world. (W.H.M.)

Echinoid. A representative of class Echinoidea of phylum Echinodermata. Echinoids are free-moving *echinoderms with exoskeletons that may be globular or shaped like discs, hearts or biscuits. Modern representatives include the sea urchins, heart urchins and sand dollars. The echinoid test (exoskeleton) is composed of many calcareous plates that enclose the animal's soft parts. The exterior of the test is typically covered with large numbers of movable spines. These are of some aid in locomotion, support the skeleton of the animal and provide a measure of protection from enemies.

The oldest known echinoids have been recovered from rocks of *Ordovician age, but it was not until the *Mesozoic that the group began to flourish. They were especially numerous during the *Cretaceous Period and have been abundant and varied from that time until today. (W.H.M.)

Echo-Sounding. A technique for determining the depth of the sea by bouncing a sound impulse off the bottom and recording the time it takes to return to the surface. The recorder is set for a speed of 1460 m per second, and most modern echo-sounders have a power source whose frequency is regulated. Any variation in the frequency or the voltage will cause changes in the time required for the impulse to travel to the bottom and back. A recording machine interprets and prints the travel time as a topographic change on the ocean floor. The operator of the machine must be able to distinguish between apparent topographic changes caused by sound-frequency variation in the impulse and genuine changes. Late-model echo-sounders, called precision depth recorders (PDR), show details of ocean-bottom topography to within 1 m of accuracy. (R.A.D.)

Eclogite. A granular *metamorphic rock composed mainly of the minerals garnet and omphacite. Eclogite has also been used to represent an assemblage of minerals (*metamorphic facies) thought to have formed within a limited range of temperature and pressure. However, recent experimental work has shown that minerals associated with eclogite may form over a much greater range. The lower portion of the temperature-pressure range overlaps another well-established metamorphic facies, and for this reason it may be eliminated as representing special metamorphic conditions.

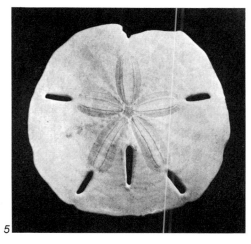

5

Earth's Planetary Motions
1 The earth's axis, tilted 23½° to the plane of its orbit, has a circular wobbling motion that completes a cycle every 26,000 years. Because of this movement, called precession, the North Pole traces a circle in the sky, pointing to different stars as it moves in its orbit.

Earth Tides
2 The moon and, to a lesser extent, the sun exert a pull on the landmasses of the earth in much the same manner as they do on the oceans. (a) The greatest rise of the crustal surface, at the time of the spring tides, occurs when the moon and sun are aligned. (b) The least rise of the earth's surface, at the time of the neap tides, occurs when the pull of the moon and sun are perpendicular to each other.

3 When the moon is north or south of the earth's equator, two successive high tides for a given latitude may have very unequal heights. For example, high tide in the Southern Hemisphere on the side toward the moon is appreciably higher than the one at the same latitude on the opposite side.

Echinoderm
4 Cross sections of the five living classes and one fossil class (cystoid) of echinoderms show the water-vascular systems (stippled) and simple digestive tracts of this phylum.

Echinoid
5 Sand dollars are flat, circular echinoderms found in ocean bottoms.

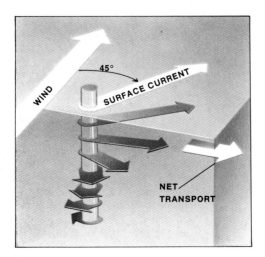

Ekman Layer

In the Northern Hemisphere the Coriolis force, caused by the earth's rotation, deflects the surface layer of a current 45° to the right of the wind direction. The angle of deflection increases with depth—due to the decrease in current speed—until the current is actually moving opposite to the direction at the surface. The net transport of water is at right angles to the wind direction.

Eddy (or Vortex). A swirling movement of turbulent water that develops wherever there is a discontinuity of flow in the stream. It may be caused by (1) velocity pulsations or accelerations, (2) deceleration of velocity, (3) separation of flow such as occurs at bends or in the lee of dunes, (4) changes in density of water resulting from change in temperature or change in load (solid or chemical) or (5) obstacles such as rocks or logs in channel beds or walls. Such eddies are important in entraining a load. The whirling vortex action lifts grains upward and downstream. Swirling eddies also cause local *abrasions, forming potholes and swales as they move particles. (M.M.)

Ekman Layer. The topmost layer of the ocean (also the layer above the earth's surface in the *atmosphere) in which the large-scale motion of the sea is theoretically represented by a balance between frictional forces and the *Coriolis effect. The effects of the frictional stress between the moving atmosphere and the sea surface are transmitted both upward into the atmosphere and downward into the ocean by internal frictional or shearing stresses. The Coriolis effect, a force due to the earth's rotation, causes a moving object to be accelerated to the right of its path in the Northern Hemisphere and to the left in the Southern Hemisphere.

As friction between the wind and the atmosphere sets the sea surface in motion, the Coriolis effect causes the water to drift to the right of the wind direction (in the Northern Hemisphere). The effect of friction, which extends downward to a depth of roughly 100 m, was first computed by the Swedish oceanographer Vagn Walfrid Ekman in 1902. According to Ekman's theory, the sea surface water flows at an angle of 45° to the wind direction (45° to its right in the Northern Hemisphere; 45° to its left in the Southern Hemisphere). The 45° angle is theoretical; observations indicate that the angle is probably considerably smaller. With increasing depth and decreasing speed the water drifts farther and farther to the right (or left) of the wind direction, its movement describing a spiral known as the Ekman spiral. Finally, at the bottom of the layer, the flow is opposite to the wind direction but quite small. The average flow of water is to the right of wind direction in the Northern Hemisphere, to the left in the Southern Hemisphere.

The theory was first applied to atmospheric motion by F. A. Akerblom in 1908. In the atmosphere the Ekman layer extends from about 10 m above the surface to the base of the free atmosphere. The free atmosphere is the region of the atmosphere beyond the zone where the wind is affected by the friction of the surface. The wind in the free atmosphere is assumed to be geostrophic; that is, its motion is a balance between Coriolis and pressure forces and is not affected by friction.

The wind speed in the Ekman layer is most retarded at the bottom of the layer because the wind there blows at an angle across the isobars (lines connecting points of equal pressure) and causes surface friction. At the top of the layer there is minimum friction because the isobars are parallel to the *geostrophic wind. Gradually, with increasing elevation, the wind picks up speed and changes direction until, at the base of the free atmosphere, the motion becomes geostrophic. (M.F.H.)

Elastic Limit. The point at which a rock will be permanently deformed due to an outside force being applied to it. This point varies for each type of rock. If a rock is subjected to an outside force of small magnitude, it will be deformed but will return to its original shape when the force is released.

Elastic Rebound. A theory developed by Harry F. Reid (1911) of Johns Hopkins University to explain the mechanism responsible for *earthquakes.

Electrical Conductivity. A quantitative measure of the ease or fluency with which an electrical current of known voltage flows through a substance. Some ore minerals have very high conductivity, which has enabled geophysicists to devise electrical prospecting methods for locating large deposits of these minerals.

Electric Log. A continuous graph (log) that shows the electrical properties of the various types of rock that may be encountered in a wellhole. Such logs are produced by lowering or by raising an electronic probe that measures the electrical properties of the rock units it traverses.

Electrum. A mineral, a natural alloy of gold and silver. Most native gold contains 10–15 percent silver in solid solution, but when silver is present in amounts of 20 percent or more, the mineral is electrum.

Element, Native. A chemical element that occurs in nature uncombined with other elements. Excluding the rare gases of the atmosphere, there are twenty pure or nearly pure elements found as minerals. Most of these are rare, and some are found so sparingly as to be considered mineralogic curiosities. Only *gold, *silver, *copper, *platinum, *iron, *sulfur, and carbon as *diamond and *graphite have historic or commercial significance. The elements can be divided into three groups: metals, semimetals and nonmetals. The native elements have representatives in each group as shown by the following list:

Metals: gold, silver, copper, mercury, zinc, platinum, palladium, iron, tin, lead, tantalum

Semimetals: arsenic, antimony, bismuth

Nonmetals: sulfur, carbon (diamond, graphite), selenium, tellurium

A mineralogic classification commonly includes natural solid solutions of metals with the native metals. Important solid solutions are *electrum (gold and silver), gold-amalgam (gold and mercury), platin-iridium (platinum and iridium) and iridos-mine (iridium and osmium). (C.S.H.)

Emerald. The variety of *beryl with a deep green color, highly prized as a gem. The color is attributed to the presence of small amounts of chromium or vanadium.

Emery. A natural fine-grained *corundum intimately admixed with *magnetite or, more rarely, with *hematite or *spinel. It is a black granular material that has been used as an abrasive since ancient times.

Enargite. A mineral, copper arsenic sulfide, and an ore mineral of copper. Although much less common than several other copper minerals, it has been an important ore at several localities, notably Butte, Mont.; Morococha and Cerro de Pasco, Peru; and Chuquicamata, Chile. See Mineral Properties.

Endogenic Processes (or Endogenous Processes). Processes that originate below the earth's surface, act on and beneath it, and include both *extrusive and *intrusive igneous activity and earth movements such as crustal warping, folding and faulting. They form crustal features such as *batholiths, thrust *faults and *anticlines, as well as geologic features on the earth's surface such as mountain ranges, volcanoes and *grabens. The sources of energy for these processes are original heat, radioactivity, gravity and, less important, exothermic chemical reactions.

Energy Resources. The raw materials available on earth that provide man with the ability to do work and to alter the land surface. Energy is required for every technological activity. It is the driving force of industry and is utilized to operate nearly every convenience known to man. Therefore a major role of earth scientists is to find raw materials that will serve as sources of energy to meet man's current and future needs. A variety of resources supply energy, most of them coming directly or indirectly from the sun. Wood, the most common fuel a century ago, is no longer a significant source. The *fossil fuels—coal, crude oil and natural gas as well as tar sand and *oil shale—are derived from organic matter once nourished by the sun's rays. When burned, these materials release stored heat and thus produce energy.

*Petroleum (crude oil and natural gas) provides over two-thirds of the world's energy supply. The search for petroleum now extends to the outer edges of the *continental shelves because the best prospects on land have been explored. The Gulf of Mexico, the North Slope of Alaska and the Norwegian Sea are examples of the development of offshore areas with substantial petroleum reserves. However, the United States, North Africa, the Soviet Union, Venezuela and especially the Middle East dominate the world's oil market, for these areas contain most of the earth's proven reserves.

The role of crude oil and natural gas in supplying energy needs in the near future is especially critical because environmental considerations limit the use of coal, and technological barriers prevent the immediate maximum utilization of nuclear fuels. Although the United States, for example, has an estimated reserve of 390 billion tons of coal, a large increase in the worldwide utilization of coal faces opposition mainly because of fear of environmental damage. Despite this, the energy crisis has made increased use of coal resources a certainty. Research has begun on reducing the environmental effects of coal burning, especially air pollution, and on developing techniques for large-scale production of gas by coal gasification.

Nuclear power holds promise of both ample and clean fuel in the future. The immediate problems include the safe location and construction of nuclear power-generating plants and the effects of thermal pollution, a by-product of the cooling process that requires large quantities of water, which become heated. Moreover, the present method of producing nuclear energy using uranium 235 is not a long-range solution because supplies of this raw material are limited.

The answer may lie in breeder reactors, which are expected to become important producers of nuclear power by the mid-1980s. These reactors utilize surplus neutrons from the fission of uranium 235 to transform other materials into fissionable products. In this process more fuel is created than is consumed. Examples of this are the transformation of thorium 232 into fissionable uranium 233 and of uranium 238 into fissionable plutonium 239. One unsolved problem in the use of breeder reactors is finding adequate methods for long-term storage or disposal of the highly radioactive wastes produced by the process.

A controlled fusion process holds even greater promise for the release of huge quantities of clean energy with negligible radioactive wastes. Fusion reactions that combine light nuclei of deuterium or deuterium and tritium are likely to prove useful. Both methods yield inactive helium as well as energy. Deuterium is ideal because ample quantities occur in the ocean; tritium is far less abundant. It is unlikely, however, that fusion plants will be operational before the year 2000.

The role of hydropower is well established. Each year more dams are constructed to control the flow of water and convert its power to electricity. Its potential is limited, however, because the most desirable locations on many rivers are already developed. Hydropower is therefore not expected to add significantly to the total available energy.

The energy potential of several other resources is being investigated. Geothermal energy, derived from the heat within the earth, is one such source. Natural steam is already proving valuable in the Geysers area of northern California, as well as in Iceland, Italy, Japan, New Zealand and several other countries. The earth's internal heat might also be utilized by tapping regions of underground hot water or by pumping surface water into subsurface zones of hot rocks, thereby producing steam.

Solar energy, captured directly, is already in use as a heat source. Solar panels mounted on buildings have proved an efficient technique for gathering the sun's rays. In several countries hot-water heaters for homes operate on energy derived in this way. Perhaps in the future, giant beacons or orbiting satellites will gather solar energy and transmit it to power-generating plants.

The oceans offer a tempting source of undeveloped power. A major tidal power plant on the Rance Estuary in Brittany furnishes electricity to France, and several similar plants are under construction elsewhere in the world. The tidal surges of Canada's Bay of Fundy make it an appealing site, although to date it has not proven economically feasible to build a tidal power plant there. Major ocean currents and well-known temperature gradients within the ocean also have potential for generating huge amounts of electricity. Estimated costs to harness these energy sources are

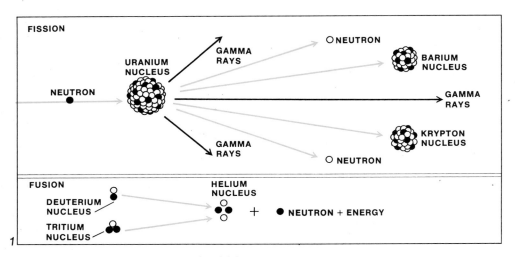

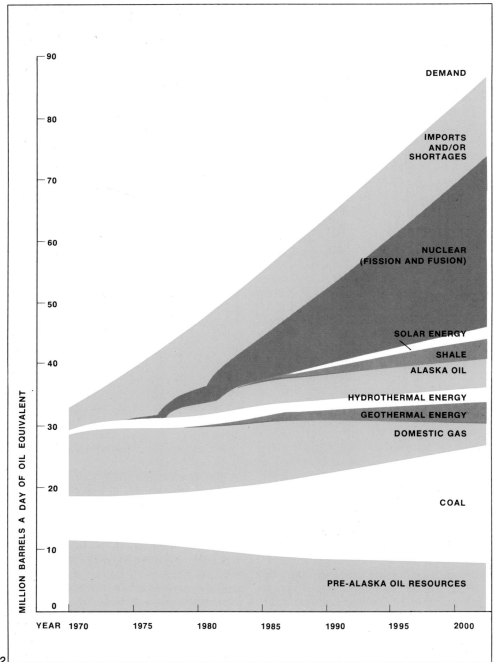

formidable, and their future usefulness to mankind is thus a matter of speculation.

Other fossil fuels in addition to coal, oil and natural gas may yield significant energy supplies. Tar sands, dense sand deposits containing tarlike petroleum, occur in western North America, Venezuela and Brazil. In the Athabascan area of Canada, an *open-pit mining operation and refining plant produce petroleum from tar sands at the rate of one barrel to every two tons of processed sand. But most tar sand is too deep to be mined, and therefore its future role is uncertain. Deposits of organic material in laminated rock, called oil shales, also hold promise as a petroleum source. Oil shale occurs most prominently in Colorado, Utah and Wyoming, with lesser deposits in Europe and Great Britain. Estimated costs for mining and refining oil shale deposits are high, but demands for additional energy are likely to make these deposits a profitable source of petroleum.

The use of solid organic wastes for fuel would satisfy a dual purpose: the demand for energy and the disposal of solid waste. Conversion techniques have been developed, but the widespread dispersal of solid wastes makes their central processing a difficult and expensive task. Although solid-waste conversion may supply some energy, it is not likely to become a major source in the near future.

It therefore appears likely that petroleum will remain our key energy source for the next decade or two. Nuclear power generation will gradually increase and, with breeders, is likely to play a major role by the year 2000. Contributions from several other sources will increase slightly, but the percentage of energy they will supply in the next few decades will probably remain small. (R.E.B.)

Enstatite. A mineral, a magnesium silicate in the orthorhombic *pyroxene group. The monoclinic *polymorph of enstatite is clinoenstatite. With the substitution of iron for magnesium, the mineral grades first into bronzite (5–13 percent iron oxide) and then into *hypersthene (13–20 percent iron oxide). Enstatite and other orthorhombic pyroxenes are common in pyroxenites, peridotites, gabbros and basalts, and are present in many meteorites, both metallic and stony. See Mineral Properties.

Eocene Epoch. The second oldest *epoch of the *Tertiary Period; it began about 54 million years ago and lasted for about 18 million years. It was named by Sir Charles Lyell in 1833 from rocks exposed in the Paris Basin of France. Its name, from

Greek words meaning "dawn of recent," refers to the relatively small number of modern species found as fossils in Eocene rocks. These rocks consist of both marine and continental deposits, which occur in northern France, southeastern England, Germany, Holland and Belgium. In the Mediterranean region they occur as a very thick, widespread deposit of marine limestone containing large numbers of the fossil *foraminifer, Nummulites. Their shells formed thick deposits of nummulitic limestones, which stretch from southern France through eastern Europe, northern Africa and across the Himalayas to the Philippines. The Egyptians quarried this fossiliferous limestone as building stones for the pyramids.

In North America, Eocene deposits of marine and nonmarine origin are exposed on the Atlantic and Gulf coastal plains from southeastern Texas to Maryland and Virginia. Eocene rocks of the Pacific Coast occur in California, Washington and Oregon and are essentially marine, although beds of coal and brackish-water deposits occur in Washington and western Oregon. Nonmarine Eocene strata cover much of the western interior of North America, including northwestern New Mexico, Colorado, Wyoming, Montana, western North and South Dakota as well as Alberta, Canada. These include the Wasatch Formation (responsible for the spectacular scenery at Bryce Canyon National Park and the "Pink Cliffs" of Utah) and the Green River Formation, consisting of ancient lake deposits that are the source of much *oil shale. Evidence of volcanic activity can be seen in Utah, northwestern Washington and elsewhere.

Eocene climates were warm, humid and almost subtropical in places as far north as Canada. Invertebrates found as fossils include foraminifers (especially Nummulites), *pelecypods, *gastropods and *echinoderms. Other marine animals included the bony fishes and the sharks (including the 24-m Carcharodon) and 24-m long, whalelike mammals known as zeuglodonts. Eocene land plants were widespread and closely related to modern forms such as beeches, elms and walnuts. Birds flourished; some, such as Diatryma (whose remains have been found in Wyoming), were as much as 2 m tall. Mammals became larger, developing more specialized feet and teeth to adapt to changing environments. Horses (which first appeared in the preceding Paleocene Epoch) were represented by fox-size Hyracotherium (also called Eohippus), and the bats, rodents, camels, rhinoceroses, monkeys and ancestral elephants made their first appearance. Among the more primitive mammals were the doglike condylarths, the meat-eating *creodonts, and the uintatheres—saber-tusked creatures, the size of a small elephant, with three pairs of blunt, bony horns.

Economic resources from Eocene rocks include petroleum (Gulf Coast area and California), coal in western Oregon and Washington, oil shale from Colorado, Utah and Wyoming, and lignite and bauxite in the Gulf Coastal Plain region. (W. H. M.)

Eolian (from the Greek for Aeolus, god of the winds). Pertaining to or caused by wind. Dust storms, sand storms, tornadoes and dune fields result from eolian activity. *Loess and dune sand are composed of material transported and deposited by wind and are therefore eolian deposits. Eolian *ripple marks are formed by wind on dunes and other deposits. *Ventifacts and *yardangs result from eolian erosion, and eolian sandstone is formed by the consolidation of sand transported and deposited by wind.

Epeirogeny. The uplift and subsidence of large portions of the earth's crust to form *arches and *basins. Generally the structures produced within the crust can be distinguished as either epeirogenic or orogenic but they may grade into each other. Epeirogenic movements at or near the end of regional faulting or folding of the crust are associated with the development of mountainous topography. A disconformity, i.e. an irregular surface between essentially parallel layers of sedimentary rock (resulting from a break in deposition of sediments and subsequent *erosion) is commonly formed by epeirogeny, whereas *folds, thrust *faults and angular *unconformities are typical of orogenic movements. See Orogeny. (J.W.S.)

Ephemeral Stream. A stream that carries water only during and shortly after a rainstorm. The channel is dry most of the year and may be referred to as a *gully, wash, *arroyo, *coulee or wadi. Even when flowing, these streams tend to lose water downstream and thus deposit debris in their channels. Because they range from flash floods to dryness, they vary greatly in amount of discharge. They are found in arid and semiarid regions but may occur in any area where the channel bed lies above the groundwater table.

Epicenter. A point on the surface of the earth directly above a source of crustal disturbance that has produced an *earthquake. The source of an earthquake is within the earth and is referred to as the *focus.

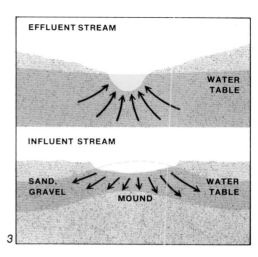

Energy Resources

1 In nuclear fission, the nucleus of an unstable element such as uranium is struck by a neutron and splits into two smaller nuclei, releasing high-energy gamma rays and more neutrons. The neutrons in turn bombard other uranium nuclei and set off a spontaneous reaction. During fusion, as in a hydrogen bomb explosion, two nuclei fuse, producing a helium nucleus, a neutron and more energy per gram of material than any other known reaction.

2 A graph of U.S. energy consumption.

Ephemeral Stream

3 In humid areas streams are perennial, or effluent. In desert areas stream channels are normally dry except after rainstorms. These temporary, or influent, streams have fluctuating water tables, losing water by percolation into highly permeable beds of coarse sand and gravel. When such streams flow from mountain watersheds to alluvial basins, they form alluvial fans.

4 A dry wash in Arizona has become a raging torrent after a sudden rain.

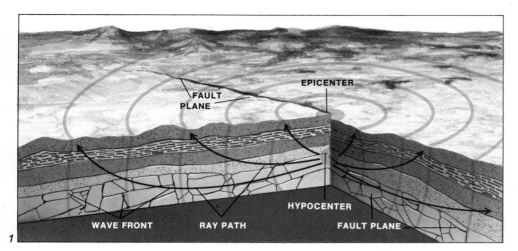

To determine the epicenter of an earth-quake requires data from several stations throughout the world—or from several *seismographs at the same observatory—that have monitored the effects of the quake, including the time it takes *seismic waves to travel from the focus to the monitoring station. Each station must record the arrival time of two types of waves, P (primary) waves, and S (shear) waves. Because these two waves travel at different velocities within the *crust, one will arrive at the monitoring station before the other. The primary wave is faster and always arrives before the shear wave. The time lapse between the P and S waves in-creases as the distance between the origin of the earthquake and the monitoring station increases. The greatest time difference occurs when the earthquake and monitoring station are 180° (longitude) apart, that is, when they are on opposite sides of the globe.

Special tables have been prepared to plot the arrival times of P and S waves against the angular difference between the epicenter and monitoring station. Using the distances in degrees from the epi-center as measured from several sta-tions, arcs are drawn on a globe cor-responding to the angular distance be-tween each station and the earthquake. When three or more of these arcs are con-structed they will cross at a point, forming a small triangle on the face of the globe. The point of intersection or center of the tri-angle marks the location of the epi-center. (J.W.S.)

Epidote. A mineral, a calcium iron alu-minum silicate, commonly found in *metamorphic rocks as an alteration of feldspar, pyroxene and amphibole. It is es-pecially characteristic of contact meta-morphic deposits in limestones. Although often black when it occurs as isolated crystals, as a rock-forming mineral epidote has a characteristic pistachio-green color. It is frequently associated with chlorite, which is also green; and the green color of many rocks results from the presence of these two minerals. Crystals usually have a prismatic aspect, striated parallel to their length. Well-developed crystals are found at many localities but the most notable are Untersulzbachthal, Austria, and the Prince of Wales Island, Alaska.

Epidote is the principal member of a mineral group which includes clinozoisite. These minerals form under similar condi-tions and have similar crystallographic properties and chemical compositions. They are calcium aluminum silicates, but in addition to these elements, epidote con-tains iron substituting for aluminum. A

variation in the physical and optical properties of epidote is the result of the amount of iron present in the mineral (5 – 17 percent). See Mineral Properties. (C.S.H.)

Epithermal Deposit. A hydrothermal vein deposit formed relatively near the earth's surface at moderate pressure and temperature. It is characteristically found in volcanic rocks of Tertiary age and is believed to be genetically related to volcanism. The chief metals of these deposits are gold, silver and mercury. The most commonly associated *gangue minerals are quartz, calcite, aragonite and fluorite.

Epoch. A division of geologic time that is shorter than a *period but longer than an *age. It is the standard subdivision of a geologic period and the time unit during which a *series of rock is formed. Epochs are especially useful in describing the subdivisions of the *Tertiary and *Quaternary periods of the *Cenozoic Era.

Epsomite. A mineral, a hydrated magnesium sulfate, the naturally occurring compound generally known as Epsom salts. The mineral is found as an efflorescence on the walls of caves and on the rocks of mine workings, and more rarely in lake deposits. Epsomite is easily soluble in water and has a bitter taste. Epsom salts can also be manufactured from other magnesium minerals and the manufactured form is often used for industrial purposes and medicines (chiefly as a cathartic). It gets its name from mineral springs at Epsom Downs, England.

Equatorial Bulge. The earth is not a perfect sphere but is an oblate spheroid, slightly flattened at the poles and bulging at the Equator. The equatorial bulge is caused by centrifugal forces induced by the earth's rotation. The dimensions of the earth have been accurately measured with the aid of satellites. The equatorial radius is 6378.16 km and the polar radius is 6356.18 km. A perfect sphere having a volume equal to that of the earth would have a radius of 6371 km. Therefore the equatorial bulge represents a departure of 7.16 km from the radius of a sphere of equal volume.

Equipotential Surface. A surface over which a body of known mass would possess an equal value of potential energy. One such equipotential surface, the *geoid, is a surface along which the pull of gravity is everywhere normal to the surface. It is used as a reference surface for astronomical observations and for geodetic surveying.

Era. The largest unit of geologic time. Each era has been given a name describing the stage of development of life at that time. Thus the life of the *Mesozoic ("middle-life") Era is intermediate in complexity between the *Paleozoic ("ancient-life") forms and the more advanced species of the *Cenozoic ("recent-life") Era. Two other eras, the *Archeozoic and *Proterozoic, represent all time before the beginning of the *Cambrian Period of the Paleozoic Era. The terms for these *Precambrian eras are less commonly used, for it is difficult to establish a workable worldwide classification of these opening chapters in earth history.

The rocks formed during an era are known as an *erathem. The end of an era is typically marked by distinct and widespread physical breaks in the geologic record. These breaks can be caused by periods of mountain-building and/or extensive uplift or depression of the land. Wholesale changes in plant and animal life normally accompany the physical change at the end of an era. Thus the dominant life forms of each era usually differ from those of preceding eras. See Color Plate "The Fossil Record" and Color Plate "The Tree of Life." (W.H.M.)

Erathem. A *time-stratigraphic unit used to designate the rocks formed during an *era of geologic time. An erathem encompasses all systems of rocks assigned to each era and is the largest *recognized time-stratigraphic unit.

Erosion. The wearing away by natural agencies of any part of the earth's surface. Material must be picked up and carried away in order for erosion to take place. The most important agents of erosion are mass wasting, wind, waves and shore currents, glacial ice, underground water and running water, which is by far the most important. All of these agents depend upon the force of gravity. The evidence for erosion is ubiquitous and varied. It includes the easily observed retreat of marine cliffs and the development of gullies on farmland as well as the cutting of the great canyons. The material carried by rivers and deposited at their mouths is evidence of erosion upstream. The vast thicknesses of the sedimentary rocks, caused by the erosion of landmasses elsewhere, bear witness to erosion of a magnitude that is difficult to conceive.

The average rate of erosion for a drainage basin can be calculated if the weight of the material the river annually carries to the ocean is known, and if the area of its drainage basin is also known. The average rate of erosion of the drainage basin of the Mississippi River and its

4

Epicenter

1 A cutaway view shows the relationship between an earthquake's focus, or hypocenter, on the fault plane, and its epicenter directly above it on the surface. Seismic waves radiate in all directions through the earth and on the surface. Ray paths (spreading circles) form at right angles to the wave fronts.

Erosion

2 The Grand Canyon, 1600 m (1 mile) deep, was gouged by the Colorado River over a period of 10 million years.

3 In Nevada's Cathedral Gorge State Park, clay and silt deposits have been sculptured by rainwash into gothic spires.

4 Weathering and wind erosion produced these sandstone columns in Skeleton Mesa, Ariz.

tributaries is about 30 cm (1 foot) in 6000 years. The rate for the Colorado River Basin is more than three times as fast.

An analysis of rocks indicates that erosion has been taking place throughout geologic time. Erosion also takes place on the moon, but it is thought that the rate is slower than on earth, and of course the agents are different. (R.L.N.)

Erosion, Differential. A difference in the rate of erosion on individual outcrops as well as over large areas. The rate varies because the rocks vary in resistance (differential resistance) and are subjected to differential attack. Such attack takes place where the blocks have fallen from cliffs into silt-laden rivers. Those parts of the fallen blocks usually below river level will be more eroded than those above river level. Pedestal rocks form in this way.

Old marble tombstones furnish excellent examples of differential erosion. Originally smooth, the marble is roughened with time through solution and the impact of wind, rain and gravity. These processes act on the surface of the stones and commonly form ridges where the marble is more resistant.

A large-scale example of differential erosion is shown in the formation of *hogbacks (ridges) and erosional valleys. Where sedimentary rocks that vary in resistance are tilted and eroded, the resistant rocks form hogbacks and the less resistant form valleys. Hogbacks may be several hundred meters high and extend for scores of kilometers. (R.L.N.)

Erratic. A rock fragment transported by either glacial or floating ice and different from the *bedrock beneath it. It is often used to refer to a glacially transported fragment.

Erythrite. A mineral of secondary origin, the hydrous arsenate of cobalt known as cobalt bloom. Although erythrite in itself has no economic importance, the presence at the surface of this crimson to pink mineral indicates that other cobalt minerals and associated native silver may be found below. See Mineral Properties.

Escarpment (or Scarp). A straight cliff or steep slope of considerable length separating a lower, gently sloping surface from a higher surface. It may be formed by either faulting or *erosion. It occurs on the margins of plateaus and is associated with *hogbacks, *cuestas and *fault block mountains. The word scarp is generally used for cliffs resulting from faulting; escarpment is more frequently used for cliffs due to erosion. See Fault; Fault Scarp.

Erratic

1 These boulders were deposited by a glacier which covered Yosemite National Park, Calif., about 10,000 years ago.

Escarpment

2 California's Sierra Nevada, culminating in Mount Whitney, 4418 m (14,495 feet) high, were formed when enormous blocks of rock slipped upward along a fault in the earth's crust. The eastern escarpment, shown here, which is more than 640 km (400 miles) long, is almost vertical; the western slope is more gradual.

Esker

3 Winding for miles across King William Island, Northwest Territories, Canada, this sand and gravel ridge was deposited by a stream that ran beneath a glacier.

Esker. A long, narrow ridge, commonly sinuous in shape, with steep sides and a gently undulating, rounded summit. Eskers result from glaciation and are composed of poorly stratified glaciofluvial sand and gravel. They range in length from about 100 m to more than 150 km, and in height from less than 3 m to more than 50 m. They are usually found in valleys, although they also cross ridges. They roughly parallel the direction of glacial motion, and where the *bedrock of a region is known, it can be demonstrated that roundstones in eskers have been transported more than 20 km. Eskers are common in glaciated regions in North America (very large ones found in Maine) and Europe.

There are three principal kinds of eskers: (1) Those formed in stagnant ice in tunnels clogged by meltwater streams with sand and gravel; on deglaciation, the tunnel fillings became eskers. Stagnant ice is thought to be necessary here; otherwise the tunnels would have had no permanence. (2) Those formed on the surface of the ice in channels that became filled with sand and gravel; on deglaciation, the sand and gravel were let down onto the ground and formed eskers. Eskers that rise out of valleys and over ridges can perhaps be best explained in this way. (3) Those formed where glaciers terminated in standing water—either in lakes or in the ocean. When meltwater streams carrying sand and gravel in tunnels in the ice entered this water, deposition took place near the mouths of the tunnels due to the reduction in the streams' velocity; as the glaciers melted, eskers formed. (R.L.N.)

Estuary. A drowned river valley where freshwater in the lower course is mixed with seawater. Estuaries are formed where a deep-cut river mouth is drowned following subsidence of the land or a rise in sea level. A shoreline composed of a system of drowned river valleys is called a ria coast. A glaciated estuary that has steep walls and a very deep bottom is a *fiord. Tidal effects are important in estuaries, not only because of the saltwater wedge that moves upstream from the ocean and gradually mixes with the fresh river water lying above it but also because of the scouring effects of the outgoing tide. Estuaries are the habitat for shellfish that live in brackish water, an important source of seafood. This is of major concern at present because river pollution is a factor in destroying this shellfish life. (M.M.)

Etch Figure. Solution pits on crystal faces. The shapes of the pits vary with the solvent and with the crystalline substance, and may also vary from face to face of a given

2

3

Etch Figures

1 *The etching of many mineral crystals, such as this topaz, results in pitted surfaces whose shapes reflect the crystal symmetry of the mineral.*

Eurypterid

2 *Closely related to the horseshoe crab, these large Paleozoic arthropods swam on their backs through brackish waters.*

3 *A fossil of the Silurian genus* Acanthoeurypterus *was preserved in dolomite at Herkimer County, N.Y.*

Evaporation

4 *In the lower latitudes evaporation is greater than precipitation because dry sinking air masses absorb vast amounts of moisture. This river in Greece is already nearly dry although it is April and the warmest months are still to come.*

crystal. The etching effect of the solvent is the same for all faces of a crystal form but is different for faces of different crystal forms. Thus the *crystal symmetry is revealed by the etch figures.

Euhedral. In crystallography, a term applied to minerals that are bounded by their own crystal faces. By contrast, *anhedral indicates a mineral fragment that is without crystal faces.

Euphotic Zone (or Photic Zone). In the ocean, a thin zone in which the light penetration is sufficient for photosynthesis to take place. It commonly extends down 60 or 70 m, but is subject to greater depths in low latitudes than in middle latitudes because there is less suspended material in the water. The zone is shallowest over the *continental shelves because of the large quantity of suspended sediment in the water there.

Eurypterid. An extinct water-dwelling *arthropod characterized by paired, broad, winglike appendages that were modified for swimming. Eurypterids resembled modern scorpions; some species apparently possessed a stinger and poison gland. At least one species, *Pterygotus*, a *Silurian species from New York State, reached a length of about 2.7 m (9 feet) and is the largest known arthropod. Eurypterid remains have been found in rocks ranging from Lower Ordovician to Permian. They are most common, however, in certain Silurian and Devonian formations.

Eustatic Change of Sea Level. A worldwide, uniform and simultaneous shift in the level of the ocean resulting from (1) changes in the volume of glacial ice, (2) changes in the volume of rivers, lakes and subsurface waters, (3) deposition of sediments in the ocean basins, (4) cooling or heating of the oceans, (5) addition of juvenile water (derived from *magma) to the oceans or (6) changes in the area and depth of the ocean basins due to uplift and subsidence of the sea floor. Such changes of sea level can be identified and measured only in geologically stable areas. Changes in sea level resulting from variation in the volume of glacial ice are known as glacio-eustatism. Changes of sea level resulting from variation in the volume of the ocean basins due to crustal movements are known as diastrophic eustatism. (R.L.N.)

Euxinic Environment. A marine or freshwater environment characterized by restricted water circulation but with a moderate amount of water runoff from adjacent land areas. The lack of circulation may be due to physical barriers such as a *sill, a constricted outlet or other topographic irregularities. It produces a general lack of oxygen and an acidic condition. This results in an accumulation of organic matter and an absence of life. Portions of the Black Sea as well as most fiords are euxinic environments. (R.A.D.)

Evaporation. The change of phase from a liquid or a solid to a gaseous state. In meteorology, the change from ice to a vapor is usually specified as sublimation. According to the kinetic theory of gases, evaporation occurs when molecules near a liquid's free surface acquire above-average, outward-directed velocities and escape into the vapor phase. Energy is always absorbed from the environment in the evaporating process; the heat removed is termed the latent heat of vaporization. At 0°C, the latent heat of vaporization of water is 597 gram calories per gram of water evaporated. The same quantity of heat is released to the environment when a gram of water condenses at 0°C. Since water is evaporated from the earth's surface and then condenses at cooler temperatures in the atmosphere several kilometers above, evaporation results in the transfer of heat to the atmosphere. Evaporation—including the conversion of ice to vapor—and transpiration are the only ways in which water can enter the atmospheric phase of the *hydrologic cycle.

Net evaporation of water into the atmosphere ceases when the air reaches saturation, that is, when the relative humidity is 100 percent. Thus evaporation takes place most rapidly when dry air moves above a wet surface. Evaporation is also speeded if the water is warmer than the air above. In this case the lowest layer of the atmosphere becomes unstable, resulting in convective and turbulent motions that transfer water vapor upward and dry air downward; this transfer permits evaporation to continue. Since turbulent motions are also favored by high wind speeds, wind is another factor controlling evaporation into the atmosphere.

The earth's climates are often characterized by a moisture index representing the difference between the moisture supplied (precipitation) and the moisture lost (evaporation). The primary factors affecting evaporation are the availability of energy, as indicated by the temperature; the availability of water; the wind speed; and the relative humidity of the air in the atmospheric boundary layer. Thus evaporation depends strongly on the surface temperature and therefore upon the latitude. All the factors favorable for high evaporation are present in the extensive belts of

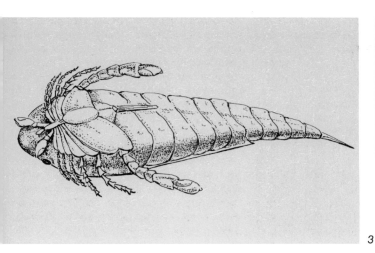

3

trade winds over the oceans on either side of the equatorial convergence zone. Precipitation, depending primarily on the upward motion of moist air in this zone, and in the belt of the polar-front cyclones, also shows a latitudinal dependence. The difference between the zonally averaged evaporation and precipitation delineates in a general way the earth's dry and moist climates. On the average, the dry climates are most extensive in subtropical latitudes where heat from the sun supplies ample energy but where the atmosphere's predominantly sinking motions and anticyclonic conditions are unfavorable for precipitation. (M.F.H.)

Evaporite. A chemical *sedimentary rock consisting of minerals precipitated from aqueous solutions concentrated by the evaporation of a solvent. They originate from precipitation of salts in water (usually seawater) during times of aridity and prolonged evaporation. Rock *gypsum, *halite (rock salt) and *anhydrite are typical evaporite deposits laid down as sediments on the bottoms of restricted bays or arms of ancient seas. They are now being formed by the evaporation of saline lakes such as the Great Salt Lake in Utah and in isolated bodies of seawater like the Rann of Kutch in northwest India. Thick beds of evaporites occur in the United States in Texas, New Mexico and Utah, and in India, Iran and Germany. The rock-salt deposits of New York State and Michigan and the *salt domes of the Gulf Coast states also originated as evaporites. (W.H.M.)

Evorsion. Erosion of *bedrock by sheer force of water under pressure, as in torrents, waterfalls or beneath glaciers. Some *potholes may be formed by evorsion. See Corrasion.

Exfoliation (from Latin *ex* "out" and *folia* "leaf"). In geology, the separation from granite and other massive rocks of concentric flakes, scales, plates, sheets or shells. These may be flat or curved and may range from less than a centimeter to several meters in thickness and from a centimeter or so up to scores of meters in length. Large-scale and small-scale exfoliation occur. In the large-scale type (sheeting) the shells separate along joints more or less parallel to the topography and may be over 100 m (330 feet) long and from 30 cm (1 foot) to 6 m (20 feet) thick. Where the *bedrock is foliated the sheets are frequently at angles to it. This type of exfoliation is thought to be produced when rocks at depth are brought to the surface by erosion. The pressure on these rocks is therefore reduced, and they expand and ultimately develop *joints more or less parallel to the topography. Large, broadly curved bedrock domes are formed in places of considerable relief. Some outstanding exfoliation domes are found in Yosemite National Park, Calif. (including the celebrated Half Dome); in Rocky Mountain National Park, Colo.; and on Sugar Loaf in the harbor of Rio de Janeiro, Brazil.

In the small-scale type of exfoliation bedrock blocks of granite and other rocks bounded by joints are changed into spheroidal cores of solid rock (spheroidal *weathering). These cores are surrounded by concentric layers of weathered rock. The core stones range from a few centimeters to a few meters in diameter. The shells, generally up to a few centimeters thick, can usually be separated from the core and from each other by a hammer blow. Spheroidal weathering is an instance of small-scale surface chemical exfoliation. Water and air enter the joints causing hydrolysis, hydration and oxidation of the silicate minerals and the formation of clay and other minerals. The formation of these new minerals involves an increase in volume, and stresses are set up that split layers and shells from the block. Since corners are attacked more than edges, and edges more than faces, the blocks gradually become spheroidal. (R.L.N.)

Exogenic Processes. Geologic processes that act at or near the earth's surface, such as *weathering, *mass wasting, running water, *subsurface water, waves and shore currents, wind, glaciers. The activities of animals, including humans, and plants are also included as exogenic processes. (Although gravity acts on and near the earth's surface, it also acts at great depth and is considered not an exogenic process but a directional force.) These processes erode, transport and deposit, producing many of the rocks and forming or modifying most landforms. Running water is the most important of these processes, but the processes vary in importance from place to place and from time to time. Thus wind is important in the Sahara, glacial activity in the Antarctic. Running water is the most important exogenic process in New England today, but 15,000 years ago glaciers were far more important there.

The main source of energy for exogenic processes is the sun; it is responsible for the heating and circulation of the oceans and atmosphere, as well as for evaporation. A much less important source is the gravitational attraction of moon and sun; these result in *tides and tidal currents. All geologic features, except those formed by meteoric impact, derive from exogenic and *endogenic processes. (R.L.N.)

2

Evaporite

1 As lakes dry up in arid regions, chemicals precipitate out to form salt deposits such as at Devil's Golf Course, Death Valley, Calif.

Exfoliation

2 Expansive pressures within the granite eventually caused the surface to crack and peel off in concentric layers, creating Royal Arches at Yosemite National Park, Calif.

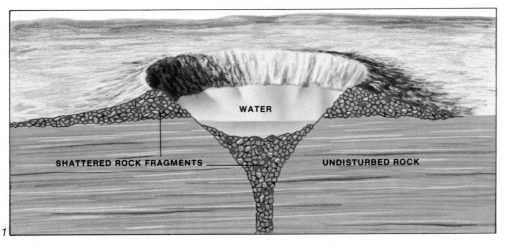

Explosion Pit (or Embryonic Volcano). A crater formed when a volcanic eruption has only enough energy to eject the material drilled out of its chimney. The pit, or crater, is surrounded by a ringed wall consisting of angular fragments of the *country rock through which the chimney was drilled. With the exception perhaps of volcanic gases, little or no magmatic material is emitted, and not enough debris is ejected to build a cone. The floor of the pit may be below the water table and may therefore contain a lake. Such a pit is called a *maar* in the Eifel volcanic district of Germany. Ubehebe Crater in Death Valley, Calif., is an explosion pit in which some magma-derived debris mixed with fragments of country rock, including *alluvium, was ejected. Due to the absence of magmatic material it is in some cases difficult to distinguish explosion pits from similarly appearing meteorite impact craters. (R. L. N.)

Exsolution. In crystallography, the separation or unmixing of two distinct phases in a disordered crystal. Minerals that form at high temperatures may accept foreign ions into their crystal structures. When cooling brings about a general contraction of the structure, these ions can no longer be tolerated and are then forced to migrate through the crystal and to group locally. They arrange themselves into a different crystalline structure that grows in the solid body of the host crystal. The formation of *perthite, thin plates of albite feldspar found in microline feldspar, is believed to result from exsolution. (C. S. H.)

Extrusive Igneous Activity. Any igneous activity at the earth's surface resulting from the movement of magma from the *mantle and hot regions in the crust up to the earth's surface. It produces extrusive rocks consisting of various kinds of lava formed by the solidification of *magma as well as *pyroclastic rocks formed by consolidation of volcanic detritus. It also adds gases to the atmosphere and water to the *hydrosphere, and forms volcanic cones and mountains of various kinds, great lava plateaus as well as *craters, *calderas, *geysers, *fumaroles and many other volcanic features. See Igneous Rock; Intrusive Igneous Activity. (R. L. N.)

Explosion Pit

1 A volcanic vent is drilled to the surface by a jet of gas, ejecting shattered rock fragments that later fill the neck and surround the crater. If volcanic activity ceases, the resulting shallow crater is called an explosion pit.

2 Pinacate craters, part of a volcanic field in northern Mexico, are the result of small eruptions that did not eject enough debris to build a cone.

F

Facies. A term describing an assemblage of mineral, rock or fossil features reflecting the environment in which a rock was formed. It refers to the nature or general appearance of one part of a rock body as distinguished from other parts of the same unit. When applied to *sedimentary rocks it describes an accumulation of deposits that exhibit distinctive characteristics and grade laterally into other sedimentary accumulations which, although formed at the same time, exhibit quite different characteristics. Sedimentary facies typically represent lateral subdivisions of a *stratigraphic unit; they include biofacies (lateral variations in the fossil record in a stratigraphic unit), which contain facies fossils representative of that particular part of the unit. A facies change is said to occur when there is a lateral or vertical change in the type of rock or fossils present in contemporaneous sedimentary deposits. Such variations are caused by and are an indication of a change in the environment of deposition.

An igneous facies denotes a physical, structural or chemical change that differentiates a part of a single mass of *igneous rock from the typical rock of the main unit. A metamorphic facies consists of a mineral assemblage that has reached chemical equilibrium during *metamorphism under restricted temperature-pressure conditions. (W. H.M.)

Fault. A break or zone of fracture within the earth's *crust along which movement of adjacent bodies of rock has occurred. Faults range in size from a few centimeters to hundreds of kilometers in length, and the magnitude of displacement between adjacent walls may also show such an extreme variation in distance.

Faults are of special interest to earth scientists because the sudden movement of crustal blocks along a fault is responsible for generating *earthquakes. Scientists believe that the more they can learn about the mechanism of faults, the more accurately they will be able to predict and possibly control earthquakes.

Faults are classified on the basis of the relative movement of two formerly adjacent points. Should movement occur in the horizontal direction and parallel to the plane of the fault, the fault is referred to as a strike-slip fault. Strike-slip faults can further be divided into right- and left-slip faults. If it appears to an observer on one side of and facing the fault that the movement of the block on the other side has been to his left, it would be considered a left-slip fault. It is so-termed because in tracing a surface feature that had been offset or displaced by the fault, one would

Fault
Strain along California's 965 km-long (600 mile) San Andreas Fault was suddenly released in 1906 in a major earthquake that displaced part of the surface as much as 6 m (20 feet).

have to turn left along the fault in order to find it on the other side of the fault. Such faults have also been called right- and left-handed faults, right- and left-lateral faults and wrench faults. The San Andreas fault in California is a right-lateral or slip fault.

Relative movement in the vertical direction along the plane of the fault results in a dip-slip fault. Dip-slip faults in which the block containing the hanging wall moves down relative to the foot wall are known as normal faults. The reverse relative motion results in a thrust fault when the dip angle is less than or equal to 45°; it is a reverse fault when the angle is greater than 45°. Motion that follows a direction intermediate between the strike-slip and the dip-slip produces a diagonal-slip or oblique-slip fault.

The displacement that has occurred between the two walls of the fault is often called the "throw and heave" of the fault. The throw is represented by the vertical displacement, while the heave is the horizontal displacement. All faults so far mentioned have involved motions in which structures within adjacent blocks have remained parallel during movement. It is also possible that adjacent blocks may twist or rotate relative to one another; these are called rotational or scissor faults. See Plate Tectonics; Seismology. (J.W.S.)

Fault Scarp. A steep slope or cliff produced by unequal vertical movements of adjacent blocks of the earth's *crust that are separated by a crack or fissure. The face of the fault scarp is generally smooth and straight and can be recognized in areas subjected to recent earthquakes. Older fault scarps may not be as readily recognized because erosion may reduce their height and smoothness.

Fauna. In paleontology, a group of fossilized animal remains representative of the animal population in a given part of the geologic or rock record.

Fayalite. A mineral, the iron end-member of the *olivine mineral group.

Feldspar. The most important group of rock-forming minerals. The importance of the feldspars rests in their abundance: They make up about 60 percent of the rocks of the earth's crust and are far more abundant than any other mineral. As essential constituents of most *igneous rocks, the kind and amount of feldspar present is used as the basis for the classification of these rocks. They are also found in many *metamorphic rocks and in some *sedimentary ones as well.

Feldspar is important economically both as an industrial and a gem mineral. In the United States alone many hundreds of thousands of tons of ground feldspar is used in the glass and ceramic industries. A small amount of transparent yellow feldspar, mostly from Madagascar, has been cut as gemstones. But far more abundant and widespread feldspar gems are the varieties of *moonstone, which have an opalescent play of colors, and *aventurine, which has internal reflections from red or green internal inclusions. The deep green variety of microcline known as *amazonite is also cut for gems and is used as an ornamental material.

Although feldspar crystallizes in both the *monoclinic and *triclinic crystal systems, all types have common properties that serve to identify them as feldspar and to distinguish them from other minerals. In crystals their forms and general appearance are similar, they all have two good cleavages at 90° or nearly 90° and a hardness of about 6. The color is more variable; most feldspar is colorless, white or light gray, but it may also be yellow, brown, red, green or black.

The average grain of feldspar as found in most rocks is small, usually less than 5 mm across. However, individual crystals in *pegmatites may be gigantic, with dimensions reaching 10 m or more. The largest crystals ever reported are feldspar from pegmatites.

The feldspars are aluminum silicated with potassium, sodium-calcium, and (rarely) barium. They can be divided into two groups: the potassium feldspars that are partly triclinic, and the sodium-calcium feldspars (*plagioclase) that are all triclinic. Barium feldspar, celsian, is rare and is unimportant as a rock-forming mineral. Potassium feldspar, also called potash feldspar, includes the monoclinic form (*orthoclase) and the triclinic form (*microline), both potassium aluminum silicates. The variety *sanidine is an orthoclase containing sodium substituting for some potassium. The plagioclase feldspars form a complete series from the pure sodium member, albite (sodium aluminum silicate) to the calcium member of the series. With compositions intermediate between these two extremes, they are given the following names in order of increasing content of calcium: albite, oligoclase, andesine, labradorite, bytownite and anorthite. (C.S.H.)

Feldspathoid. A mineral chemically similar to *feldspar but containing less silica. The feldspathoids are a group of rock-forming aluminosilicate minerals of potassium, sodium and calcium. They take the place of feldspars in *igneous rocks that crystallize from magmas too low in silica to form feldspar, or that contain more alkalies and aluminum than can be housed in the feldspar structure. The most abundant feldspathoids are nepheline and leucite; less common are sodalite, haüynite and noselite.

Felsenmeer. See Frost Action.

Felsite. A light-colored *igneous rock composed essentially of a microcrystalline aggregate of *quartz and *feldspar. Quartz, feldspar and (more rarely) dark minerals may be present as *phenocrysts; the rock is then called a felsite porphyry. With microscopic study, the rocks grouped under the term felsite can be given more specific names: rhyolite, trachyte, quartz latite, latite, dacite and andesite.

Ferberite. A mineral, the oxide of iron and tungsten. It is the iron member of the *wolframite series.

Ferromagnesian. A term applied to rock-forming minerals in which iron and magnesium are present as essential elements. The principal minerals are *amphibole, *pyroxene, *olivine and *biotite; these are also called the dark minerals.

Filter Feeder (or Suspension Feeder). Any ocean-bottom invertebrate that feeds on suspended organic material in the water. They are selective feeders, taking in only food particles rather than all suspended matter. Most of them are sessile, or attached forms, and include clams, brachiopods, barnacles, many marine worms, sponges and sea lilies. All of these organisms rely on currents to carry the suspended material to them. They then extract their food by means of special filtering devices that differ with each type of organism.

Fiord (or Fjord). A narrow, elongated deep-water arm of the ocean, found along mountainous, glaciated coasts in high latitudes. A fiord may extend inland for scores of kilometers and is commonly hundreds of meters deep. The side walls are very steep and are characterized by *truncated spurs, *hanging valleys and high waterfalls. The combined subaerial and submarine transverse cross section is generally U-shaped, and the floor of the fiord is characterized by a deep basin. A fiord results from the submergence of a glaciated valley commonly formed by a tidewater glacier. The threshold toward the ocean side of the basin may be *moraine or *bedrock. If bedrock, it marks the place where the thinning out of the glacier or its flotation in the ocean decreased glacial erosion. Fiords, which rank among the earth's great scenic features, are found in Alaska, British Columbia, Labrador, Norway, Greenland, New Zealand, Chile and Antarctica. (R.L.N.)

Fireclay. A clay capable of withstanding high temperature without disintegrating or becoming soft and pasty. Nonplastic fireclays, which are extremely hard, are called flint clays. Much fireclay is derived from underclays that occur beneath *coal beds, but not all underclays are fireclays. Fireclay is rich in hydrous aluminum silicates but is deficient in iron, calcium and alkalies; the primary clay minerals are *kaolinite and illite. Fireclays form where surface conditions are favorable for leaching out most minerals except for kaolinite and illite. Groundwater, aided by vegetation, removes the more soluble substances. Fireclay is used to manufacture firebrick (to line kilns), clay crucibles, pots for molten glass and as a binder in molding sands. (W.H.M.)

Flame Test. A method used in determinative mineralogy to detect the presence of certain elements. The test is made by heating a small mineral fragment over a Bunsen burner. If the mineral contains an element that will volatilize, it will impart a characteristic color to the flame. The yellow flame of sodium, the green flame of barium and the crimson flame of strontium are good evidence of these elements.

Flash Flood. A sudden, rapid increase of discharge in a river. A flash flood is marked by a precipitous rise in water level that forms a wall of water moving downstream. Flash floods are common in arid or in semiarid regions, where an intense local rainstorm often generates excessive runoff in an otherwise dry channel. Flash floods may occur in humid regions if conditions in the *watershed are such as to suddenly concentrate the amount of runoff during an intense storm.

Flint. A fine-grained form of *quartz that breaks with a *conchoidal fracture yielding sharp edges. Flint and *chert are similar in composition and texture, but flint is usually in dark nodules in chalk whereas chert is lighter in color and is found in bedded deposits.

Flood. The overbank discharge of a stream. Most streams overflow their banks on an average of every 1.2 years. Although looked on as a disaster by man because he builds homes and industries on river *floodplains, a flood is a natural event that often has beneficial results. The water car-

1

Flood Discharge in Cubic meters/second	Magnitude	Recurrence Interval in Years
1988	1	200
1858	2	100
1706	3	50
1580	4	25
1378	5	10
1197	6	5
924	7	2
696	8	1.25
596	9	1.11
405	10	1.01

Data adapted from U.S. geological Survey, Water Resources Branch Water record
2 from 1913—1967.

Fiord
1 An arm of the sea formed by the partial submergence of a glaciated mountain coast, a fiord such as Nordfjord, near Olden, Norway, penetrates deep inland.

Flash Flood
2 Peak flood discharges and recurrence intervals in the Susquehanna River at Conklin, N.Y.

ries sediment, which it deposits on the floodplain to renew and fertilize it; the water sinks into the soil to recharge the groundwater.

One way to minimize the disastrous effects of floods is to predict their magnitude and frequency. A flood frequency analysis is made by listing flood discharges according to magnitude.

From this data the time interval at which such discharges can be expected is calculated. It is called the recurrence interval. Each flood discharge is then plotted against its recurrence interval, and the points form the frequency curve. These curves are used by engineers in building bridges and roads. See Flash Flood. (M.M.)

Floodplain. The flat area adjacent to a river. It is subject to flooding and is composed of *alluvium deposited by the river. Although the surface is mostly flat, a floodplain does contain topography on a small scale, as evidenced by hollows, swales and low ridges. The features are bars and former channelways. In addition, the floodplain surface often has *oxbow lakes, *levees, *sloughs and *bayous. The surface features demonstrate the development of the floodplain by erosion and deposition. As a *meandering stream moves across and along the unconsolidated floodplain material, it undercuts the banks in one place and deposits material in another. Bars may be built up on the inside of meander bends while the outside bank of the bends are eroded. In a *braided stream, bar building takes place in the channel; with stabilization or abandonment these become part of the floodplain. Some alluvium, such as levees and coarse surface material, is added by flooding. Other areas of sedimentation on the floodplain include tributary junctions, abandoned channels (sloughs) and lakes. The growth of vegetation often seems to aid in floodplain growth by diminishing erosion and acting as sediment traps. Floodplain deposits are classified as (1) deposits of lateral accretion (point bars), (2) deposits of vertical accretion (overbank deposits), (3) colluvium (slump material from valley sides) and (4) channel fill (channel bars; abandoned channels). (M.M.)

Floss Ferri. A variety of *aragonite that occurs in iron-ore deposits in forms resembling coral.

Flow Regime. The turbulent state of water in a channel, which determines the form of material on the bed and transportation of the load. A tranquil regime, called stream-

ing flow, occurs when velocity is low and the *Froude number is less than one. At Froude numbers much below one, the bottom of the channel has ripples on it, the water surface is placid and the sediment transport is low. As the Froude number increases—but is still less than one—the ripples on the bottom become dunes, which generate eddies and "boils" on the surface of the water. This turbulence moves grains up and over the back of the dunes and down the front, causing sediment and dunes to move downstream as bedload.

When the Froude number is greater than one, flow is in the upper rapid flow regime and is called shooting flow. There is a transition stage from streaming to shooting flow during which the bed form is a plane. Then, as the Froude number increases beyond one, the bed changes so the bed and water surfaces are parallel. When the amplitude of water waves is greater than the amplitude of the sand bed waves, antidunes form on the bed; sediment scoured from the face of the antidune is deposited on the back of the next antidune downstream. When the amplitude of the water waves is about twice that of the antidunes, the waves break, much turbulence is created and large amounts of sand are lifted into suspension. (M.M.)

Fluorescence. The property of emitting light as a result of absorbing light. Some minerals emit visible light when exposed to radiation of short wavelengths such as ultraviolet light X rays or cathode rays. Ultraviolet light is most easily produced and is thus most commonly used. Some minerals will fluoresce only in short-wave ultraviolet light, others only in long-wave ultraviolet light. Fluorescence is an unpredictable property, for one specimen may fluoresce brilliantly whereas another of the same mineral may not react at all. The phenomenon appears to be related to small amounts of impurities. Some minerals that frequently fluoresce—but not invariably—are *scheelite, *willemite, *diamond and *autunite.

From the zinc mines at Franklin, N.J., have come the world's outstanding fluorescent mineral specimens. A score of minerals from this locality fluoresce, the most spectacular being *calcite and willemite. When subjected to ultraviolet light, the calcite glows a pinkish-red like a hot coal and the willemite becomes a vivid yellow-green. These and other selected minerals make a pleasing display, drab stones suddenly bursting into a variety of brilliant colors.

Fluorescence also has practical uses in geology. The tungsten ore scheelite and

several of the uranium minerals usually fluoresce. It is thus possible to detect their presence in a rock by examining the outcrop at night with a portable ultraviolet light. Oil also fluoresces, and its presence can be determined by examining surface cuttings in ultraviolet light. (C.S.H.)

Fluorite. Calcium fluoride, the most common and abundant mineral containing fluorine. Sold under the name fluorspar, it is an important industrial mineral; over 1 million tons are consumed annually in the United States. Fluorite is isometric and commonly occurs in well-formed cubes which in places may be as large as 30 cm (11½ inches) on an edge. The mineral may also be without crystal form, occurring in crusts as aggregates with a fibrous texture. There is perfect octahedral cleavage, and fine cubic crystals must be handled with care, for a slight blow will chip off the corners along this cleavage. Crystals exhibit a great range of colors; they may be various shades of red, yellow, green, blue, brown, purple or colorless. A single crystal may show bands of different colors parallel to the cube faces. Some fluorite, but not all, will fluoresce in ultraviolet light. Because early studies of fluorescence were carried out on fluorite, the phenomenon takes its name from the mineral.

Fluorite occurs in many different geologic environments, indicating deposition under a wide range of pressure and temperature. It is found as an accessory mineral in *igneous rocks and *pegmatites, and it is also found lining cavities in *limestone. It usually occurs in hydrothermal veins, in which it may be the principal mineral, or as a *gangue mineral associated with metallic ores. Most commercial fluorite is found in vertical veins or in horizontal bedded deposits either replacing the earlier rock (usually limestone) or filling open spaces. In southern Illinois, the major producing area of the United States, both types of deposits are found. At Rosiclare, Ill., fluorite without crystal form occurs in fissure veins that are 12 m wide in places, whereas 30 km away, at Cave-in-Rock, flat-lying open spaces are lined with coarsely crystalline aggregates.

Fluorite has been known since ancient times as an ornamental material and was used by the Romans for carved drinking cups, bowls and vases. Many other peoples, including the Chinese and the American Indians, used it for ornaments and images. It became an important industrial mineral toward the end of the 19th century when it was used as a flux in the basic open-hearth method of steel making. This use continues, but a greater amount is consumed today by the chemical industry.

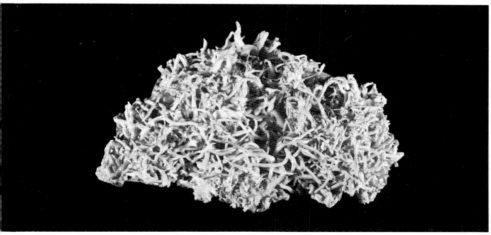

3

Floodplain

1 *Floodwaters periodically overflow the banks of this curved channel and spread across the flat countryside, depositing material to build up a floodplain.*

Flos Ferri

2 *A variety of aragonite, this mineral is deposited by groundwaters rich in calcium carbonate.*

Fluorite

3 *Interpenetrating translucent or transparent crystals of bold cubic form are characteristic of the mineral's appearance in many locations.*

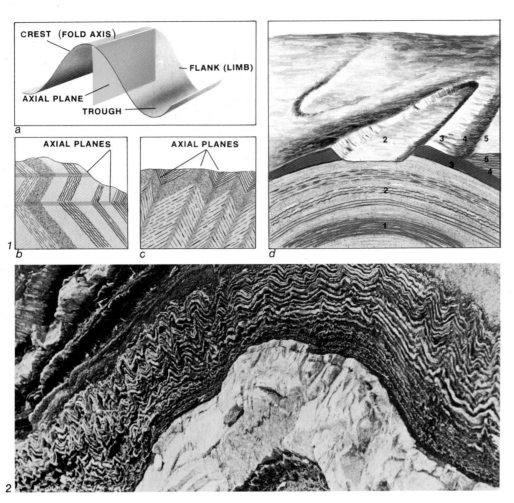

CREST (FOLD AXIS)

FLANK (LIMB)

AXIAL PLANE

TROUGH

a

AXIAL PLANES

b

AXIAL PLANES

c

d

2

3

It is the raw material used in the preparation of hydrofluoric acid, necessary for the production of metallic aluminum which is, in turn, essential in the manufacture of synthetic *cryolite. Fluorite is used in the ceramic industry to make white and colored opal glasses and enamels and in enamel coatings for such products as steel stoves, refrigerators and bathtubs. A small but important use of fluorite was as lenses and prisms in optical equipment, but today most optical material is synthetic. (C.S.H.)

Focus. A point within the earth that represents the origin of an earthquake or, more precisely, from which earthquake waves or seismic waves are generated.

Foehn. A dry, warm wind that results when strong winds passing over a mountain range are forced to descend on the lee side. See Adiabatic Process.

Fold. A structure formed by the bending or buckling of the earth's *crust. Folds vary greatly in size, with widths or wavelengths varying from a few millimeters to hundreds of kilometers. Folded structures can occur in any type of rock but are best displayed in layered rocks, such as *sedimentary ones. Because of the complexity of structural forms that folds take, they are subdivided into many types. Classification by geologists has resulted in a vast variety of terms, some of which are synonymous. Generally folds are identified by (1) the attitude of their axial plane or axis,(2) their overall appearance and (3) their mode of formation. To help classify fold structures various components are given specific names.

A simple upturned fold is known as a symmetrical *anticline. The sides of the fold are called the limbs or flanks while the highest part is the crest. Crests are important to petroleum geologists because they act as structural traps in which gas and oil accumulate. The geometry of a fold is described by the direction of trend of the axis, the angular value of its plunge measured from the horizontal, and the strike and dip of its axial plane.

The top of the fold is known as its crest, fold axis or hinge line. The hinge line represents the point of maximum curvature within the fold, and each bed or layered surface has a hinge line. In folds that are inclined away from the vertical, the hinge line and crest may not coincide. A plane that passes through or connects all the hinge lines of a fold is referred to as the axial plane of the fold. The down-curved portion of a fold (trough) represents part of a *syncline. Anticlines and synclines generally appear together in multifolded portions of the earth's crust.

When the attitude of the axial plane and therefore of the fold becomes inclined, the fold is known as asymmetrical. Further inclination of the axial plane results in an overturned fold or overfold, one in which both limbs of the fold dip in the same direction. The limb that has reversed direction has now had part of its original upper surface inverted, that is, the upper part of the surface is on the underside of the limb. Recumbent folds are inclined to such a degree that the axial plane is horizontal.

In the Alps and in parts of New England portions of the crust have been deformed into large sheetlike rock structures known as nappes that are totally or partly the result of recumbent folding. Not all folding results in the relatively simple structures described above. In many cases the hinge line is no longer horizontal and produces plunging synclines and anticlines. In other folds the axis may plunge in opposite directions from a central point (*culmination), forming domes; or the hinge line may plunge toward a central point (*depression), forming basins. (J.W.S.)

Foliation. In geology, the planar arrangement of textural or structural features in any variety of rock. Primary foliation is that which develops at the time a rock is formed; secondary foliation develops afterward. Primary foliation appears in *igneous or *sedimentary rocks. Igneous rocks develop foliation in their molten state; as *crystals form within the molten material they may align themselves into parallel planes guided by the laminar flow of the remaining liquid.

Minerals most likely to form foliate patterns are those whose crystals form plates, such as *mica, or rods, such as *hornblende. Rodlike crystals will produce foliation if their longest dimensions lie in parallel planes; when rodlike minerals in a given plane are parallel to each other the resultant texture is then called lineation. Sedimentary rock may also possess a primary foliation parallel to the *stratification of the rock. Platelike or rodlike crystals of sedimentary rock may align in parallel layers.

Secondary foliation results from the process of metamorphism of former igneous or sedimentary rocks. In secondary foliation, which is commonly referred to as *cleavage or rock cleavage, the rock tends to break along planes which may or may not coincide with the original layering of a sedimentary rock. (J.W.S.)

Food Chain. Also known as the organic cycle or the ecological pyramid, this term refers to the interdependence among all organisms. The fundamental process is

4

Fold

1 Features of a simple upright fold (a), or anticline, are flanks (or limbs), crest and trough. In recumbent folds (b), the axial planes are horizontal and the strata of one limb are upside down. In overturned folds (c), the axial planes are tilted beyond the vertical, and the limbs dip in the same direction. A cross section (d) shows the relationship of youngest (5) to oldest (1) strata.

2 A series of crimped folds in

metamorphic rock was caused by compressive stress.

3 A V-shaped syncline and two arched anticlines are exposed at a highway cut near Palmdale, Calif.

Foliation

4 Erosion has split this metamorphic rock into parallel slabs, each of different mineral composition. This type of foliation of the original sediments before metamorphism is primary foliation.

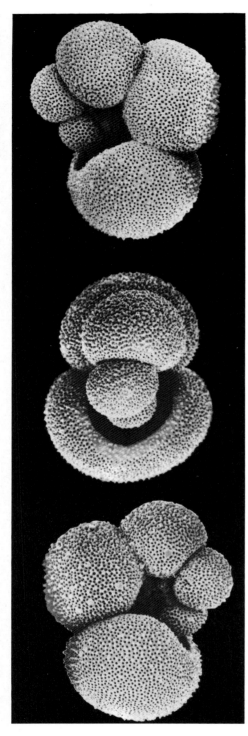

Foraminifera
Skeletons of the minute protozoan
Globigerina *occur in great abundance*
on the ocean floor.

photosynthesis, in which plants produce organic material by combining water, carbon dioxide and nutrients in the presence of sunlight. In the sea the food chain goes from algae to sharks. Small sea animals such as zooplankton feed on the plants, and are called herbivores or primary consumers. These in turn are eaten by bottom-dwelling invertebrates or small fish, which are called carnivores or secondary consumers. Depending on the composition of the environment, there may be a number of levels of carnivores, but they eventually culminate in a top carnivore, which in most areas of the ocean is a shark. During all of these stages, there is a loss of oxygen because it is utilized by animals, which also produce carbon dioxide. In addition, various nutrients result from waste products and from the decay of dead organisms; they are utilized primarily by plants. (R.A.D.)

Fool's Gold. One of several minerals that can be mistaken for gold. The most common of these is iron sulfide, *pyrite, a hard brassy mineral. Another, of a more golden appearance, is the copper iron sulfide *chalcopyrite; a third is *vermiculite, a yellow glittering mica. All of these can be proved not to be gold by scratching with a knife. Gold is soft and can be cut with a knife, and scratching leaves a smooth, shining furrow. Pyrite is too hard to be scratched, chalcopyrite scratches but leaves a greenish powder, and vermiculite splits into thin flakes. (C.S.H.)

Foot Wall. The lower wall of a fault as distinct from the upper wall, which is called the hanging wall. In mining geology the term is used to indicate the bottom of a horizontal or an inclined ore-bearing vein or bed, whether or not it lies along a fault.

Foraminifer. A member of subclass (or order) Foraminifera of the *protist phylum Protozoa. "Forams," which range from *Cambrian to *Holocene (Recent) in age, typically secrete a microscopic shell composed of many chambers pierced by minute pores or perforations. The shell is commonly made of calcium carbonate, but may consist of silica or chitin. Certain species build agglutinated or arenaceous shells by cementing together minute mineral grains or other foreign particles. Typically microscopic in size, foraminifers are especially valuable in the search for petroleum because they are not normally destroyed by the drill bit. Some are easily seen by the naked eye, and the shells of a few species may be as much as 5 cm (2 inches) in diameter.

The shells of some forams, such as *Globigerina,* have accumulated in great numbers, forming thick deposits on many present-day ocean bottoms. Known as globigerina ooze, this soft, fine-grained, limy material now covers about 130 million square kilometers of the sea floor. Similar deposits accumulated in the geologic past and have since been converted into thick beds of *chalk or *limestone. (W.H.M.)

Force. An influence that tends to set a body in motion or to change the direction of motion. Force has magnitude and direction, both represented graphically by vectors (arrows with length proportional to the magnitude of the force and with orientation representing the direction of applied force). Structural geologists use force vectors to analyze various problems associated with deformation of the *crust.

Formation, Geologic. A fundamental geologic unit used in the local classification of rocks or strata. Recognized and mapped as single units, formations are classified by distinctive physical and chemical characteristics of the rocks, not by geologic time. The names of formations are commonly derived from geographic names of places where they were first described, combined with the names of the predominant rock that makes up the bulk of the formation; for example, Beaumont Clay and St. Louis Limestone. However, if the formation consists of various types of rocks (such as alternating sands, clays and limestones) it is referred to as a formation, e.g., the San Joaquin Formation.

Geologic formations may be subdivided into members, which may also be given geographic and/or lithologic (rock type) names. Other smaller units are lentils (smaller lens-shaped rock bodies within the formation), tongues (interfingering or intertonguing bodies of different lithology) and beds (individual rock layers). If several formations have certain definite characteristics in common, they may be spoken of as a group, which is the largest recognized rock unit. (W.H.M.)

Forsterite. A mineral, the magnesium end-member of the *olivine mineral group.

Fossil. The remains of an ancient plant or animal, or evidence of this existence, preserved in the rocks of the earth's crust. The word fossil is derived from the Latin word *fossilis,* meaning "dug up," and most fossils represent the preservable hard parts of some prehistoric organism that once lived in the area where remains were collected.

During the earliest periods of recorded

history, certain Greek scholars were greatly puzzled by the occurrence of the remains of fish and seashells in desert and mountainous regions. In 450 B.C. Herodotus noticed fossils in the Egyptian desert and correctly concluded that the Mediterranean Sea had once covered that area. Aristotle in 400 B.C. declared that fossils were organic in origin but that they were embedded in the rocks as a result of mysterious plastic forces at work within the earth. One of his students, Theophrastus (about 350 B.C.), also assumed that fossils represented some form of life but thought they had developed from seeds or eggs planted in the rocks. Strabo (about 63 B.C. to A.D. 20), noting the occurrence of marine fossils well above sea level, correctly inferred that the rocks containing them had been subjected to considerable elevation.

During the "Dark Ages" fossils were variously explained as freaks of nature, the remains of attempts at special creation and devices of the devil to lead men astray. These superstitions and the opposition of religious authorities hindered the study of fossils for hundreds of years. About the middle of the 15th century the true origin of fossils was generally accepted. They were recognized as the remains of prehistoric organisms, but these remains were considered remnants of the Great Flood as recorded in Scripture. The resulting controversy between scientists and theologians lasted for about 300 years. During the Renaissance several early natural scientists, notably Leonardo da Vinci, concerned themselves with fossils. Leonardo contended that the Flood could not be responsible for all fossils nor for their occurrence in the highest mountains. He reaffirmed the belief that fossils were indisputable evidence of ancient life, and that the sea had once covered northern Italy. He explained that the remains of the animals had been buried in the sea floor, and that at some later date this ocean bottom was elevated above sea level to form the Italian peninsula. In the late 18th and early 19th centuries the study of fossils became firmly established as a science, and since then fossils have become increasingly important to geologists.

The majority of fossils are found in marine *sedimentary rocks that were formed when salt-water sediments, such as limy muds, sands or shell beds, were compressed and cemented together to form rocks. Only rarely do fossils occur in *igneous and *metamorphic rocks. Igneous rocks were once molten and had no life in them. Metamorphic rocks have been so greatly changed that any fossils in the original rock have usually been destroyed

or altered. But even in the sedimentary rocks only a minute fraction of prehistoric plants and animals have left any record. This is understandable in view of the rather rigorous requirements of fossilization.

Although many factors ultimately determine whether an organism will be fossilized, the three basic requirements are: (1) The organism must possess hard parts, whether shell, bone, teeth or the woody tissue of plants. However, under very favorable conditions it is possible for even such fragile material as an insect or a jellyfish to become fossilized. (2) The organic remains must escape immediate destruction after death. If the body parts of an organism are crushed, decayed or badly weathered, this may alter or destroy the fossil record of that particular oganism. (3) The organism must undergo rapid burial in a material capable of retarding decomposition. The type of material burying the remains usually depends upon where the organism lived. The remains of marine animals are common as fossils because they fall to the sea floor after death and are covered by soft muds, which will be the shales and limestones of later geologic periods. The finer sediments are less likely to damage the remains; certain fine-grained Jurassic limestones in Germany have faithfully preserved such delicate specimens as birds, insects and jellyfishes.

Ash falling from nearby volcanoes has been known to cover entire forests, and some of these fossil forests have been found with the trees still standing and in an excellent state of preservation. Quicksand and tar are also commonly responsible for the rapid burial of animals. The tar acts as a trap to capture the beasts and as an antiseptic to retard the decomposition of their hard parts. The Rancho La Brea tar pit at Los Angeles is famous for the many fossil bones recovered from it, including the saber-tooth cat, giant ground sloths and other now-extinct creatures. The remains of certain animals that lived during the Ice Ages have been incorporated into ice or frozen ground, and some of these frozen remains are remarkably preserved.

Although untold numbers of organisms have lived on the earth, only a minor fraction left any record of their existence. Even if the basic requirements of fossilization have been fulfilled, there are still other reasons why some fossils may never be found. For example, many have been destroyed by erosion, or their hard parts have been dissolved by underground waters. Others may have been preserved in rocks later subjected to violent physical changes, such as folding, fracturing or melting. Under these circumstances a fossiliferous marine limestone may be

changed into a marble, and any traces of organisms present in the original rock completely or almost completely obliterated. Much of the record is doubtless contained in deposits inaccessible for study. In addition, there may be well-exposed fossiliferous rocks in parts of the world that have not been studied geologically. In other areas fossils may go undetected because of poor outcrops of a formation or the scarcity of specimens. Another common problem is that organic remains may be fragmentary or poorly preserved and do not adequately represent the organism.

Gaps in the fossil record, moreover, increase as we go backward in time. Older rocks have had more opportunity to be subjected to the destructive forces that obscure the paleontological record. A further complication is the fact that many earlier organisms are difficult to classify because they are so different from organisms living today. Even so, the large number of organisms preserved as fossils have provided a good record of the past.

The many different ways in which plants and animals may become fossilized is usually dependent upon (1) the original composition of the organism, (2) where it lived and (3) the forces that affected it after death. Most paleontologists recognize four major types of preservation, each based upon the composition of the remains or the changes that they have undergone.

The original soft parts of organisms are preserved only when the organism has been buried in a medium capable of retarding decomposition of the soft parts. Such materials include frozen soil or ice, oil-saturated soils and *amber (fossil resin). It is also possible for organic remains to become so desiccated that a natural mummy is formed. This usually occurs only in arid or desert regions and when the remains have been protected from predators and scavengers.

Probably the best-known examples of preserved soft parts of fossil animals have been discovered in Alaska and Siberia. The frozen tundra of these areas has yielded the remains of large numbers of frozen woolly mammoths—a type of extinct elephant. Many of these huge beasts have been buried for as long as 25,000 years, and their bodies are exposed when the frozen earth begins to thaw. Some carcasses have been so well preserved that, when exposed, their flesh has been eaten by dogs and their tusks sold by ivory traders. Many museums now display the original hair and skin of these elephants, and some have parts of the flesh and muscle preserved in alcohol.

Original soft parts have also been

recovered from oil-saturated soils in eastern Poland. These deposits yielded the well-preserved nose-horn, a foreleg and part of the skin of an extinct rhinoceros. The natural mummies of ground sloths have been found in caves and volcanic craters in New Mexico and Arizona. The extremely dry desert atmosphere permitted thorough dehydration of the soft parts before decay set in, and specimens with portions of the original skin, hair, tendons and claws have been discovered.

One of the more interesting and unusual types of fossilization is preservation in amber. This type of preservation occurred when ancient insects were trapped in the sticky gum exuded by certain coniferous trees. In time this resin hardened, leaving the insect encased in a tomb of amber; some insects and spiders have been so well preserved that even fine hairs and muscle tissues may be studied under the microscope.

Although the preservation of original soft parts has produced some interesting and spectacular fossils, this type of fossilization is relatively rare, and the paleontologist must usually work with remains that have been preserved in stone.

The original hard parts of organisms may also be preserved. Almost all plants and animals possess some hard parts, such as the shell material of clams, oysters or snails; the teeth or bones of vertebrates; the exoskeleton (outer body covering) of crabs; or the woody tissue of plants that can become fossilized. These parts are composed of minerals capable of resisting weathering and chemical action; fossils of this sort are relatively common. Hard parts composed of calcite (calcium carbonate), such as the shells of clams, snails and corals, are very common among the invertebrates. Many of these shells have been preserved with little or no evidence of physical change.

The bones and teeth of vertebrates and the exoskeletons of many invertebrates contain large amounts of calcium phosphate. Because this compound is particularly weather resistant, many phosphatic remains such as fish teeth are found in an excellent state of preservation. Many organisms having skeletal elements composed of silica (silicon dioxide) have been preserved with little change. The siliceous hard parts of many microfossils and certain sponges have become fossilized by *silicification. Some organisms have an exoskeleton composed of chitin, a material similar to that of fingernails. The fossilized chitinous exoskeletons of *arthropods and other organisms are commonly preserved as thin films of carbon because of their chemical composition and method of

burial. Carbonization is an especially common form of preservation of the remains of plants. Because organic remains of most plants and animals undergo change after burial, the altered hard parts of organisms are especially common in the fossil record. This alteration is usually determined by the composition of the hard parts, the environment in which the plant or animal lived and the conditions under which it was buried.

*Carbonization (or distillation) takes place as organic matter slowly decays after burial. During the process of decomposition, the organic matter gradually loses its gases and liquids, leaving only a thin film of carbonaceous material. This is the same process by which coal is formed, and large numbers of carbonized plant fossils have been found in many coal deposits. In many places the carbonized remains of plants, fishes and invertebrates have been preserved in this manner, and some of these carbon residues have accurately recorded the most minute structures of these organisms.

Fossils may also be preserved by *permineralization or petrifaction. This occurs when mineral-bearing groundwaters deposit their mineral content in the empty spaces of the hard parts, making them heavier and more resistant to weathering. Some of the more common minerals so deposited are calcite, silica and various compounds of iron. *Replacement or mineralization takes place when the original hard parts of organisms are removed after being dissolved by underground water. This is accompanied by almost simultaneous deposit of other substances in the resulting voids. In some replaced fossils the original structure will be destroyed by the replacing minerals.

Fossils may be not only plant and animal remains but any evidence, or trace, of their former existence. Trace fossils normally furnish considerable information concerning the characteristics of the organism responsible for them. Many shells, bones, leaves and other forms of organic matter are preserved as *casts and molds. If a shell is pressed down into the ocean bottom before the sediment has hardened into rock, it may leave the impression, or mold, of its exterior. If this mold is later filled with another material, a cast is produced that will show the original external characteristics of the shell. External molds show the external features of the hard parts, and internal molds show the inner parts.

Some animals have left evidence of their former existence in the form of tracks, trails, borings and burrows. Some of these, such as footprints, indicate not only the type of animal that left them but often

provide information about the animal's environment. Thus dinosaur tracks not only reveal the size and shape of the foot but yield clues to the weight and length of the animal; the type of rock containing the track might help determine the conditions under which the dinosaur lived. Some of the world's most famous dinosaur tracks are found in the Lower Cretaceous limestones in Somervell County, Texas. These footprints, which are about 110 million years old, were discovered in the bed of Paluxy Creek near the town of Glen Rose. Great slabs of limestone containing these tracks were transported to museums throughout the world, mute evidence of the gigantic size of these great reptiles.

Invertebrates also leave tracks and trails, and these markings may be seen on the surface of many sandstone and limestone deposits. These may be simple tracks or they may be the burrows of crabs or other burrowing animals. Such markings provide evidence of the manner of locomotion of these organisms and of the environment they inhabited. Burrows— that is, tubes or holes in the ground, wood, rock or some other substance that have been made by an animal for shelter or in search of food—may later become filled with fine sediments and thus be preserved. On rare occasions the remains of the animal making the burrow have been found in the sediments filling the tubes. Burrows are made on soft sea bottoms by worms, *arthropods, *mollusks and other animals. The fossil burrows or borings of such species of mollusks as *Teredo,* a wood-boring clam, and *Lithodomus,* a rock-boring clam, also occur frequently. Tubular structures believed to be worm burrows are among the oldest known fossils, and many of the most ancient sandstones contain such tubes.

Borings are holes made by animals on other organisms for the sake of food, attachment or possibly shelter. Such holes frequently occur on fossil shells, wood and other organic objects, and are considered fossils. Carnivorous boring snails drill through the shells of other mollusks in order to reach the soft parts of the animal. The neatly countersunk holes bored by similar fossil snails are common on many ancient mollusk shells.

Fossils are useful in tracing the development of plants and animals because the fossils in the older rocks are usually primitive and relatively simple. Similar specimens living in later geologic time reveal that the fossils become more complex and advanced in the younger rocks.

Certain fossils are valuable as environmental indicators. For example, the reef-building corals appear to have always lived under much the same conditions as today. Hence if the geologist finds fossil *coral reefs where they were originally buried, he can be reasonably sure that the rocks containing them were formed from sediments deposited in warm, fairly shallow salt water. A study of the occurrence of such marine fossils makes it possible to outline the location and extent of prehistoric seas. The fossils present may also indicate the depth, temperature, bottom conditions and salinity of these ancient bodies of water.

One of the more important uses of fossils is in *correlation—the process of demonstrating that certain rock layers are closely related. By correlating or "matching" beds containing specific fossils, geologists can determine the distribution of a geologic formation in a given area. Certain fossils have a very limited vertical or geologic range and a wide horizontal or geographic range. In other words, they lived only a relatively short time in geologic history but were rather widely distributed. Fossils of this type are known as *guide fossils (or index fossils). They are especially useful in correlation because they are normally associated only with rocks of one particular age.

*Microfossils are especially valuable as guide fossils for petroleum geologists. The *micropaleontologist (a specialist in the study of microfossils) washes the well cuttings from the drill hole and separates the tiny fossils from the surrounding rocks. The specimens are then studied under the microscope. Information derived from these minute remains may provide valuable data on the age of the subsurface formation and the possibilities of oil production. Microfossils are so useful in the world's oil fields that some oil-producing horizons have even been named for certain key genera of *foraminifers. Other microfossils, such as fusulinids, ostracodes, spores and pollens, are also used to identify subsurface formations in many other parts of the world.

Although plant fossils are very useful as climatic indicators, they are not very reliable for purposes of correlation. They do, however, provide much information about the development of plants throughout geologic time. See Color Plate "The Fossil Record" and Color Plate "The Tree of Life." (W.H.M.)

Fossil Amphibian. The preserved remnants of a member of the class Amphibia of the phylum Chordata. Amphibians, the earliest of the developed four-legged vertebrates, now include toads, frogs and salamanders. These cold-blooded animals breathe primarily through lungs and

1

2

Fossil Amphibian

1 The bones of Ichthyostega, *one of the earliest land-going vertebrates, were found in Greenland and date from the Late Devonian Period, 340 million years ago. The tail still bears a fishlike fin.*

2 *By the Jurassic Period, 190 million years ago, frogs had developed into the present-day form, with a flat skull, wide mouth and greatly elongated hind legs.*

1

2

3

spend most of their life on land. However, during their early stages of development they live in water, where they breathe by means of gills.

Amphibians apparently evolved from *crossopterygian fishes in the Late *Devonian Period and were relatively abundant in the *Pennsylvanian, *Permian and *Triassic periods. Early species varied greatly in shape and size and have left an interesting fossil record. Most *Paleozoic amphibians were called labyrinthodonts because of the complex infolding of the teeth. They have also been called stegocephalians because of their heavily roofed skulls. Early amphibians had an opening on top of the skull behind and midway between the eyes. This marked the position of the so-called pineal eye, a vestigial structure in all higher vertebrates, including man. (W.H.M.)

Fossil Bird. The preserved remnants of a member of the class Aves of the phylum Chordata. Because of the fragile nature of their bodies, birds are not commonly found as fossils. However, some interesting and important fossil bird remains have been discovered. The oldest was found in Upper *Jurassic rocks exposed in Germany. This primitive bird, named *Archaeopteryx, is little more than a reptile with feathers, and is known from two relatively complete skeletons and a few fragments and a feather of another. These reveal that this creature was about the size of a crow, had sharp teeth, a relatively long neck, a lizard-like tail and three sharp claws on each wing. The feathers clearly indicate that Archaeopteryx was a bird and was warm-blooded. However, such characteristics as the teeth, skull, hip-girdle, hind limbs and solid bone construction are wholly reptilian. These unusual fossils furnish evidence of the close relationship between reptiles and birds, and are among the most important paleontological discoveries.

Birds of the *Cretaceous were more abundant and more specialized than those of the Jurassic. The structure of the skull, limbs and bones was definitely more birdlike, but many species still retained teeth. Cretaceous birds also show considerable evidence of specialization. Some forms, such as Hesperornis, had small, useless wings, long jaws bearing many sharp teeth, and strong feet well suited for paddling. This bird, normally about 1 m high and somewhat similar to the modern loon, appears to have been well adapted to swimming and diving. Ichthyornis, another Cretaceous bird, was smaller than Hesperornis, had well-developed wings and formed habits probably much like those of present-day sea gulls. Much is known about these birds because of unusually well-preserved skeletons collected from Cretaceous limestones in western Kansas.

*Cenozoic birds closely resembled modern species and must have been almost as varied and abundant although they are relatively rare as fossils. Fossil ducks, vultures, owls, crows, pelicans and penguins are among the *Tertiary birds. Many of these were about the same size and had the same habits as their modern counterparts. There were exceptions, however, for one extinct species of penguin attained a height of approximately 1.5 m— considerably taller than any present-day forms.

Of particular interest are the remains of large, flightless birds of the Tertiary Period. One of these, Diatryma, an *Eocene form was about 2 m tall and had strong jaws and a head almost as long as that of a horse. Dinornis, the largest known land bird, stood 3—3.6 m tall; its eggs were almost 30 cm (12 inches) in diameter. These birds lived in New Zealand within historic times and are believed to have been exterminated by the natives. (W.H.M.)

Fossil Fish. The preserved remnants of class Pisces of the phylum Chordata. A fossil of the earliest known fish was found in *Ordovician rocks; it also appears to be the oldest known vertebrate. Named *ostracoderm, these primitive, jawless fishes are related to the living lampreys and hagfishes. The head and foreparts were protected by a bony covering and some had an intricate covering of protective scales. Lacking jaws, they were incapable of biting and apparently were mud feeders. They appeared in the Middle Ordovician, increased in numbers during the *Silurian and were extinct by the end of *Devonian time.

The somewhat more advanced *placoderms were also armored and in addition had jaws. They include the large, sharklike arthrodires, which were heavily armored, had powerful jaws and jointed necks and grew to be as much as 9 m (29 feet) in length. Appearing in Late Silurian time, placoderms were the dominant vertebrates of the Devonian Period but became extinct during the *Permian.

The sharks, rays and other fishes with cartilaginous skeletons did not appear until the Devonian Period but have been abundant since that time. The true bony fishes appeared at the same time; they include trout, perch, eels, seahorses and lungfishes, and are the most highly developed and abundant of all fishes. They possess an internal bony skeleton, well-developed jaws, an airbladder and, typically, an external covering of overlapping scales.

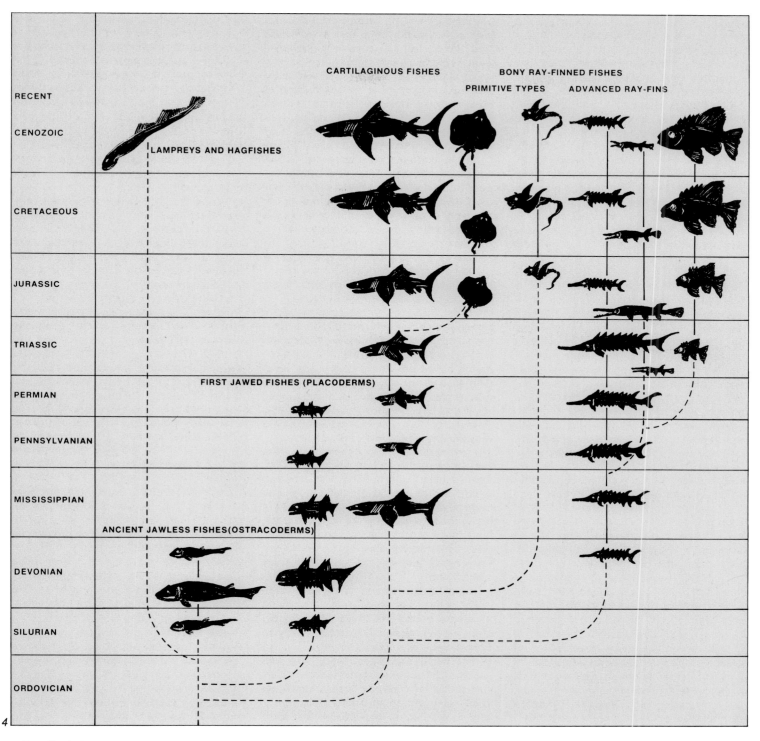

RECENT

CENOZOIC

CRETACEOUS

JURASSIC

TRIASSIC

PERMIAN

PENNSYLVANIAN

MISSISSIPPIAN

DEVONIAN

SILURIAN

ORDOVICIAN

CARTILAGINOUS FISHES

BONY RAY-FINNED FISHES

PRIMITIVE TYPES   ADVANCED RAY-FINS

LAMPREYS AND HAGFISHES

FIRST JAWED FISHES (PLACODERMS)

ANCIENT JAWLESS FISHES(OSTRACODERMS)

4

Fossil Bird

1 Diatryma, *a flightless bird 2 m (7 feet)
tall, probably preyed on small mammals
during the Eocene Epoch, some 50
million years ago.*

Fossil Fish

2 *Many placoderms, primitive-jawed
fishes such as Bothriolepis canadenisis
from the Upper Devonian, had heavily
armored head and body shields.*

3 *The fossil remains of* Acrognathus
dodgei, *a Cretaceous fish much like
modern forms, were found at Mount
Lebanon, Syria.*

4 *The evolution of fishes. The size of each
fish indicates the relative abundance of
that group in a given period.*

They include the *crossopterygians, a primitive group abundant during the Devonian and believed to be the ancestors of the air-breathing amphibians. These unusual fish were originally thought to have been extinct since the end of the *Cretaceous Period, about 70 million years ago. Then, in 1938, a crossopterygian (the *coelacanth) was dredged up by a trawler off the southeast coast of South Africa. Since then several other coelacanths, all belonging to the genus *Latimeria*, have been caught. These specimens have been extremely useful in providing information about a group known previously only as fossils. The remains of bony fish have been found in rocks ranging from Middle Devonian to Holocene in age. (W.H.M.)

Fossil Fuel. A fuel containing solar energy trapped as hydrocarbons (chemical compounds of hydrogen and carbon) in the remains of ancient plants and animals. The term generally applies to any hydrocarbon deposits that may be used for fuel, including *petroleum (oil and natural gas) and *coal, which collectively provide approximately 95 percent of the world's energy. Coal originated as plant matter in ancient swamps and is both abundant and widely distributed. It is the world's greatest single present source of industrial energy, for most of the coal that is mined is burned under boilers to make steam. In addition, liquid fuel can be derived from coal.

Petroleum may be gaseous, liquid or solid substances that occur naturally and consist chiefly of chemical compounds of carbon and hydrogen. Oil and gas probably originated as organic matter deposited with sediments on ancient sea floors and later decomposed chemically. The conversion of the organic material into petroleum is not thoroughly understood but appears to have required vast amounts of time accompanied by temperature increase in the material and compaction of the sediments. Although new petroleum discoveries are continually being made, our expanding economy and modern technology are making unprecedented demands on this valuable and nonrenewable natural resource. More than 90 percent of all petroleum production is used as fuel; the remainder goes into lubricants and into hundreds of manufactured products produced by the petrochemical industry. See Energy Resources. (W.H.M.)

Fossil, Guide. See Guide Fossil.

Fossil Mammal. The preserved remnants of a member of the class Mammalia of the phylum Chordata. Mammals first appeared during the *Jurassic Period and apparently evolved from mammal-like reptiles such as the *therapsids. Although rare during the *Mesozoic, they underwent rapid evolutionary changes during the *Cenozoic, growing extremely large and assuming unusual shapes. Most of these bizarre species became extinct, probably because they were too specialized to adapt to changing conditions. They are well known from their fossils.

The multituberculates were a group of rather small rodent-like creatures that appear to be the earliest herbivorous mammals. They originated in the Jurassic, were probably never very numerous, and became extinct during the early part of *Eocene time. They were followed by somewhat more specialized species such as the edentates, rather primitive mammals that are represented by such living forms as the anteaters, tree sloths and armadillos. They are characterized by poorly developed teeth (in some species a complete absence of teeth) and were common in the southern part of the United States during *Pliocene and *Pleistocene time. One such form, *Megatherium* (an extinct giant ground sloth), was as large as a small elephant, with a short, wide head and powerful legs. It was the ancestor of the present-day tree sloths of South America. The glyptodont also developed at about the same time as the ground sloths and is the forerunner of the armadillo. This armored mammal had a solid, turtle-like shell. A large specimen might be more than 1 m (4 feet) tall and 4.5 m (15 feet) long. Its thick, heavy tail was protected by a series of bony rings, and in some species the end of the tail was developed into a bony, heavily spiked club.

The earliest known mammal carnivores (meat-eaters) are the *creodonts, which appeared during the *Paleocene Epoch. They were followed by carnivores ranging from the size of a weasel to that of a large bear. Their claws were sharp and well developed, but their teeth were not as specialized as those of modern carnivores. Their relatively small brains indicated a very low order of intelligence, and they became extinct at the end of the Eocene. These early meat-eaters were followed by more specialized carnivores that developed throughout the Cenozoic. The sabertoothed cat (or "tiger") *Smilodon* is well known from Pleistocene deposits, as is *Canis dirus,* "the dire wolf." Both of these forms, representing the cat and dog families, have been found in large numbers at the Rancho La Brea tar pits in Los Angeles, Calif.

Among the largest of the *Tertiary mammals were the uintatheres, or Dinocerta. *Uintatherium* had three pairs of blunt horns; the males had dagger-like upper tusks and some were as much as 2 m (7 feet) tall and as large as a small elephant. The small size of the brain in relation to the body suggests that these animals were less intelligent than most mammals of that time.

The earliest proboscideans, the elephants and their relatives, were found in Upper Eocene rocks in Africa. They were about the size of a small modern elephant but had larger heads and shorter trunks. Their development is marked by an increase in size, change in tooth and skull structure and elongation of the trunk. The mammoth and the mastodon are well-known fossil forms. The latter resembled the modern elephants, but the structure of their teeth was quite different. In addition, the mastodon skull was longer than that of an elephant and the tusks were exceptionally large, some reaching a length of 2.7 m (9 feet). The true elephants, or mammoths, had several North American representatives, of which the woolly mammoth is probably best known. This animal lived until the end of the Pleistocene and is known from ancient cave paintings and frozen remains. Evidence from these sources shows that it had a coat of long black hair, with a wool covering beneath it.

The perissodactyls, or "odd-toed animals," are mammals with a greatly enlarged central toe on each limb. Modern representatives include the horses, rhinoceroses and tapirs. Extinct members of this group include the titanotheres, chalicotheres and baluchitheres, all of which grew to tremendous size and took on many unusual body forms. One of the first perissodactyls was *Hyracotherium* (also called *Eohippus),* the earliest known horse, which appeared in Paleocene time. This small animal was about 30 cm (1 foot) high, and its teeth indicate a diet of soft food. Following the first horse, there is a complete series of fossil horses, providing an excellent record of the development of this important animal group.

Titanotheres appeared in the Eocene, at which time they were about the size of a sheep. By Middle Oligocene time they had increased to gigantic proportions but still had very small, primitive brains. *Brontotherium* was a rhinoceros-like titanothere, standing 2 m (8 feet) at the shoulder. A large bony growth from the skull was extended into a flattened horn divided at the top. Despite rapid development during the Tertiary, these beasts became extinct in Middle Oligocene time. *Brontotherium* is the largest land mammal ever discovered in North America.

Although somewhat like titanotheres, the chalicotheres exhibited unique characteristics. *Moropus,* for example, had a

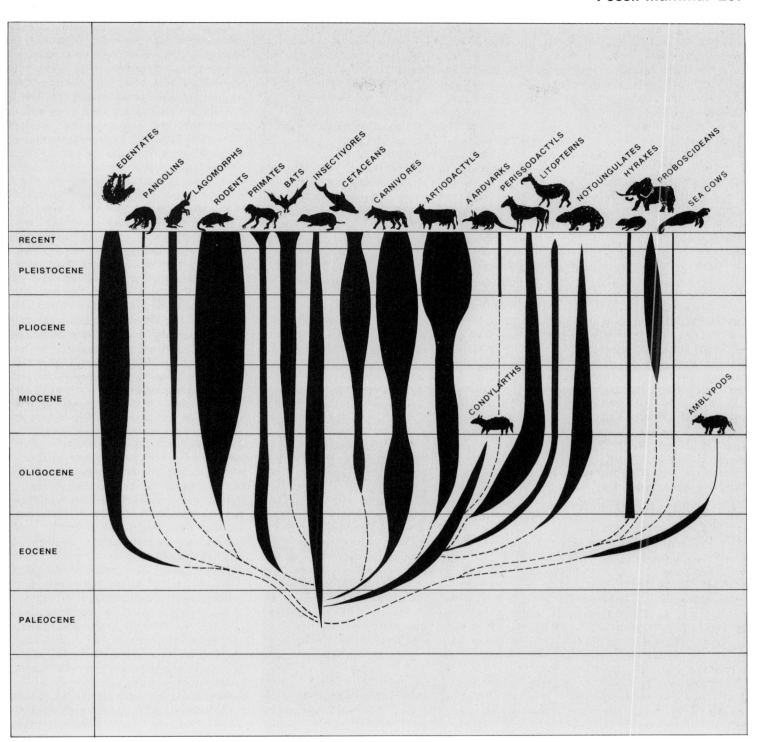

Fossil Mammal
*The evolution of mammals. The width
of the column for each order indicates
the relative abundance of that group
at a given period.*

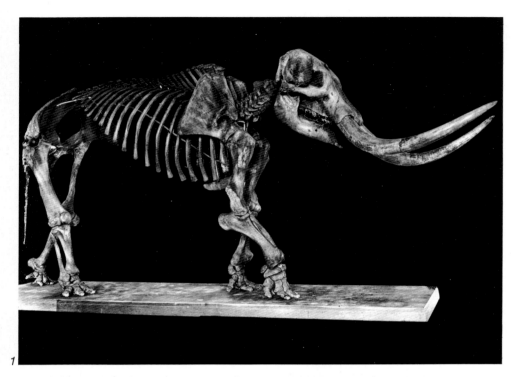

head and neck similar to those of a horse, but its front legs were longer than the hind legs and its feet resembled those of a rhinoceros except that they bore long claws instead of hoofs.

Rhinoceroses are also "odd-toed" hoofed animals; there are many interesting and well-known fossils in this group. *Baluchitherium,* the largest land mammal known, was a hornless rhinoceros that lived in Late Oligocene and Early Miocene time. This monstrous creature measured about 10 m (34 feet) from head to tail and stood almost 5.5 m (18 feet) high at the shoulder. It must have weighed many tons. Baluchitheres have not been discovered in North America and appear to have been restricted to Central Asia. The woolly rhinoceros was a two-horned species that ranged from southern France to northeastern Siberia during Pleistocene time. It is well known from complete carcasses recovered from the frozen tundra of Siberia and remains preserved in an oil seep in Poland. These unusual specimens, as well as cave paintings, have provided much information about this animal.

Artiodactyls (the "even-toed" hoofed mammals) include such familiar forms as pigs, camels, deer, goats, sheep and hippopotamuses. They are common fossils in rocks ranging from Eocene to Pleistocene, and have been found in many parts of the world. Two of the more important members of this group are the camels and the extinct entelodonts. The earliest known camels, recovered from rocks of Late Eocene age, were smaller than a lamb and had short limbs. But as these forms evolved, they underwent considerable specialization of teeth and limbs and became increasingly larger. Many camels of Middle Cenozoic time had long legs, which would have allowed them to browse among leaves of trees. The entelodonts, giant piglike artiodactyls, lived during the Oligocene and Early Miocene. They were characterized by a long, heavy skull that held a relatively small brain and a face that was marked by large knobs situated beneath the eyes and along the underside of the jaw. Certain of these giant swine attained a height of 1.8−2 m (6−7 feet) at the shoulders and were almost 3.4 m (11 feet) long.

The primates, the most advanced mammals, include the lemurs, tarsiers, monkeys, apes and man. They appear to have originated during Early Tertiary time. See Fossil Man. (W.H.M.)

Fossil Man. The preserved remnants of a member of the biological group to which modern man belongs. This group includes the genus Homo as well as other now

Fossil Mammals

1 *The American mastodon existed during the Pleistocene Ice Age and became extinct about 8000 years ago.*

2 Smilodon bonaerensis, *the giant saber-tooth cat from South America, had upper canine teeth 20 cm (8 inches) long.*

Fossil Man

3 (a) Australopithecus, *the first distinct hominid, had a gorilla-sized brain but* walked upright, lived on the ground and may have used stones for defense.
(b) *One of the first true men,* Pithecanthropus, *used primitive instruments to kill the animals he hunted.* (c) *Neanderthal Man had a large cranial capacity and made a variety of well-designed tools.* (d) *The first representative of* Homo sapiens *was Cro-Magnon Man, physically little different from modern man, with a large brain and a well-defined culture.*

extinct genera. Man's fossil history is largely confined to the Pleistocene Epoch, and consists primarily of artifacts and skeletal elements found in gravels and limy cave deposits in the Old World. The Primates are a highly developed group that includes the lemurs, tarsiers, monkeys, apes and men. Man (along with the monkeys and apes) has been placed in the suborder Anthropoidea (Greek *anthropos*, meaning "man", plus *eidos*, "resemblance"). Anthropoids possess large eyes that face forward, relatively specialized teeth and large brains. Most are able to sit in an upright position, leaving the hands free for manipulation of objects. Because of the obvious similarities between them, apes and men have been assigned to the superfamily Hominoidea (the hominoids). But man is separated from the apes by being placed in the family Hominidae (the hominids).

Dryopithecus (originally known as *Proconsul*), a hominoid primate whose remains have been found in *Miocene and *Pliocene deposits, is known from fossils collected in southern Europe and Asia. This early ape is thought to have given rise to the gorillas, chimpanzees, orangutans and also to the oldest-known hominids, *Ramapithecus,* believed to be the ancestor of all fossil men. *Ramapithecus* is known from teeth and jaw fragments found in Miocene and Pliocene strata in India, East Africa, China and southern Europe. Reconstruction of the remains suggests a small, bipedal creature, about 1 m (between 3 and 4 feet) tall, with a short face and human-like teeth. Another hominid, *Australopithecus africanus,* appeared in Early Pleistocene time; its remains were first found in South Africa in 1924. The australopithecines (or "southern apes") had apelike skulls, were bipedal, were probably less than 1.5 m (5 feet) tall, and stood and walked erect. Although their brains were only half the size of those of modern man, they were hunters who used sticks and stones as weapons and probably fashioned crude tools. Another species, *Australopithecus boisei* (formerly *Zinjanthropus boisei*), was described from remains found in Tanzania's Olduvai Gorge in 1959. Discovered by Louis S.B. Leakey, this specimen has been assigned an age of at least 1.75 million years. Recent findings suggest that these hominids may date back as much as 2–5 million years. Although the australopithecines became extinct in about Middle Pleistocene time, they were widespread, for their remains have also been found in Java, northern Israel and mainland China.

Australopithecus was succeeded by *Homo erectus,* the well-known Java man, whose remains were first discovered on the banks of the Solo River near Trinil, Java, in 1891. Originally described as *Pithecanthropus erectus,* he appeared in Early Pleistocene (perhaps during the *interglacial stage) and became extinct near the end of Middle Pleistocene time. The skull of *Homo erectus* is apelike, the forehead slopes or is flat and the lower jaw lacks a chin. But the teeth are near-human and the brain size begins to approach modern human brain capacities. In addition, a cast of the inside of the skullcap provides evidence of a near-human level of brain development, especially in relation to its speech area. Muscle attachments on the femurs indicate that Java man probably had an upright posture. Between 1927 and 1937 scientists recovered fossilized remains of more than 40 prehistoric men in the vicinity of Peking, China. These fossils, named *Sinanthropus pekinensis,* were collected from cave deposits that also contained the remains of extinct Middle Pleistocene mammals. Because the brain case and certain other skeletal parts of *Sinanthropus* closely resemble similar parts of *Homo erectus,* Peking man and Java man have now been placed in the same species. The Chinese skulls have larger brain capacity and Peking man apparently possessed cutting tools of stone and knew how to use and control fire. The Chinese hominids were probably cannibals, for the basal parts of most skulls had been broken open so that the brains could be removed. Many long bones were split to expose the marrow.

Homo sapiens, the single species that includes all present-day men, evolved from *Homo erectus* and appeared near the middle of the Pleistocene Epoch, approximately half a million years ago. One member of this species was *Homo sapiens neanderthalensis,* who inhabited Europe, Africa, China and the Near East from about 110,000 to 35,000 years ago. Among the best known fossil men, Neanderthal man was first described from a skullcap and some leg bones dug up from the floor of a limestone cave in the Neanderthal Valley near Düsseldorf, Germany. These fossils were discovered in 1856 but were not recognized as a distinct human species until 1864. Meanwhile, the bones were variously regarded as a pathologically deformed specimen, as the skull of an idiot or as belonging to primitive savages. Another skull, that of a young female, had been found in a quarry on the Rock of Gibraltar as early as 1848. However, it was not recognized as a Neanderthal until 1906. Since the original 1848 discovery, several complete skeletons and many incomplete skeletons of men, women and children

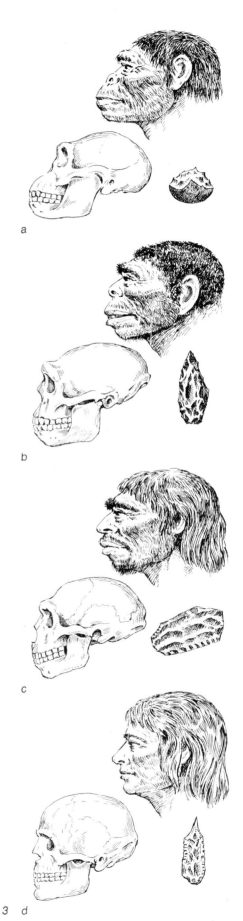

a

b

c

3 d

have been collected. These abundant remains suggest that Neanderthal man was widespread and prolific during Late Pleistocene time. They were short, stocky individuals, typically about 1.5 m in height. The head was thrust forward and the shoulders were stooped, giving the body a slouched appearance. The hands and feet were large and the knees were slightly bent. The head of Neanderthal man was large and marked by a low forehead, flat nose, receding chin and heavy brow ridges. In spite of his primitive appearance, Neanderthal man had a brain capacity approximately equal to that of modern man. They were apparently cave-dwellers and hunters, for their skeletons are often accompanied by stone implements such as scrapers, spearheads and axes. Neanderthal remains are also found associated with the bones of the animals that they hunted. These include the woolly mammoth, giant cave bear, bison, horse, and woolly rhinoceros. They also developed a primitive culture and are known to have used fire. The positions in which certain Neanderthal skeletons were found, and the arrangement of food, stone implements and other artifacts with the skeletons, indicates that the Neanderthals buried their dead. It also suggests that they had some type of religion or burial ritual.

Homo sapiens neanderthalensis was replaced by modern man—*Homo sapiens sapiens*—some 35,000 to 40,000 years ago. The first of these was Cro-Magnon man, whose remains were initially discovered in 1868 at the rock shelter of Cro-Magnon, near the village of Eyzies in the Dordogne Valley of France. Cro-Magnon man is well known from a large number of skeletons collected in western and central Europe, and appears to be modern in all respects and far advanced over Neanderthal man. He was tall (many skeletons exceeded 1.8 m in height), walked erect and had relatively long legs with straight thigh bones. The head was long with a high forehead, long nose and pointed chin. He was as highly developed mentally as he was physically, with a brain capacity equal to that of present-day man. He fashioned many well-formed tools from flint and bone. He also possessed considerable artistic ability, as evidenced by the large number of paintings and sculptured bones found in some of his caves.

The early development of man probably took place in the Old World; although recent finds in New Mexico suggest that *homo sapiens* may already have been living in the New World as much as 200,000 years ago, most scientists consider he was relatively late in arriving in the Western Hemisphere. It is generally agreed that the early inhabitants of both North and South America belong to the Mongoloid division of *Homo sapiens sapiens*, and that the ancestors of these early Americans apparently migrated from northeastern Asia (Siberia) to Alaska by way of the Bering Strait area. There were probably many such migrations, the first known occurring after the last Pleistocene glaciation about 10,000 years ago. Remains have been found in such widely separated places as Texas, Florida, Washington, Mexico and Peru. It is evident that the first men reached the New World before the extinction of such typical Pleistocene mammals as the mammoth, mastodon, ground sloth and various extinct horses and bison. In fact, there is evidence to suggest that the extinction of some of these species (especially the mastodons and mammoths) was hastened by the hunting of early man. (W.H.M.)

Fossil Reptile. The preserved remnants of a member of the class Reptilia of the phylum Chordata. Reptiles were apparently derived from labyrinthodont amphibians at some time during the *Pennsylvanian Period, and steadily increased in number to become the dominant land animals of the *Mesozoic Era—the "Age of Reptiles." Mesozoic reptiles not only lived successfully on the land but they also adapted themselves to life in the sea and in the air. The best-known and most important reptilian fossils are the extinct dinosaurs, which were plentiful during the Mesozoic.

The collective term "dinosaurs" (meaning terrible lizards) has been given to that distinctive group of reptiles prominent in Mesozoic life for some 140 million years. In size, the dinosaurs ranged from as little as 30 cm (1 foot) to as much as 26 m (85 feet) in length, and from a few pounds to perhaps 45 tons in weight. Some were bipedal (walked on their hind legs) while others were quadrupedal (walked on all fours). Although most were terrestrial in habitat, aquatic and semiaquatic forms were also present.

According to the structure of their hip bones, the dinosaurs have been divided into two orders: the Saurischia (forms with lizard-like pelvic girdle) and the Ornithischia (dinosaurs with a birdlike pelvic girdle). Members of the order Saurischia, particularly abundant during the *Jurassic, are characterized by hip bones similar to those of modern lizards. Their remains were first discovered in rocks of *Triassic age, and they did not become extinct until the end of the *Cretaceous. The lizard-hipped reptiles are divided into two rather specialized groups: the theropods (carnivorous bipedal dinosaurs that varied greatly in size) and the sauropods (herbivorous, quadrupedal, semiaquatic, usually gigantic dinosaurs).

Species assigned to suborder Theropoda walked on birdlike hind limbs and were exclusively meat-eating forms, such as *Allosaurus* of Jurassic age. Some theropods were exceptionally large and were undoubtedly ferocious beasts of prey. This is borne out by the small front limbs with long, sharp claws for holding and tearing flesh, and the large, strong jaws armed with numerous sharp, dagger-like teeth. The largest known theropod, *Tyrannosaurus rex* was almost 61 m (200 feet) tall when standing on his hind limbs. Some were as much as 15 m (50 feet) long, and they are believed to have been among the most ferocious animals to ever inhabit the earth. Members of suborder Sauropoda were the largest of all dinosaurs. Some, such as *Brontosaurus,* were as much as 26 m (85 feet) long and weighed 45 to 50 tons. They were primarily herbivorous dinosaurs that had become adapted to an aquatic or semiaquatic type of existence and probably inhabited lakes, rivers and swamps.

The order Ornithischia consists of herbivorous reptiles that were quite varied in form and size and appear to have been more highly developed than the saurischians. This order includes the duckbilled dinosaurs (ornithopods), the plate-bearing dinosaurs (stegosaurs), the armored dinosaurs (ankylosaurs) and the horned dinosaurs (ceratopsians). Species of suborder Ornithopoda were predominantly bipedal and semiaquatic; some, like the duckbilled dinosaur *Anatosaurus* (or *Trachodon*), were highly specialized. Members of suborder Stegosauria were herbivorous, quadrupedal ornithischians with large, projecting plates down the back and heavy spikes on their tails. The Jurassic dinosaur *Stegosaurus* is typical of the plate-bearing dinosaurs. This creature weighed about 10 tons, was some 9 m (30 feet) long and stood about 3 m (10 feet) at the hips. *Stegosaurus* is characterized by a double row of large, heavy, pointed plates running along its back. These plates begin at the back of the skull and stop near the end of the tail. The tail was also equipped with four or more long, curved spikes, which were probably used for defense. The animal had a very small skull that housed a brain about the size of a walnut; it is assumed that these, and all other dinosaurs, were of very limited intelligence.

Ankylosaurs (suborder Ankylosauria) were four-footed, herbivorous, Cretaceous dinosaurs with relatively flat bodies. The skull and back of the animal were protected by bony armor, and the clublike tail was armed with spikes. *Paleoscincus,* a

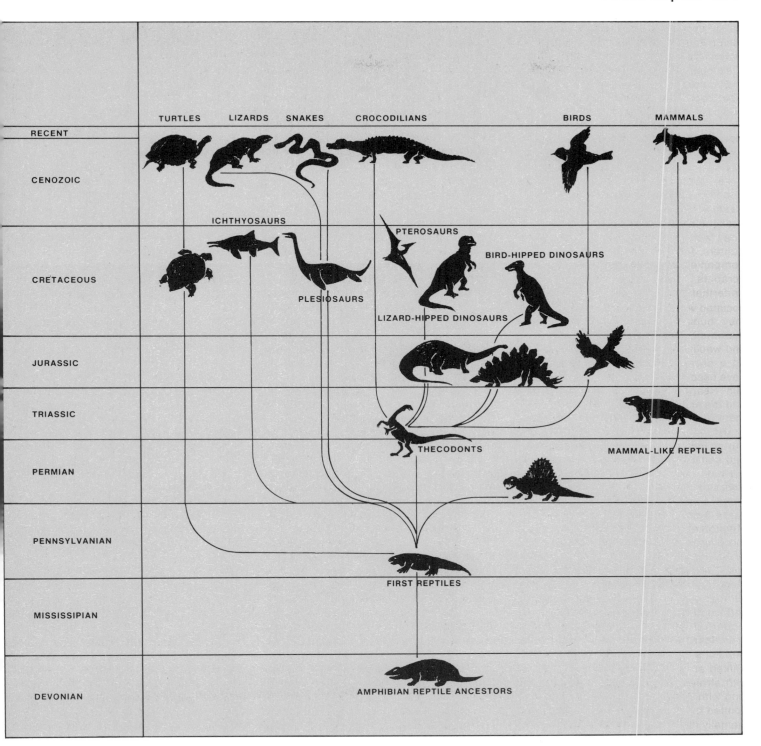

Fossil Reptile
The evolution of reptiles from their amphibian ancestors down to their bird and mammal descendants.

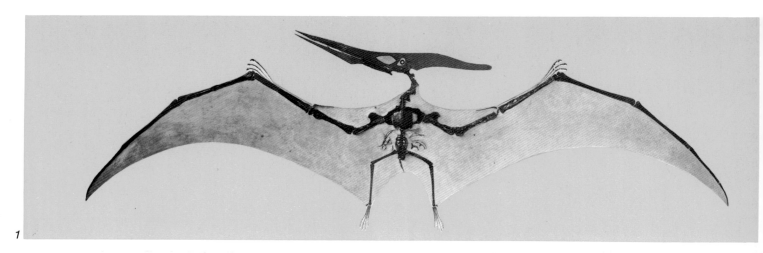

1

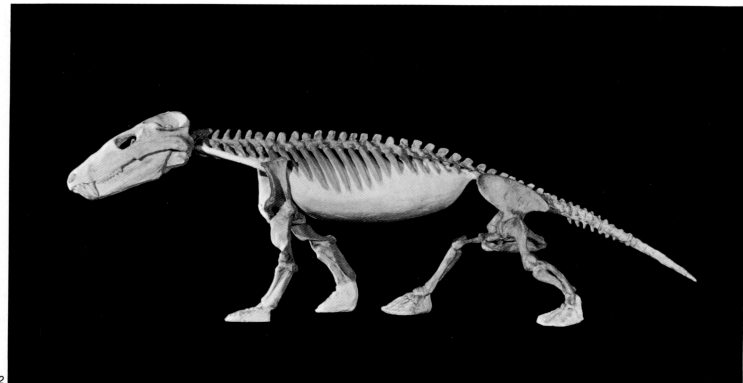

2

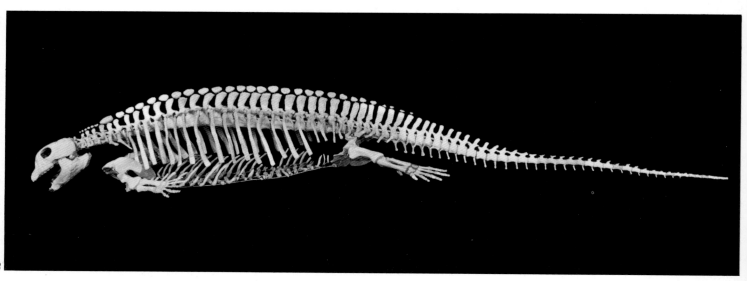

3

typical ankylosaur, had large spines projecting from the sides of the body and tail. The armored spiked back and the heavy clublike tail probably provided much-needed protection from the vicious meat-eating dinosaurs of Cretaceous time.

Species of suborder Ceratopsia are characterized by horns, and their remains are known only from rocks of Cretaceous age. These large , plant-eating dinosaurs had beaklike jaws, a bony neck frill that extended back from the skull, and one or more horns. *Triceratops* is the largest of the horned dinosaurs: some forms were as much as 9 m (30 feet) long, and the skull measured 2 m (8 feet) from the tip of the parrot-like beak to the back of the neck shield.

Extinct swimming reptiles dating from the Mesozoic Era included ichthyosaurs, mosasaurs and plesiosaurs. Ichthyosaurs were short-necked, fishlike reptiles that resembled dolphins and grew to be as much as 9 m long. Mosasaurs attained a length of 15 m, and their huge jaws were filled with many sharp, curved teeth. The plesiosaurs were characterized by a broad turtle-like body, paddle-like flippers and a long neck and tail. They were neither as streamlined nor as well equipped for swimming as ichthyosaurs or mosasaurs, but their long serpent-like neck was probably useful in catching fish and other small animals for food.

Pterosaurs (sometimes called pterodactyls) were Mesozoic reptiles with batlike wings, supported by arms and long, thin "fingers." They were well adapted to life in the air, and their lightweight bodies and wide skin-covered wings enabled them to soar or glide for great distances. The earliest known pterosaurs were found in Lower Triassic rocks, and the group became extinct by the end of the Cretaceous. During this time certain of these creatures attained a wingspread of as much as 8.2 m (29 feet) but their bodies were small and light. Recently, fossil remains of a giant winged reptile were found in Big Bend National Park, Texas. The estimated wingspan of this pterosaur is 15.5 m (51 feet), almost twice the wingspan of any pterosaur previously found, making it the largest known creature of its kind.

A number of lesser-known reptiles with evolutionary significance preceded the rise of the dinosaurs. These included the cotylosaurs, a group of primitive reptiles which, although retaining some amphibian characteristics, became adapted to an exclusively land-dwelling existence. They lived during the Pennsylvanian and the *Permian and apparently became extinct sometime during Late Permian time. The pelycosaurs were a group of Late Paleozoic reptiles, some of which were characterized by the presence of a tall fin on the back. The fossils of these unusual creatures are well known from the Permian.

The phytosaurs were an extinct group of crocodile-like reptiles whose remains are found only in Upper Triassic rocks. They had long jaws armed with sharp teeth, a cylindrical body studded with bony armor plates, short legs, and a long, deep, flattened tail for sculling through the water. Although they closely resemble crocodiles, the phytosaurs did not give rise to the crocodiles, for they are a distinct group of reptiles that underwent parallel evolution with the crocodiles. Despite their abundance during Late Triassic time, they became extinct at the end of that period. The therapsids, a mammal-like group of reptiles, were well developed for a terrestrial existence. Although the remains of these primitive reptiles are not particularly important as fossils, study of the therapsids has provided much valuable information about the origin of the mammals. Members of this group appeared first in the Middle Permian, and they persisted until the Middle Jurassic. (W.H.M.)

Fossil, Trace. See Trace Fossil.

Fracture. A parting or crack in an otherwise solid body of rock. Fractures result from forces acting from without or within solid rock. They are classified according to their mode of origin as well as their geometric relationships to other structural features. When movement occurs parallel to the fracture plane, the parting is referred to as a *fault. Fractures displaying no movement or slight movement perpendicular to the fracture surface are known as *joints.

Fracture Zone. A major crack in the earth's crust, basically at right angles to the *oceanic ridge system. These fractures produce breaks, or offsets, of up to hundreds of kilometers in oceanic ridges such as those in the central Atlantic or the eastern Pacific oceans. An example is the Mendicino Fracture Zone off the California coast. The fractures result when two adjoining plates of the earth's crust slide past each other in opposite directions. See Plate Tectonics.

Franklinite. A mineral, an oxide of zinc and iron and a member of the *spinel group of minerals. Found in Franklin, N. J. (from which it gets it name), it is an important ore of zinc. At Franklin the mineral is commonly found in black octahedral crystals, resembling magnetite, enclosed in a granular limestone associated with the other zinc minerals, *willemite and *zincite.

Fossil Reptiles

1 The giant flying reptile called Pteranodon *soared over Cretaceous seas looking for surface-swimming fish.*

2 Cynognathus *was a mammal-like reptile from South Africa, about the size of a wolf, that lived during the Triassic Period, some 200 million years ago.*

3 Placodus gigas, *1.5 m (5 feet) long, was a shell-eating marine reptile of the Triassic Period.*

4 The ceratopsians, or horned dinosaurs, were the last of the dinosaurs to evolve, appearing in Late Cretaceous times. Styracosaurus *had a large nasal horn and a series of long spikes around the borders of its head frill.*

5 In Triceratops, *another ceratopsian, the nasal horn was augmented by two large brow horns, one above each eye.*

Aside from this occurrence, franklinite is a rare mineral. See Mineral Properties.

Frasch Process. A method for removing native sulfur from the cap rock of certain salt domes along the Gulf Coast in southeast Texas and Louisiana. These cap rock sulfur deposits are typically about 450-730 m underground. At first an attempt was made to mine the sulfur by sinking shafts, but caving sands and poisonous gases made this dangerous. Finally a chemist, Herman Frasch, found a way to obtain sulfur by making use of its low melting point. When sulfur gets slightly hotter than boiling water (113° – 119°C, or 235° – 247°F) it becomes liquid. In the Frasch method a well is drilled into the cap rock of the salt dome and three pipes, one inside the other, are put into the well. Superheated water under pressure— over 100°C (212°F), at which point water ordinarily turns into steam—is sent down one of the pipes to melt the sulfur in the cap rock. Then compressed air is sent down another pipe, forcing the liquid sulfur to flow up the third pipe. At the surface, the sulfur is poured into bins, where it cools and becomes a solid again, or it may be transported molten in pipelines and tankers. (W. H. M.)

Front, Polar. See Polar Front.

Frost Action. A process of repeated freezing and thawing that plays a major role in the mechanical weathering and breakup of rock. When water freezes, the resulting expansion is responsible for frost wedging, shattering and splitting. Pressures as great as 2100 kg/cm² (15 tons per square inch) have been produced by freezing water in confined spaces in the laboratory. Although such high pressures may not be reached in nature, much less will shatter rocks. The effectiveness of the process depends upon the presence of confined spaces and the frequency of the freeze-thaw cycle. Frost action is widespread, occurring in polar and midlatitude lowlands and on mountains from the polar regions to the Equator. It is largely responsible for *talus, *rock glaciers and other features. Victorious armies used frost action to destroy the artillery of defeated armies. The gun barrels were filled with water and plugged at both ends; on freezing, the barrels split.

A block field that results from the frost shattering of bedrock and large boulders is called a felsenmeer. (R.L.N.)

Frost Heaving. The upward expansion of the ground caused by *frost action, that is, by the freezing of water in the upper part of

Frost Action

1 This barren rubble-strewn surface in Antarctica, broken up by centuries of freezing and thawing, is called a felsenmeer, a German word meaning "sea of rocks."

Frost Heaving

2 On a hillside a rock is lifted at right angles to the slope by expansion during freezing. When thawing occurs, the rock is dropped vertically and so moves downhill.

3 A pingo rises 90 m (300 feet) above Alaska's Mackenzie Delta. This giant frost heave, covered with lichens and other vegetation, has a core of solid ice.

the *regolith. If the surface is horizontal, the regolith after thawing will return to its original position. However, if the surface slopes, the regolith on freezing will rise perpendicular to the slope but on thawing will sink vertically, causing downslope creep of the regolith. Frost heaving is also responsible, as many New England farmers can attest, for bringing large boulders up to the surface, for making the ground soft and spongy in the spring and for creating bumps (frost boils) in the spring on highways and other surfaces. (R.L.N.)

Froude Number. Used to indicate the influence of gravity on surface stream flow. The formula for this is $F = V/\sqrt{gD}$. F is the Froude number, g is the gravitational constant (acceleration due to gravity and equal to the specific weight of the water divided by its density), V is velocity of flow and D is depth of flow. If F is less than 1, small surface waves can move upstream; if F is greater than 1, they will be carried downstream. When F equals 1, the velocity of flow is just equal to the velocity of surface waves. See Flow Regime.

Fulgarite. Glossy tubes of fused rock formed by lightning. They are most common on exposed mountain peaks where the lightning has not only formed small tubes but has fused the surface of the rock, giving it a varnish-like appearance. Fulgarites may form from any kind of rock, but the largest tubes have been produced by lightning striking unconsolidated sand. Under these conditions the tubes may be 6 m (20 feet) long, sometimes branching, and over 1 cm (½ inch) in diameter.

Fuller's Earth. A fine-grained, earthy, claylike substance probably formed by chemical weathering of rock in place. It contains the clay mineral *montmorillonite and resembles clay, but lacks its plasticity, contains a higher percentage of water and is rich in magnesia. Its traditional use is to "fuel," or thicken, cloth; it is now widely used in industry as a bleaching agent, to refine and decolorize oil and to process raw wool.

Fumarole. A vent at the earth's surface, usually in volcanic areas, that emits hot gases. Fumaroles are found at the surface of *lava flows, lava lakes, pyroclastic flows; in the *craters and *calderas of active volcanoes; and in areas like Yellowstone National Park where hot intrusive *igneous rock bodies occur. The gases may reach temperatures as high as 1000°C (1830°F). Steam and carbon dioxide are the principal gases, but nitrogen, argon, carbon monoxide, hydrogen and other gases are also

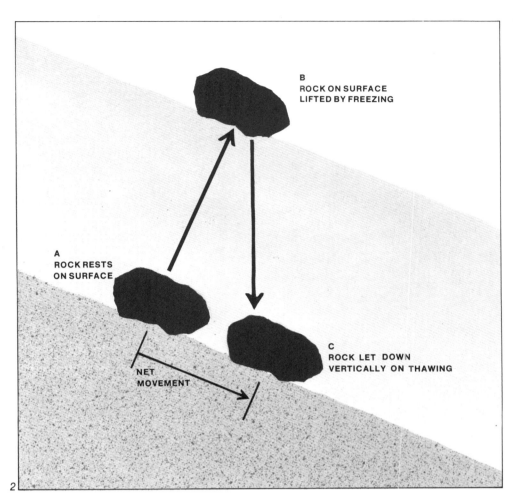

B
ROCK ON SURFACE
LIFTED BY FREEZING

A
ROCK RESTS
ON SURFACE

C
ROCK LET DOWN
VERTICALLY ON THAWING

NET
MOVEMENT

2

3

Fumarole
Evidence of the volcanic nature of Iceland, fumaroles emit steam and gases from vents in a valley near Namaskard.

present. The gases vary greatly not only from fumarole to fumarole but even in the same fumarole.

Fumaroles also emit compounds that are gaseous only at high temperatures and collect as solids around the vents. It is not known exactly how much of these fumarole gases came from *magma and how much from the atmosphere, groundwater and other sources. Carbon dioxide given off from fumaroles may collect at the bottom of depressions when there is little wind and prove dangerous to animals. At Death Gulch in Yellowstone National Park, grizzly bears and other animals have been asphyxiated by such concentration. Fumaroles that emit sulfurous gases are called *solfataras* (from the Italian word for sulfur). Those that emit large quantities of carbon dioxide are called *mofettes*. In the United States fumaroles are found in Yellowstone and Lassen National Parks, on Mount Rainier and at Geysers, Calif. The heat of the fumaroles in the crater at the summit of Mount Rainier has saved the lives of climbers trapped on the summit by blizzards. The "smokes" in the Valley of Ten Thousand Smokes, Alaska, are fumaroles. (R.L.N.)

G

Gabbro. A coarse-grained *plutonic rock containing plagioclase feldspar, most commonly labradorite. Although the classification is based on the composition of the feldspar, *dark minerals are usually present. The most common of these is pyroxene, which may make up 50 percent or more of the rock; olivine may also be present. Although some gabbros are composed almost wholly of feldspar and thus have a light color, in most the dark minerals are sufficiently abundant to give the rock a dark color. The most common accessory minerals in gabbro are magnetite, ilmenite and apatite.

Galena. A mineral, lead sulfide, the principal ore of lead. Silver is usually present, probably as admixtures of silver minerals, and thus galena is also an important ore of silver. It is characterized by its cubic crystals, cubic cleavage, black color and high specific gravity. It is one of the most common metallic sulfides and is found in veins, open space fillings and in irregular bodies replacing limestone. Commonly associated minerals are: sphalerite, pyrite, marcasite, chalcopyrite, quartz, calcite, barite and fluorite. It is also found in *contact metamorphic deposits, *sedimentary rocks and *pegmatites. As the only important source of lead, galena is extensively mined in many countries on every continent. See Mineral Properties. (C.S.H.)

Gangue. The worthless rock or minerals associated with the valuable minerals in ore.

Garnet. A mineral group in which the members have a common crystal structure but vary in occurrence and also in chemical and physical properties. Although it has been known for thousands of years as a dark-red gemstone and red is its most common color, garnet is found in all colors except blue. Some garnet breaks with angular fractures; this, coupled with a high hardness, makes it a valuable abrasive. At Gore Mountain, N.Y., for example, thousands of tons of garnet are mined annually for abrasive purposes.

Although chemical variation within the garnet group is great, all members conform to the same basic formula, which may be written as $A_3B_2(SiO_4)_3$. The letter A in the formula may be magnesium, iron, manganese or calcium; B may be aluminium, iron or chromium. In a given garnet it is possible that only one of the A elements and one of the B will be present. In this case the garnet would correspond to one of the members whose names are given below. However, most garnets contain two or more A elements and at least two B elements, which gives rise to a great range of chemical composition. In spite of this variation the following names are given to the subspecies: pyrope (magnesium, aluminum); almandite (iron, aluminum); spessartite (manganese, aluminum); grossularite (calcium, aluminum); andradite (calcium, iron); and uvarovite (calcium, chromium).

Since all garnets are isometric and have the same crystal structure, they have similar crystal forms. Most common are the twelve-sided dodecahedron and the twenty-four-sided trapezohedron, or combinations of these two. However, the other properties—including color—vary with the chemistry.

All garnets except green uvarovite have been used as gems. The most familiar is the deep-red pyrope, which has been used in jewelry for centuries. Much of this has come from Czechoslovakia and is known as Bohemian garnet. The red to violet-red almandite is called carbuncle and has been used as a gem since Biblical times. It is believed that carbuncle was one of the stones in the breastplates of the high priests in the Bible. Grossularite is found in several colors—white, green, yellow, brown and red. But the stones most highly prized as gems are the orange to golden variety called hessonite. Although not well known, spessartite, which has an orange-red color and comes from the gravels of Ceylon, makes a most pleasing and attractive cut stone. Andradite is a common garnet but most are worthless as a gem; however, a green variety known as demantoid, which means diamond-like, is the most valuable of any of the garnets.

Garnet forms under a wide range of geologic conditions. It is common in metamorphic rocks, gneisses and schists and in *contact metamorphic deposits in crystalline limestone. It is an accessory mineral in igneous rocks and occurs in large crystals in pegmatites. Because of its high hardness and resistance to chemical weathering it is a common constituent of sand, particularly black sand made up of minerals with high specific gravity. See Asterism; Mineral Properties. (C.S.H.)

Garnierite. A mineral, the apple-green nickel-bearing serpentine formed as an alteration of nickel-bearing peridotites. In a few places, notably New Caledonia, it is an important ore of nickel. See Mineral Properties.

Gastrolith (from the Greek *gastros,* meaning "stomach," and *lithos,* "stone"). A highly polished, rounded stone believed to have been used by certain extinct reptiles in grinding their stomach contents. "Stom-

Galena

1 An unusual combination of large cubic crystals spotted with smaller cubo-octahedrons, this specimen comes from a lead mine in Kansas.

Garnet

2 The larger, more complex garnet crystal has the dodecahedron as its dominant form, modified by faces of the trapezohedron. The smaller garnet only displays trapezohedral faces.

ach stones" are found in abundance in the body cavities of certain reptiles. They almost invariably accompany the remains of *plesiosaurs (extinct long-necked, small-headed swimming reptiles). The remains of one large plesiosaur contained half a bushel of these stones, the largest being 10.2 cm (4 inches) in diameter.

Gastropod. A member of Gastropoda (the largest class of phylum Mollusca), including such forms as snails, slugs, whelks, limpets and abalones. The living animal has a distinct head with one or two pairs of tentacles and a broad, flat foot. It is typically enclosed in a coiled, single-valve shell that is secreted by the mantle and carried on the animal's back. Most gastropods have gills and live in shallow marine waters, but some inhabit fresh water. Others (the pulmonates) are terrestrial and breathe by means of lungs. Gastropod shells, both living and fossil, exhibit great variety: spirally coiled, ear-shaped, cap-shaped or wormlike, cone-shaped, flat, turreted or cylindrical. The external shell surface may be marked by ribs, grooves, nodes, spines or other ornamentation.

Most gastropod shells are composed of *aragonite, an unstable form of calcium carbonate. The shell is therefore rather easily dissolved, and fossil gastropods are frequently preserved as *casts and molds.

Indirect fossil evidence of gastropods is sometimes seen in the form of fossil snail trails made as the animal moved over soft sediments. In addition, the neatly drilled, countersunk holes made by carnivorous snails are found on many recent and fossil *mollusk shells. These are valid evidence of the presence of these animals even though their shells or molds are not to be found.

The earliest known gastropods were of the Early Cambrian, at which time they must have been relatively rare. Their remains become increasingly abundant from Ordovician until Holocene (Recent) time, and are found in both marine and freshwater deposits. (W.H.M.)

Geanticline. A broad upwarping or *anticline of regional proportions, composed of crustal rock that originally formed in a subsided portion of the earth's crust (or *geosyncline). The Green Mountains of Vermont form a geanticline.

Gem. A substance that because of its beauty is cut and polished for ornamental purposes. Traditionally gems are minerals and thus natural products. But today, with the synthesis of many gem minerals, the definition must be expanded to include the man-made "minerals" as well as sub-

stances that have no natural counterparts.

The primary attributes of gem mineral or other gem substance are beauty, durability and rarity. Some gems have all of these qualities, others only one. The principal qualification is beauty, which depends upon transparency, brilliancy, luster and color. Durability is a function of *hardness. A soft mineral makes a poor gem because it scratches easily and will soon lose its attractiveness. To maintain a polish a gem should be harder than the dust of the air, which invariably contains quartz particles. Consequently, the most valued stones are those harder than quartz; these include diamond, emerald, ruby, sapphire, zircon, spinel, chrysoberyl and topaz. The rarity of a mineral also has considerable effect on its value. A gem with desirable properties, so scarce that only a few can possess it, brings a higher price than an abundant stone with comparable qualities.

The gems of greatest value, sometimes called precious stones, are diamond, ruby, sapphire and emerald. Of these, only some diamonds have all of the requisites of a fine gem: rarity, high hardness, transparency, high luster, brilliance and color. However, most gem diamonds are colorless, and it is the exceptional stone that also has color. Although ruby, sapphire and emerald lack the high luster and brilliance of diamond, they have the other desirable properties plus their pleasing colors. Some gems, such as turquoise and lapis lazuli, are valued only for their color, for they are nontransparent and of a comparatively low hardness. Opal, which to many is one of the most pleasing gems, has as its only gem attribute internal reflections that give it a play of colors. (C.S.H.)

Geochemistry. The science dealing with the chemical composition of the earth and the chemical changes that have taken place within it. In a more restricted sense it refers to the abundance of elements and their distribution and migration in various parts of the earth, including the atmosphere and hydrosphere as well as the rocks of the crust.

Geochronology. The study and classification of time as applied to the history of the earth. Geologists mark the passing of time by establishing a sequence of geologic events. Time sequences make it possible to establish relative age (the age of one geologic event or object in relation to another). Time sequences based on relative time have been used to establish a *geologic time scale consisting of named units of geologic time such as *eras, *periods, *epochs and *ages.

Relative time simply relates one event to

another, but measured (absolute) time can be expressed quantitatively in years. Natural evidence of measured time can be observed in the growth rings of certain shells and trees and in seasonal rock layers called *varves. Early attempts to measure the earth's age were based on studies of the internal heat of the earth, the salinity of the sea and the rate of deposition of sediments. These unreliable methods have been replaced by *radiometric (or atomic) dating, a complex technique based on measurements of the rate of decay of radioactive elements such as carbon-14 or uranium-238. Radiometric dating suggests that the earth formed approximately 4.5 billion years ago, a figure supported by the measured ages of meteorites and lunar samples returned by Apollo astronauts. (W.H.M.)

Geode. A hollow, usually spheroidal body commonly consisting of an outer shell of *chalcedony covering an inner shell of well-formed crystals that project inward toward the hollow center. It ranges in diameter from a few centimeters to more than 30 cm (1 foot) and can easily be identified by its light weight and the hollow sound when tapped with a hammer. It may contain quartz and other minerals, including calcite, dolomite, barite, sphalerite and galena. Although a geode may form in many kinds of *sedimentary rocks, it is generally found in *limestone or *shale. The geode minerals are usually more resistant to weathering and erosion than the enclosing rock. Its removal frees the geodes, and they accumulate in the soil mantling the *bedrock. The origin of the cavity in which the geode minerals are deposited is sometimes difficult to determine. It is usually the result of decay of organic material or the solution of soluble rocks, and there is good evidence that some cavities have been enlarged by osmotic pressure.

A geode on the earth's surface generally results from the following sequence of events: (1) the formation of marine limestone or shale, (2) the initiation and growth of a spheroidal or ellipsoidal cavity in the rock, (3) the precipitation in the cavity of mineral matter from subsurface solutions, (4) erosion that brings the geode to the earth's surface. Geodes are prized by mineral collectors for their perfectly formed crystals. During World War II they were sought because of industrial demand for quartz crystals. (R.L.N.)

Geoid. The surface of the earth as it would appear if it coincided at every point with mean sea level. Such a hypothetical continuous surface would be perpendicular at every point to the direction of gravity. It is represented on the surface of the earth by (1) mean sea level over the oceans and (2) an imaginary surface through the continents, at which mean sea level would prevail if the continents were cut by channels that allowed seawater to flow through them. The concept of the geoid is used by both surveyors and geophysicists as a reference surface for mapping elevations and for the reduction of gravity data to comparative values.

Geologic Column. The total succession of rocks, from the oldest to the most recent, found either locally or in the whole earth. Thus the geologic column of California includes all rock units known to be present in that state. By referring to the geologic column worked out for any given area, a geologist can determine the type of rocks he may expect to find there. This should not be confused with the *geologic time scale, which deals with units of geologic time.

Geologic Range. The total duration of a group of organisms through geologic time. Certain species that lived but a very brief time in geologic history are valuable as *guide fossils because of their short geologic range. Certain *trilobites, for example, lived only during Early Cambrian time, making them especially useful as Lower Cambrian guide fossils.

Geologic Time Scale. A "geological calendar" composed of named intervals of geologic time during which the rocks represented in the *geologic column were deposited. The need to subdivide earth history into manageable "chapters" was recognized early in the development of geology. One of the first attempts at subdivision was made in 1756 by Johann Lehmann, who developed a time scale on the basis of rocks exposed in central Europe. Lehmann recognized three classes of mountains and three classes of rocks that formed them. This pioneer work was followed by a fourfold classification of rock and time proposed by Abraham Werner, an 18th century mineralogist who believed that all rocks had been precipitated from seawater and could be assigned to one of four major subdivisions. Although Werner's theory of rock formation did not hold up, his idea of establishing distinctive time units persisted. With advances in geologic thought during the 19th century, geologists were increasingly in need of a geologic time scale. Their efforts resulted in the geologic column and time scale in universal use today.

A major problem in devising a time scale was the establishment of criteria to

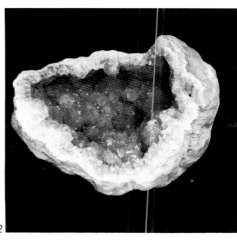

GEOLOGIC SUCCESSION			ABSOLUTE OR RADIOMETRIC DATING USED	APPROX. AGE IN 10⁶ YEARS
ERA	SYSTEM-PERIOD	SERIES-EPOCH		
CENOZOIC	QUATERNARY	RECENT		
		PLEISTOCENE		3
	TERTIARY — NEOGENE (UPPER)	PLIOCENE		
		MIOCENE		22
	TERTIARY — PALEOGENE (LOWER)	OLIGOCENE		
		EOCENE		62
		PALEOCENE		
MESOZOIC	CRETACEOUS			130
				180
	JURASSIC			230
	TRIASSIC			280
PALEOZOIC	PERMIAN			
	CARBONIFEROUS — PENNSYLVANIAN			340
	CARBONIFEROUS — MISSISSIPPIAN			400
	DEVONIAN			450
	SILURIAN			500
	ORDOVICIAN			570
	CAMBRIAN			640
PROTEROZOIC	UPPER			950
				1350
				1650
	MIDDLE			1800
	LOWER			2600
ARCHEAN				3600
NO RECORD				4700

Dating methods shown (vertical columns): URANIUM 238, URANIUM 235, LEAD 206, LEAD 207, RUBIDIUM 87, STRONTIUM 87, POTASSIUM 40, ARGON 40, CARBON 14

Gastropod

1 *This fossil is the remains of a spiral-shelled snail that lived in the sediment at the bottom of Cenozoic seas.*

Geode

2 *Crystals of quartz growing within a cavity formed by solution helped create this geode from Keokuk, Iowa.*

Geologic Time Scale

3 *The sequence of geologic time units. The most recent units are at the top of the column and the earliest are at the bottom.*

separate the major episodes in earth history. These natural breaks were eventually based on geologic events supposedly of such magnitude as to be clearly discernible in the rock record. Thus intervals of mountain-building activity and times when the seas changed their positions were generally recognized as of sufficient importance to delimit eras and periods of the time scale. However, we now know that mountain-building movements may be restricted to a single continent and that the seas did not go "in and out" with determinable regularity. Consequently, *correlation of the succession of fossil assemblages in the rocks and the basic tenets of *superposition are the basis for the standard rock column and geologic time scale in use today.

The *time units of the geologic time scale bear the same names that were originally used to distinguish the rock units in the geologic column. Thus one can speak of Cambrian time (referring to the geologic time scale) or Cambrian rocks (referring to the geologic column). The largest time unit is an *era. Eras are divided into *periods, and periods, into *epochs.

Unlike the days and months on a standard calendar, the units of geologic time are arbitrary divisions of unequal duration, for there is no way of knowing precisely how much time was involved in each era, period or epoch. Nevertheless, the geologic time scale does permit earth scientists to say, for example, that a certain rock is of Cretaceous age, meaning that it was formed during the Cretaceous Period about 65 million years ago.

Each of the five eras of geologic time has a name that describes the degree of life development characterizing the era. Thus, Paleozoic, which literally means "ancient life," refers to the relatively simple and ancient stage of life development during this era. The eras and the literal translation of their names, beginning with the most recent, are as follows:

Cenozoic—Recent life
Mesozoic—Middle life
Paleozoic—Ancient life
Proterozoic—Fore life
Archeozoic—Beginning life

Archeozoic and Proterozoic rocks are commonly grouped together and referred to as Precambrian in age. Precambrian rocks have been greatly deformed and are very old; hence the record of this portion of earth history is difficult to interpret. Precambrian time represents the portion of geologic time from the beginning of earth's history until the deposition of the earliest fossiliferous Cambrian rocks, and therefore may represent as much as 85 percent of all geologic time.

Most of the periods in each era are named after the regions in which the rocks of each were first studied.

The Paleozoic Era has been divided into seven periods (with the oldest at the bottom of the list). These periods and the sources from which their names are derived are as follows:

*Permian—from the province of Perm in the Soviet Union
*Pennsylvanian—from the State of Pennsylvania
*Mississippian—from the Upper Mississippi River Valley
(The term *Carboniferous is used by European geologists to denote a geologic period that is equivalent to the Pennsylvanian and Mississipian Periods of North America.)
*Devonian—from Devonshire, England
*Silurian—for the Silures, ancient tribe of Britain
*Ordovician—for the Ordovices, an ancient tribe of Britain
*Cambrian—from Latin *Cambria*, meaning Wales

The periods of the Mesozoic Era and the sources of their names are:

*Cretaceous—from Latin *Creta*, meaning chalky
*Jurassic—from the Jura Mountains between France and Switzerland.
*Triassic—from Latin *trias*, meaning three

The Cenozoic Periods get their names from an outdated classification system that divided all of the earth's rocks into four groups. The two divisions below are the only names of this system still in use:

*Quaternary—meaning fourth derivation
*Tertiary—meaning third derivation
(W. H. M.)

Geology. The science that deals with the origin, composition, structure and history of the earth and its inhabitants. Derived from the Greek *geo* (earth) plus *logos* (discourse), the scope of geology is so broad that it has been divided into two major divisions: physical and historical geology. Physical geology treats the earth's composition, its structure, the movements in and upon the earth's crust and the geologic processes by which the earth's surface is, or has been, changed. Its more important specialized branches include *mineralogy, the study of minerals; *petrology, the study of rocks; *structural geology, the study of earth structures; *geochemistry, the chemistry of earth materials; *geophysics, the study of the physical behavior of earth materials; and *economic geology, the study of economic

products of the earth's crust and their commercial and industrial application. Historical geology deals with the origin and evolution of the earth and its inhabitants. Its subdivisions include *paleontology, the study of prehistoric organisms as revealed by their fossils; *stratigraphy, which is concerned with the origin, composition, proper sequences and correlation of rock strata; and *paleogeography, the study of the distribution and relationships of ancient seas and landmasses.

Geology relies heavily on other basic sciences. Astronomy has provided information about the earth's origin and its place in the universe. Chemistry is used to analyze and study earth's rocks and minerals, and the principles of physics are used to explain the physical forces that affect the earth and the reaction of earth materials to these forces. Botany and zoology have provided a better understanding of prehistoric plants and animals and how they developed throughout *geologic time. The term "earth science" is commonly used in conjuction with the study of the earth and includes the science of geology. However, it also encompasses the sciences of *meteorology, the study of the atmosphere, and *oceanography, the study of the oceans; it also overlaps with astronomy, the study of celestial objects. (W. H. M.)

Geology, Economic. The branch of geology that deals with the useful natural materials. It is concerned with geographical distribution, methods of location and the genesis of many substances. These substances fall readily into two groups: the metallic and nonmetallic. In the metallic are the ores of such metals as gold, silver, tin, copper, lead and zinc. The nonmetallic, the larger group, includes substances such as sand, clay, building stone and all industrial minerals, of which asbestos, sulfur, salt and borax are but a few examples. The nonmetallic group also includes coal and petroleum, two of the most important natural resources. (C. S. H.)

Geology, Environmental. The science of earth processes and resources and their relationships to man. It deals especially with the problems resulting from the use of earth materials and the way our habitat and the planet in general reacts to that use.

Man is an important geologic agent whose very existence alters the environment in which he lives. Furthermore, the rapid increase in population and in the concentration of population compounds the environmental problems. For example, two-thirds of the people of the United States live in or around cities, and it is probable that by the year 2000 nearly five-sixths of the population will be urbanized. This pattern is being repeated throughout the world. The urban areas therefore feel the greatest environmental stress, especially as necessary commodities—such as fresh water, food, energy and land—become ever more scarce. For this reason urban geology, the study of problems in environmental geology in highly populated areas, has become particularly significant.

The continued vitality of every city demands a steady consumption of mineral resources of all kinds. In turn, an outflow of wastes and contamination spreads from the city to the land and sea. The problem is how to utilize the resources efficiently and how to dispose of the wastes with minimal harm to the environment. Planning urban growth and the ultimate use of land must be largely determined by the resources needed and their availability. Multiple land use is one essential of efficient resource management. For example, sites for residential and industrial development should first be utilized for their mineral wealth. Deposits useful for construction material such as certain rock types or sand and gravel can be removed first. The sites may then be well suited for schools, playgrounds, churches and other necessary facilities. Natural drainage patterns should also be carefully developed, perhaps protected by green belts bordering streams, thereby saving natural runoff and reducing flash flooding and stream pollution as well as providing recreational facilities.

An important aspect of environmental geology is the cooperation it requires among many types of experts. Architects, realtors, engineers, city planners, construction company owners and investors, to name a few, are concerned with geology. Environmental geologic maps include assessments of physical properties of the soil and bedrock, the distribution of mineral deposits, surface runoff and groundwater flow pattern, and the identification of natural hazards (fault zones, cavernous areas, regions of subsidence, potential land slides and the like). Such maps are essential to planning for the full development of available resources and minimum harm to the environment.

Although concentrated in urban areas, environmental geology applies to the oceans as well. Nearly all resources, and especially food and water, the most critical ones, originate outside the urban zone. The unwise exploitation of less populated areas and of the oceans will quickly have adverse affects on many people. Thus environmental interplay extends beyond national boundaries. The oceans exemplify this since their resources, such as oil and protein-rich seafoods, as well as their pollution by practices such as the dumping of radioactive wastes, are the concern of all nations. In some cases international control is essential.

Some degradation of the environment is inherent in the use of most earth resources. Consequently, a basic conflict has developed between economic activities and environmental protection. This conflict is illustrated by the energy problem facing the world today. Coal presents a good example: extensive deposits are known and coal could serve as a major source for world energy needs for two centuries or more. However, its use causes significant environmental problems, including sulfur contamination of the air, land surface devastation by strip mining, and surface and groundwater pollution by acid-rich leaching substances.

The conflict is difficult to resolve. Some balance between availability of needed resources and controlled alteration of the environment will be necessary. Care must be taken to avoid unnecessary consumption of resources, and a great effort must be made to find alternatives less harmful to the environment. (R.E.B.).

Geology, Historical. The record of the movement of continents, the formation and destruction of landforms, the invasion and retreat of the seas, the deposition of vast amounts of sedimentary rocks, the injection of igneous masses into the earth's crust and the eruptions of long-extinct volcanoes, the erosion of the lithosphere, the growth of deltas and coral reefs, and the swing of climate between the tropical and glacial and between wet and dry. Historical geology also deals with the study of prehistoric life in relation to these physical developments and changes.

Geology, Lunar. See Lunar Geology.

Geology of Mars. See Mars, Geology of.

Geology, Physical. A science dealing with the minerals, rocks, rock bodies and geologic structures in the earth's crust. It considers the characteristics, origin and development of the features of the landscape, and the processes and forces acting on, above and beneath the earth's surface. It theorizes on the early history of the continents and ocean basins, on the origin of ocean and atmosphere, and on the nature of the deep interior. Among its many subdivisions are volcanology, seismology, glacial geology, hydrology and structural geology. It makes use of chemistry, physics and mathematics, and helps to make available the raw materials for industry. (R.L.N.)

Geology, Structural. The geometric classification and analysis of features found on or within the earth, and the study of rock subjected to outside forces. The structural geologist not only must be able to recognize various features but must also be able to interpret the processes responsible for their formation and develop a chronological order in which geologic events occurred.

Geomagnetic Reversal. The theory that the earth's *magnetic field and the polarity of the field have not remained constant throughout geologic time. Evidence of geomagnetic reversals is found in the orientation of the *natural remanent magnetism associated with some types of rock. In 1906 the French physicist Bernard Brunher, studying the direction of magnetism in volcanic rocks, found that the rock had been magnetized in a direction opposite to that of the earth's present field. As a result of analyses of rocks from a range of ages, scientists have been able to construct a history of magnetic pole reversals. (J.W.S.)

Geomorphology. A geologic science dealing with the surface features of the earth, including those on land and under the oceans, and, in particular, with the characteristics, origin and geologic history of all landforms and of the *regolith. It also deals with the physical and chemical processes that act at or near the earth's surface, such as weathering, mass wasting, rivers, groundwater, glaciers, wind, waves and shore currents.

Geomorphography is the description of geomorphic features. Geomorphogeny deals with the origin of geomorphic features. Paleogeomorphology deals with ancient erosion surfaces and other kinds of ancient topography, whether buried or at the earth's surface. Physiography as originally defined includes not only geomorphology but also climatology and oceanography. (R.L.N.)

Geophysics. The study of physical phenomena that act on or originate within the earth. It embraces three basic areas of study—the atmosphere and hydrosphere, the solid earth and the magnetosphere, and is concerned with the study and understanding of such aspects as magnetism, gravity, radioactivity, composition and seismic activity. Moreover, the geophysicist has been able to use what he has learned about these subjects to develop methods to search for a variety of

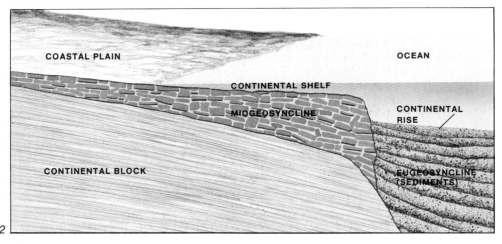

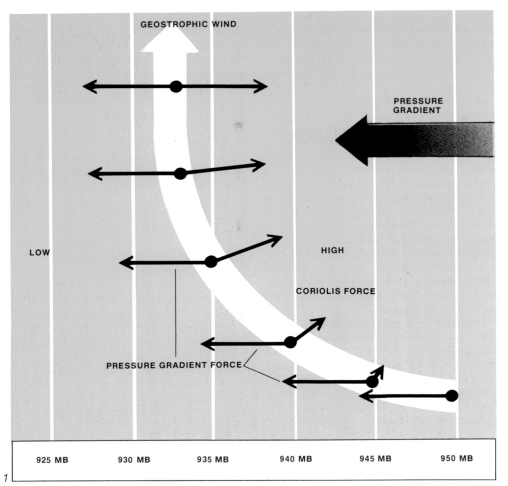

natural resources. Studies of earthquakes and their causes provide information helpful to engineers and architects in reducing seismic risk in the construction of man-made structures. The branch of geophysics called *seismology is beginning to make an important contribution to earthquake prediction. (J. W. S.)

Geostatic Pressure. The pressure exerted on a segment of the earth's crust by a vertical column of material. See Lithostatic Pressure.

Geostrophic Wind. The horizontal wind velocity arising from an exact balance between the Coriolis force (an acceleration or apparent force resulting from the earth's rotation) and the pressure-gradient force due to differences of pressure within the atmosphere. The geostrophic wind blows in a direction parallel to the isobars—the lines connecting points of equal pressure on a weather map. In the Northern Hemisphere, as one faces downstream in the direction in which the wind is blowing, low pressure is to the left and high pressure to the right. In the Southern Hemisphere this relationship is reversed: low pressure is to the right of the flow instead of to the left. This qualitative statement of the geostrophic wind and pressure relation is known as Buys Ballot's Law, after the Dutch meteorologist who formulated it in 1857. At a given latitude the speed of the geostrophic wind is inversely proportional to the distance between the isobars—the closer the isobars for a given pressure interval, the higher the wind speed.

In any fluid, such as air, the difference in pressure represents a force that drives the fluid from high to low pressure; this force, proportional to the pressure difference divided by the distance between the two points, is termed the pressure-gradient force. On a weather map, the pressure-gradient force is directed from high to low pressure and is perpendicular to the direction of the isobars. The geostrophic wind thus blows at right angles to the pressure-gradient force. See Vorticity. (M. F. H.)

Geostrophic Wind

1 As a pressure gradient force moves a parcel of air from high to low pressure, the Coriolis force, caused by the earth's rotation, pulls it to the right. Eventually the air mass achieves a balance between the forces and travels at a constant speed parallel with the isobars.

Geosyncline

2 A cross section of a typical geosyncline consists of shallow water sediments on a continental shelf which grade into deposits typical of the continental slope, rise and abyss.

Geosyncline. An elongate or basin-like subsiding portion of the earth's crust. It may extend for several thousand kilometers, and sediments accumulated within it may attain thicknesses of thousands of meters, representing millions of years of deposition. Not all geosynclines have similar morphologic features, and scientists have therefore divided them into several subclasses based on their relationship to the stable interiors of continents (*cratons), type and derivation of their rocks, and their form.

Probably the most important geosynclines are those between the craton and ocean basin, since it is now believed that their evolution is directly related to the process of *plate tectonics and *sea floor spreading. These geosynclines, called orthogeosynclines, are linear troughs of crustal subsidence divided along their length into two distinct zones. The zone nearest the craton, the miogeosyncline, is characterized by lesser subsidence and a lack of volcanic activity; it may chiefly represent deposits laid down on a *continental shelf. The rocks associated with miogeosynclines are sandstones and limestones derived from erosion of the adjacent craton and deposits of skeletal remains of marine life. Both types of sediments indicate that they were deposited in relatively shallow water and that the rate of subsidence was nearly equal to the rate of sediment deposit. The second zone, eugeosyncline, lies between the miogeosyncline and the deep ocean floor. Within this zone subsidence has been relatively rapid and is associated with volcanic activity. The rocks found in the eugeosynclinal area are derived from: (1) volcanic activity, (2) very fine sediment eroded from the craton and transported beyond the miogeosyncline and (3) heterogeneous rocks known as *turbidites (suspensions of mud and coarser materials that cascade down the steep slope of the continental shelf). In cross section the sedimentary deposits of an orthogeosyncline form a wedge, thin along the craton and thickening seaward until it ends abruptly in the eugeosyncline.

A good illustration of a presently developing orthogeosyncline is the contintental shelf and *continental rise along the east coast of North America. It must be noted, however, that the portion of the rise representing the eugeosyncline is not known to be volcanically active at present. A series of *seamounts off the Atlantic coast of North America may represent ancient volcanoes. Submarine eruptions may produce finely divided volcanic material, characteristic of the eugeosyncline, which may be transported into the area of the rise and slope. Scientists now believe that ancient orthogeosynclines are preserved in areas occupied by belts of fold mountains such as the Appalachian Mountains of eastern North America (*continental accretion). According to this theory, the ocean bordering the Atlantic coast may very well be the site of a future mountain range. The Swiss Alps are the result of uplifting and buckling of an ancient geosyncline.

Other areas of crustal subsidence are known to exist within the otherwise stable craton. Originally all areas of cratonic subsidence were classified as parageosynclines and defined as areas in which the thickness of sediment deposited is greater than that in parts adjacent to them. (Parageosynclines are considered areas of less active subsidence than orthogeosynclines.) As more information became available, parageosynclines were subdivided into three classes: (1) exogeosynclines, or transverse basins, (2) autogeosynclines, or isolated basins, and (3) zeugogeosynclines, or yoked basins. Exogeosynclines are characterized by areas of subsidence within the craton (but near its edge) that are being supplied with sediments eroded from an uplifted orthogeosynclinal area adjacent to the craton. Autogeosynclines are basins essentially filled with carbonate sediments (limestones), including some sands. The minor amount of sandy material filling the basins comes from nearby lowlands or from distant uplifted orthogeosynclines. The last group of geosynclines, zeugogeosynclines, are found in the interior of the craton and contain sandy material derived from nearby uplifted portions of the craton. The Denver Basin adjacent to the Front Range of the Rocky Mountains in Colorado exemplifies this type of geosyncline. (J.W.S.)

Geothermal Energy. Energy from naturally occurring steam and hot water, found in hot springs, geysers and fumaroles, or from the heat of the earth's interior. *Fumarole fields have been developed at many localities to use volcanic steam for generation of power. At Larderello, Italy, boric acid has been recovered from steam vents since 1818; as early as 1904, electric energy was developed there. Wells have been drilled in which the temperature and volume of the steam increase with depth. Geothermal energy is now utilized in Iceland, Italy, Japan, Mexico, New Zealand, the United States and the Soviet Union, and new installations are under construction or planned in several other countries. The 1969 production of geothermal energy in Italy was 390,000 kw, and an additional unit of 50,000 kw capacity is under construction. At The Geysers, Sonoma County, in northern California, the 1969 capacity was 83,000 kw, but the total projected capacity will be 633,000 kw. The United Nations presently supports five exploratory projects. It is likely that many countries poor in the usual sources of energy will rely more and more on geothermal energy. The total estimated outflow of heat from the interior of the earth is more than 3000 times that from areas of volcanic activity. If this energy can be utilized economically by deep drilling, man will have a new and important source of energy.

Brine is associated with steam vents in some areas. The production of freshwater from these brines, by the utilization of the steam, is under consideration. (R.L.N.)

Geothermal Gradient. The rate of rise in temperature with increasing depth in the earth. The actual geothermal gradient is known only for those depths (less than 16 km) to which boreholes and mine shafts have been driven. Even at these depths there are variations caused by local conditions of heat flow and thermal conductivity. An average of 25°C per kilometer (about 78°F per mile) represents the geothermal gradient near the earth's surface. Determination of gradients deeper in the earth depends on many factors, including assumptions earth scientists make about the conditions prevailing at the time of origin and development of the earth, its composition and physical properties. Based on these assumptions several different curves have been formulated for estimating the gradients in deeper portions of the earth. (J.W.S.)

Geyser. A tubular channel in the earth's crust from which boiling water and steam are intermittently ejected. A major geyser throws water 30 m or more in the air; a minor geyser less than 30 m. The time interval between eruptions varies not only from geyser to geyser but also for any one geyser. The geyser cone or mound built up around the orifice is composed of geyserite (siliceous sinter), a hydrated oxide of silicon precipitated when the geyser water evaporates.

Geysers are found in Iceland, where they were first observed, and in New Zealand as well as in America's Yellowstone National Park. There are approximately 200 geysers in Yellowstone, more than in all the rest of the world. Old Faithful, in Yellowstone, is the world's most famous geyser. Its interval between eruptions is about 65 minutes; Castle Geyser, also in Yellowstone, has an interval of about 9 hours. The duration of the eruption itself also varies: for Old Faithful it is 2 – 5 minutes and for the Castle Geyser 45 – 60 minutes. Old Faithful throws a column of water 32 – 56 m (106 – 184 feet) in the air, depending on the violence of the eruption and the wind velocity. During an eruption it discharges approximately 40 cubic meters (1400 cubic feet) of water at a temperature above 95°C (200°F). Most of this water does not drain back into the tube but starts on a long journey toward the Atlantic Ocean.

Many partially petrified stumps, logs and pine needles embedded in the cone of Old

Faithful prove that after the cone reached its present size, hundreds of years of thermal inactivity ensued, during which trees grew on the cone. Then renewed thermal activity killed the trees. The size of the cone, coupled with measurements of the rate of geyserite deposition, suggests that it is thousands of years old.

Geysers owe their existence to irregularly shaped tubes extending down from the surface, a source of water generally of surface origin (in Yellowstone, magmatic water amounts to less than 5 percent) and a source of heat sufficient to boil the water. It is now thought that heat is supplied when deeply circulating surface water is brought close to a slowly cooling *intrusive rock body.

If the tube is wide, has no constrictions and the source of heat is adequate, a boiling pool results.

The intermittent activity of geysers is explained by the fact that the greater the pressure on water the higher its boiling point. Thus the boiling temperature of water in a tube increases with depth. If a tube is long, crooked and constricted in places, the water may first boil in only one part of the tube, forming steam with an increase in volume. Water above the point where the boiling occurs will be pushed upward and some will be ejected, forming the preliminary eruption that often precedes a main eruption. The pressure on the water remaining in the tube is then reduced and its boiling point is lowered. More water, after perhaps additional heating, will flash into steam, more water will be ejected, the pressure will be further reduced, more boiling will ensue and finally the rest of the water in the tube, together with steam, will form the main eruption. The next eruption will take place when the tube has filled with water and has again been brought to its boiling point. (R. L. N.)

Geyserite. An opaline material deposited by *hot springs. It is also called *siliceous sinter.

Gibbsite. A mineral, an aluminum hydroxide, that in association with boehmite and diaspore makes up *bauxite, the ore of aluminum. See Mineral Properties.

Ginkgo. The maidenhair tree. Although abundant in the geologic past, only one species, *Ginkgo biloba,* is living today. These plants were widespread during the Mesozoic and probably originated during Late Permian time. They have been considered sacred in Japan and are planted near temples. Because of their primitive characteristics they are called "living fossils." Their known *geologic range is from Permian to Holocene (Recent), and they attained their maximum abundance during the middle portion of the Mesozoic Era.

Glacial Drift. Sediments (commonly called drift) deposited by glaciers and glacial *meltwater. Stratified drift is deposited by meltwater, unstratified drift (*till) by glaciers; both are common in glaciated regions. Glaciofluvial sediments are deposited on land by meltwater streams; glaciomarine and *glaciolacustrine sediments are brought to the ocean and to lakes by meltwater streams. (R. L. N.)

Glacially Transported Fragment. A block or boulder transported and deposited by glacial ice. In general, continental glaciers do not carry fragments far, but many fragments in both North America and Europe are known to have been carried approximately 500 km and some more than 1000 km. Fragments of copper from ledges in Michigan have been carried almost 1000 km to Illinois. Boulders have been lifted up mountain slopes as much as 1200 m. The fragments vary in size from microscopic particles to such a gigantic mass as the Madison Boulder near Conway, N.H., which is about 27 m by 12 m by 12 m (90 by 40 by 40 feet) and weighs approximately 9000 metric tons.

Glacial-Marine Deposit. A deposit consisting of *rock flour brought to the ocean by *meltwater streams and usually containing ice-rafted fragments. The fragments are generally angular or almost angular, may be striated, and range in size from granules up to fragments weighing several metric tons. The matrix commonly contains cold-water shells and other remains of cold-water organisms, and may be stratified if the ice-rafted fragments are not too abundant. During the Pleistocene Epoch, this type of sedimentation was common in the high latitudes of both hemispheres. Glacial-marine deposits are being formed today off the coast of Alaska, around Greenland and Antarctica, and wherever tidewater glaciers occur. These deposits are not yet covered by modern *pelagic sediment. (R. L. N.)

Glacial Stage. A subdivision of a time of extensive glaciation, such as the Ice Ages of the *Pleistocene Epoch of the *Quaternary Period. Glacial stages mark times during an ice age when glaciation was at its maximum (glacial maxima) and climates were much colder in both the Northern and Southern Hemispheres. The stages are separated by longer and warmer intervals of time called *interglacial stages (glacial

2

Geyser

1 *Old Faithful, at Yellowstone National Park, Wyo., periodically ejects a column of boiling water more than 30 m (100 feet) in the air. Each eruption lasts two to five minutes.*

Ginkgo

2 *Lone survivor of an ancient group of softwoods that once covered much of the Northern Hemisphere, the maidenhair tree,* Ginkgo biloba, *has remained virtually unchanged for 150 million years.*

1

Glacial Striation

1 *The shape and alignment of grooves and scratches etched on rock surfaces by a glacier show the direction of flow.*

Glaciated Valley

2 *A stream eroding a mountain region forms a curving, V-shaped valley (a) with many side-wall spurs. As the climate cools, a glacier is formed (b) filling the valley with ice. As the glacier moves downward it widens and deepens its valley by plucking and abrasion. The ridges create depressions called cirques. Rubble that accumulates along both sides of the glacier is called a lateral moraine. When two glaciers merge, their lateral moraines join to form a medial moraine. As the climate warms, the glacier melts away, revealing a wide, straight, U-shaped valley (c) with a hanging tributary whose stream drops to the main valley as a waterfall. The glaciers sheared two sides of a ridge to form a horn, and one cirque has become a meltwater lake.*

3 *The trough of Clinton Canyon, in Otago, New Zealand, was gouged out by a large glacier during the Ice Age.*

minima). Four glacial stages (sometimes called ages) have been described in the Upper Mississippi Valley of the United States, named for the states where their deposits are best exposed: beginning with the oldest, the Nebraskan, Kansan, Illinoian and Wisconsin. A similar series of glacial stages in Europe—the Günz, Mindel, Riss and Würm—are believed to be essentially equivalent to those in North America. (W. H. M.)

Glacial Striation. One of a series of straight or slightly curved, parallel or nearly parallel scratches that may be less than 1 mm (1/25 inch) in depth and only a few centimeters in length. A glacial groove may be 25 mm (1 inch) or more in depth, 50 – 100 mm (2 – 4 inches) in width and many meters in length; a glacial trench may be 30 m (100 feet) deep and 1.5 km (1 mile) in length. These striations, grooves and trenches are best developed on soft massive rocks. They occur on steep slopes, in valleys and on mountain tops, at the surface or deeply buried by *regolith or *bedrock, and beneath modern glaciers as well as far removed from them. They cover great areas in northern North America, Europe and elsewhere, and are found in the high valleys of all the great mountain ranges. They are usually associated with Pleistocene deposits, but they also occur in the Permo-Carboniferous, the Precambrian and other parts of the stratigraphic column.

Many agents can form striated, grooved and polished bedrock surfaces. Among them are volcanic gas blasts, faulting, eolian abrasion, snow avalanches and landslides, grounded icebergs, sea and river ice, and glaciers. Those considered here, however, have been formed entirely by the scratching, scouring and planing of bedrock by glaciers. This is proved by their widespread or even continental distribution, association with *till and other glacial deposits, uniform alignment over large areas and occurrence in areas from which glaciers have recently retreated.

The striations are formed by rock fragments in the bottom of glacial ice dragged slowly across the bedrock by the glacial motion. The abrasion resulting from their formation produces an abundance of very fine-grained particles called *rock flour. Silt and sand form the finer striations; larger fragments form the grooves and trenches. Striations end where fragments are either destroyed by abrasion or are retracted up into the ice and out of contact with the bedrock. Those formed on slow-weathering rocks such as quartzite have considerable permanence. However, well-developed striae and polish may disappear

completely in 25 years or so on rocks that weather rapidly. Many striations now at the surface have probably been buried for most of post-glacial time and have only recently been exposed.

Cross striae (the crossing of an earlier striation by a later one) are common. They are formed by a change in the direction of glacial motion or perhaps by the rotation of fragments held in the ice. Most of the striations found in northern North America were formed during the last glacial period (the Wisconsin) 8000 – 18,000 years ago. (R. L. N.)

Glacial Surge. A relatively sudden, rapid movement of a glacier. While most valley glaciers move less than 1 m (3 feet) per day, surges with velocities of more than 120 m (400 feet) a day have been reported. Such movement causes looped and contorted medial *moraines, broken and crevassed ice, a displacement of surface features down valley, a lowering of the surface of the headward part of a glacier, and other features. It is thought that basal *meltwater lubrication is a primary cause of surging. The meltwater may result from friction or high geothermal gradients, but more important, perhaps, from increased accumulation (possibly due to climatic fluctuations triggered in places by earthquakes), which lowers the melting point of the ice and induces basal melting. When the film of meltwater reaches a critical thickness, the attachment of the glacier to *bedrock is drastically weakened and the glacier surges forward. Surges have been observed in hundreds of glaciers in Alaska, India, Spitsbergen and elsewhere. A dozen major glaciers in Alaska surged in 1966. (R. L. N.)

Glaciated Valley. A fluvial valley modified by glaciation. It has steep walls that may be vertical, and its transverse *bedrock cross section is U-shaped as a result of glacial action. It is commonly characterized by bedrock basins that may contain lakes, and by steplike irregularities. It is usually straighter and more open than a fluvial valley because the spurs between tributaries have been cut off by the glacier. Unlike a fluvial valley, where the tributary streams meet the main stream on grade, a glaciated valley has tributary streams that fall from *hanging valleys. Thus glaciated valleys are marked by *waterfalls, including some of the highest and most impressive in the world. The bottom of the valley is usually characterized by terminal, recessional and ground *moraines, and the valley walls by lateral moraines. The terminal and recessional moraines often dam up the drainage and form beautiful lakes such as Lake

Louise in the Canadian Rockies and St. Mary Lake in Glacier National Park. The bedrock floor and the lower side walls are usually polished, striated and characterized by *roches mountonnées unless they were destroyed by post-glacial erosion. If occupied by a *valley train, the valley is generally flat-bottomed. The upper end of a glaciated valley commonly terminates in one or more *cirques, the lower end in a V-shaped fluvial valley. Yosemite Valley is perhaps the most famous glaciated valley in the United States, and Lauterbrunnen Valley is a well-known example in Switzerland. (R.L.N.)

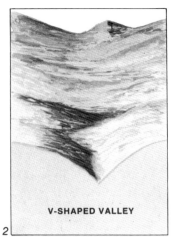

V-SHAPED VALLEY

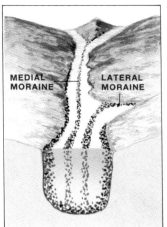

MEDIAL MORAINE   LATERAL MORAINE

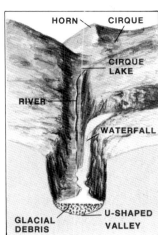

HORN   CIRQUE

CIRQUE LAKE

RIVER

WATERFALL

GLACIAL DEBRIS   U-SHAPED VALLEY

Glaciation. The covering and modification of the earth's surface by glaciers. Landscapes resulting from mountain glaciation differ greatly from those produced by continental glaciation. Mountain glaciation forms erosional features such as *glaciated valleys, *hanging valleys, *truncated spurs, *cirques and *horns; and depositional features such as *moraines, *valley trains and *kame terraces. Areas that have undergone continental glaciation are characterized by *drumlins, *eskers, *outwash plains and moraines. Lakes are common in both areas, as are *erratics, *roches moutonnées, *glacial striations, *kettle holes and *kames. In general, mountain glaciation increases the sharpness and steepness of topography, whereas continental glaciation produces a flatter and more subdued landscape.

Early in the 19th century, Swiss naturalists began to realize that the Swiss glaciers had previously been more extensive. Later it was shown that northern Europe and the British Isles had also been glaciated, and by the middle of the 19th century it was demonstrated that large areas in North America had once been covered with glacial ice. Some time later, glacial geologists realized that the *Pleistocene Ice Age had been more complex than a single glacial advance followed by a retreat. They concluded that during the Pleistocene the glaciers advanced and retreated four times, with the climate during interglacial episodes similar to that at present. In the United States the four advances are named the Nebraskan, Kansan, Illinoian and Wisconsin glacial stages, after the states where the glacial deposits that had been laid down were first studied or were well exposed. The European advances are called the Günz, Mindel, Riss and Würm glacial stages. Today even greater complexity has been demonstrated; potassium-argon age determinations indicate that the glacial period started a few million years age, and carbon-14 measurements show that the ice most recently

Glacier
Fed by a mountain snowfield, Alaska's North Sawyer Glacier carves a valley as it creeps toward the sea.

retreated from its most advanced position in Illinois, Ohio, New England and adjacent states less than 15,000 years ago.

During the Wisconsin maximum approximately 30 percent of the earth's land surface was covered with glacial ice. A continental glacier centered in Canada extended down into northern United States and covered approximately 13 million square kilometers (5 million square miles). Musk-oxen were living in Arkansas, which was then about as cold as northern Canada is now, and walrus inhabited the ice-strewn waters off the coast of New Jersey. Continental glaciers also covered Europe, Antarctica and Greenland, and huge alpine glaciers occurred on all high mountains, including those of Australia.

Not only was glaciation widespread during the Pleistocene, but extensive large-scale pre-Pleistocene glaciation also occurred. The presence of *tillites, glacial pavements and other glacial features in the stratigraphic record has demonstrated that parts of Africa, India, South America, Australia and Antarctica were glaciated in the Late *Paleozoic, approximately 230 million years ago. Still other periods of glaciation occurred in the Late *Precambrian, more than 600 million years ago.

Erosion and deposition during Pleistocene glaciation not only modified the areas it covered but also had other effects, some global in extent. During the glacial periods vast quantities of water were taken from the oceans and locked up in the glaciers on land. This caused a lowering of sea level, amounting to over 100 m during the Wisconsin. A similar lowering occurred during the other glacial stages, but sea level was high during the interglacial stages. The evidence for the lowering is found in the presence of features such as valleys, fluvial deposits, peat and beach ridges now over 100 m underwater. During the low stages large areas of the ocean bottom were laid bare, and streams deposited diamonds and tin ores in these areas. Also, islands were joined to mainlands and to other islands, facilitating animal migrations that now would be impossible.

The weight of the Pleistocene continental ice sheets depressed the earth's crust as much as hundreds of meters in some places. This is shown by the uplift due to gradual rebound that is still taking place in recently deglaciated areas such as northern Canada and Scandinavia.

During the Pleistocene glaciation the climatic zones in North America were pushed southward by the continental ice sheet that covered Canada and northern United States. The southwestern states, which today are arid or semiarid, were then colder and wetter. As a consequence, numerous *pluvial lakes occupied the undrained basins of the area; Lake Bonneville was the largest. At its maximum it was more than 50,000 square kilometers in area, and over 300 m deep; it covered parts of Utah, Nevada and Idaho. Its shoreline features are conspicuous on the lower slopes of the Wasatch Mountains near Salt Lake City.

There is no generally accepted explanation of the Pleistocene and early glaciations. It is certain, however, that they are not due to a progressive cooling of the earth, since the Precambrian and Late Paleozoic glaciations were followed by prolonged warmer periods. Among the theories are: (1) the radiant energy of the sun has varied through time. (2) Atmospheric temperature drops 1°C for every 100 m increase (5.5°F for every 1000 feet) in altitude, and mountain-building periods when the earth's crust was high should therefore favor glaciation. It is known that the crust was high when the earlier glaciations took place and this may have been an important cause for these older glaciations. Moreover, a progressive lowering of the temperature, known to have taken place during the Tertiary, may have resulted from progressive mountain building. It seems very unlikely, however, that recurring crustal uplift and depression could have been responsible for the fourfold glaciation of the Pleistocene, or for still shorter post-glacial fluctuations in the size of glaciers. (3) Dust clouds are scattered through the empty spaces in our galaxy; if the solar system should pass through such a cloud, there would be a decrease in the solar energy received by earth. However, there is no evidence of this. (R.L.N.)

Glacier. A mass of ice derived from snow, which moves or has moved in the past. Glaciers are found on high mountains and in high latitudes where more snow falls in the cold season than melts in the warm season. Glacial ice now covers approximately 15.5 million square kilometers or about 10 percent of the earth's land surface. The Antarctic and Greenland ice sheets contain most of it, but it is also found on all continents except Australia. Most modern glaciers are dwindled remnants from the *Pleistocene (the glacial period), when about 30 percent of the earth's land surface was covered with glacial ice. If the ice now in existence should melt due to an amelioration of climate, sea level would rise perhaps 60 m, much of the wealth of the world would be destroyed and London, New York, Sidney, Honolulu and many other coastal cities would be partly or completely inundated.

When freshly fallen snow above the snowline is progressively buried, it slowly changes into *névé and finally into ice. When the accumulation gets thick enough—usually tens of meters, depending on the steepness of the slope and the temperature of the ice—the mass begins to move downslope and a glacier is formed. That part of it that moves below the snowline will slowly waste away. Thus a glacier is composed of two zones: (1) the zone of accumulation above the snowline, where more snow falls than is lost by *ablation and (2) the zone of wastage below the snowline, where more snow and ice is lost than is replenished by snowfall. A glacier is in equilibrium if its terminus remains more or less stationary. If the climate becomes colder, less ablation takes place, perhaps more snow falls, and the terminus of the glacier advances farther down the valley until it reaches a new equilibrium. If the climate becomes warmer, the terminus will retreat up the valley until it again reaches equilibrium. Many of the glaciers in both the Northern and Southern hemispheres have been in retreat during the first half of this century, reflecting a warming trend.

The movement of glaciers can be demonstrated in many ways: (1) glaciers have moved into areas formerly free of ice; (2) huts built on glaciers have been gradually carried down a valley; (3) bore holes drilled into glaciers become deformed; (4) movement has been detected in glacier tunnels; (5) a straight row of stakes across a glacier moves down valley and loses its alignment; (6) rocks that have fallen on glaciers from the valley walls may be unlike those in the walls above them but resemble rocks in the walls farther up the valley; (7) the striations and grooves on *bedrock once covered by glaciers can have resulted only from the movement across the bedrock of ice containing rock fragments.

Studies show that glaciers move more rapidly in the center than on the margins, near the surface than at depth, when the temperature of the ice is higher, where the valleys are narrow and where the ice is thicker. Velocities of over 45 m per day have been observed on a few glaciers, but most move only a meter or so per day. The mechanics of flow is still under investigation, but it is known that one part of a glacier can move past another part along a shear plane. Moreover, an entire glacier may slide along bedrock (basal slip). The *meltwater between bedrock and ice is thought to facilitate this movement. In addition, ice under pressure is a plastic substance; glaciers therefore also flow.

Although Antarctic glaciers may have temperatures as low as -21°C (-7°F), many mid-latitude glaciers are at about 0°C (32°F). Since most snow falls at subfreezing temperatures, the source of heat for this higher temperature is of interest to glaciologists. Heat is developed by friction where one part of a glacier shears past another and where it slides on the bedrock beneath it. The temperature of a glacier can also be raised by the geothermal heat supplied to its base. Calculations have shown that in the Greenland ice sheet the heat generated by shearing may equal the geothermal heat. Meltwater formed on a glacier during the warm season may percolate downward and freeze, giving up its latent heat (80 calories per gram) and thus increasing the temperature of the ice. In addition, if the climate becomes warmer, heat from the atmosphere can be conducted down into the glacier.

There are several major types of glaciers. A river of ice that moves down a valley and is fed by a snowfield higher up is known as an alpine or mountain glacier. It originates in one or more *cirques and terminates either in a valley below or in the ocean. Many glaciers which exist today are dwindled remnants of larger Pleistocene glaciers. Alpine glaciers may be as much as 100 km long, several kilometers wide and over 600 m thick, with several tributaries. They usually move only a meter or so a day. They are characterized by ice falls, seracs, *crevasses, lateral and medial *moraines and tunnels from which meltwater issues. The mountain scenery they created in the Himalayas, Andes and Caucasus, and in Alaska, New Zealand, Scandinavia and elsewhere is among the grandest on earth. The fifty-odd alpine glaciers in the continental United States (excluding Alaska) are found in the Sierra Nevada, Cascade Range, Olympic Mountains, Northern Rockies and in Rainier and Glacier National Parks.

A piedmont glacier is a large, gently-sloping, slow-moving sheet of ice at the foot of mountains. It is fed by valley glaciers that originate in the mountains and coalesce and expand on reaching the lowlands where they are no longer confined by valley walls. If alpine glaciers resemble rivers of ice and continental ice sheets resemble oceans of ice, then piedmont glaciers are like lakes of ice. They are found in Alaska, Greenland, Iceland, Spitsbergen and Antarctica. During the *Pleistocene, piedmont glaciers were found on the plains that border the Alps and the Rocky Mountains. The best known piedmont glacier and the first to be recognized is the Malaspina in Alaska. Fed by numerous valley glaciers originating in the Mount St. Elias Range, it covers about 2200 square kilometers (850 square miles) and is at least 600 m (2000 feet) thick. It has a gentle slope of about 20 m (70 feet) per mile. White, crevassed, active ice occurs in its central portion, and an ablation moraine as much as 5 m (15 feet) thick, on which a spruce forest grows, is found on stagnant ice at its margin. Of the many piedmont glaciers in Antarctica, the Wilson is perhaps the best known; it is about 60 km (37 miles) long and 15 km (9 miles) wide.

When a single valley glacier issues from the mountains and expands on reaching the lowlands, it is known as an expanded-foot glacier. Good examples are found in the McMurdo Sound region of Antarctica.

An ice sheet or ice cap is a sea of glacial ice that moves slowly outward from the center and may cover hundreds of thousands of square kilometers. It moves outward because of greater accumulation of snow at the center. Ice sheets are found in Iceland, Spitsbergen, the archipelago north of Canada and elsewhere, and continental ice sheets cover most of Greenland and Antarctica. The Greenland ice sheet has an area of roughly 1.7 million square kilometers (650,000 square miles), and much of it is more than a thousand meters thick. In places there is a narrow strip of land between it and the ocean; elsewhere the ice reaches the ocean and forms icebergs by calving. The Antarctic ice sheet covers an area of more than 11.5 million square kilometers (4,500,000 square miles)—more than the combined areas of the United States and Mexico—and has thicknesses greater than 4500 m (14,850 feet). It contains about 90 percent of the world's ice and 75 percent of its fresh water. In most places it terminates in the ocean and forms many icebergs. During the Pleistocene a continental ice sheet covered about 15 million square kilometers (approximately 6 million square miles) of Canada and northern United States; another covered about 5.2 million square kilometers (2 million square miles) of Europe.

A cirque glacier is restricted to a cirque basin and is therefore small. If the snowline becomes progressively lower, the ice may move out of the basin. An outlet glacier usually occupies a valley, but instead of having its own snowfield in one or more cirques it is fed by an ice sheet. It is found where ice escapes through valleys in the mountain ranges that dam the ice sheets. Such glaciers are common in Iceland, Norway, Greenland and Antarctica. See Color Plate "Rivers of Ice." (R.L.N.)

Glaciolacustrine Deposits. Sediments deposited in ice-dammed lakes. The two most important kinds are beaches on the margins of such lakes and sediments carried into the lakes mainly by meltwater streams. The beaches are important because they outline the former extent of these lakes; their regional tilt records the nature of the crustal movements that have occurred since their formation; and a series of beaches, one lower than another, signifies a lowering of lake outlets resulting from glacial retreat. They were early recognized in Glen Roy, Scotland, and have been identified in many other places, notably at the Great Lakes, where they were formed around the lakes that immediately preceded the present Great Lakes.

The sediments carried into ice-dammed lakes are rhythmically laminated, with alternating coarse and fine layers. They are formed as follows: During the summer, when melting was at a maximum, sand, silt and *rock flour were carried into the lakes by meltwater streams. The coarse material soon settled to the bottom but the rock flour remained in suspension. During the cold season melting stopped and the lakes froze over. In the absence of waves and currents under the ice, the rock flour settled to the bottom and buried the summer layer. Thus a coarse and a fine-grained layer were deposited each year. By counting the pairs, the duration of ice-dammed lakes can be determined; this method is also used to measure the rate of glacial retreat. A pair of layers is called a *varve; the bottom sediments are sometimes called varved clays. The glaciolacustrine deposits formed during the close of the *Pleistocene glaciation are usually devoid of fossils. (R.L.N.)

Glaciology. The branch of geology concerned with the snow and ice that occur naturally on the earth's surface. Emphasis is on glaciers, but sea, lake and river ice are also studied. Thus glaciology is closely related to meteorology, climatology and hydrology.

Glaucophane. A mineral, sodium magnesium silicate of the *amphibole group (a series extended from glaucophane to riebeckite with the substitution of iron for aluminum). It is found only in *metamorphic rocks; in a few places it is a major constituent, as in the glaucophane schists of the California Coast Ranges. See Mineral Properties.

Glomar Challenger. A specially designed research ship used for the Deep-sea Drilling Project, a program sponsored by the National Science Foundation of the United States and begun in 1968. Its objective is to explore the suboceanic crust of the earth by obtaining sediment cores up to several hundred meters long from the ocean bottom. The *Glomar Challenger* is 120 m long and has 10,500 tons displacement. Amidship is a drilling derrick that rises 60 m above the waterline. It has more than 6 km of drill pipe and contains such advanced drilling equipment as a gyroscopically controlled roll-stabilizing system; satellite navigation to fix its position accurately; a dynamic positioning system that utilizes computer-controlled thrusters to keep the ship on station; a funnel on the drill string that remains on the ocean floor when the pipe is removed (the funnel can then be located with a high-resolution scanning sonar probe system, thus enabling re-entry into the hole); and a laboratory that permits specialists to examine core samples aboard ship.

The extraordinary drilling capabilities of the *Glomar Challenger* have led to such accomplishments as the confirmation of the *sea floor spreading, *continental drift and the Northwest Pacific as an area of very old oceanic sedimentation, proof that much of the deep Gulf of Mexico is underlain by salt, recovery of sediments very rich in metals (primarily iron and manganese oxides) just above the original igneous rock floor of the ocean, and evidence that deep ocean water of the past contained oxygen much as it does today. (R.E.B.)

Gneiss. A coarse-grained *metamorphic rock commonly composed of quartz and feldspar, with lesser amounts of mica. The segregation of quartz and feldspar from mica produces an irregular banding in the rock. Since gneiss may be composed of many minerals besides quartz, feldspar and mica, the term refers more to the texture and structure of the rock than to its chemical-mineralogic composition. Gneisses generally are labeled according to the most abundant mineral or minerals in the rock; for example, a gneiss with a large percentage of hornblende is called a hornblende gneiss. They may also be classified by texture or mode of origin. Gneiss is considered to be the product of regional metamorphism associated with moderate to high temperatures and pressures. Gneiss may represent the recrystallization of *igneous rock types (orthogneiss) or of *sedimentary rocks (paragneiss).

Gneisses are found throughout New England, in the Adirondack Mountains of New York, in large areas of Scandinavia, in the British Isles and the Alps, and generally as the core rocks or older rocks of mountain chains. Gneiss, and especially the variety known as granite gneiss, is used as a building stone. The striped pink, Stony Creek granite of southern Connecticut is such a building stone. Granite gneiss differs from most other gneisses in that it contains more quartz and feldspar and less platy micaceous minerals. The lower the percentage of mica in the stone, the greater its strength, for mica grains normally form bands that constitute planes of weakness within the rock. (J.W.S.)

Goethite. A widespread mineral, a hydrated iron oxide that makes up most of the material known as brown iron ore. It was named in 1806 in honor of Goethe. In many localities it is an important ore of iron; major occurrence is in Alsace-Lorraine.

Goethite is a crystalline mineral with definite properties and composition, but in many places it is impossible to distinguish it from *limonite, an impure amorphous iron oxide with which it commonly occurs. A well-crystallized specimen can be easily determined as goethite, but an earthy amorphous-appearing specimen cannot be identified by observation alone. Much of what was formerly called limonite is now known to be goethite. Both are characteristically formed under oxidizing conditions as weathering products of iron-bearing minerals. They are common natural pigments and impart their yellow and brown colors to the rocks in which they occur. Goethite forms extensive bedded deposits as a direct inorganic or biogenic precipitate from water and is common as a deposit in bogs. It also is the major constituent in the *gossan overlying metalliferous veins, and in residual lateritic mantles formed on the weathering of *serpentine. See Bog Iron Ore; Mineral Properties. (C.S.H.)

Gold. An element that occurs in nature principally as a native metal, that is, uncombined with other elements. However, it is commonly alloyed with silver, and the rich yellow color of pure gold becomes paler with an increasing proportion of silver. When the amount of silver equals or exceeds 20 percent, the alloy is called electrum.

Although gold is a rare element, it is widely distributed in nature and in a variety of geologic environments. It is most commonly found in hydrothermal gold-quartz veins that are related to granitic types of igneous rocks. Sulfide minerals, of which *pyrite is most common, are usually present. The gold at and near the surface is liberated because it is quite inert to the chemical weathering that breaks down the associated minerals. It may either remain deposited in the soil mantle or be washed into neighboring streams to form a *placer deposit. Because of its high specific gravity, 19.3, gold works its way downward through the sand and gravels to the

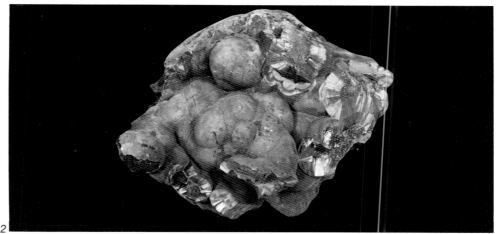

1

2

Glomar Challenger
1 Used by the Deep-sea Drilling Project, an expedition sponsored by the National Science Foundation, this ship has drilled and cored bottom sediment in all oceans of the world.

Goethite
2 Radiating fibrous crystals of brownish-black goethite are packed together tightly to form kidney-like masses.

bedrock of the stream bottom, where it lodges in cracks or behind irregularities. It was the discovery of flakes and nuggets of placer gold that resulted in the great gold rush to California in 1849. In fact, the first evidence of gold in most great gold-producing areas of the world has been in placers.

The formation of gold placers has been going on throughout geologic time, and the gravels containing the gold may be buried beneath younger rocks and consolidated into conglomerates. Such is the Precambrian Witwatersrand conglomerate of the Transvaal, South Africa. Gold was discovered in these ancient placers in 1886; since that time they, and similar conglomerates to the south in the Orange Free State, have been the major source of the world's gold. See Mineral Properties. (C.S.H.)

Gondwanaland. The name given by the German meteorologist Alfred Wegener (1880-1930) to a hypothetical landmass of the Mesozoic Era that included most of the landmasses of the present Southern Hemisphere. As *continental drift took place, Gondwanaland broke up and formed what is now South America, Africa, peninsular India, Australia and Antarctica. See Pangaea. See also Color Plate "An Ancient Supercontinent Breaks Up."

Gorge. A narrow, steep-walled valley or passage, usually carved by stream abrasion. A gorge is steeper than a ravine and smaller than a canyon. Although it may be the only passage through a mountain, the deep gorge of a major river is often a barrier to movement because there is no room for a roadbed between the stream and the sheer walls.

Gossan. A surface deposit composed in large part of hydrated iron oxides overlying metal-bearing veins. The surface oxidation of veins rich in sulfide minerals, particularly *pyrite, results in the removal of sulfur and the formation of a cellular mass of brownish-yellow *limonite and/or *goethite. The oxidation may also cause the removal of copper, lead, zinc and silver. However, in many places secondary minerals of these metals form near the surface surrounded by the more abundant iron oxides. Thus the gossan, or "iron hat" as it is sometimes called, is the home of many rare, interesting minerals. (C.S.H.)

Gouge. Pulverized, claylike rock found along a *fault. It is produced by the crushing and grinding of rock fragments caught between the walls of large fault blocks. Although there is no general rule, in some cases the amount of gouge indicates the magnitude of displacement that has taken place along the fault. The term is also used by miners to describe the soft rock adjacent to an ore vein because a miner can easily "gouge out" the soft material with his pick.

Graben. A narrow trenchlike portion of the earth's crust that has been depressed relative to the land on either side of it. Graben are bordered by steeply dipping planes that represent fractures within the crust. It is thought that they result from tensional forces acting on the underlying rock parallel to the earth's surface. The forces are created by uplifting of the crust into a broad arch or by a stretching of the crust. In either case forces are generated greater than the cohesive strength of the rock, causing development of steeply dipping to nearly vertical fracturing with subsequent lengthening of the crust. The lengthening allows some of the fracture-bounded, wedge-shaped blocks to become depressed relative to adjacent blocks. This process may be compared to the movement of a keystone in an arch if the sides of the arch were pulled apart. Graben are responsible for the topography in the Basin and Range states of the southwest United States, in the Rhine Graben of Germany and in East Africa's Rift Valley.(J.W.S.)

Graded Stream. A river that neither erodes nor deposits material along its channel; a condition of balance whereby the river adjusts its ability to degrade or aggrade its channel to maintain an equilibrium profile, that is, a *longitudinal profile where the slope at each point is just that needed to continuously transport the load downstream. If, over a long period of time, the slope is adjusted to discharge and hydraulic channel characteristics so that the load the stream has to carry can be transported, the river is graded. In modern terms, the stream is an open system in a steady state where discharge and load entering the system are balanced by the discharge and load leaving. In this state the system is self-regulatory; any change in the system's environment will result in an automatic readjustment of gradient, hydraulic characteristics (channel form, roughness) and channel pattern to absorb the effects of the change. (M.M.)

Gradient. In hydrology, the slope of a stream, measured by the fall between two points on the channel bed (or water surface) divided by the distance between the points, expressed in meters per kilometer. The gradient is generally steep near the head and gentle near the mouth. The slope

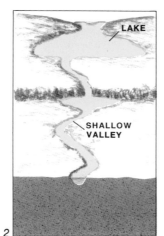
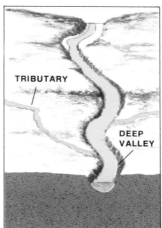
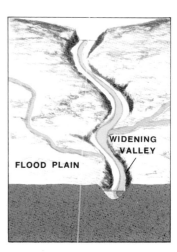

2

3

Gorge

1 The steep slopes of the Urubamba River Valley in Peru were carved by rivers flowing toward the Amazon basin.

Graded Stream

2 The stages in the development of a graded stream are: (a) a stream forms on the surface of recently elevated land; (b) the valley deepens, lakes and marshes drain and tributary valleys develop; (c) downcutting diminishes and the stream widens the valley, creating a floodplain in the process.

Graben

3 An elongate rock mass slipping down between two faults caused this deep trench along the base of California's Panamint Range.

is intimately interrelated to caliber and amount of load carried, discharge, channel form, bed roughness and channel pattern. If the stream has excess energy for the amount and caliber of the load, it may lower its gradient by eroding the channel. If the stream is unable to move the load provided, it may increase its gradient by depositing in the channel.

Rock type and lithology affect gradient since it may be easier for the stream to adjust some factor other than slope on resistant *bedrock. (M.M.)

Granite. A coarse-grained igneous rock composed essentially of *feldspar and quartz. Potash feldspar (orthoclase or microcline) is the dominant mineral but some oligoclase feldspar is usually present. In addition to quartz and feldspar, small amounts (about 10 percent) of *mica and *hornblende are common. Biotite is the principal mica, but muscovite may also be present. The accessory minerals are zircon, sphene, apatite, magnetite and ilmenite. See Granodiorite.

Granite, Graphic. See Graphic Granite.

Granitization. A process, other than melting, whereby any preexisting rock is converted into granite. The process most likely results in the addition of material to form *feldspars and in the removal of elements of iron and magnesium from the original rock to an outside source, thus forming granite. Some geoscientists apply the term to the formation of granitic rock from sedimentary rock regardless of whether melting has occurred. The exact operation and extent of the process is still not clearly understood, but many geologists believe that large granite masses of the world formed not from *magma but from granitization.

Granodiorite. A coarse-grained igneous rock composed essentially of *feldspar and *quartz. It is thus similar to a *granite except that in granodiorite the feldspar is dominantly *plagioclase with minor potash feldspar; in granite the reverse is true. Without microscopic examination one cannot distinguish between these two rocks.

Granophyre. A fine-grained *granite characterized by a graphic intergrowth of *quartz and *feldspar. The relationship of quartz to feldspar is the same as in *graphic granite except on a smaller scale.

Granular Disintegration. The separation of mineral grains from coarse-grained rocks such as *granite, *gabbro and *marble. The grains separate along their boundaries, forming particles suffering little chemical alteration and retaining essentially their original size and shape. Granular disintegration operates most rapidly on corners, less rapidly on edges and least rapidly on faces, resulting in rounded *outcrops and boulders. It occurs in Antarctica, in hot deserts and in temperate climates. It probably results from several processes, including solution, *frost action, expansion due to chemical alteration, and changes in temperature. An accumulation of mineral grains, common in deserts, is called grus. (R.L.N.)

Granulite. Any *metamorphic rock composed of equal-sized interlocking grains. Granulites may contain flat lenses of quartz and/or feldspar that impart a crude layered structure to the rock. North American geologists use the term in reference to texture, while most European geologists refer to a granulite rock as composed of specific minerals resulting from restricted conditions of metamorphism, regardless of texture or structure. The American term can be applied to any rock without regard to chemical composition. An example of a granulite in the latter sense is *marble, a metamorphosed limestone used as an architectural and ornamental stone.

Graphic Granite. *Quartz and *feldspar intergrown in such a manner that in certain sections dark quartz appears as areas resembling cuneiform characters on a background of lighter colored feldspar. Because of this resemblance to writing the intergrowth is called graphic granite. The feldspar, usually *microcline, may be in crystals as large as 1 m across, and the quartz rods within the feldspar may all have a common crystallographic orientation. Quartz in other feldspar crystals may have several different orientations. It has been proposed that graphic textures represent a simultaneous crystallization of the two minerals; however, evidence seems to point to the replacement of the microcline by quartz.

Graphite (from the Greek meaning "to write," because of its use in pencils). A mineral, a black form of the element carbon. Graphite is dimorphous with diamond; that is, the two minerals have the same composition, native carbon. The mineral also is called plumbago and black lead because for centuries it was confused with *galena, the common black sulfide of lead, although aside from the black color the properties of the two minerals are quite different.

Graphite most commonly occurs as small scales disseminated through *metamorphic rocks such as marbles, schists and gneisses. In some of these rocks it is found in beds or veins large enough to permit mining. Such graphite was probably derived from carbonaceous material of organic origin. In places such as Rhode Island and Sonora, Mexico, coal beds have metamorphosed to graphite. Elsewhere, hydrothermal veins of graphite have formed from hydrocarbons during the *metamorphism of the regions.

Graphite has various important uses. Mixed with clay it serves as the "lead" of lead pencils; mixed with oil it is used as a lubricant. Because it is chemically inert at high temperatures it is manufactured into crucibles for handling molten metals. It is also used in electroplating, foundry facings, batteries and electrodes. See Mineral Properties. (C.S.H.)

Graptolites. A group of extinct colonial animals very abundant during Early Paleozoic time. The colony consisted of a chitinous exoskeleton composed of a series of tubes or cuplike structures that housed the living animals. The cups grew along single or branching stalks, some of which were attached to seaweeds, rocks, shells or other objects. The stalks of other graptolites were attached to bladderlike floats. Certain of these floating species attained worldwide dispersal and are especially useful in the long-range correlation of Cambrian, Ordovician and Silurian rocks. Because graptolites have no living counterparts, there has been much uncertainty as to their proper place in the animal kingdom. Earlier classifications treated them as Coelenterata, but they are now considered an extinct chordate. (W.H.M.)

Gravimeter. An instrument used by geophysicists to determine the variation in intensity of the force of gravity at any point on the earth's surface. The instrument can measure changes in the gravitational potential in the order of 0.01 milligals (1 milligal equals 1/1000 of the standard acceleration of gravity, 980 cm per second squared). Gravimeters may be used to locate ore and oil-bearing structures beneath the surface of the earth, since such bodies may affect the *gravitational potential over the area. Although the gravimeter has many variations, the principle of operation is essentially the same in all of them: A mass suspended by a spring or level system is displaced from a reference position by variations in the pull of gravity. The change in displacement is very small and must be magnified by optical, mechanical or electrical means before it is transmitted to the reading dial of the instrument. (J.W.S.)

Gravitational Constant. In Newton's Law of Gravitational Attraction, $F = G\left(\frac{m_1 m_2}{T_2}\right)$, G represents what is known as the gravitational constant. This law states that a force of attraction (F) between two bodies equals the product of their masses $(m_1 m_2)$ divided by the distance between their centers (r) squared, times the gravitational constant (G). G equals $6.670 \pm 0.005 \times 10^{-8}$ cubic centimeters per gram-second squared. This is the force exerted by two bodies, each with the mass of 1 gram separated by a distance of 1 cm.

The constant was first measured by Lord Cavendish in 1791. He used a horizontal beam suspended at the center by a delicate torsional fiber, that is, a wire subjected to a torque of known magnitude. At each end of the beam were fixed two small bodies of equal mass. Two large external masses were placed near the ends of the beam in such manner that the attractive force between the small fixed weights and the large weights would cause the beam to rotate. Knowing the torsional constant of the fiber, the weight of the four masses and their separation, and the angle through which the beam was deflected, he was able to calculate the gravitational constant. His value of 6.754×10^{-8} is not greatly in error when compared with the present calculation of the constant.

To give some idea of the force of attraction between two objects, visualize two billiard balls touching each other. The attractive force between the two balls equals less than one thirty-billionth of the weight of one ball. (J.W.S.)

Gravitational Potential. Any mass lying within the influence of a gravitational field has potential energy caused by the attractive force of gravity acting on the body. The reference or zero potential energy of a mass lies at an infinite distance from the center of the gravitational field. As the mass moves closer to the center of gravitational force, its potential energy increases in a negative direction. The gravitational potential then can be defined as $V = -GM/t$, where V is the gravitational potential, G is the universal gravitational constant, M is the mass of the body acted upon by the gravitational force and t is the distance between the body and the center of the gravitational field. (J.W.S.)

Gravity Anomaly. An observed departure from the theoretical value of gravity calculated for any point on the earth's surface. Gravity anomalies are best displayed when equal values of gravity are plotted on a map and connected by lines, producing a surface like that of a topographic map. The map allows a geoscientist to locate areas of

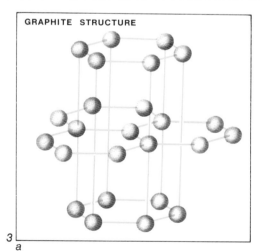

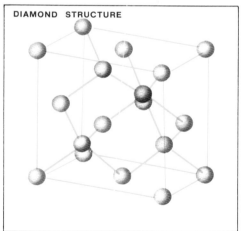

Granite

1 South Dakota's Mount Rushmore Memorial to Presidents Washington, Jefferson, Lincoln and Theodore Roosevelt was sculpted in granite by Gutzon Borglum.

Graphic Granite

2 This piece of feldspar, intergrown with dark quartz rods came from Bradbury Mountain, Maine.

Graphite

3 Carbon can crystallize in two forms: graphite and diamond. (a) The structure of graphite, a soft mineral, shows sheetlike layers of carbon atoms, with only weak bonds holding the sheets together. (b) The structure of diamond, a very hard mineral, shows each carbon atom sharing a bond with four other carbon atoms at the corners of a tetrahedron, a very difficult structure to disrupt.

Greenhouse Effect

1 *Water vapor and carbon dioxide in the atmosphere act like the glass of a greenhouse. Short-wave solar rays can pass through the atmosphere warming the earth's land and water. But the atmosphere absorbs the long-wave heat rays emitted by the earth's surface, thus warming it, rather than allowing the long-wave radiation to escape back into space.*

Guide Fossil

2 *Widespread fossils that are typical of a particular period of earth history enable the geologist to correlate the age of rock strata around the world. The best index fossils are swimming or floating organisms that evolved rapidly and were distributed widely, such as the ammonites illustrated here in an early book about fossils.*

Gully

3 *Water runoff, concentrated into narrow channels by surface irregularities, eventually erodes a trench in a sparsely vegetated slope.*

Guyot

4 *A submarine volcanic mountain called a seamount (a) is uplifted and truncated by wave action (b), then subsides to form a flat-topped guyot (c).*

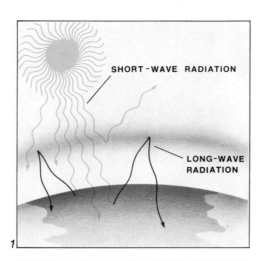

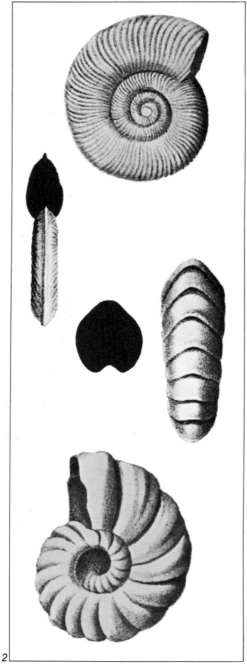

gravity highs and lows. Departure of measured gravity values from a theoretical value results from a segment of the earth's *crust being composed of material having a greater or lesser density than the average crust. If the crust of the earth were homogeneous in its composition and thickness, gravity anomalies would not exist. Gravity values measured at sea and on land have been made for many years, and more recently by satellites orbiting the earth. From the extensive gravity data that has been gathered, geoscientists have been able to determine that the crust under the oceans is thinner and denser than the crust forming the continents. On a smaller scale local variations in gravity are used by exploration geophysicists to locate underground reservoirs of oil and bodies of some economically important materials. It was the variation in the attractive force of gravity that led J. H. Pratt and George B. Airy to develop their hypotheses of *isostasy. (J.W.S.)

Graywacke (or Lithic Arenite). A hard, coarse-grained sandstone characterized by angular particles of quartz, feldspar and other rock fragments embedded in a matrix of clay-sized particles. This cemented "dirty sand" is typically dark green, greenish gray or black. Like *arkose, graywackes represent environments in which erosion, transportation, deposition and burial of sediments were rapid enough to prevent complete chemical weathering of the original material.

Greenhouse Effect. A heating effect in which the atmosphere serves the earth's surface much as the glass panes serve a greenhouse. Just as the glass permits shortwave solar radiation to pass inside but absorbs and re-emits outgoing infrared radiation, so atmospheric water vapor and, to a lesser extent, carbon dioxide transmit much of the incoming solar radiation to the earth. Most of the radiation reaching the earth is absorbed; morever, the long-wave radiation emitted by the earth's surface is, to a great extent, absorbed by water vapor in the troposphere just above, and some is re-emitted to the earth's surface. In this way the earth's surface temperature is kept higher than it would be without the infrared absorption of water vapor and carbon dioxide.

The average surface temperature of the earth, about 15°C (59°F), is roughly 40°C (72°F) higher than that required for radiative equilibrium of a black body at the earth's mean distance from the sun. A black body is a hypothetical body that absorbs and emits all of the radiation striking it; its temperature depends upon the amount of energy radiated per unit time from a unit area of its surface, hence upon the rate at which energy is received. The earth's black-body temperature may therefore be calculated if the radiation striking it is known. The earth's actual surface temperature represents a balance between the energy emitted and the energy received both from the sun and from the atmosphere. (M.F.H.)

Greensand. A sand having a greenish color, usually due to the presence of the mineral *glauconite but sometimes caused by algae growing along a beach.

Greisen. An altered granitic rock composed largely of *quartz, *mica and *topaz. Other minerals formed during the alteration process are tourmaline, fluorite, wolframite and cassiterite. The presence of cassiterite, tin oxide, is characteristic, and greisen has been defined by some as a tin-bearing rock.

Grossularite. A mineral, a calcium aluminum silicate, a member of the *garnet series. It is found principally in crystalline limestone as a product of regional or contact *metamorphism.

Groundmass. The fine-grained material enclosing larger crystals (*phenocrysts) in an igneous rock.

Guide Fossil. A fossil so characteristic of certain strata that it can serve as a guide or index to the age of those strata. These are the remains of rapidly evolving species that lived only a short time in geologic history but while alive attained widespread distribution. The usefulness of a guide fossil is enhanced if the species is abundant and possesses easily identifiable features.

Gully. A small, generally dry stream channel. It is cut by *runoff from extreme or intense rainfall on a land surface covered by little or no vegetation. When very small and narrow, such channels are called shoe-string gullies or rills. They grow headward by vertical cutting, by sapping (undercutting) and collapse, or by piping (washing out of fine particles in sandy materials). Some gullies are discontinuous: They occur in a linear series down valley but are not connected. See Arroyo.

Guyot (or Tablemount). Submarine volcanoes that have a flat crest. Guyots were discovered in the Pacific Ocean and identified during World War II. They have since been found throughout the world, but are most common in the Pacific. The flat upper surface of a guyot is as much as 1 km below

3

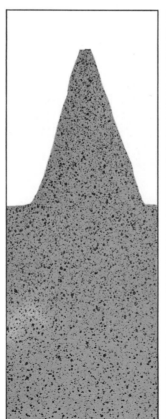
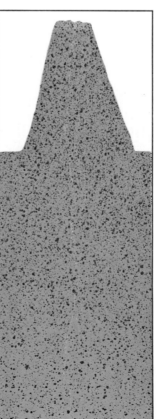
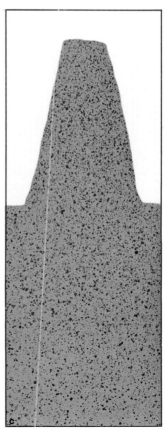

4

Gypsum
Although gypsum commonly forms crystals with smooth flat faces, it may grow from the walls of caves in curved fibrous aggregates, such as this unusual specimen from Texas.

sea level. Wave activity has been suggested as the cause of the flat crest, and this is supported by the presence of remains of reef-type corals and other shallow-water organisms. The depth of the upper surface is due to subsidence of the guyot. Although most flat crests of guyots are nearly horizontal, some are tilted as a result of conditions related to subsidence. (R.A.D.)

Gymnosperm. One of the seed plants; typically evergreen trees with frondlike or scalelike leaves. Most species have a dominant main trunk supporting much smaller branches. They are found as fossils in rocks ranging from Devonian to Holocene (Recent) in age, and include the seed ferns and *cordaites* abundant in coal deposits of the Mississippian and Pennsylvanian periods. The living sago palms, or *cycads and cycadeoids, of Mesozoic age, as well as the conifers such as pines, firs and cedars, are gymnosperms. The *ginkgo, or maidenhair, was common during Mesozoic time, but only one species is living today.

Gypsum. A mineral, hydrous calcium sulfate, of commercial importance as the raw material used in the production of plaster of Paris. This common mineral has other uses as well; the massive, fine-grained variety, alabaster, has been cut and polished for ornamental purposes for centuries. Clear transparent crystals known as *selenite can be cleaved into thin sheets for use in optical instruments. Satin spar is a fibrous variety with a silky luster which, when cut as a smooth rounded cabochon stone, shows a beautiful, changeable luster called *chatoyancy. Were it not so soft (hardness 2) such stones would make lovely gems.

Gypsum is a widespread mineral most commonly found as a sedimentary rock interstratified with limestone and shale and usually underlying beds of rock salt. Such bedded deposits result from the evaporation of seawater, which causes the crystallization and precipitation of gypsum and other dissolved salts. See Evaporite; Mineral Properties. (C.S.H.)

Habit, Crystal. See Crystal Habit.

Hadal Zone. The deepest zone in the ocean, exceeding 6000 m. It is essentially restricted to the oceanic *trenches. Organisms are relatively scarce because of the intense pressure, low temperature, absence of light and lack of food. Sediments in this environment accumulate very slowly and are quite fine-grained.

Half-Life. A measurement of the spontaneous disintegration of radioactive substances; specifically, it is the time required for half of a radioactive element's atoms to disintegrate to a stable daughter (derivative) element. The rate of disintegration is constant for each element. However, half-lives of the different radioactive elements vary greatly: the half-life of uranium is 4500 million years, while the half-life of thorium is 13,900 million years and potassium-40 is only 1300 million years.

Halides. The chemical class in which the electronegative element is a halogen. It thus includes the chlorides, bromides, fluorides and iodides. The most common mineral halides are *halite, which is sodium chloride; sylvite, which is potassium chloride; and *fluorite, which is calcium fluoride.

Halite. The mineral sodium chloride, also known as common salt and *rock salt. Because salt is a physical necessity for animal life, it was one of the first minerals to be sought and traded by early man. Human consumption, originally the principal or only commercial use of salt, is today small compared with industrial uses. A major raw material of the chemical industry, salt is the source of the sodium and chlorine used in countless products for home and industry. Large amounts are also used in food processing in the canning, fishing and meat-packing industries.

Seawater contains nearly 3 percent sodium chloride in solution. Man purposely evaporates seawater to recover salt, but throughout the geologic past it has been evaporated by natural causes and the salt deposited as a sedimentary rock over vast areas on the earth's surface. Halite is the most abundant of the *evaporites. It follows gypsum and anhydrite in the sequence of precipitation of salts from seawater. It is common, therefore, to find rock salt overlying beds of gypsum or anhydrite. In places the salt is impure, interbedded with these and other precipitated minerals such as calcite and sylvite as well as sand and clay.

In over 75 countries salt is recovered on a commercial basis. If the salt beds are not too far below the surface, shafts are sunk to them and the mineral is mined underground. Elsewhere a common method of extraction is to pump water to the water-bearing layers and pump the resulting brine to the surface. The salt is then recovered by evaporation. About one-third of the salt produced today is recovered by evaporation of seawater in man-made ponds (salt pans) into which seawater is admitted and evaporated by solar heat until salt is precipitated.

Although halite most commonly is in bedded deposits, it is found in well-formed cubic crystals with perfect cubic cleavage. The crystal's habit and salty taste make identification of the mineral easy. See Mineral Properties. (C. S. H.)

Hanging Valley. A tributary valley whose floor hangs above that of the main valley where they join. Hanging valleys are common in glaciated mountains. They can be formed in many ways but perhaps occur most often where a main valley was deeply eroded by its glacier while a tributary valley, containing a smaller glacier, was less deeply eroded. The absence of a glacier in a tributary valley and the presence of one in the main valley can also result in a hanging valley. It is usually not until the area is deglaciated that the tributary valley is left hanging. Well-known examples are found in Yosemite National Park in California and in Lauterbrunnen Valley in Switzerland. (R.L.N.)

Hanging Wall. The upper wall of a *fault; it is distinct from a lower wall, which is known as the foot wall. The term is commonly used in mining geology to indicate the upper boundary of a horizontal or inclined ore-bearing vein or bed, whether or not it lies along a fault.

Hardness. The resistance a mineral offers to scratching. Hardness is related to the bonding forces between atoms in the crystal structure: the stronger the bonding, the harder the mineral. Some minerals, such as talc, are so soft that rubbing will tear off sheets of its atoms, giving a slippery feeling. At the other extreme is diamond, whose atoms are so firmly bonded that nothing save another diamond will be able to scratch it.

The degree of hardness of a substance is determined by the minerals it will scratch on the Mohs Scale of Hardness. This scale, arranged by the German mineralogist Friedrich Mohs (1773–1839), consists of the ten minerals listed below in order of decreasing hardness. For example, a substance that scratches orthoclase but not quartz has a hardness of 6½.

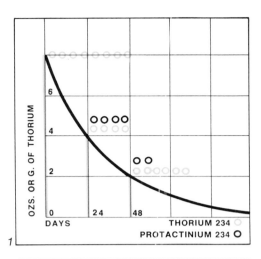

Half-Life

1 Radioactive decay occurs when the nuclei of certain elements throw off particles, transforming themselves into other elements. The rate is measured in terms of half-life, the time necessary for a radioactive substance to lose half its radioactivity during decay. This process occurs at a constant rate and provides an extremely accurate measurement of the age of rocks and fossils. Eight ounces of radioactive thorium 234 has a half-life of about 24 days. During each time unit, half of the atoms remaining in the parent isotope decay into daughter isotopes, eventually forming a new element, protactinium 234.

Halite

2 These salt crystals are called "hopper-shaped" because material added to the edges left a depression in each cube face that looks like an angular funnel.

Heat Balance

1 At the Equator the earth's surface is perpendicular to the sun's rays and receives the most direct radiation. At higher latitudes solar beams spread over more surface area, decreasing the amount of heat radiation.

Heat Flow Provinces

2 The western United States is divided into four zones based on the quantity of heat flowing through the crust of the earth. The boundaries of the various zones generally coincide with physiographic boundaries. The Basin and Range and Columbia Plateau provinces are areas of high heat flow. The Rocky Mountains and Colorado Plateau provinces have intermediate values of heat flow. The coastal region consists of areas of both high and low values, while the Sierra Nevada is an area of low heat flow.

Hematite

3 This specimen from Cumberland, England, called kidney iron ore, is composed of radiating crystalline aggregates terminating in shiny black kidney-shaped masses. The distinctive test for hematite is the red color of its powder, or streak.

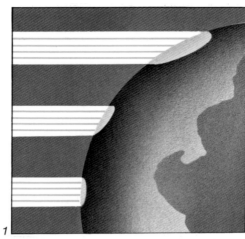

1

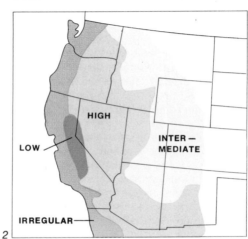

2

3

1.	Talc	6.	Orthoclase
2.	Gypsum	7.	Quartz
3.	Calcite	8.	Topaz
4.	Fluorite	9.	Corundum
5.	Apatite	10.	Diamond

Heat Balance. The balance between incoming solar radiation (insolation) and the energy emitted by the earth-atmosphere system. Such a balance is demonstrated by the relative constancy of the earth's surface temperature. On one hand the heat balance involves the reflection or absorption of the solar radiation, and on the other, the emission of the absorbed energy in the form of infrared terrestrial and atmospheric radiation.

Of the incoming solar radiation, about 30 percent—called the earth's *albedo—is reflected back into space. If the incoming energy is counted as 100 units, the earth and atmosphere together absorb about 70 units of the incoming energy, the amount not reflected back into space; the air absorbs about 20 units and the earth's surface the remaining 50 units. To maintain a balance, the earth-atmosphere system must emit 70 units to space. Since by far the greater part of the terrestrial radiation is absorbed in the atmosphere, the earth's surface itself emits only a small fraction of the absorbed energy directly to space. Most of the energy emitted to space is radiated from the atmosphere, and the major portion of this comes from the *troposphere, where the main absorbing gas, water vapor, is concentrated.

The energy exchange between the earth's surface and the atmosphere is accomplished primarily by radiative transfer, but the transport of latent heat energy by the evaporation of water into the atmosphere is also of great importance. The turbulent transfer of "sensible" (in contrast to "latent") heat to and from the earth's surface is another mechanism for this exchange. For example, the surface air may be warmed by conduction through contact with the earth's surface and then transported bodily upward by turbulent mixing with the air above.

The strong radiative exchange of heat between the atmosphere and the earth's surface is one of the important aspects of the *greenhouse effect. The other important aspect is the relative transparency of the air to shortwave solar radiation, which allows most incoming energy to reach the earth's surface.

The earth as a whole constantly absorbs and emits the same amount of energy. However, the lower latitudes absorb more energy than they emit, while the high latitudes emit more energy than they absorb. This radiative imbalance between low and high latitudes is a primary factor in the atmosphere's general circulation. To maintain the earth-atmosphere heat balance, the excess heat at low latitudes must be transported poleward by the circulation of the atmosphere and of the oceans. (M.F.H.)

Heat Capacity. The amount of heat required to raise the temperature of a body one degree centigrade. Heat capacity differs for various earth materials and is usually expressed in calories per degree centigrade. The value of heat capacity when determined for a single gram mass of material is referred to as the *specific heat of that material.

Heat Flow. The flow of heat from the hotter to the cooler regions of a substance. The greater the temperature difference, or thermal gradient, the greater the amount of heat flow. Thus heat produced in the interior of the earth, as measured by the heat production of certain types of rocks, flows to the cooler surface regions. Calculations based on the distribution of various kinds of rock indicate that over 8×10^{20} joules of heat per year, or 30-40 calories per square centimeter per year, reach the earth's surface. The speed at which this heat flows to the surface is governed not only by the *geothermal gradient (the earth's thermal gradient) but also by the conductivity of the rock between the source of heat and the surface. Rock is such a poor conductor that most heat now reaching the surface has probably come from a maximum depth of several hundred kilometers and has probably taken all of geologic time to reach the surface. (J.W.S.)

Heat Flow Provinces. Areas in continents and oceans having a defined *heat flow. Their boundaries can commonly be correlated with regions of typical landforms (physiographic provinces) or particular mountain-building characteristics (tectonic provinces). Within oceans, ridges are characteristically provinces of high but variable heat flow; basins are typically of moderate, relatively uniform flow; and trenches have low heat flow. The old nuclei of continents, the *Precambrian shields, have low heat flow values; regions of much younger mountain-building activity have variable heat flow. The Basin and Range provinces and the Sierra Nevada of North America, formed during the Mesozoic-Cenozoic cycle of mountain-building activity, show high and low heat flow characteristics respectively. Observed heat flow values are generally compatible with hypotheses of sea floor spreading and *plate tectonics. Heat flow is high along oceanic

ridges and low near ocean trenches where a cold slab of oceanic crust is thought to be thrusting under the *continental crust. (J.W.S.)

Heavy Liquid. A liquid of high *density used to determine the *specific gravity of mineral grains and also to separate minerals of different specific gravities. The density of the various heavy liquids ranges from about 2.5 g/cm³ to about 4.25 g/cm³, and many different solutions and compounds have been used. The three most widely used and their maximum densities in g/cm³ at room temperature are bromoform, 2.89; methylene iodide, 3.31; and Clerici solution (an aqueous solution of thallium malonate-formate), 4.25.

Specific gravity is determined by immersing a mineral fragment in the heavy liquid and noting whether the fragment rises or sinks. If it floats, the liquid has the greater specific gravity and is diluted until the fragment stays suspended, neither rising nor sinking. The mineral and liquid then have equal density, and the density can be determined by means of a Westphal balance. Bromoform and methylene iodide must be diluted with an organic solvent such as acetone or benzol; Clerici solution must be diluted with water.

Separations of mechanical mixtures of mineral grains can also be easily accomplished by heavy liquids. If the densities of two components are known, the heavy liquid is adjusted to an intermediate density and the mineral mixture is immersed in it. The higher density material immediately sinks while that with lower density rises. (C.S.H.)

Heavy Mineral. A rock-forming mineral with high *specific gravity (greater than 2.8) as opposed to a *light mineral with lesser specific gravity. In general, dark minerals such as amphiboles, pyroxenes, olivine, biotite mica and most of the accessory minerals of rocks are heavy minerals. The dark and light minerals can be separated by means of *heavy liquids.

Hematite (from the Greek for "blood," alluding to the many red varieties). A mineral, an oxide of iron, and the principal ore mineral of iron. Although other metals are gradually taking its place, iron still remains the most important structural metal. For this reason hematite is among the most valued *economic minerals. In a few places hematite occurs in beautifully formed rhombohedral crystals with a brilliant metallic luster. But more commonly the crystals are flattened and form as extremely thin plates, which in some specimens are grouped in rosette forms known as iron

roses. Elsewhere the plates are found arranged in a foliated micaceous aggregate called *specularite (or specular hematite). All these well-crystallized varieties are black, but their "streak," that is, the color of the powdered mineral, is red. This is the color of the variety known as kidney ore, radiating aggregates with a reniform (kidney-shaped) surface. Red is also the color of the most abundant hematite, an earthy material showing no evidence of crystal form. This kind is the red ochre once used as a paint pigment.

Hematite is a widespread mineral found in all types of rocks. It occurs as small grains scattered through many igneous rocks; as the specular variety it is found in enormous masses in regionally metamorphosed rocks. In red sandstones it is the material that cements the quartz grains together and colors the rock.

To be mined economically, a hematite deposit must contain tens of millions of tons. Accumulations of this magnitude are largely sedimentary, most of Precambrian age, and they generally have silica as a major impurity. Most of the high-grade ores, containing at least 50 percent iron, have been enriched by the removal of silica by meteoric waters. These rich ores have been the source of the world's iron, but the reserves are dwindling. To supplement them, mining companies are turning to the primitive iron formation called *taconite, which contains only 25 to 30 percent iron but is present in essentially inexhaustible amounts. By mechanical means the iron minerals of this low-grade material are concentrated, and thus taconite can be considered a permanent iron reserve.

Large deposits of hematite are found and mined on every continent. In 1961 the Soviet Union took the lead from the United States as the largest producer; following the United States are France, Canada, China, Sweden and Australia. In the United States the major production of iron ore since the late 19th century has been in the form of hematite from the Precambrian rocks of the Great Lakes region. See Mineral Properties. (C.S.H.)

Hemimorphite. A mineral zinc silicate, a secondary product found in the oxidized portions of zinc deposits. It crystallizes in a low symmetry class of the *orthorhombic system with different forms at the top and the bottom of crystals; hence the name hemimorphite, meaning half-form. Single crystals are rare, however, and the mineral usually occurs in colorless mammillary or *botryoidal masses with crystal faces encrusting the surfaces. Hemimorphite contains 54 percent zinc and thus is a valuable ore wherever it is abundant.

Heulandite. A mineral in the *zeolite group, found in cavities of basic igneous rocks associated with other zeolites and calcite.

Hexagonal System. In *crystallography, the crystal system encompassing all crystals having a sixfold symmetry axis, or a threefold symmetry axis. All crystals in the system are referred to four crystallographic axes: three of equal length lie in a plane at 120° to each other; the fourth is perpendicular to the plane of the other three.

Hiatus. A gap or break in the continuity of the geologic record. It is represented by the absence of rock units that would normally be present but were never deposited or were removed by erosion before the beds immediately overlying the break were deposited. A geologic hiatus represents the elapsed time represented by strata missing at a physical break or *unconformity in a *geologic column.

Hill. A small and isolated eminence, usually with a rounded summit and distinct slopes, that rises above the surrounding country. It is smaller than a mountain and is generally defined as less than 300 m from base to summit, but this is arbitrary. Thus the Normandy Hills (Collines de Normandie) in France rise some 400 m while the Back Mountains (Montagnes Noires) of Brittany rise only about 300 m.

Hogback. A linear, sharp-created ridge formed on resistant, steeply dipping sediments. Such ridges are left by the erosion of upfolded layers of rock. Because of high dips, a hogback may be symmetrical in shape, with its back slope as steep as the eroded front (cliff). The sandstone hogback on the east side of the Black Hills, S.D., is an excellent example of a hogback. Hogback ridges can also be found on the east side of the Bighorn Mountains, Wyo., and in the Weald region of England. Hogbacks are similar to *cuestas but have steeper dip slopes.

Holocene Epoch (or Recent Epoch; Postglacial Epoch). The last *epoch of the *Quaternary Period; it began at the end of the Pleistocene Epoch, about 10 million years ago, and continues to the present. It is the shortest known period of geologic time—the name literally means "wholly recent"—and includes all time since the last *Ice Age.

Holocene Transgression. The rise in sea level since the melting of the *glaciers formed during the last Ice Age. This rise began about 18,000 years ago and continues today. During the last glaciation (the Wisconsinan), sea level was more than 100 m below its present level and shorelines were close to the outer edge of the *continental shelves. Between 18,000 and 7000 years ago, sea level rose at a rate of 10 m per 1000 years. Between 7000 and 3000 years ago the rate of increase slowed down to 2 m per 1000 years. There is considerable disagreement about the rate of rise during the past few thousand years, but it is thought to be only about 1 m per 1000 years. (R.A.D.)

Homeomorph. In mineralogy, a mineral similar to another in crystal structure and crystal form but differing in chemical composition. For example, calcite ($CaCO_3$) and soda niter ($NaNO_3$) are homeomorphs with similar crystal constants, cleavage and optical properties.

Hook. A split whose terminus is curved landward. Sandy Hook, extending from the New Jersey coast into New York Harbor, and the tip of Cape Cod, Mass., are well-known hooks.

Horizons, A-, B-, and C-. The three layers in a soil profile. The uppermost, A- horizon (also called the zone of leaching or the zone of eluviation), extends 1 m or more below the earth's surface and is characterized by humus, which is mainly decomposed vegetation. The parent material has been profoundly decomposed and physically altered. In a humid climate, leaching by water containing dissolved carbon dioxide removes the soluble salts formed by decomposition and groundwater removes the clay particles formed by chemical alteration of the parent material.

B-horizon lies below A-horizon and is poorer in humus. In a humid climate the iron and clay minerals formed by the decomposition of the parent material in A-horizon move downward into B-horizon. In arid climates calcium carbonate is brought up from below by capillarity and deposited in B-horizon by evaporation; some also percolates down from A-horizon. B-horizon is sometimes called the zone of accumulation because so much material is added to it.

C-horizon is below B-horizon and grades downward into a zone of unaltered parent rock. It contains both unaltered and altered parent material. Some disintegration and decomposition takes place and new materials are formed. See Soil; Soil Profile. (R.L.N.)

Horn. A steep-sided, pyramidal peak with three or more ridges radiating down from its summit. The highest and the most

3

Hogback
1 These sharp ridges along the east coast of Colorado's Rocky Mountains are the eroded edges of rock layers tilted by a major uplift.

Hook
2 Longshore currents and wave action in Icy Bay, Alaska, curved this spit of land into a hook.

Horn
3 Towering 1500 m (5000 feet) above a Swiss valley, the Matterhorn was formed by the headward erosion of a ring of hollow basins, called cirques, from which glaciers flowed.

Hornblende

1 Well-developed crystals of the dark green-black amphibole cluster together in an unusually fine specimen of this mineral.

Hot-spring Deposits

2 Layers of calcium carbonate that precipitated from the cascading hot water formed Minerva Terrace at Mammoth Hot Springs, Yellowstone National Park, Wyo.

3 Mud mixed with pulverized lava seethes at the surface of a spring in New Zealand.

Hydraulic Geometry

4 Stream dimensions and flow rates are determined by a set of hydraulic measurements. Depth (D) is measured at any specific point in the stream as the vertical distance from water surface to bed. Width (W) is the distance across the stream from one edge to another. Wetted perimeter (P) is the length of contact between the water and the channel, half of which is shown in this cross section. Slope or gradient (S) is the angle between the water surface and the horizontal plane. Velocity distribution is indicated by the arrows, each tracing a given drop of water. The line of maximum velocity (V) is in mid-stream and one-third of the distance down from the surface to the stream bed, where friction between the water and the floor and sides of the channel is least. The line of maximum shear occurs along the stream bottom, where friction between the overlying water and the floor and sides of the channel is greatest.

notable feature in a glaciated mountain range, a horn is a relatively unreduced portion of a preglacial upland formed by the headward erosion and intersection of three or more *cirques. The Matterhorn and the Weisshorn in the Swiss Alps, Mount Assiniboine in the Canadian Rockies and Grand Teton in the American Rockies are well-known horns. (R.L.N.)

Hornblende. A mineral, the most common member of the *amphibole group. It is a dark green to black rock-forming mineral found in both igneous and metamorphic rocks. See Mineral Properties.

Hornfels. A metamorphic rock composed of fine equidimensional mineral grains and produced by *contact metamorphism. Some hornfels contain isolated large crystals that are known to represent either *porphyroblasts or relics of large crystals formed with the original rock. Others may show structural features related to the original rock, such as the *bedding of sedimentary rocks. Such relict features are preserved because rock undergoing transformation to hornfels, unlike many other metamorphic rocks, is not subjected to deformation. Hornfels results from muds, sands and their lithified counterparts, or fine-grained, quartz-rich volcanic rocks subjected to temperatures from 200° to 800°C (392° to 1472°F) and to pressure within the limits encountered below 10 km beneath the earth's surface. The heat necessary to form hornfels comes from molten rock transported from within the earth's depths and injected into rock in the outer portions of the earth's crust. (J.W.S.)

Horse Latitudes. A zone of winds coinciding with the subtropical high-pressure belts between 30° and 40° N and S latitudes. The horse latitudes were once called the subtropical belts of winds and calms. Calms prevail as much as a quarter of the time in the horse latitudes, and the weather is generally fair and dry. It is said that sailors on trading ships becalmed in this zone threw their horses overboard when the drinking water supplies ran low; hence the term.

Horst. A raised block of crustal material generally longer than it is wide. The sides of a horst represent planes of normal *faults. The Basin and Range landforms of the southwestern United States developed in part from horst structures. The Ruwenzori Mountains of east-central Africa are carved from a horst that is an uplifted block of the crust.

Hot Spring. See Thermal Spring.

Hot-spring Deposits. Mounds, cones and terraces commonly formed where hot water reaches the earth's surface. If groundwater circulates deeply enough, it may be heated by the earth's internal heat, by contact with hot igneous rocks and by mixing with magmatic gases and liquids. Since hot water is a more powerful solvent than cold water, it dissolves calcium carbonate, *silica and other substances in the rocks and deposits the dissolved materials at the surface. Deposition of calcium carbonate (*travertine) takes place because of the cooling and evaporation of the hot water after it reaches the surface, because of the loss of carbon dioxide from the hot water due to the reduction of pressure as the water moves upward, and because of the metabolic activity of algae living in the hot water at the surface.

The hot-spring terraces at Mammoth in Yellowstone National Park, which are among the largest and most beautiful in the United States, are composed of travertine. Those now active and growing are brilliantly colored by the algae in the hot waters, while the inactive are generally pure white. That these deposits are quite old is indicated by the glacial deposits on the inactive terraces and the travertine in nearby glacial deposits. See Mineral Spring; Thermal Spring. (R.L.N.)

Huebnerite. A mineral, the oxide of tungsten and manganese. It is the manganese member of the *wolframite series.

Hurricane. See Cyclone, Tropical.

Hyalite. A clear, colorless variety of opal with globular or botryoidal surface that commonly forms as crusts in rock cavities. In some specimens the glassy globules look like drops of water.

Hydraulic Geometry. The configuration of the cross section of a stream channel. The channel form is closely related to the discharge of the stream and is interrelated with its velocity gradient, roughness and the load it carries. The width, depth and velocity of a river changes with the amount of water flowing in it in a regular way. In general, width changes more rapidly than depth as the discharge increases downstream. Velocity usually changes least as the discharge downstream increases and, in fact, tends to remain fairly constant.

Hydraulic Gradient. The slope of a *water table. The hydraulic gradient is equal to the difference in elevation between any two points on the water table divided by the horizontal distance between them. It may be expressed as $G = H/L$, where G is the hydraulic gradient, H is the difference in elevation and L is the distance between the two points.

Hydraulicking. Removal of loose material by the sheer force of impact of water. It may occur naturally in alluvial channels or soil, and is used by man when mining in loose material, especially old stream terraces or floodplains.

Hydraulic Mining. A method of mining in which high-pressure jets of water are used to wash away earth or gravels containing a valuable mineral. It has been used chiefly in gold mining to break down banks of unconsolidated material and carry it into sluices, where the gold separates from the other minerals because of its high specific gravity. In many mining districts hydraulic mining is prohibited by law because the large amount of debris it discharges into streams fouls the water, obstructs navigation and damages adjacent property.

Hydraulic Radius. The effective depth of a channel, equal to the cross-sectional area of the channel divided by the perimeter of the channel that is wetted by the water. Hydraulic radius is used as mean depth in all hydraulic equations.

Hydrocarbon. An organic compound consisting solely of the elements hydrogen and carbon. Each molecule of a hydrocarbon is distinguished by the number of carbon atoms and the associated hydrogen atoms it contains. For example, one carbon atom combined with four hydrogen atoms forms one molecule of methane gas (CH_4). Ethane (C_2H_6) consists of two carbon atoms combined with six atoms of hydrogen. Hydrocarbons may occur in a liquid, gaseous or solid state. Many of them, such as petroleum, coal, asphalt and natural waxes, are of great commercial value.

Hydrograph. A graph showing the variation in streamflow with time. It may show changes in yearly, monthly, daily or instantaneous discharge. It is useful in determining periods of low or high flow or variability of flow and in supplying data for the design of irrigation and power developments, flood control, available water supply and sanitation.

Hydrologic Cycle (or Water Cycle). The interchange of water substance between the ocean, the atmosphere and the land areas of the earth. The continuous sequence, affecting the supply and distribution of water on land, is central to the study of *hydrology and plays a part in many aspects of meteorology and oceanography.

2

3

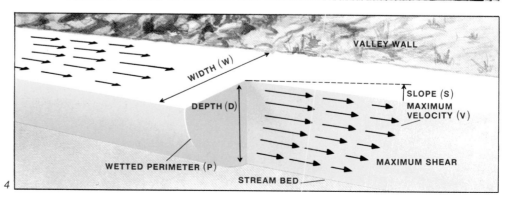

4

WIDTH (W)

DEPTH (D)

VALLEY WALL

SLOPE (S)

MAXIMUM VELOCITY (V)

MAXIMUM SHEAR

WETTED PERIMETER (P)

STREAM BED

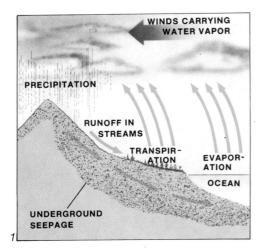

WINDS CARRYING WATER VAPOR

PRECIPITATION

RUNOFF IN STREAMS

TRANSPIR-ATION

EVAPOR-ATION

OCEAN

UNDERGROUND SEEPAGE

1

WATER	CUBIC KILOMETERS
Oceans	1,330,000,000
Glaciers, ice fields	29,400,000
Lakes, inland seas, rivers	230,000
Subsurface	8,500,000
Total	1,368,130,000

2

Hydrologic Cycle

1 *Water continually evaporates into the atmosphere from land areas, from ocean, lake and stream surfaces, and from plant transpiration. Winds carry the water vapor until it eventually precipitates to earth again as rain or snow. The water is temporarily stored underground and in glaciers and lakes before returning to the sea by surface runoff and underground seepage.*

Hydrosphere

2 *Distribution of the earth's water.*

The hydrologic cycle in its simplest form may be considered a three-way exchange: (1) from the ocean, the primary source, into the atmosphere by means of *evaporation; (2) from the atmosphere to the land as *precipitation; and (3) from the land back to the sea as runoff via surface and groundwater flow into streams and rivers. Actually, the cycle is considerably more complex. It is estimated that about 6 percent of the total flux of water is involved in the direct ocean/atmosphere—land/ocean circuit. (M.F.H.)

Hydrology. The science of the distribution, quantity and quality of the snow, ice and water on and beneath the land areas of the earth, and the laws of their geologic activity. Where oceanography deals with oceans and seas, and meteorology deals with water in the atmosphere, hydrology is concerned with snowfields and glaciers, subsurface waters, rivers and lakes. The hydrologist studies the velocity, height and flow rate of rivers, the response of flow rate to rainfall, the quantity of sediments carried by rivers, the dissolved salts in water and the magnitude and probability of floods and droughts. He is concerned with the volume of water locked in snowfields and glaciers, and their variation in volume due to climatic changes. He studies the position of the *water table, the nature and movement of groundwater, the characteristics of *aquifers and the changes in storage in groundwater basins. Hydrology is invaluable in flood control, reduction of soil erosion, development of water power and finding sources of water for irrigation, industrial and domestic uses. (R.L.N.)

Hydrometeor. Any liquid or solid product resulting from the change of phase of atmospheric water vapor, whether formed within the atmosphere or at the earth's surface, or from water particles blown by the wind from the earth's surface. Hydrometeors appear in a variety of forms, including (1) particles that are formed and remain suspended in the air, such as clouds, fog, damp haze, mist and ice fog; (2) the various forms of liquid, freezing rain or drizzle, hail and snow; (3) virga, the falling particles that evaporate before reaching the ground; (4) particles lifted by the wind from the earth's surface, such as drifting snow or blowing spray; and (5) deposits of water or ice on exposed surfaces, exemplified by dew, frost and glaze. (M.F.H.)

Hydrosphere. All the water, both liquid and frozen, found in the oceans, on land and underground as well as the water vapor in the atmosphere. It is an envelope of irreg-ular thickness around the planet. Along with the atmosphere, lithosphere (earth's crust), biosphere and centrosphere, it is one of five spheres of earth. It includes surface water, *subsurface water, snow, ice fields, glaciers, ice in the ground, and the gases, liquids and solids suspended or dissolved in water. It was formerly believed that the hydrosphere condensed out of a primitive atmosphere as it slowly cooled. However, evidence now suggests that it was derived from the interior of the earth and reached the surface by way of volcanic vents and *fumaroles. (R.L.N.)

Hydrostatic Pressure. The pressure exerted by a homogeneous fluid acting on an object. It is nondirectional and equal in magnitude on all parts of the surface. An increase in hydrostatic pressure tends to decrease the volume of an object but will not change its shape. Rock deep within the earth's *crust is subjected to a similar pressure, referred to as *lithostatic pressure, produced by overlying rock.

Hydrothermal Alteration. The mineralogic changes brought about by hot aqueous hydrothermal solutions reacting with pre-existing minerals as they pass from a deep-seated source toward the earth's surface. Deposition of hydrothermal vein minerals is frequently accompanied by alteration of the minerals of the wall rock.

Hypabyssal Rock. An igneous body resulting from crystallization of *magma injected at shallow depths in the earth's crust. Such rocks are exemplified by comparatively small intrusives such as *dikes, *sills and *laccoliths. Hypabyssal rocks are commonly porphyritic and have a texture finer than the deep-seated *plutonic rocks but coarser than extrusive ones.

Hypersthene. A mineral, magnesium-iron silicate, a member of the series of orthorhombic *pyroxenes. *Enstatite, the pure magnesium end-member, contains 40 percent magnesium oxide. Iron can substitute for magnesium in all proportions; when the mineral contains between 13 and 20 percent iron oxide it is called hypersthene. See Mineral Properties.

Hypothermal Deposits. An ore deposit formed from hydrothermal solutions at high pressures and high temperatures. The ore minerals of greatest importance in these deposits are cassiterite, wolframite, scheelite, molybdenite, native gold, chalcopyrite and galena. The most common *gangue mineral is quartz, but it is commonly accompanied by fluorite, tourmaline, topaz and axinite.

Ice Age. A nontechnical term describing a time of extensive glaciation. It is more specifically used to describe the *Pleistocene Epoch of the *Quaternary Period.

Iceberg. A mass of glacial ice detached from the end of a tidewater glacier or, less often, from a glacier terminating in a lake. Icebergs are common in the waters along the Greenland, Alaskan and Antarctic coasts, where tidewater glaciers occur. They float in deep water usually with about 80 percent of the ice under water and are sometimes larger than a liner. They may drift hundreds of kilometers, becoming progressively smaller with increasing distance from their source. They are a menace to North Atlantic marine transportation and have accounted for such marine disasters as the sinking of the *Titanic.* The breaking away of ice from glaciers terminating in water is called calving. (R.L.N.)

Ice-contact Slope. A slope formed by the slump and collapse of *outwash (glaciofluvial sediments) when the stagnant glacial ice against which it was deposited melts away. An ice-contact slope inclines downward toward the former position of the ice front and reflects its alignment. The deposits that terminate against an ice-contact slope are characterized by deformed bedding resulting from collapse due to the melting of the ice. *Eskers, *kames, *kame terraces and *moraines, and *kettle holes and proglacial *deltas are usually characterized by ice-contact slopes. (R.L.N.)

Iceland Spar. The chemically pure, optically clear, colorless variety of *calcite, so called because it was early found in Iceland as large crystals in basalt cavities.

Ice-rafted Fragment. A fragment transported by icebergs, sea or river ice, common in glaciomarine and *glaciolacustrine deposits. Abundant in the marine deposits that surround the Antarctic, such fragments are often found many hundreds of kilometers from their source. See Erratic.

Ichthyosaur. An extinct, short-necked, marine reptile, fishlike in appearance. They resembled modern dolphins, and some attained lengths of 7.5 – 12 m (25 – 40 feet), although the average length was much less. The group is known from rocks ranging from Middle *Triassic to Late *Cretaceous. See Fossil Reptiles.

Idioblastic. A texture found in metamorphic rocks in which minerals have formed characteristic crystal faces. See Crystalloblastic Series.

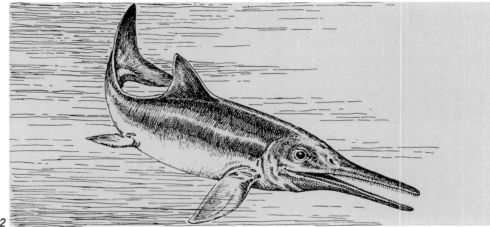

Iceberg

1 Arctic icebergs, usually with jagged tops, are formed mainly from the Greenland and Alaska glaciers and from those along the shores of the Arctic Ocean. Flat-topped Antarctic icebergs break off from the shelf of ice surrounding the continent.

Ichthyosaur

2 These carnivorous "fish-lizards," 3 m (10 feet) or more in length, were the most completely adapted to marine life of all seagoing reptiles.

3 Eurhinosaurus was a porpoise-shaped carnivorous reptile that lived during the Lower Jurassic Period, some 190 million years ago.

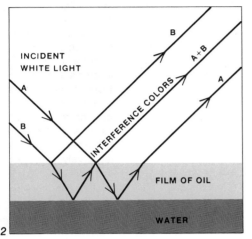

1

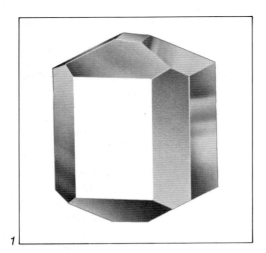

2

Idocrase

1 Tetragonal prisms with truncated pyramids are characteristic of this mineral.

Interference Color

2 Incident white light rays A and B striking a thin layer of oil are partly reflected on the surface and partly reflected at the interface between the oil and water. The portion of ray B that penetrates the oil film is slowed because it travels a longer distance, and its vibrations are out of phase with the reflected portion of ray A. The resulting combination of wavelengths differs from the original mixture, and is seen as color.

Idocrase. A mineral, a silicate with complex chemical composition. It is usually found as a *contact metamorphic mineral in impure limestones. Although frequently in well-formed tetragonal crystals with square cross section, it is more common as striated columnar aggregates. Idocrase is usually green or brown but may also be yellow, red or blue. A compact green variety from California used as an ornamental material is known as californite. See Mineral Properties.

Igneous Rock. A rock that has formed by the cooling and consequent solidification of *magma, a hot molten mass of rock material. These rocks are divided into two large groups: *plutonic rocks crystallized within the earth's crust, and *volcanic rocks formed from magma that has reached the earth's surface. Many different names have been given to rocks of both types, depending on chemical and mineral composition, texture and occurrence. See Extrusive Igneous Activity; Intrusive Igneous Activity; Magmatic Differentiation.

Ignimbrite. The extrusive *igneous rock formed by the consolidation of a type of volcanic deposit known as *nuée ardente. Typically, a nuée ardente is a mass of glowing hot, solid particles formed by frothing of the magma and the bursting of the bubbles of froth. It spreads laterally along the land surface, sometimes attaining hurricane velocity in a short distance. Ignimbrite may also refer to deposits of nuée ardente that form a stratigraphic layer. Ignimbrites occur extensively in the Great Basin of the western United States, some of them tens of meters thick over an area of many thousand square kilometers. They were deposited as essentially horizontal sheets and, as such, serve as ideal reference planes for measuring deformation during the Tertiary intervals (the last 70 million years of geologic time) of their extrusion. See Welded Tuff. (R.E.B.)

Ilmenite. A mineral, iron-titanium oxide, the principal ore mineral of titanium. Ilmenite is common as an *accessory mineral in igneous rocks, but in some *gabbros and *anorthosites it is present in extensive bodies usually intermixed with magnetite. These large masses are mined chiefly for the titanium in them, but the magnetite, separated from the ilmenite magnetically, is used as an iron ore. Ilmenite is resistant to chemical attack; during the weathering of rocks in which it occurs it is liberated and washed into neighboring streams. There it may collect as a *placer mineral or may continue downstream to the sea to be incorporated in beach sands. Large accumulations of ilmenite — with other *heavy minerals such as rutile, monazite and zircon — are mined in beach sands and coastal dunes in the United States on the east coast of Florida, and in India, Brazil and Australia. See Mineral Properties. (C.S.H.)

Incised Meander. See Meander, Incised.

Index Mineral. A mineral selected to represent a grade of *metamorphism. See Isograd.

Index of Refraction. A number expressing the ratio of the velocity of light in a vacuum (or in air) to the velocity of light in a given transparent substance. When a ray of light passes obliquely from air into a denser medium it is bent or refracted. The greater the obliquity of the incident ray, and the greater the density of the medium, the greater is the refraction. The relation of the angle of incidence, i, to the angle of refraction, r, both measured from a normal angle to the interface, is expressed by Snell's law. This states that for the same two media, the ratio of $\sin i : \sin r$ is constant. The ratio is usually expressed as $\sin i/\sin r = n$, where n is the index of refraction. Every nonopaque mineral is characterized by one, two or three indices of refraction. Their determination is an important means of mineral identification. See Becke Test. (C.S.H.)

Indicator. A *glacially transported fragment of rock so distinctive that its *bedrock source can be located and the distance and the direction it has traveled can be measured. Some indicators have been carried more than 1100 km. See Erratic.

Infiltration. The movement of rain and *meltwater from the surface down into the *regolith and the *bedrock through voids, cracks and joints. It provides water for terrestrial plants and furnishes the groundwater for irrigation and for domestic and industrial use. Percolation, on the other hand, is the slow movement of groundwater through interconnected openings from one part of the regolith or bedrock to another.

Inosilicates. A silicate group in which the oxygen-silicon tetrahedra are linked into chains by sharing oxygen ions. There are two types of chain: (1) the simple chain in which two of the four oxygens in the SiO_4 tetrahedra are shared, giving rise to the ratio $Si:O = 1:3$, and (2) the double chain made up of two single chains joined side by side. In this double chain or band structure half of the tetrahedra share three oxygens and the other half share two oxygens, yielding the ratio Si:O = 4:11. The inosilicates comprise two important groups of rock-forming minerals: the *pyroxenes as single-chain members, and the *amphiboles as double-chain members. See Silicate Structure and Classification. (C.S.H.)

Insequent River. A waterway whose course cannot be ascribed to control by any evident surface slope or weakness in the earth's surface or to rock type. The course of the stream and its tributaries is random, and thus its basin develops a treelike, or dendritic, *drainage pattern.

Interference Color. A color that may be observed when an *anisotropic crystal is viewed in white light between crossed polarizers. The light from the first polarizer is broken into two rays traveling at different velocities as it passes through the crystal. Because of their different velocities these rays interfere with each other when they enter the second polarizer, eliminating certain wavelengths and appearing as complementary colors. The colors so produced are called interference colors. They play an important role in mineral identification with a polarizing microscope.

Certain minerals show another type of interference color, an internal iridescence, caused by light diffracted and reflected from closely spaced parallel cleavages, fractures, twin lamellae or foreign inclusions. This type of interference is shown well by some specimens of labradorite feldspar. A surface color or iridescence similar to that produced by thin films of oil on water is caused by interference of light as it is reflected from the thin surface films. This phenomenon is shown most commonly on metallic minerals, particularly on limonite and sphalerite.

Interference Figure. An image composed of colored rings or curves and dark bars or curves seen when certain sections of *anisotropic crystals are viewed in convergent, polarized light. These figures are useful in mineral identification with a polarizing microscope.

Interfluve. See Drainage Basin.

Interglacial Stage. A subdivision of a time of extensive glaciation, such as the Ice Age of the *Pleistocene Epoch of the Quaternary Period. Interglacial stages, sometimes called ages, mark times during an ice age when glaciation was at a minimum (glacial minima) and much of the ice melted, permitting temperatures to rise to those of the present day. These stages are separated by shorter and much colder intervals of time called *glacial stages. Three interglacial stages have been described in the United States, named for the localities where their deposits are well displayed. These are, beginning with the oldest, the Aftonian (from Afton, Iowa), Yarmouth (from Yarmouth, Iowa) and the Sangamon (from Sangamon County, Illinois). Three interglacial stages are known in Europe, the Günz-Mindel, Mindel-Riss and the Riss-Würm; they are probably the time equivalents of those in North America. (W.H.M.)

Intermittent Stream. A watercourse that carries water only part of the year. It receives water from the groundwater table when it is high enough but runs dry when the water table declines below the channel bed during the dry season. Dryness may also occur if the channel flows on highly porous rock such as limestone.

Intertidal Zone (or Littoral Zone). A narrow area along a coast that extends from the low-tide position to the high-tide position. Its width varies according to the slope of the bottom and the tidal range. Changes in width are also affected by the moon's position; thus the zone is at maximum width during spring *tides and at minimum during neap tides. Areas of high tidal ranges and gently sloping coasts, as in the Bay of Fundy on the Atlantic Coast of Canada or the Bay of Saint-Malo on the northern coast of France, may have an intertidal zone several kilometers wide. Such a zone is characterized by special fauna and flora. Because of the alternate exposure and flooding in the zone, any organism that lives there must be able to tolerate alternating wet and dry conditions. (R.A.D.)

Intrusion. Injection or implacement of magma into pre-existing rock. The process may be violent, forcing the magma by the development of fractures, or passive, as when the magma flows into fractures already present.

Intrusive Igneous Activity. Any kind of igneous activity that takes place beneath the earth's surface. It is responsible for some ore deposits and for thermal metamorphism, and it forms the intrusive igneous rock bodies that result from the solidification of *magma beneath the earth's surface. These rock bodies are of various sizes and shapes, such as *dikes, *sills and *batholiths, and are usually characterized by coarse-grained textures because they cool slowly. *Granite, *diorite, *gabbro and other coarse-grained igneous rocks are intrusive rocks. Erosion subsequent to solidification exposes them at the surface. They are found on all the continents and were formed at various periods throughout geologic time. Precambrian intrusive rocks are found in the Inner Gorge of the Grand Canyon in Arizona. Jurassic intrusive rocks including granites and gabbros occur in Yosemite Valley National Park, California. Cenozoic intrusives are found at Devil's Tower, Wyoming. See Extrusive Igneous Activity. (R.L.N.)

Ion. An atom with an electric charge. Atoms are the smallest units of a chemical element that retain the characteristics of the element. They are made up of a very small nucleus composed of protons and neutrons surrounded by a much larger region sparsely occupied by fast-moving electrons. Each proton has a positive electric charge and each electron carries a negative electric charge. Since atoms are electrically neutral, the number of protons equals the number of electrons. If an atom loses one or more electrons, it assumes a positive charge and is called a cation. An atom that has gained one or more electrons has a negative charge and is called an anion. Ions are thus either positively charged cations or negatively charged anions.

With the exception of the *native elements, the building units of minerals are ions. As the positive pole of one magnet attracts the negative pole of another magnet, the attraction of unlike electrostatic charges of cations and anions joins them together in a crystal. This attractive force, the ionic bond, is the principal type of chemical bonding in minerals.

The space surrounding the tiny highly charged nucleus of an ion is so thinly populated by electrons there is no definite surface. The size of an ion or its ionic radius can be defined only in terms of the interaction with other ions. As a pair of oppositely charged ions are brought together as a result of the attractive electrostatic force, repulsive forces are set up. The distance at which attractive and repulsive forces balance is designated as the characteristic interionic distance for this pair of ions and equals the sum of their radii. In general cations have smaller ionic radii than anions.

Ions with the same electrical charge and the same or nearly the same ionic radius can substitute for one another during the crystallization of a mineral. For example, olivine may form as a pure magnesium silicate. If iron is present, it can enter into the crystal structure in place of some of the magnesium to form a magnesium-iron silicate. This ionic substitution is common and gives rise to mineral series.

Some minerals, notably the *zeolites,

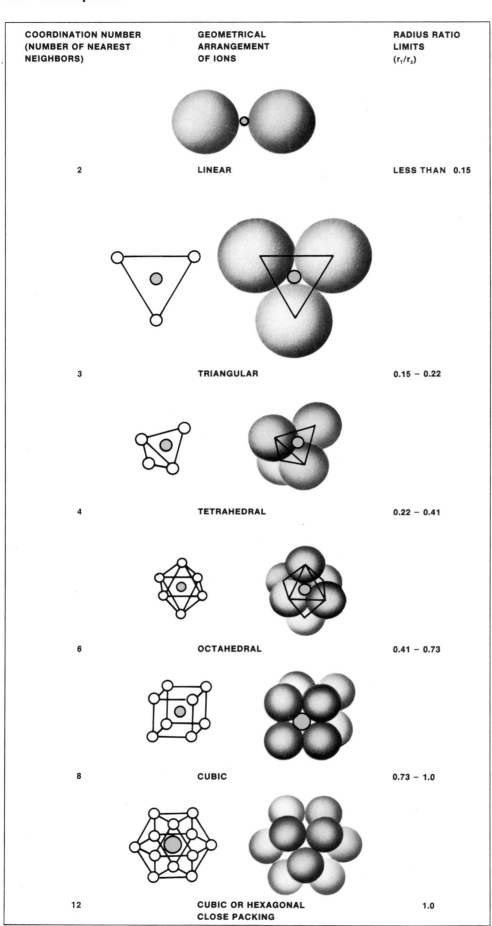

COORDINATION NUMBER (NUMBER OF NEAREST NEIGHBORS)	GEOMETRICAL ARRANGEMENT OF IONS	RADIUS RATIO LIMITS (r_1/r_2)
2	LINEAR	LESS THAN 0.15
3	TRIANGULAR	0.15 – 0.22
4	TETRAHEDRAL	0.22 – 0.41
6	OCTAHEDRAL	0.41 – 0.73
8	CUBIC	0.73 – 1.0
12	CUBIC OR HEXAGONAL CLOSE PACKING	1.0

exhibit the property of ion exchange. Cations only weakly held to the structural framework of the crystal may be replaced by cations from solution. This is the principle on which water softeners operate. See Bonds, Chemical. (C.S.H.)

Ionosphere. A region of the atmosphere characterized by a relatively high number of ions (electrically charged particles); more specifically, the region of high electrical conductivity in the earth's upper atmosphere—extending from about 50 km to the edge of the magnetosphere—which contains enough free electrons to noticeably affect the propagation of radio waves. Its existence was first deduced by the German mathematician Karl Friedrich Gauss (1777–1855) from studies of variations in the earth's *magnetic field. When the Italian physicist Guglielmo Marconi succeeded in sending electromagnetic waves over long distances, it was not at first understood how the waves could bend round the earth. Unaware of the suggestions by Gauss and later physicists, Professor A. E. Kennelly of Harvard University and Oliver Heaviside, a British engineer, in 1902 independently proposed the existence of an electrically conducting layer that would reflect the radio waves back to the earth's surface.

Ultraviolet radiation from the sun is the principal source of ionization in the upper atmosphere, but X rays, cosmic rays and other particle radiation also contribute. The ionosphere as a whole is assumed to be electrically neutral, that is, the number of its positively charged ions is equal to the number of its electrons plus the number of negatively charged ions. The importance of solar radiation in maintaining the ionized layer is demonstrated by the large differences in electron concentration between day and night and by its variation during the solar cycle and during magnetic storms.

Within the ionosphere there is a tendency toward concentration of ions in several distinct layers: the D region (50–85 km above the earth's surface), the E region (85–140 km), the F region (140–600 km), and the high ionosphere extending from 600 km to the edge of the magnetosphere. The daytime electron concentration increases from about 100 electrons per cubic centimeter in the D region to about 1000 electrons per cubic centimeter at a distance of three earth radii. The number of charged particles in the ionosphere is still small compared with the number of neutral particles; for example, at 300 km the concentration of neutral particles is about 1 million times the daytime concentration of electrons. (M.F.H.)

Iridescence. In mineralogy, the interference of light either at the surface of or within a mineral thereby producing a series of colors as the angle of incident light changes. Surface iridescence, similar to that produced by soap bubbles, results from interference of light as it is reflected from a thin surface film. Internal iridescence is caused by the reflection and refraction of light from closely spaced fractures, cleavage planes or inclusions. Some specimens of labradorite feldspar show a beautiful internal iridescence, the color therein changing from blue to green to yellow. (C.S.H.)

Iron. Usually alloyed with nickel, iron is found sparingly as a native element. It has two types of occurrence: as terrestrial iron and in meteorites. Some terrestrial iron is regarded as a primary magmatic mineral, but most of it was probably formed by the reduction of iron compounds through assimilation of carbonaceous material. Iron meteorites always contain from 7 percent to 75 percent nickel. See Mineral Properties; Appendix 3.

Iron Meteorite. A *meteorite composed essentially of iron and nickel-iron, as opposed to a stony meteorite which is made up of silicate minerals.

Island. A body of land smaller than a continent and surrounded by water. Islands range in size from under 100 square meters to hundreds of thousands of square kilometers. Some are so low as to be awash during storms; others, such as New Zealand, have mountains thousands of meters high. The world's largest islands, in descending size, are Greenland, New Guinea, Borneo and Madagascar.

Islands are formed in many ways. Volcanic islands, such as the Hawaiian and Canary islands, are initiated by submarine eruptions and are built up above sea level by subaerial eruptions. Islands may also form where a rugged coastal area has been submerged and peaks and ridges remain above water, as along the coast of Maine, the Finnish Archipelago, the Dalmatian Coast and the west coast of Scotland. *Barrier beach islands are built by waves and shore currents off the shore of a mainland; they parallel the coast, are separated from it by a lagoon or marsh and are low-lying. *Stacks are small islands formed when *marine erosion detaches headlands from the mainland; they are found on many coasts and are common along the California and Oregon coasts. Finally, corals and other marine organisms build *barrier reefs off the shore of a mainland as well as the circular islands, or *atolls, found in

the Pacific. A group of islands is called an archipelago. (R.L.N.)

Island Arc. A chain of volcanic islands commonly associated with deep-sea *trenches. Island arcs occur on the landward side of the trenches and are curved, with their convex side oceanward. Many such island chains are around the margins of the Pacific Ocean, as, for example, the Aleutians, the Japanese islands and the Marianas.

Isograd. A line drawn on a map to delimit zones on the earth's surface containing rock subjected to a particular range of temperature and pressure during *metamorphism. Each isograd marks the boundary at which a particular mineral or mineral assemblage begins to appear in the rock. Minerals chosen to represent grades of metamorphism are those that form under certain limited conditions of temperature and pressure. They are called index minerals and are, in order of increasing degree of metamorphism: chlorite, brown biotite, almandine, staurolite, kyanite and sillimanite.

Isometric System. In crystallography, a system of crystals in which the forms are referred to three equal and mutually perpendicular axes. Isometric crystals are usually equidimensional and transmit light with a uniform velocity in all crystallographic directions. Many common minerals, such as halite, fluorite, garnet and galena, crystallize in the isometric system.

Isomorphous. In crystallography, the term originally referred to minerals having the same crystal form. It is now generally used synonymously with solid solution.

Isostasy. A condition of equilibrium whereby the earth's *crust is buoyantly supported by the plastic flow of material in the earth's *mantle. The theory of isostasy evolved during the middle of the 19th century to explain large discrepancies in measurements made by surveyors while mapping India. A near north-south line between the cities of Kalianpur (at the south end of the line) and Kaliana (lying near the foothills of the Himalaya Mountains) was measured by means of a surveying method known as triangulation and checked by astronomical determination of longitude and latitude. When the results were computed and compared, a discrepancy of about 150 m (500 feet) along the 603-km (375-mile) line between the two cities was observed.

To explain the discrepancy, a British cleric, John Henry Pratt, suggested that

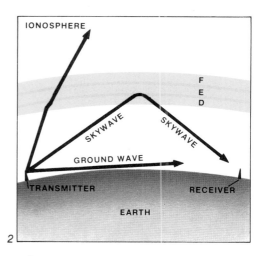

Ion

1 The geometric arrangement of ions in a crystal structure is greatly determined by the relative size of the ions that are bonded together. The type of packing depends on the ratio of the radii, with the radius of the anion (larger ion) taken as unity.

Ionosphere

2 Ground waves travel in a straight line and can only be transmitted as far as the horizon. Radio waves reflected by the ionosphere's layers of ionized air can be transmitted beyond the horizon, making long-distance radio communication possible.

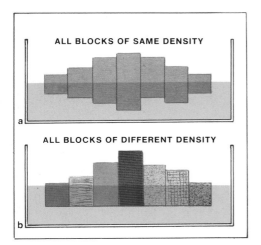

ALL BLOCKS OF SAME DENSITY

a

ALL BLOCKS OF DIFFERENT DENSITY

b

Isostasy

According to the principle of crustal flotation, the continents, floating in equilibrium over a soft mantle, are thicker than the oceanic crust. Two theories have been proposed to explain this phenomenon: (a) In one model, all blocks have the same density and are floating in a denser material. Each block floats at a depth where the proportion of volume above the mantle level to that lying below the level is the same for all blocks, even though they are of different lengths. If part of a block were removed, the remaining section would rise to a new position of equilibrium. (b) In the second theory, the crust is composd of different materials with different densities. Blocks rising higher have proportionally lower densities than surrounding blocks

there was an attraction between the Himalayas and the plumb bob (the weight that hangs beneath the surveyor's tripod). The plumb bob and the line to which it is attached were used as a reference from which all angular measurements were made. Pratt believed that the plumb line was deflected by the mass of the Himalayas at both locations, but that the deflection was greatest at Kaliana since it was closer to the mountain range. He calculated that the amount of error due to the horizontal attraction of the mountains was three times greater than the observed error. In 1855 he reported his findings to the Royal Society of London. Soon afterward, George B. Airy, Royal Astronomer, undertook to explain the differences. Airy declared that the earth's crust floated on a substratum of material of higher density. To maintain equilibrium between the crust and the substratum, the roots of the crustal segments forming mountains would have to be sunk deeper into the substratum than would segments of the crust containing plains or ocean bottoms. The concept can best be understood by viewing two icebergs floating in the sea; both are of the same density but they float in a medium of greater density. If one iceberg rises twice as far above the water as the other, the larger iceberg will be found to lie twice as far below the surface of the sea so as to remain in a state of isostatic equilibrium.

Similarly, Airy suggested that the Himalaya Mountains, composed of crustal material, penetrate deeper into the denser substratum. Because the roots are of lesser density, they represent a deficiency of mass beneath the mountains and would have less of an attractive force on the surveyor's plumb bob than that assumed by Pratt. A recalculation of the deflection based on Airy's theory produces a deflection that is nearly equal to the observed deflection.

Four years later Pratt presented another explanation. He maintained that the substratum on which the crust floats is of an equal density but that various segments of the earth's crust differ in density; that mountainous segments are of a lesser density than, say, crustal segments forming ocean floors. The crust everywhere would then extend equally into the denser substratum without the mountain roots postulated by Airy. Nonetheless, just as Airy's theory suggested, the mass associated with mountain crust would exert less of an attractive force on the plumb bob.

With the information that has been learned about the crust since Pratt and Airy presented their views on isostatic equilibrium, it appears that either theory can be applied over various portions of the earth's crust. The amount of crustal area maintained in isostatic balance in Airy's theory as compared to Pratt's theory remains a subject of debate. (J.W.S.)

Isotherm. A hypothetical line connecting points of equal temperature within the earth. It is commonly used on weather maps.

Isotope. One of two or more kinds of the same chemical element having an identical number of protons but a different number of neutrons in its nucleus. Isotopes of the same element have the same atomic number but have different atomic weights and slightly different physical properties.

Itacolumite. A flexible sandstone that will sag under its own weight. Its flexibility is the result of porosity, a combination of three flexible minerals (mica, chlorite and talc) and an interlocking arrangement of quartz grains.

J

Jade. Either of two minerals, *jadeite or *nephrite, used for ornamental purposes and highly regarded as a material for carving. Although jadeite and nephrite belong to different mineral groups (jadeite is a pyroxene and nephrite an amphibole) and differ chemically, they are both extremely tough and are usually green. Under the microscope, jadeite is seen to be composed of minute grains, whereas nephrite is a felted mass of fine fibers. Nephrite is not as hard as jadeite, but the interlocking fibers give it a greater cohesive strength and make it difficult to break with a hammer. In carved and polished pieces it is difficult to distinguish one kind from the other. However, when viewed together a slight difference in luster can be noted: jadeite is vitreous, nephrite oily. The expert can also detect a difference in color: jadeite is commonly mottled whereas the color of nephrite is more uniform. A fine emerald-green material, sometimes translucent, is jadeite; a spinach-green material with darker spots is nephrite. Nephrite mineralogically is tremolite, a white or colorless calcium magnesium silicate. However, iron may be present, coloring it green — the more iron the deeper the color. Consequently, nephrite can vary from white through deepening shades of green to nearly black. Although jade is usually green, it may be blue, brown, yellow, mauve, gray or black.

For many centuries jade has been valued by the Chinese above all other minerals, and its delicate and elaborate carving is an ancient art. Undoubtedly nephrite from Chinese Turkestan was first used, and only comparatively recently has jadeite from Burma been available. Today jadeite is the more valued variety.

Jade was known by primitive peoples and worked by them into implements as well as ornamental pieces. Ancient jade artifacts have been found in Mexico and Central America, and when Captain Cook arrived in New Zealand in 1769, he found that the native Maoris on the South Island used jade not only for ornamental purposes but for knives, axes and war clubs. (C.S.H.)

Jadeite. A mineral of the *pyroxene group. Jadeite and the amphibole *nephrite comprise *jade. See Mineral Properties.

Jasper. A compact, granular, microcrystalline variety of quartz. It is usually red because of the inclusion of finely divided hematite.

Jet. A dense black coal capable of taking a high polish. It is used for beads, jewelry and other ornaments.

Joides. A term made up of the initials of the Joint Oceanographic Investigation for Deep Earth Sampling, an organization that consists of five large American marine research institutions. It is devoted to studying the history of the ocean basins. The group's first effort in the early 1960s was the drilling of several holes in the *continental margin of the southeastern coast of the United States. Presently the group is conducting a worldwide drilling program, called the Deep-sea Drilling Project, in ocean basins. JOIDES institutions include Woods Hole Oceanographic Institution, Lamont Geological Observatory of Columbia University, the University of Miami, Scripps Institution of Oceanography (University of California) and the University of Washington. (R.A.D.)

Joint. A potential parting surface or crack through a solid body of rock. There is little or no movement parallel to the joint surface. *Weathering or movement perpendicular to the joint plane may form an open joint, or fissure. Because joints are planes of weakness, they must be closely examined when selecting sites for large construction projects such as dams, tunnels and highway road cuts. Geologists and engineers have devoted much time in the field and in the laboratory to the study of jointing. Field and structural geologists use what has been thus learned to understand the patterns of forces that produce structural features on and in the earth's *crust.

Generally, joints have patterns of distribution and are rarely found as a single crack in a solid mass of rock. Based on their geometric orientation or relationships to each other or other structural features, joint systems have been classified as follows: Joints that form essentially parallel to one another in map view and may or may not show the same relationship in cross section are known as systematic joints. These also form plane surfaces and may intersect other joint planes. Several systematic joints constitute a joint set; when two or more sets are recognized in an area, the overall pattern is called a joint system. Nonsystematic joints form curved nonparallel surfaces and may meet but not cross other joint surfaces.

When joints occur in rock displaying some layered feature — such as *foliation, *schistosity or *bedding — they are classified on their orientation relative to that feature. Joints striking parallel to the planar feature are called strike joints; those that strike parallel to the *dip of a planar feature are called dip joints. Oblique or diagonal joints strike at some random orientation to the planar feature. In sandstones and other sedimentary rocks or-

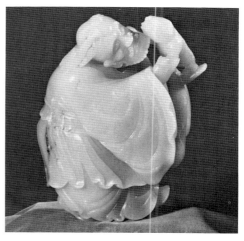

Jade
Jade has long been esteemed for its beauty by many peoples and for centuries has been fashioned into delicate and elaborate works of art such as this Chinese figurine.

dinarily displaying *bedding, joints parallel to bedding are classed as bedding joints.

A special kind of jointing, known as columnar jointing, can be seen in certain kinds of igneous rock. Outstanding columnar jointing is well displayed at Devil's Tower National Monument in northeastern Wyoming, in the Palisades along the Hudson River in New Jersey and in the Giants Causeway in Northern Ireland. (J.W.S.)

Jointing, Columnar. See under Columnar Jointing.

Jolly Balance. An instrument for obtaining the *specific gravity of minerals in which relative weights are determined by stretching a spiral spring. Two pans are carried at the lower end of the spring, the upper one in air and the lower in water. The specimen is placed in the upper pan, and the amount *(a)* that this causes the spring to stretch can be read on a scale. The specimen is then placed in the lower pan and the amount *(w)* that the spring thus stretches with the specimen immersed in water is noted. The stretching *(a)* is proportional to the weight in air and the stretching *(w)* proportional to the weight in water. Thus $a/a-w$ = specific gravity. (C.S.H.)

Jurassic Period. The second oldest period of the *Mesozoic Era; it began about 190 to 195 million years ago and lasted some 54 to 59 million years. It was named from the Jura Mountains located between France and Switzerland, where its richly fossiliferous rocks were first described by Alexander von Humboldt in 1795. Jurassic deposits are widespread in England, Germany, France, Scotland, northern Russia and parts of the Mediterranean region. They also occur in Siberia, northern China, India and other parts of Asia and are quite widespread in the Southern Hemisphere. In North America, Jurassic rocks are unknown east of South Dakota, but there are marine deposits in the West. These are not abundant but do occur in the western interior of the continent, including eastern British Columbia and Alberta, Canada; Idaho, Wyoming, Montana, Colorado, Utah, Arizona, New Mexico, southern Nevada and South Dakota. These rocks consist mostly of continental deposits believed to represent windblown desert sands and sediment laid down by streams; in places gypsum and fossiliferous marine rocks can also be found. Most of these formations are rather colorful and have been eroded into spectacular scenery in areas such as Zion Canyon National Park and Rainbow Bridge and Arches National Monuments in Utah. In western North America thick strata (mostly marine) ex-

3

Joint

1   Tension within the rock created these rectangular fractures along joint planes in diabase in Fort Lee, N.J.

2   This polygonal paving is the top of a great pile of basaltic columns at Devil's Postpile National Monument, Calif., one of several such outstanding features in the world. It was formed 4 million years ago when a deep pool of lava, cooling and shrinking, developed vertical cracks of unusual symmetry.

Jurassic Period

3   Named from the Jura Mountains between Switzerland and France, the Jurassic Period began about 190 million years ago and lasted over 50 million years. Among the most abundant dinosaurs in this period were the carnivorous theropods, which walked upright on powerful but birdlike hind limbs. Antrodemus (top right), a 12-m (40-foot) giant with a skull 60 cm (2 feet) in length, was capable of attacking the mightiest dinosaurs as well as smaller prey like Ornitholestes (top left), a light, nimble theropod only about 2 m (6 feet) long. The Jurassic dinosaurs reached their greatest size with the massive sauropods, such as Brontosaurus (bottom left), which was 24 m (80 feet) long and weighed over 30 tons. These reptiles fed on plants in swamps and lakes, where the buoyancy of the water took some of the enormous weight off their limbs. Stegosaurus (bottom right), was a plated dinosaur as big as a modern elephant.

tend from Alaska along the western coast of Canada and into Washington, Oregon, California and Nevada. Igneous rocks in some of these formations are evidence of volcanic activity during Middle and Late Jurassic time.

Jurassic rocks have also been encountered in deep wells in the Gulf Coast region of the United States. These marine limestones are hundreds of meters thick and have been the source of much petroleum. At the end of the period North America and the other continents were uplifted, causing withdrawal of the continental seas. There was mountain building in the western part of North America, as the Nevadan *orogeny elevated mountains occupying a position equivalent to the present Sierra Nevada, Coast Range, Alaskan Range, Cascade and Klamath mountains. This disturbance was accompanied by much igneous activity, and great masses of granite were intruded into the severely deformed Jurassic rocks. These intrusive rocks can be seen in the sheer granite cliffs and rounded domes of Yosemite National Park in northern California.

The Jurassic seems to have been a time of rather uniform warm to temperate climates. However, windblown deposits believed to represent ancient desert sands suggest arid conditions for parts of the southwestern United States.

Jurassic plants were varied and abundant. They resembled those of the *Triassic, with the notable exception of an increase of the conifers and the *cycads. The latter were so numerous in the Triassic and Jurassic that this portion of the Mesozoic has been called the "Age of Cycads." The *ginkgo, or maidenhair tree, which had appeared during Late Permian time, became increasingly abundant. Ginkgos occurred worldwide during the Jurassic but today are represented by a single living species. Modern flowering plants and hardwood trees had not yet appeared. Marine invertebrates were very abundant, including reef-building corals, *pelecypods, *gastropods, *foraminifers, *bryozoans and *echinoderms. *Ammonoids and *belemnoids were especially numerous; both of these groups reached the peak of their development during this period.

Land-dwelling invertebrates included land snails and a variety of insects and other *arthropods. The vertebrates were predominantly reptilian, including many strange and unusual forms. *Reptiles of all shapes and sizes successfully occupied a variety of environments on the land, in the sea and in the air. Some typical Jurassic forms were Brontosaurus and Diplodocus, huge quadrupedal plant-eating dinosaurs. Some of these were as much as 25 m (82

feet) long and weighed 35 or 40 tons. Stegosaurus was another quadrupedal herbivorous dinosaur; it had a double row of large, bony, triangular plates along its back, which served as protection from large dinosaurs. For additional protection Stegosaurus had a long, flexible tail bearing four long, sharp-pointed spikes. Although this animal weighed as much as ten tons and reached a length of 4.5–7.5 m (15–25 feet), its ridiculous-looking small head contained a brain about the size of a walnut. Allosaurus was the largest of the Jurassic carnivorous dinosaurs. This great beast was 10.5 m (35 feet) long, walked on its hind legs and had powerful jaws with large, sharp teeth. The tail was long and heavy, undoubtedly serving to balance its body. The forelimbs were relatively small, with sharp claws that were used in tearing prey. Another group of reptiles became adapted to flying. Known as pterosaurs, these unusual creatures had hollow bones (as birds do) and relatively small bodies. Some Cretaceous forms had wingspreads of as much as 7.5 m (25 feet), and one recently discovered pterosaur fossil shows a wingspan of 15.5 m (51 feet), but most species were much smaller. The wings of these creatures consisted of a thin, leathery membrane stretched between the limbs. Gliding was probably the principal mode of flight. Their relatively large skull indicates that they may have attained a higher degree of intelligence than did their marine and terrestrial counterparts. Reptiles also continued to expand in the sea, and ichthyosaurs and plesiosaurs were common.

Although small and primitive, mammals were definitely present in Jurassic time. They are known from fragmental tooth and jaw remains which indicate that they ranged from the size of a rat to that of a small dog. Their teeth suggest that one group was adapted to plant eating while another, more sharp-toothed group was apparently carnivorous. The scarcity of Jurassic mammalian fossils makes it difficult to compare these early forms with the typical forms of today, and their early development is not clearly understood.

A major event of the Jurassic was the appearance of the first birds. They are known from the skeletons of two specimens and the single feather and a few fragments of a third. Named *Archaeopteryx, these primitive birds had teeth, claws on their wings, and other distinct reptilian characteristics. Indeed, only the presence of feathers distinguishes their fossils from the remains of certain of the smaller dinosaurs. All of the remains were found in fine-grained limestone at Solnhofen, Bavaria.

The igneous activity that accompanied

the Nevadan orogeny gave rise to mineral-bearing veins of *quartz that have yielded millions of dollars in gold. Jurassic coal has been mined in China, Siberia, Japan, Hungary and Australia. Oil has been produced from Jurassic rocks in Wyoming, Arkansas and the Gulf Coast states of Texas and Louisiana. (W.H.M.)

Juvenile Water. Water on the earth's surface that is known to be derived directly from *magma and is appearing on the surface for the first time.

K

Kame. An irregular mound generally composed of coarse glaciofluvial gravel. Kames are formed when the sediments deposited in the depressions on stagnant glaciers or against their margins are let down onto the ground when ice melts. They are characterized by *ice-contact slopes and by the deformed bedding that results from collapse and slump. The precise way in which these features are formed is still uncertain.

Kame Moraine. A large accumulation of glaciofluvial sediments deposited by *meltwater streams that are on or downstream from the terminus of a glacier. An *alluvial fan is commonly formed where each meltwater stream leaves a glacier; where the fans coalesce, a kame moraine is developed. An ice-contact slope facing up-valley is formed when the sediments next to the ice are left unsupported as the glacier retreats. In additon, a kame moraine containing numerous *kettle holes will form when the ice melts away if sediments are deposited on the marginal zone of a glacier with an irregular surface. Kame moraines owe their origin to such factors as an abundance of meltwater, to the morainal material on and in the ice, and to a stagnant ice front. See Kame Terrace. (R.L.N.)

Kame Terrace. A deposit composed of fluvial, lacustrine and mass-wastage sediments left between the edge of a tongue of glacial ice and a valley wall after the ice has melted. Kame terraces are commonly found on both sides of a valley; a series of terraces, one below the other, may result from a succession of glacial stillstands (periods when the glacier neither advances nor retreats) occurring at lower and lower positions of the ice during its thinning. They slope down-valley, usually have *kettle holes and may be continuous with a *valley train farther down the valley. They form next to valley glaciers and the dwindled remnants of ice sheets. They may be identified by their association with *eskers and *kames, the presence of an *ice-contact slope, the absence of *meander scars on the face of the terrace and the deformation of the sediments resulting from the collapse that follows melting of the adjacent ice. (R.L.N.)

Kaolin. A rock composed essentially of clay minerals. It is low in iron and is usually white. The minerals present are all hydrous aluminum silicates, of which *kaolinite is most abundant. Kaolin is used in manufacturing china and porcelain and is thus sometimes called china clay or porcelain clay.

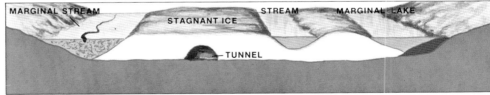

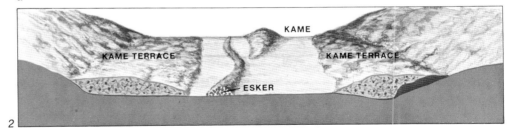

Kame

1 The region around Windsor, Vt., has many irregular, rounded hillocks of outwash deposited by meltwater.

Kame Terrace

2 Stream and lake sediments deposited between a stagnant ice mass and the sides of a valley (a) remain after the ice has melted and form kame terraces (b) A debris-filled stream tunnel beneath the ice remains as a long, rounded ridge called an esker.

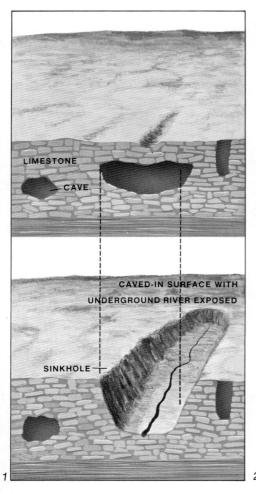

Kaolinite. A mineral, hydrous aluminum silicate, the commonest clay mineral and the chief constituent of *kaolin or clay. Kaolinite is a secondary mineral formed by alteration of aluminum silicates, particularly *feldspars. In places it is found in large deposits uncontaminated by other minerals; more commonly, as a constituent of soils or when transported by water and deposited in beds of clay, it is found mixed with quartz and with other minerals. See Mineral Properties; Appendix 3; Kaolin.

Karst Topography. Named after Karst, an area on the Dalmatian coast of Yugoslavia characterized by a scarcity of surface drainage and by sinkholes, caves and caverns, natural bridges, sinking streams, dry valleys and an abundance of subsurface drainage. This topography results from the presence of limestone dissolved by subsurface water, plus other factors such as highly jointed and thinly bedded limestone, moderate or abundant rainfall, and deep valleys cut into the limestone. The joints, bedding planes and valleys facilitate the movement of groundwater. Karst topography is widespread; it is well developed in parts of Indiana, Arizona and Kentucky, where there are over 500,000 sinkholes, and in the Causse region of southern France, the Yucatan Peninsula, Jamaica, Cuba, Puerto Rico and the Kwangsi area of China. (R.L.N.)

Kernite. A mineral, a hydrated sodium borate similar in composition to borax but containing less water. It was discovered in 1926 beneath the Mohave Desert near the present town of Boron, Calif. Millions of tons occur here associated with other borates in a bedded series of *Tertiary clays. The kernite is found near the bottom of the deposit and is believed to have formed by the dehydration of borax caused by increasing temperature and pressure. The mineral is also found associated with borax at Tincalayu, Argentina. Kernite is characterized by two good cleavages that cause it to break into long, splinterlike fragments. See Mineral Properties; Appendix 3. (C.S.H.)

Kettle Hole. A pit or basin formed by the melting of a block of ice partly or completely buried by *glacial drift. Kettle holes are found on *valley trains and on *outwash plains. Many contain ponds, lakes or swamps. They range from a few meters to several kilometers in diameter, and from about 1 m to more than 30 m in depth. Small kettles may be formed by the burial of blocks of ice detached from glaciers and transported away from them by melt-

Karst Topography

1 *In the development of a karst landscape, limestone leached out by subsurface water forms underground caves (a), some of which produce sinkholes on the surface when their roofs collapse (b).*

2 *The numerous sinkholes at Chevelon Fork, Ariz., display karst topography, which is named after a plateau in Yugoslavia with a similar landscape.*

Kettle Hole

3 *Deep hollows formed by glacial action and filled with water, such as this one near Amqui, Quebec, are common in the northern United States and in Canada.*

4 *A block of stagnant ice is almost buried by outwash from a glacier (a). As the block melts, a depression, or kettle, is formed that fills with water (b). In (c), the sides of the kettle hole have collapsed.*

3

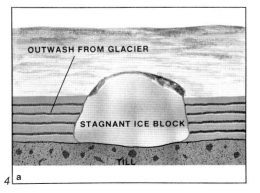

4 a

OUTWASH FROM GLACIER

STAGNANT ICE BLOCK

TILL

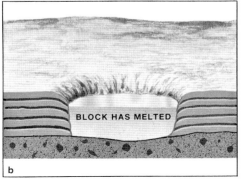

b

BLOCK HAS MELTED

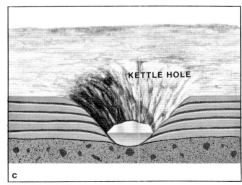

c

KETTLE HOLE

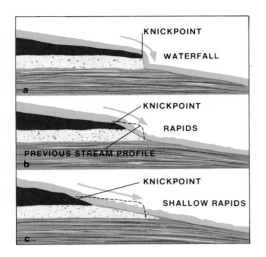

Knickpoint

At a sudden drop, or knickpoint, in a river's longitudinal profile (a) the stream breaks into a waterfall. The constant force of water plunging over the edge eventually erodes the bedrock to a more gentle slope (b) and the waterfall becomes rapids. As the bedrock continues to erode, the river's profile becomes smoother (c) and the rapids and knickpoint retreat upstream.

water streams. Large kettles, however, are formed by the burial of extensive masses of stagnant glacial ice. Pitted valley trains and outwash plains contain numerous kettle holes. (R. L. N.)

Kimberlite. A variety of the rock type *peridotite; it is composed essentially of *olivine and *phlogopite mica. Kimberlite is the primary rock type in which diamonds occur. The name comes from Kimberley, South Africa, where diamonds were first recovered from the rock.

Kinetic Metamorphism (or Dislocation; Cataclastic Metamorphism). The process by which brittle rock is shattered by external forces and subsequently reduced in particle size by crushing or grinding. The result is the formation of a new rock without alteration of the original chemical and mineralogic composition. The process occurs near the surface of the earth's *crust where temperature and pressure are too low for chemical reactions to take place. See Cataclasite; Mylonite.

Kipuka (A Hawaiian term). An area ranging from a few square meters to several square kilometers where older rocks, either volcanic or nonvolcanic, are surrounded but not buried by younger lava. Unlike *steptoes, kipukas may be lower or higher than the surface of the surrounding lava. They are common on Hawaii, where some can be identified even from a distance because the older rocks are covered with vegetation whereas the younger flows are bare. Some kipukas are low areas between two flows; others result from the bifurcation of one flow and the subsequent union of the two streams. The bifurcation results from a slight rise in the older topography or from irregularities in the advance of the front of the flow. Kipukas yield information about the older rocks that otherwise might not be available. (R.L.N.)

Knickpoint (or Nickpoint). A break in the smooth concave profile of a river, denoted by rapids, cascades or a *waterfall. The break may be caused by a resistant layer of rock that the stream cannot erode. Knickpoints may also occur from anything that causes renewed downcutting of the channel. A lowering of base level, either locally or regionally, increases the stream gradient and enables the river to lower its channel, causing a knickpoint that moves upstream. Additional discharge caused by *stream capture or climatic change may also give the river renewed downcutting energy. All knickpoints tend to disappear with time as a result of their retreat upstream.

Kyanite. A mineral, aluminum silicate, having the same chemical composition as andalusite and sillimanite. It is a metamorphic mineral characteristically in long-bladed crystals found in *gneiss and *mica schist, and often associated with garnet and staurolite. Kyanite shows better than any other mineral a variation of hardness with crystallographic direction; its hardness on *Mohs scale is 5 parallel to the length of crystals but 7 across the length.

Important commercial deposits are in India, Kenya and, in the United States, in North Carolina and Georgia. Kyanite is used to manufacture spark plugs and other highly refractive porcelains. See Mineral Properties. (C.S.H.)

Labradorite (named for its occurrence in Labrador). This *plagioclase feldspar is the major constituent of *gabbro and *basalt. Its composition in terms of percentages of albite and anorthite varies from 30 percent albite and 70 percent anorthite to 50 percent each. In Labrador it is found in large cleavable masses that show a beautiful series of blue, green and yellow as the specimen is turned.

Laccolith. A roughly circular body (as viewed from above) of igneous rock intrusion that has an upper surface parallel to the layering of the rock into which it intrudes. Laccoliths have diameters ranging from 1.5 to 8 km (1 to 5 miles). In cross section they are dome-shaped and generally, but not always, have a flat floor. A sill is similar to a laccolith but larger in extent and thinner in cross section. It has been suggested that these two be distinguished according to the ratio of their diameter to their average thickness, with laccoliths having ratios of less than 10 and sills having ratios of 10 or greater. Laccoliths most commonly occur where the bedding is mainly flat, as in the Colorado Plateau of the western United States. (J.W.S.)

Lagoon. A coastal bay that does not have significant exchange of water with the open ocean and does not receive significant freshwater runoff from the land. A lagoon is commonly formed by an elongated sand peninsula (*spit) that closes off the mouth of a bay, or by a *barrier beach. The barrier may be built up of coral reef. Lagoons are low-energy environments in that large waves and strong currents are typically absent. Wind serves as the primary factor in circulating lagoonal waters. High salinity conditions may result in the formation of evaporate minerals such as gypsum and halite (salt). Organisms living in this environment must tolerate the hypersalinity and general absence of water movement. As a result, the diversity of species is rather low but the population density of adapted species is high. (R.A.D.)

Lake. A sizeable inland body of standing water. The source of the water is precipitation that falls into the lake as rain or snow, or is fed into it by streams and springs. Lakes without outlets are usually found in arid regions and tend to be saline; some, such as Great Salt Lake and the Dead Sea, are more salty than an ocean. The salts in most saline lakes result from weathering and volcanic activity; they were carried into the lakes by water and wind, usually when the lakes were larger. The concentration of salts is due to evaporation. Some saline lakes, like the Caspian and Aral seas, are

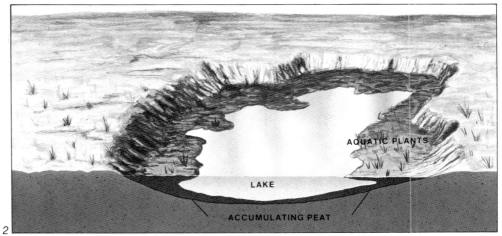

Lake

1 Moraine Lake, in Banff National Park, Alberta, Canada, was formed by a dam made of deposits from a retreating glacier.

2 As a small lake ages, it slowly fills with decaying plants until it becomes a peat bog.

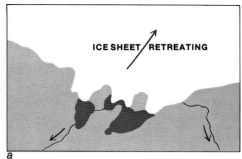

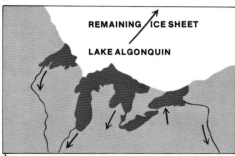

Lake

1 *The development of the Great Lakes began about 15,000 years ago during the last ice age. (a) Sufficient retreat of the ice sheet had taken place by some 13,500 years ago to leave sizeable ice-dammed lakes along the ice margin. (b) By 12,000 years ago the ice had retreated further and the early Michigan and Huron lakes were joined together to form Lake Algonquin (center). (c) Some 4500 years ago the Great Lakes looked much as they do today.*

Landslide

2 *A slide of thick, laminated clay loosened by unstable conditions blocks a highway near Redcliffburg, Ohio.*

landlocked arms of the ocean. Lakes can be permanent or ephemeral, and those in cold regions may be perennially frozen.

Lake Superior, located between the United States and Canada, is the largest freshwater (non-saline) lake (82,414 square kilometers, or 31,820 square miles); Lake Baikal, southern Siberia, has the greatest volume (23,000 cubic kilometers, or 5,520 cubic miles) and depth (1,621 meters, or 5,315 feet); Lake Titicaca, South America, 3,809 m (12,497 feet) above sea level, is the highest large lake; and the Dead Sea, Israel, is 397 m (1,302 feet) below the Mediterranean. Canada is richest in lakes, South America poorest; Finland has 55,000 lakes; Minnesota is said to have 10,000; lakes occupy 75 percent of certain provinces in Canada. They are more numerous in high latitudes and high altitudes.

Lakes are formed in many ways. Faulting and gentle upwarping of the earth's crust are responsible for some of the greatest lakes. Faulting produced those of the African Rift Valley and of southwestern United States. Upwarping formed Lake Victoria in Africa, and the Caspian and Aral seas in Asia. Lakes are also formed by volcanic activity; they occupy *craters and *calderas (as in Crater Lake, Oregon) and occur where lava flows and volcanoes obstruct drainage. Landslides form lakes by damming up valleys, especially in mountainous regions. A lake 70 km long and over 300 m deep was formed in 1840 when a landslide blocked the Indus River, Kashmir; when the dam failed, a flood raced down the valley causing destruction for hundreds of kilometers.

Lakes of glacial origin are far more numerous than all other types put together. Some are dammed by ice, as in Greenland and in other regions now glaciated. Many are located in glacially-formed bedrock basins. Others result from the irregular deposition of *till. Ponds and small lakes occur where *kettle holes extend below the water table, as on Cape Cod, Mass. *Sinkholes formed by the solution of limestone may contain ponds and small lakes; they are common in Kentucky and in the Karst region of Yugoslavia. Rivers form lakes in many different ways, as, for example, the *oxbow lakes that occupy abandoned channels on *floodplains or the lakes found on *deltas. Lakes are formed behind *baymouth beaches between *tombolos, between *dunes and where wind-blown sand dams small streams.

Lakes are short-lived geologic features; their destruction commences as soon as they are formed. In humid regions the outlet stream lowers the barrier, and the deposition of detrital sediments and of

organic matter helps fill the basin. In arid regions the deposition of chemical precipitates and sediments borne by wind and streams help fill the basin, and evaporation reduces the size of the lake.

Lake deposits are found in buried layers of the earth. The best known in the United States is the Green River formation, which accumulated in an Eocene lake that covered parts of Utah, Colorado and Wyoming. In places this formation is 1500 m thick, and someday its organically rich shales may be a source of petroleum. (R.L.N.)

Laminar Flow. The theoretical flow of water in a smooth, straight channel at low velocities. The water moves in parallel layers that slide over one another with no mixing. The *shear stress of this kind of flow is proportional to the resistance of the layers of water to movement. The resistance depends on the viscosity (stickiness) of the water and on the change in velocity from layer to layer. The velocity of the layers increases from the bottom upward, with the layer of maximum velocity just below the water surface. Theoretically the boundary layer of stream flow comprising layers close to the bed and banks is laminar because friction reduces the velocity and turbulence. Laminar flow cannot transport a solid debris load. (M.M.)

Lamprophyre. A dark dike rock composed chiefly of *dark minerals that occur both as *phenocrysts and in the groundmass. Any one of the ferromagnesian minerals biotite, hornblende, pyroxene and more rarely olivine may be dominant. Feldspar or feldspathoids are also present.

Land Bridge. A land area, typically narrow and subject to submergence, that forms a connecting link between continents. Some of these bridges, such as the Bering land bridge now beneath the waters of the Bering Strait, served as dispersal routes for migrating plants and animals.

Landslide. The downward movement of masses of *bedrock and/or *regolith on steep, unstable slopes and cliffs. The material can fall, slide or flow; it may move tens of meters or several kilometers; and the rate may be hardly perceptible or very rapid. Landslides can take place on land or beneath the ocean. Circumstances favoring their occurrence are steep slopes, moisture, *foliated rocks, closely spaced *joints, *bedding planes parallel to the slopes, unconsolidated or poorly consolidated material having little cohesion, rocks such as clay and shale that are slippery when wet, and earthquakes.

Depending on these various factors, the landslides may be rock and debris falls, rock and debris slides, debris flows, and *mudflows, earthflows or *slump. The great mobility of some kinds of landslides has been attributed to a cushion of air between the slide and the ground. Such landslides can be very destructive to life and property: They can generate huge waves when they fall into *fiords, lakes or reservoirs, and they can loose catastrophic floods when they dam up water in valleys and the dams fail. Landslide scars with deposits below them are common on the sides of mountains. Some of these deposits are characterized by hummocky topography with small ponds occupying undrained depressions. (R. L. N.)

Land-use Planning. See Environmental Geology.

Langley. In meteorology, a unit of energy per unit area equal to one gram-calorie per square centimeter. It is named after the American solar radiation physicist Samuel P. Langley. (The 15° gram-calorie commonly used in meteorology is the amount of heat required to raise the temperature of one gram of water from 14.5°C to 15.5°C.) It is often used in conjunction with a time unit to express the rate of flow of radiant energy. For example, the solar constant — that is, the solar energy received outside the earth's atmosphere on a surface perpendicular to the incident radiation at the earth's mean distance from the sun — is about 2.00 langleys per minute. (M.F.H.)

Lapis Lazuli. A gem and ornamental material composed of several minerals, of which deep blue *lazurite is the chief constituent. In addition to lazurite, lapis (as it is sometimes called) is composed of *augite, *hornblende and *pyrite embedded in a matrix of white *calcite. Northeastern Afghanistan is the principal source of lapis lazuli today, as it has been for centuries. Other notable occurrences are near the southern shore of Lake Baikal in Siberia and at Ovalle in the Chilean Andes.

Lateral Secretion. A theory postulating that the elements forming the minerals of an ore deposit were derived by solution and diffusion from the neighboring country rock.

Laterite. A highly weathered, reddish-brown to blackish deposit that develops in moist, tropical regions. The name (derived from the Latin *later* for brick) refers to the bricklike color and hardness of dried laterite, which can be used as building material. An end product of *weathering, it

varies widely in composition but typically contains aluminum hydroxides, iron hydroxides and oxides, residual quartz and varying amounts of clay minerals such as kaolinite. Neither a mineral nor a soil, laterite more properly refers only to the soil zone in which the iron and aluminum have accumulated. Prolonged weathering may produce rather sticky lateritic soils, or latosols, that contain nodules or *concretions of mineral material. Certain laterites of Australia, Brazil, Cuba and southwest Asia contain concentrations of hydrous aluminum oxides sufficient to form *bauxite, an ore of aluminum, and some Cuban laterites contain such large volumes of hydrous iron oxides that they are useful as iron ore. (W. H. M.)

Latite. An extrusive rock, the fine-grained equivalent of *monzonite. Plagioclase exceeds potash feldspar in latite, but frequently the texture is so fine that determination of amount and kind of feldspar is impossible even microscopically. Augite or hornblende is commonly present.

Laurasia. The name given by the German meteorologist Alfred Wegener (1880-1930) to a hypothetical landmass of the Mesozoic Era that included most of the landmasses of the present Northern Hemisphere. As *continental drift took place, Laurasia broke up and formed the continents of the Northern Hemisphere. See Pangaea. See also Color Plate, "The Dividing Supercontinent."

Lava. Molten rock poured out of vents at the earth's surface or on the floor of the ocean; also the rock formed when such molten material cools and solidifies. The vents may be located in *craters or in *calderas, on the flanks of volcanoes or elsewhere. *Lava flows are formed where lava moves as a river; lava falls drop over cliffs; and lava lakes are formed where drainage is obstructed. The temperature of molten lava on reaching the earth's surface ranges from about 750°C (1382°F) to 1200°C (2192°F). Still higher temperatures have been reported from Kilauea Crater on Hawaii, where escaping gases may have been burning. Hawaiian lava when extruded commonly has a yellowish-red color, indicating a temperature of about 1100°C (2012°F). Lava cools very slowly because it is a poor conductor of heat. The greater the thickness of the lava, the slower the cooling. It is estimated that a lava flow 9 m (30 feet) thick extruded at a temperature of about 1100°C (2012°F) would take more than 10 years to cool to air temperature. A lava flow about 9 m thick that was extruded at the Paricutin volcano in Mexico in 1944

was still steaming 12 years later. It was reported that a cigar could be lighted in cracks in the lava 21 years after the Jorullo volcano in Mexico erupted in 1759, and that vapor was still rising from it 87 years later.

Lava varies in composition from volcano to volcano; also, lava extruded early from a volcano may differ from that extruded later. Lava generally contains varying proportions of SiO_2, Al_2O_3, Na_2O, K_2O, CaO, MgO, Fe_2O_3, FeO and H_2O, together with smaller quantities of many other elements. Based on chemical composition, lava is of three principal types: sialic lava (rhyolite, obsidian) contains 66 percent or more silica; mafic lava (basalt) contains less than 52 percent silica; intermediate lava (andesite) contains between 52 and 66 percent silica. The composition of lava partly determines the type of volcanic eruption. Sialic lava is much more viscous than mafic; the gases cannot escape as easily as the gases of mafic lava, and the volcanic eruptions are generally much more violent.

The viscosity of lava depends mainly on its chemical composition and temperature. Lavas with high silica content have much higher viscosities than those with low silica content. Thus the rhyolite flows are much more viscous than the basaltic kind. The viscosity of lava also increases greatly with decreasing temperature. The viscosity of Hawaiian lavas ranges from about 1000 poises (100,000 times that of water) where they are extruded from vents, to 10,000 poises several kilometers from the vents. Lava viscosities as high as 10 million poises (like shoemaker's wax at 8°C, or 46°F) have been measured in Iceland. (R.L.N.)

Lava Curtain (or Fire Curtain). See Lava Fountain.

Lava Flow. A stream of liquid rock from a vent during a relatively quiet eruption; also the rock body formed by its solidification. The dimensions of the flow depend essentially upon viscosity. Thus fluid basaltic lava tends to form fast-moving, thin, long flows; viscous sialic lava such as obsidian forms slow-moving, thick, short flows. Flows 80 km (50 miles) long have been reported from Iceland, and the 1859 flank flow on Mauna Loa, Hawaii, flowed 50 km (33 miles) before it reached the ocean. Flows vary greatly in thickness. In Hawaii they range from about 30 cm (1 foot) to about 20 m (60 feet). The average of about 400 measured flows in the Columbia River basalts is 23.5 m (78 feet). The largest historic flow, from the Laki fissure in southern Iceland in 1784, was calculated to have a volume of about 12 cubic kilometers

(9 cubic miles). With the high viscosities that occur in lava, high velocities are not expected. Flows as fast as 65 km an hour have been reported from Surtsey, Iceland, but most flows move much more slowly. Lava flows cool more rapidly than intrusive rocks, which are usually coarse-grained. Lava flows have a glassy or fine-grained texture, generally being composed of basalt, andesite, rhyolite, obsidian, scoria or pumice.

There are three principal kinds of lava flow. The ferromagnesian, dark colored lavas form flows known by their Hawaiian names pahoehoe and aa. Lavas of andesitic or intermediate composition form the type called block because of their regularly shaped fragments. Pahoehoe flows, which are usually basaltic, have smooth, undulating, continuous surfaces; closely spaced, arc-shaped, concentrically aligned ropes; *lava tunnels and *collapse depressions formed by the incomplete collapse of their roofs; abundant spheroidal vesicles, or gas bubbles; in thick flows a vesicular zone at the top, a thinner zone at the bottom and no vesicles in the center.

Aa flows, which are also generally basaltic and may have the same chemical composition as pahoehoe, have a surface covered with clinkery fragments (generally less than 1 m in diameter) and with jagged spines; surface channels that formed when lava rivers that fed the flows drained away at the end of eruptions; vesicles generally very irregular and elongated in the direction of flow; a clinkery zone at the top and bottom of a flow and a zone of massive lava in the center.

Block flows are usually more sialic than pahoehoe and aa flows. They have a surface covered with irregularly shaped fragments with relatively smooth sides; a larger proportion of glass than in pahoehoe and aa flows; spines similar to those on *cumulo-domes but smaller; a blocky zone at the top and bottom of a flow and massive lava in the center; and irregularly shaped vesicles.

When aa lava enters the ocean, as it often does on Hawaii, billowy banks of steam and black clouds of tephra, formed from liquid lava fragmented by the steam, may rise hundreds of meters in the air. The ocean boils dramatically, marine organisms are killed, tephra accumulates on the sea bottom and the land is extended out into the ocean. Littoral cones, similar to *cinder cones, may be built by the tephra, and when reworked by waves and shore currents may form the black sand beaches found along the coast. During the 1960 eruption on Hawaii, about 3 km (2 miles) of shoreline was widened as much as 800 m (2650 feet) in places by lava flows.

Lava Flow
Streams of glowing lava covered 2000 acres in three weeks during the 1959 eruption of Kilauea Iki, Hawaii.

1

In historic time lava flows have been observed on the flanks of Mauna Loa and Kilauea in Hawaii, on Vesuvius and Etna in Italy, and in Iceland, Mexico, Japan and the Azores. Since towns and farms have been buried, efforts have been made to divert flows from populated areas. The first recorded attempt occurred on the slopes of Etna in 1669, when several dozen men with iron bars broke the crust on the side of a flow and temporarily diverted a lava stream from the city of Catina below. Walls have been built in Sicily and on Hawaii to divert flows, also with limited success. In 1935 a lava flow on Mauna Loa threatened the city of Hilo and its harbor. The U.S. Army, under the guidance of a volcanologist, dropped twenty 272-kg (600-pound) bombs in an attempt to destroy the roof of the tunnel in which the lava was flowing. It was hoped that freeing and thereby cooling the lava in the tunnel would slow it down, and that the collapse of the tunnel roof would block and divert the flow. The bombing blasted liquid lava into the air, and where the tunnel roof collapsed a thread of lava could be seen. On the day of the bombing the front of the lava was advancing 250 m (800 feet) per hour; thirty hours later, forward motion had stopped. While some authorities believed the bombing had stopped the flow, others thought it would have stopped anyway.

The only incontrovertible success in mastering a lava flow of such immense proportions occurred in 1973 on the Icelandic island of Heimaey. A new volcano formed on Heimaey on January 23, and after considerable activity and damage for two months a lava lake rose in the crater and overflowed, advancing at a rate of 3 to 8 meters (10 to 26 miles) a day toward the most important fishing port in Iceland. United States cargo planes brought in 47 powerful pumps with a total output of 4500 tons of water per hour. Three large pipes, several hundred meters long, conveyed the pumped seawater to the lava front. The discharge of the seawater lowered the temperature of the front, and as the flow cooled bulldozers mounted the lava, leveling it so that the three pipes could be brought into the interior of the flow. Simultaneously, trucks and a steamroller built barriers of ashes at the front. After two weeks of continuous watering, the lava had cooled from 1000°C (1800°F) to 100°C (180°F). The wall of lava, now 40 meters (132 feet) high, solidified, and the molten lava behind it was arrested and diverted.

*Shield volcanoes and lava plateaus and plains, such as the Deccan Plateau of India and the Columbia Plateau of the United States, are composed almost entirely of lava flows; composite cones consist partly of such flows. Lava flows have occurred in every period of geologic time. (R.L.N.)

Lava Fountain (or Fire Fountain). Clots of incandescent lava thrown as much as 600 m (2000 feet) into the air by the extrusion of frothy lava and the bursting of gas bubbles at the surface of liquid lava. They are analogous to the droplets of water thrown 30 cm (1 foot) or more above the surface of the water in a glass in which seltzer tablets are effervescing. The lava cooling in the air may land as solid fragments or as plastic clots. Spatter cones, cinder cones, or spatter-and-cinder cones may be formed by the accumulation. A fire curtain or lava curtain is formed when the incandescent clots are emitted from elongated fissures. Cinder or spatter ridges may result from the accumulation of the ejecta along the fissures. Lava fountains and curtains, such as appear when Hawaii's Mauna Loa and Kilauea are active, are among the most spectacular and awesome of volcanic phenomena. (R.L.N.)

Lava, Pillow. See Pillow Lava.

Lava Tree. A hollow cylindrical sheath of *lava that projects above the surface of a *lava flow as much as several meters. It is formed when fluid lava solidifies around a tree and later the surface of the flow subsides. Sometimes the pattern of the bark is preserved in the lava, but more commonly the wood is charcoalized. Lava trees can be used to determine the direction of flow because they are higher on the upstream side (just as the surface of a swiftly flowing river is higher on the upstream side of a post standing in it) and because material from the trees alters the lava downstream from them. Lava trees are common on the Island of Hawaii, in the Lava Cast Forest near Bend, Oreg., and in other parts of the world these features are found wherever youthful flows have invaded forests.

Lava Tunnel (or Lava Cave, Lava Tube). A tunnel inside a solidified *lava flow. The front of a pahoehoe lava flow advances because *lava is supplied to it from one or more pipelike zones of very fluid lava inside the flow. The fluid lava is bounded by more viscous lava. When lava no longer issues from the source of the flow, the fluid lava may drain away, leaving a tunnel as much as 8 m wide and several kilometers long. Lava *stalactites and stalagmites may form in the tunnels. The partial collapse of the roofs of these tunnels results in *collapse depressions on the surface of the flows. Lava tunnels are found in the United States, in New Mexico, California, Oregon, Idaho and Hawaii, and in New Zealand,

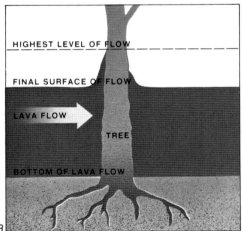

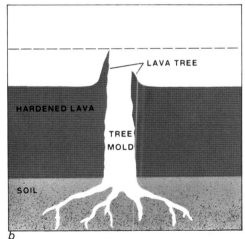

Lava Fountain
1 Shooting as high as 300 m (1000 feet), a fiery geyser of molten rock and gases bursts from Kilauea Iki.

Lava Tree
2 At Lava Tree State Park in Hawaii the lava trees project more than 3 m (10 feet) above the flow surface.

3 A lava flow surrounds the trunk of a tree, the surface of the lava lowering as the flow advances. (a) When the tree burns out, a hollow pillar of lava called a lava tree projects above the surface of the flow. (b) The vertical cylindrical opening below the surface of the flow is called a tree mold.

Lava Tunnel
Liquid lava drained away beneath the crust, creating this lava cave at Valley of Fires State Park, N.M.

Samoa and many other places where young flows occur. (R.L.N.)

Lazurite. A mineral, a silicate of aluminum, sodium and calcium with some sulfur. Deep aqua blue, it is the principal mineral in the ornamental stone *lapis lazuli. Lazurite is a rare mineral formed in crystalline limestones by *contact metamorphism.

Leaching. The transportation, usually downward, of soluble materials from the *regolith and *bedrock by water and other chemicals. The soluble material may be deposited at depth or can be carried to rivers and thence to the ocean. Leaching is an important soil-forming and *weathering process. It enriches ore bodies at depth by adding metallic minerals to them (secondary enrichment), and it removes the products of chemical weathering from near the surface, leaving bauxite and other valuable substances (residual concentration). It is also important in metallurgy.

Lepidolite. A mineral of the *mica group containing lithium. Its occurrence is essentially confined to *pegmatites where it is associated with other lithium-bearing minerals such as spodumene, amblygonite and petalite. Although usually pink to lilac, lepidolite may be pale yellow to grayish white. Where sufficiently abundant and pure, it is mined as a source of lithium.

Leucite. A mineral belonging to the *feldspathoid group. It is characteristically found in white trapezohedral crystals embedded in a fine-grained matrix. These crystals are isometric, indicating that they formed at temperatures higher than 605°C (1120°F); below that temperature leucite inverts to a tetragonal form. Although comparatively rare, it is abundant in certain recent lavas. Its most notable occurrence is in the lavas of Mount Vesuvius.

Levee. A natural levee is a low ridge formed at the edge of a bank by material dropped by an overflowing river. Rivers on a *floodplain will often overflow; as the debris-laden waters leave the channel, the depth of flow decreases. This results in a loss of ability to carry the load, which piles up on or near the bank. Natural levees are too small to be noticed on most rivers. However, large rivers such as the Mississippi often have prominent levees.

Artificial levees are erected by man along many rivers to protect the surrounding countryside from floods. If a river restrained by such a levee cannot overflow its banks and deposit its load on the floodplain, it will lay down the debris in its own channel. This sediment tends to raise the level of the riverbed. To prevent the next flood from overflowing, man must raise the levee ever higher.

Light Mineral. A rock-forming mineral that has a *specific gravity less than 2.8, as contrasted with a heavy mineral which has greater specific gravity. The term is also applied to minerals light in color, including feldspars, quartz, feldspathoids and calcite. The minerals light on the basis of specific gravity are essentially the same as those light in color.

Lightning. The visible electrical discharge that occurs during *thunderstorms along a channel of highly ionized air particles between electrical charge centers of opposite sign. It occurs within a thundercloud, between a thundercloud and the earth's surface, between different clouds and between a cloud and the air below. The spark discharge consists of an irregularly shaped main stroke with many branches; it may be up to 6 km long in large thunderstorms. Within the channel of the main stroke, only a few centimeters wide, the air is heated to perhaps 15,000°C (27,000°F) within a few ten-millionths of a second. The resulting explosive expansion of the heated air creates shock waves that degenerate into sound waves, heard as thunder.

Lightning discharges within thunderclouds indicate that electrical charges of opposite sign exist within the cloud itself. The cloud top is apparently positively charged, while the bottom part has an excess of negative ions. Laboratory experiments have shown that there are a number of ways *convection currents, moving raindrops and the ice-and-water mixture within most thunder clouds can lead to this separation of charges. However, it has not yet been established which explanation accounts for thunderstorm electricity, if indeed any one process is dominant. See Atmosphere Electricity; Saint Elmo's Fire. (M.F.H.)

Lignite. A brownish-black, low-grade *coal, intermediate in composition between *peat and *bituminous coal. It is rich in volatile material, ignites easily and burns smokily. Known also as brown coal, it is used as fuel for residential and industrial heating and for generating electric power. Carbonized lignite has been made into briquettes for cooking and heating; it is also a source of activated carbon used for water treatment, recovering gold and extracting iodine. In some parts of the world lignite is a primary raw material for heating and for the production of chemicals, gases and liquid fuels. In the United States, lignite occurs primarily in the northern Great Plains (western North Dakota, eastern Montana, northeastern Wyoming and northwestern South Dakota); the Gulf coastal region (Alabama, Arkansas, Kansas, Louisiana, Oklahoma and Texas); and the Pacific coast states—California, Oregon and Washington). (W.H.M.)

Limestone. A sedimentary rock composed largely of *calcite, a calcium-carbonate mineral. It also commonly contains grains of quartz, clay minerals, dolomite or other materials. If a large amount of dolomite is present, the rock is called dolomitic limestone. In some limestones the mineral grains are too small to be distinguished without a magnifying glass; in others they can be observed easily. Because limestone contains calcite, a simple test can identify it: A drop or two of dilute hydrochloric acid will fizz and bubble when placed on limestone. Although pure limestone is white, it may contain clay or plant or animal matter, causing it to be gray or black; it may also be yellow, brown or red. It is fairly soft and can be scratched with a knife.

Most limestones are marine in origin, formed from the calcareous sediments deposited in ancient seas; other limestones formed in fresh water such as lakes. As some earlier-formed rocks are weathered their calcium minerals are dissolved. Streams may carry this dissolved material to the sea, where small animals such as corals, crinoids, sponges and foraminifers take the dissolved material out of the water to build their calcium carbonate shells. Plants such as algae can also take calcium carbonate out of solutions, and it collects on them. Shells, shell fragments and plant remains accumulate on the sea floor, forming limy deposits that may later form organic limestone.

Limestones also originate from chemical reactions. When the temperature and chemical composition of the water permit, calcium carbonate will precipitate as countless minute grains of calcite. This may form a limy mud that is later converted to limestone. Many limestones also contain shell or plant fragments.

Limonite. An amorphous hydrated iron oxide, brown to yellow in color. *Goethite forms under similar conditions and may have a similar appearance, but is crystalline and has a definite chemical composition. Most amorphous-appearing hydrated iron oxide formerly called limonite is now known to be goethite. However, the name limonite is retained for the truly amorphous substance and is used to refer to natural brown iron oxides of uncertain identity.

Lineation. A rock feature produced by the parallel alignment of any linear constituent of the rock. It may result from the parallelism of rodlike (prismatic) crystals or the alignment of the axes of many small folds or from other structural and textural features. Like *foliation, lineation is divided into two major categories based on the time at which the linear feature appeared. Primary lineation forms in conjunction with the development of the rock in which it is found. Any lineation formed after the original development of the rock is considered of secondary origin. A rock may show more than one form of lineation if it has been subjected to repeated lineation-producing processes.

Primary lineation, when represented in *igneous rock by parallel arrangement of rodlike mineral crystals, results from the direction of flow of liquid components within molten bodies of rock. Secondary lineation may result in the formation of such structural features as *boudinage, *slickensides, *mullion structure or the parallelism of fold axes. Other secondary lineations may result from the orientation of deformed pebbles in conglomerates or it may result from the intersection of planar features such as *bedding, *jointing, and *cleavage.

Lineation, like other structural features, can help geologists reconstruct the history of the forces that formed the crustal structure of an area. (J.W.S.)

Lithification. The process whereby unconsolidated *sediments are converted into *sedimentary rock. Lithification is accomplished by cementation of newly deposited, loose, individual rock-forming particles and may be accompanied by other processes such as compression, compaction, crystallization, recrystallization and desiccation. Loose sand grains, for example, may undergo lithification and become bonded together to form the sedimentary rock known as sandstone.

Lithology. The description and study of rocks, especially those of sedimentary origin, as seen in a hand specimen or as exposed on the earth's surface. Lithologic descriptions are made on the basis of distinctive characteristics such as grain size, rock type, mineralogic composition, color and structure.

Lithosphere. The outermost portion of the earth. It consists of the *crust of the earth and the cooler rigid part of the upper *mantle. The term can also be used synonymously with oxysphere, that is, the zone of the earth in which rock consists of at least 60 percent oxygen.

Lithostatic Pressure (or Geostatic Pressure). The pressure exerted on a segment of material within the earth's *crust by a vertical column of material lying above it. At a depth of 1.6 km (1 mile) within a crust consisting entirely of granite, the lithostatic pressure on a square centimeter of material is greater than 350 kilograms (over three tons per square inch).

Littoral Zone. The portion of a coast which lies between the spring high tide and spring low tide. See Intertidal Zone.

Lodestone. A variety of the mineral *magnetite. It is so strongly magnetic that it acts as a natural magnet.

Loess. An unconsolidated, generally unstratified, silt-size windborne deposit that may contain minor amounts of fine sand and clay. Loess is usually composed of unweathered angular particles of *quartz, *feldspar and other rock-forming minerals. It is generally yellow or buff-colored, although where unweathered it may be gray. In some places, such as the Loess Plateau of China, it is a few hundred meters thick. Loess may form vertical cliffs, which may be due to the presence of clay and vertical tubes resulting from root casts, the interlocking of the angular grains and cementation from the precipitation of calcium carbonate.

Geologists now believe that most loess is wind-deposited because (1) it blankets the topography irrespective of elevation, (2) it contains the remains of various terrestrial organisms, (3) some of the shells that it contains are too fragile to have been transported by water; they must have lived on the surface and have been buried by wind-blown dust, (4) it has much the same grain-size distribution as wind-blown dust, (5) in places it merges into coarser deposits that are definitely eolian and (6) when wetted and loaded it shrinks, suggesting that it was laid down dry.

Loess is being formed today in two different environments: silt deflated from deserts and deposited to the leeward (desert loess) and silt deflated from sediments deposited by *meltwater streams from modern glaciers. These sediments containing silt are laid down during the melting season. Later, during the cold months, when meltwater is absent and these deposits are dry because of the absence of meltwater, silt is deflated. Considerable thicknesses of loess can be built up where this sequence is repeated year after year.

Deposits of loess and loesslike material cover about one-tenth of all land areas. Some of the best agricultural soils are

Lodestone

1 *A form of magnetite, this mineral is a natural magnet and will deflect a compass needle.*

Loess

2 *A wind-deposited layer which in places may be 30 m (100 feet) thick, covers much of the central United States. The exposed bank above is at Wynne, Ark.*

Longshore Current

3 *Because waves usually strike the shore at an angle, there is a net longshore transport of sediment caused by upwash and backwash along the beach margin.*

4 *Flowing parallel to the shore, currents deposited a sandbar at the south tip of Nauset Beach, Cape Cod, Mass.*

developed on loess. It covers large areas in midwestern United States, Washington, Idaho and Alaska; Europe; northern China; and elsewhere. In the United States it is interbedded with glacial deposits and is also found at the surface. The age of these deposits is shown by their stratigraphic position, by the weathering profiles developed on them and by the fact that in some places they rest on unweathered *till. (R.L.N.)

Longitudinal Profile. In hydrology, a sketch of the *gradient of a stream from head to mouth. Since a stream is steeper at the head and gentler at the mouth, the gradient is generally concave upward. However, the longitudinal profile of an aggrading stream may be convex. The profile may be smooth or irregular depending on the presence of *knickpoints. See Graded Stream; Profile of Equilibrium.

Longshore Current. A shallow movement of water parallel to the shore and generally limited to the surf zone. It is caused by waves approaching the beach at an angle. As the waves reach shallow water they are bent (refracted) so as to become parallel to the shore. As this happens, some of the wave energy is transmitted in a direction parallel to the shore, thus setting up a current. Such longshore currents may exceed a velocity of 1 m per second during storms. The angle of approach and the size of the waves are important in determining the velocity of the current.

Lopolith. A large bowl-shaped body originating from the intrusion of molten rock into the earth's *crust. Lopoliths cover large regions, their diameters ranging from tens to hundreds of kilometers and attaining thicknesses of several kilometers. The average ratio of thickness to diameter is between 1:10 and 1:20. The molten rock that forms the lopolith is intruded into the crust parallel to structural features such as layered rocks. The exact mechanism of the formation of lopoliths — or even whether they actually form within the crust or on the surface—is still unclear. The Sudbury Lopolith in Canada is considered a classic example although it has also been interpreted by some geologists as a crater-like feature produced by the collision of a large meteor with the earth's surface. Another example, the Bushveld Lopolith in the Republic of South Africa, is also well known. (J.W.S.)

Low Velocity Zone (LVZ). The zone in the earth's upper mantle in which seismic waves begin a gradual decrease in speed. See Mantle.

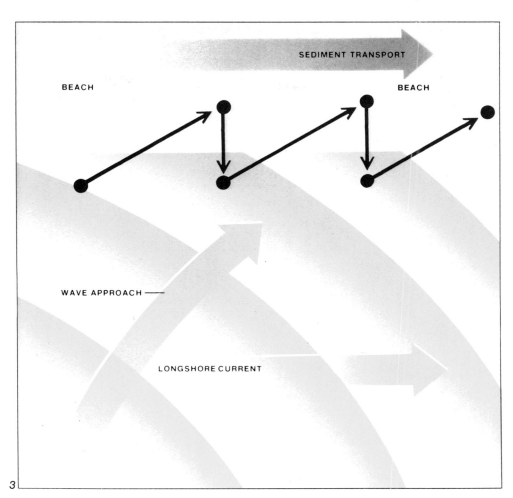

3

4

	Sun	Earth	Moon
Mass	1.99×10^{30} kg	5.976×10^{27} gm	7.35×10^{25} gm
Radius — Equatorial		6378.16 km	1737.7 km
— Polar	avg. 696,000 km	6356.18 km	1736.4 km
Surface gravity	274 m/sec²	9.78 m/sec²	1.62 m/sec²
Average density	1.41	5.52	3.34
Period of Rotation	25.4 days equator 33 days 75°N or X	24 hours	28 days

Lunar Geology

1 The moon's surface is dotted with craters and dark, smooth mare or "seas," the result of meteor impact and a series of volcanic eruptions.

2 Physical constants of the sun, earth and moon.

3 Albedo, or reflectivity, differences are clear in this photograph. The mare are dark, the highlands are light and the ejecta rays from craters are very light.

Luna Program. The Soviet unmanned series of lunar exploration vehicles, known by the prefix "Luna" followed by a number. Luna 1 was launched in January, 1959; the vehicle, which missed the moon and went into orbit around the sun, carried a small number of geophysical experiments on its flight. Luna 2, launched in September, 1959, crashed into the moon; it was the first man-made object to strike another planet and it produced the first evidence of the moon's weak magnetic field. In October, 1959, Luna 3 transmitted a series of low-resolution television pictures of the hidden face of the moon. These showed that the moon's farside, like its frontside, is cratered, but that the farside is almost free of dark *mare areas except for two regions named Mare Moscorviense and Mare Tsiokovsky, the latter named for a leading Soviet space scientist. Lunas 4, 5, 6, 7 and 8 were unsuccessful launches that produced few scientific results. Luna 9 (January, 1966) was the first space vehicle to survive landing on another planet. It produced a series of panoramic television pictures of the lunar surface comparable with those of the early U.S. Surveyor missions launched at the same time. The pictures indicated the rocky, barren nature of the lunar surface. In March, 1966, Luna 10, a lunar orbiting mission, carried a meteorite particle recorder, a gamma spectrometer, a magnetometer and other scientific equipment. The mission produced a great deal of valuable data about the moon and its reaction to the solar wind and other phenomena of interplanetary space. The gamma ray emissions of lunar rocks were correctly recognized as being like terrestrial basalt, and the extremely weak magnetism of lunar rocks was confirmed. Luna 11 (August, 1966) was unsuccessful. Luna 12, launched in October, 1966, resembled the U.S. Lunar Orbiter series in that it took pictures of the lunar surface and transmitted them back to earth. Luna 13 (December, 1966) had a more sophisticated landing apparatus than Luna 9. In addition to taking pictures, it measured the relatively high bearing strength of lunar soil as well as its low density. Luna 14 (April, 1968) was an orbital mission like Luna 10, but none of its results has been released. Luna 15 (July, 1969) flew to the moon at the same time as Apollo 11; the exact mission of the Soviet flight was never made public after the craft crashed in an attempted landing. The trajectory and landing site chosen suggest that it probably contained a roving vehicle and was not intended to return lunar soil to earth. Luna 16, launched in September, 1970, did return about 3 ounces of lunar soil to earth from an area not sampled by the Apollo series. In November, 1970, Luna 17 carried a sophisticated roving vehicle called Lunakhod, which took pictures of the surface, analyzed the surface with an X-ray spectrometer and carried a laser reflector for precise measurement of the earth-moon distance and location of the Lunakhod. Little data on this mission has been released. Luna 18, launched in September, 1971, was an unsuccessful landing mission. Luna 19, launched the same month, was a lunar orbiting vehicle that made geophysical measurements and took photographs, few of which have been released. Luna 20 (February, 1972) was a successful sample-return mission like Luna 16, and Luna 21, launched in January, 1973, was a second successful Lunakhod. (C.H.S. and J.L.W.)

Lunar Geology. The moon, the earth's only natural satellite, can be seen on most clear nights as a white disc in the sky. Like the sun, it rises in the east and sets in the west. The scientific study of the moon may be divided into three periods. First is the astronomical study that deciphered the moon's physical constants and motions. Second came the photogeology phase. Based on photographs of the moon taken through telescopes and those relayed to earth by Ranger, Surveyor and Lunar Orbiter spacecraft, lunar morphology was studied. Finally, *Apollo and *Luna missions brought back samples of moon rock and dust as well as many new lunar photographs.

The earth, moon and sun all lie in a plane (along with the other planets) known as the ecliptic. The earth revolves about the sun in 365.26 days (one year), and rotates about its axis in 23 hours 56.07 minutes (one sidereal day) relative to fixed stars and in 24 hours relative to the sun. The moon in turn revolves about the earth, and rotates on its axis, doing both in 27 days 7 hours 43 minutes relative to the earth and in 29 days 12 hours 44 minutes relative to fixed stars. This coupling of the period of the moon's rotation and revolution causes one side of the moon to face toward the earth at all times. The moon's geography is thus divided into nearside and farside. The average earth-moon distance is 384,400 km (238,900 miles).

Photographs of the moon reveal two basic types of terrains, one dark and the other light. The dark regions (called *"mare" [plural "maria"] after the Latin word for sea) are smooth compared to the hilly light regions (called "highlands," or "terra" after the Latin word for land). The main lunar topographic division into mare and highlands is analogous to the division of the earth into ocean basins and conti-

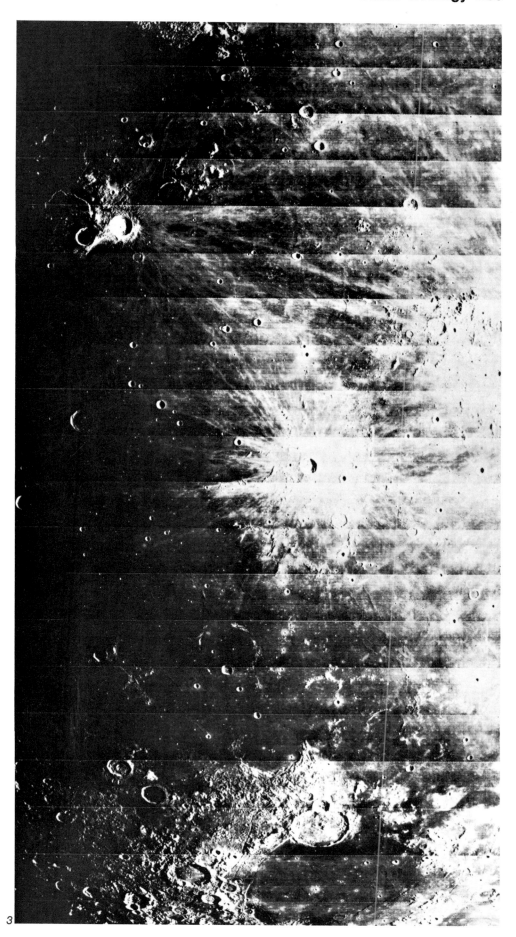

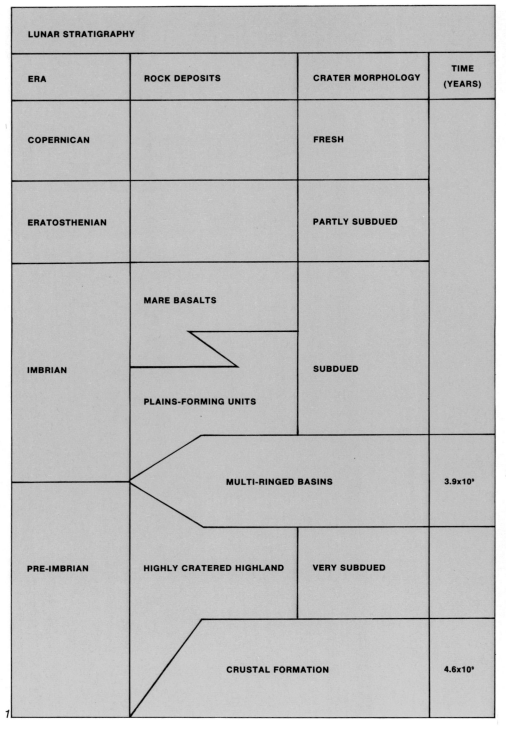

LUNAR STRATIGRAPHY			
ERA	ROCK DEPOSITS	CRATER MORPHOLOGY	TIME (YEARS)
COPERNICAN		FRESH	
ERATOSTHENIAN		PARTLY SUBDUED	
IMBRIAN	MARE BASALTS / PLAINS-FORMING UNITS	SUBDUED	
		MULTI-RINGED BASINS	3.9x10⁹
PRE-IMBRIAN	HIGHLY CRATERED HIGHLAND	VERY SUBDUED	
		CRUSTAL FORMATION	4.6x10⁹

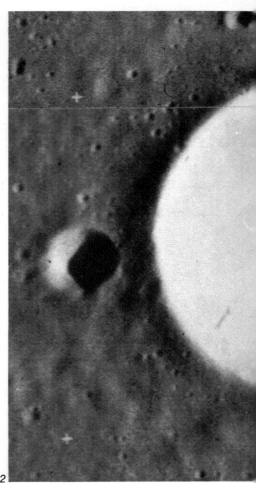

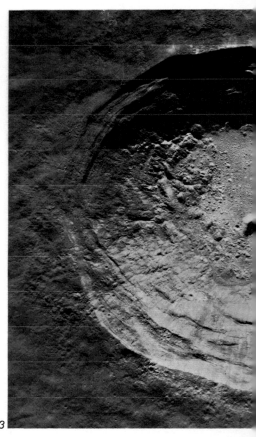

Lunar Geology

1 A lunar time scale shows general rock units and typical crater morphology. Only two dates have been well established.

2 These photographs of lunar craters taken from orbiting spacecraft illustrate the change in crater shape as a function of size. Lansberg D, a small crater with a diameter of about 10 km (6 miles), displays a smooth, bowl-shaped form and ejecta deposits.

3 Aristarchus, a medium-sized crater with a diameter of about 40 km (25 miles), displays a flat bottom, central uplift, slump terraces on the inside walls and ejecta deposits.

4 Mare Orientale, a large crater with a diameter of several hundred kilometers, displays several sets of rings, mare basalt filling and ejecta deposits.

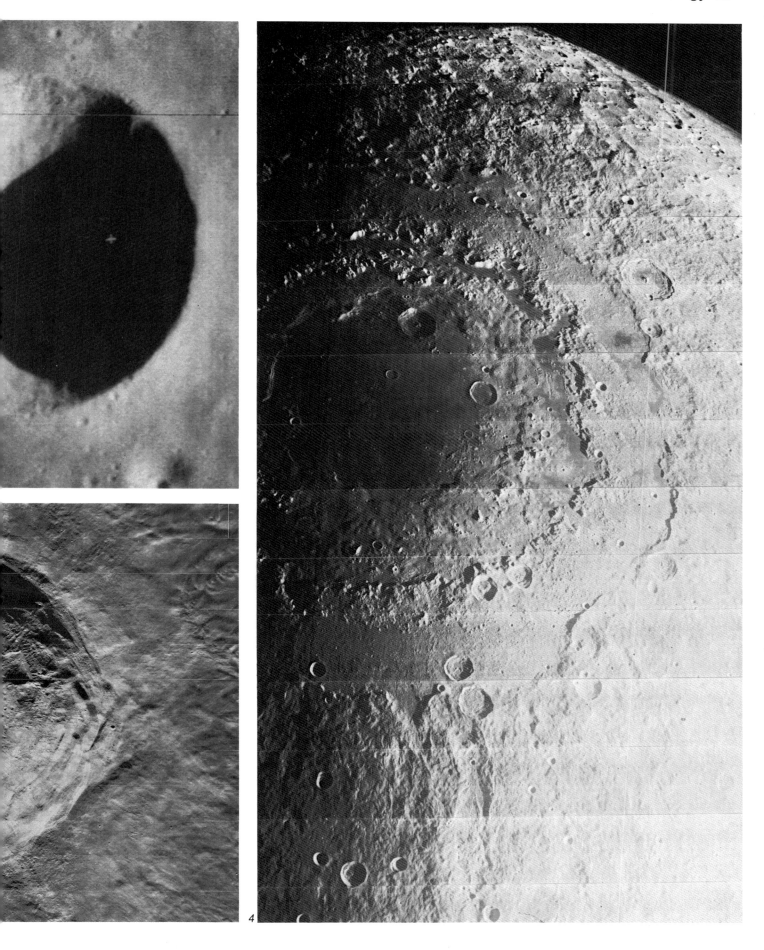

4

nents. Both the lunar mare and the earth's oceans are low with respect to the highlands and continents. The topographic and geologic features of the mare and ocean basins are also young with respect to the highlands and continents. The circular forms on the mare surfaces are craters. Craters are also present in the highlands, but in greater abundance; in fact, the hills of the highlands are due to overlapping craters of many sizes. Thus a knowledge of the formation of craters, gained from studying the earth's craters, is critical to understanding the moon's geology.

There are three types of natural craters on the earth: volcanic, meteorite impact, and subsidence. All three are undoubtedly present on the moon. The question of which process is responsible for the majority of lunar craters was solved by comparing the morphology of lunar craters and the rocks near them with the morphology and rocks of three types of terrestrial craters. Virtually all lunar craters are the result of meteorite impact.

The changes in morphology with crater size are characteristic of lunar craters. Tiny lunar craters, less than a millimeter across (known as zap pits), are found on lunar rocks brought back from the moon by the Apollo project. These consist of a central pit that is glass-lined and a surrounding spall (splinter) zone that is about two pit-diameters across. There is no earth equivalent of these microcraters. Since man cannot send a small projectile fast enough to produce a glass-lined zap pit, even though velocities of up to 7 km per second have been obtained, the impacting particles that produced lunar zap pits must have had velocities of tens of kilometers per second.

Lunar craters with diameters from several centimeters to about 15 to 20 km (10 to 12 miles) are bowl-shaped, with the bottom of the crater lower than the surrounding ground level. The crater rim is raised above the surrounding level and consists of fine-grained material and large blocks. Outward from the rim for several crater diameters is series of tangential and then radial ridges. The material underlying these ridges once filled the crater; it was ejected during the formation of the crater and is known as the continuous ejecta blanket. The material in this blanket came from a debris-laden cloud of gases generated during the impact. (Terrestrial impact craters and explosive volcanoes have similar features.) Loops of small secondary craters extend farther out from the main crater. Narrow, linear deposits of light dust and blocks stretch far in all directions; these make craters look like sunbursts and are known as *rays.

Medium-size lunar craters, 20 to 100 km (12 to 62 miles) in diameter, display the same extra-crater features as the smaller craters, but there are major differences in the crater itself. They are not bowl-shaped but have a flat floor. Slumping and terraces are common on the crater walls, and there is a central peak — a small cluster of hills in the center of the crater floor. Like the medium-size craters, large lunar craters more than 100 km (62 miles) across also have extra-crater deposits, flat floors, slumped and terraced walls, and central peaks. But the basin has many rings, and instead of one crater wall, there are several bands of mountains around the basin.

Evidently most lunar craters are not similar to terrestrial volcanic craters or to subsidence craters such as *sinkholes and *calderas, but are like terrestrial-impact craters such as Meteor Crater in Arizona, Gosses Bluff in Australia and Ries Crater in Germany.

Stratigraphy (that is, the study of the layers of rock) on the moon is unlike terrestrial stratigraphy because the moon lacks a series of fossil-bearing sedimentary rocks such as permit us to correlate terrestrial rocks. Lunar stratigraphy is based on photogeologic mapping of material ejected from craters, the interpretation of crater age and the abundance of craters on specific surfaces. The principle of superposition holds, that is, the overlying ejected material and the associated crater is younger than the underlying material and its crater. Relative crater age is judged by the sharpness of the crater and the structure of the ejecta. The abundance of craters is an indication of relative age: the more craters present, the older the surface. Crater abundances range from mare basins, which have few craters, to highlands, which are saturated with craters. A surface is saturated when there are so many craters present that any new craters will destroy old ones of the same size.

Much information has been gathered on the lunar time scale. The oldest rocks are of the Pre-Imbrian system, that is, older than the Mare Imbrium basin. These form the high-albedo, highly cratered highlands, a region saturated with craters as large as 50 km across. About 75 percent of the nearside of the moon is underlain with Pre-Imbrian material. The craters are very subdued and have no rays, secondary craters or ejecta deposits; the rims are low, rounded and cut by later craters; the floors are shallow. Most Pre-Imbrian craters are so subdued and so overlain by younger craters that they are not obviously craters.

Pre-Imbrian time ended with the formation of the multiringed circular basins. Mare Orientale is the last of the multiringed

3

Lunar Geology

1 *Photographs of lunar craters taken from the Lunar Orbiter spacecraft illustrate the erosion of lunar craters. Young craters are fresh in appearance.*

2 *Old craters are degraded with a rounded appearance and have been impacted by later craters.*

3 *These craters are intermediate in age.*

basins and was immediately preceded by Mare Imbrium. Based on radiometric dating of rocks formed by the Imbrian impact and brought back to earth in the *Apollo program, these events took place about 3.9 to 4 billion years ago. Large areas of the moon are covered with ejecta from these events; the Fra Mauro formation sampled by the Apollo 14 mission consists of such ejecta.

There are two widespread types of deposits of Imbrian age. The older is the "plains-forming" material such as the Cayley Formation sampled by the Apollo 16 mission. "Plains" material is intermediate in *albedo and in crater density, being darker and less cratered than the Pre-Imbrian highlands and lighter and more cratered than the basalts that fill the mare basins. "Plains" units are more or less level deposits of reworked ejecta material that fills craters in the highlands and are found along the inner margins of some of the multiringed basins. "Plains" material, which makes up about 10 percent of the moon's nearside, tends to be found near the multiringed basins. Based on radiometric dating of rocks on a "plains" unit, the Early Imbrian period ranged from 3.85 to 4.05 billion years ago.

The low-albedo, lightly cratered mare filling deposited in the later part of the Imbrian system is a layered series of basalts making up about 15 percent of the moon's nearside. Mare basalts fill most of the nearside multiringed basins and also flood large low-lying areas. Radiometric dating for the Late Imbrian period ranges from 3.15 to 3.85 billion years.

Craters of the Imbrian period are subdued but, unlike Pre-Imbrian craters, are clearly recognizable. Archimedes and Gassendi are typical Imbrian craters. Ejecta deposits and secondary craters are subdued; crater rims are low, broken and rounded; floors are shallow or filled with later material. No rays have been recognized.

The Eratosthenian period marks the beginning of greatly reduced rates of cratering and the virtual cessation of volcanism. The period has left partly subdued craters with accompanying ejecta, and minor features that are interpreted as volcanic. The craters (such as Eratosthenes and Aristoteles) display well-developed, but partly subdued, ejecta deposits and secondary craters. No rays have been recognized. Crater rims are ragged, and floors are not filled with later material.

The time when the Eratosthenian period ended and the Copernican period started is not known. However, material from a Copernican ray sampled by the Apollo 12 mission suggests that the ray formed about one billion years ago. The Copernican-age craters (such as Copernicus, Tyco and Kepler) are fresh and have well-defined ejecta, secondary craters and rays. The crater rims are sharp; the floors are deep and are not filled with deposits or mare basalts.

The lunar highlands have a high albedo and are saturated with large impact craters. The rocks forming the surface of the highlands have all experienced multiple impact events. The following data are based on highland samples brought back by Apollo and Luna missions:

1. The highlands are old, probably between 3.8 and 4.2 billion years.
2. The average highlands composition is equivalent to a terrestrial rock of *anorthositic *gabbro composition that is depleted in alkali elements.
3. Eighty-five percent of the highland rocks are *breccias, many of which have been shocked, crushed, metamorphosed, and/or melted or partially melted by impact events.
4. The breccias range in composition from gabbroic anorthosite to a basalt that is enriched in trace elements and contains low calcium pyroxene.
5. Fourteen percent of the highland rocks are anorthosites.
6. One percent of the highland rocks are magnesium-rich cumulates.
7. The anorthosites and cumulates are coarse-grained, low in trace elements and have been shocked, crushed and annealed by impact events.

Most highland rocks are breccias; that is, they are rocks with a fine-grained matrix containing larger clasts. (Clasts are more or less rounded mineral, glass and lithic fragments.) The matrix of breccias consists of small fragments similar to the clasts, with two exceptions. First, the smaller particles contain a lower percentage of mineral and lithic fragments and a higher percentage of glass. Second, the smaller particles are more angular. Breccias are formed by annealing at about 700°C (1300°F). Breccias heated to higher temperatures, 800° to 1000°C (1475° to 1800°F), are metamorphosed, and those heated to still higher temperatures are partly melted. Breccias heated by impact to above 1100°C (2000°F) are partly or totally melted. The Apollo 16 breccias from the Cayley Formation are partly melted. When the partial melt is separated from the residue, a basalt rich in trace elements is formed. This rock type is called KREEP, an acronym for potassium (K), rare earth elements (REE) and phosphorus (P).

Highland rocks that are not breccias are cataclastic anorthosites, that is, coarse-grained rocks (with crystals of several cen- timeters) that have gone through repeated cycles of crushing and annealing.

Chemically the lunar highlands reveal a diverse series of compositions that range from basalt to anorthosite. A number of magnesium-rich cumulate rocks have been found in the highlands that do not fall into the above series, but the amount of these is small. Nevertheless, these cumulates, along with the cataclastic anorthosites and gabbros that fit in the main highland chemical series, suggest that plutonic processes have some role, albeit minor, in the evolution of the lunar highlands.

Because only about 50 different highland rocks have been analyzed to date, it is difficult to calculate the average composition of the lunar highlands. One technique used to estimate the average composition is the orbital X ray fluorescence experiment that analyzed the surface of the moon for aluminum, silicon and magnesium from lunar orbit during the Apollo 15 and 16 missions. The results of the experiment show that the lunar highlands possess a high aluminum-silicon ratio and a low magnesium-silicon ratio and that both ratios are approximately equal to those in anorthositic gabbro.

A second method of calculating the average lunar highland composition is by statistically analyzing the glass particles in lunar soil. This method assumes that glass particles found in lunar soils result from the total melting of preexisting lunar rock and soil. Some soils from each Apollo and Luna missions have been studied with this technique, and a preferred composition of anorthositic gabbro has been found in each soil. These data, taken with the orbital X ray fluorescence data and the rock analyses, suggest that the average lunar highlands composition must be in the center of the series of rock analyses, near anorthositic gabbro. These data also suggest that the lunar highlands originally had a composition near anorthositic gabbro, but with slight differences from area to area. During a period of meteorite impact, the original crust was crushed, partly melted, annealed and recombined into the breccias and cataclastic anorthosites found today.

Radiometric datings of lunar highland rocks yield ages between 4.2 and 3.8 billion years. This means that the intense impacting of the highlands was essentially complete in the first 700 million years of lunar history.

The lunar maria consist of iron-rich basalts and breccias of similar composition. Mare material is found in all nearside multi-ringed circular basins and in large areas adjacent to these basins. Thus Mare

Imbrium is in a multi-ringed basin, and the mare Oceanus Procellarum is adjacent, but not in a circular basin. A similar relation holds between Mare Serenitatus and Mare Tranquillitatus. The mare basalts display a wide range of volcanic textures and chemically form several well-defined groups at each landing site, which nevertheless contrast strongly with highland rocks, being richer in FeO and TiO_2 and poorer in Al_2O_3.

High-temperature, high-pressure experiments performed on mare basalts indicate that mare lavas are derived from a pyroxenitic source rock by partial melting at a depth between 150 and 400 km (100 and 250 miles). Most chemical groups of mare basalts have shown some evidence of near-surface fractionation. The Apollo 12 and, to some extent, the Apollo 15 basalts display evidence of olivine accumulation, whereas the Apollo 11 and 17 basalts underwent olivine, spinel and ilmenite accumulation.

The repeated meteoritic impact has crushed and ground the surface of the moon, resulting in a loose material known as the lunar regolith. Over large areas of the mare the regolith is from 3 to 10 m (10 to 33 feet) thick, whereas in heavily impacted parts of the highlands the regolith may be as thick as 25 m (83 feet). Below the regolith the rocks are still broken up but not ground up. It is well to note that small meteorites are as important as large ones. Although small impacts do not do as much work as large ones individually, there are a far greater number of these smaller impacts, so the net effect of small impacts is large. Phenomena directly attributed to meteorite impact include: (1) glass beads in the regolith, (2) glass splashes on rock surfaces, (3) rounding of rocks on the lunar surface and (4) microcraters (zap pits) on rock surfaces.

The regolith consists of ground-up rock, mineral and glass particles that are angular to subrounded. About 5 percent of the material disturbed by an impact gets melted and quenched to a glass. If the melt is quenched while in flight, a glass bead is formed. If the glass is quenched when it hits a rock surface, a glass splash is formed. If the glass is quenched when it strikes regolith, an *agglutinate (a glass-cemented aggregate) is formed. The percentage of agglutinates is used as a measure of regolith maturity. Freshly ground regolith will have no agglutinates, whereas mature regolith may have up to 50 percent agglutinates.

Repeated meteorite impact causes a phenomenon known as "gardening," that is, a slow overturning of the regolith. An impact will create a crater about 1000 times

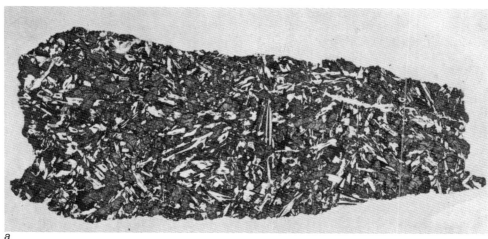

a

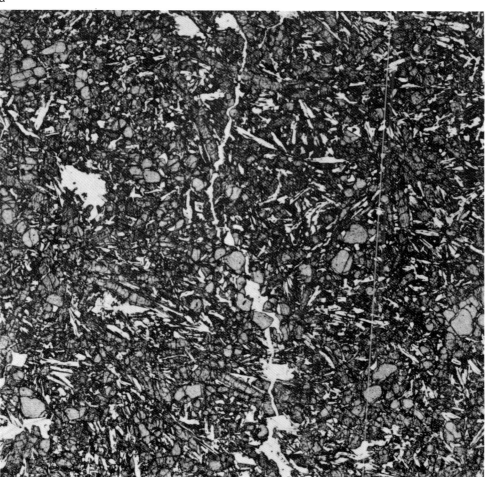

b

Lunar Geology.
*Photographs (a and b) of thin sections
of mare basalts illustrate the range in
textures observed.*

the volume of the impacting body. The excavated material is deposited in a thin layer over an area about four times the crater area. Repeated impacts turn over the regolith, and each impact produces a distinct layer. Sequences of these layers have been observed in the core tubes taken during the Apollo missions; for example, one Apollo 15 core is 9.5 m long and has over 50 layers representing the last 600 million years of lunar history.

Several lines of evidence yield an understanding of the moon's interior. Astronomical observations give the moon's average density as 3.34 grams per cubic centimeter (gm/cc). If we add to these data the densities of mare basalts (about 3.3 gm/cc) and highland rocks (about 3.1 gm/cc), it would seem that the moon cannot have any major internal layers of different density and, hence different composition. At the most there can be a metal core of about 5 percent of the moon's mass or 20 percent of the moon's radius.

The upper 25 km (16 miles) of the moon's interior are interpreted as fractured mare basalts. The region from 25 to 65 km (16 to 40 miles) is interpreted as being anorthositic crust, and the region below 65 km is interpreted as being pyroxenitic material (which is probably the source of the mare basalts). Recent seismic data suggest that the moon may have a liquid or partly liquid core with a radius of about 20 percent of the moon's radius.

Moonquakes are rare and of low magnitude — less than 2 on the Richter scale. They occur at depths of between 800 and 1000 km (500 and 620 miles), that is, deeper than any terrestrial earthquake, suggesting that the moon has a thick, rigid crust. The timing of the moonquakes correlates with lunar tides, suggesting that, unlike the earth, the moon does not have an active global tectonic system.

Lunar gravity on a global scale has been calculated by careful tracking of spacecraft in lunar orbit. The most striking feature of the lunar gravity field is the excess mass concentrations (mascons) of about 800 kilograms (1760 pounds) per square centimeter over all circular mare basins that are more than 200 km (125 miles) in diameter. These *mascons are best explained by disc-shaped bodies beneath but near the surface of the basins. Since the mare rocks are over 3 billion years old, the moon must have supported the mascons out of isostatic equilibrium for billions of years, suggesting that the lunar crust must have enormous strength or a very high viscosity.

Lunar gravity observations have shown that the center of mass of the moon is about 2 km (1.3 miles) closer to the earth than the moon's physical center. Varying crustal thickness has been suggested as a cause of this phenomenon.

Heat flow determinations have been made on the lunar surface at the Apollo 15 and Apollo 17 landing sites. Both determinations are about 30 erg/cm²/sec, about half the earth's average heat flow, but higher than predicted. The distribution of radioactive elements (potassium, uranium, and thorium) needed to produce these fluxes is not known, but it does indicate that the interior of the moon has some heat today even though there has been little surface volcanism for the last 3 billion years.

Magnetic measurements by lunar orbiting satellites showed that at a distance of a few hundred kilometers the moon has no planet-wide magnetic field. However, magnetometers on the lunar surface detected fields as high as hundreds of gammas, and lunar samples from all missions have natural remanent magnetism (*NRM). The local surface fields are due to the NRM in the rocks, but if there is no lunar field, how did the rock acquire their NRM? Apparently the answer is that when the rocks formed, there must have been a lunar field of 100 to 1000 gammas (the earth's field is about 50,000 gammas). Perhaps the lunar core acted as a dynamo billions of years ago.

The present hypotheses on the origin of the moon are: (1) it was formed by fission from the earth; (2) it was formed elsewhere in the solar system and was captured by the earth; (3) it was formed as a binary planet together with the earth; and (4) it was formed from a sediment ring of earth-orbiting planetesimals.

The fission theory was first suggested by George Darwin (a son of Charles Darwin) in 1878. However, returned lunar samples, although chemically similar to terrestrial rocks, differ to such an extent in detail that the fission origin has lost favor. In addition, if the moon were formed by fission its orbit would have to lie in or near the earth's equatorial plane, and the moon's orbit actually varies between 18 and 28 degrees from that plane.

The capture theory has enjoyed popularity partly because it raises no chemical problems. But it does pose a celestial mechanics problem. It requires that the moon never be closer to the earth than 2.89 earth radii (the Roche limit) or else the moon will spontaneously break up, and that the capture must be prior to 3.5 billion years. Many detailed models have been proposed, including capturing the moon in retrograde orbit. The main problem with all capture theories is that the probability of capture is very low.

The twin-planet hypothesis suffers from the same problems as the fission theory: the chemical differences between the moon and earth and the inclination of the moon's orbit from the earth's equatorial plane. In addition, the overall volatile-element content of the moon is lower than earth's. This is very difficult to explain in a twin-planet hypothesis.

The theory that the moon was formed by accretion of earth-orbiting planetesimals (similar to the rings of Saturn but more massive) has the same chemical problems as the fission and twin-planet theories. However, complex schemes have been proposed to account for the chemical disparity.

It is, in short, considerably easier to unravel the evolution of the moon than to determine its origin.(J. L. W. and C. H. S.)

Lunar Orbiter. A series of unmanned satellites designed to take pictures of the lunar surface. Five were launched between August, 1966, and August, 1967. An orbiter consists of an aluminum frame built around several cameras. The cameras take photographs, process the film, scan the resulting pictures with a television scanner and then transmit the pictures to earth. Because the orbiters were placed in lunar polar orbits, they could photograph all of the lunar surface. The result was complete coverage of the moon with the exception of a small area on the farside near the lunar south pole. The area on the moon photographed by the Apollo program was much more restricted.

M

Macrofossil (or Megafossil). A fossil that in its typical form is readily visible to the naked eye. These include such invertebrate forms as fossil sponges, corals, snails and clams, and vertebrate remains.

Magma. Molten or fluid material generated from rock deep within the earth that may force its way upward into the crust (as plutonic rock) or onto the surface (as *lava.) When cooled and solidified, magmas become a great variety of *igneous rocks. See Magmatic Deposit; Magmatic Differentiation; Magmatic Stoping.

Magmatic Deposit. A mineral deposit that has formed as an integral part of an *igneous rock mass. It indicates that the minerals forming the deposit crystallized from a molten *magma. Most magmatic deposits are made up of minerals more commonly found as accessories, that is, as isolated grains scattered through large volumes of rock. Important among these minerals are magnetite, ilmenite, chromite, pyrrhotite, pentlandite, corundum, platinum and diamond. It is through some process such as *magmatic differentiation that the minerals are concentrated in the form of dikes, layers or irregular masses to form an ore body. In some magmatic deposits, such as *peridotites carrying disseminated diamond, the ore body is the whole rock mass itself. (C.S.H.)

Magmatic Differentiation. A process by which rocks of varying mineral and chemical composition derive from a common *magma. As a magma of basaltic composition cools, olivine and the plagioclase feldspar bytownite are the first to crystallize. If they remain essentially in place, they react with the surrounding melt, olivine forming pyroxene and bytownite forming labradorite. The resulting rock will be gabbro or basalt. However, if the early-formed crystals are removed, the remaining melt will crystallize to form a different rock. If other minerals are removed before complete crystallization, still other rock types develop. Thus, a large variety of igneous rocks may result from a common parent magma.

The accumulation of certain minerals to form large ore deposits is also attributed to magmatic differentiation. Minerals such as magnetite, ilmenite and chromite, formed early in the crystallization of a magma, may settle from the melt and accumulate in big masses. See Reaction Series. (C.S.H.)

Magmatic Stoping. A process that enables molten material called *magma to work its way upward into overlying solid crustal rock. As the magma comes in contact with the cooler rocks of the earth's *crust, rapid heating causes these rocks to expand rapidly, resulting in cracks from thermal stress. The mechanical stress produced by the magma pushing against the overlying rock is also responsible for much *fracturing. Magma then flows into the cracks and, with further fracturing, is able to surround and isolate large segments of the crustal rock. Because the blocks of solid rock are denser than the magma, they sink into the magmatic body and are partly or completely melted. As blocks are removed from the overlying crust, new surfaces are in turn exposed to the magma and subjected to thermal stress. As the process continues, the magma is able to eat, or stope, its way upward into the crust. This process is verified when workers encounter a block of foreign rock (called a *xenolith) within a granite being quarried for building stone.

The process of magmatic stoping can only work effectively when magma is not as dense as the rock it intrudes or when the rock undergoes compression. For instance, a magma of granitic composition has a density of about 10 percent less than its solidified counterpart. Therefore, blocks of granite will sink in a magma of the same composition, allowing exposure of new surfaces on which the intruding magma may work. However, in a situation where the magma is as dense as or more dense than the intruded solid rock, the fractured blocks will not sink into the magma and will shield the overlying rock from further fracturing. An example of this situation would be a magma of gabbro trying to invade a granite, where the magma has the same or greater density compared to the solid granite. When this process involves the removal of blocks of *country rock measured in meters or hundreds of meters it is called piecemeal stoping; when the blocks are kilometers in size it is called cauldron subsidence.(J.W.S.)

Magnesite. A mineral, magnesium carbonate, belonging to the calcite group of carbonates. It usually occurs in white, compact, earthy masses or in cleavable aggregates. The compact variety is associated with serpentine or peridotite and is formed when these rocks are altered by waters containing carbonic acid. The cleavable magnesite originates either as a product of *metamorphism associated with talc schists, or by the replacement of limestone rocks by magnesium-containing solutions.

Both types of magnesite are used as a source of magnesia. The magnesite $MgCO_3$ is heated until the carbon dioxide is driven off, and the resulting magnesia (MgO) is used for a variety of purposes, including the making of firebricks and refractory mortars. Major deposits of the sedimentary type are in Manchuria, the Soviet Union, Austria and in the United States in Washington and Nevada. The principal deposit of the compact variety is on the island of Euboea, Greece. It is also found in irregular masses in the Coast Ranges of California. See Mineral Properties. (C.S.H.)

Magnetic Anomaly (or Nondipole Field). A deviation from the uniform or idealized spherical *magnetic field of the earth. The earth generates a magnetic field similar to that displayed by a bar magnet when sprinkled with iron filings. Earth scientists record the direction and intensity of the field; values of equal intensity are connected by lines to form contours in much the same way that mapmakers connect points of equal elevation to produce topographic maps. Compared with a field produced by a homogeneous spherical magnet, the earth's field is very irregular, the contours forming a series of hills and valleys. On a small scale, anomalies are produced by compositional irregularities within the earth's crust in the form of concentrations of magnetic minerals. Geologists use magnetic prospecting techniques to locate magnetic anomalies in order to find ore deposits. (J.W.S.)

Magnetic Declination (or Magnetic Variation). The angle in degrees between true north and magnetic north.

Magnetic Field of Earth. A field of influence surrounding the earth in which magnetic forces operate. The earth is a huge magnet, producing a field not unlike that of a bar magnet. The lines of force representing the field start at its south geomagnetic pole, curve around the earth's surface and enter the north geomagnetic pole, where they pass through the earth to form a closed loop. That the earth has a magnetic field has been known for many centuries. William Gilbert, physician to Queen Elizabeth I, used a spherical magnet to study the form of the field. Results from his work agreed with what was then known of the earth's magnetic field, leading him to conclude that the earth itself is a huge magnet. Much has since been learned about the distribution and shape of the field. Although the field generally conforms to that of a bar magnet, there are many areas within it where the lines of force deviate from the idealized form. These are called *magnetic anomalies. The deviations are attributed to forces in the earth's crust and and deeper interior.

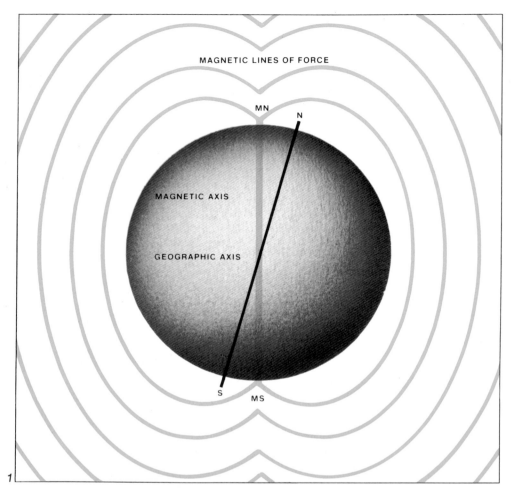

MAGNETIC LINES OF FORCE

MN

N

MAGNETIC AXIS

GEOGRAPHIC AXIS

S

MS

1

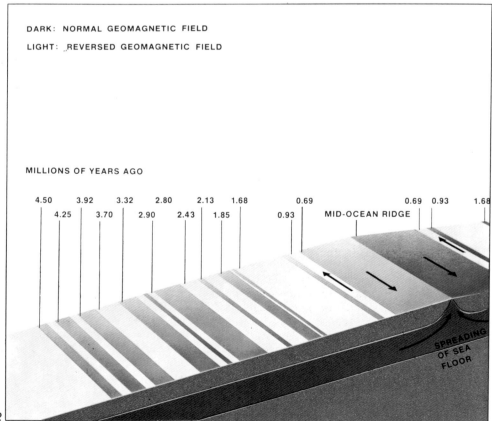

DARK: NORMAL GEOMAGNETIC FIELD

LIGHT: REVERSED GEOMAGNETIC FIELD

MILLIONS OF YEARS AGO

4.50   3.92   3.32   2.80   2.13   1.68   0.69   0.69   0.93   1.68

4.25   3.70   2.90   2.43   1.85   0.93   MID-OCEAN RIDGE

SPREADING
OF SEA
FLOOR

2

At any point in the earth's field the intensity and direction may be represented by a vector (a line drawn to represent magnitude and direction in three-dimensional space). At most localities the vector representing the total field is broken down into the following components: (1) a horizontal and vertical component representing the magnetic intensity of the field (measured in oersted units, or gammas, which equal 10^{-5} oersted). The total field of the earth generally measures approximately 0.3 to 0.7 oersted, depending upon location with respect to the earth's geographic coordinates; (2) the declination or angle by which the magnetic field deviates either east or west of true geographic north.

Records on the magnitude and direction of the total field in various places have been kept for many centuries. The information is generally recorded on maps in the form of lines that connect points of equal intensity (isogams). Such records indicate that the total field has been constantly drifting westward. A point representing zero declination has drifted along the Equator from 20° east longitude to 65° west longitude in approximately four centuries.

Changes in the earth's magnetic pattern are recorded daily (diurnal variation) and yearly or longer (secular variations). Daily variations amount to changes in declination of a few minutes of arc, while intensity fluctuates on the order of 10^{-4} oersteds. Such changes are periodic and constant and can be attributed to the effect of the sun and moon on electric currents in the earth's outer atmosphere. Other changes in the field occur on a daily basis, are irregular and affect the magnetic field to a great extent. Rapid changes of relatively large magnitude are called *magnetic storms and are believed to be caused by bursts of energy from the sun. They generally affect the magnetic field for several days, may affect radio communications and may produce auroras in polar regions. Secular variations such as the westward drift mentioned previously are atributed to causes originating within the earth. (J.W.S.)

Magnetic Iron Ore. A common synonym for *magnetite.

Magnetic Poles of Earth. In a truly physical sense the earth's south magnetic pole lies in the geographic north while the north magnetic pole is located in the southern region. The explanation of this interchange of poles is apparent when we consider that the earth's magnetic field is similar to that shown by the alignment of iron filings sprinkled around a small bar magnet; magnetic lines of force leave the north pole of a bar magnet and enter its south pole. Opposite poles attract, which explains why the North-seeking pole of a compass points north. However, by convention the magnetic pole in the Northern Hemisphere is referred to as the geomagnetic north pole or North-seeking pole.

The axis of the earth's magnetic field (a line representing the north-south magnetic poles and extending through the center of the earth) does not coincide with the earth's geographic poles (those on the rotational axis) but is inclined approximately 11.5° to the rotational axis. This inclination places the earth's geomagnetic north pole at 78.50° North latitude and 69° West longitude, while the south magnetic pole is located at 78.50° South latitude and 111° East longitude. Scientists have also shown that the geographic locations of the magnetic poles are not stationary but have changed (a movement called *polar wandering) throughout geologic time and have also exchanged positions, reversing the direction of the magnetic field. (J.W.S.)

Magnetic Reversal. A periodic shift, for reasons not completely understood, in which the magnetic polarity of the earth reverses itself so that the north and south magnetic poles are interchanged. The magnetic reversals are very useful in understanding *plate tectonics and *continental drift because they result in strips of normal and reversed magnetic fields produced in the earth's crust; matching these magnetic strips can demonstrate sea-floor spreading and movement along transform faults. A total of 130 magnetic reversals since the Cretaceous have been detected.

Magnetic Storm. A disturbance of the earth's *magnetic field associated with solar disturbances and caused by streams of electrically charged particles from the sun. Magnetic storms were first observed by ground-based *magnetometers. Satellites have made it possible to study the reaction of the earth's magnetosphere to solar particle radiation. *Auroras, the colorful, shimmering displays arising from radiation emitted by the energized gases of the upper atmosphere, are most intense and are observed at lower latitudes during magnetic storms. The more intense magnetic disturbances often seriously disrupt radio communication. Like the auroras, magnetic storms show periodicities related to the 27-day rotation period of the sun and the 11-year sunspot cycle. (M.F.H.)

Magnetic Susceptibility (or Susceptibility). The ratio of the strength to which a substance has been magnetized by an external magnetic force (of known strength) to the strength of the external magnetic force. Not all rocks or minerals have the same magnetic susceptibility. Substances such as iron that have positive susceptibilities are known as paramagnetic, whereas materials of negative susceptibility, for example, quartz or feldspar, are referred to as diamagnetic. The susceptibility of rocks is their most significant magnetic property and can be readily measured. Geophysicists have devised methods to detect changes in magnetic susceptibilities within the earth caused by concentrations of paramagnetic or diamagnetic substances. By plotting such changes on a map they are able to locate areas that may contain bodies of important ore minerals. (J.W.S.)

Magnetism. A physical phenomenon that results in forces of attraction or repulsion between atomic particles or between magnets or electric currents. The generation of a magnetic field within the earth allows certain natural substances (ferrimagnetic minerals) such as magnetite (generally known as lodestone) to form natural magnets. Other minerals capable of being magnetized, but to a lesser degree, are ulvospinel, hematite, limonite and pyrrhotite. Some metals found in nature such as iron, nickel, and cobalt make extremely good magnets and are capable of producing magnetic strengths thousands of times greater than the ferrimagnetic minerals. (J.W.S.)

Magnetite. A mineral, an iron oxide, one of the major ores or iron. Its most notable physical property — and the one for which it is named — is its magnetism. Small grains are attracted to a magnet, and one variety, lodestone, acts as a natural magnet. Octahedral crystals are common, but the great mass of magnetite occurs in granular massive aggregates.

Magnetite is a common accessory mineral, disseminated through many igneous rocks. In certain rocks it is segregated by *magmatic differentiation to form large ore bodies. Such deposits frequently contain *ilmenite. Magnetite also occurs in *metamorphic rocks as beds and lenses of immense size. The world's largest magnetite deposits at Kiruna and Gellivare, Sweden, are believed to have formed by magmatic segregation. Other important deposits are in Norway, Rumania and the Soviet Union. In the United States minable quantities are found in New York, Utah, California, New Jersey and Pennsylvania. In the low-grade iron ore called *taconite, mined in northern Minnesota, magnetite is the principal ore mineral. See Mineral Properties. (C.S.H.)

3

Magnetic Pole
1 Like a bar magnet, the earth produces lines of magnetic force. These lines leave the south magnetic pole and sweep around the earth's surface to enter the north magnetic pole.

Magnetic Reversal Correlation
2 Periodic changes in the direction of the earth's geomagnetic field are recorded in new sea floor created by spreading of fissures at the crest of mid-ocean ridges. The magnetized field reversals match on either side of the crest.

Magnetite
3 A cluster of partly intergrown crystals shows the characteristic octahedral form of this magnetic mineral.

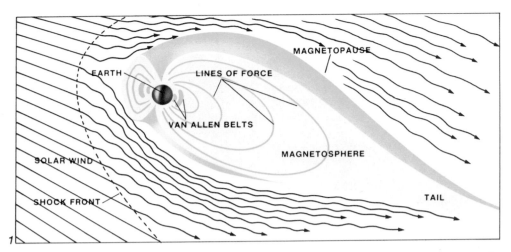

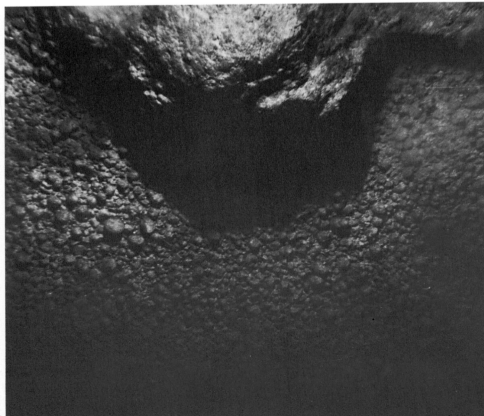

Magnetometer. An instrument that measures the magnitude of the earth's *magnetic field or of changes within the field. Magnetometers of many types have been developed to help in exploring for mineral resources. They are designed to be used on land or to be towed behind an airplane or a ship. Magnetometers help geoscientists locate *magnetic anomalies that may be produced as the result of large concentrations of iron-rich mineral deposits.

Magnetosphere. The outer region of the earth's *magnetic field. Its lower boundary is near the base of the F region (the top layer of the ionosphere, about 150 km above the surface of the earth), in which magnetic forces predominate over gravity in the motions of electrically charged gases (plasma). It has long been known that the earth is a relatively strong magnet with an extensive *magnetic field, but it is only recently that its magnetic field in space has been mapped with the aid of satellite observations. The magnetosphere is not symmetrical, as was once assumed, but is compressed on the side of the earth facing the sun and trails far outward on the opposite side. The compression on the sunward side and its extension in a comet-like tail on the side opposite the sun are caused by the impact of the solar wind on the magnetic field. The solar wind, a continuous stream of charged particles from the sun, had been predicted in 1951 by the German astrophysicist Ludwig Biermann, who suggested that such a flow of particles could explain why comet tails are observed to point away from the sun. A standing shock wave is created on the upstream side of the magnetospheric boundary by this stream of solar particles. The bounded region of space in which the magnetic field is confined by the solar wind plasma is often referred to as the earth's magnetic cavity.

The outer boundary of the magnetosphere, called the magnetopause, is compressed along the earth-sun line to a distance of about 10 earth radii; at the earth's poles its distance is about 14 radii. The long cylindrical tail of the magnetic cavity, on the side of the earth opposite the sun, extends outward several earth-moon distances. (M.F.H.)

Malachite. A mineral, green hydrated copper carbonate, a minor ore of copper. When cut out and polished it is used for decorative and ornamental purposes. Malachite is a secondary mineral formed in the oxidized zone over primary copper deposits. It is associated with other secondary minerals, particularly the blue copper carbonate *azurite, from which it may form as an alteration. Malachite is commonly botryoidal; when cut, such masses show concentric rings in varying shades of green. The best specimens come from the Ural Mountains in the Soviet Union; Katanga, Zaire; Tsumeb, Namibia and Bisbee, Ariz. See Mineral Properties; Appendix 3. (C.S.H.)

Manganese Nodule. A potato-shaped lump rich in manganese and iron and distributed irregularly on the ocean floor. The largest recorded quantities occur in patches in the central deep-water basins of the North Pacific Ocean, but concentrations of potential commercial value are found in other areas as well. Surface crusts and pavements rich in manganese have also been sampled. The nodules apparently form by precipitation, commonly around a nucleus containing a sand grain, a particle of volcanic material or even a fossil fragment. Low rates of sedimentation are prerequisite for the accumulation of the nodules. Concentrations are also clearly related to areas of red clay deposits with very fine sediment. The abyssal plains and low-lying *abyssal hills, located between near-shore areas of high sediment dispersal and the flanks of major mid-ocean ridge systems, contain the highest accumulation of nodules.

Manganese nodules have attracted attention as potential commercial sources not only of manganese but also of metals such as cobalt, copper and nickel, as well as several rare elements including titanium, vanadium and zirconium. Recovery technology presents serious problems; several companies have experimental programs to develop methods for mining the deep-sea floors. Various dredge and suction-pump techniques are being tested. The incentive continues, for although the precise value of these manganese deposits is unknown, estimates indicate perhaps 1.6×10^{12} metric tons of nodules on the Pacific Ocean floor alone, and an average formation rate of 6×10^6 metric tons per year in that area. (R.E.B)

Manganite. A mineral, an oxide of manganese, formed as a low-temperature hydrothermal mineral and as a secondary mineral deposited by meteoric waters. It is commonly associated with other manganese oxides, particularly *pyrolusite, to which it may alter. Manganite is a minor ore of manganese; pyrolusite is the major source. See Mineral Properties.

Manning Equation. Used to calculate the mean velocity of flow in a stream. The average flow velocity (V) is related to the slope of the water (S), the depth (R for hydraulic radius) and a coefficient of roughness (n): $V = 1.5 / n \, R^{2/3} S^{1/2}$. Once the slope and depth of a stream are known, the roughness can be estimated by comparison with the known roughness of other streams (ranging from 0.01 for smooth to 0.06 for extremely rough).

When equated with the *Chézy equation, the Chézy discharge coefficient is seen to be a function of the Manning roughness coefficient and the hydraulic radius.

Mantle. The zone within the earth that extends below the *crust to the earth's *core. The upper limit of the mantle is marked by the *Mohorovicic Discontinuity, located about 40 km below the land surface and 10 km beneath the surface of the ocean floor. The lower limit of the mantle is about 3000 km beneath the surface of the earth at the mantle-core boundary. The lower boundary can be recognized by the abrupt decrease in the velocity of earthquake waves as they enter the core.

The mantle is further divided into an upper and lower mantle. The boundary between the two zones does not appear to be sharp, and the depth assigned it ranges from about 700 km to 1000 km. Another boundary within the upper zone has been placed at about 60 km beneath the oceans and 150 km beneath the continents. The boundary represents the beginning of a zone in which seismic waves begin a gradual decrease in speed. The existence of this low velocity zone (LVZ) was first proposed by Beno Gutenberg in 1926. Today the existence of the LVZ is well established and is known to extend to depths greater than 350 km, depending on its relationship to overlying crustal features. Beyond that boundary there is a rapid increase in seismic wave velocity and rock density down to the lower mantle boundary. Within the lower mantle, seismic waves and rock density again increase gradually.

The mantle forms 83 percent of the earth's volume and 67 percent of its mass. *Convection currents in the mantle are the source of energy which thrust the *crustal plates over the earth's surface and a source of intrusive *igneous rock in which iron-rich and magnesium-rich minerals predominate. (J.W.S.)

Map. In geology, a graphic representation generally showing selected physical features of the earth and using colors, lines, symbols and words to depict these features. Although maps commonly pertain to the earth's surface, they may relate to its interior or to the moon or other planets. Characteristic of most maps is a means of orientation that enables the

4

Magnetosphere
1 *Powerful radiation from space is trapped in the magnetosphere, the magnetic field that surrounds the earth. Its outer edge, the magnetopause, is compressed on one side and drawn out on the other by the solar wind, a continuous flow of electrons and protons from the sun.*

Malachite
2 *This deep green hydrous carbonate of copper is frequently found in nodular masses built up layer by layer. Malachite makes a pleasing ornamental material, for when the nodules are cut through the layering appears as concentric banding.*

Manganese Nodule
3 *Photographed 4754 m (2600 fathoms) deep on the ocean floor, these potato-sized lumps form around bits of clay, sharks' teeth and whales' ear bones. Manganese nodules are a rich future source of the metals cobalt, copper and nickel as well as manganese.*

Manganite.
4 *Magnificent crystals of manganite, an ore of manganese, are found near Ilfeld, Germany.*

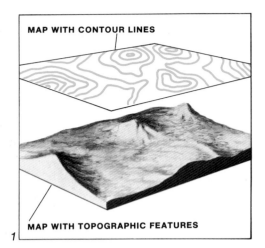

MAP WITH CONTOUR LINES

MAP WITH TOPOGRAPHIC FEATURES

1

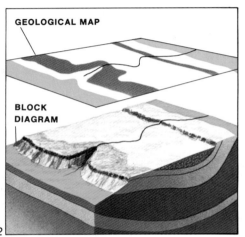

GEOLOGICAL MAP

BLOCK
DIAGRAM

2

Maps

1 *A flat map, with lines (contours) passing through points having the same elevation above sea level, is used as the basis for a three-dimensional model showing actual topographic features.*

2 *A geologic map showing outcrop patterns of gently folded strata is translated into a three-dimensional block diagram below.*

Marble

3 *Horizontal lines in a marble quarry near Poughkeepsie, N.Y., result from the quarrying operation. Evidence of the bedding planes disappeared when the original limestone was metamorphosed into marble.*

Marcasite

4 *A stalactitic mass of this iron sulfide is coated with small individual crystals of marcasite and a few crystals of white calcite.*

viewer to locate all points, both on the map and on the surface represented. Maps of the earth use the North Pole as a reference, and every map has a scale that establishes the relative size and position of features.

Several kinds of maps are important to the earth sciences. The most frequently used are topographic maps: they depict the configuration of the land surface as well as water features, vegetation and the works of man. Surface shape on such a map is generally shown by a series of contour lines, each connecting a number of points at the same elevation. These lines are established in uniform increments of elevation, with sea level as zero elevation.

Making such maps was once a tedious task requiring the field measurement of numerous elevation points by surveying techniques. Topographic maps of most terrains can now be made speedily and accurately by means of photographs taken by precision aerial cameras. Pairs of these overlapping aerial photographs are then scanned with an optical stereo-plotting instrument that allows precise measurement of the three-dimensional characteristics of the land surface.

The U.S. Geological Survey and several other federal agencies publish topographic maps in a number of series. Maps in each series have a designated areal size and map scale; for example, common Survey series have maps showing 7½-minute areas (that is, covering 7½ minutes of latitude and 7½ minutes of longitude). On these maps a scale of 1 inch represents 2000 feet of ground surface; this is shown by a bar scale and the ratio 1:24,000 (that is, 1:24,000 inches). Another series covers 15-minute areas at a scale of 1 inch to nearly 1 mile (1:62,500). These maps play an essential part in the planning and construction of highways, transmission lines, pipelines, municipal development, dams and river maintenance and industrial growth. They facilitate mineral exploration, hydrologic research, flood control, soil conservation and forest and wildlife preservation. They also serve as a base on which other types of information can be placed and thus become an integral part of many geologic maps.

A geologic map is a graphic portrayal of the relationship of rocks at or near earth's surface. Most such maps are based largely on information obtained from careful inspection of the rocks. However, large areas may lack exposed rock because they are covered with soil, vegetation or manmade objects. The mapmaker must project the rock data for such covered areas. Geologic maps are therefore not restricted to documented data, and in this way differ from topographic maps.

Geologic maps generally benefit from insights into the nature of rock bodies below the surface. Types of soils, the shape of surface features and even vegetation may offer clues to the underlying rocks. Especially helpful in getting a view well below a land surface are the natural cuts made in the upper crustal layers by rivers and glaciers. A classic example is the Grand Canyon of the Colorado River, which provides an unparalleled exposure of layers below the surface. Data from wellborings, deep drilling and subsurface mining also furnish useful information. In addition, techniques have been developed to measure the properties of buried rocks, such as variations in density and in magnetic attraction. Collectively, this information gives scientists a three-dimensional understanding of the rocks that make up the earth.

Like topographic maps, geologic maps use a series of symbols and often add colors coded to signify certain rock types and their shapes. A full explanation of all symbols and colors is included on each map. Generally, one or more vertical views, called cross sections, of the shallow upper crust fill out the three-dimensional picture.

Geologic maps have many uses. They have proved valuable in unraveling the 4.5 billion years of earth history and especially of the past 600 million years. They also are a great help in the search for natural resources. Many such resources have a close affinity with certain rock types or occur in specific rock body configurations. Examples of this are crude oil and natural gas, coal, limestone, sand and gravel, sulfur, salt, precious metals such as silver and gold as well as numerous other metals like nickel, copper, chromite and zinc. Even the distribution of groundwater, which is becoming more important as the demand for freshwater increases, is largely controlled by the nature of shallow subsurface rocks.

Application of geologic information to environmental problems has resulted in environmental geologic maps. They show land features that have environmental significance, including slope stability, runoff water, sewage infiltration, actual or potential subsidence and the distribution of resources in relation to urban development and land use. Such maps are invaluable aids to efficient land-use planning, resource management and pollution containment.

Other types of maps are widely used in weather forecasting, and, utilizing information provided by weather satellites, in monitoring storms. Hydrologic maps of various types are increasingly in demand in the growing search for water resources.

Oceanographic exploration has led to many revisions in oceanographic mapping. Lunar maps are undergoing extensive refinement as a result of the space probes. Most recently, space flights have provided a unique vantage point for surveying earth's surface, and new maps reflecting high altitude views of large areas of the earth are now being developed. (R.E.B.)

Marble. A fine to coarse-grained *metamorphic rock consisting chiefly of recrystallized *calcite and/or *dolomite. The stone usually has a granular, sugary texture capable of being highly polished. It may feature colorful streaks and swirls produced by inclusions of minerals such as mica, talc, pyrite, actinolite, epidote, quartz, feldspar or iron oxide. Marble has long been used as an architectural and ornamental stone. Sculptors prize it because of its intermediate hardness, which makes it soft enough to be readily worked and hard enough and of such a composition as to be fairly resistant to weathering and breakage. Purer grades of marble composed of granular calcite or dolomite, without inclusions, are used in sculpture, while the less pure varieties are used as ornamental stone and decorative facing on buildings and monuments. Centers of marble production are located throughout the world. Pure varieties are quarried in Europe; the quarries at Carrara, Italy, have furnished blocks of high-quality marble to such sculptors as Michelangelo. Several grades come from the Appalachian district of the United States, especially Vermont, Georgia and Tennessee. New sources of marble are being found in the Rocky Mountains and the Coast Ranges. (J.W.S.)

Marcasite. A mineral, iron sulfide; it is dimorphous with the more abundant mineral *pyrite. Marcasite is orthorhombic whereas pyrite is isometric, and since both minerals are commonly in crystals, they frequently can be distinguished by crystal form. Marcasite is pale bronze-yellow, lighter than the brass-yellow of pyrite, and hence sometimes called white iron pyrites. Marcasite is a near-surface mineral often found in metalliferous veins deposited at low temperature. See Mineral Properties.

Mare (plural: maria). A large, relatively flat region on the surfaces of Mars and the moon that has a low *albedo or reflectivity and therefore appears darker than the surrounding area. The first observers of the moon thought that the maria were oceans. (Hence the name mare, meaning "sea" in Latin.) The eyes of the so-called man-in-the-moon are maria, his right eye being

3

4

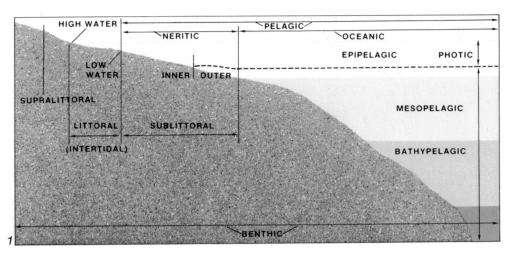

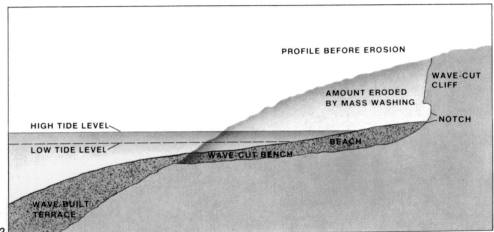

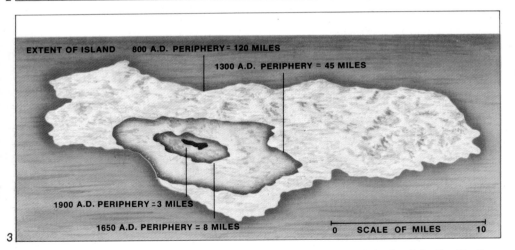

Marine Environment

1 *The various layers of the marine environment are known as biozones.*

Marine Erosion

2 *Profile of a coastline showing a cliff, notch and bench cut by waves and a wave-built terrace, all typical of coastlines worn back by marine erosion.*

3 *The island of Helgoland, in the North Sea, was reduced by waves from an original periphery of 193 km (120 miles) in A.D. 800 to 6 km (3 miles) in 1900.*

Mare Imbrium and his left Mare Serenitatus. The mare basins are filled with iron and titanium-rich basalts deposited between 3.8 and 3.1 billion years ago.

Margarite. A mineral, hydrous calcium aluminum silicate. It is platy and has a single perfect cleavage. Like the micas it is a *phyllosilicate but it is harder, and thin plates of it break rather than bend. It is therefore called a brittle mica. Margarite usually occurs with corundum and is characteristic of *emery deposits. In this association it is found on the island of Naxos, Greece, and in the United States at Chester, Mass. See Mineral Properties.

Marginal Plateau. Rather steeply terrace-shaped features on the seaward side of the *continental margin. The depth is greater than the continental shelf, and the plateau is separated from the shelf by a low continental slope. The Blake Plateau off the southeastern United States and similar plateaux off eastern South America and western Africa are good examples of this feature.

Marine Environment (or Biozone). A subdivision or zone within the *oceans based on depth and characterized by a particular set of conditions and occupied by a particular community of organisms. There are two broad categories, which can be subdivided into smaller zones: the *benthic, or bottom, and the *pelagic, or water environment. Both plankton (floating life) and nekton (swimming life) occupy a pelagic environment, whereas benthos (organisms that live on the ocean floor) move about or are actually attached to the bottom. The zonation of these various communities is based on such environmental conditions as depth of water, availability of light, food, temperature and pressure.

The initial attempt at zoning the marine environment was made by the pioneer oceanographer Edward Forbes in 1834 in the North Sea area. Many classifications and zonations of the marine environment followed. In 1940 the National Research Council of the U.S. National Academy of Sciences formed the Committee on Marine Ecology and Paleoecology and its efforts have provided a generally accepted zonation of the oceans. (R.A.D.)

Marine Erosion. The removal of material from the land by waves and shore currents. The waves erode by means of solution, hydraulic action and abrasion. Solution is of minor importance but does operate where limestone and other soluble rocks are present. The hydraulic action of storm waves may exert a pressure of over 29,000 kilograms (64,000 pounds) per square meter. *Joints and cracks are widened and extended, and as the waves retreat the compressed air expands, loosening and detaching fragments. Coastal rocks are abraded by fragments hurled against them by waves. The combined action of these processes, aided by *longshore currents and *mass wasting, forms marine cliffs, which usually increase in height as they retreat inland and which furnish the waves with a continuous supply of fragments. Where coastal rocks are less resistant, chasms and coves are formed; where more resistant, headlands. In addition to the marine cliff, wave-cut platforms, stacks, caves and arches are among the features formed by such erosion. The sand and gravel resulting from the erosion are moved along shore by currents; when deposited they may form spits, tombolos and other such coastal features. The finer-grained material is moved seaward and forms offshore deposits.

A remarkable example of marine erosion is illustrated by Helgoland, a small island in the North Sea that is about 1600 m long and 500 m at its widest part and is surrounded by marine cliffs. Historical and geological evidence indicates that its coastline was approximately 200 km long in A.D. 800, 75 km in 1300, 15 km in 1649 and only 5 km by 1900. The Yorkshire coast south of Flamborough Head, composed of unconsolidated deposits, is another example of extreme erosion: the average rate of coastal retreat during the last 100 years has been between 1 and 2 m per year; since Roman times the coast has been worn back 3 to 5 km and more than 30 villages have been destroyed. (R.L.N.)

Marl. A sedimentary rock consisting of an earthy mixture of *clay, *calcite and *dolomite. It commonly contains fossil shells and may form in either fresh or salt water. Marl may become indurated to form marlstone or calcareous mudstone.

Mars, Geology of. All four terrestrial planets (i.e., the earth, Mercury, Venus and Mars) have diameters between 4800 and 12,500 km and densities between 3.9 and 5.5 gm/cm³ (grams per cubic centimeter). In contrast, the giant or Jovian Planets (Jupiter, Saturn, Uranus and Neptune) have diameters between 43,000 and 137,000 km and densities of 0.71 to 2.47 gm/cm³. Of the terrestrial planets, Mars is most like the earth.

Mars is about half the diameter of the earth. It has a density of 3.94 gm/cm³, which is considerably less than the earth's 5.52 gm/cm³. Typical rocks at the surface of the earth have densities of 2.7÷3.2;

therefore the earth must have denser material in its interior. The greater density of the earth's interior is caused mainly by the fact that the rocks in the interior are richer in iron and magnesium than those at the surface, with a metallic iron core in the center. Another reason is that the weight of rock compresses material in the interior and raises its density slightly. The surface of Mars appears to have rocks that cannot be much denser than those at the surface of the earth or the moon. By simply assuming that Mars consists of rocks like those of the earth or the moon, and that they are compressed at the interior of the planet but less than they would be at the center of the earth, we can account for the observed density of Mars.

Another indication that Mars does not have a large metallic core is the moment of inertia of the planet, that is, the measure of the ability of the planet to store kinetic energy by spinning on its axis. (The moment of inertia of a homogeneous sphere is 0.4 times the mass of the sphere, m, times the square of its radius, R^2, times the angular velocity, w.) The moment of inertia of the earth is $.33\ mR^2w$, because much of the earth's mass is a metal core deep in its interior. For Mars the value is $.375\ mR^2w$, which, after we allow for compression of surface rocks, indicates that Mars does not have a large metallic core.

Mars lies about 1.52 times farther away from the sun than does the earth, so each square meter of Martian surface receives about 43 percent as much solar energy as an equivalent area on earth. Thus Mars is much colder than the earth. The axis of rotation of Mars is inclined to the plane of its orbit by 25° and thus Mars, like the earth, has pronounced climatic seasons. Because the orbit of Mars is more elliptical, the intensity and duration of seasons in its northern hemisphere is different from those in its southern hemisphere. The Martian north pole is tilted away from the sun when Mars is nearest the sun and moving the fastest. Thus its northern winter lasts 156 days, its northern summer 177 days.

The Martian atmosphere is much less dense than the earth's and its composition is very different. The total atmospheric pressure at the surface of Mars is about one percent of that at the surface of the earth. The composition of the Martian atmosphere has been inferred from infrared spectroscopic observations, which have detected substantial amounts of carbon dioxide and small amounts of water. Other gases possibly present include nitrogen, the most abundant gas in the earth's atmosphere, and argon; however, these are not easily detected by spectroscopic techniques.

The total pressure at the surface of Mars ranges from 3 to 8 millibars except at the tops of the highest peaks, where the pressures are on the order of 1 millibar. In contrast, the atmospheric pressure at the surface of the earth is about 1000 millibars (14.7 lbs/in.²), falling to about half that at an elevation of 3 km.

The Martian atmosphere has a small but significant amount of water. If all of it were condensed, it would form a film only 0.03 mm thick over the Martian surface. The abundance of water is strongly dependent on season. During the winter no atmospheric water is detectable over the poles, but the amount increases to a maximum in late summer.

Because Mars is far from the sun its surface is cold. The highest afternoon summer temperatures at the equator are about 22°C (70°F), cooling to −70°C (−94°F) just before sunrise. The minimum temperatures on the planet occur at the poles during winter, and temperatures as low as −138°C (−216°F) have been measured. These great temperature variations cause extremely strong winds, with velocities of hundreds of meters per second. And even though the Martian atmosphere is very tenuous, the high winds commonly cause major dust storms that completely cover the planet.

The cratered terrain on Mars looks very much like the highland areas on the moon except that most craters appear to be more subdued, possibly because they have been filled with substantial amounts of windblown dust and the rims of the craters have been eroded by wind. The large smooth areas of Mars (which have no counterpart on the moon) are believed to have resulted from a combination of volcanic flows and of windblown sedimentary deposits. A number of huge shield volcanoes are concentrated between 80° and 220° longitude. The largest of these, Nix Olympica, is estimated to be as much as 23 km high, over twice the height of Hawaii's Mauna Loa, the highest volcano on earth. The polar regions of Mars are covered with an ice cap that advances and retreats with the seasons. The residual summer cap is believed to be solid carbon dioxide, while the winter extensions consist of water ice. A number of channel-like features are recognized and interpreted as evidence of liquid water in the Martian past.

The Martian highlands appear to be covered with flat-floored craters over 20 km in diameter; however, the density of smaller bowl-shaped craters appears to fall well short of saturation by a factor of 10. The bulk of the cratered terrain falls in the southern hemisphere, extending to 70° south latitude. Areas between the

PHYSICAL PROPERTIES OF MARS

	MARS	EARTH
Mean diameter (km)	6775.0	12,742.1
Equatorial diameter (km)	6786.8	12,756.3
Polar diameter (km)	6751.6	12,713.6
Density (gm/cm³)	3.94	5.52
Moment of inertia	$.375mR^2$	$.33078mR^2$
Mean distance from sun (km)	227,941,000	149,600,000
Maximum distance from sun (km)	249,226,000	152,100,000
Minimun distance from sun (km)	206,656,000	147,100,000
Eccentiricity orbit	.0933	.0167
Inclination of equator to ecliptic (plane of earth's orbit)	1.85°	0.0°
Inclination of equator to orbit	25.2°	23.45°
Period of revolution about sun (days)	687.00	365.26
Period of axial rotation	24:37:22	23:56:04
One day	24:39:35	24:00:00
Escape velocity (km/sec)	4.0	11.2
Gravity at surface (cm/sec/sec)	380	978

1

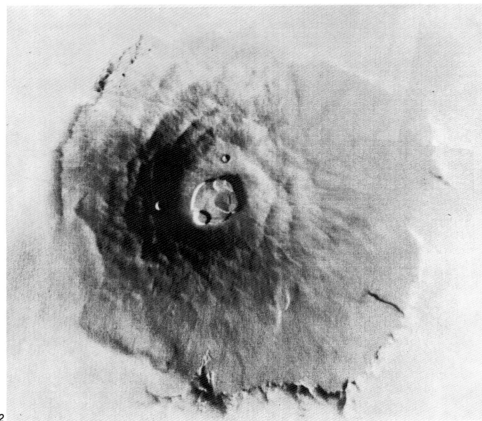

2

large craters are relatively flat and featureless. The intercrater areas are believed to be made of material ejected from craters and possibly volcanic and windblown deposits. This material would hide small craters and may account for the paucity of such craters. The mantling material appears to thicken toward the south pole regions, and features interpreted as volcanic flow fronts are recognized.

Ringed basins have been identified on Mars, which are comparable to similar features on the moon. As with lunar basins, the larger the basin, the greater the number of rings. It is generally agreed that the ringed basins result from the impact of meteorites measuring from tens of kilometers to over a hundred kilometers across. Six ringed basins have been described. The largest is Helles, centered at 295° longitude, 45° south latitude. The outer ring of this basin is over 2000 km across and thus is much larger than the largest lunar ringed basin, Mare Imbrium, whose comparable diameter is 1500 km.

The Martian volcanoes are much larger than most terrestrial volcanoes. (Although the moon has several large areas flooded with extrusive flows of basalt, unequivocal volcanoes are absent.) The largest of the Martian volcanoes, Nix Olympica, is over 600 km across, with a series of summit calderas tens of kilometers across. The edge of the volcano is approximately 2 km high.

Mars has a major rift system extending more than 4000 km between 30° and 100° longitude close to the equator. This giant canyon is clearly a tectonic feature, not an erosional one. The eastern end of the canyon merges into an area of large, irregular, flat-topped blocks forming hills and depressions in an area surrounded by cratered highlands. The western end of the giant chasm branches into a series of much smaller troughs.

Wind erosion is much more important on Mars than on either the moon or earth. The moon has no substantial atmosphere, and thus no wind erosion. The effects of wind erosion on earth in contrast to Mars are less significant than those of water erosion. A giant dust storm in late 1971 carried great amounts of very fine dust particles to such altitudes that virtually all Martian topography was blocked out. The dust appears to have had a high silica content, indicating that Mars, like the earth and moon, is basically composed of silicates. Wind erosion appears to erode topographic heights and to deposit material in low areas. In general, such erosion seems to remove material from equatorial regions and deposit dunelike features in the polar regions. Numerous smooth plains areas are covered with light and dark streaks that changed position and shape during the Mariner 9 observations. These streaks are believed analogous to the areas in terrestrial desert that have relatively little sand covering. (C.H.S. and J.L.W.)

Mascon. A term coined for a concentration in the moon with a mass in excess of that expected for gravitational equilibrium. Mascons cause positive *gravity anomalies over all of the circular lunar mare basins that are more than 200 km in diameter.

Mascons are important since they indicate that isostatic equilibrium has not been attained since the mascon formed. Because the rocks in the mare basins are over 3 billion years old, we assume that the mascon also formed more than 3 billion years ago. These relations suggest that the lunar crust is strong enough to have preserved major features like these over billions of years. The earth's crust, in contrast, returns to isostatic equilibrium in less than 1 million years. See Lunar Geology.

Mass Wasting. The downslope movement of surficial material due to gravity. The material may fall, roll, flow or creep. The movement may be very rapid or imperceptibly slow; it may take place on land or beneath the ocean, and on both steep and gentle slopes. Small or large masses of both regolith and bedrock may be moved. *Avalanches, *landslides, *mudflows and *creep are examples of mass wasting. Although mass wasting is ultimately dependent on gravity, it is aided by steep slopes, jointed and dipping rocks, water, freezing and thawing, wetting and drying, burrowing of organisms, decay of vegetation, earthquake vibrations and other factors. It occurs almost everywhere because of the many slopes. It widens valleys, denudes land, and furnishes sediment for forming sedimentary rocks. (R.L.N.)

Matrix. In geology, the small particles of sediment or rock material that occupy the spaces between the larger fragments forming the framework of a rock. The matrix of certain igneous rocks is called its *groundmass. The term is also applied to the rock that naturally surrounds fossils, concretions, minerals, etc.

Meander. A bend in a river channel. Streams with many bends are said to be meandering or sinuous. Most streams meander over some of their lengths. Meandering is a result of the adjustment of a stream to its environment in order to carry its load most efficiently.

Both the *thalweg (the line joining the deepest points) and streamlines (lines con-

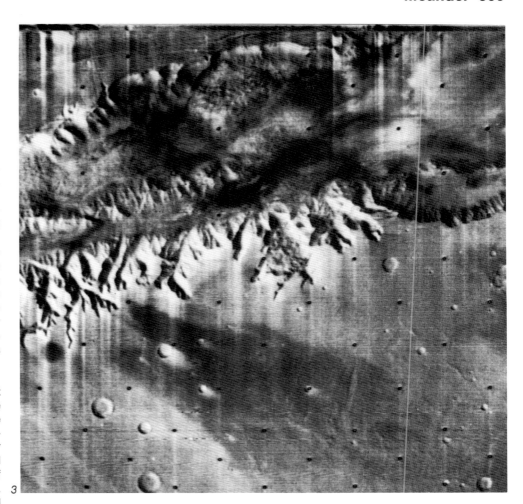

3

4

Mars, Geology of
1 Physical properties of Mars.

2 Nix Olympica, shown here in a photograph taken by Mariner 9, is over 20 km (12 miles) high and is the largest volcano known on earth or Mars. Note the cliff surrounding the volcano and the complex caldron at the summit.

3 The equatorial rift on Mars, as photographed by Mariner 9, is a valley of massive proportions. It extends more than 3000 km (over 1850 miles) in length, almost the width of the United States, and is several kilometers deep. This canyon is located just south of the Martian equator and east of Nix Olympica. The rift is probably a tectonic feature, and the side canyons suggest water erosion.

Mass Wasting
4 Loosened by weathering, tons of rock fragments flowed down a slope forming this talus cone in Alberta, Canada.

necting a series of fluid particles so that the velocity of every particle is tangent to it) of the maximum velocity of a meandering river move close to the outer side of each bend, crossing over at the point of inflection between bends. As the water moves in this winding motion, an excess is piled up on the outside of the meander bend. Pressure from the excess water causes a strong downward movement on the outside of the bend, resulting in erosion. At the same time there is a deficiency of water on the inner side of the bend. This deficiency, plus separation of flow of water from the inner bank, causes a backwater eddy and the deposition of a point bar. The combination of erosion on the outside bank and deposition on the inner bank causes meander bends to migrate.

The gradient of a meandering reach is uneven, marked by pools and shoals. Deeper pools are near the axis of bends, and shoals occur at crossovers. The channel cross section at bends is typically asymmetric, whereas that of crossover points is less so.

Relationships have been established showing that meander wavelength depends upon the width of the channel and the radius of curvature, and that the breadth of meander bends depends upon the width of the channel.

Numerous causes have been suggested for meandering. Traditionally, mature streams were thought to meander when they no longer had the energy to downcut. Another theory suggested that streams meandered in order to lower a gradient that provided more velocity than needed to transport its load. Meanders lengthen the course and hence decrease the gradient. Meandering has also been related to the size of the material carried in a channel and to channel morphology. See Meander, Entrenched. (M.M.)

Meander, Entrenched (or Incised). A riverbed with one or both banks vertical. When a meandering river cuts down deeply into its channel it becomes entrenched or incised. This occurs when the stream's environment changes so that downcutting is renewed while the stream maintains the same channel. Entrenched meanders may be first cycle or may be formed during multiple cycles. First-cycle meanders are formed from an original winding pattern on a land surface that is continuously uplifted (or where the base level is lowered) as the stream cuts down. The multiple cycle meanders are formed when a stream meandering on a *floodplain is *rejuvenated so that the meanders are incised.

If the river is cutting down as uplift occurs, the bends migrate. This leaves

prominent *slipoff slopes or spurs on the inner sides of the bends and undercut banks on the outside of the bends (ingrown or asymmetric meanders). Whether such meanders are ingrown or show little difference between the slopes on opposite sides of the bend depends upon the rate of uplift in relation to sediment load and rock resistance. Some incised meanders are controlled by rock type such as those of the Shenandoah River, Va., where elongation of meanders has occurred because of preferential erosion of weak rock. (M.M.)

Mediterranean. A term generally applied to very large bodies of saltwater or inland seas surrounded by land, but with openings to the ocean proper. Examples are the Gulf of Mexico, the Arctic Sea and the Mediterranean Sea itself.

Meerschaum. The popular name for the claylike mineral sepiolite used to carve meerschaum pipes.

Meltwater. Water from melted snow and ice. In some areas it has great geologic effect, depositing all kinds of stratified drift, such as *outwash plains, *valley trains, *eskers, *kames, *kame moraines and *kame terraces, and *glaciolacustrine materials. It forms marginal channels, *potholes and other erosional features, and helps to form ephemeral features such as ice-dammed lakes and channels in and beneath glaciers. Meltwater streams issuing from glaciers are commonly grayish-white because they contain *rock flour resulting from glacial abrasion.

Meltwater poses a genuine threat. If all the glacial ice now on the earth's surface should melt because of a warming-up of climate, sea level the world around would rise approximately 70 m. Many of the great coastal cities would be partly or completely inundated, and a large part of the world's wealth would be destroyed. Some authorities believe that the amount of carbon dioxide currently being added to the atmosphere by the burning of fossil fuels will increase global temperatures and bring about this melting. (R.L.N.)

Member. A minor *rock stratigraphic unit consisting of part of a *geological formation. It can be distinguished from the rest of the formation by distinctive characteristics such as rock type, hardness or color. Since it is usually of considerable geographic extent, members may be given formal names.

Mesa. A flat-topped, tablelike, isolated hill or mountain bounded by cliffs and steep *bedrock slopes, with an extensive summit area, composed generally of horizontal

1

2

Meander

1 "Goosenecks"—entrenched, or incised, meanders—are cut deep into the rock by the rejuvenated San Juan River in Utah.

2 Alternately cutting and filling its channel, a stream makes smooth, looping curves as it wanders across a valley floor in Yugoslavia.

Meltwater

3 A stream of water from a melting glacier pours toward the sea in Norway.

Mesa

1 Weather-resistant cap rock may have prevented this table mountain, or mesa, at Springdale, Utah, from eroding to the level of the surrounding countryside.

Metamorphic Rock

2 Formed under intense heat and pressure, gneiss is typical of metamorphic rocks, with a coarse-grained texture and minerals arranged in streaks or bands.

*sedimentary rocks. Mesas range from 30 m to 600 m in height and from a few hundred meters to several kilometers in length. They are erosional remnants that persist because they are located between main drainage lines, or are capped by rocks such as *lava, *sandstone or conglomerate that resist erosion, or because of a combination of both. In the United States they are found in Arizona, Colorado, New Mexico and Utah. Mesa Verde, Colo., is the site of the most notable cliff dwellings in the United States. Monument Valley, Ariz., is dominated by prominent, colorful mesas, by buttes, which have smaller summits than mesas; and by chimneys, which have little or no summit areas. Masada, a mesa in Israel, is famous because of the heroic defense there by the Jews in A.D. 73 against the Romans. (R.L.N.)

Mesosiderite. A rather rare type of *stony-iron meteorite. Nickel-iron as a matrix encloses fairly large crystals of olivine, hypersthene and anorthite.

Mesothermal Deposit. A variety of ore deposit formed at intermediate depth and temperature; that is, the depth is less than in *hypothermal deposits but greater than in *epithermal deposits. The principal ore minerals in such deposits are pyrite, chalcopyrite, arsenopyrite, galena, sphalerite, tetrahedrite and native gold; the chief metals recovered are gold, silver, copper, lead and zinc.

Mesozoic Era. The *era of geologic time from the end of the Paleozoic, 225 million years ago, to the beginning of the Cenozoic, about 70 million years ago. The Mesozoic is composed of the *Triassic, *Jurassic and *Cretaceous periods, and has been called the "Age of Reptiles" because of the numerous and diverse reptilian fossils (especially dinosaurs) found in Mesozoic rocks. In addition to the dinosaurs—which appeared in the Triassic and were extinct by the end of the Cretaceous—flying reptiles (pterosaurs) and marine reptiles (ichthyosaurs, plesiosaurs and mosasaurs) were common throughout Mesozoic time. The first birds, mammals and flowering plants appeared during this era, and invertebrates (especially mollusks) were numerous and diverse.

In North America, Mesozoic formations are exposed from Nova Scotia to North Carolina near the Atlantic coast of North America, in the inner border of the Atlantic and Gulf Coastal Plains, in the Great Plains and in the Colorado Plateau and Rocky Mountain Regions. In westernmost North America igneous and sedimentary rocks of Mesozoic age occur in a belt from Alaska to Lower California. Mesozoic rocks are widely exposed in England, Germany and Russia. They are also found in western South America, eastern Australia, Africa, and Asia. Toward the close of the Mesozoic many of the earth's largest mountain ranges (including the Andes and the Rocky Mountains) underwent maximum uplift. (W.H.M.)

Metacryst. A large crystal in a metamorphic rock formed during the process of *metamorphism. Examples are *garnet, *staurolite and *andalusite in mica schist.

Metal, Native. A metal that occurs in nature uncombined with other elements. See Element, Native.

Metallogenetic Province. An area in which conditions at a given time were favorable for the formation of metallic ore deposits. Examples of two metallogenetic provinces are the gold-quartz veins of California and the lead-zinc deposits of the Mississippi Valley region.

Metamict Mineral. In crystallography, a mineral, originally crystalline, whose crystal structure has been destroyed by radiation from radioactive elements and is thus amorphous. Consequently, such a mineral has no cleavage, is optically isotropic and does not diffract X-rays. Many metamict minerals are bounded by crystal faces and are *pseudomorphs after an earlier crystalline phase. On heating a metamict mineral, the crystal structure is reconstituted and its density increased. All metamict minerals are radioactive and the structural breakdown results from the bombardment of alpha particles from the contained uranium and thorium. (C.S.H.)

Metamorphic Facies. A concept that was developed by the Finnish petrologist Pentti Elias Eskola in 1915 and applied systematically to group and classify the many varieties of *metamorphic rocks. His formal definition of a metamorphic facies appeared in 1939 in German; freely translated it reads: "A certain metamorphic facies comprises all rocks exhibiting a unique and characteristic correlation between chemical and mineralogic composition, in such a way that rocks of a given chemical composition have always the same mineralogic composition, and differences in chemical composition from rock to rock are reflected in systematic differences of this mineralogic composition." Each facies develops within a limited range of temperature and pressure, and as one or both of these conditions exceed the limits of the range, a new facies will develop.

Since 1939 other geoscientists in the field of metamorphic petrology have somewhat modified the original definition. Nevertheless, the basic concepts established by Eskola remain valid. Thirteen facies have been recognized to date, each named for a particular mineral or mineral assemblage peculiar to that facies. Although there are slight differences of opinion as to the exact number of facies, the majority are used by all geologists. Each facies is produced by one of two major types of metamorphism: local metamorphism (*contact metamorphism) or *regional metamorphism. The second type is divided into two subgroups, dynamothermal metamorphism and burial metamorphism. The former is associated with bending and buckling of the earth's crust whereas burial metamorphism occurs simply by rock being buried deep within the earth. (J.W.S.)

Metamorphic Rock. A rock produced in the solid state from preexisting rock that has been subjected to striking changes in one or more of the following conditions: temperature, pressure, shearing stress or chemical activity. Such physical and chemical changes produce new minerals that are stable in the new environment, along with changes in the texture and structure of the rock. See Cataclasite; Eclogite; Gneiss; Granulite; Hornfels; Marble; Mylonite; Phyllite; Phyllonite; Quartzite; Schist; Skarn; Slate; and Soapstone.

Metamorphism. A process that produces a change in the chemistry, structure or mineralogic composition of solid rock. The most important agents of metamorphism are changes in temperature and pressure. Minerals in any rock are generally stable under a limited range of temperature and pressure. When new physical conditions are imposed, they become unstable and react with each other to form new stable minerals. The lower limits of temperature and pressure necessary for the process of metamorphism are not clearly defined. An arbitrary lower limit used by some geologists is a temperature greater than 100°C (212°F) and a pressure equal to that encountered at a depth of approximately 3 km below the surface of the earth. Upper limits at which metamorphism can occur are reached when temperature and pressure conditions approach 800°C (1472°F) —about 35 km beneath the surface. Above these upper limits rock begins to melt to form *magma, the molten material associated with volcanoes.

Although temperature and pressure control the metamorphic process, they are not solely responsible for reactions

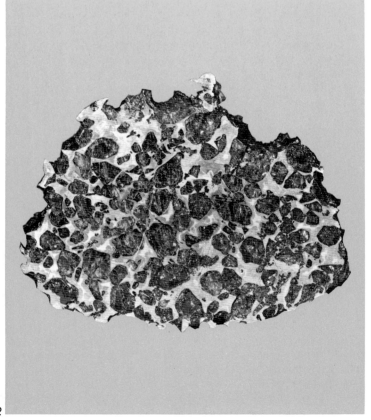

between mineral constitutents. Laboratory experiments with combinations of minerals similar to these occurring naturally in rock show that the presence of water or water vapor or other gases greatly increases the rate of chemical reactions. Without these substances some reactions could not occur at all, or would take millions of years to complete. The water necessary to bring about chemical reactions exists in premetamorphic rock in pores or cracks, or as part of the atomic structure of some minerals. Should temperature and pressure conditions become lower after initial metamorphism, the minerals present in the rock will tend to react with one another to attain a new state of equilibrium (*retrograde metamorphism). Because much of the water has escaped or is unable to react chemically at lower temperatures, the chemical reaction between minerals may not occur or at best may occur at very slow rates. Thus the mineral assemblage found in a metamorphic rock generally reflects the highest temperature and pressure reached during metamorphism. See Metamorphic Rock. (J.W.S.)

Metamorphism, Contact. See Contact Metamorphism.

Metamorphism, Kinetic. See Kinetic Metamorphism.

Metamorphism, Retrograde. See Retrograde Metamorphism.

Metasomatism. A process whereby existing minerals are transformed in part or totally into new minerals by the replacement of their chemical constituents. This occurs with the introduction of migrating fluids, originating either outside or inside the rock, that can react with existing minerals. Metasomatism is important to economic geologists since it is responsible for the existence of many localized ore deposits. The lead-zinc deposits in the Pioche district, Lincoln County, Nev., and the sulfide ores of Bisbee, Ariz., are of this type, as well as the sulfide deposits at Rio Tinto, Spain. See Granitization.

Meteor Crater. A circular basin formed by meteoritic impact and the resulting explosion. Such craters are continually being discovered: In 1930 less than 10 were known, in 1966 more than 33; an estimated 250 may ultimately be found. Craters buried or destroyed by erosion number in the thousands, and those not formed because the meteorites fell into the oceans are even more numerous. Among the many criteria for the identification of such craters are: (1) the presence of meteorites;

(2) the presence of coesite and stishovite, as both require pressures very much higher than those resulting from volcanic activity; (3) the presence of shatter cones associated with craters; (4) the presence of fused quartz sandstone, which requires for its formation temperatures higher than those of volcanic activity; (5) the absence of lava, pyroclastic deposits and hydrothermal activity in and around the crater; (6) the occurrence of tilted and faulted beds.

Perhaps the best-known impact crater is Meteor Crater, on the Coconino Plateau of northern Arizona. It is approximately 1200 m (4000 feet) in diameter, 150 m (500 feet) deep and has a rim about 50 m (165 feet) above the surrounding area. The *sedimentary rocks under the rim have been tilted and faulted. Meteoric fragments have been collected in great quantities from the surrounding area. Fused quartz sandstone, coesite, stishovite and shattered and pulverized rock are present, but no volcanic rocks. The Wabar Craters in Saudi Arabia, the Ashanti Crater in Ghana and the Chubb Crater in Quebec, Canada, are thought to be of impact origin, as are others in Europe, Africa, Australia, Argentina and Tasmania. (R.L.N.)

Meteoric Dust. Minute particles formed by the melting and oxidation of meteors when they enter the atmosphere of the earth Meteoric dust has been identified in the Greenland and Antarctic ice sheets and in deep-water marine deposits. See Cosmic Sediment.

Meteoric Iron. Native iron occurring in meteorites. It is alloyed with nickel. Small amounts of other metals are also generally present, as are small inclusions of *graphite and *silicates.

Meteoric Water. Water derived from the atmosphere. It constitutes almost all of the atmospheric, surface and subsurface water. Juvenile water is that recently derived from *magma and *lava flows and from the oxidation of volcanic hydrogen. Cosmic water is that almost negligible amount which comes from space with meteorites.

Meteorite. An extraterrestrial rock that has fallen to earth from outer space. Meteorites may be placed in three groups: *stony meteorites (sometimes called stones or aerolites) composed essentially of silicate minerals; *iron meteorites (irons) made up largely of nickel-iron; and stony-irons, composed of both nickel-iron and silicate minerals. Stony meteorites are further divided into two groups, the *chondrites and *achondrites.

4

Meteor Crater

1 A meteorite smashing into the earth a few tens of thousands of years ago near present-day Flagstaff, Ariz., formed a crater about 1200 m (4000 feet) wide and more than 150 m (500 feet) deep.

Meteorite

2 This rare stony-iron meteorite from Bolivia is composed of rounded olivine crystals embedded in a nickel-iron matrix.

3 Composed chiefly of silicate minerals, this stony meteorite weighing 8320 grams (about 18 pounds) was found in Richardton, N.D.

4 The pits called "thumb marks" on the surface of this iron meteorite are the result of friction, which burned out weak portions of the rock as it entered the earth's atmosphere.

Microfossil
Skeletons of minute plankton called radiolarians make up much of the ooze on the ocean bottom.

Meteorology. As used synonymously with atmospheric science, the science of the atmosphere of the earth and of other planets. According to the goals sometimes ascribed to it, meteorology seeks to understand, predict and eventually control atmospheric phenomena. A distinction is often drawn between meteorology and climatology, the latter being mainly concerned with average — as opposed to actual—weather conditions. However, under the definition given above, climatology must be considered one of the major branches of meteorology.

Meteorology is subdivided into many other specialized disciplines, depending on the method of approach, the particular area of study and the application to human activities. Physical meteorology, or atmospheric physics, traditionally deals with the optical, electrical and acoustical phenomena of the atmosphere, its chemical composition, the laws of radiation and the physics of clouds and precipitation. The thermodynamics of the atmosphere are sometimes treated in physical meteorology and often in dynamic meteorology, which is the study of *atmospheric motions as solutions of the basic hydrodynamic equations. Radar meteorology and satellite meteorology deal with the analysis and interpretation of atmospheric phenomena as observed by radar and other electromagnetic techniques. Synoptic meteorology and numerical weather analysis and prediction concentrate on the analysis and interpretation of meteorologic observations obtained simultaneously over a wide area for the purpose of predicting weather changes. Still other branches of the science, such as biometeorology and agricultural meteorology, deal with the relations of meteorology with other disciplines. (M.F.H.)

Miarolitic Cavity. A small angular opening in *plutonic rock, particularly common in granites. The cavity is frequently lined with crystals of the rock-forming minerals.

Mica. A group of phyllosilicate minerals characterized by their platy habit and excellent cleavage. Although the micas are monoclinic they are pseudohexagonal, and the prism faces that bound the tabular crystals intersect at nearly 60°. This gives rise to diamond-shaped or hexagonal plates. The thin sheets into which mica can be cleaved are both flexible and elastic; that is, after being bent they will return to their initial flat condition. This distinguishes them from other platy minerals such as talc, which is flexible but inelastic, and margarite, one of the so-called brittle micas, which breaks on bending.

The principal members of the mica group are: muscovite and paragonite (white mica), phlogopite (brown mica), biotite (black mica) and lepidolite (lithia mica). The chief occurrence of the micas, with the exception of lepidolite, is as small grains in metamorphic and igneous rock. Lepidolite is found almost exclusively in *pegmatites, and from these bodies have also come the largest and best crystals of all other members of the group. (C.S.H.)

Microcline. A mineral, potassium aluminum silicate, a triclinic *feldspar. This low-temperature form and its high-temperature polymorphs, *orthoclase and *sanidine, are known as the potash feldspars. Microcline occurs as a rockforming mineral in igneous and metamorphic rocks. It is the common potash feldspar of *pegmatites, where it may form gigantic crystals and be intergrown with quartz in graphic granite. Such pegmatite occurrences yield most of the feldspars used for industrial purposes.

Microfossil. A fossil that in its typical form is microscopic in size. They are especially useful to the petroleum geologist since their small size makes it unlikely that they will be destroyed by the drill bit. They usually consist of the remains of invertebrate animals such as foraminifers, radiolarians, conodonts and ostracods. Among the common plant microfossils are diatoms, algae, spores and pollen. When small enough, fish scales, bones and teeth may also occur as microfossils. Although most microfossils must be studied under the microscope, certain of them, such as the larger species of foraminifers, may have a diameter of 2.5 – 5 cm.

Micropaleontology. The branch of *paleontology that deals with microfossils— fossils too small to be studied with the unaided eye. These minute remains typically represent parts of microscopic plants or animals incorporated into the rocks that enclose them. The remains of microscopic organisms such as foraminifers and diatoms are so small that they can be brought to the surface intact and undamaged by the mechanics of drilling. Consequently, microfossils are particularly useful as *index fossils in the search for petroleum.

Microseism. A vibratory motion within the earth of very small magnitude. In many instances microseisms are probably unrelated to earthquakes. They are created by many sources, such as the waving of trees, motor vehicles, or more important, the action of the ocean waves breaking along a shoreline, the interference of waves in deep ocean water or tidal movements within the rocks of the earth's *crust.

Mid-Oceanic Ridge. See Oceanic Ridge.

Millerite. A mineral, a nickel sulfide, a minor ore of nickel. It is brass-yellow but usually can be distinguished from similarly colored pyrite because of its slender hairlike crystals. Millerite is a low-temperature mineral often formed in cavities and as crystal inclusions in other minerals. See Mineral Properties.

Mimetite. A mineral, lead arsenate, formed as a secondary mineral similar to *pyromorphite in appearance, occurrence and physical properties. It was so named because it "mimics" pyromorphite. See Mineral Properties.

Mineral. A naturally occurring homogeneous solid with a definite (but generally not fixed) chemical composition and an ordered atomic arrangement. It is usually formed by inorganic processes.

Mineral, Cleavage. The property by which some minerals break along plane surfaces. The direction of breaking is controlled by the crystal structure and is always parallel to a crystal face or a possible crystal face. Cleavage is defined by the quality of the break, and its direction in the mineral. The quality is expressed as excellent, good, etc., and the direction by giving the name or indices of the form it parallels. For example,*mica has an excellent basal cleavage, whereas *augite has an imperfect, prismatic cleavage.

Mineral, Color. For some minerals, color is a fundamental, never-changing property that serves as a major criterion in identification. Thus malachite is always green, azurite and lazurite always blue, and rhodochrosite always pink, because of a major chemical constituent. Most metallic minerals also have a constant color, as shown in the brass yellow of chalcopyrite, the silver-white of arsenopyrite and the bronze of pyrrhotite.

Most nonmetallic minerals are composed essentially of elements that produce no characteristic color and are colorless. However, color varieties of these minerals may be common due to the presence of small amounts of strongly light-absorbing elements known as chromophores. Thus the presence of chromium produces the red of *ruby in *corundum and the green of *emerald in *beryl. Trace amounts of *iron in *quartz are responsible for the purple of *amethyst. The color in other minerals results from selective absorption of light due to imperfections in the crystal structure. Such may be the reason for green, blue and yellow crystals in usually colorless *diamond. During their growth, minerals frequently incorporate, and are colored by, small amounts of other minerals. *Hematite (red), manganese oxide (black) and *chlorite (green) are common pigmenting agents that impart their colors to otherwise colorless minerals. (C.S.H.)

Mineral, Dark. See Dark Mineral.

Mineral Deposit. Any valuable concentration of a mineral or group of minerals. In general, a twofold division is made: (1) metallic mineral deposits, which include ores of the metals and (2) nonmetallic mineral deposits, which include all other useful natural materials. The term is commonly used as a synonym for ore deposit but with no implication as to occurrence, size or shape.

Mineral Hardness. See Hardness.

Mineral, Heavy. See Heavy Mineral.

Mineral, Light. See Light Mineral.

Mineral, Luster. The quality or character of light reflected by a mineral. In general, minerals are divided into two groups on the basis of luster: metallic and nonmetallic. The metallic minerals have the brilliant appearance of a metal; all others are nonmetallic. When the division to which a mineral should be assigned is in doubt, it is sometimes classified as submetallic. Other terms are used to further characterize the luster of nonmetallic minerals. These are *adamantine, the luster of a *diamond; vitreous, the luster of broken glass; resinous, the luster of resin, as seen in *sphalerite; pearly, an iridescent pearl-like luster seen parallel to cleavage planes on the surfaces of some minerals, such as *apophyllite; greasy, as covered with a thin film of oil, as in nepheline; or silky, like silk as the satin spar variety of *gypsum. (C.S.H.)

Mineral, Parting. Planes of structural weakness resulting from pressure or twinning along which minerals tend to break. This property resembles *cleavage, since the breaking takes place parallel to crystallographic planes. It differs from cleavage in that it is present only in those specimens that are twinned or have been subjected to the proper pressure. Common examples are the rhombohedral parting in corundum and hematite and the octahedral parting in magnetite. See Twin Crystal.

Mineralogy. The science of the study of minerals. It includes a consideration of the crystallographic, physical and chemical properties of minerals as well as their origin, occurrence and association.

Mineraloid. A naturally occurring substance that is amorphous and thus is not included in any rigorous definition of *mineral. Mineraloids form by solidification of colloidal solutions that have the power to absorb other substances; thus they have a wide variation in chemical composition. The commonest examples are *opal and *limonite.

Mineral, Primary. See Primary Mineral.

Mineral Properties (See Table of Mineral Properties; Appendix 3). The basic characteristics of a given mineral. These properties depend on the chemical composition and internal structure of a mineral and are the factors that determine its identification. Among the major properties of minerals are chemical composition, *crystal system, *mineral luster, *specific gravity, *mineral color and *mineral cleavage.

Mineral Resources. The known and potential reserves of materials that can be extracted from the earth and that have some actual or potential economic value. Unlike some other natural resources such as forests and wildlife, mineral resources are nonliving. They are commonly naturally occurring, inorganic, crystalline substances, although they may be fluid (like petroleum), solid but noncrystalline (like coal) or derived from organic materials (like the fossil fuels).

Mineral resources are commonly classified into four broad types: mineral fuels, metals, nonmetals and water. Mineral fuels include coal, oil, natural gas and related products used chiefly as a source of energy. Metals are elements that carry a positive charge and are good conductors of heat and electricity. They typically have a shiny luster, called metallic, and can be hammered into sheets or drawn out as wire. Metals are traditionally subdivided into iron and ferro (iron) alloy metals, nonferrous metals, precious metals and minor metals, as follows:
1. Precious metals—gold, platinum, rhenium and silver
2. Nonferrous metals — aluminum, copper, lead, tin and zinc
3. Iron and the ferroalloy metals — chromium, cobalt, iron, manganese, molybdenum, nickel, tungsten and vanadium
4. Minor metals — antimony, arsenic, barium, beryllium, bismuth, cadmium, calcium, lithium, magnesium, mercury, radium and uranium, rare-earth metals, selenium and tellurium, tantalum and columbium, titanium, and zirconium

The elements radium and uranium are sources of nuclear energy and are therefore properly considered mineral fuels, although they are commonly listed as minor metals. Nonmetals include elements that lack the characteristics of metals and a broad grouping of other earth materials, as follows:
1. Construction materials — aggregates, binders, cement materials, crushed stone, dimension stone, gypsum, insulators, lime and pigments
2. Ceramic materials — clay, feldspar, talc and pyrophyllite, wollastonite and minor ceramic materials (barite, bauxite, borax, fluorspar, lithium minerals)
3. Metallurgical, chemical and refractory materials — bauxite, brucite, chrome, diaspore, dolomite, fire clays, fluorspar, foundry sands, limestone, magnesite, phosphorite, quartz and quartzite, salts and brines, sillimanite-andalusite-kyanite-dumortierite, spinel, sulfur and zircon
4. Industrial and manufacturing minerals — abrasives, asbestos, barite, bentonite, diatomite, graphite, mica, silica sand, talc, zeolites and other mineral fillers and filters
5. Fertilizer materials — agricultural limestone, nitrate, phosphate, potash and sulfur
6. Gemstones

Water is generally placed in a separate category, although it may be included as a nonmetal.

Several classifications of mineral resources have been devised, some based on origin of the resources and others on descriptive aspects of the occurrences. These groupings, especially those implying origins, are useful in understanding the formation of mineral resources and thus can aid in the exploration for additional deposits.

An important characteristic of most mineral resources is that they are nonrenewable (wasting) assets. They exist in finite amounts; once removed from the earth and consumed, they are gone forever. In this respect they differ markedly from certain types of energy resources (running water and sun's rays, for example) and other natural resources (agricultural products, timber and food from the oceans, for example) that are renewable. Only a few mineral resources are known to be "growing," that is, to some degree replenishing the supply being consumed. Examples include aluminum-rich bauxite deposits formed by *weathering and alteration of source rocks near earth's surface, salt-rich brines in desert valleys and phosphate deposits accumulating by leaching of bat guano in some caves. Even these deposits are "growing" at a slow rate, generally far less than present consumption rates. Careful consideration must thus be given to the total available supply and rate of use of each mineral resource.

A factor in the supply of mineral resources is their distribution and whether their value depends on where they are found. Most materials used in construction, as well as water, are low-value resources; their worth depends largely on their proximity to use (for example, deposits of sand and gravel). Although vital to the construction industry, haulage costs become prohibitive if the deposit occurs far from the site of its use. Construction companies desire local sources of the materials they consume in bulk quantities. Thus a deposit on the fringe of an urban center is eagerly sought after, whereas one 30 miles away holds little appeal. In contrast, high-value resources are sought and developed wherever they occur and are then transported to markets. Such is the case of petroleum, which is literally moved around the world in tankers and pipelines; nearly all metals are likewise considered high-value resources.

Some countries have much more extensive high-value mineral resources than others. World trade is controlled by this distribution of mineral wealth. Petroleum serves as one example; Iran, Iraq, Kuwait and Saudi Arabia in the Middle East contain the world's largest known reserves. The economy of these countries centers on the trade value of petroleum, needed by many other nations throughout the world. Another example is tin, a mineral largely concentrated in the Far East countries of Burma, Indonesia, Malaysia and Thailand. Some nations, especially those that are highly industrialized, consume quantities of mineral resources far in excess of their own production. Trade is a peaceful solution, but at times it has proved unsatisfactory. Conflicts between nations are commonly an outgrowth of these needs and the desire of an aggressor nation to acquire additional mineral resources.

Demands for resources increase as world population grows and more nations seek to improve their standard of living. Increased industrialization requires a continually greater consumption of the exhaustible supply of earth's mineral deposits. Partial solutions to this dilemma include: technological discoveries to produce substitute materials for those in critical supply; recycling used products to yield some minerals for reuse; conservation to reduce consumption; exploration of undeveloped areas such as the oceans to produce new supplies; and planning for future development in society to increase efficiency in the use of materials.

A growing awareness of environmental protection also reflects the concern for mineral consumption. Industry has been identified as a major polluter. A basic conflict therefore exists between demands for industrial products on the one hand and environmental protection on the other. One result has been a significant reduction in the use of some mineral resources identified as particularly harmful to the environment: for example, mercury and coal. The search for clean energy (such as solar radiation and geothermal sources rather than coal) is an example of the desire for environmentally cleaner resources. But the growing need for mineral resources despite their exhaustibility and the problems arising from their use continues to challenge mankind. (R.E.B.)

Mineral Spring. A hot or cold spring containing various salts in solution in appreciable quantities. *Carbonates, *sulfates, *phosphates and chlorides of calcium, magnesium, potassium, sodium and lithium are commonly present; so too are dissolved gases, including carbon dioxide and hydrogen sulfide. The mineral matter comes from the *bedrock and *regolith through which the water passed. Therapeutic properties are commonly attributed to the heat of the waters and the dissolved salts, but are perhaps due as much to the rest and relaxation that patients who take the cure generally get. Among the famous springs are those at Aix-les-Bains, France; Baden, Switzerland; Carlsbad, Czechoslovakia; and in the United States, Hot Springs, Ark., and Saratoga Springs, N.Y. See Spring; Thermal Spring. (R.L.N.)

Minerals, Silicate. See Silicate, Structure and Classification.

Miocene, Epoch. The fourth oldest *epoch of the *Tertiary Period; it began about 26 million years ago and lasted for approximately 14 million years. Its name, meaning "less recent," was introduced by Sir Charles Lyell in 1833. Miocene marine deposits are exposed in Italy, southern France, Austria, Rumania, Belgium and Holland. They are also well known in the West Indies and the Antilles. In North America, marine strata of Miocene age occur in the northern Atlantic Coastal Plain from North Carolina to Long Island and are especially well developed in the

Chesapeake Bay area. Scattered Miocene deposits occur in the southern Atlantic Coastal Plain of Florida. Both marine and nonmarine beds are known in the Gulf Coast region of western Florida, southern Mississippi, Alabama, southern Louisiana and southeastern Texas. The Miocene of the Pacific Coast area consists largely of a thick section of marine rocks in California. Nonmarine deposits of Late Tertiary age occur in the continental interior between the Rocky Mountains and the Mississippi River and from South Dakota to Texas. The Miocene of the northwestern United States is characterized by large quantities of volcanic ash, cinders and lava. Structural disturbances affected parts of North America near the close of the period, and there were mountain-building movements in the Alps-Himalaya region of Eurasia.

General uplift of the northern continents and localized mountain building produced cooler and drier climates in much of the world, and salt deposits in southern Nevada suggest desert-like conditions for that part of North America. Conditions were ideal for land life, with grasses and grains becoming more predominant and forests considerably reduced. The earliest redwoods appeared in Colorado, U.S.A., Greenland, and Europe. Marine invertebrates and fish were abundant, and many resembled modern species. Mammals were at the peak of their development and expanded with the grasslands. They were much more advanced than their *Oligocene ancestors, and included three-toed horses, camels, deer and antelopes. The entelodonts, tall, piglike creatures, about 2 m (6 feet) tall, were common in North America, as were rhinoceroses and the mastodons, which represented the early elephant family. Carnivores, some of them quite large, were abundant; primates (including the earliest apes) continued to advance.

Petroleum has been found in Miocene strata in California, the Gulf Coast region, the Soviet Union, the Middle East, the East Indies and parts of South America. Other valuable products associated with rocks of Miocene age include salt, clays, diatomaceous earth and building stone. (W.H.M.)

Misfit Stream. A stream occupying a valley that seems much too large or too small for it. Most commonly these are underfit rivers where the meanders of a stream are smaller and more intricate than those of the valley walls. Underfit streams have been ascribed to erosion by floodwaters, the influence of rock structure, and stream capture. Misfit streams are commonly found in northern Europe, northern North America

and other regions glaciated during the Pleistocene, when melting glaciers provided great volumes of water that excavated large valleys. With the disappearance of the ice, the discharge decreased, leaving the misfit streams.

Overfit rivers may occur where stream capture has swollen the discharge of a captor river. (M.M.)

Mississippian Period. The fifth period of the *Paleozoic Era; it began about 345 million years ago and lasted for approximately 25 million years. It is named from the Upper Mississippi River valley of the United States, where the system is represented by extensive limestone deposits. The Mississippian is not recognized as a separate period in Europe but is the approximate equivalent of the lower part of the *Carboniferous. Rocks of this age are widespread in Europe, where they consist of thick limestone beds containing numerous marine fossils. In the Soviet Union's Donetz Basin there are valuable coal deposits of Mississippian age, and sedimentary deposits occur in China, Siberia, Asia Minor, Australia, Africa and South America.

Most of the Mississippi Valley was inundated by the sea during this period, and the Appalachian *geosyncline received marine sediments along the eastern margin of the continent. Some of these Mississippian rocks form certain of the more prominent ridges in the eastern United States. There are widespread deposits of Mississippian marine limestones exposed in the western United States and Canada. Continental deposits, some containing the fossils of land plants and freshwater fishes, occur in eastern Quebec, Nova Scotia and New Brunswick. The end of the Mississippian was a time of widespread crustal unrest, and there was a general elevation of the North American continent. There is no evidence of extensive mountain building in the United States and Canada, but mountains were formed in western Europe.

Mississippian climates were probably warm and humid; the widespread seas gave rise to warm, temperate climates in much of the world. The presence of salt and gypsum in Michigan and Newfoundland suggests fairly arid conditions for those areas.

Mississippian life forms appear to be closely related to those of the preceding *Devonian Period. *Bryozoans (especially in corkscrewlike *Archimedes*), *foraminifers, and spine-bearing *brachiopods were common. *Crinoids and *blastoids were very abundant, and their remains are found as fossils in many Mississippian

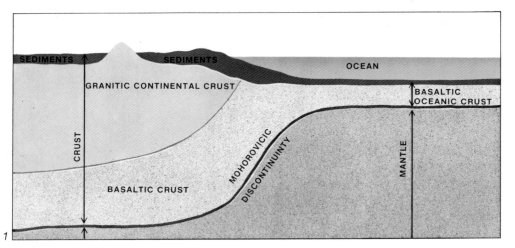

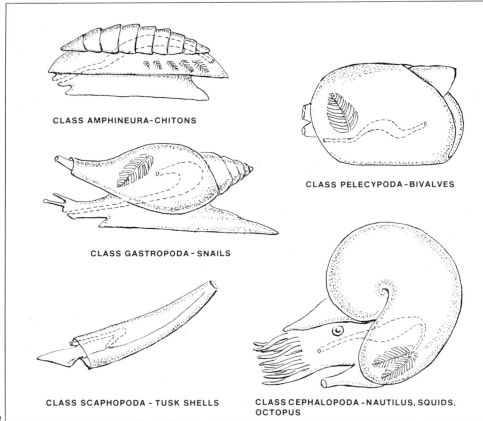

CLASS AMPHINEURA-CHITONS

CLASS GASTROPODA-SNAILS

CLASS SCAPHOPODA - TUSK SHELLS

CLASS PELECYPODA-BIVALVES

CLASS CEPHALOPODA -NAUTILUS, SQUIDS, OCTOPUS

limestones. There were dense swamp-forests on the land, and *amphibians lived among the ferns, rushes and other water-loving plant species. There were also many fishes, including about 300 species of shark.

Economic products from Mississippian rocks include coal from the Soviet Union and from Virginia and the Appalachian region; petroleum from the mid-continental and Appalachian areas of the United States; lead and zinc ores from the Tri-State District (parts of Missouri, Kansas and Oklahoma); and building stone from south-central Indiana and southwestern Missouri. (W. H. M.)

Mistral. A local name in southern France for a regional type of *wind that occurs where cold air flows from inland centers of high pressure through gaps in mountain ranges. Along the northern Adriatic coast, this wind is called the *bora*.

Modulus of Elasticity. The ability of a solid to resist deformation.

Mohorovicic Discontinuity (or Moho). The abrupt boundary between the relatively rigid outer shell of the earth (the *crust) and the hotter plastic material comprising much of the rest of the earth (the *mantle). The depth at which the Moho is encountered ranges from about 10 km beneath the surface of the ocean to 30 or 40 km below the continents. The existence of the Moho is known from the behavior of earthquake waves as they pass through it. As a result of the greater density of the mantle rock as compared to the crust above it, the waves undergo an abrupt increase in velocity as they pass from the crust into the mantle.

The discontinuity was named after its discoverer, Andruya Mohorovicic (1857-1936), a Croatian geophysicist. He is most widely known for his study of earthquake wave patterns produced by the Balkan earthquake of 1909 and his resulting conclusions that the earth consists of a layered structure.

A drilling operation to penetrate the earth's crust and the upper mantle was proposed in the early 1960s. Entitled "Project Mohole," the project was abandoned due to lack of funds after preliminary operations conducted by a specially constructed drilling ship off the western coast of North America. (J. W. S.)

Mohs Scale of Hardness. See Hardness.

Mold. In paleontology, the preserved impression of a prehistoric organism. See Casts and Molds.

Mollusk. A member of phylum Mollusca, a large group of aquatic and terrestrial invertebrates that includes such familiar forms as the snails, clams, oysters, squids and octopuses. Most mollusks possess a calcareous shell that serves as an exoskeleton and is well suited for preservation as a fossil. However, some mollusks, the slugs, have no shells; others, the squids, have an internal calcareous shell. Because of their relative abundance and great variety, fossil mollusks are particularly useful to paleontologists. Moreover, the remains of certain mollusks, such as the oysters, are important rock builders.

The phylum Mollusca has been divided into five classes:
1. Amphineura — the chitons or seamice, not a common fossil. Shell composed of eight valves or plates; range from Ordovician to Holocene (Recent) in age.
2. Scaphopoda — the various tusk-shells, generally not a common fossil but locally abundant in certain Cenozoic formations. Shell composed of a single tusklike valve; found in rocks ranging from Devonian to Holocene in age.
3. Gastropoda — the snails and slugs, common fossils in Paleozoic, Mesozoic and Cenozoic rocks. Slugs lack shells, whereas snails have a single-valve shell that is typically coiled; first appeared in Cambrian time and is very abundant today.
4. Pelecypoda — the clams, mussels, oysters and scallops, common fossils especially in Mesozoic and Cenozoic rocks. Shells composed of two valves, usually of equal size; range from Cambrian to Holocene (Recent) in age.
5. Cephalopoda — the squids, octopuses, the pearly nautilus and the extinct ammonoids, valuable fossils especially in Paleozoic and Mesozoic rocks. Shell composed of one valve, usually coiled and partitioned by septa; have been collected from rocks as old as Cambrian, and the squids, octopuses and cuttlefishes are common in the seas today.

(W.H.M.)

Molybdenite. A mineral, molybdenum sulfide, the principal ore of molybdenum. It is found in black, hexagonal-shaped plates with perfect cleavage, and thus closely resembles *graphite, a mineral with which it has long been confused. The bluish tone of molybdenite compared to the brown tinge of graphite helps distinguish the two. A positive differentiation can be made on the basis of specific gravity: molybdenite is 4.65, graphite is 2.23. Molybdenite commonly occurs in high-temperature vein deposits associated with cassiterite, wolframite and fluorite. It is also found in

*contact metamorphic deposits and as an accessory mineral in certain granites and pegmatites. The world's most important commercial deposit is at Climax, Colo., where molybdenite occurs in veinlets in a silicified granite. See Mineral Properties. (C.S.H.)

Monadnock. An isolated hill or mountain rising above a plain that has been leveled by stream erosion. The ultimate stage of the fluvial cycle of erosion is a *peneplain; however, denudation is never complete, and there are residual peaks (monadnocks) composed either of more resistant rock or farther away from base level. The classic example is Mount Monadnock, New Hampshire.

Monocline. A unit of bedded rock that changes dip in one direction, with the bedding on either side of the dipping layers horizontal or nearly flat lying. In cross section a monocline resembles a stair step with identical beds on either side of the *fold being separated vertically by hundreds or even thousands of meters.

Monoclinic System. In crystallography, the *crystal system in which the forms are referred to three unequal axes, one of which is perpendicular to the other two. If the three axes are lettered a, b and c, this definition implies that b is perpendicular to both a, and c; it does not impose a restriction on the angle between a and c. This latter angle in a few monoclinic crystals is 90° but in most it is oblique.

Monsoon. A flow of surface *winds that changes in dominant direction with the seasons because of great seasonal extremes of barometric pressure. The monsoon is particularly characteristic of southern Asia. In summer, warm, humid air moves northward and northeastward from the Indian Ocean and the southern Pacific, passing India, Indochina, and China. This air flow, which persists for several months, is called the summer monsoon and is accompanied by heavy rainfall. In winter, the air flow is reversed; it blows south and southeastward from north central Asia toward the equatorial oceans. This winter monsoon brings months of dry, clear weather.

Montmorillonite. A group of clay minerals whose outstanding characteristic is the capacity to absorb water to produce a marked swelling. Montmorillonite is the dominant material in *bentonite, altered volcanic ash that expands several times its original volume when placed in water. See Mineral Properties.

4

Mohorovicic Discontinuity
1 The earth's mantle is separated from the crust by an abrupt change in density known as the Mohorovicic Discontinuity, or Moho, which may be 32 km (20 miles) deep beneath the continental plains but only 6÷8 km (4÷5 miles) deep beneath the oceans.

Molluska
2 These bivalved marine mollusks are extremely varied and abundant, from the small, edible common clam or sea scallop to the astonishing giant clam of the South Pacific, which may reach 2 m (over 6 feet) in diameter.

3 Mollusks include five classes of animals with similar internal structure but different external appearance. Dashed lines indicate the digestive system. Leaflike structures represent the gills.

Monadnock
4 This rolling hill, known as Big Cobbler, is a conspicuous solitary feature on the worn piedmont plain near Sperryville, Va.

Monzonite. A granular *plutonic rock in which plagioclase exceeds potash feldspar and quartz is present in amounts no greater than 5 percent. It is thus intermediate in composition between *syenite and *diorite. It is commonly intermediate in color as well; the dark minerals, which may be hornblende, biotite or pyroxene, are usually more abundant than in syenite but less abundant than in diorite. Accessory minerals are apatite, sphene, zircon and magnetite.

Moonstone. Several varieties of translucent feldspar, orthoclase, albite or oligoclase, which show an opalescent play of colors.

Moraine. A landform generally composed of *till and found on glaciers and in areas once occupied by glaciers. It includes lateral, medial, terminal, recessional, end and ground moraines.

Where a valley glacier moves between cliffs, *frost action detaches fragments from the cliffs and they fall onto the glacier below. The blanket of debris thus developed on both sides of the glacier is called a lateral moraine. When valley glaciers disappear due to warming of the climate, the lateral moraines are lowered as the glaciers thin out and are finally deposited along the valley walls. Ridges are formed ranging from a meter to hundreds of meters high. The height of a lateral moraine above the bottom of its valley is a measure of the glacier's former thickness.

A medial moraine is a ridge of debris extending along the middle of a glacier. It is formed when two inner lateral moraines coalesce where two valley glaciers flow together. A region's principal valley glacier will generally have as many medial moraines as it has tributary glaciers. The terminus of a glacier remains stationary if the glacier's forward motion equals the rate at which its terminus melts back. A glacier, like a giant escalator, carries material on, within and beneath itself down to its terminus, where it deposits the material. Where the terminus of a glacier is stationary, an end moraine is formed.

A terminal moraine is an end moraine formed at the front of the most advanced position of a glacier. A recessional moraine is formed when the front of a glacier retreats and remains stationary for a time at a new position. When the terminus of a glacier retreats at a uniform rate and the debris it carried is deposited over the deglaciated region, ground moraine is formed. The height of terminal or recessional moraines depends on how long the terminus of the glacier remains stationary, the quantity of debris the glacier carries

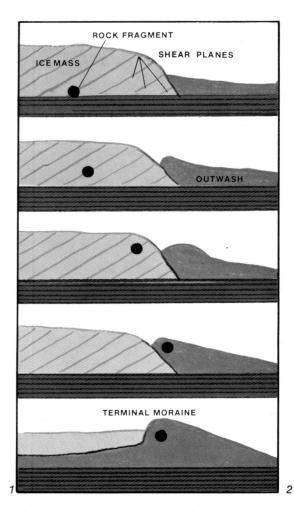

1

2

3

LATERAL MORAINE

...ESSIONAL MORAINE

TERMINAL MORAINE

Moraine

1 In this sequence of diagrams, the progressive movement of a rock fragment is shown as it is moved by ice from the bedrock floor to the ice margin, and finally deposited as part of the moraine.

2 Terminal and recessional moraines record the stages in the retreat of a valley glacier. Debris falling on the glacier from the valley walls creates lateral moraines.

3 Medial moraines on the Barnard Glacier, Alaska are ridges of debris formed where valley glaciers coalesce.

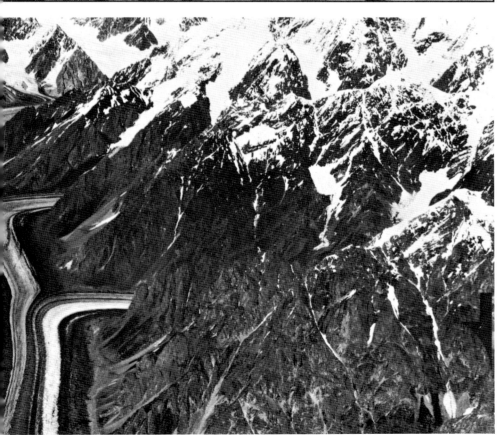

1

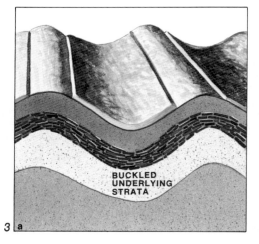

2

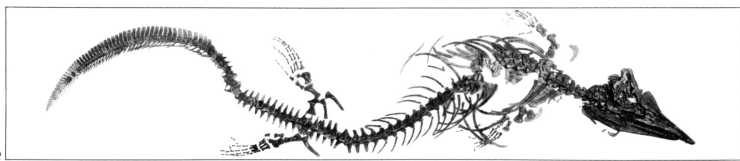

3 a

ERODED SEDIMENTARY ROCK STRATA

SOLIDIFIED
IGNEOUS
ROCK

b

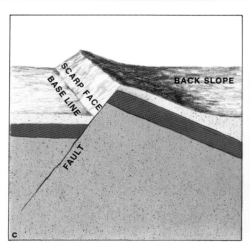

SCARP FACE

BASE LINE

BACK SLOPE

FAULT

c

Mosasaur

1 These Cretaceous marine reptiles, evolved from lizard-like ancestors, were found in oceans throughout the world.

2 *Tylosaurus,* more than 9 m (30 feet) long, had the powerful body and strong jaws that marked the culmination in the development of the mosasaurs.

Mountains

3 Folded mountains (a) are formed when two areas of the earth's crust are

compressed, causing the land between to buckle. Domes (b) result from an upwelling of molten rock that collects in a huge pocket beneath the surface and pushes the ground up in a bulge. Faulted mountains (c) are created when underground pressures force a mass of rock to rupture, rising on one side of the fault and subsiding on the other.

4 The face of Half Dome, Yosemite Valley, Calif., was sheared off by a glacier during the Ice Age.

5 The steep fault scarp of the Cathedral Group, in Wyoming's Teton Range, culminates in Grand Teton, 417 m (13,766 feet) high.

and the glacier's rate of motion. They range in height from a few meters to hundreds of meters. They commonly dam up their valeys, forming long, finger-shaped lakes that are among the most beautiful in the world: for example, Lakes Como and Garda in northern Italy, Lake Louise in the Canadian Rockies and St. Mary Lake in Glacier National Park, Mont. The terminal and recessional moraines of continental glaciers may extend for scores of kilometers and are often hundreds of meters wide and tens of meters high. They are conspicuous in Michigan, Ohio and Indiana, as well as in various parts of Europe.

Terminal moraines separate regions that differ greatly geologically. Recessional moraines, *kame terraces, till, ground moraines, *glacial striations, *roches moutonnees, an abundance of lakes and many other glacial features are found inside terminal moraines. Beyond them these features are absent and fluvial features usually predominate. (R.L.N.)

Mosasaur. One of a streamlined group of extinct marine lizards that lived in Cretaceous seas. Some of these great reptiles grew to be as much as 15 m (50 feet) long, and their great gaping jaws were filled with many teeth that are sharp and curve backward. See Fossil Reptiles.

Mountain. Topographically, a steep-sided eminence composed of *bedrock exposed at the surface or covered with vegetation, which rises more than 300 m (1000 feet) above the surrounding country. The summit is restricted in area and may be sharp or rounded. It is differentiated from a *hill by its greater altitude and from a *plateau by its smaller summit area. Mountains are formed by volcanic activity (mountain of accumulation), *folding and *faulting (mountain of dislocation) and *erosion. Their height, according to geographers, is the difference in altitude between sea level and their summits and for mountaineers, the difference in altitude between base and summit. High mountains are usually characterized by higher rainfall than the surrounding country, and the windward side is commonly wetter than the leeward. As the temperature of the atmosphere decreases with increasing height, the summits of high mountains, even those near the Equator, may be covered with snowfields that support glaciers. They have been effective barriers to migration and have profoundly affected the course of history.

The process responsible for the formation of mountainous topography is *epeirogeny, a larger scale process than the more localized process of *orogeny, which

4

5

Mud Cracks
Shrinkage cracks formed in the parched silt and mud deposits of the Colorado River at the head of Lake Mead, Ariz.

is responsible for the development of the interior structures of mountains such as folds, faults, and metamorphic features. (R.L.N.)

Mountain Cork (or Mountain Leather). A variety of tremolite asbestos occurring in felted aggregates resembling cork or leather. The "leather" type is found in thin, tough, flexible sheets; the "cork" is found in thicker but light, porous masses with a corklike quality.

Mud Cracks (or Sun Cracks, Desiccation Cracks). A system of polygonal cracks that form when mud, silt or clay shrinks in the course of drying. They occur in certain *sedimentary rocks and are typically wedge-shaped in cross section, tending to thin out downward. They develop in places where long exposure and a dry, warm climate prevail. They are largely continental in origin and indicate that the water in which the mud or clay accumulated was relatively shallow.

Mudflow. A flow of predominantly fine-grained, water-saturated, fluid *regolith. The water content ranges from 10 to 60 percent, and more than 50 percent by weight of the solid material is smaller than sand size. Mudflows range in length from 1 m to more than 80 km, and in thickness, from a few centimeters to tens of meters. The material is commonly unsorted, lacks stratification and contains angular fragments, some of which may be striated. It is therefore similar to and sometimes confused with glacial *till. The fluidity of mudflows ranges from that of freshly poured concrete to that of thick soup. Velocity ranges from 1 km per hour to more than 80 km per hour, depending on the steepness of the slope and the amount of water mixed with the debris. Their high density enables them to carry house-size fragments, and they sometimes emit a dull roar due to the collision of fragments.

The conditions necessary for the formation of mudflows are (1) steep slopes, as are found in mountainous areas, (2) fine-grained regolith veneering the slopes, (3) little or no vegetation and (4) intermittent heavy rains. These conditions are found in semi-arid and arid regions and on volcanoes mantled with fine-grained, unconsolidated tephra. Mudflows are common in the fault-block mountains of Nevada, California and Arizona, and in many other parts of the world. Material saturated by torrential rains moves down the steep slopes into valleys and sometimes out onto the *alluvial fans at the foot of mountains.

Volcanic mudflows are formed when

fine-grained, unconsolidated tephra is saturated by rain, melting snow, condensed volcanic steam or water from the bursting of crater lakes. They can be dangerous. Herculaneum, on the slopes of Vesuvius, was buried by a mudflow during the eruption of Vesuvius in A. D. 79. One of the most catastrophic mudflows occurred on the slopes of the volcano Gunong Keloet in Java in 1919, destroying 104 villages and killing more than 5000 people. (R.L.N.)

Muscovite. A mineral, hydrous potassium aluminum silicate, a member of the *mica group of minerals and commonly known as white mica. However, in thick crystals the mineral may be green, light brown or red. Muscovite is the common light-colored mica of granites, gneisses and schists. In some pegmatites it is found in crystals known as "book mica," which can be cleaved into thin, transparent sheets. Before the advent of glass these sheets, coming from Muscovy (an old name for Russia), were used as windowpanes under the name of Muscovy glass; hence the name of the mineral. Because of its heat-resisting nature, transparent muscovite is today used as windows in furnaces and ovens. This sheet mica is also used in the electrical industry in punched forms in radio tubes and as a base for winding electrical resistance elements. Finely ground muscovite used in roofing material, paint and wallpaper imparts to them a glitter or a sheen because of light reflected from the perfect cleavage. See Mineral Properties.

Mylonite. A fine-grained *metamorphic rock produced by the grinding and crushing of larger rock fragments. Development of mylonites occurs along planes of movement, called *faults, between large masses of solid rock. Mylonites form under conditions of low temperature and pressure found near the surface of the earth and do not undergo chemical or mineral transformations. It is believed that such conditions prevail along such features as the San Andreas Fault Zone in California or the Moine Thrust Zone in the Assynt region of northwest Scotland.

N

Nansen Bottle. A water-sampling bottle named after its developer, the Norwegian scientist Fridtjoff Nansen (1861÷1930). Water may be collected at any depth by means of Nansen bottles attached to a cable. Each bottle consists of a metal tube with valves at both ends. Bottles are lowered in the open position. When the first bottle is at the desired depth, a messenger weight is slid down the cable; on impact the bottle turns end over end, closing the valves and trapping the water sample inside. Each bottle may carry a weight to close the bottle below it; in this manner Nansen bottles may be used as the cable is played out at various depths. (R.A.D.)

Native Element. See Element, Native.

Natrolite. A mineral, hydrous sodium aluminum silicate. A *zeolite, it is found lining cavities in basaltic rocks associated with other zeolites and calcite. It characteristically occurs in acicular (needle-shaped) crystals, frequently in radiating groups. See Mineral Properties.

Nappe. Any large allochthonous (having been transported from place of origin) mass of rock that has been thrust or folded over an adjacent portion of the earth's surface. The distance over which the folded or faulted crustal rock must travel is on the order of a kilometer or more. Any structures that have moved a shorter distance are not classified as nappes. The Swiss Alps, for example, are the eroded remnants of large nappe structures.

Natural Bridge. An arch-like feature, usually spanning a river or a valley, that was formed by natural agents. Natural bridges excite interest because of their beauty, size and rarity. They are of little geologic importance, however, because they are short-lived, incidental features doomed by the processes that created them. They result from the actions of waves, shore currents and underground water, and from river erosion and volcanic activity. Bridges are commonly formed by waves and currents along coasts undergoing *marine erosion. Waves attacking a narrow promontory from both sides may break through it, forming a bridge between the mainland and the outer part of the promontory. Continued erosion will eventually destroy the bridge, forming a *stack, as seen along the coasts of California, France and Australia, at Perce Rock on the Gaspe Peninsula and at many other places.

Narrow, elongated caves or tunnels formed by the solution of limestone by underground water are often developed

Natrolite
This radiating group of needlelike crystals growing in a cavity in basaltic lava is typical of zeolite minerals.

along vertical joints. When all but a small part of the roof of such a cave collapses, a natural bridge is formed. The most famous example in the United States is the Natural Bridge in Virginia, which rises about 60 m (200 feet) above Cedar Creek and is about 45 m (150 feet) wide. A major highway passes across the bridge.

A bridge may also be formed when *lateral planation by a stream enables it to cut through the neck of a deeply entrenched *meander. If the upper level of the neck does not collapse, a natural bridge will be formed, and the river will then flow beneath the bridge. Rainbow Bridge in southeastern Utah, the world's largest natural bridge, was formed in this way. It rises 94 m (309 feet) above Bridge Creek and has a span of 85 m (278 feet).

A lava bridge is formed when the liquid lava in the central part of a lava flow drains away and all but a small part of the tunnel roof collapses. Examples are found in the Hawaiian Islands and the Columbia River Lava Plateau. The arches and bridges in the Arches National Monument in Utah were formed in still another way: Deformation of the crust resulted in vertical cracks; erosion widened the cracks into slots or canyons, and narrow vertical ridges or fins were formed between them. Where the rock in the fins was less resistant, groundwater, *weathering, *frost action, wind and gravity perforated the fins, forming windows and eventually bridges and arches. (R. L. N.)

Natural Gas. A naturally occurring, gaseous hydrocarbon (a chemical compound of hydrogen and carbon) commonly associated with deposits of oil. The natural gas content of a *petroleum reservoir may range from minute quantities dissolved in the oil to as much as 100 percent of the petroleum content. The gases range from natural hydrocarbons of low boiling point to those that consist almost entirely of hydrogen, nitrogen or carbon dioxide. The latter gases are of little commercial value. However, helium, which may form as much as 2 or 3 percent of some gas, may be commercially important. Methane (CH_4), the most stable of all petroleum hydrocarbons, makes up the greater part of the hydrocarbon content of natural gas. Variable but small amounts of other paraffin hydrocarbons — such as ethane (C_2H_6), propane (C_3H_8), butane (C_4H_{10}), pentane (C_5H_{12}) and hexane (C_6H_{14}) — are often present. Free hydrogen occurs only rarely in natural gas, as in some volcanic regions and around certain salt mines in Germany. Carbon monoxide and unsaturated gases occur only in minor amounts in most gases. The total amount of carbon in a typical natural

gas is approximately 35 pounds per 1000 cubic feet. About one-third of this may be recovered as carbon black, which consists of extremely fine particles of carbon from 10 to 150 microns in diameter, when the gas is burned in a limited supply of air. Air is taken to have a density of 1.0 and the density of natural gas ranges from that of methane, which is 0.554 relative to air, up to densities greater than air for certain gases. Most natural gases have a relative density between 0.65 and 0.90.

Natural gas issuing from a well may be classified according to the amount of natural gas liquid vapors that it contains. A dry gas contains less than 0.1 gallon natural gas liquid vapors per 1000 cubic feet, and a wet gas 0.3 or more. Natural gas from which the vapors of natural gas liquids have been extracted is called residue gas. The terms sweet gas and sour gas are used to describe natural gases that are low or high, respectively, in hydrogen sulfide. The principal gaseous impurities in natural gas are nitrogen, carbon dioxide, hydrogen sulfide and helium. Excessive quantities of carbon dioxide and nitrogen will decrease the flammability of natural gas and thus lower its heating value. Nitrogen and carbon dioxide also raise the temperature necessary for the combustion of natural gas.

Although its origin is not clearly understood, natural gas, like oil, is believed to have originated from the remains of microscopic plants and animals buried in sediments of ancient prehistoric seas, where they eventually underwent bacterial decomposition. The natural gas in a petroleum reservoir may occur as free gas, which occupies the upper part of the reservoir; as gas dissolved in oil; as gas dissolved in water; or as liquefied gas. Liquefied gas may occur at depths below 1800 m under reservoir conditions of high temperature and high pressure—conditions that make natural gas and crude oil physically indistinguishable.

Natural gas, like oil and coal, is a most important fossil fuel and energy source and is found in most petroleum-producing areas. (W. H. M.)

Natural Remanent Magnetism (NRM). A magnetism that is acquired by rocks at the time of their formation and that remains stable throughout their existence. NRM is used in the study of *paleo magnetism, which is the determination of the earth's *magnetic field throughout geologic time. Some rocks are more susceptible to magnetization than others because they contain certain minerals with magnetic properties. These minerals are iron oxides such as magnetite and hematite, or sul-

Natural Bridge
Lateral erosion through a neck of rejuvenated meandering stream created the Owachomo rock arch, Natural Bridges National Monument, Utah. The arch is 60 m (200 feet) wide and more than 30 m (100 feet) high.

fides such as pyrrhotite. Basaltic lavas and red sandstones show the greatest capacity to retain their NRM because they possess the greatest percentage of minerals capable of being magnetized.

When a rock such as basaltic lava forms from molten material, the minerals capable of being magnetized crystallize and become magnetized when their temperature drops to a level known as the Curie point. This point is about 500°C (932°F) for the magnetic minerals associated with lavas. Lavas have a greater capacity to become strongly magnetized at these very high temperatures than if magnetization were induced at room temperature. For this reason the NRM of lavas remains stable even though the rock may be subjected to other magnetic fields of differing intensity and direction at some later period.

During the formation of red sandstone, small fragments of magnetic minerals align themselves with whatever magnetic field prevails at the time. Several factors may cause the NRM produced in this fashion to be of less magnitude than that found in lavas. The mineral particles may be shifted by water currents after they have settled into position, or the sediments may be deformed by *compaction, thus decreasing the total effect of the magnetic particles. (J.W.S.)

Nautilus. The only living externally shelled *cephalopod. It is of great scientific interest. A living connection between the nautiloids and the *ammonoids, it has been termed a "living fossil." The comparatively simple nautiloids have undergone relatively little change since early Paleozoic time. The more complex ammonoids, however, have been extinct for some 70 million years. The earliest nautiloids were not coiled but had rather long, straight, slightly tapering shells. They were especially common during Ordovician and Silurian time. *Endoceras*, an uncoiled Ordovician form, grew to be almost 5 m long; most, however, were much shorter. (W.H.M.)

Nekton. A subdivision of marine life that includes any organism able to swim well enough to control its movement and position. This excludes invertebrates or juvenile fish that have some swimming abilities but are at the mercy of ocean currents. Nekton includes a wide variety of organisms, with fishes constituting the bulk. Invertebrates included among the nekton are squid, the pearly nautilus and some shrimp. The mobility of the nekton permits it to avoid many predators, seek food and shelter and in general move to more suitable locations.

Neogene Period. A unit of geologic time of Late *Cenozoic age, incorporating the *Miocene and *Pliocene epochs. In classifications that designate the *Tertiary Period as an era, the Neogene is considered the youngest period of the Tertiary Era and follows the *Paleogene Period.

Nepheline. A mineral, sodium aluminum silicate. This rock-forming mineral is the most abundant *feldspathoid. It is a major constituent of *nepheline syenite, *phonolite and other intrusive and extrusive rocks deficient in silica. Small crystals in rocks may resemble quartz but can be distinguished by the inferior hardness and greasy luster of nepheline. See Mineral Properties.

Nepheline Syenite. A variety of syenite that contains 5 percent or more *nepheline. The *potash and *plagioclase feldspars are in roughly the same proportions as in an ordinary *syenite, and dark minerals are usually present. In addition, some nepheline syenites contain sodalite (blue) and cancrinite (yellow), whereas others carry corundum.

Nephelinite. A fine-grained rock composed essentially of *pyroxene and *nepheline. Minor minerals include leucite or other feldspathoids, plagioclase and olivine. With increase of plagioclase and olivine, nephelinite grades into nepheline basalt. Common accessory minerals are sphene, apatite and iron oxides.

Nephrite. A tough, compact variety of tremolite-actinolite that constitutes part of the material known as *jade. Its color varies from white through deepening shades of green to nearly black.

Neptunism. An obsolete theory holding that all rocks were formed in or by water. It was proposed in 1787 by Abraham Gottlob Werner (1750 – 1817), a famous German mineralogist who assumed that all *bedrock (including basalt and granite) had been precipitated from a universal sea that once covered the entire earth. The Neptunists supported Scriptural interpretation, especially Noah's Flood. They explained fossils as victims of the Deluge and granite and basalt as the oldest deposits in the original sea. Neptunism was countered by the equally obsolete theory of *Plutonism, which assumed that all rocks solidified from an original molten mass. (W.H.M.)

Neritic Province. The *pelagic environment covering the *continental shelves from the shoreline to the outer shelf edge. Most of this environment receives light, and nutrients are abundant. As a result, there is a great abundance and diversity of both plants and animals. Much of the food obtained from the sea comes from this environment. See Marine Environment.

Nesosilicate. A group of silicates in which the silicon-oxygen tetrahedra occur as isolated units; that is, they are not linked together by sharing oxygen ions as they are in all other silicate types. In these structures the tetrahedra are bound together by other ions. The ratio of silicon to oxygen is thus 1:4, as in the mineral zircon, $ZrSiO_4$. See Silicate Structure and Classification.

Névé. A form of snow intermediate between freshly fallen snow and ice. It is formed near the surface of perennial snowfields and glaciers. Whereas snow is a light, fluffy aggregate of hexagonal crystals that may have a porosity as high as 95 percent and a specific gravity as low as 0.1, névé, with a porosity of around 50 percent and a specific gravity of about 0.5, is composed of compacted, nearly circular granules that may be as much as 1 cm or more in diameter. Snow is changed into névé by alternate melting and freezing, sublimation, condensation and pressure. The same processes change névé into ice. Névé is considered by some to be snow more than a year old. Skiers call it corn snow; it is also called *firn* (from the German for "of last year"). (R.L.N.)

Niccolite. A mineral, a nickel arsenide mined as a minor ore of nickel. Until the Swedish chemist Axel Fredrik Cronstedt isolated nickel from it in 1751, it was believed to be a copper mineral because of its copper-red color, and was called copper nickel. Niccolite occurs with other nickel arsenides and sulfides in vein deposits with silver and cobalt minerals. In this association it was found and described in the 12th century in the silver mines of Saxony, in what is now East Germany. It also occurs with *pyrrhotite, pentiandite and chalcopyrite, as in the nickel mines at Sudbury, Ontario. See Mineral Properties.

Nickpoint. See Knickpoint.

Nicol Prism. A device made of optical calcite for producing plane-polarized light. In the construction of the prism a cleavage rhombohedron is cut at a specified angle and then recemented with Canada balsam. Light entering the prism is double refracted, and the ordinary ray, because of its greater refraction, is totally reflected by the Canada balsam at the interface. The extraordinary light ray emerges from the prism as plane-polarized light. This prism, devised by William Nicol in 1828, was the first efficient polarizer. Today, because of the high cost, its use has largely been replaced by sheet polarizers such as Polaroid. (C.S.H.)

Niter. A mineral, the potassium nitrate that commonly goes under the name of saltpeter. As a mineral it is found in arid regions in white crusts or as a surface efflorescence in caves and other dry, sheltered places. It also occurs as crusts or impregnations in soils rich in organic matter. Niter is found in the desert regions of Chile associated with more abundant *soda niter. See Mineral Properties.

Nitrates. A group of compounds that contain the NO_3 radical. Only two natural nitrates can be considered common: *soda niter ($NaNO_3$) and *niter (KNO_3).

Nivation. Processes operating on slopes in polar regions and on high mountains that enable a snow patch to dig itself into either regolith or bedrock. Nivation includes *frost action, *creep, *solifluction and *sheet and *rill wash, all of which take place beneath or on the margin of the snow patch. They form nivation hollows, which may be several meters deep. When the snow patch becomes thick enough to move downslope as glacial ice, *glaciation is initiated.

Nondipole Field. Areas within the earth's *magnetic field that deviate from the idealized form. See Magnetic Anomaly.

Norite. A type of *gabbro with two pyroxene minerals, *augite and *hypersthene; hypersthene occurs in greater abundance.

Nuclear Breeder Reactor. See Energy Resources.

Nuclear Energy. See Energy Resources.

Nuée Ardente (or Glowing Cloud, Glowing Avalanche). Named from the French for "cloud" and "fiery," a hot cloud or mass of superheated gases and gas-charged fragments, sometimes incandescent, that erupts from a volcano. It originates in gas-rich acidic *magma and moves with hurricane velocity due to gravity and sometimes to the explosive force of the eruption. Nuées ardentes are also called pyroclastic, sand, ash and pumice flows. One of the most famous nuées ardentes rolled down the slopes of Mount Pelée in 1902 at about 160 km per hour and completely destroyed St. Pierre on Martinique, West Indies. Approximately 30,000 persons were killed; only two men survived — one a pris-

Nautilus

1 An X-ray photograph of a nautilus shows the curved partitions separating the gas-filled chambers of the shell, and reveals the living animal in the outermost chamber of the spiral.

Nekton

2 Organisms able to swim without relying on wave and current action are known as nekton (from the Greek for "swimming"). The group includes the majority of fishes and a few mammals, such as the harbor seal shown here.

oner in a dungeon. Glass bottles were melted by the blast, and a statue of the Virgin Mary weighing about 3 metric tons was carried 15 m. The cloud was said to be black, but in later eruptions a dull red color was reported.

The great mobility of a nuée ardente is thought to result from gases emitted by the fragments. The gases cushion the fragments, greatly reducing internal as well as basal friction, and enable the cloud to move swiftly and silently. Some nuées ardentes are explosively ejected from the sides of *cumulo-domes; others roll over the crater rim like boiling milk or foaming beer.

*Welded tuff (ignimbrite), which occurs in many places, is thought to result from the accumulation of nuée ardente deposits. Nuées ardentes have been observed at Soufrière Volcano, St. Vincent Island, West Indies; Merapi in Java; Santa Maria Volcano, Guatemala; and elsewhere. (R.L.N.)

Nunatak (from the Eskimo term). A hill or peak rising above and completely surrounded by glacial ice. Common in the Greenland and Antarctic ice sheets, nunataks are generally found near the margins of ice sheets where bedrock hills or the tops of mountains project above the surface of the ice. When the ice is thinning during deglaciation, nunataks increase in size; nunataks adjacent to each other may coalesce to form larger ones. They have the same relation to ice as islands to water and *steptoes to lava.

Nutation, Earth's Free (also known as the Chandler Wobble, named in 1891 after its discoverer, Seth C. Chandler, an American amateur astronomer). A wobble of the earth about its spin axis. It is the result of a displacement between the earth's spin axis and its axis of rotational inertia, that is, the axis around which there is an equal distribution of mass. To an observer in space it would appear that the earth's spin axis would describe a circle around the axis of rotational inertia. Since latitude equals the complement of the angle between the spin axis and the vertical at the point of observation, a change in this angle occurs because of the wobble. The angular change results in a cyclic variation in latitude. The cycle or period of time necessary for a point on the surface of the earth of known latitude to reach the maximum variation in its latitude and return to its original value is 435 days. (J.W.S.)

Nuée Ardente

1 The base of this tremendous glowing cloud that accompanied the 1902 eruption of Mt. Pelée rolled down the mountainous slopes like a fiery avalanche, killing tens of thousands of people within a few minutes.

Nunatak

2 An immense ice cap covers most of Antarctica, burying all but the tallest peaks, such as this bedrock peak, or nunatak, in McMurdo Sound.

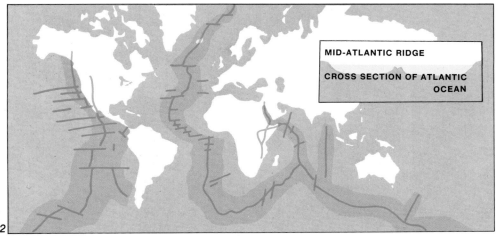

MID-ATLANTIC RIDGE

CROSS SECTION OF ATLANTIC OCEAN

Oceanic Circulation

1 *The general patterns of major surface currents in the world's oceans are primarily produced by two factors. Variations in density, created by differential heating and evaporation, cause the waters to move both horizontally and vertically. The friction of wind on the surface layers of the ocean creates surface currents. Once the water mass begins to move, it is deflected by the Coriolis force, caused by the earth's rotation.*

Oceanic Ridge

2 *The mid-oceanic ridge system winds around the world for more than 50,000 km (over 30,000 miles). At the ridge crest, rock cools as it wells up through the central rift zone and spreads outward, forming new sea floor. Insert shows a cross section of the Mid-Atlantic Ridge. The lines intersecting the ridges are fractures in the crust known as faults.*

Oceanography

3 *Marine technicians working off Barbers Point, Hawaii, install a long-period wave recording transducer that records small pressure changes and thereby permits measurement of waves.*

4 *A diver completes inspection of the research submarine* Deep Quest *before its descent 2518 m (8310 feet) in the Pacific.*

prevailing winds. This type includes the familiar current systems of the world such as the Gulf Stream and the Alaskan and Japanese currents. Surface currents move in large elliptical paths called gyres, whose patterns are formed by the interaction of prevailing winds, landmasses and the earth's rotation. Each of the large ocean basins of both the Northern and the Southern Hemisphere has a large central gyre caused by the prevailing westerly and trade wind belts. Smaller gyres may develop in the lower latitudes and in the higher latitudes of the Northern Hemisphere. In the Southern Hemisphere a belt of water called the West Wind Drift moves eastward around the globe. It can do so because no landmass obstructs its path as it moves around the Antarctic Continent. Surface currents are rapid and are swiftest on the west side of ocean basins because of the earth's west-to-east rotation and the *Coriolis effect.

*Thermohaline or deep-water currents dominate circulation elsewhere in the ocean. These currents may be quite slow and may move vertically as well as horizontally. They are generated by density differences that result primarily from temperature changes; the denser water seeks to move "downhill" to the site of less dense water, causing the density gradients that result in deep-water circulation. See Longshore Current; Upwelling. (R.A.D.)

Oceanic Crust. The thin, continuous portion of the outer zone of the earth. It has a composition similar to that of *basalt and its average density is 2.9. It is also referred to as the simatic crust because of its high silicon and magnesium content. The oceanic crust comprises the entire *crust under the oceans, where it is less than 8.0 km (5.0 miles) thick, and under the continents, where it may be somewhat thicker. Although the entire crust has not been penetrated by drilling, its thickness can be determined by the path and velocity of *seismic waves transmitted through it.

Under the oceans the crust is covered by about a kilometer of sediment and sedimentary rock. The thickness of sedimentary material over the oceanic crust reaches several kilometers along the continental margins where most of the sediment carried from the land is accumulated. Volcanic activity on the continents may cause some basaltic crust material to reach the surface on the continents. See Continental Crust. (R.A.D.)

Oceanic Ridge (or Mid-Oceanic Ridge). The largest single feature on the earth's surface, extending as a volcanic mountain range for more than 50,000 km (31,000 miles). It is present in all ocean basins and nearly bisects the Atlantic and Indian oceans; hence the commonly used term mid-oceanic ridge. In some areas volcanic peaks rise above sea level, forming islands such as Iceland, the Azores, and Ascension Island in the Atlantic. Although the central high-relief zone of the ridge is narrow, the ridge province itself is several hundred kilometers wide and may, as in the Atlantic, comprise over one-third of the ocean basin. Although the ridge is globally continuous, there are numerous dislocations caused by strike-slip faults along its extent. In addition, the ridge has a central elongated rift valley formed by a *graben, that is, a central block with a high-angle fault on each side.

Oceanic ridges are characterized by much earthquake activity, by very young rocks of a basaltic or mafic (ferromagnesian) composition, high heat flow, thick crust and higher than normal gravity values (see Isostasy). It has been shown generally that the crust of the ocean basin becomes older from the ridge toward the continents. This supports the belief of most geologists that the oceanic ridge system serves as the center of spreading in the *continental drift and *sea floor spreading process. Sediments on the ridges are largely pelagic carbonate *ooze, which is accumulating at only 2−4 mm per 1000 years. Clay is present on the lower ridge areas, but its rate of accumulation is less than 1 cm per 1000 years. In very steep areas of the ridge these fine oozes and clays may move downslope and accumulate in low areas. (R.A.D.)

Oceanic Rise. Generally a large area of the ocean basin elevated above the surrounding abyssal hills. It is distinctly separated from the *continental margin and *oceanic ridges and usually encompasses thousands of square kilometers. It is thought that rises originate as gentle upliftings of the ocean floor. They are at present seismically inactive, in contrast to oceanic ridges, where earthquakes are common. Examples are the Bermuda Rise in the Atlantic and the Darwin Rise in the Pacific. (R.A.D.)

Oceanography. The application of all sciences, including biology, chemistry, geology, meteorology and physics, to the study of the marine world. Oceanography is a relatively young discipline, the first systematic study of the oceans being that of the *Challenger Expedition of 1872-76. The title of "first oceanographer" is commonly bestowed upon Lt. Matthew F. Maury (1806-1873) of the U.S. Navy; while attached to the U.S. Hydrographic Office, he wrote the first book on the subject,

3

4

Physical Geography of the Sea (1855), and he was the first person to devote full-time duties to the study of the oceans.

Most of the early studies of the oceans were made by biologists interested in the abundant and diverse life of the marine realm. These studies were accompanied and followed by investigations of ocean currents and water chemistry. Because of the great amount of travel by sail in former times a knowledge of currents and weather was of great importance. It became common practice for ships' logs to include valuable information on these matters. In time, more detailed information was needed for the purposes of transportation, naval defense and commercial fishing.

Among the specialists in the field of oceanography are biological oceanographers, who are concerned with the nature and distribution of life in the ocean; chemical oceanographers, who study the chemical composition of, and reactions in, the oceans; and geological oceanographers, who study the topography, history and sediments of the ocean floor. Physical oceanographers and marine meteorologists are interested in the movements of fluids (air and water) within and above the marine world. Because of the interactions among these disciplines, it is of course important for a specialist to have a general knowledge of all aspects of oceanography.

In comparison to what we know of the continents, we know very little about the 70 percent of the earth that is covered by marine waters. Because of the ever-increasing need for new sources of food and natural resources as well as the continuing importance of the seas for transportation and defense, the ocean environment is receiving more and more attention from scientists. Most of this intensive investigation began after World War II. At present the bulk of all oceanographic research is funded by various federal agencies in the United States and in other countries where active research is in progress, such as Canada, Great Britain, Germany, Japan, Monaco, and the Soviet Union. In the United States the primary support comes from such sources as the National Science Foundation, the National Oceanic and Atmospheric Administration (NOAA), the U.S. Bureau of Commercial Fisheries, and the U.S. Geological Survey.

Oceanography, or marine sciences as it is sometimes called, is fast becoming a part of the curriculum at many universities. In the United States, the universities of California, Delaware, Miami, Michigan, Rhode Island, Washington and Wisconsin, as well as Oregon State University, Texas A & M University and Woods Hole Ocean-

ographic Institution have comprehensive programs. Because students need a general background in science prior to study in oceanography, the vast majority of these institutions give degree programs only on the graduate level. (R.A.D.)

Octahedron. A form of the *isometric crystal system composed of eight symmetrically equivalent faces. In the geometrically perfect octahedron, each face is an equilateral triangle.

Oil. See Petroleum.

Oil Shale. A very fine-grained *sedimentary rock that when processed contains sufficient organic matter (hydrocarbon) to yield oil. Most oil shale is dark (black, brown or dark gray). However, some layers are yellow, red or green because they contain waxy organic materials; very thin slices of these rocks viewed through a microscope reveal small mineral grains surrounded by dark, shapeless masses of organic material speckled with brilliant yellow, red or green particles. These particles are probably fossil remains of spores or pollen or filaments of algae, a single-celled plant. The remains of other plants or animals may also be conspicuous and abundant, although most of the organic matter in oil shale is generally so finely pulverized that the fossils in it cannot be identified. The shale also contains large amounts of incombustible minerals such as calcite, dolomite, clay, quartz and feldspar, which generally make up more than half the rock.

Oil shale accumulated as layers of organic ooze and mud on the bottoms of ancient lakes, ponds, lagoons and shallow seas, where plant or animal life was abundant. As such organisms died and sank to the bottom, their remains were partly preserved, particularly where the water was stagnant and lacked sufficient oxygen for complete decay of organic matter. The organic-rich layers of mud were gradually buried as younger sediments accumulated above them, and the weight of the overlying material slowly turned the mud into hard shale. Oil shale is present in some of the earliest known fossiliferous rocks (about 600 million years old) and also in many younger rocks, including a few small deposits that formed in modern lakes. It is rather widely distributed in certain sedimentary rocks throughout the world.

When oil shale is heated to a temperature of about 480°C (900°F), much of its organic material melts, and gases are released when this liquid is heated to higher temperatures. Some of these gases can be condensed to form oil, and part

remains as a hydrocarbon gas. From 25 to 75 percent of the organic matter in oil shale can thus be converted to oil and combustible gas. Oil shale can be burned as a solid fuel or processed to produce oil or combustible gas. Although the major item of commercial interest in such deposits generally has been oil, other products can also be obtained from them. Combustible hydrocarbon gas, for example, has been a major product of some European shale industries. After the oil is removed, the spent shale can be used for lightweight aggregate for construction purposes.

The known deposits of oil shale are estimated to be capable of yielding 10 gallons or more oil per ton of rock or a total of more than 2 trillion barrels. Thus, the oil shale deposits of the world can be considered collectively as an enormous low-grade source of oil, hydrocarbon gas, or solid fuel. The most extensive high-grade oil shale deposits known are in North America. They lie in the United States in the Rocky Mountain region, extending almost 8000 square kilometers in parts of Colorado, Utah and Wyoming. These deposits are part of sedimentary materials that accumulated in the bottoms of two vast shallow lakes during the Eocene Epoch of the Tertiary Period. These lake-basin sediments, named the Green River Formation from exposures near the Green River in Wyoming, were deposited about 50 million years ago. Deposits of high-grade oil shale in the Piceance Basin of northwestern Colorado are estimated to contain 1.3 trillion barrels of oil.

Large deposits of oil shale that formed during the Jurassic and Cretaceous periods occur near the Brooks Range in northern Alaska, and deposits of Triassic age occur in eastern Alaska. They also occur in New Zealand.

Only the highest grade and more accessible oil shales are of commercial interest. About 80 billion barrels of oil from deposits in the Green River Formation can be considered available at costs approaching the costs of petroleum of comparable quality. Such potential resources are more than double the proved reserves of petroleum in the United States. Experiments to extract oil or gas by heating the shale in place have been made on some deposits. If the extraction of oil from shale eventually proves to be economic on an industrial scale, the potential supply of oil from shale will be enormous. (W.H.M.)

Oligocene Epoch. The *epoch of the *Tertiary Period that began about 38 million years ago and lasted for approximately 12 million years. Originally part of the *Eocene Series as described by Sir Charles Lyell in

1833, the Oligocene (literally, "few recent") was named in 1854. The rocks were first studied in the vicinity of the Paris and London basins where they typically consist of fossiliferous clays, sandstone, limestone and lignite. Oligocene rocks include marine, brackish water deposits and continental deposits in France, Germany, Russia, Belgium, Spain, the British Isles, western Asia, South America, North Africa and the West Indies. Lava flows in Iceland and the northwestern British Isles suggest volcanic activity, and crustal deformation apparently occurred in the Alps, Pyrenees, Apennines and Himalayas near the end of Oligocene time. In North America, Oligocene strata (mostly marine) are exposed in the Gulf Coast region and southern Atlantic Coastal Plain. Similar deposits occur in the Pacific Coast area (California, Oregon and Washington) and in southern Alaska. Continental deposits are exposed in north-central Oregon and the plains area of Wyoming, Colorado, Nebraska and South Dakota. These include the White River beds, which have been eroded into the famous "Badlands" area southeast of South Dakota's Black Hills. Consisting largely of sands and clay, these stream-laid deposits contain the remains of countless perfectly preserved fossil mammals.

Oligocene climates were somewhat cooler and drier than in the preceding Eocene Epoch, but fossil palms in northern Germany and the remains of alligators and palms in the Dakotas suggest warmer climates for these areas.

Life of this epoch resembled that of the Eocene, and marine invertebrates included foraminifers, gastropods, pelecypods, corals and echinoderms. Rays, sharks and bony fishes were also numerous. Plant life was marked by an expansion of the grasses to form vast prairies, which were inhabited by a rapidly expanding mammalian population. Many of the more primitive mammals had become extinct, and more modern groups such as horses, deer, antelopes, camels, rhinoceroses and members of the cat and dog families were abundant. Giant mammals such as the elephant-size titanotheres and the baluchitheres were also present. *Baluchitherium*, whose remains were found in central Asia, was a rhinoceros-like beast about 10 m (33 feet) long and standing 5.5 m (18 feet) at the shoulder—believed to be the largest land mammal of all time. Proboscideans (elephant-like animals), apes and monkeys were also present.

Economic resources from the Oligocene are rather limited, but they include Middle Eastern and North American petroleum and European lignite. (W.H.M.)

Oligoclase. A mineral, one of the *plagioclase feldspars, present as a major constituent in *granodiorite and *monzonite rocks. Its composition in terms of percentages of the end-members of the plagioclase series, *albite (ab) and *anorthite (an), varies from 10ab, 90an to 30ab, 70an. When oligoclase contains hematite inclusions which give it a golden sparkle it is called aventurine or sunstone.

Olivine. A mineral series extending from *forsterite, magnesium silicate, to *fayalite, iron silicate. The pure or nearly pure end-members are rare; the most common olivines are richer in magnesium than in iron. Olivine is a rock-forming mineral varying in amount from that of accessory to that of a main constituent. Its principal occurrence is in dark-colored rocks such as gabbro, peridotite and basalt; in the rock known as dunite it is the only major constituent. In all of these rocks it occurs as small cracked and clouded grains, but in a few places larger clear crystals have been found that are cut as the gemstone *peridot. The name olivine comes from its characteristic olive-green color. See Mineral Properties. (C.S.H.)

Onyx. A microcrystalline variety of quartz with different colored layers of *chalcedony arranged in parallel planes. It is similar to *agate except for the arrangement of the layers, which are concentric in agate. Onyx with alternating white and black layers is used for cameos. The name onyx is also used in the building-stone industry to describe a variety of banded marble, which does resemble true onyx.

Oolite. A *sedimentary rock, usually limestone, consisting of small spheroidal or ellipsoidal particles called ooliths. The ooliths resemble fish roe and range from 0.25 to 2 mm (0.01 to 0.08 inch) in diameter. They are typically composed of concentric layers of calcium carbonate, but some consist of silica, phosphate or iron compounds. They are formed as accretionary growths from chemical precipitation of minerals around a nucleus of foreign matter such as a shell fragment or a grain of sand. This occurs when warm marine water is agitated, producing evaporation that increases the concentration point at which calcium carbonate can be precipitated. The rounded shapes are the result of ooliths continually being rolled about on the ocean bottom. Ooliths are being formed today in the shallow seawater of the lagoons and tidal channels of Florida and the Bahamas. Rocks with an oolitic texture are typical of certain marine sedimentary series. (W.H.M.)

Ooze. A type of *pelagic or deep-sea sediment made up of at least 30 percent sediment derived from the skeletons of microscopic floating organisms. These may be either plants, most commonly *diatoms and *coccoliths, or animals such as *foraminifera and *radiolarians. Its composition may be calcium carbonate (foraminifera and coccoliths) or silica (radiolaria and diatoms). Foraminiferal oozes are the most widespread of all deep-sea sediments. They occur throughout the Atlantic, Indian and southern Pacific oceans. Diatomaceous oozes are most abundant in the Southern Hemisphere. (R.A.D.)

Opal. A mineraloid, some varieties of which are highly prized as gemstones. It is classed as a mineraloid rather than as a mineral because it is noncrystalline, and its chemical composition ($SiO_2 \cdot nH_2O$) shows that it is silica containing an indefinite amount of water. The specific gravity varies from 2.0 to 2.25 depending on the water content; the less water, the higher the specific gravity.

Gem or precious opals have long been valued for their subtle beauty resulting from an internal play of colors. One can scarcely improve on the description of Pliny nearly 2000 years ago: "There is in them a softer fire than in the carbuncle, there is the brilliant purple of the amethyst; there is the sea-green of the emerald—all shining together in incredible union." Today, various names are given to precious opal: fire opal is a translucent gem emitting flashes of red to orange colors; black opal a dark body that emits a brilliant play of colors. At the time of Pliny and for centuries thereafter opals rivaled the diamond, but early in the 19th century opals became unpopular because some people believed they were unlucky. Today opal has regained its place of high esteem.

Australia is the principal source of gem opals, from widely separated localities in Queensland, New South Wales, South Australia and Western Australia. Other important sources are Querltaro, Mexico, and the Virgin Valley, Nev.

Although attention through the years has focused on precious opal, other varieties are far more plentiful. Common opal—which may be white, blue, yellow, green or red—has the characteristic milky luster or "opalescence" but lacks the internal reflections. Hyalite occurs on the walls of cavities in clear, colorless globules resembling drops of water. Diatomite, a fine-grained deposit resembling chalk, is composed of countless tiny opaline shells of diatoms that accumulate on the sea floor or lake bottom.

Opal forms under conditions of low pressure and low temperature and is thus deposited in rock cracks and cavities near the earth's surface. It is also a common petrifying material in petrified wood in recent volcanic ash and sedimentary rocks. See Mineral Properties. (C.S.H.)

Open Pit Mining. A process in which valuable minerals are recovered by surface operations. This method, sometimes called open cut or open cast, is less expensive than underground mining and is usually employed in deposits with large volumes of low-grade or low-value ore, for example, in mining iron ore, phosphates and porphyry copper.

Order of Streams. A ranking or classification of streams in a *drainage basin. Unbranched, beginning tributaries are first-order; two or more first-order streams unite to form a second-order; two or more second-order streams join to form a third-order, and so on. It takes at least two streams of a given order to form the next higher order, but rivers of a given order may have any number of lower-order streams.

Ordinary Ray. Light entering optically uniaxial crystals is refracted into two rays: the ordinary and extraordinary. The ordinary ray has a constant index of refraction regardless of the crystallographic direction traversed.

Ordovician Period. The second oldest period of the *Paleozoic Era; it began about 500 million years ago and lasted for about 75 million years. Ordovician rocks, first studied in northern Wales, derived their name from the Ordovices, a Celtic tribe that once inhabited this region. The system was named in 1879 by British geologist Charles Lapworth (1842-1920). His studies of Lower Paleozoic strata in Wales clearly separated the Ordovician rocks from the underlying Cambrian and from the Silurian formations lying above the Ordovician. In establishing the boundaries of the Cambrian, Ordovician and Silurian systems, Charles Lapworth settled one of the most celebrated controversies in the history of geology.

The Ordovician Period was a time of widespread marine sedimentation. Shallow seas flooded what is now the Mediterranean region, the British Isles, Scandinavia, the Baltic area and much of Siberia. In North America the sea covered as much as 70 percent of the continent, and the surrounding landmasses were considerably lower than they had been during Cambrian time. Thousands of meters of

Open Pit Mining
At Bingham, Utah, one of the world's largest copper producers exploits low-grade ore by excavating it at the surface.

sedimentary rocks have formed from sediments deposited in the apparently warm Ordovician seas. Some of these—graywackes, black shales and volcanic ash—indicate deposition in deep trenches or subsiding basins. Others, such as dolomites, limestones and calcareous shales, represent sediments deposited in relatively shallow seas that covered the continents. Many of the black shales contain large numbers of *graptolites, floating colonial animals that attained worldwide distribution by drifting with the ocean currents. When they died their remains settled to the ocean floor and were incorporated into the sediments. Certain Early Ordovician graptolites of North America are similar to those found in Lower Ordovician rocks in Wales. These guide fossils permit intercontinental *correlation of many Ordovician strata. Graptolites were so numerous that the Ordovician has been called the "Age of Graptolites."

Near the end of Ordovician time there was uplift in eastern North America along a line extending from New Jersey to Newfoundland. Known as the *Taconic orogeny, this mountain-building disturbance was accompanied by much volcanic activity and rock deformation. The Taconic Mountains of eastern New York State are the eroded remnants of the ancient Ordovician landmass produced by this uplift.

Ordovician climates were probably mild and equable over much of the world, and there is no evidence of well-defined climatic zones. The climate and the widespread Ordovician sea were favorable to the development of life. The oceans teemed with large numbers of marine plants and invertebrates, and their fossilized remains are abundant worldwide. These organisms were more complex and varied than those of Cambrian time, and the trilobites and brachiopods that had dominated Cambrian life were even more numerous. They were joined by distinctive species of corals, cystoids, blastoids, crinoids, gastropods and pelecypods. Especially significant was the development of exceptionally large cephalopods with straight, cone-shaped shells. Some of these, such as *Endoceras*, were as much as 4.5 m (15 feet) long. The bryozoans, which made their earliest known appearance in Early Ordovician time, underwent rapid and extensive development.

The most noteworthy event of the Ordovician was the appearance of the *ostracoderms, the earliest known vertebrates. Known from fragmental bony plates and scales collected in the Rocky Mountain region, these small, armored, jawless fish are related to the modern hagfish or lamprey eel.

Ordovician rocks have yielded a variety of products. The well-sorted, pure quartz sand of the St. Peter Sandstone formation in northern Illinois and Missouri is an important source of glass sand. Marble of Ordovician age is quarried in Tennessee and Vermont, and limestone for the manufacture of cement, whitewash, fertilizer and building stone is produced from the northern and central Appalachian region. Lead and zinc ores are mined in southern Wisconsin, northern Illinois and Iowa. Subsurface Ordovician formations have produced much oil and gas in the mid-continent region of the United States. (W.H.M)

Ore. A mineral or an aggregate of minerals from which one or more metals can be profitably extracted. In an ore the ore minerals contain the valuable metals, but these are usually intimately mixed with worthless minerals termed gangue. The word "ore" implies that the mineral assemblage can be mined and the metal extracted at a profit. Thus the ore mined when the price of the metal is high is no longer considered ore when the price falls so low as to make mining uneconomic.

Traditionally the word was used to refer to metalliferous minerals or aggregates, but it is now expanded to include any natural material used as a source of a nonmetallic substance. Thus pyrite is an ore of sulfur and colemanite is an ore of borax. (C.S.H.)

Ore Body. A mass of rock containing one or more minerals of value. Ore bodies vary greatly both in size and shape, and are characterized by mineralogy or structure that distinguishes them from the surrounding country rock. An ore body rarely contains only the valuable ore minerals, and commonly a high percentage of it is composed of waste material.

Ore Deposit. A localized, natural occurrence of a mineral or group of minerals in sufficient concentration to have economic value. See Mineral Deposit.

Ore Dressing. The process by which valuable ore minerals are separated from gangue (the waste material). The method that is used depends on the nature of the ore minerals and gangue. Methods include mechanical separations by gravity, magnetism, flotation and static electricity as well as by the more primitive procedures of hand-cobbing and hand-picking.

Ore Shoots. Those parts of a deposit in which the valuable minerals are sufficiently concentrated to permit their economic extraction. For example, in a gold-quartz vein, quartz alone may be present in most of the vein and the gold confined to pipelike masses (ore shoots) within it.

Orogeny. Literally meaning "formation of mountains," the term was originally applied to the process responsible for the development of mountainous terrain and the deformation of rock within mountains. Today it refers chiefly to the latter process. The formation of mountain systems is in part related to vertical movements within the earth's crust and is termed *epeiorogeny. Orogeny also involves the bending and fracturing of the outer layers of the crust in mountainous areas to form *folds and *faults. Deep inside mountain chains, where high temperatures and pressures occur, orogenic structures are formed by the plastic flowage of rocks into folds, the transformation of original rock into new rocks (*metamorphism) and the introduction of large masses of molten rock into mountain cores (*plutonism). Geologists generally, but not universally, agree that orogeny results from compressional forces acting on segments of the crust. The stress resulting from these forces is relieved by folding and faulting with subsequent shortening of the crust.

Possibly the most notable episodes of orogeny that have occurred in North America throughout geologic time are the Taconic, the Acadian and the Alleghenian of eastern North America and the Nevadan and the Laramide of western North America. In Europe the Caledonian of Ordovician and Silurian age produced great mountain ranges that extended from Ireland to Scandinavia. Although detailed correlations with the Taconic orogeny of the northern Appalachians are not yet possible, the available information suggests that the Caledonian and the Taconic may have been essentially contemporaneous. Some geoscientists have theorized that the mountain chains of eastern North America and western Europe were originally continuous.

The Acadian and Alleghenian orogenies occur in what is now the linear chain of mountains extending from Newfoundland to Alabama; the Taconic occurs in the northern Appalachians. The Acadian orogeny can be related to a relatively short-lived crustal movement, metamorphism and/or plutonism occurring from 410 to 380 million years ago. Most of the metamorphic and plutonic structural features related to this orogeny are present in large segments of the Appalachians. Its climax is considered to have occurred in New England in the Middle Devonian (approximately 380 million years ago). The Taconic orogeny affected large parts of the

Orogeny
The Sierra Nevada Mountains of northern California were originally thrust up between 180 million and 80 million years ago during a period of deformation, igneous intrusion and metamorphism known as the Nevada Orogeny. They were then eroded almost to a peneplain, but later faulting uplifted a great tilted block from which the present mountains have been eroded.

1

2

northern Appalachians, deformations taking place with varying degrees of intensity in different parts of the mountain belt. This orogenic activity spanned the interval from late Ordovician to late Silurian time. The elleghanian orogeny cannot be as well dated as the Acadian, but it is believed that it affected crustal rocks in parts of the northern, central and southern Appalachians in the Pennsylvanian to Early Permian and possibly into the Early Triassic periods (approximately 320 to 220 million years ago). It is characterized by strong folding in the southern and central section. The present topography of the Appalachian Mountains is related not to these events alone but also to epeiorogenic movements that occurred considerably after the close of the elleghanian orogeny.

In the western United States a period of deformation and plutonism known as the Nevadan orogeny occurred in parts of California, Nevada, Oregon, Washington and Idaho as well as British Columbia. It was characterized by widespread and complex folding and faulting and the intrusion of massive granitic *batholiths. It began in the Jurassic and lasted through much of the Cretaceous, ranging in different areas from 180 to 80 million years ago. The episode known as the Laramide orogeny lasted roughly from the late Cretaceous to the Early Tertiary (approximately 80 to 60 million years ago). In the broadest sense, the Laramide orogeny covers many separate orogenic events generally occurring in the area now occupied by the Rocky Mountains. It elevated a long chain of mountains as compressional forces folded and faulted hundreds of meters of sedimentary rocks. The copper deposits of Butte, Montana, and the gold, silver and lead ores of Colorado are related to more or less contemporaneous igneous activity that produced zones of mineralization in these areas. (J.W.S.)

Orthoclase

1 Prismatic crystals of this feldspar mineral are found lining cavities in a granite at Baveno, Italy. In the pegmatite environment in which the mineral commonly forms, crystals may grow to a length of several feet.

Ostracoderm

2 Birkenia (A) was a small, primitive, jawless fish of Late Silurian time whose head was covered with an armor of small scales. It probably fed on

plankton near the surface of the water. Hemicyclaspis (B),which had a flattened bony head shield and a scale-covered body, lived along the bottom of shallow streams, sucking up small food particles with its wide mouth.

Outcrop

3 The tilted bedrock exposed at Vasquez Rocks, Calif., is sandstone and conglomerate.

Orpiment. A mineral, arsenic sulfide, with a bright lemon-yellow color. It is found in veins of lead, silver and gold ores associated with another arsenic sulfide, *realgar. In the United States it is found at Mercer, Utah, and Manhattan, Nevada. Orpiment was formerly used as a pigment but this use has been discontinued because of its poisonous nature. See Mineral Properties; Appendix 3.

Orthoclase. A mineral, potassium aluminum silicate, a member of the feldspar group of minerals. It is monoclinic, dimorphous with triclinic microline; the two minerals are known as the *potash feldspars. Orthoclase is a common constituent of granitic rocks. See Mineral Properties.

Orthorhombic System. The crystal system in which the forms are referred to three unequal axes that are at right angles to each other. The unit of structure, the unit cell, of the orthorhombic system has three mutually perpendicular edges of unequal length.

Ostracod. A small, bivalved, aquatic crustacean that has the external appearance of a minute clam. The shell is normally microscopic in size and houses a typical joint-legged, segmented *arthropod body. When fossilized, the remains of certain of these minute creatures are valuable as *guide fossils. Ostracods range from Early Ordovician to Holocene in age.

Ostracoderm. A primitive, jawless fish, a member of the class Agnatha. Relatives of the living lamprey eel and hagfishes, the ostracoderms were covered with bony plates and scales and reached a maximum length of about 30 cm. First discovered in an Ordovician sandstone near Canon City, Colo., ostracoderm remains have also been found in Norway, eastern Greenland and the British Isles. They are the earliest recorded fishes and the first known vertebrate animals. The ostracoderms first appeared in the late Ordovician, increased in numbers during Silurian time and were extinct by the end of the Devonian Period. Fragments of the teeth, spines and bony armor plates of these primitive fishes are especially abundant in certain Devonian rocks. (W. H. M.)

Outcrop. Any crustal rock exposed on the earth's surface. Outcrops are important to geologists in observing as well as extrapolating, on the basis of surface observations, the kinds of rock and types of structure that form the outer layer of the earth. In humid regions the bedrock is usually covered by vegetation, soil and mantle rock. In desert and mountainous regions, vegetation and mantle rock are scarce and outcrops are common. In areas where outcrops are limited or absent, geologists must turn to drilling or geophysical methods to determine the composition and structure of the crustal rock.

Outwash. Glaciofluvial sediments, composed mainly of stratified, cross-bedded sand and gravel, washed out from glaciers by meltwater streams and deposited on land. Outwash becomes finer grained, better sorted and stratified, and the fragments become more rounded with increasing distance from the glaciers. Outwash from modern glaciers may contain the remains of plants and animals. The outwash from Pleistocene ice sheets, however, is

generally devoid of fossils because few plants and animals lived close to the ice. Most outwash is deposited during the retreating stage of glaciation, when the climate is warm enough to cause glacial retreat and produce abundant *meltwater. Less outwash is deposited during the advancing stage of glaciation since the colder climate prevents abundant meltwater from forming. Moreover, some of the outwash deposited during the advancing stage is overridden by the ice and incorporated in its *moraine. *Valley trains, *outwash plains, *kames, *kame moraines and *kame terraces, and proglacial deltas are composed of outwash.

Outwash deposits have considerable economic value since they are used in making concrete and in highway construction. They are also excellent *aquifers—conductors of water—and much water is pumped from them for industrial and municipal use. (R.L.N.)

Outwash Plain. A broad plain consisting of coalescing *alluvial fans formed in front of a glacier by glaciofluvial sand and gravel deposited by *meltwater streams. Such a plain is formed where the land in front of a glacier is more or less flat and slopes away from it and where meltwater streams flow off the ice or issue from tunnels within it. An outwash plain may extend for many kilometers in front of a glacier and for scores of kilometers along it. These plains are usually associated with continental ice sheets and form in front of either terminal or recessional *moraines. Those associated with recessional moraines commonly have an abundance of *kettle holes. Excellent examples are found on Cape Cod, Martha's Vineyard and Long Island; in Minnesota, Wisconsin, Michigan and Illinois; and in western Denmark, northern Germany and New Zealand. See Outwash; Valley Train. (R.L.N.)

Overland Flow. The nonchannel flow of water over the surface of a drainage basin, from the divide to the nearest channel. It often occurs as a thin sheet-flow on smooth surfaces and as intertwining flow lines on grassy slopes. When rain first falls on a land surface, it may be intercepted by vegetation. Precipitation that reaches the ground may infiltrate the soil or be detained by hollows or obstructions. Only when the ground has reached its infiltration capacity, and surface detention and vegetative interception have been fulfilled, does water begin to run down a slope as overland flow. Overland flow is generally mixed flow, partly turbulent and partly *laminar, with increasing turbulence downslope.

Erosion by overland flow depends upon infiltration capacity and resistance of the soil or rock, intensity of rainfall, velocity of flow, length of slope and vegetative cover. The steeper part of slopes are eroded while the crest undergoes little or no erosion; the base is the site of deposition. (M.M.)

Overthrust. A low-angle *fault (less than 10°) that involves the movement of a large sheet of crustal rock over horizontal distances measured in miles. The term overthrust is somewhat misleading because the hanging wall may also slide down along the fault surface as well as be thrust up over it.

Oxbow Lake. A lake formed in an abandoned meander loop, common on *floodplains. It forms when the *meander of a channel is cut off; since the new channel is shorter and steeper than the old one, it is maintained by the river and the old channel is bypassed. The ends of the old channel are rapidly silted up and become separated from the stream, and the old channel becomes a crescent-shaped lake. An oxbow lake may eventually fill in and become a marsh.

Oxide. A chemical compound containing oxygen and one or more other elements. In the mineralogic classification the oxides are commonly divided into three groups. The first group is the simple oxides, that is, compounds of one metal and oxygen; rutile, titanium dioxide (TiO_2), is an example. The second group is the multiple oxides in which two nonequivalent metal atoms occupy different positions in the crystal structure; an example is spinel, magnesium aluminum oxide ($MgAl_2O_4$). The third group is the hydroxides, minerals containing the hydroxyl ion (OH); such as brucite or magnesium hydroxide, $Mg(OH)_2$.

Oxidized Zone. The upper portion of an *ore body that has been subjected to surface waters carrying oxygen, carbon dioxide and soil acids. The zone of oxidation extends from the surface to the water table where the downward movement of the solution terminates. Within this zone the primary ore minerals react with the solution to produce a large number of secondary minerals containing oxygen. Common examples are goethite and limonite from pyrite, iron sulfide; cuprite, malachite and azurite from copper sulfides; cerussite and anglesite from galena, lead sulfide; and smithsonite and hemimorphite from sphalerite, zinc sulfide. (C.S.H.)

Oxysphere. The outermost portion of the earth. See Lithosphere.

Ozone. A nearly colorless but faintly blue gas composed of molecules in which oxygen atoms are combined. It has an odor like that of weak chlorine. Ozone is produced mainly by photochemical processes involving ultraviolet radiation. Trace quantities are found in the atmosphere, but O_3 is largely concentrated in the ozonosphere at heights between 10 and 50 km. Ozone screens out most of the sun's potentially damaging ultraviolet radiation, which, even in the small amounts received at the earth's surface, can cause severe sunburn. It has been estimated that without the shield of ozone, sunburn would occur 50 times faster than it does under the hottest summer sun. The maximum concentrations of ozone normally occur between heights of 20 and 25 km, reaching values as high as 5 to 10 parts per million.

Ozone production near the earth's surface contributes to *air pollution. It is a significant threat to leafy plants, shrubs, and fruit and forest trees. Tobacco, for example, is sensitive to ozone at a level of 0.03 parts per million. Ozone also damages textiles and accelerates the cracking of rubber. Ozone has been used as a "tracer" of *atmospheric motions, particularly within the stratosphere. By mapping the distribution of ozone in relation to temperature and wind fields, and studying its transport, meteorologists have been able to deduce features of the stratospheric circulation.

Meteorologically, ozone is important because of the strong absorption of solar radiation at the top of the ozonosphere. As a result of this absorption, the temperature is typically high near 50 km. This temperature peak marks the boundary between the stratosphere and the mesosphere. (M.F.H.)

P

Paleobotany. The study of fossil plants to obtain a record of the history of the plant kingdom. Most paleobotanical studies are approached from the viewpoint of botany; however, a geologic background is also necessary. One of the more recent practical applications of paleobotany has been in the field of palynology, the study of fossil spores and pollen. These microscopic plant elements are especially valuable *microfossils and are used by petroleum geologists in the search for oil and natural gas.

Paleocene Epoch. The oldest *epoch of the *Tertiary Period; it began about 65 million years ago and lasted for some 11 million years. Originally part of the *Eocene Series described by Sir Charles Lyell in 1833, the Paleocene (literally, "ancient recent") was established by Wilhelm P. Schimper in 1874. The rocks were first studied in the vicinity of Paris and London, where they consist largely of fossiliferous clays, limestone, sandstone, and lignite. In North America, Paleocene deposits occur in the San Juan Basin of New Mexico, the Big Horn Basin of Wyoming, in Montana and in the Dakotas. They are also exposed in the Pacific coast area and the Atlantic and Gulf Coastal Plain. In Europe they are present in western Germany, Belgium and southeastern England. Paleocene climates were generally warm and humid, and the life forms resembled those of the Eocene.

Marine organisms included foraminifers, gastropods, pelecypods and echinoderms, plus many sharks and bony fishes. Land mammals were abundant although unspecialized and relatively small, and included primitive meat eaters, insectivores, primates, uintatheres, and the first horses. The similarity of European and North American land animals suggests the presence of a land bridge between the two continents. However, South American species differ considerably, suggesting that this region was geographically isolated. Plant life was modern in appearance: palms, ferns, conifers, cycads and flowering plants were numerous. Deposits of lava and volcanic ash suggest that Late Paleocene time was marked by considerable volcanic activity in New Mexico, Colorado and Montana, but the epoch ended quietly elsewhere. Paleocene economic resources are rather limited but include clay, lignite, bentonite and some bauxite. (W.H.M.)

Paleoclimatology. The study of world climates throughout past time, helpful in reconstructing paleogeographic maps and studying earth history. The interpretation of ancient climates assumes the validity of the principle of *uniformitarianism, for such inferences are based on analogy with known climatic indicators of today. Paleoclimatology requires an understanding of modern climatology, life forms and ecology. For example, the fossil remains of corals (which live mostly in warm, shallow sea water) and tropical plants (which have been found in Antarctica) are useful paleoclimatological indicators. Coal beds typically indicate lush plant growth under tropical to semitropical conditions; other coal formations, such as those in the Antarctic, formed under more temperate conditions and suggest that the climate there was once much warmer. *Cross-bedding in certain wind-deposited sandstones and evaporites suggest arid conditons. *Tillites, glacial markings and other ice-associated features indicate glacial conditions of the geologic past. (W.H.M.)

Paleoecology. The study of ancient associations of both plants and animals in relationship to their environment. Interpreting the ecology of the geologic past makes possible a better understanding of past physical and biological environment and the reconstruction of natural communities and their surroundings.

Paleogene Period. A unit of geologic time incorporating the *Paleocene, *Eocene and *Oligocene epochs of the *Cenozoic Era. In classifications that designate the Tertiary Period as an era, it is considered the oldest period of the Tertiary and is followed by the Neogene Period.

Paleogeographic Map. A map that is an inferred restoration of the major geographic features of a region at a given time in the geologic past. Such maps are based on inferences assembled from the physical characteristics of rocks and fossils that may be present. The interpretations are based on the postulated environment in which the sediments were deposited and the ancient animals lived. These inferences in turn are based on an analogy with modern depositional and biologic environments. Such maps may vary greatly in accuracy depending on the age and type of rocks exposed in the area and the nature of the geologic record. (W.H.M.)

Paleogeography. The study of the distribution and relationship of ancient seas and land masses. "Ancient geography" is reconstructed through the interpretation of *sedimentary rocks and the fossils they contain. *Paleogeographic maps have been prepared for the major periods and epochs of geologic time and may provide clues to the location of valuable mineral resources.

Paleogeology. The study of the geologic conditions and events of an earlier period of the earth's history. Such studies are used to prepare paleogeologic maps, which reveal the geology of an ancient but now-buried surface representing some former portion of geologic history.

Paleomagnetism (or Fossil Magnetism). The study of the magnitude and direction of the earth's *magnetic field during the geologic past. The rocks most suitable as indicators of paleomagnetic fields are basaltic lavas and well-bedded red sandstones, for they possess a high degree of magnetic stability. The capacity of these rocks to become magnetic indicators results from the presence in them of small amounts (usually less than 10 percent) of magnetic iron oxide and certain sulfide minerals. Lava is stronger magnetically than sandstone because it contains magnetite, a mineral highly susceptible to magnetization, whereas sandstone contains the mineral hematite, which is less magnetic.

Lava and sandstone become paleomagnetic indicators in the following manner. Lava begins to cool as it is intruded into the upper portion of the earth's *crust or poured out onto its surface. As it cools down through a temperature below 575°C (1067°F), the small grains of magnetite, acting as small bar magnets, begin to align themselves parallel to the direction of the earth's magnetic field. This is analogous to the way iron filings on a piece of paper lying on a bar magnet align themselves with the magnet's lines of force. As the rock solidifies, the grains of magnetite or other magnetic minerals are frozen in such a position that the entire rock is magnetized parallel to the earth's field. In sandstones the magnetizing process is caused by the alignment of individual mineral particles with the earth's magnetic field as they are deposited in sedimentary environments. As the earth's field varies in intensity and direction relative to the newly formed magnetized rocks, a small part of the original magnetization remains. This residue is known as the *natural remanent magnetism (NRM).

Paleomagnetism has proved useful in strengthening the theory of *continental drift and is providing evidence that the earth's magnetic field is not permanently stable but capable of change, including the reversal of its magnetic poles. (J.W.S.)

Paleontology (from Greek *palaios,* "ancient"; *ontos,* "a being"; and *logos,* "word" or "discourse"). The study of *fossils. Information gathered with the help of paleontology has greatly increased knowledge about ancient plants and animals and the world in which they lived. Fossils represent the remains of so many types of organisms that paleontologists have established four main divisions of their science:

*Paleobotany: The study of fossil plants and the record of the changes they have undergone. Palynology, a specialized branch of paleobotany, deals with the study of spores and pollens and their dispersal through time.

Invertebrate Paleontology: The study of fossil animals lacking a backbone or spinal column, including such forms as protozoans (tiny one-celled animals), snails, clams, starfish and worms, and the remains of spineless animals that lived in prehistoric seas.

Vertebrate Paleontology: The study of the fossils of the animals that had a backbone or spinal column. The remains of fishes, amphibians, reptiles, birds and mammals are typical vertebrate fossils.

Micropaleontology: The study of microscopic fossils. They are called microfossils and usually represent the shells or fragments of minute plants or animals. Because of their small size, microfossils can be brought out of wells without being damaged by the mechanics of drilling or coring, and are therefore particularly valuable to the petroleum geologist in identifying rock formations far below the surface. (W.H.M.)

Paleozoic Era. The *era of geologic time from the end of the *Precambrian, 600 million years ago, to the beginning of the *Mesozoic, about 225 million years ago. It has been divided into seven units of geologic time: the *Cambrian, *Ordovician, *Silurian, *Devonian, *Mississippian, *Pennsylvanian and *Permian periods. The beginning of Paleozoic time, which marks the start of the first accurate records in geologic history, is characterized by the appearance and development of the major types of invertebrates. Their fossils, and the relatively undisturbed Paleozoic rocks, make possible a fairly accurate reconstruction of the historical events of this time. Paleozoic rocks cover much of eastern and central North America; there are more limited exposures in the northern and western parts of the continent. They are also exposed in northern and western Europe, South America, Australia, Africa and Asia. Most of these rocks are formed of sediments deposited in ancient seas, and many are extremely fossiliferous. (W.H.M.)

Paleozoology. The study of *fossil animal remains. The subdivisions of paleozoology (commonly referred to by the general term *paleontology) are invertebrate paleontology and vertebrate paleontology.

Invertebrate paleontology is the study of fossil animals lacking a spinal column or backbone, including such diverse forms as protozoans (tiny one-celled animals), snails, corals and starfish. Vertebrate paleontology is the study of ancient animals possessing a backbone or spinal column, including organisms such as fishes, reptiles, birds and mammals.

Palynology. The study of fossil spores and pollens.

Pangaea. According to the Wegener theory of *continental drift, Pangaea was a primitive landmass from which all our present continents originated. In 1912 Alfred Wegener, a German meteorologist, suggested that the landmasses of the world are not static but are large drifting islands floating on a substratum of the earth. Not all geologists agree today that there was one supercontinent. Some have proposed that two subcontinents existed, Laurasia, composed of North America, Europe and Asia, and Gondwanaland, consisting of South America, Africa, India, Australia and Antarctica. Others consider the supercontinents as one configuration among many which the earth's landmasses have occupied in the geologic past. The theory that continents drifted remained controversial until the 1960s, when convincing evidence led to widespread acceptance.

Wegener himself was the first to try to reconstruct the shape of Pangaea. His work, and that of later scientists, was based on matching the present outline of the continents to produce the best fit, in much the same way as we assemble a jigsaw puzzle. As more became known about *sea floor spreading and *plate tectonics, scientists were able to retrace the paths along which the continents have been drifting. Their reconstructions, however, were lacking in reference to where Pangaea lay with respect to the poles and the coordinate system defined by latitude and longitude.

In 1970 Robert S. Dietz and John C. Holden, scientists with the Atlantic Oceanographic and Meteorological Laboratories in Miami, Fla., worked out a reconstruction with absolute geographic coordinates for the supercontinent as it existed at the beginning of the Mesozoic Era, about 225 million years ago. They also established coordinates for the present continents as they appeared at the end of various geologic periods following the breakup of Pangaea. They described Pangaea as an irregular landmass covering about 40 percent (200 million square kilometers or 80 million square miles) of the earth's surface. It was surrounded by an ancient sea called Panthalassa, part of which is today the Pacific

Ocean. The Atlantic and Indian oceans are considered new ocean basins formed after the breakup. The reconstruction shows that present continents moved north and west of their original position. Before the breakup of Pangaea the continents were nearly balanced on both sides of the Equator, with the Equator passing through the city of New York and traversing the southern portion of North America. New York City would have been on a longitude 84° farther east than it is today, or approximately on the same meridian as Hamburg, Germany. India (its location is not well documented) and Australia bordered the Antarctic, far south of their present location. Other geologists, while accepting the former existence of large landmasses such as Laurasia and Gondwanaland, are reluctant to concede the existence of an original Pangaea, noting that early Paleozoic and Precambrian continental displacements remain largely undeciphered. (J.W.S.)

Paragonite. A mineral, hydrous sodium aluminum silicate, a mica that occurs with and is indistinguishable from *muscovite in its physical properties. The difference between the two micas lies in their chemistry. In paragonite, sodium ions occupy the structural sites that in muscovite are occupied by potassium.

Parallel Drainage. See Drainage Patterns.

Paramorph. A type of *pseudomorph with the same chemical composition as the original crystal. A common example is calcite after aragonite, for both minerals have the composition $CaCO_3$.

Parasitic Cone (or Lateral Cone). *Cinder cones and other small volcanoes on the flanks of large composite cones and *shield volcanoes. They are called parasitic because they are fed from the same chimney as a main cone and drain volcanic material from it. They form after a volcano reaches a certain height, when it is easier for *magma to break out on the flanks of the volcano rather than move up to the summit crater. Sicily's Etna, essentially a shield volcano, has more than 200 parasitic cones on its flanks; Hawaii's Mauna Loa and Kilauea are also dotted with parasitic cones; Shasta, in the Cascade Range, has a large parasitic cone called Shastina. (R.L.N.)

Parting. See Mineral Parting.

Peat. A soft, spongy, brownish deposit of partially reduced plant material containing approximately 60 percent carbon and 30 percent oxygen. It resembles tightly packed, rotted wood and forms when vegetable matter decomposes in waterlogged environments in the absence of oxygen. An intermediate material in the formation of *coal, peat is used as fuel in some parts of the world.

Pectolite. A mineral, hydrous calcium sodium silicate, similar in its occurrence to *zeolites. It is found lining cavities in basalt associated with zeolites and with prehnite, datolite and calcite. Pectolite is characteristically in aggregates of white, sharp-pointed crystals that can puncture the skin. See Mineral Properties.

Pedalfer (from *pedon,* meaning "soil," and *al* and *fer,* abbreviations of "aluminum" and "iron"). An acid soil that contains large amounts of humus and has undergone much *leaching. The iron present in the A-horizon is removed by leaching and is deposited below in the B-horizon. Na, K, Mg and Ca are removed by leaching from the A-horizon as ions and bicarbonates, some of which move in groundwater to streams and eventually reach the ocean. The secondary clay formed by the decomposition of the parent material in the A-horizon is removed mechanically and deposited in the B-horizon. The upper part of the soil profile is generally light gray. Pedalfers develop best under a forest cover and in a temperate climate where the rainfall is greater than 60 cm per year. They are common in the northeastern part of the United States and in northern Europe. See Horizons, A-, B-, and C-. (R.L.N.)

Pediment. A gently sloping, nearly flat bedrock surface, usually having a thin cover of loose gravel. A pediment is a landform typical of arid regions where it is found at the foot of mountain ranges merging basinward with the *bahada. Some pediment surfaces rise gently upward to the mountain crest, forming a pediment pass.

The formation of pediments has been, and still is, a matter of controversy. One proposed origin is by sheetwash — or erosion by sheets of rainwater running over the surface. This process not only removes the weathered material but as the loose particles are transported they wear away the bedrock surface. Gradually after a long period of time, a pediment could be formed through this process. Another way in which a pediment might originate is by sideward erosion of a river flowing out from the mountains into the basin. As the river shifts its channel sideward from time to time it could erode the broad pediment surface. Finally, there is a theory of composite origin, which proposes that pediments are formed as a result of the combined forces of weathering, sheetwash and sideward erosion and shifting of a stream. (M.M.)

Pedocal (from the Greek *pedon,* "soil," and cal, an abbreviation for calcium). An alkaline soil rich in calcium carbonate. A pedocal is formed in regions of low rainfall, where the vegetation is grass or bushes and the soil is dry and hot. Much of the calcium carbonate has been precipitated near the surface by the evaporation of solutions that moved upward by capillarity. Because of insufficient rainfall and humus, chemical weathering and therefore the formation of clay minerals take place more slowly than in humid regions. Pedocals are common in the western United States and in other regions of low rainfall. (R.L.N.)

Pedology. The science of the characteristics, constitution, origin, use and conservation of the soil. It is concerned with the chemistry, physics, geology, climatology and biology of soil formation.

Pegmatite. An extremely coarse-grained *igneous body closely related genetically to large masses of fine-grained *plutonic rocks. It may be present as a vein or a dike in the granular igneous rock, but more commonly is found completely enclosed within the neighboring country rock. Although pegmatites vary considerably in size, they are generally small; the length of the largest is over 3 km, whereas some have a length of only 1 m or so.

Although pegmatites may be related to several types of rocks, they are most commonly associated with *granites and, unless otherwise specified, pegmatite refers to granite pegmatite. These small, coarse-grained bodies represent the end phase of the crystallization of a granitic *magma. As minerals slowly separate from the molten rock material, water and other volatiles are concentrated in the still-fluid portion. This highly mobile material, out of which pegmatites form, is expelled from the magma chamber. With it go many rare elements whose atoms were either too large or too small to be housed in the crystal structure of the rock-forming minerals. The eventual crystallization of this fluid magma results in large crystals of the common minerals of a granite: quartz, feldspar and mica. The largest known crystals have been found in pegmatites: a 5-ton quartz crystal from Brazil, a 100-ton phlogopite crystal from Ontario and hundreds of tons of feldspar from a single crystal in the Soviet Union.

In addition to the rock-forming minerals, some pegmatites house many unusual

Pectolite
The ball-shaped forms of this mineral are tightly packed bundles of radiating needle-like crystals.

Pelecypod

1 *The genus of bivalved mollusks called*
Pecten *includes this scallop, a form that
lived during the Mississippian Period
some 330 million years ago.*

Pelagic Sediment

2 *Two chains of diatoms and several other
marine organisms called dinoflagellates
are here magnified 69 times. The shells
of diatoms sink to the bottom to form a
thin cover of diatomaceous ooze over
great areas of the sea floor.*

3 *Fragmented remains of foraminifera,
one-celled animals with skeletons of
calcium carbonate, are among the most
common particles in deep-sea
sediments.*

minerals, because of the rare elements
present in the fluid pegmatite magma.
Beryl, tourmaline, lepidolite, amblygonite,
spodumene, topaz, chrysoberyl and many
less common minerals come almost exclusively from pegmatites. Although these
are usually flawed and imperfect, most
of them have transparent gem varieties
that occasionally occur in cavities in pegmatites. Thus pegmatites are the source of
a greater variety of gemstones than any
other geologic occurrence. (C.S.H.)

Pelagic Environment. Sometimes called
the water environment, this includes all of
the marine water above the ocean bottom.
The *neritic (shelf environment) and oceanic (open-ocean environment) provinces
are its two primary subdivisions. All
floating and swimming organisms in the
sea occupy this environment.

Pelagic Sediment. The fine-grained material that settles very slowly out of the pelagic environment with essentially no contribution from the land or from currents.
These sediments may be composed of
skeletal particles from plants or animals
(biogenic oozes), clay particles or *authigenic sediments. The rate of accumulation is slow, ranging from a few millimeters
to 1 cm per 1000 years. Because of the way
these particles accumulate, they tend to
form rather thin and uniform layers over
the ocean floor without regard to topography; this contrasts with land-derived sediments, which fill in low areas. (R.A.D.)

Pelecypod. A member of class Pelecypoda
of phylum Mollusca, the pelecypod is
characterized by a shell composed of two
calcareous valves that enclose the soft
parts of the animal. Members of this class
live exclusively in an aquatic habitat and
are abundant in marine environments.
Most pelecypods are slow-moving, bottom-dwelling forms, but some are attached (the oyster, for example) and others
are swimmers (the scallop, for example).
The Pelecypoda includes such familiar
saltwater forms as the clam as well as the
common freshwater mussel.

Most of the pelecypod shell is composed
of calcium carbonate, but the outer layer of
each valve is a horny material. The inner
surface of the shell is lined with a calcareous layer of pearly or porcelaneous
material. The outline of the shell may vary
greatly, but most pelecypods are clamlike.
The majority have a beak that represents
the oldest part of the shell. Pelecypods
range in size from tiny clams a fraction of a
centimeter to the great *Tridacna* of the
South Seas, which may attain a length of
almost 2 m. Some forms crawl over bottom

sediments; others burrow into mud, rock,
wood or even other shells. (W.H.M.)

Pele's Tears. Named after the Hawaiian
goddess of fire and volcanoes, Pele's
tears are spherical, cylindrical, dumbbell-
shaped or pear-shaped drops of glassy
basaltic lava from 6 mm to 13 mm (¼ to ½
inch) long. They erupt as small blobs of liquid lava and solidify during flight. Pele's
hair is threads of solid lava, sometimes
over 2 m long. When the drops of liquid
lava that form Pele's tears are erupted, they
usually have liquid threads of lava attached
to them. These threads solidify in flight,
break off from the tears and may drift with
the wind for many miles. Pele's hair is commonly found on Hawaii during eruptions of
Mauna Loa and Kilauea. (R.L.N.)

Pelycosaur. A member of the order Pely-
cosauria of class Reptilia. These extinct
mammal-like reptiles, which lived only during Late *Pennsylvanian and *Permian
times, derive their name from Greek *pely,*
meaning "basin," from the shape of their
pelvis. They are normally characterized by
a large fin or sail-like structure on the back,
supported by elongated spines of the
vertebral column. A study of their teeth indicates that both herbivorous and carnivorous forms existed.

Peneplain. A low, almost featureless surface, the ultimate stage in the humid cycle
of fluvial erosion. It is developed primarily
by stream erosion and *mass wasting of
*divides. The peneplain may cut across
various rock types and structures and is
controlled only by the *base level of erosion. Ancient peneplains are identified by
the wide, regional extent of the low undulatory surface that truncates rocks of
varying resistance, by deep *soil profiles
and by the presence of upland *alluvium.
Streams in an old peneplain should show
evidence of *rejuvenation: their headwater
tributaries may begin in wide, shallow
valleys; their longitudinal profiles show
*knickpoints; and upraised alluvial ter-
races line the valleys. If the peneplain has
been eroded, the old surface should be
recognized by equal elevation of the present summit levels. Such ancient peneplains have been identified in the Appalachian and Rocky mountains in the United
States, in the central Massif of France and
in the Alps.

Identification of such peneplains and
even the validity of the peneplain concept
is in dispute, however. It is argued that
continents are not stable long enough
for streams to attain peneplanation, as
evidenced by the fact that there are no
identifiable, present-day peneplains at sea

level. The accord in summit levels may be the result of rock resistance. Likewise, level surfaces may be formed as a result of: (1) structural stripping, that is, the erosional stripping away of horizontal sediments to a resistant layer, as in the Colorado Plateau; (2) marine planation by wave action; many such level surfaces are found near the sea and may have been formed inland when the oceans covered more of the continents; (3) pediplanation, or reduction of a desert region to a gently sloping rock surface or pediment; so-called peneplains in the Uinta mountains of Utah and Bighorn Mountains of Wyoming have been reinterpreted as pediment surfaces; (4) panplanation, or reduction to a level surface by lateral corrosion of streams.

Peneplains that have been buried and later exposed by erosion (morvans) have been identified. Examples are the Morvan Plateau (type example) of northeastern France, the Fall Zone peneplain on the eastern edge of the Piedmont Plateau of southeastern United States, and the pre-Cambrian rock surface in parts of Canada. (M.M.)

Pennsylvanian Period. The period of geologic time that began about 320 million years ago; it comes after the *Mississippian and before the *Permian periods of the *Paleozoic Era. It was named for the state of Pennsylvania and is the equivalent of the upper part of the Carboniferous System (or "Coal Measures") of Europe. Rocks of this age are exposed in eastern North America; the mid-continent region of the United States, including west Texas; and parts of western North America, including the Northern Rocky Mountains, New Mexico, Colorado, the Great Basin region (especially Utah and Nevada) and along the Pacific Border from California to British Columbia and Alaska. In Europe, Pennsylvanian (Upper Carboniferous) rocks are widespread in Great Britain, northern France, Germany, Belgium, the Soviet Union, Spain and the eastern Alps. They also occur in South America, Australia, South Africa, China, Korea and northern India.

The interior of North America was a broad low-lying area of slight relief during most of Pennsylvanian time. The warm, swampy lowlands were covered with extensive forests of ferns, rushes and other moisture-loving plants. The remains of these plants form the bulk of the *coal taken from the coal fields of West Virginia, Ohio, Indiana, Illinois, Pennsylvania and Kentucky. These fields cover approximately 520,000 square kilometers (200,000 square miles). Continental deposits and extensive coal beds suggest similar conditions for western Europe. *Cyclothems developed in the coal beds of central and eastern United States (especially in Illinois), and these alternating layers of marine and nonmarine sediments suggest repeated changes in sea level for these areas. Although the period closed relatively quietly in North America, there were episodes of mountain building in several areas of the Southwest.

Pennsylvanian climates are generally assumed to have been mild, humid and tropical to subtropical in nature. Locally, however, there is evidence of aridity and cooler temperatures.

The Pennsylvanian seas teemed with a variety of life, and corals, brachiopods, bryozoans, crinoids, and mollusks (especially clams and snails) were common. The distinctive spindle-shaped foraminifers called fusulinids were very abundant; some rocks consist almost exclusively of the remains of these tiny bottom-dwelling organisms that resemble grains of wheat. Amphibians were the dominant vertebrates; some were highly specialized and as much as 3 m (10 feet) long. Although their remains are rare, the earliest known *fossil reptiles have been found in Pennsylvanian rocks. *Fossil plants are very abundant and reveal much about the vegetation of this period. Ferns, scouring rushes and giant scale trees were particularly numerous, and their carbonized remains made possible the rich coal deposits for which the Pennsylvanian is noted. The dense jungle-like forests were inhabited by myriad spiders, scorpions, centipedes and insects, including cockroaches as much as 10 cm (4 inches) long and dragonflies with a wingspread of 74 cm (29 inches).

Eighty percent of the world's coal comes from Pennsylvanian rocks in Europe and North America. They also yield much petroleum (especially in the mid-continent region of Texas, Oklahoma and Kansas); iron ore in northern West Virginia, eastern Ohio and Pennsylvania; as well as limestone and clay for Portland cement, ceramics and other clay products in the eastern United States. (W.H.M.)

Pentlandite. A mineral, an iron nickel sulfide, the principal ore mineral of nickel. It usually occurs in basic *igneous rocks where it is associated with other nickel minerals, *chalcopyrite and *pyrrhotite. In this association pyrrhotite is dominant and the other minerals are included within it. Since pentlandite and pyrrhotite have the same yellowish-bronze color, it is difficult to distinguish them by inspection. But pyrrhotite is magnetic and pentlandite nonmagnetic, so they can be separated

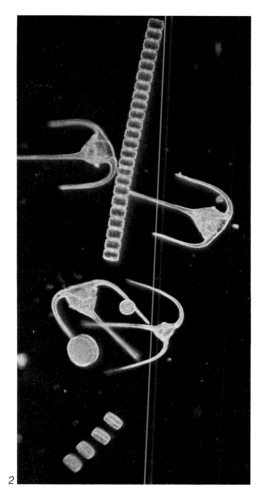

2

3

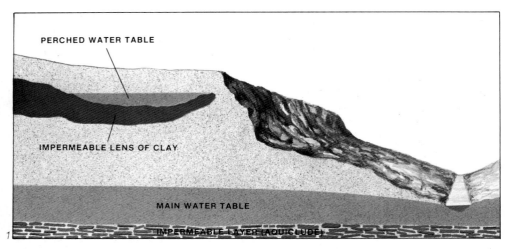

PERCHED WATER TABLE

IMPERMEABLE LENS OF CLAY

MAIN WATER TABLE

IMPERMEABLE LAYER (AQUICLUDE)

1

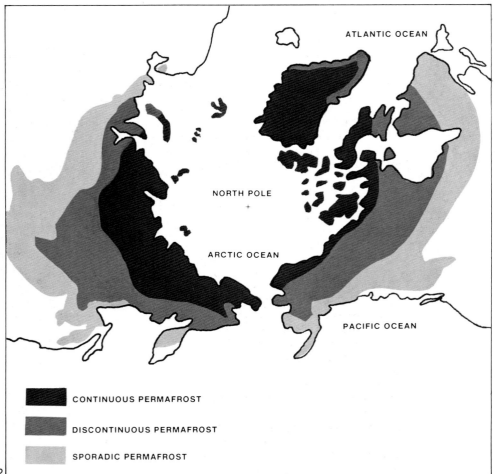

ATLANTIC OCEAN

NORTH POLE
+

ARCTIC OCEAN

PACIFIC OCEAN

CONTINUOUS PERMAFROST

DISCONTINUOUS PERMAFROST

SPORADIC PERMAFROST

2

Perched Water Table

1 *A bed of impermeable rock may block the downward percolation of water and create an isolated pool suspended above the main water table.*

Permafrost

2 *Perennially frozen ground occupies about 20 percent of the land area in the Northern Hemisphere.*

magnetically from a powdered aggregate. At the world's largest nickel mines, pentlandite is the chief source of nickel. Sudbury, Ontario, the Lynn Lake district of Manitoba and the Petsamo district of the Soviet Union are the major producing areas. See Mineral Properties. (C.S.H.)

Perched Water Table. The top of a zone of saturation, held above the main *water table by an impervious bed. It is formed when water moving downward is prevented, commonly by a bed of clay or shale, from reaching the main water table.

Perennial Stream. A waterway that flows year-round and is fed by stable groundwater flow, that is, which has a channel below the water table.

Peridot. The gem variety of the mineral *olivine. Although peridot has a rather low hardness (6½−7), its olivine-green color makes it an attractive gem. The best peridot has come from Saint John's Island in the Red Sea. Gem-quality crystals are found in the Navajo Indian country.

Peridotite. A granular igneous rock composed essentially of dark minerals, which are usually olivine and pyroxene in varying proportions; hornblende and, more rarely, mica may be present. When the rock is composed essentially of pyroxene it is called a *pyroxenite; and when it is composed essentially of olivine it is a *dunite.

Period. A division of geologic time that is of shorter duration than an *era and longer than an *epoch. Each geologic period refers to a specific interval, and most of these *time units derive their names from the areas in which the rocks were first studied and described. Thus the Pennsylvania Period is named for the State of Pennsylvania, where there are many exposures of rocks of Pennsylvanian age. The rocks formed during a given period of geologic time are said to be a *system. Most geologists recognize 12 periods: 7 in the *Paleozoic Era, 3 in the *Mesozoic Era and 2 in the *Cenozoic Era. See Geologic Time Scale. (W.H.M.)

Perlite. A volcanic glass with many small spheroids formed by concentric cracks. The spheroids vary from the size of small shot to the size of peas; they result from cracking during cooling of the glass. Most perlites have the composition of *rhyolite.

Permafrost. *Regolith or *bedrock in which the temperature is permanently below freezing. If the material contains water, frozen ground results. The permafrost is found below the "active zone," which extends downward from the surface for 1 m or so and is frozen in winter and thaws out in summer. Below the permafrost is unfrozen ground that increases in temperature with increasing depth. In places the permafrost is as much as 580 m (1900 feet) thick. It underlies approximately 22 percent of the land area in the Northern Hemisphere, including large expanses in Alaska, Canada and Siberia. A mean annual temperature of at most -2°C (28°F) is necessary for its formation. It is not found under large rivers or lakes. Prehistoric woolly mammoth and woolly rhinoceros carcasses have been preserved in the permafrost of Siberia and Alaska for thousands of years. A building constructed on permafrost will slowly sink into the ground unless the permafrost is insulated from the heat of the building. Permafrost is therefore of vital interest in planning land usage in northern areas. (R.L.N.)

Permeability. A measure of the ability of a porous substance to transmit water and other fluids. A permeable solid is one that can transmit fluids. Permeability depends upon porosity as well as the size of the interstices and the size and continuity of their connections. Thus clay, which may have a greater porosity than sand, has a lower permeability because its openings are so small that capillarity inhibits the flow of liquids through it. Gravel generally has a greater permeability than sand because its openings are larger. Permeability is of economic significance because it partly determines the quantity of water, oil or gas that can be recovered from a reservoir and the speed with which it can be obtained. (R.L.N.)

Permian Period. The youngest period of the *Paleozoic Era; it began about 280 million years ago and lasted some 55 million years. It was named in 1841 by Sir Roderick Murchison, from the province of Perm near the Ural Mountains in northeastern Russia. Permian rocks also occur in Germany, England, Asia (especially the Salt Range of India), Africa, South America, Australia and the East Indies. In North America they are best exposed in the mid-continent region, the northern Appalachian area, and in most of the western United States, Alaska and Canada. Marine *sedimentary rocks of Permian age are quite rare east of the Mississippi Valley region. Continental deposits (including extensive *evaporites and redbeds) are also common in the far western United States. In New Mexico the famous Carlsbad Caverns formed as groundwater dissolved underground chambers in the soluble Capitan Limestone.

The Paleozoic Era ended with the close of the Permian Period. This event was marked by the folding and uplift of the old Appalachian *geosyncline to form a great mountain range. This marked the last phase of a sequence of mountain-building events that had occurred throughout much of Paleozoic time in the Appalachian region.

Drastic climatic changes occurred during the Permian due to the restricted seas and general continental uplift. The presence of *tillite in Brazil, Madagascar, Australia, Africa and Antarctica is considered proof of *glaciation and extremely cold climates in the Southern Hemisphere during Permian and *Pennsylvanian time. One such formation, the Dwyka Tillite of South Africa, is considered a strong argument in favor of the theory of *continental drift. Evaporite deposits are well developed in Kansas, southeastern New Mexico and western Oklahoma. These salt and gypsum beds suggest conditions of local aridity when landlocked arms of sea slowly dried up, leaving the salts as a residue.

These drastic climatic and geographic changes coincided with changes in Permian life, and mass extinctions occurred in major groups of animals such as the trilobites, blastoids, cystoids, fusulinids and certain species of highly specialized brachiopods. Other species such as the ammonoids and certain species of pelecypods and gastropods became much more abundant. Amphibians and reptiles became more numerous and the reptiles became highly specialized. Pelycosaurs were common and were characterized by long, bony spines growing out of their backs: the spines were joined together by skin that made the animals appear as if they had sails on their backs. Other reptiles more closely resembled the large lizards of today and were the forerunners of the great dinosaurs of the Mesozoic Era. Another group, the mammal-like therapsids, were probably the ancestors of the mammals. Permian plant life differed greatly from that of the Pennsylvanian, for many of the coal-forming plants became extinct and were replaced by the more advanced conifers (cone-bearing plants).

Permian rocks have yielded oil and gas in the mid-continent region and western United States; salt in Texas, Kansas, New Mexico and Oklahoma; gypsum in New Mexico, Texas, Colorado, Kansas, South Dakota, Iowa and Oklahoma. Calcium phosphate (a mineral fertilizer) has been produced in Idaho, Utah and Wyoming; potash salts in Germany, Texas and New Mexico; and Permian coal has been mined in Europe, India, China, Australia and other parts of the world. (W.H.M.)

Permineralization (or Petrifaction). The process whereby the hard parts of many ancient organisms have been preserved by mineral-bearing solutions after burial in sediment. These solutions, usually percolating groundwater, infiltrate porous bones and shells, depositing their mineral content in the pores and open spaces of the skeletal parts. A fossil thus produced is referred to as a petrifaction, because it has literally turned to stone. The addition of these minerals tends to make the hard parts of the organism even harder and more resistant to weathering or other destructive agents. Fossils that have undergone such preservation may be correctly referred to as "petrified wood," "petrified bone," and so forth.

Silica, calcite and various iron compounds are the usual permineralizers, although many others are known. Petrified wood is the most common example of this type of preservation. After a piece of wood has been buried by sediments, circulating groundwaters tend to saturate its woody tissues, filling all spaces and some of the surrounding material with minerals, usually silica. The result is increased weight and durability in the permineralized plant material. (W.H.M.)

Perthite. An intergrowth of two feldspars in which the host *potash feldspar, usually microcline, is traversed by thin lamellae of *albite. The intergrowth is interpreted as a product of *exsolution during cooling. At high temperature there was but a single feldspar whose structure accommodated both potassium and sodium. When the temperature was lowered the sodium could not be tolerated in the structure of the microcline and it migrated in the solid state to form the thin lamellae composed of albite.

Petrifaction. A process whereby wood or other organic matter is converted into a stony substance by impregnation with dissolved minerals; it has preserved the hard parts of many ancient organisms. See Permineralization.

Petrified Wood. Wood that has been fossilized by the process of petrifaction or *permineralization. The original woody tissue of the plant has been preserved by mineral-bearing groundwater after burial of the plant in sediment. The groundwater infiltrated the buried wood, depositing its mineral content in the pores and open spaces of the plant remains. Petrified wood is common in certain Cenozoic and Mesozoic rocks and is especially abundant at Arizona's Petrified Forest National Park. There hundreds of tree trunks and

stumps have been exposed by the weathering away of the shale in which they were once buried. See Permineralization; Silicified Wood; Replacement.

Petrogenesis. The branch of *petrology concerned with the origin of rocks. The term is commonly used more narrowly to refer to the origin of *igneous rocks.

Petrography. The branch of *petrology that deals with the systematic study of rocks, particularly by microscope.

Petroleum. A complex mixture of naturally-occurring chemical compounds of hydrogen and carbon (hydrocarbons) that may contain impurities (nitrogen, oxygen and sulfur). It is found in the earth in a liquid, solid or gaseous state. Petroleum contains oil and *natural gas. It is believed to have originated from the remains of microscopic marine plants and animals that were buried in the mud and sand of shallow prehistoric seas and underwent slow decomposition by bacteria, leaving a residue of hydrocarbons. The processes by which the organic material was finally converted to petroleum are not thoroughly understood; they appear to have required vast amounts of time plus temperature increases and compression of the sediments. Most shales are too dense and solid to permit oil or gas to remain in them, but sandstones and the limestones have empty spaces (called pore space and measured as *porosity), in which oil and gas generated in the earth may accumulate. If the empty spaces are connected by passages, the reservoir rock is permeable; ideally it is both porous and permeable.

Very little, if any, of the oil and gas found in a petroleum reservoir originated in the reservoir rock. Originally the pores in the reservoir rock were filled with the ancient seawater in which the sediment was deposited; later the petroleum moved in and displaced the water. However, not all of the water was forced out, and in practically all petroleum reservoirs water is present in the pore spaces along with oil and gas. In some instances migrating oil has seeped out onto the surface of the earth. However, if the oil encountered obstacles that it could not penetrate or bypass, migration stopped and oil and gas accumulated. *Anticlines, *faults, *salt domes and changes in porosity and permeability of the reservoir rock may block migration, forming traps (or geologic structures) that force the oil to accumulate in economically recoverable concentrations of petroleum called reservoirs. Accumulation of oil in the reservoir itself evidently occurred as a result of the buoyancy of oil and gas, the

Petrified Wood

1 *These 180-million-year-old fossil trees at Petrified Forest National Monument, Arizona, have been so completely silicified by groundwater that they have remained long after the shale in which they were buried weathered away.*

Petroleum

2 *In a typical petroleum-bearing anticline (a), oil and gas accumulate over the water table in a permeable reservoir rock, such as sandstone, which is overlain by an impermeable cap rock, such as shale. (b) An oil pool, formed in*

a permeable sandstone bed, is prevented from escaping by the impermeable shales. (c) Surrounding strata are bent and faulted against the sides of a salt plug, creating traps for petroleum. Oil is also trapped in the limestone cap rock above.

3 *Offshore oil drilling platforms extract petroleum from huge reserves beneath the Gulf of Mexico and the North Sea.*

4 *The West Coyote Hills oil field near La Habra, Calif., is a typical oil-producing anticline.*

propulsive force of moving gas and the circulation of underground water.

Modern techniques of exploring for petroleum utilize our understanding of its origin and how it accumulated within certain portions of the earth's crust. It is believed that oil formed in the ancient sediments of prehistoric seas first existed in the form of tiny droplets of petroleum dispersed throughout the sediments. Salt water also occurred in the sediments, and gas apparently formed from the same organic matter as the oil. Because oil is lighter than water, and gas is lighter than oil or water, the three tend to separate. In the process, the oil and gas normally work their way above the water, seeping through porous rock toward the earth's surface. Much oil and gas surfaced thousands of years ago and was dissipated into the atmosphere. But not all of the petroleum escaped, for some was caught in underground traps formed by the buckling and folding of the earth's crust.

In the search for oil, petroleum (or exploration) geologists study and map rocks exposed at the surface, or examine rock fragments brought to the surface when exploratory (or so-called wildcat) wells are drilled. Numerous oil companies use the science of geophysics in their search for oil and employ a *seismograph similar to that used to record earthquakes. Such seismic prospecting is carried on by producing small artificial "earthquakes" with explosives. The seismic record of the shock waves as they travel through the earth can suggest the type of rocks, their depth and whether a suitable trap is present.

Although the major source of oil and gas is rocks deeply buried within the earth's crust, increasing demands placed on this rapidly dwindling nonrenewable resource have kindled interest in other sources. An especially promising source of new oil is *oil shale. These fine-grained rocks represent sediments deposited on the floors of ancient lakes and oceans that covered parts of the earth during prehistoric time. The remains of the organisms that lived in these ancient bodies of water were trapped in the sediments after they died; they sank to the bottom and were later converted into petroleum. Oil shale can be burned as a solid fuel or processed to produce oil or combustible gas. Known deposits of oil shale are estimated to be capable of yielding more than 2 trillion barrels of oil. The oil-bearing tar sands that occur in certain parts of the world are another potential source of petroleum from surface rocks. One such formation, the Athabasca Tar Sands of northern Alberta, Canada, is a hydrocarbon-bearing formation estimated to hold more than 600 million barrels of

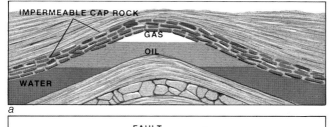

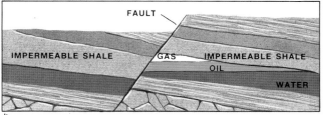

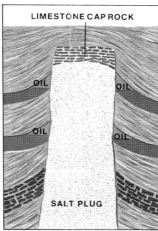

bitumen—a tarlike petroleum substance. If large quantities of this black, gummy material can be released from the sandstone and reduced to a liquid state, it may eventually prove to be a vital source of petroleum. As with oil shales, however, the major problem with releasing petroleum from tar sand is that it is a relatively expensive process and a large amount of waste material is generated after the oil has been extracted, ruining large areas of land.

Petroleum is found in many parts of the world, in rocks ranging from Cambrian to Late Tertiary in age. Oil occurs all over the world, but the United States produces more than any other country. Texas is the leading oil-producing state; Louisiana, California, Oklahoma, Kansas, Illinois and Wyoming also produce large quantities. The Soviet Union, Rumania, the Middle East (Kuwait, Saudi Arabia, Iran and Iraq), the East Indies (Borneo and Sumatra), certain South American countries (such as Venezuela and Colombia), Canada and Mexico are among the other important oil-producing areas. (W.H.M.)

Petrology. The branch of the earth sciences concerned with the study of rocks, including their origin, occurrence, composition, texture and structure. Petrology includes petrogenesis, which deals with the origin of rocks, and petrography, which is concerned with the description and classification of rocks, especially by means of microscopic examination.

Phacolith. A body of *igneous rock associated with layered rocks that have been deformed into folds. Phacoliths are normally found at the crests of *anticlines or at the troughs of *synclines. In cross section they appear crescent shaped and some of them measure thousands of meters at the thickest part of the crescent. Some geologists believe that the *magma that forms phacoliths does not force its way between the layers of rock but passively flows into zones of weakness or open spaces left between the folded layers.

Phanerozoic. A broad division of geologic time that includes the part of the geologic record for which there is much evidence of plant and animal life. It includes about 10 percent of earth history and chiefly comprises the *Paleozoic, *Mesozoic and *Cenozoic eras and their corresponding rocks.

Phenocryst. In an *igneous rock, a crystal that is larger than the crystals of the matrix in which it is embedded. The general term for rock with such large conspicuous crystals is *porphyry. Phenocrysts vary

greatly in size; in a fine-grained rock they may be only 1—2 mm across, whereas in coarse-grained rock they may be as much as 10 cm in length.

Phlogopite. A mineral, a member of the mica group. It is brown and magnesium-rich, characterized by one highly perfect cleavage along which it can be split into thin folia. Cleavage sheets of phlogopite are used like those of *muscovite, chiefly as an electrical insulator. The mineral is found in metamorphosed magnesium limestones and ultrabasic rocks. See Asterism; Mineral Properties.

Phonolite. A volcanic rock, the fine-grained equivalent of a *nepheline syenite. The principal minerals are sodium-rich orthoclase and nepheline. In addition, pyroxenes or amphiboles and other feldspathoids, particularly sodalite, are usually present.

Phosphate. One of a group of chemical compounds containing the phosphate ion, the radical PO_4^{-3}. This group is extremely large, embracing well over 250 distinct species. Many are secondary minerals and all but a few are very rare. The most abundant primary phosphate is *apatite, and the phosphorus that enters into the secondary minerals is to a great extent derived from its alteration.

Phosphate Rock. A *sedimentary rock consisting largely of calcium phosphate. See Phosphorite.

Phosphorescence. Luminescence, that is, the emission of visible light, that continues to be emitted from a substance after removal of the radiation that causes the luminescence. This should be distinguished from *fluorescence, where continued exposure to the stimulating radiation source is necessary. The duration of phosphorescence varies greatly; in some minerals it lasts for only a fraction of a second, whereas in others it continues for several minutes.

Phosphorite (or Phosphate Rock). A *sedimentary rock consisting largely of calcium phosphate. Although it may occur as pellets or nodules, it more commonly resembles layered masses of limestone. Most phosphorite was probably originally derived from the bones of vertebrates or the shells of certain arthropods and brachiopods, but much of it was altered and enriched after deposition. Animal excrement, such as guano and the coprolites of ancient organisms, may also contribute to the formation of phosphorite. It is used

to manufacture fertilizers. In the United States commercial deposits occur in Florida, Idaho, Montana, Wyoming, Utah, Tennessee and South Carolina; elsewhere they occur in Canada, Tunisia, Algeria, France and Russia. (W.H.M.)

Phyllite. A *metamorphic rock essentially composed of the minerals white mica, chlorite and quartz. The grains of chlorite and mica are sufficiently large to impart a silky sheen to the rock. Some phyllites may possess thin but recognizable alternating bands of chlorite and mica with bands of quartz. Phyllites are derived from rocks formed by deposits of muds and clays on sea floors and subsequently buried under accumulations of sediments. The subsiding *sedimentary rocks reach depths where temperature and pressure are great enough for chemical reaction to occur. The chemical activity is responsible for generating new minerals and converting the existing rock into phyllite. (J.W.S.)

Phyllonite. *Metamorphic rock of a characteristic texture, produced close to the earth's surface by the crushing of solid rock. The process occurs along surfaces in which massive bodies of rock are in motion relative to one another. Such surfaces are called fault planes or shear zones. Crushing or granulation and the recrystallization of some minerals and the metamorphic growth of others, such as mica, impart a silky sheen to the rock. In many cases phyllonites resemble another rock, *phyllite, which is formed by a different metamorphic process.

Phyllosilicates. The group of *silicates in which the silicon-oxygen tetrahedra are linked together by sharing oxygen ions to form sheets of indefinite extent. The structural unit, sometimes called the "siloxene sheet," imparts to all members of the group a platy habit and one prominent cleavage. In the phyllosilicates the ratio of silicon to oxygen, $Si:O = 2:5$, is reflected in the formulas of the minerals, as, for example, in talc, $Mg_3(Si_4O_{10})(OH)_2$, and serpentine, $Mg_6(Si_4O_{10})(OH)_2$. Other important members of the group are the micas, the clay minerals and the chlorites.

Physiography. See Geomorphology.

Phytosaur. An extinct crocodile-like reptile 1.5 to 7.5 m (6 to 25 feet) in length. It resembled the crocodile both in appearance and in mode of life. This similarity is only superficial, however, for phytosaurs and crocodiles are two distinct groups of reptiles. The phytosaurs are exclusively Triassic in age. See Fossil Reptiles.

Piezoelectricity. Electricity, or electrical polarity, produced on the surface of a crystal by mechanical stress. Not all crystals are piezoelectric; in those that are, the electricity is developed only where pressure is exerted parallel to certain crystallographic directions. To show this phenomenon a crystal must belong to one of 20 symmetry classes that lack a center of symmetry and have at least one polar axis. If pressure is exerted at the ends of a polar axis, a flow of electrons toward one end produces a negative charge while a positive charge is induced at the opposite end.

Piezoelectricity was first detected in 1881 by Pierre and Jacques Curie in the mineral quartz, but it had no practical application until the end of World War I. At that time it was discovered that sound waves produced by a submerged submarine could be detected by the piezoelectric current generated when they impinged on a quartz plate. In 1921 it was discovered that quartz through its piezoelectric property could be used to control radio frequencies. When a properly oriented and dimensioned quartz slice is placed in a circuit of alternating current, it mechanically vibrates when the frequency of the quartz plate coincides with the frequency of the circuit. Millions of quartz oscillators are in use in radio today to control and maintain a constant frequency for radio transmission and reception.

Tourmaline also is piezoelectric and because of this property is used to manufacture gauges to measure high transient pressures. (C.S.H.)

Pillow Lava. Basaltic or andesitic lava composed of ellipsoidal, spherical or pillow-shaped structures piled on top of each other, with the upper pillows often shaped against the lower. Pillows are commonly about 1 m in diameter, but elongated pillows may be as much as 4 m in length. The margins are composed of glass (if it has not been altered to palagonite), the interiors are of more crystalline material. The voids between pillows may contain glassy angular fragments (commonly altered to palagonite) spalled off from the pillows, together with marine deposits.

A flow in the Samoan Islands is said to have formed pillows as it moved along the ocean floor. Underwater photography has shown that recent submarine lava flows off Hawaii are characterized by pillows. In many localities they are interbedded with marine deposits. Pillow lavas associated with lake deposits are common in the Miocene basaltic flows of the Columbia River lava plateau. They are common in the Precambrian of Canada and have been

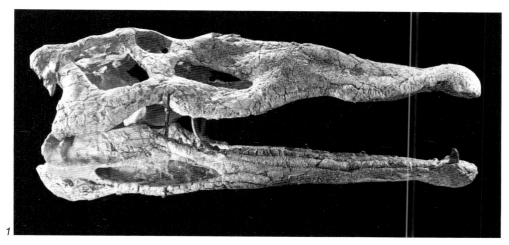

Phytosaur

1 *This skull and jaws of* Machaeroprosopus *was found in Arizona's Painted Desert.*

Pillow Lava

2 *The bulbous masses in this igneous rock at Great Notch, N.J., formed millions of years ago when hot lava flowed underwater.*

found in every part of the geologic column in many parts of the world. For the following reasons it is generally agreed that they are formed in oceans, lakes, rivers or swamps, and that they were either erupted beneath water or flowed into it: (1) They have been observed to form on the ocean floor; (2) recent pillow lavas occur in deep water; (3) quick chilling, such as would be expected from contact with water, is indicated by glassy margins; (4) marine deposits are present in the interstices between pillows; (5) they are associated with lake deposits; and (6) they are not found in subaerial flows. (R.L.N.)

Pinacoid. A crystal form composed of two parallel faces. Pinacoids are found on crystals of all crystal systems, except the isometric system.

Piping. Removal of fine solid particles by water percolating through the ground, generally in unconsolidated permeable material. It produces underground openings or conduits and is a common phenomenon in badlands.

Piracy (or Stream Capture). Interception of the channel of one stream by another and the taking over of its drainage. The stream whose waters are intercepted is a *beheaded stream. Capture commonly occurs when one stream has a steeper gradient than another on the opposite side of a drainage divide, or flows on less resistant rock and can erode more quickly. Capture may also be made by lateral swinging of a *meander. In the early development of an expanding drainage system, one tributary may capture another by *abstraction, that is, by expanding its drainage area to take in that of another tributary, thus diverting part of its *overland flow.

Stream capture is evidenced by rivers that seem too large or too small for valleys (*misfit stream), by *barbed tributaries or by terraces that slope upstream rather than down. There is often a waterfall at the point of capture. Stream capture causes migration of the divide since the captor drains part of the beheaded stream's basin. It is thought that in this way the whole drainage divide of the east coast of the United States is gradually shifting westward, with eastward-flowing streams slowly capturing drainage areas of streams flowing to the Mississippi. (M.M.)

Piston Corer. The most widely used of many types of devices that take cores of *deep-sea sediment. It consists of a hollow steel tube containing a piston. As the core barrel falls, the piston remains fixed and creates a suction which facilitates entry of

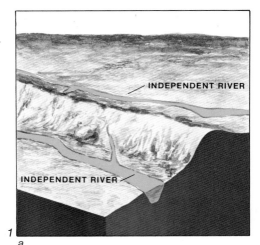

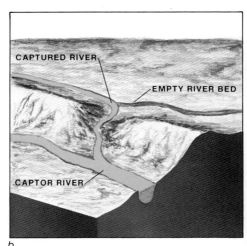

1 a   b

2

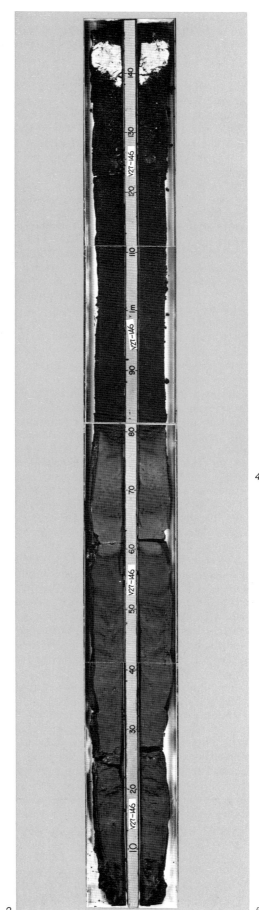

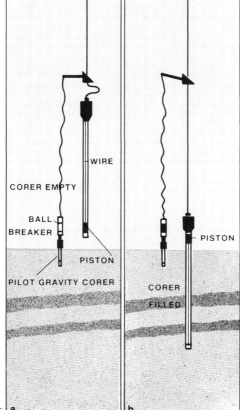

Piracy

1 As the ridge separating two independent rivers gets worn away by the headward erosion of tributary streams (a), one of the rivers may capture the other and divert the water into its own channel (b).

2 The drainage pattern for these stream valleys near San Juan Capistrano, Calif., was changed when a more rapidly eroding neighboring stream captured and diverted their headwaters.

Piston Corer

3 The cylinder of a core sample represents hundreds of thousands of years of undisturbed geologic history.

4 A piston corer is lowered over the side of an oceanographic research vessel.

5 In the Kullenberg type of piston corer (a), the pilot corer, upon hitting the ocean bottom, trips a release mechanism and the main corer falls. (b) Steel coring tube penetrates deep into the sediment, creating a vacuum that helps push the sample into the tube.

the sediment core into the barrel. The barrel and its sample are then hauled up to the ship. The apparatus can remove sediment cores up to 30 m in length.

Pitchstone. A volcanic glass having the luster of a pitch or resin. Pitchstones vary in color and in composition and may be the glassy equivalent of a rhyolite, dacite or andesite.

Placer. A superficial deposit, such as sand or gravel, containing grains or particles of one or more valuable minerals. Although most commonly thought of as gold deposits, many other minerals of economic importance are found in placers. For a mineral to accumulate in placers it must have a high *specific gravity (greater than 3) and be chemically inert and mechanically strong. Cassiterite, ilmenite, monazite and many gem minerals including diamond, zircon, corundum, topaz, garnet and spinel meet these requirements. In many places in the world they, as well as gold, are taken from placers. (C.S.H.)

Placer Mining. A type of mining in which the unconsolidated surface material is mined to recover the valuable minerals. Equipment for this varies in size and complexity from a small pan to a gigantic dredge. In one type of placer mining, called hydraulicking, water under high pressure is used to break down the deposits of unconsolidated sand and gravel.

Placoderm. A member of the extinct class of fishes called the Placodermi. Unlike the more primitive *ostracoderms, the placoderms had well-developed articulating jaws. They were plate-skinned and they were protected by strong bony armor in the head region. There were several different types of placoderms, but the largest were the huge sharklike arthrodires, which had massive, bony jaws armed with pointed toothlike projections and a sharp cutting edge. *Dinichthys,* a Devonian arthrodire, grew to be as much as 9 m (30 feet) long. Placoderms have been found in rocks ranging from Late Silurian to Permian, but were especially abundant during the Devonian Period. (W.H.M.)

Plagioclase. A group of triclinic *feldspars forming a complete solid solution series between pure albite ($NaAlSi_3O_8$) and anorthite ($CaAl_2Si_2O_8$). Calcium substitutes for sodium in all proportions with a concomitant substitution of aluminum for silicon.

The series is divided arbitrarily into six divisions on the basis of the relative amounts of albite (ab) and anorthite (an).

The names of the divisions and their compositions in terms of proportions of albite and anorthite are:

	% ab	% an
Albite, $NaAlSi_3O_8$	100 — 90	0 — 10
Oligoclase	90 — 70	10 — 30
Andesine	70 — 50	30 — 50
Labradorite	50 — 30	50 — 70
Bytownite	30 — 10	70 — 90
Anorthite, $CaAl_2Si_2O_8$	10 — 0	90 — 100

A common feature of all plagioclase is *albite twinning that is evidenced by striations crossing the basal cleavage. The presence of the striations makes plagioclase distinguishable from other feldspars. The specific gravity increases uniformly with the increase of calcium from albite, 2.62, reaching a maximum of 2.76 in anorthite.

The plagioclase feldspars, the most abundant of the rock-forming minerals, are found in *igneous, *metamorphic and, more rarely, *sedimentary rocks. The classification of igneous rocks is based largely on the amount and kind of plagioclase feldspar. As a rule the plagioclase in the light-colored rocks is near the albite end of the series; the darker the rock, the closer the plagioclase approaches anorthite in composition. See Twin Law. (C.S.H.)

Plankton. Aquatic organisms that float, drift or have only feeble swimming capabilities, and are therefore unable to cope with currents or control their location. Most of them are microscopic, but a few are large forms. A great variety of both the plant type (phytoplankton) and the animal type (zooplankton) are located in the photic zone, that is, where light penetrates. Included in the phytoplankton are many types of algae, *diatoms and *coccoliths. Zooplankton include protozoans (single-celled animals), insects, worms and some large types such as jellyfishes. Most bottom-dwelling animals have planktonic larval or juvenile forms. This allows widespread distribution even though adult forms are not very mobile. (R.A.D.)

Plate Tectonics. A branch of earth science dealing with the formation, destruction and motion of segmented sections of the earth's crust, called plates. The concept of plate tectonics draws heavily upon the now accepted concepts of *continental drift and *sea floor spreading to help explain the cause of earthquakes (seismic activity), volcanism and fold mountain systems (linear mountain ranges produced by bending and buckling of the earth's crust).

A crustal plate is made up of material representing the outermost portion of the earth's surface, known as the lithosphere. It may be composed of oceanic crust (that portion underlying the oceans and made up of dense rocks that are rich in iron and magnesium), continental crust (composed mainly of silica and alumina, characteristic of landmasses) and the uppermost zone of the *mantle. Plate boundaries need not coincide with continental margins; many of the major plates, for example the American Plate, are composed of both continental and oceanic crustal material. The number and size of the plates vary. Geologists agree that there are between six and eight major plates and several smaller plates covering the earth's surface. Estimates of the number of smaller plates change as earth scientists continue to discover new plate boundaries.

Plate thickness varies between 70 and 230 km, the thicker portions of a plate coinciding with continental landmasses. They are believed to ride on a weaker, viscous layer known as the *asthenosphere (mantle). The mechanism by which the plates are driven along is still in doubt, but one reasonable explanation is that *convection currents generate in the material below the lithosphere.

From evidence so far gathered there appear to be three types of boundaries associated with individual plates: (1) a boundary where adjacent plates are moving away from each other, (2) a linear zone where two plates are able to slide past each other without either being deformed, and (3) a zone where two plates are colliding. It is also known that volcanic and seismic activity occurs mainly along plate boundaries, with the interior of the plates remaining relatively rigid and stable.

The first kind of boundary is found along the mid-ocean ridges that form a nearly continuous mountain range dividing the ocean basins. Ridges are zones of tension within the lithosphere. The tensional forces create fractures along the axis of the ridge and the adjacent plates move apart. New crust is formed along the fracture by the upwelling of molten material from deeper parts of the earth. Along with volcanic activity, mid-ocean ridges are the sites of seismic activity, with earthquakes originating from shallow depths within the crust (less than 70 km). Since most scientists today believe that the earth has not expanded to any great degree (about 2 percent in the past 200 million years), some explanation must be found for the large amounts of new crust that have been generated without resultant global expansion. The answer lies along the leading edge of the growing plate, the boundary opposite the ridge axis. This junction corresponds to the third class of plate boundary. To relieve the stress created by the colliding plates, one plate is deformed and slips below the other at approximately a 45° angle. In areas where an oceanic plate collides with a continental plate, the thinner and denser oceanic crust will be the downgoing plate. The continental rocks, being lighter than the oceanic crust, are unable to sink into the denser plate and therefore override it.

The zone in which the downgoing plate dips under the overriding plate is called a *Benioff zone or subduction zone. Physiographic features associated with a subduction zone are deep ocean trenches and volcanic *island arc systems such as those found around the western edge of the Pacific Ocean. As the downgoing plate continues to be overridden, earthquakes are produced. They can range from shallow to deep within the earth's interior(down to 700 km). The earthquakes associated with subduction zones are produced in two ways. The shallow earthquakes originate from the downgoing plate breaking as it bends to dive under the leading edge of the overriding plate. Farther down on the subsiding plate, deeper earthquakes are generated by slippage between the downgoing oceanic crust and the overriding plate. Another feature associated with colliding plates is the formation of fold mountains on the overriding plate. They are caused by the buckling of the leading edge of the overriding plate. Along with buckling, *magma is generated when the downgoing plate melts. The molten material rises forcefully into fractures in the continental plate.

The second type of boundary is found on the sides of the plates adjacent to the mid-ocean ridges. These boundaries are fractures trending essentially perpendicular to the ridge system and are known as *transform faults. The San Andreas Fault in California is now considered by many geoscientists to be a transform fault, associated with the mid-ocean ridge located on the eastern edge of the Pacific Ocean. Along these boundaries, plates are able to slip past each other without being greatly deformed. See Color Plates, "The Shifting Plates of the Crust," "Drifting Continents" and "The Dynamic Crust of the Earth."

Plateau. An elevated region having a summit that is large and flat or gently inclined, bounded on one or more sides by steep cliffs separating the summit from the lower adjacent area. More specifically, a broad elevated area more than 300 m above adjacent country. Plateaus are differentiated from *mesas by their greater size and from mountains by their larger summit areas. They may be dissected by canyons, and *volcanoes, *laccoliths and *fault scarps may occur on their summits. Some authors

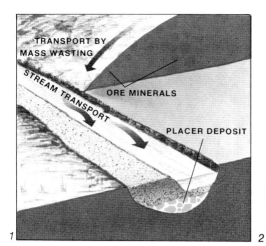

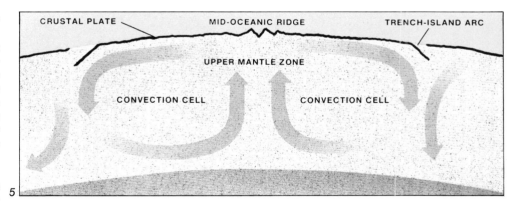

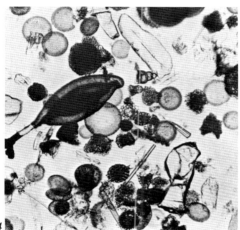

Placer

1 Heavy ore minerals weathered from fissure veins are washed downslope and become concentrated in stream gravels as placer deposits.

Placoderm

2 The giant Upper Devonian Dinichthys, 9 m (30 feet) long, was one of the earliest of the jawed vertebrates.

Plagioclase

3 This large crystal of plagioclase feldspar shows a face with numerous parallel striations formed by its repetitive albite twinning characteristic.

Plankton

4 Skeletons of aquatic organisms of the Miocene Age, 20 million years ago, are seen in this microphotograph taken over 3600 m (12,000 feet) below the surface.

Plate Tectonics

5 Currents of molten material are believed to move in convection cells within the earth's mantle, rising through rifts in the mid-ocean ridges, cooling as they flow outward as plates along the ocean floor, then sinking into deep trenches along the perimeter of the ocean basin.

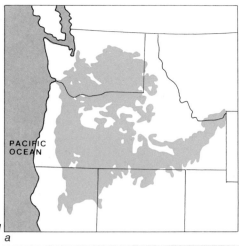

PACIFIC
OCEAN

ARABIAN
SEA

1
a

b

2

restrict the term to those flat-top elevated regions composed of horizontal *sedimentary rocks. Others include areas formed by the successive extrusion of *lava flows, as in the case of the Deccan Plateau of India and the Columbia River Plateau of northwestern United States, and by processes which have resulted in regional uplift, as in the case of the Colorado Plateau of western United States. (R.L.N.)

Plateau Basalt (or Lava Plateau, Flood Basalt). Formed by great floods of basaltic lava, a plateau basalt consists essentially of horizontal and overlapping flows that may cover hundreds of thousands of square kilometers. These flows came out from elongated fissures rather than circular vents, and their flatness and wide distribution resulted from their fluidity and a shifting of the feeding fissures. Subordinate interbedded *pyroclastic material and fluvial and lacustrine deposits are not uncommon. The presence of interbedded residual soils and other features indicates that in places considerable time elapsed between successive flows. Where the terrain that preceded the extrusion of the flows was rugged, the higher eminences may project above the flows as *steptoes. Plateau basalts range in age from *Precambrian to Recent.

Two well-known areas of plateau basalts are the Columbia Plateau in northwestern United States and the Deccan Plateau in India. Others are found in South America, Scotland, Iceland and elsewhere. Lavas blanket approximately 520,000 square kilometers (200,000 square miles) in northern California, Oregon, Washington and Idaho. The maximum thickness of these flows is not known, but the deep canyons of the Columbia and Snake rivers expose more than 1200 m (4000 feet) of lava, and a drill hole penetrated more than 3000 m (10,000 feet) without reaching bottom. The estimated volume of these flows is approximately 250,000 cubic kilometers (60,000 cubic miles). The average thickness of over 400 flows is 23.5 m (77 feet). The great bulk of the flows are Miocene, and in places they have been *folded and *faulted. The Eocene lavas of the Deccan Plateau in India are nearly continuous over an area of 650,000 square kilometers (250,000 square miles) and erosional remnants nearby suggest that the basaltic lavas originally covered at least 1,300,000 square kilometers (500,000 square miles). (R.L.N.)

Platform. A continental area consisting of basement rocks covered by essentially flat-lying layers mainly of sedimentary rock. See Craton.

Plateau Basalt

1 The dark shading indicates areas in (a) the northwest United States and (b) India that are thickly covered with lava flows. India's vast basaltic plateau covers nearly 647,500 square kilometers (250,000 square miles).

Playa

2 Rocks propelled by winds have furrowed trails across Race Track Playa in Death Valley, Calif. Playas composed of fine-grained sediments are the beds of ephemeral, shallow lakes and are found in dry regions of the western United States.

Platinum. A metal, the principal member of the platinum group. Although it occurs in nature in compounds with other elements, its mineralogic occurrence is chiefly as the native metal. Because it has a high specific gravity (21.45 when pure) and is chemically inert, it accumulates in *placers. It was in just such a deposit in Colombia, South America, that platinum was discovered in 1735. It is still mined as a placer mineral, but most of today's production comes from primary deposits in basic *igneous rocks. The principal platinum-producing countries are the Soviet Union, South Africa and Canada. See Mineral Properties; Native Elements; Appendix 3. (C.S.H.)

Playa. A flat, sunbaked, barren plain at the bottom of a desert basin, floored with silt, clay and *evaporites. The surface may be very hard if dry, soft and spongy if wet. It is the bed of an ephemeral lake and may be covered with white salts. It is usually surrounded by *alluvial fans and, at greater distances, by pediments and mountain ranges. *Dunes and *loess may be found on the leeward side. Some playas cover thousands of square meters, others hundreds of square kilometers. Many playas are the sites of former *Pleistocene lakes. An extensive playa is found in Death Valley, and playas are common in California, Nevada and New Mexico and in the Arabian, Atacama and Sahara deserts. (R.L.N.)

Playa Lake. A shallow, ephemeral lake that covers a *playa after infrequent rains. It is usually undrained and therefore salty. It may cover scores of square kilometers, but nevertheless disappears a few weeks after the rains.

Playfair's Law. The law of accordant stream junctions, elucidated by J. Playfair in 1802: "Every river appears to consist of a main trunk, fed from a variety of branches, each running in a valley proportioned to its size, and all of them together forming a system of valleys, communicating with one another, and having such a nice adjustment of their declivities, that none of them join the principal valley, either on too high or too low a level." The significance of this law is that the *gradients of tributaries merge smoothly with that of the main stream.

Exceptions to Playfair's Law occur in (1) *hanging valleys, where glaciation has interfered with fluvial processes; (2) *piracy (stream capture), where the newly captured stream may not yet be adjusted to the gradient of its captor; (3) *rejuvenation, where the stream has not adjusted to uplift throughout its course.

Pleistocene Epoch. The earliest *epoch of the *Quaternary Period; it began 2 to 3 million years ago and lasted until the start of the *Holocene Epoch, approximately 10,000 years ago. The name (which means "most recent") was introduced by Sir Charles Lyell in 1839. The Pleistocene was a time of widespread continental glaciation during which ice blanketed about 30 percent of the land. At the height of the *Ice Age in North America, a vast ice sheet covered all of Canada and reached as far south as St. Louis, Mo. Similar continental glaciers flowed southward from Scandinavia to cover eastern Britain, and ice from Scotland covered northern Ireland. In Europe the ice reached into southern France, central Germany and Poland. Elsewhere mountain *glaciers extended well below their present limits in the Himalayas, the Andes, the Soviet mountain ranges, in New Zealand, the Rocky Mountains and Coast Ranges of North America and on the Equator in East Africa. Glaciers were also found on mountains in southeastern Australia and New Guinea that are free of ice today. The exact extent of Pleistocene ice is not known, but it may have covered almost 34 million square kilometers (13 million square miles), about one-fifth of the earth's land surface. This included 13 million square kilometers (5 million square miles) in Antarctica, about 3.25 million square kilometers (1.25 million square miles) in Europe and Asia, and some 2 million square kilometers (760,000 square miles) in Greenland. Today less than 16 million square kilometers (6 million square miles) are covered by glaciers, and most of this is in Greenland and Antarctica. The Pleistocene was not a single, prolonged time of glaciation; instead, multiple glaciations during at least four relatively short *glacial stages — the Nebraskan (the oldest), Kansan, Illinoian and Wisconsin — have been recognized in the states that bear their names. Corresponding glacial stages in Europe include (from the oldest to youngest) the Günz, Mindel, Riss and Würm. These were separated by warmer and much longer *interglacial stages during which the ice retreated. In the United States three interglacial stages named (from oldest to youngest) the Sangamon, Yarmouth and Aftonian have been described. In Europe, the corresponding interglacials are (from oldest to youngest) the Günz-Mindel, Mindel-Riss and Riss-Würm. The duration of the Pleistocene is not clear, but glaciation was probably at its maximum 20,000 years ago and may have ended in Canada as recently as 6,000 years ago. Some geologists believe that the Holocene (or Recent) Epoch is an interglacial stage and that

Pleistocene time has continued without interruption to the present day.

The extent of the ice can be documented by the type of sediment deposited and the landscape produced by the glaciers. Pleistocene deposits of *till and *outwash occur in Finland, Scandinavia, Russia, Poland, northern Germany and the British Isles. The Alps and other European mountains show the effects of glacial *erosion, and marine terraces along the Mediterranean Sea reveal the effects of the ice sheets on sea level. Extensive Pleistocene glaciation is evident in the Himalayas, Caucasus and other great mountain ranges of Asia, and an ice sheet is thought to have blanketed parts of northern Siberia. In North America the beginning of Pleistocene time is marked by the advance of the first continental glacier, which eventually covered as much as 12.5 million square kilometers (4.5 million square miles) of the continent. Glaciated regions include Greenland, Newfoundland and most of Canada. The ice advanced almost to the mouth of the Ohio River in the Mississippi Valley region, and glacial deposits occur as far south as the Missouri and Ohio rivers. Much of this region is a land of lakes, irregular rolling topography and poorly drained depressions. Much soil has been removed by glacial erosion, exposing bedrock that has been scratched or polished by the advancing ice. The melting glaciers deposited sediments that produced features such as moraines, drumlins, eskers, kames, outwash plains and huge boulders called erratics. Soil removed from eastern Canada and New England was transported south, where it was deposited on land or in the sea. Long Island, New York, consists almost entirely of sediments laid down by a melting glacier. Sea level was lowered during glacial stages because the normal return of water from the land to the oceans was reduced. But sea level rose as Ice Age glaciers melted, permitting the meltwaters to flow into the ocean. Shoreline fluctuations were also produced through elevation or depression of the land. The great weight of the ice slowly depressed the earth's mobile crust; subsequent removal of the weight through glacier melting allowed the crust to assume relative equilibrium. Such movement, common in glaciated areas, is best documented in Scandinavia and Finland. Evidence of similar uplift can be observed in the region of the Great Lakes and Lake Champlain, where old shorelines, originally horizontal, are now raised and tilted so that the greatest uplift is to the north.

The Pleistocene ice sheets caused major changes in stream and drainage patterns in some regions. When the ice retreated, stream valleys were typically aligned in the direction that the ice flowed. The ice also gouged out long, narrow, linear troughs that later became filled with water to form lakes. The Finger Lakes of New York State and countless small lakes in Minnesota and northwestern Canada were created in this fashion. The Great Lakes were also formed during the retreat of the last great ice sheet (the Wisconsin). They are the result of drainage changes and the ponding of water in great depressions that were formed in part by glacial erosion. Lake Agassiz, which no longer exists, was the largest Pleistocene lake in North America and once covered parts of Minnesota, North Dakota and Manitoba. The fertile soils of these areas have been derived from sediments deposited in this ancient lake. Lake Bonneville was a similar Ice Age lake in the western United States and was the result of increased rainfall produced by climatic changes associated with glaciation. Utah's Great Salt Lake is a small, salty remnant of this huge Pleistocene lake. Extensive glaciation also occurred in the mountain ranges of the western United States and Canada. Large valley glaciers played an important role in carving the spectacular scenery of the Teton Range, Sierra Nevada, Rocky Mountains, Cascade and Coast ranges. Evidence of mountain glaciation can be seen as far south as Arizona and southern California. Although the basic cause of the Pleistocene Ice Age is not completely understood, a number of explanations have been proposed, relating to one or more of the following: (1) variations in solar radiation, (2) changes in atmospheric composition, (3) shifting in the geographic position of earth's magnetic poles, (4) changes in elevation of the continents, (5) shifting of the relative positions of continents with respect to the ocean basins and (6) drastic changes in both ocean and atmospheric currents.

The Pleistocene was also a time of much volcanic activity and gave rise to many of the larger volcanic mountains of the Cascade Range. Numerous *cinder cones, *lava flows and other evidence of violent *extrusive igneous activity can be seen in Arizona, Colorado, California, New Mexico, Utah, Nevada, Washington, Oregon, Idaho, western Canada and Mexico.

The fluctuating climates, mountain-building activities and advances and retreats of the ice sheets had a profound effect on Pleistocene life. Many plants and animals were able to migrate or adapt to changing climates, but others could not do so and became extinct. Many large animals inhabited Europe and North America; these and other Ice Age creatures are discussed under the *Cenozoic Era. Although manlike primates began their early development in Late Tertiary time just before the earliest Pleistocene glaciation, true man probably did not appear until Early Pleistocene time — between 1 and 2 million years ago. He probably underwent relatively rapid physical and cultural evolution, and his artifacts and skeletal remains have been found in Pleistocene deposits in Europe, Asia and Africa. Some of these occur in association with the remains of extinct Ice Age animals such as the woolly mammoth. See Fossil Man. (W.H.M.)

Pleochroism. A property of some crystals of showing different colors when polarized light is transmitted along different crystallographic directions. It is seen only in *anisotropic crystals and results from the selective absorption of certain wavelengths of light; the resulting color is a combination of those wavelengths that reach the eye. In tetragonal and hexagonal crystals there are only two directions in which there is differential absorption, and the phenomenon is called *dichroism. In orthorhombic, monoclinic and triclinic crystals there are three such directions along which light may be transmitted to yield three different and distinct colors. In many pleochroic minerals the color differences are slight and seen only as subtle variations of the same color. But in a few the differences are striking, as in "tanzanite," the gem variety of *zoisite, which exhibits deep shades of red, blue and yellow. (C.S.H.)

Plesiosaur. One of a group of marine reptiles characterized by a broad turtle-like body, paddle-like flippers and a long neck and tail. Plesiosaur remains range from Middle Triassic to Late Cretaceous in age. See Fossil Reptiles.

Pliocene Epoch. The last *epoch of the *Tertiary Period; it began about 12 million years ago and lasted for about 9 million years. Its name (which translates as "more recent") was established by Sir Charles Lyell in 1833. Pliocene marine deposits are exposed in northwestern Germany, Belgium, Holland, southeastern England, southern France and Italy. Continental deposits occur in southeastern Europe and western Asia, and volcanic rocks (including volcanoes such as Mount Etna and Mount Vesuvius) are found in Sicily and Italy. In North America both marine and continental deposits are known along the Atlantic Coast in North and South Carolina and Florida and in the Gulf Coast states of Louisiana, Texas, Mississippi and Alabama. Pliocene strata of marine and nonmarine origin occur in Oregon but are particularly well developed in California. Igneous rocks of Late Tertiary age cover large areas of the southwestern and northwestern United States, and volcanic mountains such as Mount Shasta and Mount Rainier were probably associated with the general uplift and mountain-building activities of this time. Continental beds of at least part Pliocene age occur with Miocene deposits in the Great Plains area and in part of the Colorado Plateau.

Pliocene climates were increasingly cool and dry, and life forms of this time show the effect of adaptation to the changing environment. Mammals were widespread, and camels, horses, rhinoceroses, deer, antelopes and giraffes were common. Mastodons underwent a high degree of specialization, and primates (including monkeys, apes and gibbons) made significant advances. Carnivores were abundant and included hyena-like dogs, saber-toothed cats and bears. Plant fossils are generally poorly preserved but appear to have resembled those of the Miocene. Marine invertebrates were dominated by the mollusks, foraminifers, echinoderms and corals. Sharks and bony fishes were numerous and resembled those of today. Economic resources of the Pliocene include petroleum, phosphate rock (used for fertilizer), sand, gravel and clay for ceramics. (W.H.M.)

Plug Dome. A very steep-sided *cumulodome.

Plunge. The angle between a fold axis or other linear geologic structure and a horizontal plane. See Fold.

Pluton. A body of igneous rock that has been intruded into the earth's crust. Plutons are classified according to their size, shape and structural relationships to the intruded crustal rocks. The size of plutons ranges from sills or dikes only centimeters thick to batholiths covering many square kilometers. See Batholith, Dike, Laccolith, Lopolith, Phacolith, Sill.

Plutonic Rock. Any rock of molten origin (*magma) that forms a large body within the earth's *crust when it solidifies. Plutonic rocks are characterized by relatively large grain size, such as the crystals normally present in granite. They are the rocks that comprise such plutonic bodies as *batholiths, *stocks and *phacoliths.

Plutonism. An obsolete theory that assumed all the earth's rocks solidified from an original molten mass. It was proposed by the Scottish Geologist James Hutton

(1726–97) and derives its name from Pluto, Greek god of the lower world and keeper of the earth's eternal fires. The Plutonists (or Vulcanists) conducted extensive studies of volcanoes and igneous rocks and disproved the opposing concept of *Neptunism (advocated by geologists who believed all rocks were formed in or on water). (W. H. M.)

Pluvial Lake. A lake existing during the cold, wet pluvial (glacial) periods during the *Pleistocene in areas now dry and hot. Such lakes were common in Nevada, California, Utah and adjacent states; in this area approximately 120 freshwater lakes occupied intermontane basins, which are now dry or occupied by saline lakes. Lacustrine cliffs, *wave-cut platforms, tufa deposits, lake deposits and deltas formed by rivers entering the lakes testify to the existence of these lakes. Stratigraphic evidence and carbon-14 measurements indicate that these lakes were contemporaneous with the younger glacial episodes.

Lake Bonneville was the largest pluvial lake. At its maximum it was more than 50,000 square kilometers (20,000 square miles) in area and more than 300 m (1,000 feet) deep. Great Salt, Provo and Sevier lakes are its saline remnants. As Lake Bonneville grew, the earth's crust was depressed by the weight of the water; it was domed upward as the water evaporated following the pluvial period. This is indicated by the Bonneville shorelines, which are not horizontal. Basin deposits indicate that Lake Bonneville was preceded by at least three other pluvial lakes, all related to earlier glacial events. Another pluvial lake approximately 200 m (600 feet) deep occupied Death Valley, Calif. Pluvial lakes also occurred in the desert areas of Bolivia, Chile and Argentina, and in the Saharan region of North Africa. The Dead Sea during the pluvial period was more than 300 m (1000 feet) deeper than it is now. (R. L. N.)

Pluvial Period. An episode during the *Pleistocene when areas that today are arid or semiarid were colder and wetter. During the Pleistocene glaciation the climatic zones in North America were pushed southward by the continental ice sheet that covered Canada and northern United States. The southwestern part of the United States was colder and wetter, and there was much less evaporation. As a consequence, numerous *pluvial lakes occupied the undrained basins of the Southwest. The shoreline features of these lakes are still conspicuous on the lower slopes of the fault-block mountains. Data are accumulating that indicate there were

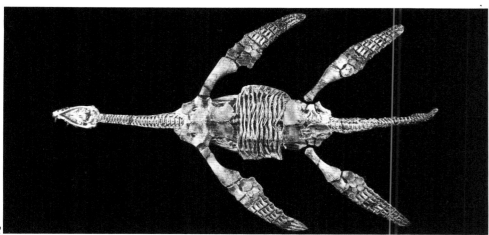

Plesiosaur

1 Elasmosaurus, *more than 12 m (40 feet) long, had a neck almost twice the length of its body. As it swam through shoals of fish, it darted its snakelike head from side to side to catch prey.*

2 Thaumalosaurus, *from the Lower Jurassic, had a strong central girdle and large "paddles" that enabled it to row both forward and backward with great speed.*

Pluvial Lake

3 *Today the southwestern United States is an arid region with few lakes (a). During the Ice Age, when temperatures were lower and rainfall in the region was greater than now, vast areas were covered by pluvial lakes (b) such as the gigantic Lake Bonneville, which was more than 300 m (1000 feet) deep and had a volume comparable to that of Lake Michigan.*

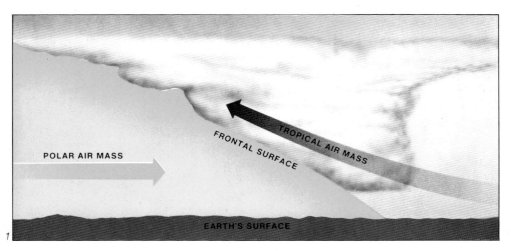

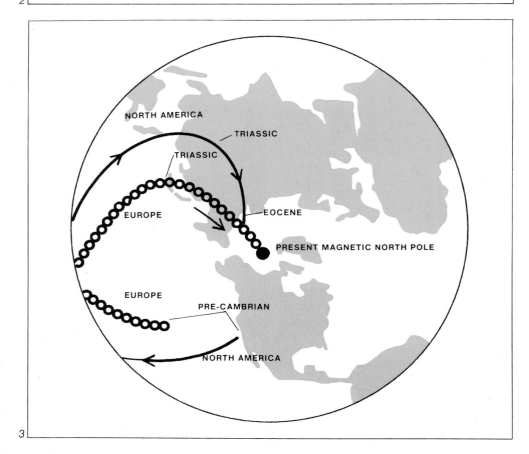

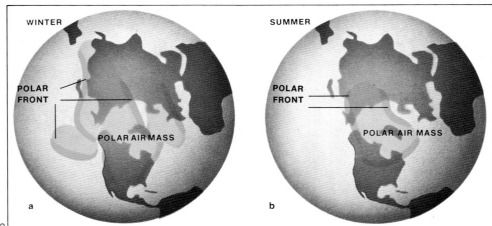

as many pluvial as glacial periods in the Pleistocene. Evidence of pluvial periods is also found in South America, Asia and Africa. (R.L.N.)

Podzol. See Pedalfer.

Poikiloblastic. A texture associated with some *metamorphic rocks. It results from the growth of a new mineral around pre-existing grains, which produces a sievelike texture.

Polar Front. In the atmosphere, the major front, or transition zone, between air masses of different density. The polar front, separating air masses of tropical and polar origin, is a nearly continuous, semipermanent feature of the general circulation, marking the zone of most frequent cyclone development.

According to the polar-front theory proposed by the Norwegian school of meteorologists early in the 20th century, frontal cyclones originate as disturbances on the surface or interface that separates the warm and cold air masses. However, the modern theory of dynamic meteorology explains the major features of *cyclones, and of *anticyclones (high pressure areas with counterclockwise circulation in the Northern Hemisphere), without reference to fronts. But the newer theory does invoke a zone of marked temperature contrast to explain the characteristics of the waves associated with frontal disturbances and anticyclones in the lower troposphere.

Fronts do exist in the atmosphere, nevertheless, and the polar-front theory offers a useful description of the growth and decay of frontal cyclones and of the accompanying weather. The cold, dense air mass on the poleward side of the front is initially a wedge underlying the warmer air of the westerlies. This would seem to be an unstable situation — the dense air only partly underlying the warmer air mass — but it is a balanced condition made possible by the earth's rotation. When the cyclonic circulation begins, the polar front is distorted. A portion of it is swept rapidly toward the Equator as a cold front. The other portion, the warm front, moves less rapidly in the cyclonic circulation because the warmer air tends to glide up over the colder air on the poleward side of the warm front. Eventually the cold front will overtake the surface warm front. This indicates that the growth stage of the cyclone is nearing completion. When the cyclone is fully occluded, that is, when the cold air mass in the lower troposphere entirely surrounds the cyclone, the storm ordinarily begins to decay.

The typical weather accompanying the passage of a frontal cyclone is largely the result of the vertical air motions that accompany it. Clouds and precipitation develop as a result of the mechanical lifting of the warm, usually rather humid air mass. Clear, dry weather is typical of the subsiding polar air mass that follows the cold front. The abrupt lifting of the air by the rapidly advancing cold front frequently produces large cumuliform *clouds and *thunderstorms in the warm air mass. Stratiform or layer-like clouds are more often typical of the moist air that ascends slowly over a warm-front surface. (M.F.H.)

Polarization of Light. The process by which ordinary light, vibrating in all directions perpendicular to its direction of propagation, is constrained to vibrate in a single plane. Light reflected from a smooth nonmetallic surface is partially polarized with the vibration direction parallel to the reflecting surface. Complete polarization is produced by the *Nicol prism by eliminating one of the two polarized rays into which the light is broken on entering the prism. Polarized light is also effectively produced by differential absorption in a birefringent crystal. In sheet filters, such as Polaroid, light entering is separated into two polarized rays; one suffers nearly complete absorption, the other very little and emerges as plane polarized light. (C.S.H.)

Polar Wandering. Changes in the position of the magnetic and geographic poles. These movements were discovered during studies of the evolution of the earth's *magnetic field (*paleomagnetism) and its ancient climates. When paths of polar wandering at various periods of geologic time (plotted from data gathered on the North American and European continents) were compared, they were found to be roughly parallel but not to coincide except near the present pole position. To explain the difference in polar paths — equal to approximately 30 degrees of longitude — scientists have applied the concept of *continental drift. When the separation between North America and Europe is taken into account, it is found that the two paths of polar wandering coincide. Thus paleomagnetic evidence on the wandering of the pole positions simultaneously supports the theory of continental drift. (J.W.S.)

Pollution, Air. See Air Pollution.

Polymorph. One of two or more distinct forms — that is, having different crystal structures — that may be assumed by an element or a compound. There are many examples of polymorphism in minerals. Diamond and graphite are polymorphs; although their physical properties are vastly different, they are both composed of the element carbon. Likewise FeS_2 crystallizes as two distinct minerals, pyrite and marcasite. When, as in these examples, only two substances are involved, they are sometimes referred to as dimorphs. An example of three minerals having the same composition (Al_2SiO_5) is andalusite, sillimanite and kyanite. (C.S.H.)

Porosity The property of a rock of containing voids, spaces and openings. A rock with few openings has low porosity; one with many has high porosity. It is expressed quantitatively as the ratio of the volume of the openings to the total volume. Porosity ranges from practically zero to more than 50 percent. The porosity may be less than 1 percent in igneous and metamorphic rocks; it averages 20 percent in sediments but much less in most sedimentary rocks. Porosity decreases with depth because of increasing compaction, cementation and recrystallization; at great depth the weight of overlying rocks crushes all openings and the rocks no longer have porosity. The openings may result from (1) unfilled spaces between the grains of sediments and sedimentary rocks, (2) fracturing of rocks, (3) solution in limestone and other soluble rocks or (4) gas cavities and other openings in lava. The porosity of rocks has a considerable amount of economic significance because it sets a limit on the quantity of gas, oil and water they can contain. (R.L.N.)

Porphyroblast. A well-developed crystal in a fine-grained matrix of a *metamorphic rock. The crystals were formed during the process of metamorphism.

Porphyroblastic. A characteristic texture in *metamorphic rocks produced by noticeably large crystals of one or more minerals within a finer-grained mass of other minerals. The large grains (porphyroblasts) may show their entire crystal outline or only a few crystal faces. The ability of some minerals to attain large size depends partly on the internal structure of the mineral and to a considerable extent on the availability of elements essential to mineral growth.

Porphyry. An igneous rock in which coarse mineral grains, *phenocrysts, are enclosed in a finer-grained groundmass. The texture is known as porphyritic. Rocks of any composition may have a porphyritic variety, and are then described as a porphyry — for example, granite porphyry.

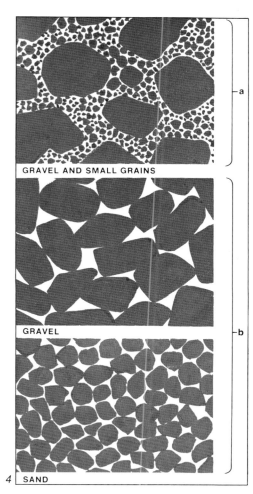

GRAVEL AND SMALL GRAINS — a

GRAVEL — b

4 SAND

Polar Front

1 A dense, cold air mass slides under tropical air like a wedge. As the lighter warm air rises, it cools and clouds and precipitation follow.

2 During winter (a) the continental polar air mass, which originates over Alaska and northern Canada, expands over much of the Northern Hemisphere. In summer (b) the sun melts northern snowfields and the polar air mass shrinks to half its winter size.

Polar Wandering

3 Curved lines trace the path of the north magnetic pole through time, based on paleomagnetic measurements in North America and in Europe. Some 50 million years ago the two curves merged.

Porosity

4 When the pore spaces between large fragments in gravel are filled by fine particles (a), the material is poorly sorted and its capacity to hold water is low. Although the particle sizes differ, water retention of well-sorted sand and gravel (b) is equally high because both contain the same volume of space.

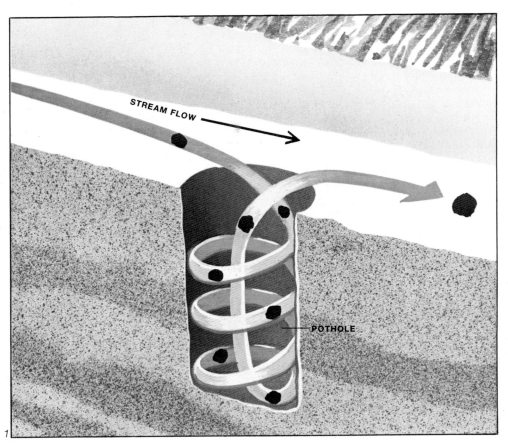

STREAM FLOW

POTHOLE

1

2

Pothole

1 *Driven by eddies of water, rocks spiraling downward into depressions in the bedrock eventually carve deep holes.*

2 *A cluster of stones in a pothole is exposed at the low-water stage in the bed of West River, South Londonderry, Vermont.*

Porphyry copper. The term as originally used referred to a porphyritic rock with various copper ore minerals disseminated throughout. Present usage is not restricted to porphyries but to any copper deposit having huge size, uniformly disseminated copper minerals and a low percentage of copper. They are sometimes called low-grade copper ores because the copper content is usually less than 1 percent and may be as low as 0.5 percent. With the rapid depletion of high-grade ores, an increasingly large amount of the world's copper comes from porphyry copper deposits. The largest open-pit mine in North America, at Bingham Canyon, Utah, is such a deposit. Although the ore averages only 0.7 percent copper, the annual production is about one-quarter million tons. (C.S.H.)

Postglacial Epoch. A synonym for *Holocene Epoch, the last epoch of the Quaternary Period.

Potash Feldspar. A *feldspar that contains potassium as an essential element, that is, orthoclase, microline and sanidine. All are potassium aluminum silicates.

Pothole. A circular depression eroded in *bedrock by the rotational movement of sand, gravel or boulders spun around by a river current. A special kind of pothole formed in bedrock beneath a glacier by *meltwater plunging down under pressure is known as a moulin. Large potholes at the base of falls are termed plunge pools.

Precambrian. Geologic time preceding the *Cambrian Period, and all rocks formed during this part of earth history. It encompasses that time between the origin of the earth and the appearance of complex forms of life about 600 million years ago, and is believed to be equivalent to as much as 90 percent of the earth's 4.5-billion-year history. The oldest known Precambrian rocks in North America are from the Morton Gneiss of southwestern Minnesota and have been dated at 3.6 billion years. The most ancient rocks (2.7 billion years) in the Baltic region are from the Karelian area of eastern Finland. Rocks from Tanzania are as much as 3.6 billion years old, and rocks 3.8 million years old are found in Greenland.

The Precambrian has been divided into the *Archeozoic Era (Early Precambrian) and *Proterozoic Era (Late Precambrian). The terms Prepaleozoic, Azoic and Cryptozoic have also been applied to all or certain portions of this part of geologic time.

Precambrian rocks generally lack fossils and have undergone much *metamorphism; thus the record of this part of earth history is difficult to decipher. Rocks of this age occur on all continents including Antarctica. They are also presumed to be present under the younger rocks in many areas of the earth's continental crust but are exposed over less than 20 percent of the earth's surface. They generally occur in (1) the cores of folded mountain ranges such as the Rocky Mountains, (2) the bottom of deeply eroded canyons (the Grand Canyon of Arizona) and (3) large stable areas known as shields. The term shield derives from a gently arched, shield-shaped area of relatively low relief in Canada that covers more than 4.5 million square kilometers (3 million square miles). It has been explored in some detail because of the presence of valuable ore deposits such as iron, copper, zinc, nickel, gold, silver and platinum. Other extensive exposures of Precambrian rocks occur on the Antarctic, Angara, Amazonian, Australian, Baltic, China-Korean, Ethiopian, Guianan, Indian and Platian Shields.

The lack of fossils in Precambrian rocks has long puzzled geologists. The older Precambrian formations consist mostly of highly metamorphosed volcanic and sedimentary rocks intruded by granite. Any fossils present may have been destroyed, but certain carbon-bearing deposits may be of organic origin and are considered indirect evidence of life. Rocks formed during Late Precambrian time contain the oldest known evidence of life. These consist largely of worm burrows, *radiolarians, algae and bacteria. The oldest known Precambrian fossils and earliest known direct evidence of life are 3.2 billion years old and consist of bacteria-like structures found in the Onverwacht Series of Swaziland in southern Africa. The oldest known North American fossils are the remains of microorganisms resembling bacteria and algae. They occur in the Gunflint Formation of Southern Ontario, Canada, and are approximately 1.9 billion years old. The first known evidence of multicellular animals was found in 1947 in the Edicara Hills north of Adelaide, Australia. These extraordinary fossils occur as impressions in fine-grained Precambrian sandstone and apparently represent the remains of soft-bodied marine animals such as segmented worms, jellyfish and other *invertebrates. (W.H.M.)

Precipitation. Water particles, whether liquid or solid, that fall from the atmosphere and reach the ground. Nearly all precipitation falls from clouds, but precipitation particles are many times larger than the droplets suspended in clouds, and the formation of such droplets requires rather special conditions.

Precipitation in liquid form includes drizzle and rain. Drizzle consists of very small drops that often seem to float along on air currents. Raindrops are larger than drizzle, and drops with a diameter of at least 0.5 mm (2/100 inch) are classed as rain. The largest drops are highly distorted as they fall, being flattened at the bottom; those with diameters larger than 5 to 8 mm (2/10-3/10 inch) are unstable and break up in falling. Freezing drizzle and freezing rain fall as liquid drops that freeze upon impact and form a coating of glaze upon exposed objects. Frozen precipitation takes a number of forms. Snow is composed of white or translucent ice crystals, usually in complex branched hexagonal forms normally clustered into larger flakes, although snow may occasionally fall as single crystals. In very calm air the clusters may be as much as 25 cm (10 inches) in diameter. Perfectly symmetrical crystals are uncommon, since single crystals are broken as they collide and cluster with each other.

The type of precipitation called ice crystals consists of very small, unbranched crystals. Seeming to float in the air, and visible only in direct sunlight, these are sometimes called "diamond dust" or "frost in the air." The crystals may fall either from clouds or from a clear sky.

The frozen equivalent of drizzle consists of very small, white, opaque particles of ice called snow grains. Snow pellets are somewhat similar in appearance but are larger, and they tend to rebound and shatter on striking a hard surface. Sleet consists of transparent globules of ice formed when raindrops freeze upon falling through a cold layer of air next to the earth's surface. Small hail is usually made up of snow pellets that have acquired a thin layer of ice.

Hail is precipitation taking the form of spheres, cones or irregular lumps of ice more than 5 mm (2/10 inch) in diameter. It is always produced in clouds that have strong updrafts, such as cumulonimbus or thunderheads, which contain supercooled water. Hailstones larger than baseballs have been reported. Hailstones are formed by the cumulative accretion of ice as the initial small ice particles are slowed in their descent, or perhaps lifted by upward air currents, through layers of supercooled liquid drops.

Showery precipitation — for example, rain showers, hail and snow flurries — falls from convective clouds of the cumuliform type. Steady precipitation — such as drizzle, steady rain and snow grains — is produced in the more stable clouds of the stratiform type.

Precipitation, which completes the atmospheric phase of the *hydrologic cycle, is the source of practically all of the fresh water stored in lakes and rivers and as groundwater. Thus temperature, evaporation and precipitation constitute the basic elements of the earth's climate.

The worldwide distribution of precipitation is primarily zonal. It is controlled essentially by the general circulation of the atmosphere, although mountain ranges, continents and oceans strongly modify the zonal distribution. On the average, precipitation is ample in the two zones of predominantly upward motion — the intertropical convergence zone and the belt of westerly winds traversed by polar-front cyclones. Regions of predominantly sinking air — the subtropical and polar high-pressure areas — are marked by deficient rainfall or snowfall and by some of the world's greatest deserts. The differential heating of land and water creates the winds known as monsoons, which exert a seasonal control on rainfall. Summer monsoons usually bring ample rainfall, as in India, southeast Asia and southeastern North America. Drier conditions occur with the reversal to the winter monsoon with its cold, dry air and typically anticyclonic flow. Within extensive land masses such as Eurasia, the great distance of the continental interior from the ocean is often sufficient to limit rainfall, and mountain ranges may increase this effect. As moisture-bearing winds ascend a mountain range, the water vapor condenses and precipitation occurs along the windward slope and atop the range. The winds descending the lee slope are dry, and deserts created in the "rain shadow" of the mountains may extend far into the continental interior in the absence of another moisture source. See Atmosphere; Clouds. (M.F.H.)

Prehnite. A mineral, hydrous calcium aluminum silicate, characteristically associated with *zeolites in cavities in *basaltic rocks. Although it may occur in colorless crystals, it is usually green and grows most frequently in *botryoidal aggregates. See Mineral Properties.

Primary Mineral. A mineral deposited during the original formation of a rock or an *ore body. It is in contrast to a secondary mineral formed by alteration of pre-existing minerals. For example, feldspar in a granite that crystallized from a magma is primary, whereas kaolinite that formed on alteration of the feldspar is secondary.

Prism. In crystallography, a prism is a form composed of three or more equivalent faces parallel to a single axis. The shape of the cross section is usually used as a modifier to designate the *crystal system

TRIGONAL PRISM

HEXAGONAL PRISM

1

Prism

1 *Crystals of many minerals exhibit prisms. Simple trigonal and hexagonal prisms with basal terminations are shown here.*

Productivity

2 *Food chains such as the organic cycle in the ocean are the pathways by which living things obtain, use and transfer energy.*

3 *An ecological pyramid shows the amount of food available at each level in an ocean community. Large numbers of organisms and food are necessary for the lowest level, but the quantity of organisms and nutrients decreases at each successive level, shaping the pyramid. Its broad base represents the food producers (diatoms and other plants), which trap solar energy while manufacturing food by photosynthesis. The primary consumers are the herbivores (such as snails) that feed exclusively on plants. The secondary consumers are the carnivores (such as fish) that prey on plant-eating animals and on each other. At the top of the pyramid are the tertiary consumers, made up of the larger animals (such as sharks, seals and polar bears) that feed on other carnivores.*

and number of faces: e.g., hexagonal prism (six faces), ditetragonal prism (eight faces), rhombic prism (four faces).

Productivity. The rate at which photosynthesis occurs in an environment. In areas of shallow, clear water and abundant nutrients, productivity is generally quite high. Even though the productivity of an area is high, the rapid consumption of plants by animals may keep down the total mass of plant material. The rate of photosynthesis, and thus productivity, may be slowed by turbid water, a choppy surface that reflects light, seasonal changes in length of day and lack of nutrient materials. In general, shallow bays, estuaries and reefs are among the most productive marine areas. (R.A.D.)

Profile of Equilibrium. The longitudinal profile of a *graded stream, essentially a smooth, concave profile steep at the head and gentle near the mouth. The profile reflects the balance of *capacity and *competency of the river to its load and discharge, with the concavity resulting from an increased capacity and competence downstream. These powers will increase downstream because discharge increases, cross-section area (width and depth) increases and the flow resistance (relative roughness) decreases. These factors, plus a general decrease in grain size downstream, means the gradient can be lowered and the load can still be carried.

The stream may make adjustments in the equilibrium profile to offset changes in its environment. If a change prevents the river from moving its load below a given point, it will deposit material at that point in order to increase the gradient and thus its transportive power downstream. However, such deposits decrease the gradient above that point, so the stream continues to aggrade upstream until a new equilibrium profile is established that enables the load to be transported along. If a change occurs in the watershed environment so that the river has excess transporting power below a given point, it will scour its bed at that point. This lowers the slope below, causing a velocity decrease that reduces the ability to transport the load. However, the gradient above the point is increased, so the river continues to degrade its channel upstream to establish a new profile of equilibrium consistent with a reduced load. The stream thus attempts to maintain a profile of equilibrium. (M.M.)

Proterozoic Era. The second oldest *era, corresponding to the most recent part of *Precambrian time. Proterozoic rocks sometimes contain sedimentary rocks and

fewer igneous and metamorphic rocks than the underlying *Archeozoic, but volcanic activity can be seen in many places. Invertebrate fossils (fragments of sponges, radiolarians and algae) found in certain sedimentary rocks are believed to be the oldest direct evidence of life. Late Precambrian rocks occur in western Ontario, Canada; the Lake Superior region of North America; as well as the Appalachian, Rocky Mountain and Grand Canyon regions of the United States. They are also well exposed in Scotland and in the northern parts of Finland and Sweden, and have been widely recognized in South America, Africa, India, Australia and China. Proterozoic rocks contain rich deposits of iron, copper, nickel, silver, gold and cobalt, as in occurrences in Canada and the United States. (W.H.M.)

Protist. A single-celled organism that may possess both plant and animal characteristics, assigned to the kingdom Protista. They include protozoans, bacteria, fungi, viruses and some algae. Because of their small size, certain protists (for example, *foraminifers and *radiolarians) are useful as *microfossils.

Protore. The mineral assemblage of a deposit that is too low-grade to be mined at a profit under existing conditions. It may become *ore with increase in price or with technological advances.

Proustite. A mineral, silver arsenic sulfide, frequently associated with *pyrargyrite, silver antimony sulfide. Both minerals are red and are known as the ruby silvers; proustite is light ruby silver whereas pyrargyrite is dark ruby silver. See Mineral Properties.

Pseudofossil. An object of inorganic origin that resembles the remains of some form of plant or animal life. Objects of this type commonly occur in certain geologic formations and may be mistaken for true fossils. Concretions may develop in the shape of plants or animals, and rocks may weather to form patterns that resemble sponges, algae and other organisms. The mosslike mineral incrustations called *dendrites superficially resemble ferns, mosses or similar types of plants. *Slickensides that develop along certain rock fractures sometimes look much like palm leaves. (W.H.M.)

Pseudomorph. In mineralogy, a mineral crystal that is altered or removed but with the original external form preserved. Most commonly a pseudomorph has the chemical composition and crystal structure of

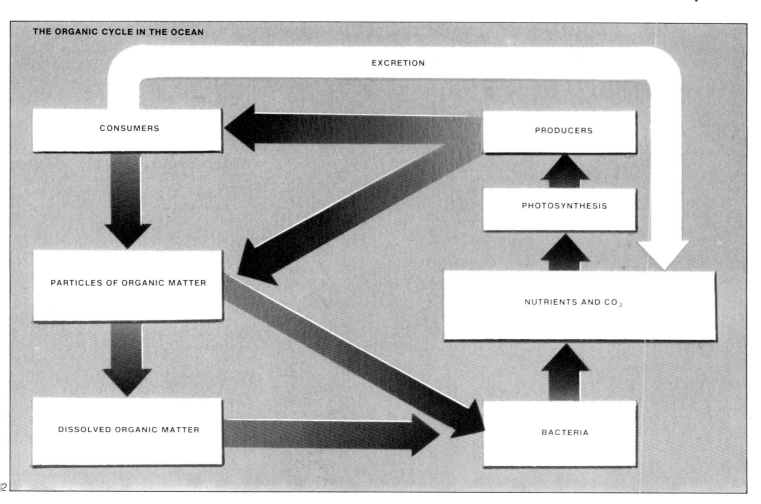

THE ORGANIC CYCLE IN THE OCEAN

EXCRETION

CONSUMERS

PRODUCERS

PHOTOSYNTHESIS

PARTICLES OF ORGANIC MATTER

NUTRIENTS AND CO$_2$

DISSOLVED ORGANIC MATTER

BACTERIA

2

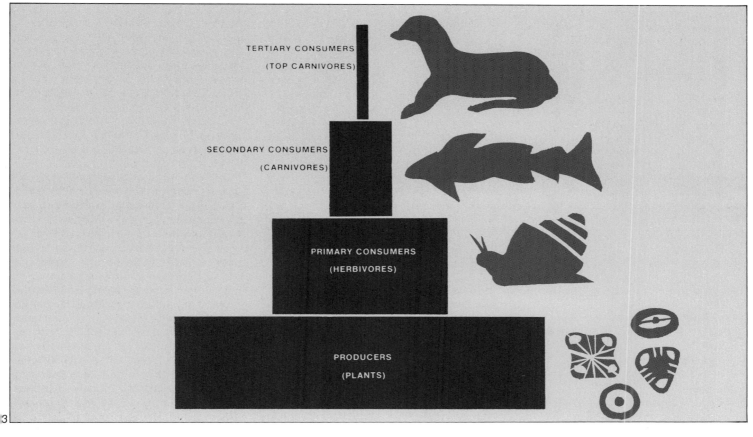

TERTIARY CONSUMERS
(TOP CARNIVORES)

SECONDARY CONSUMERS
(CARNIVORES)

PRIMARY CONSUMERS
(HERBIVORES)

PRODUCERS
(PLANTS)

3

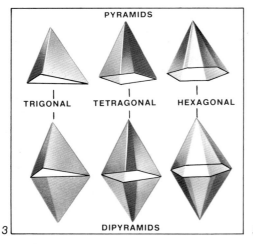

PYRAMIDS

TRIGONAL   TETRAGONAL   HEXAGONAL

DIPYRAMIDS

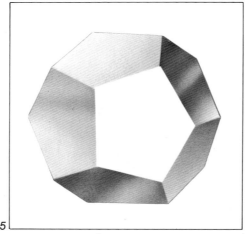

one mineral and the outward form of another. Frequently pyrite alters to limonite, preserving all the external features of the pyrite; it is then spoken of as "limonite after pyrite." Formed in analogous manner are pseudomorphs of malachite after azurite and kaolinite after feldspar. Less commonly, pseudomorphs form by incrustation in which a crust of one mineral is deposited over crystals of another: for example, quartz incrusting cubes of fluorite. If the fluorite is subsequently removed by solution, casts of its crystals left in the quartz indicate its former presence. (C.S.H.)

Psilomelane. A mineral, hydrous manganese oxide containing barium, an ore of manganese. It is a black secondary mineral usually found in *botryoidal or *stalactitic masses. Pyrolusite and manganite are also black manganese oxides, but psilomelane can be distinguished from them by its somewhat greater hardness (5–6) and its botryoidal form. See Mineral Properties; Appendix 3.

Pterosaur (or Pterodactyl). One of an extinct group of flying reptiles, with batlike wings, supported by long, thin fingers. Their very wide, skin-covered wings and lightweight bodies enabled them to glide or soar for great distances. The earliest known pterosaurs were found in Lower *Triassic rocks; they became extinct by the end of the *Cretaceous. During this time certain of these creatures attained a wingspread of as much as 8.2 m (29 feet) but their bodies were small and light. Recently, fossil remains of a giant winged reptile were found in Big Bend National Park, Texas. The estimated wingspan of this pterosaur is 15.5 m (51 feet), almost twice the wingspan of any pterosaur previously found, making it the largest known creature of its kind. (W.H.M.)

Pumice. A white or gray, lightweight, extrusive *igneous rock composed of natural glass. The abundance of *vesicles (gas cavities) accounts for its light weight. It is found on the surface of obsidian flows, as an important constituent of some *cinder cones and as fragments in *pyroclastic rocks. Pumice will float on water. Great rafts of pumice occasionally encountered at sea result from submarine or coastal eruptions; they are sometimes so numerous and large that they impede the passage of ships. After the eruption of Krakatoa in the East Indies, sailors left their ships and walked to shore over 3 km of floating pumice. It is used as an abrasive, an insulator, and in stucco, plaster and cement. (R.L.N.)

Pterosaur

1 Pteranodon was a turkey-size flying reptile of the Cretaceous Period, with a wingspan of almost 8 m (26 feet).

Psilomelane

2 Deposited by manganiferous groundwater, this specimen shows some of the characteristic stalactitic forms.

Pyramid

3 Many prismatic crystals are terminated by pyramids, crystal faces meeting in a point. Several types of pyramids and dipyramids are shown here.

Pyrite

4 A bold cubic shape, metallic luster and brassy metallic color identifies the mineral pyrite, also known as "fool's gold."

Pyritohedron

5 This 12-sided crystal form of the isometric crystal system is so named because pyrite assumes this form.

Pyramid. In crystallography, a crystal form composed of 3, 4, 6, 8 or 12 nonparallel faces that meet at a point. In specifying a pyramid we designate it as trigonal, tetragonal, hexagonal, etc. The crystal form composed of two pyramids base to base is called a dipyramid.

Pyrargyrite. A mineral, silver antimony sulfide, important as an ore of silver. Pyrargyrite and the arsenic analogue *proustite are commonly associated, and because of their color go jointly under the name of ruby silver; proustite is light ruby silver whereas pyrargyrite is dark ruby silver. Both minerals are valuable ores of silver, but pyrargyrite is more abundant. They have been mined in Germany, Mexico, Chile, Bolivia and the United States.

Pyrite. A mineral, an iron sulfide and the most common and abundant mineral sulfide. Because of its brass-yellow color it was early mistaken for gold and called *fool's gold, a name it now shares with other yellow minerals. A distinction can be quickly made between the two, for gold is easily scratched whereas pyrite—with a hardness greater than 6—cannot be scratched.

Pyrite forms under a wide range of temperature-pressure conditions and thus is found in many geologic environments. At high temperatures it forms as an accessory mineral in *igneous rocks, in *contact metamorphic deposits and in hydrothermal veins. It is deposited under atmospheric conditions in caves and forms nodules on the sea floor at temperatures only a few degrees above the freezing point of water. Pyrite is associated with many minerals in ore veins but is found mainly with galena, sphalerite and chalcopyrite.

Large and extensive deposits of nearly pure pyrite are mined as a source of sulfur. The sulfur, which makes up 53.4 percent of the pyrite, is liberated and collected by roasting. Sulfur is also collected from pyrite and other sulfides during the smelting of metallic ores. See Mineral Properties. (C.S.H.)

Pyritization. A type of fossilization whereby the original and hard parts of an organism are replaced by sulfides of iron. Such replacement is common in calcareous remains. The replacing mineral may be *pyrite or *marcasite, the latter transforming to pyrite under certain conditions. The iron sulfide is probably formed from the sulfur of the decomposing organisms and the iron present in the sediments. The remains of many brachiopods and crustaceans, and some mollusks, are found in a pyritized condition. (W.H.M.)

Pyritohedron. A form of the *isometric crystal system composed of 12 pentagonal faces. It is sometimes called the pentagonal dodecahedron.

Pyroclastic Material (or Tephra). Liquid drops and blobs, most of which solidify in the air, and fragments of solid rock of various sizes thrown up by volcanic eruptions. Pyroclastic material is classified according to the size, source and composition of the fragments, and on their physical condition when erupted. The size classification is as follows: (1) Fragments between 64 mm and a few meters (2½ inches to several feet) in diameter are blocks if erupted as a solid, bombs if erupted in a plastic condition. (2) Fragments between 64 mm and 2 mm (2½ inches to 4/5 inch) are called lapilli (from the Italian for "little stones"). (3) Volcanic ash is composed of fragments having diameters less than 2 mm but more than 0.25 mm (4/5 inch to 1/100 inch). Medium-size fragments that generally erupted as a liquid are called cinders. (4) Volcanic dust is the finest material erupted, the individual fragment having a diameter of less than 0.25 mm (1/100 inch).

Volcanic dust, ash and lapilli erupt in a solid or, more often, a liquid state. When violent gas explosions occur, solid rock can be broken into dust, ash, lapilli and blocks that may be blown high into the air. Frequently, fragments of sedimentary, metamorphic and plutonic igneous rocks are erupted. These accidental ejecta were broken (by volcanic gases and *magma) from the walls of the chimney leading up from the volcanic hearth (magmatic chamber) to the earth's surface. Fragments of volcanic rocks formed in earlier volcanic episodes and called accessory ejecta are also erupted. The material derived from the volcanic hearth responsible for the eruption is called essential ejecta.

When frothy lava reaches the surface it may be sprayed (*lava fountains) hundreds of meters into the air. Much of the material solidifies in the air, forming dust, ash and lapilli composed of volcanic glass.

Pyroclastic material is also classified on its petrographic composition. Thus it can have a rhyolitic, andesitic or basaltic composition. Rhyolitic material is light in color, generally white or light gray; basaltic material is darker, commonly black or red, and andesitic material is intermediate in color. *Cinder cones and *spatter cones are formed by the accumulation of pyroclastic material around volcanic vents. Composite volcanoes (see Volcano, Composite) are formed by the accumulation of pyroclastic material and by the extrusion of

lava flows. Much fine-grained pyroclastic material is commonly carried long distances by the wind during eruptions. When Parícutin, in Mexico, was active between 1943 and 1952, volcanic dust fell in considerable quantities 50 km (30 miles) away and small quantities fell in Mexico City 300 km (180 miles) away. It is estimated that in the eruption of Krakatoa in the East Indies in 1883 four cubic kilometers of pyroclastic material was blown 27 km (17 miles) into the air. Volcanic dust was carried around the world, causing beautiful sunrises and sunsets in the fall of 1883 and the spring of 1884. Solar radiation reaching the earth's surface was only 87 percent of normal during the year.

Since fine-grained pyroclastic material is deposited in greater quantity and at greater distances on the downwind side of a volcano during an eruption, the distribution and thickness of fine-grained pyroclastic material around a volcano will indicate the wind direction during an eruption.

If pyroclastic material falls on land it may later be picked up and transported by running water, and some may ultimately reach the ocean. Rocks formed from the consolidation of pyroclastic material may give clues about its geologic history following eruption. *Tuff, pyroclastic *breccia, *agglomerate, ignimbrite (or *welded tuff) and other pyroclastic rocks are formed by the consolidation of pyroclastic material. See Pele's Tears. (R.L.N.)

Pyroclastic Rocks. Consolidated rocks consisting of fragments formed either by the solidification of a spray of liquid droplets and clots erupted from volcanic vents or by the shattering and disintegration of older volcanic and nonvolcanic rocks. Pyroclastic rocks are usually classified according to the size of the fragments. Thus *tuff consists of consolidated volcanic dust and sand whereas *breccia consists of consolidated angular fragments ranging in size from a few centimeters to blocks two or three meters in diameter.

Pyroclastic rocks bridge the gap between *igneous rocks and *sedimentary rocks. Although the material of pyroclastic rocks has an igneous source, it is commonly deposited in the same way as are other sediments. Pyroclastic rocks have been formed throughout the earth's history. Cenozoic pyroclastic rocks are very common in western United States. After the volcanic landforms in an area are completely destroyed, the presence of pyroclastic rocks is often the only evidence of earlier volcanic activity. The following characteristics aid in their identification: (1)

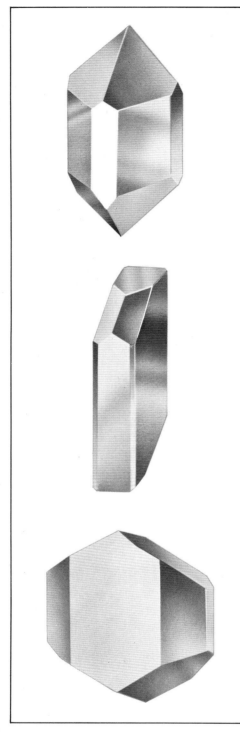

Pyroxene
Both orthorhombic and monoclinic crystals occur among minerals in the pyroxene group. Shown are some combinations of forms commonly found in pyroxene crystals.

They are commonly interbedded with *lava flows; (2) their chemical composition is similar to that of *extrusive igneous rocks; (3) they are commonly composed in part of glassy material; (4) they may contain bits of frothy *pumice and the broken walls of glassy bubbles. (R.L.N.)

Pyrolusite. A mineral, a manganese oxide, the most important ore of manganese. It is used principally in the production of spiegeleisen and ferromanganese to be utilized in the production of steel. About 13 pounds of manganese are consumed in the production of one ton of steel. Smaller amounts of manganese are used in various alloys with copper, zinc, aluminum, tin and lead, and to manufacture dry cells and chemicals.

Pyrolusite is one of three black manganese oxides, but it can be distinguished from the others, *plilomelane and *manganite, by its hardness. It is soft (H = 1-2), sometimes sooty and soils the fingers. All of these minerals are secondary and form under near-surface conditions. Present in small amounts in most crystalline rocks, manganese may be dissolved from the rocks and redeposited as the oxides, chiefly as pyrolusite. Some of the secondary minerals also result from the decay of manganiferous limestones and form beds of manganese ore in residual clays. Nodular deposits, mostly of pyrolusite, are found in swamps, on lake bottoms and in nodules on the sea floor. See Mineral Properties; Appendix 3. (C.S.H.)

Pyromorphite. A mineral, lead phosphate, a *secondary mineral found in the oxidized zone of lead deposits. It is characterized by its *adamantine luster and high specific gravity (7.04). Although pyromorphite contains 82 percent lead, it is considered only a minor ore of the metal because deposits of it are small. See Mineral Properties.

Pyrope. A mineral, magnesium aluminum garnet. It varies from deep red to nearly black; when transparent, it is used as a gemstone. See Mineral Properties.

Pyrophyllite. A mineral, a hydrous aluminum silicate, identical with *talc in appearance, in many physical properties and in uses. The two minerals can be distinguished by X ray or chemical tests. Like talc, pyrophyllite is found in foliated and radiating lamellar aggregates in *metamorphic rocks. See Mineral Properties.

Pyroxene. One of a group of minerals closely related in structure, chemical composition and physical properties. Although these minerals crystallize in both the *orthorhombic and *monoclinic systems, all species have similar cleavage: two cleavage directions intersecting at angles of about 87° and 93°. It is by means of cleavage that one can distinguish pyroxenes from similar-appearing *amphiboles. In the amphiboles the two cleavages intersect at about 56° and 124°.

The pyroxenes are inosilicates in which the SiO_4 tetrahedrons are linked into chains by sharing oxygens, with the ratio of Si:O = 1:3. The parallel chains are bound together by ionic bonds through X and Y cations. The general formula can be written as $XY(Si_2O_6)$, in which X is magnesium, iron, calcium, sodium or lithium and Y is magnesium, iron, manganese, aluminum or titanium. There is thus considerable substitution in both X and Y positions, giving rise to a large variation in chemical composition. The principal orthorhombic members of the group are *enstatite and *hypersthene; the monoclinic members are *diopside, *augite, *spodumene, *jadeite and *aegirine. See Silicate Structure and Classification. (C.S.H.)

Pyroxenite. Coarse-grained *igneous rock composed essentially of *pyroxene.

Pyrrhotite. A mineral, an iron sulfide. Because it is attracted to a magnet it is known as magnetic pyrites. Its chemical formula, $Fe_{1-x}S_x(x = 0-0.2)$, indicates a deficiency of iron with respect to sulfur; the greater the amount of iron, the weaker the magnetism. The mineral troilite, with composition close to FeS, is nonmagnetic. Pyrrhotite is commonly found in basic *igneous rocks, usually as small disseminated grains. But in some places, as at Sudbury, Ontario, and in Western Australia, it is present in large masses and is mined for the associated copper, nickel and platinum minerals. Pyrrhotite is also a minor source of sulfur. See Mineral Properties. (C.S.H.)

Q

Quartz. A mineral, an oxide of silicon. Abundant and widespread, it occurs as an important constituent in many igneous, sedimentary and metamorphic rocks. It is the principal mineral in sandstones and quartzites as well as in unconsolidated sand and gravel. Quartz is found in well-formed crystals more commonly than any other mineral. As such it occurs either by itself or associated with other minerals in veins, in rock cavities and geodes, and in pegmatites. A single quartz crystal weighing five tons has been reported from a *pegmatite in Brazil. Most often quartz is colorless and transparent; if found in good crystals it is known as *rock crystal. Other coarsely crystalline varieties that are used as gemstones are named according to the color: amethyst is purple to bluish violet, citrine is orange-brown, rose quartz is pink and smoky quartz varies from pale yellow to deep brown.

In addition to the coarsely crystalline quartz, there are many varieties in which the individual grains or fibers are not visible to the unaided eye. These may be divided into two types on the basis of the shape of the tiny particles that compose them. Flint, chert and jasper are granular, whereas carnelian, agate and other color varieties of chalcedony are fibrous. All of these have a dull to wavy luster; some are translucent even in thick slices, but others transmit light only on the thinnest edges.

Because of the clarity of many of its crystals, coupled with its unusual crystallographic properties, quartz is used in several interesting scientific devices. Its ability to transmit light in the ultraviolet and infrared portions of the spectrum makes it an ideal material to be cut into prisms and lenses for optical spectrographs. Quartz is enantiomorphous, that is, it is right-handed or left-handed depending on whether the plane of transmitted polarized light is rotated to the right or left. This property, known as optical activity, has been utilized in an instrument to produce monochromatic light. Quartz is also *piezoelectric and is therefore used to manufacture oscillator plates to control radio frequencies. The principal source of large natural quartz crystals for such uses is Brazil. The heavy demand for them has led to the increasing use of synthetic quartz.

The quartz that we see and handle, sometimes called low quartz, is one of seven polymorphic forms of SiO_2. It is the stable form of silica below 573°C (1063°F). When heated above this temperature it inverts to high quartz with different physical and crystallographic properties. The inversion is reversible; thus it again becomes low quartz when the temperature falls below 573°C. (C.S.H.)

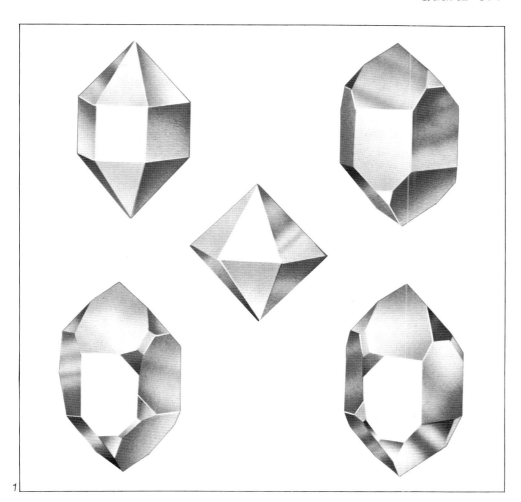

Quartz

1 Normally found in distinctive hexagonal prisms with pyramidal terminations, quartz may also occur in a variety of complex forms modifying its crystal shape.

2 Colorless to milky prismatic crystals of quartz are found in many places. These are from Hot Springs, Ark.

Quartz, Smoky
This dark, lustrous smoky quartz crystal stands out prominently on a matrix of microline feldspar.

Quartz Diorite. A coarse-grained *igneous rock composed essentially of *plagioclase feldspar with usually a small amount of *orthoclase. The common dark minerals are biotite and hornblende; the principal accessory minerals are apatite, sphene and magnetite. (C.S.H.)

Quartzite. A *metamorphic rock consisting chiefly of quartz grains. These grains result from the recrystallization of rock composed of sand.

Quartz, Smoky. A variety of coarsely crystallized *quartz with color ranging from pale yellow to smoky brown to almost black. It was early discovered in the Cairngorm Mountain in Scotland and is therefore also called cairngorm stone.

Quaternary Period. The second and the youngest period of the Cenozoic Era; it began 2 to 3 million years ago and includes all time since the end of the Tertiary Period. Its name means "fourth derivation" and originally was proposed as an era in an earlier classification of geologic time. It consists of two epochs, the Pleistocene and the Holocene (known also as Recent). Another classification places the Quaternary in the Neogene Period along with the Miocene and Pliocene Epochs of the Tertiary. Early Quaternary time (Pleistocene) was marked by extensive glaciation, for great ice sheets spread over much of North America and Eurasia and smaller glaciers were present in many other regions. During this Ice Age, there were four major glacial stages as the ice advanced across the land. These were separated by three warmer, much longer interglacial stages when the ice retreated and the temperature rose to that of today. Although glaciers blanketed but a fraction of the earth's land surface, the ice had a profound effect on the geography, climates, and plant and animal life of Quaternary time. The continental glaciation of the Late Cenozoic time was also important in shaping the modern landscape.

Nonglacial sediments of Quaternary age are present in many parts of the world and consist largely of wind-blown sand and dust, deposits laid down by streams and lakes, volcanic products (ash, lava, and cinders), and marine sediments. Marine terraces (which record higher relative positions of the sea) occur near the Mediterranean Sea, and nonglacial deposits are also found in other parts of Europe. Thick accumulations of *loess are present in Mongolia and north-central China, and fertile deposits of alluvium have been laid down by large rivers such as the Brahmaputra, Irrawaddy, Yangtze-kiang and the Hwang-Ho. Similar nonglacial materials occur in Africa, South America, and other continents. In North America, there are marine terraces of Quaternary age along the Atlantic and Gulf coastal plains; some of these are more than 250 feet above sea level. Meltwater from the Pleistocene glaciers increased stream flow in nonglaciated areas and readjustments in drainage patterns resulted in increased erosion and stream deposits in the Appalachian region and in the south-central and southeastern United States. There is also evidence of considerable erosion and stream deposition in the Central and High Plains east of the Rocky Mountains. Erosion predominated in the Colorado Plateau where the Colorado River carved the Grand Canyon in Late Quaternary time. Many lakes were formed in the Great Basin region as climatic changes associated with glaciation caused increased rainfall in the western United States. One of these, the Great Salt Lake of Utah, is a very small remnant of Lake Bonneville, a Pleistocene lake that originally covered about 50,000 square kilometers (20,000 square miles) and was as much as 300 m (1000 feet) deep. Along the Pacific coast from California to the Cascade Range there are thick deposits of stream-deposited sand, gravel, and clays. Marine deposits, now raised above sea level, are also present.

During Pleistocene time the proboscideans (elephants and their kin) underwent considerable development. The mastodon, woolly mammoth and the woolly rhinoceros were common in North America. Carnivores were well represented by *Smilodon*, the sabertooth cat, and *Canis dirus*, the dire wolf. *Smilodon* was about the size of a lion, had powerful jaws and highly developed, dagger-like upper canine teeth. The dire wolf was much larger than any living canine, and, along with *Smilodon*, was apparently quite common in southern California in Pleistocene time. The remains of both have been found in great numbers in the La Brea tar pits in Los Angeles. But by far the most noteworthy of all Pleistocene events was the introduction of the earliest known men. (W.H.M.)

Quicksand. Sand so supersaturated with water that it yields readily to pressure and can therefore engulf persons, animals or objects. However, the density of quicksand is heavier than that of a body and the quicksand will support it; drowning can only be caused by struggling and consequent loss of balance. Quicksand occurs when sand grains are separated by water that is generally moving slowly upward from a *spring. It is found in the beds of rivers and along coasts.

R

Radial Drainage. See Drainage Patterns.

Radiation. See Atmospheric Radiation; Solar Radiation; Thermal Radiation.

Radioactive Age Determination. A complex laboratory analysis of earth materials to determine when they were formed. The method involves the ratio of parent radioactive atoms to their decay products (radiogenic daughter atoms) and comparison of this ratio to the known half-life of the radioactive element. (A radioactive element will transform half of its nuclei into daughter elements in a calculable number of years.) Such tests can be used to determine the age of crystallization by *metamorphism of a preexisting rock; the age at which a molten *magma crystallized; the age of fold mountains; and the age of *sedimentary rocks in which new minerals formed during sediment deposition.

The accuracy of these calculations depends on (1) how accurately the half-life of the radioactive element is known, (2) whether any of the radioactive parent elements and/or their daughter elements have been added or removed from the rock after formation, (3) the amount of the daughter element in the rock at the time of formation and (4) the accuracy of the assumption that the time span for rock formation is short compared to the total age of the rock.

Several radioactive elements found in rock are now used as geologic clocks. Of all the methods used, the uranium-lead method, based on the radioactive decay of uranium 238 and 235 (U-238 and U-235) to lead 206 and 207 (Pb-206 and Pb-207) respectively, is the most useful. It can be employed to date rock formed over the greater part of geologic time but is most useful for rocks more than 100 million years old. The younger the rock, the less accurate the method, mainly because of the limited amount of the daughter element lead 207 present in rocks. Uranium-lead determinations are usually made on the minerals zircon and sphene, since these contain uranium and are associated with many *igneous and *metamorphic rocks in amounts up to about 1 percent of the rock.

The potassium-argon (K-Ar) method depends on the disintegration of the radioactive isotope potassium 40 (K-40) to argon 40 (Ar-40). Potassium is a very common element, occurring in many rocks and minerals. This method can be used to date rocks ranging in age from the very beginning of earth history to as recently as 30,000 years ago. Here, too, the method is less accurate for younger rocks because the amount of Ar-40 that has formed is very small in relatively young rocks. Min-

Parent Isotope	Half-life (in years)	Stable Daughter	Minerals and Rocks Commonly dated
Uranium 238	4.5 billion	Lead 206	Zircon
			Uraninite
			Pitchblende
Uranium 235	0.7 billion	Lead 207	Zircon
			Uraninite
			Pitchblende
Potassium 40	1.3 billion	Argon 40	Muscovite
			Biotite
			Hornblende
			Glauconite
			Sanidine
			Whole volcanic rock
Rubidium 87	47.0 billion	Strontium 87	Muscovite
			Biotite
			Lepidolite
			Microcline
			Glauconite
			Whole metamorphic rock

Radioactive Age Determination
Uranium, thorium, rubidium and potassium are the four principal elements used to determine age by methods based on radioactive properties. Shown here are the half-life and stable daughter isotopes of each of these elements and the rocks commonly dated with each.

Radiolarian

1 The silica shells of these minute protozoans, such as Eycyrtidium calvertense, *form an ooze common in deep parts of the Pacific and Indian oceans.*

Ray

2 This young crater, approximately 2 km (1¼ miles) in diameter, displays well-developed rays.

Reaction Series

3 As magma cools, various minerals form as their temperature of crystallization is reached. In the series on the left, each silicate is re-formed into the next in a chain of discontinuous reactions. The series on the right shows a continuous reaction of feldspar from one end to the other.

erals such as hornblende, nepheline and the micas, muscovite and biotite are used in K-Ar dating since one of these is found in most igneous and metamorphic rocks. Some marine sedimentary rocks may also be dated by this method provided they contain the potassium-bearing mineral glauconite. The usefulness of the K-Ar method is somewhat diminished by the possibility that Ar-40 was lost into the atmosphere after the mineral or rock was formed. This can result in errors of millions of years.

The rubidium-strontium method utilizes the decay of the rubidium isotope rubidium 87 (Rb-87) to strontium 87 (Sr-87). Any rock older than 20 million years may be dated by this method, and so may younger rocks if they contain enough Rb-87. (Rubidium does not form its own mineral but occurs in minerals such as muscovite, biotite and all potassium-bearing feldspars.) Accurate dating using the rubidium-strontium method is difficult, however, because the half-life of Rb-87 is not well established; half-life values of 4.7 billion or 5.0 billion years are commonly used.

Relatively recent events cannot be dated by any of these methods. The carbon 14 method partly fills this gap; it requires the study of C-14 neutrons produced by cosmic-ray bombardment of atmospheric nitrogen (N). C-14 has a half-life of 5730 years. The radioactive carbon in substances develops as follows: The newly formed C-14 combines with oxygen to form CO_2, which is metabolized by plants, which in turn are eaten by animals and man. Thus all living organisms contain some amount of C-14. The amount remains constant so long as the organism is alive, but as metabolism ceases the intensity of radioactive emission of C-14 is reduced, the amount of reduction being a function of the half-life of C-14. Dating using the C-14 method can be performed on substances that once had life provided they are not over 50,000 years old. The method has been widely used in dating glacial events and archaeological finds. (J W.S.)

Radiogenic Heat. Heat produced by the decay of radioactive matter. It is generally believed to be a major source of heat energy in the earth's interior, where its chief sources are uranium, thorium, and an isotope of potassium, potassium 40. The heat released by the disintegration of one atom of uranium 238 to lead 206 is 1.85×10^{-2} calories. It has been calculated that during the disintegration of one gram of U-238 into stable lead, enough heat is produced to burn 800 kg of coal. Granitic rock contains 4 grams per ton of uranium, 13 grams per ton of thorium and 4.1 grams per ton of potassium 40. Therefore the amount of heat produced by granitic rock alone can be quite substantial. Although basaltic and other more mafic rocks contain lesser amounts of radioactive elements, it is evident that radiogenic heat is a major source of heat energy within the earth. (J.W.S.)

Radiolarian. A member of the subclass (or order) Radiolaria of the *protist phylum Protozoa. These exclusively marine organisms range in age from Cambrian to the Holocene (Recent). They secrete a delicate, spine-covered shell of silica; the shells have formed substantial, siliceous deposits of radiolarian ooze in certain parts of the Indian and Pacific oceans. Rocks composed of fossil radiolarian ooze, called radiolarian earth, are used to manufacture filtering and insulating materials. Radiolarians may be responsible for certain siliceous deposits such as *chert or *flint. The Franciscan Chert of California, for example, is composed of the hardened radiolarian ooze that is known as radiolarite. (W.H.M.)

Rain. See Precipitation.

Ray. Lunar surface feature consisting of crater ejecta deposits with a high *albedo. They form long, narrow streaks that extend out from young craters for tens to hundreds of crater diameters. The ray material was ejected along trajectories during crater formation. These features are in clear contrast to deposits that are closer to the craters in that the deposits near the craters were transported in surge clouds that consisted mainly of gas and dust.

Reaction Series. A series of minerals crystallizing from a magma in which an early-formed mineral reacts with the melt to form a new mineral further down in the series. This concept, originally proposed by the American petrologist N. L. Bowen (1887-1956) and sometimes called Bowen's reaction series, suggests that the ferromagnesium minerals form a discontinuous reaction series and the feldspars form a continuous reaction series. Thus olivine, the first mineral to crystallize, reacts with the surrounding melt to form *pyroxene; later the pyroxene reacts to form amphibole, and so on. At the same time a continuous reaction is taking place in the feldspar series. Early-formed bytownite reacts to form labradorite, and later labradorite reacts to form a more sodic feldspar.

If at any stage in the reaction series crystals are extracted from the melt, the resulting rock will have a composition different from that of the parent magma. See Magmatic Differentiation. (C.S.H.)

Realgar. A mineral, an arsenic sulfide with a characteristic orange-red color. It is found associated with the yellow arsenic sulfide, *orpiment, in lead, silver and gold veins formed near the surface of the earth. Realgar also occurs as a volcanic sublimate, as at Pozzuoli, Italy, and is deposited from hot springs, as in the Norris Geyser Basin, Yellowstone National Park.

Recent Epoch. A synonym for *Holocene Epoch, the last epoch of the *Quaternary Period.

Red Beds. Red *sedimentary rocks of any age formed in a highly oxidizing, sedimentary environment. Consisting of sandstone, siltstone and shale, their color is due to the presence of *hematite (ferric oxide). They are generally considered to indicate an arid, continental environment of sedimentation and are commonly associated with *evaporite deposits.

Regional Metamorphism. The metamorphism (alteration of rock) of large areas of the earth's *crust. As the term was originally used, the conditions necessary for regional metamorphism are those of moderate temperature and high pressure attained only by the deep burial of rock that was once close to or at the surface of the earth. The term now includes a wider range of temperature and pressure associated with large-scale bending and buckling of the earth's crust. Regional metamorphism therefore may include rock alteration occurring within a temperature range of 400° to about 800°C (752° to 1472°F) and a wide range of pressures encountered at depths of about 10 km to 35 km beneath the surface of the earth. Rock undergoing alteration over this wider range of temperature and pressure, and including deformation by *folding, is generally called dynamothermal metamorphism, a term now used almost synonymously with regional metamorphism. The Canadian Shield as well as other *cratons are areas that contain regionally metamorphosed rock. The most common rock types produced in such areas by regional metamorphism are *gneiss, *schist and *amphibolite. (J.W.S.)

Regolith (or Mantle Rock). Unconsolidated material that is found at and near the earth's surface. It rests on *bedrock but, unlike bedrock, can usually be removed

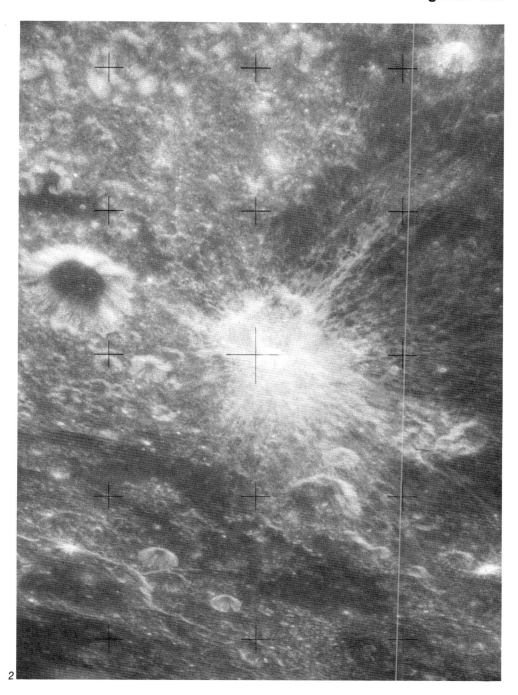

2

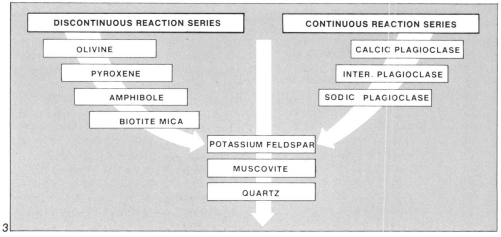

3

DISCONTINUOUS REACTION SERIES	CONTINUOUS REACTION SERIES
OLIVINE	CALCIC PLAGIOCLASE
PYROXENE	INTER. PLAGIOCLASE
AMPHIBOLE	SODIC PLAGIOCLASE
BIOTITE MICA	
POTASSIUM FELDSPAR	
MUSCOVITE	
QUARTZ	

Regolith

1 Deposits of unconsolidated loamlike material called loess form a bluff in Arizona. This fine, wind-blown dust is a type of regolith, the term for all loose depositional material that covers much of the surface of the earth.

2 An alluvial fan of fragmented rocky material carried by glacial meltwater, has been deposited in a semi-circular foam known as a cone at the foot of a mountain in the Alps.

Rejuvenation

3 Accelerated downcutting by a stream that had slowed its downward erosion (a) often results from uplift or tilting of the land (b), which increases the stream's energy.

Remote Sensing

4 This scale plots the electromagnetic spectrum, from short gamma rays to long radio waves, showing the continuum of frequency (cycle per second) and wavelength. The various sensors used to detect the different wavelength bands are indicated. The visible spectrum occupies a relatively narrow band.

with a shovel. There are two principal kinds of regolith: residual (also called saprolite) and transported. Residual regolith is formed by the mechanical and chemical *weathering of bedrock. Near the surface it may be profoundly altered, both physically and chemically, but at greater depths the intensity of weathering decreases and the regolith becomes more and more like the bedrock from which it was derived. It has undergone little or no transportation, and thicknesses of tens of meters are common. The principal factors responsible for its thickness and other characteristics are the kind of bedrock and drainage, the steepness of the topography, the climate under which it was formed, and time. *Laterite is a residual regolith with high iron content; it develops in humid, subtropical and tropical regions. *Bauxite is a residual regolith with high aluminum content.

Transported regolith is moved and deposited by processes that act at or near the earth's surface. It includes *alluvium (which is a fluvial regolith), *till (a glacial regolith), *loess and dune sand transported by the wind, beach deposits, *talus and volcanic dust. Transported regolith may be hundreds of meters thick in basins between mountains. Just as solid bedrock may change into unconsolidated regolith through weathering, so the regolith can be changed into bedrock by cementation, compaction, recrystallization and other processes. Nearly all sedimentary rocks have been formed from regolith.

The regolith has economic significance: gold, platinum, diamonds, cassiterite (an important tin ore) and other *placer minerals, certain types of iron, manganese and nickel ores, clay and bauxite and great quantities of sand and gravel important in construction are obtained from it. (R.L.N.)

Rejuvenation. Renewal of a stream's ability to downcut in its channel, caused by (1) uplift, tilting or warping of the land, (2) *eustatic lowering of sea level, (3) an hydrologic change that increases runoff or (4) a decreased load. Uplift, uptilting or upwarping of the land, as well as eustatic lowering of sea level, result in lowering the base level, which gives the stream renewed energy. A change in climate that increases runoff (without increasing load) or capture of another stream also gives the stream increased ability to erode by increasing discharge. Finally, a decrease in load allows the stream to lower its gradient by eroding its channel. Rejuvenation of a stream is reflected by a change in profile (*knickpoint), the presence of uplifted *terraces and incisement of the channel (incised *meanders), and the appearance of wide, flat valleys at its source. (M.M.)

Relict Sediments (or Palimpsest Sediments). Sediments occupying a location in which the overall environment has changed significantly since the time they were deposited. The term is commonly applied to sediments that currently occupy the bottom of the outer *continental shelf. These deposits originally accumulated in the beach or nearshore areas during the periods of Pleistocene glaciation, when the shoreline was far out on the continental shelf. Deposited in a shallow environment of high wave and current energy, they were covered by relatively deep water when sea level rose as the glaciers melted. Subsequently these sediments have either been uncovered by wave and current activity or they were never covered by modern sediments and therefore remain on the bottom surface. (R.A.D.)

Remote Sensing. The detection of an object or phenomenon without direct contact by an information-gathering device (sensor). Various remote-sensing techniques are now commonly employed in the earth sciences. The transfer of information is generally accomplished through the use of electromagnetic radiation, although sensing by means of gravity and magnetic force fields as well as acoustics is important. Remote sensing involves conversion of the data received into a stimulus or image that human beings can perceive.

The eye is a prime sensor. Electromagnetic energy reflected from an object is received by the eye, transmitted as a nerve impulse to the brain and recorded as an image. Conventional cameras record a similar wavelength span of electromagnetic energy on film. However, a great deal of radiation is not perceived by these common sensors, for they utilize only visible light, which is a very small portion of the electromagnetic spectrum.

The electromagnetic spectrum represents a wide range of radiation, subdivided by the nature of energy propagation, called wavelength and frequency. Various remote sensors are designed to detect and record radiation at different wavelengths. These range from the shorter waves of cosmic, gamma and X rays to very long radio waves. The diverse wavelengths of energy register different aspects of the objects viewed. Thus remote sensing provides a fascinating tool to record a variety of conditions about the earth that would otherwise remain undetected.

At slightly longer wavelengths than visible light in the spectrum lies the region of radiation known as the near-infrared. Reflectance of vegetation is especially critical within such a wavelength region. Different plant species as well as the vigor

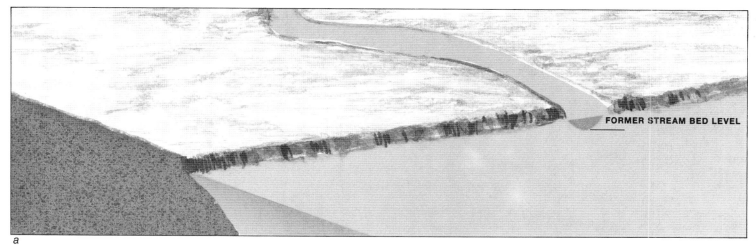

a

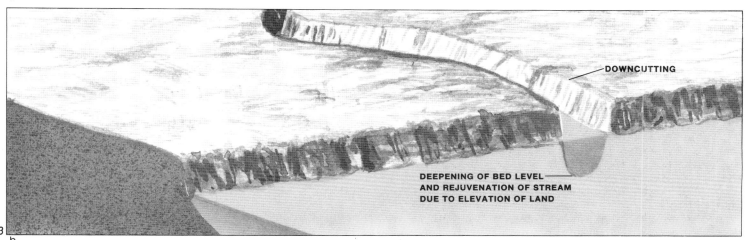

FORMER STREAM BED LEVEL

b

3

DOWNCUTTING

DEEPENING OF BED LEVEL
AND REJUVENATION OF STREAM
DUE TO ELEVATION OF LAND

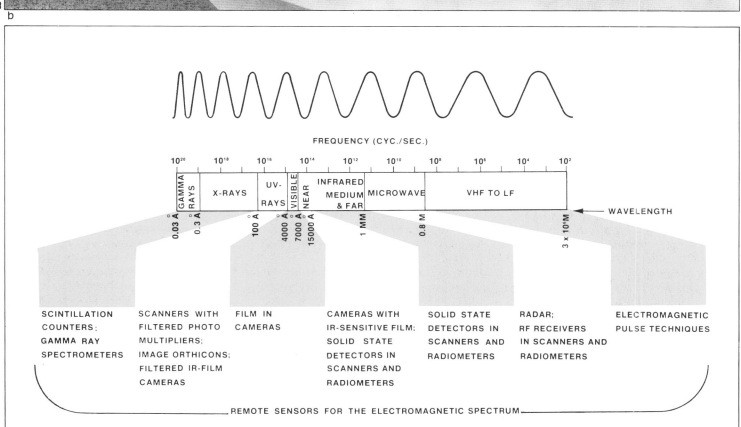

FREQUENCY (CYC./SEC.)

10^{20}	10^{18}	10^{16}	10^{14}	10^{12}	10^{10}	10^{8}	10^{6}	10^{4}	10^{2}

GAMMA RAYS | X-RAYS | UV-RAYS | VISIBLE | NEAR | INFRARED MEDIUM & FAR | MICROWAVE | VHF TO LF

WAVELENGTH

$0.03\,\text{Å}$ $0.3\,\text{Å}$ $100\,\text{Å}$ $4000\,\text{Å}$ $7000\,\text{Å}$ $15000\,\text{Å}$ $1\,\text{MM}$ $0.8\,\text{M}$ $3 \times 10^{6}\,\text{M}$

SCINTILLATION
COUNTERS;
GAMMA RAY
SPECTROMETERS

SCANNERS WITH
FILTERED PHOTO
MULTIPLIERS;
IMAGE ORTHICONS;
FILTERED IR-FILM
CAMERAS

FILM IN
CAMERAS

CAMERAS WITH
IR-SENSITIVE FILM;
SOLID STATE
DETECTORS IN
SCANNERS AND
RADIOMETERS

SOLID STATE
DETECTORS IN
SCANNERS AND
RADIOMETERS

RADAR;
RF RECEIVERS
IN SCANNERS AND
RADIOMETERS

ELECTROMAGNETIC
PULSE TECHNIQUES

REMOTE SENSORS FOR THE ELECTROMAGNETIC SPECTRUM

4

of the plants, commonly indicating the health of the vegetation, can be detected by the color contrasts recorded. In this way near-infrared photography proves useful in agricultural studies and enables a rapid inventory of crop production. It also allows prompt assessment of the extent and intensity of plant disease.

Near-infrared is effective in monitoring water bodies for certain types of pollution, especially when algal blooms and other patterns of water surface and shoreline plant growth result from the pollutants. Other applications of near-infrared photographs to the study of the earth include discerning soil and bedrock types for mineral exploration, snow and ice surveillance and snow-thickness mapping, categorizing and assessing land use, forest and wildlife management, soil-moisture content analyses, differentiation of ecological patterns and identification of urban stress and urban-growth patterns.

Thermal radiation (also known as far-infrared) lies at still longer wavelengths beyond visible light. Special recording detectors receive the thermal radiation and convert it to electrical energy, which can then be amplified and recorded as an image. This reflects differences in temperature and therefore registers thermally contrasting features of the earth's surface.

Applications of thermal infrared to our knowledge about the earth's surface are many: monitoring volcanoes, hot springs and other areas of high heat flow may lead to the accurate prediction of volcanic eruptions as well as the discovery of sources of valuable geothermal energy; discerning temperature contrasts within the ocean aids detection of cold-water upwellings and delineation of ocean currents, thus improving fishing capabilities by tracing paths of schools of fish and other seafood; surveillance of circulation and mixing patterns of heated waters, a general problem in water pollution and of special concern in nuclear power generation; quickly spotting forest fires and even identifying the location of underground mine fires; pinpointing the location of domestic animals on the open range and wild game in desolate country in times of heavy snowfall.

The microwave region occurs at wavelengths beyond infrared. In this wavelength span it is common to generate waves by energy pulses, rather than sensing naturally-occurring radiation. Radar waves are emitted in this way. A particular advantage of radar waves is their ability to penetrate clouds and to "see through" hindrances such as smog, heat shimmer, rain, smoke and haze. Thus radar provides the only accurate maps of some parts of the earth's surface, such as the Darien

province of Panama, where adverse weather is common.

Because the radar wavelengths emitted can be controlled, a variety of data about the earth's surface can be obtained. Radar images provide distinct definition of landform configuration valuable to geoscientists. With radar, geologic features such as gross bedrock types, faults, joint (fracture) distribution, drainage patterns, stream terraces and glacial deposits can be more readily identified. Accurate surface terrain maps are useful in highway construction, watershed studies and reservoir planning, and urban and rural land management.

Space-age opportunities to view large segments of the earth's surface from great heights has added another dimension to remote sensing. Photographing the earth from space began with unmanned satelites and grew in importance during the manned American spaceflights of the Gemini and Apollo series. Launching of the first Earth Resources Technology Satellite, ERTS-1, in July, 1972, was another step toward systematic worldwide surveillance from space. The regional view provided by space photographs allows a larger-scale study of earth processes and an appreciation of the association of major features, many previously considered unrelated. Single-space photographs cover areas that required more than 500 conventional aircraft photos. Thus an entire mountain range, a significant part of a coastline, a system of lakes or a province of mineralization is photographed and studied in one view without the distracting lines present on an assembled photo mosaic. This regional scale of understanding is particularly important in modern concepts of sea-floor spreading and in the movements of crustal segments over the surface of the earth.

Added to the broad-scale look are the advantages of sequential coverage, permitting the systematic monitoring of the earth's dynamic features. Seasonal variations can be readily identified and assessed, and changes dependent on climate or that otherwise fluctuate can be recognized. Satellite photographs contain a minimal lateral distortion because each point is viewed from directly overhead. This greatly facilitates mapmaking for land-use analysis, resource inventory, city planning and environmental studies. The combination of accurate mapping and repetitive coverage permits the first effective surveillance of many parts of the earth's surface on a uniform basis. In this way the diverse activities of mankind can be studied and their effects on the environment evaluated.

The manned Skylab Project in 1973, a major step toward a space shuttle-bus

system, allows further testing and refinement of remote-sensing techniques. A better resolution of photographs and images obtained from an array of sensors utilizing different portions of the electromagnetic spectrum resulted. Additional space ventures, both manned and unmanned, will improve current techniques and offer a continual "eye in the sky" to monitor our resources with more efficiency. (R.E.B.)

Replacement. The process by which material is dissolved and replaced with mineral matter by the action of groundwater and hydrothermal solutions. In the formation of *petrified wood, the open spaces are filled and the cellulose in the wood is replaced by silica. The process is frequently so delicate that minute features are faithfully preserved. In the well-known petrified forests of Yellowstone National Park and in Petrified Forest National Park, Ariz., the bark, growth rings, cells and other structures have been replaced and are preserved. The chitinous structures of insects and spiders in a Miocene lake deposit near Yermo, Calif., have been replaced with such fidelity that the muscles, heart, alimentary canal and other structures are still discernible. In addition to silica, such minerals as calcite, dolomite and pyrite may act as the replacement. Zinc, lead and copper ore deposits have been formed when hydrothermal solutions that issue from magma have replaced limestone and other rocks with these ore minerals. (R.L.N.)

Resequent Stream. A waterway that flows in the same direction as a *consequent stream but at a lower level and on a surface lower than the original surface.

Retrograde Metamorphism. Certain minerals or mineral assemblages produced by *metamorphic processes involving high temperature and pressure may become unstable if the rock that contains them is subjected to reduced temperature and pressure. The minerals may react to form new combinations capable of stability within the adjusted physical conditions. The process of mineral transformation is very slow, and evidence of its having taken place is generally easy to recognize under microscopic observation. When the process has not been completed, the magnified grains usually show cores of the original high-metamorphic-grade minerals surrounded by the rims of newly formed, low-metamorphic-grade minerals. The minerals that are being altered help geologists determine the highest temperature and pressure attained by a rock during metamorphism. (J.W.S.)

Reversing Thermometer. The most common means of measuring the temperature of seawater. It permits accuracy to within 0.02°C, which is necessary for detecting the small temperature variations that produce density gradients in the sea and generate deep-water currents. The thermometer is designed so that it can record the temperature at any depth and not change as it is brought back to the surface. The reversing thermometer is attached to a *Nansen bottle and lowered to the desired depth. A weight sent down the cable releases the bottle, causing it to reverse its position; this also reverses the position of the thermometer. Upon reversal, the mercury-filled tube is broken at a narrow point in the glass tube, causing some of the liquid to be trapped at one end. The remaining mercury falls to the other end of the tube. The volume of mercury at either end is dependent upon the water temperature. (R.A.D.)

Reynolds Number. A formula used to distinguish *laminar flow of water from *turbulent flow. The Reynolds number is a dimensionless parameter that is the product of the velocity, density, and hydraulic radius (depth) divided by the viscosity. Flow is laminar for small values of the Reynolds number and turbulent for high values. For streamflow the number varies from 300 to 600.

Rhodochrosite. A mineral, a manganese carbonate characterized by its pink color. Well-formed crystals of the mineral are rare; when deep-colored, as are those from Alam, Colo., they are highly prized. More commonly, rhodochrosite lacks a crystal outline and is massive or columnar. Beautifully banded rhodochrosite with deep-colored layers alternating with those of lighter color is mined for ornamental and decorative purposes in Catamarca, Argentina. The principal occurrence of rhodochrosite is in hydrothermal veins accompanying ores of silver, lead and copper, as at Kapnik, Rumania. Although most manganese used in industry is derived from manganese oxide minerals, in a few places, as at Butte, Mont., rhodochrosite is mined as an ore of the element. See Mineral Properties. (C.S.H.)

Rhodolite. A pale rose-red to purple variety of *garnet, intermediate in composition between *pyrope and *almandite.

Rhodonite. A mineral, manganese silicate, characterized by its rose-red color. It occurs in manganese ore bodies associated with pink rhodochrosite and black manganese oxides. Excellent crystals of rho-

Rhodochrosite
1 This rose red mineral, a manganese carbonate, is generally associated with silver ores. Shown here is a specimen from South America.

Rhodonite
2 This specimen is characteristic of the large well-formed crystals found in Franklin, N. J.; elsewhere it is usually fine-grained and associated with black manganese oxides.

1

donite have been found with the zinc ores at Franklin, N.J., although the mineral is usually massive and without crystal form. Because of its attractive color it has long been used as an ornamental stone for both large and small carvings. The outstanding source is near Sverdlovsk in the Soviet Union. The sarcophagus of Alexander II of Russia was hewn from a single block of rhodonite from this locality. See Mineral Properties. (C.S.H.)

Rhombohedron. In crystallography, a crystal form of the hexagonal system bounded by six rhomb-shaped faces. The carbonate minerals of the calcite group commonly show this crystal form.

Rhynchocephalian. A member of the order Rhynchocephia of class Reptilia. These reptiles first appeared in the Early Triassic, at which time they were abundant, attaining almost worldwide distribution. Today the group is represented by a single species, *Sphenodon punctatum*, more commonly called the tuatara. This primitive, lizard-like reptile has remained virtually unchanged for approximately 180 million years, becoming so important to zoology that it is now protected by law. Found only on a few small islands off the coast of New Zealand, the tuatara is 60-90 cm (2 to 3 feet) in length; it has a loose-fitting, scaly skin and a fringed crest extending from the head to the tip of the tail. One theory suggests that the tuatara has survived through its migration from Australia to remote islands where it faces a minimum of competition, and that its burrowing habit removes it still further from competition. See also Fossil Reptiles. (W.H.M.)

Rhyolite. A volcanic rock, the extrusive equivalent of a granite. The principal minerals are *orthoclase and *albite feldspar and one or more of the *silica minerals quartz, crostobalite or tridymite), usually quartz. Biotite, and less commonly hornblende, may be present. Many rhyolites are fine-grained and others are composed wholly or in part of glass, making identification of individual minerals difficult or impossible. However, phenocrysts of feldspar, quartz and biotite are generally present, even in fine-grained varieties.

Rift (or Taphrogeosyncline). A sediment filled, elongate trough created by the sinking of a portion of the earth's *crust. Rifts result from a downward movement along one or more steeply dipping *faults bounding the depression. During formation, rifts are characterized by abundant earthquake activity and volcanicity.

The narrow depression on the central

crest of the worldwide oceanic ridge system, the Red Sea basin, and the Triassic basins along the east coast of the United States are considered rifts. The term also implies a mode of genesis since rifts are believed to have been formed by a stretching of the earth's crust. The Triassic basins of the Appalachians were produced when the earth's crust began to break apart to form the individual continental landmasses now present on the earth's surface. See Continental Drift. (J.W.S.)

Rift Valley. A narrow cleft with a mountainous floor that marks the central crest of the oceanic ridge system. It is usually only a few tens of kilometers wide and extends throughout the ridge system and also underlies portions of the continents. The best known and most prominent example of the latter is the African Rift Valley, which runs from Ethiopia down to South Africa. Its high relief is due to numerous high-angle *faults, and the region is characterized by much earthquake activity. Rift valleys are also the site of active volcanoes, the most extensive being found in the Iceland area, which is a part of the oceanic ridge system.

The nature of a rift valley changes in character from place to place along the ridge. For example, the relief of the valley floor is quite high in the north Atlantic Ocean but is minimal in the East Pacific Rise area in the central Pacific. See Oceanic Ridge. (R.A.D.)

Rill. A very thin, elongate channel sometimes called a shoestring *gully. It is often found on steep, unvegetated slopes in badlands or road cuts.

Rille. A straight or sinuous V-shaped valley of unknown origin on the lunar surface. Rilles tend to be found near margins of *mare basins and extend for as much as hundreds of kilometers. One such formation, Hadley Rille, was visited by members of the Apollo 15 mission.

Ring Dike. A sheetlike body of *igneous rock that forms a circular or oval ring when exposed on the earth's surface and forms a cylinder with nearly vertical sides when exposed in cross section. The cylinder may be continuous, forming a solid ring, or it may be segmented, leaving gaps within the ring structure. The diameter of ring dikes varies considerably, some measuring a few hundred meters while others reach several kilometers.

It is known that ring dikes form by injection of molten rock into cylindrical fractures produced within the crust, but how the fractures are formed is still speculative. One possibility is that fracture formation

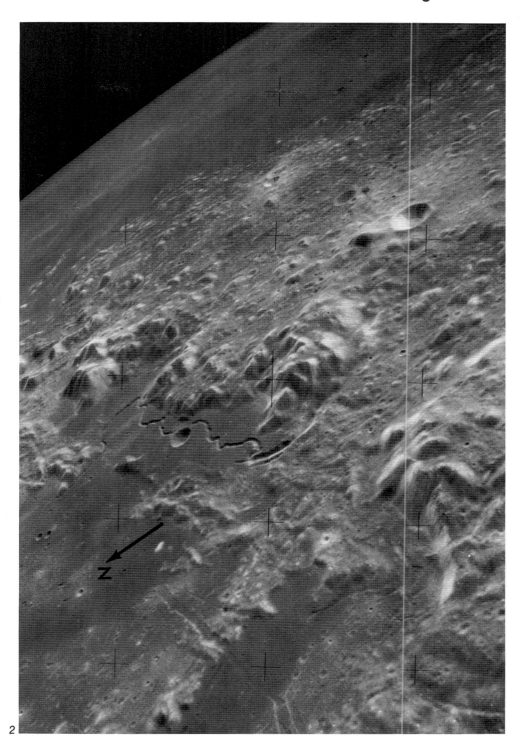

2

Rift

1 A miles-long fissure near Myvatn, Iceland, may actually be a portion of the Mid-Atlantic Ridge that stands above sea level.

Rille

2 This photograph of Hadley Rille was taken from lunar orbit during the Apollo 17 mission. Due west of the north arrow is an unnamed straight rille.

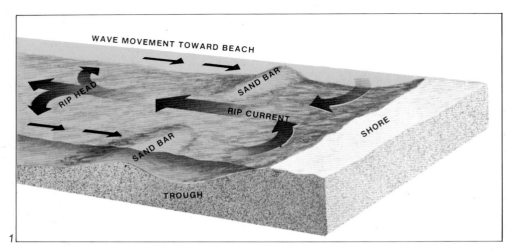

WAVE MOVEMENT TOWARD BEACH

RIP HEAD

SAND BAR

RIP CURRENT

SHORE

SAND BAR

TROUGH

1

2

3

results from a body of *magma pushing upward on the overlying crust like a punch. Ring dikes have been recognized in many parts of the world, notably in Norway, Nigeria, Australia and in the state of New Hampshire in the United States. (J.W.S.)

Rip Current. A strong, narrow flow of water that moves seaward from the shore. Such a current is often called an undertow because it moves away from shore. Rip currents are caused by waves that pile up water behind shallow sandbars as they seek the path of least resistance in returning to deeper water. Rip currents can generally be recognized by seaward moving bubbles or plumes of suspended sediment. They are known to be a significant mechanism for transporting sediment. (R.A.D.)

Ripple Mark. A wavelike surface sculpture consisting of a series of small ridges and depressions produced in unconsolidated, granular materials such as sand. They may be caused by wind, water currents or the agitation of water in wave action. A current ripple mark has an asymmetric form and is caused by air or water moving more or less continuously in one direction. A ripple mark of oscillation is symmetrical in outline and is produced by the oscillating movement of water, as along a shore outside the surf zone. Ripple marks are often found in sedimentary rocks, demonstrating that the processes that formed them have been active for much of geologic time. (W.H.M.)

River. A flow of water in a natural channel. The term generally refers to the large, main trunks of a drainage system; smaller channels may be called streams, creeks, brooks, or rills, in descending order of size. Rivers may carry water all the time, and are then called perennial streams or rivers, or they may carry water part of the time, in which case they are called intermittent streams. If the base flow is from a fluctuating groundwater table, a river will carry water only when the table is above or at the level of the river bottom, and will be dry when the groundwater table lies below the bed of the channel. If the stream has a bed which is always above the water table, water will be lost by infiltration as it moves downstream; it will then be called an *ephemeral stream. Rivers are extremely important agents of erosion. They carry material to the sea as *bed load, *suspended load and in solution. (M.M.)

Roche Moutonnée (or Sheepback, Sheep Rock, Sheepback Rock). A streamlined *bedrock knob resulting from glacial erosion. As a result of glacial *abrasion, the up-glacier side (stoss side) slopes gently and is usually rounded, polished and striated. The down-glacier side (lee side), which results from glacial plucking, aided perhaps by the freezing and thawing of *meltwater, is steep and jagged. Roches moutonnées are elongated and asymmetrical in the direction of glacial motion but are more or less symmetrical along their transverse axes. Although usually less than a few scores of meters long, large ones are found in Maine and elsewhere, including small hills sloping gently on the up-glacier side and steeply on the down-glacier side. (R.L.N.)

Rock. A mineral aggregate constituting an appreciable part of the earth's crust. In the broad sense rock, when of considerable extent, includes beds of sand, gravel, clay and other noncohering masses of mineral materials. However, a more restricted usage implies a consolidated and relatively hard mineral aggregate. The expression "hard as a rock" reflects the latter and more common usage. Rocks have been given many different names depending on their origin, texture and mineralogy, but in general they are grouped into three main divisions: *igneous, *sedimentary and *metamorphic. Igneous rocks are formed from the solification of magma by intrusion, extrusion, or volcanic activity. Sedimentary rocks are formed from fragments of other rocks or organisms by transportation or settling. Metamorphic rocks are rocks that underwent profound changes effected by heat, pressure and water. See Rock Cycle. (C.S.H.)

Rock, Basic. An *igneous rock low in silica (22-45 percent) and high in iron and magnesium. It is dark colored due to the abundance of *biotite, *amphiboles, *pyroxene and/or *olivine. The term is used in contrast to acid rock, which is rich in silica and lighter in color. The terms "basic" and "acid" are misleading and are discouraged by modern petrologists, who prefer to use the terms "mafic" and "sialic."

Rock Crystal. An early term—but still in use—applied to clear, colorless crystals of *quartz.

Rock Cycle. A basic geologic concept of the sequences concerned with the creation, alteration and destruction of earth materials when subjected to processes such as erosion, transportation, deposition, heat, pressure and volcanic activity. The interrelationships of the three major rock types—igneous, metamorphic and sedimentary—the forces that place them in contact with surface forces and the effect of

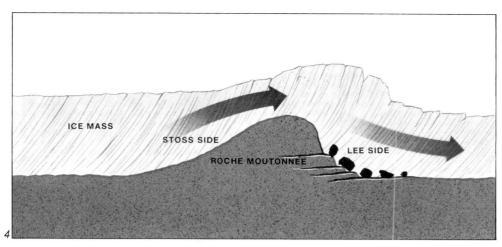

Rip Current

1 At intervals along the shore, surface currents flow seaward through the line of onrushing breakers, carrying water from longshore currents. Rip currents form where there is a path of low resistance, such as a topographic low or a walking path in the nearshore sandbar.

2 This aerial photo taken over the California coast shows offshore rip currents.

Ripple Mark

3 Currents near the shore of an ancient sea created a corrugated surface that is now preserved in shale near Denver, Colo.

Roche Moutonnée

4 As a glacier moves over an outcrop it smooths and rounds the upstream surface (stoss side) but leaves the lee side sharp and angular where rocks have been plucked away.

5 Groups of roches moutonnées, such as this knob on Tom Jones Mountain, N.Y., were believed to be named by the 18th-century Swiss naturalist Saussure after the curls on sheepskin wigs (perruques moutonnées) worn in France in the 1800s. Much later the term was used to refer to individual glaciated outcrops because they resembled sheeps' backs both in form and surface texture.

these forces on the rocks is very complex, but may be considered as starting with molten rock, called *magma, deep below the earth's surface. This igneous material may be extruded on the surface as *lava (extrusive rock) or may crystallize beneath the surface as *plutonic (or intrusive) rock. Crustal movements such as mountain-building may bring plutonic rocks to the surface, where they will be subjected to disintegration and chemical decay. Geologic agents (wind, water, ice) may transport the products of weathering and erosion, depositing them as sediments that may eventually be consolidated to form *sedimentary rocks. Deformation may later produce intense heat and pressure that will convert the sedimentary rock to *metamorphic rock to form magma. When this magma cools and solidifies to form an *igneous rock the theoretical or "ideal" cycle has been completed.

The rock cycle represents a response of earth materials to various forms of energy, and the cycle is not always completed. There may be many interruptions or short-circuits as geologic events are introduced out of sequence with respect to the ideal cycle. For example, igneous rocks may be metamorphosed before being weathered to form sediments. See Color Plate "The Rock Cycle." (W.H.M.)

Rock Flour. Pulverized rock consisting of fine sand, silt and clay resulting from debris frozen in the bottom of a moving glacier scraping across *bedrock. Microscopic examination shows that the grains are angular, unweathered and composed of many minerals, but without the clay minerals formed by chemical *weathering. Rock flour is an important constituent of *loess, *glaciomarine and *glaciolacustrine deposits. It gives meltwater streams their milky appearance.

Rock Glacier. A glacier-like body with a steep front composed of angular rock fragments that moves (or has moved) downslope away from the foot of a high cliff in a mountainous region. Rock glaciers are composed of fragments detached from a cliff by *frost action; they may be 1.5 km (1 mile) long, 3 km (2 miles) wide and 45 m (150 feet) thick, and they may move as much as 4 m (13 feet) per year at the front. The deeper interstices between the rock fragments are filled with ice. Their surfaces show concentric wavelike wrinkles that are convex downslope (somewhat similar to but much larger than those on *lava flows), indicating that the rock glaciers move more rapidly in the center than at the margins. Some observers attribute their flow to the interstitial ice. Others suggest that the

movement is a type of *creep resulting from *frost heaving. It is also possible that all gradations between true glaciers and rock glaciers exist and that they have been derived from small glaciers. Rock glaciers are commonly found in mountainous regions such as the Andes, Alps, Pyrenees, Rockies and Sierra Nevada, and in Scandinavia and Alaska. (R.L.N.)

Rock Salt. The term applied to the mineral *halite when found in granular masses.

Rockslide. A rapid movement of rock that occurs where *bedding, *joint, *cleavage or *fault planes are more or less parallel to steep slopes; also the deposit formed by such a movement. The conditions that favor development of a rockslide are (1) massive beds overlying clay or shale so that resistance to sliding is not great, (2) planes of weakness in rocks dipping toward a steep cliff and (3) heavy rains that lubricate the clay or shale and decrease resistance to sliding. A well-known example of a rockslide, which was reported to have traveled about 80 km (50 miles) an hour, occurred in 1925 in Wyoming's Gros Ventre River Valley. On the south wall of the valley the rocks dipped approximately 20° toward the river. A bed of sandstone at the surface rested on shale. Erosion by the river had cut a cliff in the sandstone, which was therefore unsupported. After heavy rains had diminished resistance to sliding, 40 million cubic meters (50 million cubic yards) of debris slid into the valley, dammed up the river and formed a lake that was 6 km (4 miles) long and 60 m (200 feet) deep. (R.L.N.)

Rock-Stratigraphic Unit. Rock strata grouped together on the basis of their physical, chemical and structural characteristics as rocks rather than on their fossil content or relative age. The basic rock-stratigraphic unit is the formation, a recognizable unit of similar rocks possessing distinctive features useful for description and mapping. Such units may be subdivided into members and beds. Formations that have certain features in common may be placed in the same group.

Rock, Ultramafic (or Ultrabasic). Igneous rocks containing little or no feldspar. Such rocks are composed essentially of one or more dark minerals, such as pyroxene, amphibole and olivine. Metallic oxides and sulfides are frequently present.

Ropy Lava (or Corded Lava). One of the characteristic surface features of pahoehoe *lava flows, these are festooned, concentric arrangements that resemble

Rock Glacier

1 A mass of large rock fragments, lubricated by ice and water, slowly creeps downhill in the San Juan Mountains, Colo.

Rockslide

2 A sudden slide loosened from bedrock along planes of weakness plunged down a mountainside at Banff National Park, Alberta, Canada.

curved pieces of rope laid next to each other or the wrinkles that develop in flowing hot tar. Ropy lava is commonly found on small tongues on the surface of flows. Individual "ropes" are generally a centimeter to many centimeters (¼−4 inches) in width but may be 6 or more meters (20 feet) in length. The thickness of the ropy zone ranges generally from a few centimeters to a meter (an inch to a few feet). The larger ropes are often composed of twisted smaller ones. Lava ropes are formed by the drag of fluid lava below the plastic crust of a flow. The curvature of the ropes is convex downstream because the fluid subcrustal lava moves faster and creates more drag in the center of the flow than on the margins. Ropes are therefore reliable indicators of the direction of lava flow. They have been described on lavas as old as Precambrian. (R.L.N.)

Rose Quartz. The rose-pink *quartz used as a gem or ornamental stone. It occurs in *pegmatites, sometimes forming large masses, but only rarely in distinct crystals. The best known source of the crystals is in Minas Gerais, Brazil. The pink coloring is believed to be due to the presence of a small amount of titanium.

Rubellite. The pink to red gem variety of *tourmaline. It contains lithium and is found in pegmatites containing other lithium-bearing minerals.

Ruby. The red variety of *corundum, used as a gem. The red color is attributed to the presence of a small amount of chromium. Some rubies show *asterism and are called star rubies.

Ruby Silver. A term used for the red silver minerals *pyrargyrite and *proustite. Pyrargyrite is called "dark ruby silver" and proustite "light ruby silver."

Runoff. The total amount of rainfall which reaches a stream channel and flows out of the watershed. Runoff includes water flowing over the surface to the channel as well as that which reaches the stream by infiltration. Runoff is the residual amount of rainfall that remains after evaporation and transpiration takes place. Thus: runoff = rainfall - evapo-transpiration. It has been estimated that approximately 80 percent of the world's rainfall is lost to rivers and only the residual 20 percent becomes runoff.

Rainfall reaches a river channel in several different ways. Some rainfall travels over the land surface to the channel as surface runoff. This occurs only after interception of the rain by plants, surface

detention in depressions, and infiltration into the soil has been exceeded. Surface runoff is the most important contribution of rain to flood flows. A second component of runoff is interflow—the water which infiltrates the soil and moves laterally toward a stream channel. The amount of interflow depends upon the type of soil and its ability to transmit water. Interflow is often the largest contributor to discharge. A third component is flow from the groundwater table. Although this part of the runoff is negligible during a storm, it is the most important component of year-round runoff since it supplies the deep seepage and steady flow of rivers between rains. Direct precipitation into the channel is a minor constituent of runoff. (M.M.)

Rutile. A mineral, a titanium oxide of a dark red color, most often found as an accessory mineral in granites, gneisses and schists. It is the most abundant of the three titanium dioxide polymorphs, the others being anatase and brookite. Rutile is a resistant mineral both chemically and mechanically, and has a high specific gravity (4.2). Thus the tiny crystals in which it occurs are liberated intact and find their way into *placers or, with other heavy minerals, into beach sands. As a commercial product, rutile is recovered from beach sands, where it is associated with *ilmenite, magnetite, zircon and monazite. The most notable such occurrence, in northern New South Wales and southern Queensland, makes Australia the world's largest rutile producer.

Most rutile is used as a coating on welding rods, but some is a source of titanium used in alloys. The titanium used to manufacture the oxide for paint pigments comes largely from the mineral ilmenite. Natural rutile is too dark for a gem; however, manufactured rutile is transparent with only a slight yellow tinge and is cut into beautiful gemstones. See Mineral Properties. (C.S.H.)

Ropy Lava

1 The surface of lava flows solidify into long, ropy lengths (called pahoehoe) or rough, pitted fragments (called aa), depending on viscosity.

Rutile

2 A specimen of rutile illustrates its characteristic twinned crystal form.

3 The mineral has pyramidal terminations on tetragonal prisms, and frequently shows this characteristic twinning.

S

Saint Elmo's Fire. A corona discharge characterized by a luminous, often audible manifestation of atmospheric electricity intermediate between the spark discharge and the point discharge. St. Elmo's fire is frequently seen as a glow from the masts and rigging of ships and on high mountain peaks when thunderstorms pass overhead. It emanates from objects, especially from pointed ones, when the electric field strength near their surfaces reaches a value of about 1000 volts per centimeter. The name was given to the phenomenon long ago by Mediterranean sailors who regarded it as a visitation from their patron saint, Saint Elmo. Its appearance was regarded as a good omen, since it tends to occur during the latter phase of a violent thunderstorm. (M.F.H.)

Salina (or Alkali Flat). Generally, a salt-encrusted land area formed in arid regions by salts precipitated from evaporating lakes or the evaporation of solutions moving upward by capillarity. Halite, gypsum, borax, mirabilite and other minerals may be present. The Bonneville Salt Flats west of Salt Lake City, on which automobile speed records have been made, are an excellent example. Salinas are common in Nevada, California and New Mexico and in the Arabian, Atacama and Sahara deserts. The term is also applied to a marsh or pond having a high concentration of salts.

Salinity. The total amount of dissolved solids in one kilogram of seawater. Although sodium (Na) and chlorine (Cl) comprise the bulk of these dissolved solids, there are more than 60 other elements in seawater, most of them in very small quantities. In 1884 a Scottish chemist, William Dittmar, while analyzing seawater samples collected by the *Challenger expedition, found that regardless of the total amount of dissolved material, the constituents invariably occurred in the same proportion. Salinity can therefore be determined from the concentration of only one element, commonly chlorine; this is multiplied by 1.80655 to determine salinity. Originally, salinity determinations were made by chemical analysis (titration); today they are made with a salinometer, which measures the electrical conductivity of seawater as an index of its concentration of dissolved solids. Salinity may also be determined from the density, refractive index, velocity of sound or freezing point of the seawater, but these are cumbersome methods and are rarely used.

Open marine waters have a salinity of about 35 parts per thousand. However, there may be considerable variation in areas of freshwater runoff where salinity is low, or where evaporation is high and causes hypersalinity. (R.A.D.)

Salt. The mineral *halite, common salt.

Salt Dome. A roughly circular structure produced by the upward movement of a vertical, pipelike plug of salt. Most of these plugs are about one kilometer in diameter but may have a vertical extent of hundreds of meters. They occur in certain areas underlain by thick layers of sedimentary rocks and are derived from deposits of deeply buried *halite. Halite is a mineral with a relatively low specific gravity (2.1) that deforms readily under unequal pressure. The halite, or salt, column is thrust toward the surface by plastic flow due to the pressure of the overlying strata. Thus the enclosing rocks are arched upward to produce a dome-shaped feature on the earth's surface. Domes in which the salt plug has broken through the enclosing rocks are known as piercement domes or diapirs.

A *cap rock of gypsum, anhydrite and varying amounts of sulfur and calcite is found on top of many salt plugs, and some contain commercial deposits of sulfur and petroleum. Oil and gas may also accumulate in traps created by the deformation of the rocks above and surrounding the salt plug. Valuable reservoirs of this type are associated with salt domes in the Gulf Coast Plain of North America and the North German Plain in Europe. (W.H.M.)

Salt Marsh. A low, heavily vegetated area within reach of ocean tides. Such marshes are extremely abundant along coasts with many bays or barrier islands. Fine sediment carried by runoff from the land is trapped along the margins of bays or behind barrier islands and often forms extensive salt marshes there. The marshes are dominated by grasses and support abundant and diverse communities of organisms, including fish, shellfish, crabs and birds. The Atlantic coast of the United States has numerous salt marsh areas, many of which, because of their rich wildlife, have been made into wildlife refuges or similarly protected areas. Marshes are also common along embayed coasts of Great Britain, Northern Europe and the Gulf Coast of the United States. (R.L.N.)

Sand. A loose, uncemented, sedimentary deposit made up of fragments of weathered rocks and minerals. The fragments must be between 1/16 mm and 2 mm—less than 8/100 inch—in diameter to be called sand grains. The largest are about the size of a pinhead. Sand grains

Salt Marsh
The Camargue in southern France is a lowland plain and a delta composed of mud brought down by the dwindling branches of the Petit and the Grand Rhone. Freshwater from the rivers mixes with saltwater from the Mediterranean Sea to form a sanctuary of extensive wetlands inhabited by many kinds of wildlife.

398 Sand Crystal

Sandstone
A sheer sandstone cliff, weathered along vertical joint planes, rises above the ruins of Betatakin in Navajo National Monument, Ariz.

are smaller than the fragments called granules but larger than silt. Although most sands consist largely of small grains of quartz, many other minerals—for example, feldspar, mica, gypsum, magnetite and garnet—are found as sand grains. The coral sands of certain tropical islands consist of sand-size particles of calcareous material of organic origin. Hawaii's famous "black sand" beaches are composed of tiny sand-size fragments derived from eroded lava flows.

A useful mineral resource, sand is used for mortar and concrete in the construction industry. Blast sand is a coarse-grained sand used with compressed air to clean the walls of brick and stone buildings. It is also used to etch designs on building stones such as granite and marble. Finer-grained sands are used in making sandpaper and other abrasives. Glass sand is an exceptionally pure quartz sand used in manufacturing glass. (W.H.M.)

Sand Crystal. A crystal with rough external form that contains a considerable amount of quartz sand. Calcite, gypsum or barite sand crystals develop by growing within a deposit of unconsolidated sand. As crystallization proceeds, the growing crystals enclose sand grains that may amount to as much as 60 percent of the volume.

Sandstone. A *sedimentary rock consisting of naturally cemented sand-size sediments. Certain sandstones are formed when mineral-laden underground water percolates through loose sand. As the dissolved mineral matter comes out of solution, it forms a cement that binds the sand grains together. The cement may consist of material such as calcite, quartz, and limonite and hematite. Clay, deposited with the sand or derived from sand-size fragments of weathered feldspar, also may serve as a cement.

The color of a sandstone is typically determined by the cementing agent. Iron oxide cement, for example, gives the sandstone a red, yellow or brown color. Sandstones are also white, black, gray, green or cream colored. Sandstones normally break through the cementing material rather than through the individual sand grains; consequently the broken surface of the rock feels rough and gritty. However, some quartz sand grains are so tightly cemented with silica that they form an extremely hard and compact rock. If this rock breaks smoothly through the grains instead of between them, it is known as *quartzite. In addition to this sedimentary quartzite, there is another type of quartzite described under *metamorphic rocks. (W.H.M.)

Sand Wave. A feature that develops on the sandy bottom of water areas subjected to rapid bottom currents. Such currents cause a movement of sediment, which commonly forms an undulating surface on the sand. In shape, sand waves closely resemble a sinusoidal (S-shaped) wave, with smoothly curved crests and troughs. The length of the waves commonly ranges between a few meters and a few hundred meters, while the height may reach as much as 4 m or 5 m. Because strong currents are needed to generate such waves, their distribution is limited to streams, estuaries and inlets that have much tidal exchange of water. Rapid tidal currents in the constricted mouth of the Chesapeake Bay have created well-developed sand waves on the bottom in that area. See Bed Load. (R.A.D.)

Sanidine. A mineral, potassium aluminum silicate, the high-temperature potassium feldspar, of which orthoclase and microcline are low-temperature forms. It occurs in lavas in the form of glassy, often transparent crystals.

Sapphire. Specifically, blue *corundum used as a gem. However, the word is also used with the appropriate color modifier (with the exception of red) to refer to other gem-quality corundum, such as green sapphire, yellow sapphire. The color of blue sapphire is attributed to the presence of small amounts of titanium. The phenomenon of *asterism is strikingly shown in the star sapphire.

Satin Spar. A variety of *gypsum composed of closely packed parallel fibers that give the specimen a satin-like appearance.

Scapolite. A group of sodium and calcium aluminum silicate minerals that form a solid-solution series between marialite, the sodium member, and meionite, the calcium member. The scapolites are metamorphic minerals occurring in schists, gneisses and crystalline limestone. See Mineral Properties.

Scheelite. A mineral, calcium tungstate, one of the ore minerals of tungsten. It occurs in granite *pegmatites, *contact metamorphic deposits and high-temperature hydrothermal veins associated with cassiterite, molybdenite and wolframite. Scheelite may be yellow, green or brown; it also is white in many deposits and resembles quartz in appearance. Usually, however, scheelite can be distinguished from quartz by its characteristic pale blue fluorescence in ultraviolet light. See Mineral Properties.

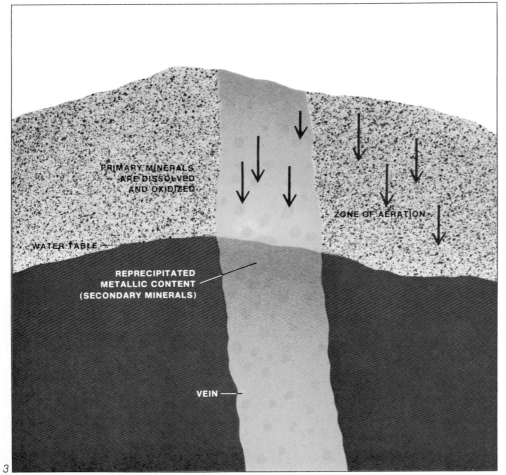

Schist. A *metamorphic rock coarse enough to allow identification of individual mineral grains. Micas are generally the dominant minerals in the rock and tend to orient themselves in such a way that their thin plates lie essentially parallel to each other, imparting a layered structure to the rock. The term "schist" relates more to the texture or structure of a rock than to its mineral constituents. When one or more minerals predominate in the rock, their names precede the term schist, as quartz-mica schist. Schist is the product of recrystallization of a wide variety of pre-existing types of rock: metamorphosed shales or clayey sandstone, impure limestones and *igneous rock rich in iron or magnesium. Schists develop under a wide range of temperature and pressure and are associated with bending or buckling of the earth's *crust over large areas. (J.W.S.)

Schistosity. A layered or planar structure in coarse-grained *metamorphic rocks. It is produced by the parallel alignment of platy minerals such as mica or by grains shaped like ellipsoids. This alignment is the result of external forces acting on the rock while mineral growth is taking place. Platelike minerals grow most readily when their largest cross-sectional area is perpendicular to the maximum compressive stress acting on the rock.

Scoria. A black, dark gray or red *igneous rock with many *vesicles (gas cavities), a glassy or fine-grained texture and a basaltic or andesitic composition. Since the vesicles are not as numerous as in *pumice, scoria is heavier and generally will not float. It commonly occurs as fragments in basaltic *cinder cones and in the upper 1−3 m of basaltic *lava flows.

Scuba. A term coined from the initials of Self-Contained Underwater Breathing Apparatus. This apparatus, developed in 1943 by Emile Gagnan and Jacques Y. Cousteau in the Mediterranean Sea, permits man to move about with relative ease underwater without any type of·connection to the surface. The invention of scuba is a major contribution to the marine sciences, for it allows direct investigation of the shallow areas of the oceans. Although descent with the scuba apparatus has been made to depths over 150 m, most scientific use takes place at less than 50 m. (R.A.D.)

Sea Floor Spreading. A hypothesis, formulated by Harry H. Hess at Princeton University in 1960, that the sea floor was growing outward from a central ridge system located on the floor of the oceans. Since 1960 enough evidence has been

Schist

1 A section of Manhattan schist, of Middle Ordovician age, reveals the foliation of minerals, formed under intense heat and pressure, that characterizes metamorphic rock.

Schistosity

2 The minerals in schist are separated into discontinuous layers that readily split and flake off into thin sheets.

Secondary Enrichment

3 Primary minerals are oxidized and dissolved as they are carried down through a vein by percolating groundwater. Their metallic content is reprecipitated and deposited at the water table, enriching the ore already there. Iron oxides with concentrations of insoluble minerals such as gold and oxidized compounds of silver or lead may remain at or near the surface, forming a deposit called a gossan.

gathered to gain wide acceptance for this theory. This concept helped establish the validity of the controversial theory of *continental drift formulated by the German meteorologist Alfred Wegener in 1912, and to lay the foundations for a new branch of earth science, *plate tectonics. Until the 1950s scientists believed that the ocean floors represented the oldest parts of the earth, but as a result of advances in technology and instrumentation during World War II, they made two important discoveries: (1) The earth's *crust forming the ocean floors is of uniform thickness and much thinner than the crust forming the continents (6 to 7 km for oceanic crust compared with 30 to 40 km for continental crust). (2) A nearly continuous mountain range, or *mid-ocean ridge system, dividing all the major oceans, extends for more than 40,000 km in length, and volcanic and earthquake activity is associated with the ridge system.

The dating of fossils and rocks from the ocean floor has shown that ocean basins are relatively youthful features, no older than the Cretaceous Period, some 130 million years. Using this information, Hess theorized that the mid-ocean ridges represented zones of weakness within the earth's crust. These zones were being constantly fractured and filled with molten material derived from deeper within the crust. The ocean floor on either side of the ridges moved farther away from the fracture zone with each new injection of material. Hess also suggested that the material underlying the crust (*mantle) was capable of plastic flow, and that the flow was caused by deep-seated *convection currents forming large cells of circular movement within the mantle. The mid-ocean ridges represented the confluence or junction of rising convection currents from adjacent cells, and the site for the development of a new oceanic crust. Based on the age of the oldest rocks from the floor of the Atlantic Ocean and the distance between continents on either side, Hess suggested that the sea floor was spreading away from the Mid-Atlantic Ridge at approximately 2 cm per year. This theory supports the concept of continental drift, offering a mechanism by which continents could move apart. It also offers a logical explanation for the absence of sediments on mid-ocean ridges as well as the youthful age of the rock composing them and the volcanic and earthquake activity associated with them.

Using *paleomagnetism (the intensity and direction of the earth's *magnetic field during geologic history), geoscientists made two important discoveries. First, the ocean floor is segmented into long parallel strips of rock that vary in magnetic intensity. Some of these strips parallel the mid-ocean ridges and display a symmetrical pattern; that is, a strip of ocean floor with a specific magnetic intensity on one side of a ridge has a counterpart on the other side, with the center of the ridge acting as a plane of symmetry. Second, the earth's magnetic field has reversed direction several times during geologic history, causing the north and south magnetic poles, as well as the polarity of magnetism in the rock, to change position a number of times.

In 1963 a way of testing the Hess hypothesis was proposed by Fred J. Vine and D. H. Matthews at Cambridge University. They suggested that if new material was being injected upward through a central fissure of a mid-ocean ridge, the magnetic particles within the solidifying rock would be aligned with the earth's magnetic field and have the same direction of polarization. After solidification, forces produced by convection cells within the mantle would cause the newly formed rock to fracture in equal segments along its length. This would allow another pulse of molten rock to be injected into the newly formed fissure. As this process continued through millions of years, strips of rock of equal magnetic intensity and direction should be found parallel to both sides of a ridge if the sea floor were actually spreading. A magnetic study made across the Mid-Atlantic Ridge south of Iceland bore out the assumptions and gave further support to the theory of sea floor spreading. See Color Plate "The Shifting Plates of the Crust." (J.W.S.)

Sea Ice. Ice that results when *seawater freezes. Normal seawater has a salinity of 35 parts per thousand and therefore will not freeze until the temperature drops to -1.9°C (29°F). As the ice begins to form, the salt brine separates from the water. The salt is not accepted into the crystal structure but is trapped in pockets within the ice. The salinity of sea ice itself is quite low and depends on the rate at which the ice freezes. In the Arctic, sea ice generally accumulates to a thickness of 2 m to 3 m (6 to 10 feet), whereas it is only about 1.5 m (5 feet) thick in the Antarctic. The Weddell Sea in the Antarctic has a perennial cover of sea ice.

Seamount. A submarine volcano that does not rise to sea level. Seamounts are scattered throughout the oceans but are particularly common along and near *fracture zones in the *oceanic crust. They are a kilometer or more in height and somewhat circular. See Guyot.

Seawater Density. The mass per unit of volume, which depends on salinity (the amount of dissolved substances in the water), temperature, and pressure. The latter depends in turn on the depth of water. Fresh water has a density of essentially 1.00000 gram per cubic centimeter, whereas seawater density commonly ranges between 1.02400 and 1.03000 grams per cubic centimeter. Although this range may seem insignificant, differences of only a few ten-thousandths of a gram per cubic centimeter can be important in deep currents and water masses. Because seawater density is always expressed to the fifth decimal place and because the first two numbers are always the same (that is, 1.0), the shorthand notation σ STP is commonly used. It is derived as follows: σ STP = (density-1) x 10^3 gm/cm³. (R.A.D.)

Secondary Enrichment. A process by which a primary low-grade *ore body is later enriched to a higher-grade ore. The enrichment process is often related to the oxidation of overlying masses of ore. As a result of the oxidation of the primary minerals near the surface, some of the metals are taken into solution and move downward toward the groundwater table. The lack of oxygen encountered at lower levels may cause the precipitation of new metallic minerals that enrich the low-grade primary ore.

Secondary enrichment has been a major factor in the formation of many copper deposits. Oxidation of copper sulfides and associated pyrite produces ferric sulfate and sulfuric acid, which react with copper to form soluble copper sulfates. Below the zone of oxidation the solutions react with the primary sulfides, either coating them or replacing them by secondary copper sulfides, usually chalcocite. As oxidation proceeds deeper, the new minerals in turn react, and there is a downward migration of the copper. The water table marks essentially the lower limit of oxidation, and near it there may be a zone of secondary enrichment that contains much of the copper originally in the primary minerals above. (C.S.H.)

Secondary Mineral. A mineral that formed by alteration of pre-existing minerals.

Sediment. Solid material of organic or inorganic origin that has settled from suspension and been deposited by water, wind or ice (the process of sedimentation). It may consist of fragments of preexisting rocks, the products of chemical action or of evaporation, the remains or secretions of organisms, or a mixture of these. Sediments are typically described in terms of their origin (sedimentary environment) or mode of transportation. The latter includes marine (in the sea), terrestrial or continental (on land), fluvial (by streams), lacustrine (in lakes), eolian (by wind), glacial (by ice) and organic (associated with the remains or products of plants or animals). Loose sediments may become consolidated to form coherent, solid *sedimentary rock such as *conglomerate, *sandstone or *limestone.

Sedimentary Rock. A rock formed by the consolidation of loose *sediment that has settled out of water, ice or air and accumulated on dry land or in water. It is typically stratified or layered as the result of variations in the velocity of the sediment-transporting currents or changes in the supply of sediment during deposition. Sediments are converted to a coherent rock mass by lithification, a process that may involve cementation (when spaces between individual grains of sediment are filled by a binding material such as quartz or calcite), compaction by the weight of overlying sediments or earth movements, desiccation as sediments dry out, and crystallization or recrystallization in certain chemical sediments.

Sedimentary rocks are classified according to the nature of the sediments from which they are formed. Clay, quartz and calcite are the most common minerals in sedimentary rocks and usually consist of particles directly derived from the erosion of pre-existing rocks, or of material precipitated from solution by organic or inorganic means, or a mixture of these. The nature of the constituent particles and the way in which they are bound together determines the texture of a sedimentary rock and is related to the conditions under which the sediment was deposited. Clastic (or detrital) sedimentary rocks make up at least two-thirds of all sedimentary deposits and consist of rock particles, minerals or shells bound together by calcite, quartz, clay or some other mineral cement. The names of clastic rocks refer principally to the size of their sedimentary particles. Typical clastic rocks (arranged in order of decreasing size of constituent particles) are conglomerate, sandstone and shale.

Chemical sedimentary rocks form from deposits composed mostly of material produced by chemical processes, especially precipitation of material dissolved in water. Precipitation may be direct, as when evaporites such as halite (rock salt) and gypsum are formed by the evaporation of seawater or a saltwater lake. Dolomite, travertine and certain types of limestone are also produced by chemical means.

Biochemical (or organic) sedimentary rocks consist of sediments produced either directly or indirectly by the chemical processes of plants or animals. Organic limestones such as chalk and coquina, peat and coal are derived from plant or animal fragments lithified into rock layers. Organic sediments can also be produced by organisms such as corals, foraminifers and radiolarians. These creatures extract chemicals from seawater and use them to build hard exoskeletons made of calcium carbonate or silicon dioxide. When the animals die, their remains accumulate as biochemical sediment that may, over a long period of time, form a biochemical sedimentary rock.

*Stratification is the most distinctive feature of sedimentary rocks, but they may contain other characteristic sedimentary structures such as *fossils, *concretions and *geodes, *cross-bedding, sedimentary *facies, *ripple marks and mud cracks. These may occur within a rock layer or on its surface, and may provide information as to conditions of sedimentation at the time they were formed. The color of sedimentary rocks varies greatly and is primarily due to small amounts of coloring matter such as iron oxide, carbon and organic material. Such colors are especially noticeable in arid climates where there is little vegetation and minimal discoloration from atmospheric weathering. The colorful sedimentary rocks that have been exposed in such arid areas as Grand Canyon, Bryce Canyon and Petrified Forest National Parks contain hematite and other iron minerals.

Sedimentary rocks extend over approximately 75 percent of the earth's surface but form only about 5 percent by volume of the outer 16 km of the earth. Of these, about 99 percent are sandstone, mudstone and shale. Because sedimentary rocks form at or very near the surface of the earth, they provide evidence of past events on the earth's surface and are thus valuable in reconstructing geologic history. They are also an important source of such valuable natural resources as soil, water, coal, petroleum, building stone, construction materials and of certain mineral ores. (W.H.M.)

Sedimentation. The act or process of depositing *sediment. In geology this also includes the process whereby rock particles are separated from the material from which the sediment is derived, the transportation of these particles to their place of deposition, the actual settling out (or deposition) of the particles, *diagenesis and other changes in the sediment, and the eventual consolidation of the sediment into solid *sedimentary rock.

Sediment Coring. In oceanography, the process of extracting elongate vertical sections or cores of deep-sea sediment in order to interpret the recent history of the ocean floor. These cores are taken by means of coring devices manipulated by scientists on research vessels. Most corers take a column of sediment from 5 to 10 cm in diameter and up to 30 m in length. The length depends largely on the type of coring apparatus and the type of bottom sediment. Piston corers are the most efficient and most common, but gravity and explosion corers are also used.

Because of the slow rate of accumulation on the ocean floor (a few centimeters per 1000 years) it is possible to find a record of up to a few million years in a core. The type of sediment and fossils included in the core will reveal the conditions that prevailed in the ocean during that interval.

Recent advances in drilling techniques have enabled investigators to drill in the oceanic sediment and rock to depths of hundreds of meters from ships operating in water thousands of meters deep. These advances have resulted largely from the Deep Sea Drilling Project (DSDP), which began in 1968 and has been drilling holes in the ocean floor around the entire globe in order to reconstruct the history of ocean basins. (R.A.D.)

Sediment, Cosmic. See Cosmic Sediment.

Sediment, Relict. See Relict Sediment.

Sediment Transport. A term that generally refers to the movement of the solid debris load of streams, wherein material is carried on the bottom as *bed load or moved in suspension (*suspended load). The amount of material transported varies from time to time and place to place along a river. The bed load is difficult to measure, but the suspended load can be measured.

The force required to move a given grain as bed load is the critical *tractive force, which for small grains depends primarily upon depth-slope product (*Duboys equation). For large grains, stream velocity is critical for movement, regardless of depth or slope. The lowest velocity at which given grains will move is called the critical erosion velocity. This, along with the settling velocity of the particle and the turbulence of the water, determines the suspended load. See Flow Regime. (M.M.)

Seismic Belts. The zones of most intense earthquake activity, as plotted on a world map. The belts are located mainly in three regions: (1) along the nearly continuous submerged mountain range (*mid-ocean ridges) that separates all the major oceans,

(2) along the western coast of South and North America, the Aleutian Islands and the western margin of the Pacific Ocean and (3) a broader zone extending from the western margin of the Pacific Ocean up the Malaysian Peninsula, across the top of India and westward along the northern coast of the Mediterranean Sea. Scientists now know that these belts are related to the lateral growth (*sea floor spreading) of the ocean basins along the mid-ocean ridges and the collision of the large plates that make up the outer shell of the earth. See Continental Drift; Plate Tectonics; Seismology. (J.W.S.)

Seismicity. A term that pertains to the phenomenon of movements within the earth. Included also is the study of the geographic distribution of earthquakes and the amount of energy released by them.

Seismic Reflection. A geophysical method based on the measurement of travel time of waves generated by explosions. As measured by sensitive instruments, the waves travel down through the earth's subsurface and are reflected back to the surface from buried structures. Seismic reflection methods are now extensively used in determining the detailed structure lying below the earth's surface. One important application is in exploring for oil. Unlike most other geophysical prospecting methods, seismic reflection maintains its reliability with depth of penetration.

Seismic Reflection Profiling. A geophysical technique for determining the characteristics of strata below the floor of the sea. An electric sparker, a boomer that produces an acoustic impulse, or an air gun that emits a bubble impulse is towed behind a ship. These impulses are sent to and beneath the ocean floor. They are reflected off the various layers and transmitted back to the research vessel, where they are received by hydrophone and automatically recorded on a chart.

Seismic Refraction. A method of geophysical prospecting that uses seismic waves generated by explosions to determine crustal structure. Sensitive instruments measure the time required for waves to travel down through the earth, where they are refracted along various layer boundaries and then refracted upward again to the earth's surface. Refractive methods were the first seismic techniques employed by geologists to determine probable sites of oil deposits. Although no longer as extensively used as it once was in petroleum exploration, but

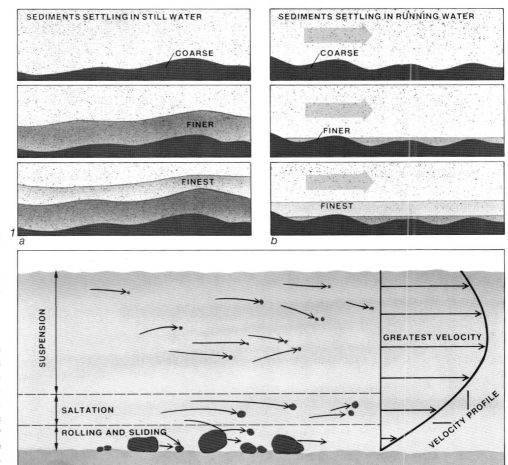

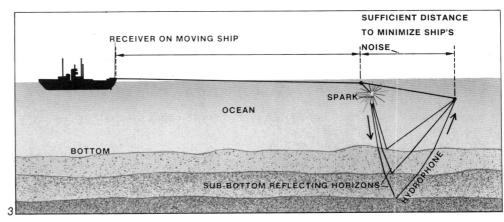

Sediment

1 In still waters (a), sediments tend to settle in layers, with the largest, heaviest particles on the bottom. (b) In moving water, the largest particles are deposited in ripples and finer sediments settle in even, horizontal layers.

Sediment Transport

2 In a turbulent stream large rocks tend to roll and slide along while smaller rock particles jump forward in a process called saltation. Fine particles move at greatest speed one-third of the distance down from the surface, following irregular paths and suspended in the water by upward-moving currents.

Seismic Reflection Profiling

3 The continuous-recording reflection instrument traces a cross section of the ocean bottom to a depth of as much as 100 m (330 feet), which provides information about the distribution of sediment and rock layers. Sound impulses emitted by an energy source are reflected from the bottom and sub-bottom horizons, picked up by a hydrophone and converted to electrical signals that are recorded on a chart.

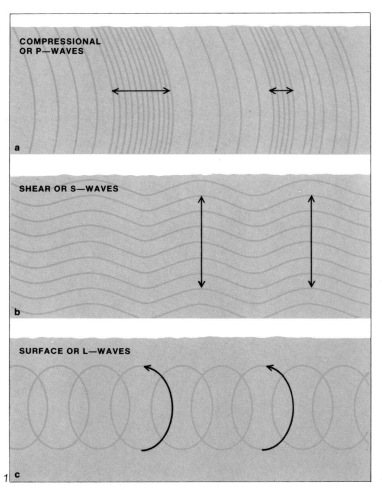

COMPRESSIONAL OR P—WAVES

a

SHEAR OR S—WAVES

b

SURFACE OR L—WAVES

c

1

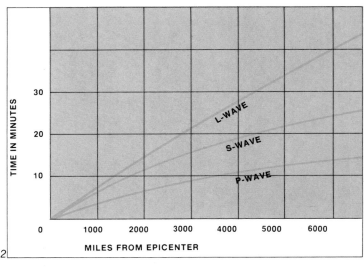

TIME IN MINUTES

L-WAVE

S-WAVE

P-WAVE

MILES FROM EPICENTER

2

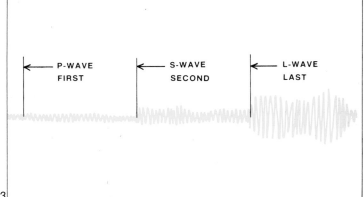

P-WAVE
FIRST

S-WAVE
SECOND

L-WAVE
LAST

3

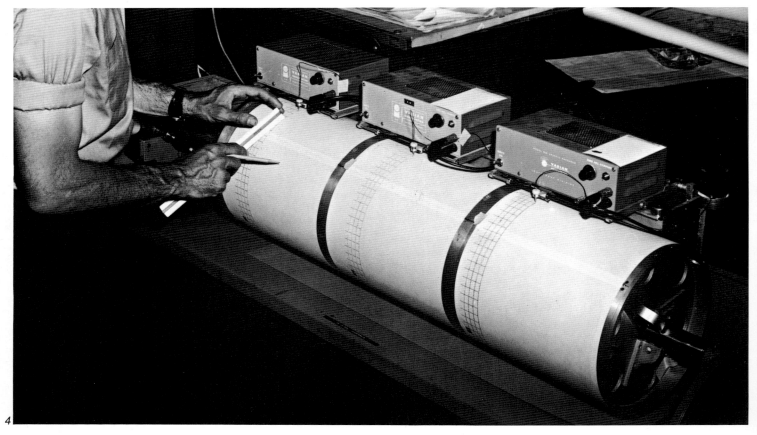

4

very heavily used in engineering operations, the refraction method still has the advantage of providing data used in determining not only the structure of various subsurface layers but also their composition. The method is used by engineering geologists to determine the depth of soil cover in areas where large buildings, highways, dams and bridge abutments are to be constructed.

In seismic wave refraction, waves radiate out from a source in concentric wave fronts. The waves propagate through the earth until they encounter a boundary between two layers of different physical characteristics. At the boundary the waves are refracted into the lower layer but only where the velocity in the lower layer is greater than in the upper layer. This is similar to the way light waves are refracted upon entering a body of water. At some point along the interface between the two layers (a) a concentric wave front will strike the boundary at an angle (called the critical angle θ) such that a refracted wave front perpendicular to the boundary is formed. At this point the waves begin to travel along the boundary. As they travel through the lower medium they produce oscillations along the boundary that act as point sources for the generation of wave fronts in both upper and lower layers. The new fronts in the upper layer will return to the surface at the same critical angle in which the original wave front was refracted into the lower layer. (J.W.S.)

Seismic Waves. The form taken by the energy released when rocks within the earth break apart or slip rapidly along fault planes during an earthquake. Seismologists record these waves to determine the location of the focus and the magnitude of the earthquake. One type of seismic wave, called the compressional or primary *(P)* wave, is similar to a sound wave. As a compressional wave passes through a substance, the behavior of individual particles is comparable to the expansion and contraction of a spring. *P* waves can travel through solids, liquids or gases.

Another type of seismic wave is known as a shear or secondary *(S)* wave. As it passes through a substance, individual particles vibrate back and forth at right angles to the direction of motion of the wave. Because of this difference, shear waves cannot travel through liquids or gases. As some compressional and shear waves travel through the interior of the earth, they can be detected on seismographic records. Surface *(L)* waves, also produced by earthquakes, travel only on the earth's surface. (J.W.S.)

Seismograph. An instrument used to detect vibrations of the earth caused by earthquakes or nuclear explosions. There are many types of seismographs, but their basic principle of operation is the same. A simple seismograph consists of two main elements, a frame or supporting structure and a free-swinging weight (pendulum). The frame is rigidly fastened to a body of rock capable of transmitting earth movements; the frame will thus move in conjunction with its rock support. Because the pendulum is free-swinging and has a large inertia (the capacity to resist motion), it remains stationary. Attached to the pendulum is a scribing device such as a pen. In contact with the pen and attached to the seismograph frame is a chart on which the motion of the pendulum, and thus of the earth, is recorded. The earth may move vertically as well as horizontally as a result of *seismic waves, and so seismographs are designed to record vertical as well as horizontal north-south and east-west motions. (J.W.S.)

Seismology. A branch of earth science concerned with earthquakes and the waves of energy they generate. Through the study of *seismic waves earth scientists have been able to construct a model of the earth's interior. Seismological studies have long been concerned with the location of the *epicenter (that is, the point from which the earthquake originates), and the causes of earthquakes.

Progress in recent earthquake study makes it likely that scientists will soon discover widely applicable methods of earthquake prediction, and thus reduce the hazards of major earthquakes near centers of population. Moreover, an understanding of earthquake vibrations allows engineers to design buildings that minimize earthquake damage. Mapping the location of major earthquakes over a period of years has lent much support to the concept that the *lithosphere is composed of moving plates since the earthquakes so mapped tend to cluster near the inferred plate boundaries. (J.W.S.)

Selenite. A variety of *gypsum that can be cleaved into broad, colorless, transparent plates.

Septarium. A roughly spheroidal *concretion of limestone (or clay-limestone) characterized by a network of intersecting mineral-filled cracks. Internally it consists of irregular, many-sided blocks cemented together by veins of minerals such as *calcite. Septaria occasionally weather so as to leave the veins standing in relief; some resemble fossil turtles.

5

Seismic Waves
1 *Compressional waves, or P-waves (a), can travel through any earth material, alternately compressing and relaxing molecules in their paths. Shear waves, or S-waves (b),which can pass through solids but not liquids or gases, vibrate molecules perpendicular to their direction of travel. Surface waves, or L-waves (c), cause a circular motion of earth particles in their path.*

2 *The first earthquake waves to arrive at a recording station are called primary (P) waves, or compressional waves, and are of low intensity. The next set of waves to arrive are called secondary (S) waves, or shear waves, and are stronger. The last to arrive are called surface (L) waves and are the longest, or most intense of all.*

3 *A seismogram indicates both the arrival time and intensity of seismic waves.*

Seismograph
4 *A seismologist at Cornell University measures horizontal components of motion recorded on a revolving drum, which vibrates from side to side during an earthquake.*

Selenite
5 *Displaying the characteristic perfect cleavage of gypsum, this often colorless, transparent variety of gypsum is named after Selena, goddess of the moon.*

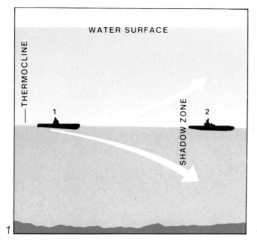

WATER SURFACE

THERMOCLINE

SHADOW ZONE

1

2

1

Shadow Zone
1 *Sound waves in water are refracted in both vertical directions at the thermocline, the layer of maximum temperature change, creating a zone where no sound waves are travelling. The first submarine cannot detect the second, since the latter is within the shadow zone.*

Shale
2 *A tilted bed of laminated shale, composed mainly of hardened clays, is exposed at Lost River, Va.*

Sheeting
3 *These thick layers of parallel plates at a Vermont granite quarry are believed to have been caused by the upward expansion of the rock as surface pressure decreased.*

Sericite. A fine-grained mica, usually *muscovite, occurring in small flakes. It is especially common in schists but is also found as an alteration of the wall rock adjacent to hydrothermal ore veins.

Series. A *time-stratigraphic unit smaller than a *system and representing the rocks formed during an *epoch of geologic time. Hence the Eocene Series represents a sequence of rocks formed during the Eocene Epoch.

Serpentine. A hydrous magnesium *silicate mineral, usually formed as an alteration product of preexisting magnesium silicates, especially olivine and pyroxene. There are two distinct forms of serpentine: *antigorite, a platy variety with a waxlike luster mottled in various shades of green; and *chrysotile, a fibrous variety with a silky luster, varying in color from yellow to dark green. Most *asbestos is chrysotile.

Serpentine as a rock name is applied to rock masses composed almost entirely of antigorite. Serpentine rocks are considered metamorphic, having formed from the alteration of *peridotites. Chrysotile asbestos forms within the massive serpentine rock. Serpentines may also be a source of associated chromite and platinum, as in the Ural Mountains, or a source of nickel from associated garnierite, as in New Caledonia. See Mineral Properties. (C.S.H.)

Serpentinite. A rock composed almost entirely of the mineral *serpentine. It is considered a metamorphic rock, having formed from the alteration of *peridotite and similar iron-rich and magnesium-rich rock types. In addition serpentinite may have veins of *chrysotile asbestos formed within it. It may also be a source of associated chromite and platinum, or a source of nickel from associated garnierite.

Shadow Zone. An area on the earth's surface in which primary and secondary shock waves originating from an earthquake or nuclear explosion are rarely observed. The area lies between 100° and 140° of arc away from the origin of the waves. Primary waves are not observed in this region because they are reflected off or refracted below the boundary between the earth's *core and *mantle. Primary waves reflected from the core intersect the earth's surface at an angle of less than 100°, while those that are refracted through the core travel in such a manner that they reach the earth's surface at angles greater than 140° from their source. Secondary waves are not observed beyond 100° from their source because their properties are such

that they cannot travel through a liquid. It was the observed behavior of secondary waves that led scientists to conclude that the earth's core was in part liquid. (J.W.S.)

Shale. A fine-grained clastic *sedimentary rock consisting of tightly packed sediments less than 1/16 mm in diameter. It is rich in the hydrous aluminum silicates (clay minerals), but other minerals and organic matter may be present. Shale is also fissile—it splits readily along closely spaced partings. This distinguishes it from clay which breaks into blocky fragments. The most abundant of all sedimentary rocks, shale may be formed from mud that accumulates in deltas, on river flood plains and on the bottoms of lakes and oceans. Shales rich in organic matter may be black or dark gray; red shales owe their color to various iron minerals. *Oil shale, from which petroleum can be extracted, is rich in hydrocarbon derivatives such as kerogen.

Shear. A force that causes parts of a solid to slip past one another, like cards in a deck. The shear per unit area is called shear stress — it is measured in pounds per square inch or newtons per square meter. Shear stress acts tangentially or parallel to the plane along which a solid body will ultimately break or deform by shear. Theoretically, the maximum shear stress within a solid will occur along a plane inclined at an angle of 45° to the surface on which an external force is being applied. Laboratory tests of rock subjected to external forces that cause the solid to fail by shearing show that the shear planes that develop are always oriented at an angle greater than 45° to the surface of the rock. This deviation in the angle of shear from the theoretical value is caused by internal friction and by the strength of the material. (J.W.S.)

Sheeting. A phenomenon that produces fracture planes in otherwise unfractured massive bodies of rock. (This phenomenon is known as *exfoliation or *spheroidal weathering when it occurs on a small scale.) It is exemplified by a system of fractures that roughly parallels the topography of the land surface. The sheets are closely spaced near the surface and increase in thickness with depth. Sheeting is best displayed in granite or other coarse-grained crystalline rock, although it also occurs in sandstone. It is believed to result from the increase in compressive forces parallel to the land surface as erosion removes overburden from the surface of massive rock. Compressive stress increases until the breaking strength of the rock is exceeded

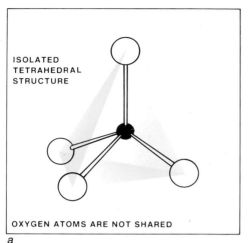

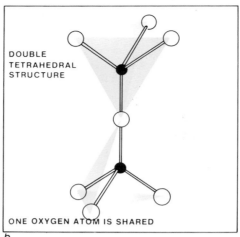

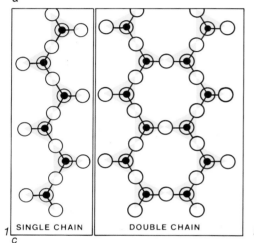

and fractures develop. The formation of sheeting has been observed in quarries and may occur with such violence that large plates of rock shoot upward from the quarry floor. (J.W.S.)

Sheetwash (or Rainwash). Flow of sheets of unchannelized water down a slope, usually occurring on unvegetated surfaces under intense rainfall. It is generally characteristic of semiarid regions but may also occur in humid climates if vegetation is low or sparse. See Overland Flow.

Shield. A continental area in which a large, gently convex surface of basement rock is exposed. See Craton.

Shield Volcano. Named because of their similarity to the circular shields of ancient warriors, shield volcanoes are broad convex domes with gentle slopes, composed mainly of superimposed *lava flows and subordinate *pyroclastic rocks. They may range in height from under 30 m to nearly 9200 m. The slopes near the base vary between 2° and 4° in steepness and often increase to more than 5° near the summit. The great fluidity of the lava is responsible for the gentle slopes. If the principal vent is circular the base is circular; but if the vent is elongated the base will be elliptical, as in the two volcanoes that constitute the island of Oahu. Although most shield volcanoes are composed of basaltic lava flows, some are of andesitic flows; other kinds of lava may be present in very small amounts. At Mauna Loa the flows at present do not issue from vents at the summit but from radial fissures on the flanks. Where shield volcanoes have been considerably dissected, a great concentration of dikes is found near the summit and on the slopes where the flank flows issued.

Large, steep-walled, flat-floored basins called *calderas are found at the summits. *Parasitic cinder cones and ridges, *spatter cones and ridges, pit craters and radial and peripheral *fault scarps are found on the flanks. A shield volcano range results when several shields are aligned, and a shield volcano cluster occurs when the shields are irregularly distributed. Shield volcanoes are also found in Iceland, Mexico, Ross Island (Antarctica) and elsewhere. Etna, in Sicily, is essentially a shield volcano. (R.L.N.)

Shock Wave. A compressional wave traveling through a medium at a velocity greater than the speed of sound in the same medium. Shock waves acting upon earth materials are capable of deforming their structure or changing their mineral composition. In some cases, shock waves

Silicate Structure and Classification
1 The unit of structure of silicate minerals is the SiO_4 tetrahedron in which each silicon atom is bonded to four oxygen atoms. There are six types of silicates; three are shown here. In the nesosilicates (a) isolated tetrahedra are joined together by atoms of other elements. In the sorosilicates (b) two tetrahedra share one oxygen atom. In the inosilicates (c) all the tetrahedra in the single chain share two. In the double chain half of the tetrahedra share two

oxygen atoms; the other half, three.

Silicified Wood
2 Silicified wood is formed when silicate minerals, carried by groundwater, replace the woody tissues in tree stumps or fallen logs.

Sill
3 A thick sheet of Precambrian diabase was intruded between ancient sedimentary rocks to form this sill on Banks Island, Northwest Territories, Canada.

may even melt or vaporize rock. The meteorite that formed the giant crater near Winslow, Arizona, produced a shock wave that transformed some earth material into stishovite (a high-pressure form of quartz).

Sial. A term combining the first two letters of "silica" and "alumina"; it is applied to the uppermost layer of the earth's *crust. Rock found within the sial contains relatively large amounts of silica and alumina as compared to rock in other parts of the earth. The sial is confined to that portion of the crust that forms the continental landmasses; it lies above a layer of crustal material known as *sima. The boundary between the two layers is called the *Conrad Discontinuity.

Siderite. A mineral, iron carbonate, belonging to the *calcite group of minerals and similar to calcite in crystal structure. As a well-crystallized mineral it is common in veins associated with silver ores. More abundantly it is found as clay ironstone in concretions, impure because of admixed clay. And as impure earthy material, blackband ore, it occurs in stratified formations associated with coal beds that have been worked as an iron ore in Great Britain. More important as an ore of iron are deposits formed by the replacement of limestone by siderite, as at Styria, Austria.

The name siderite is also used to refer to an *iron meteorite. See Mineral Properties. (C.S.H.)

Silica. Silicon dioxide, SiO_2. Silica has many polymorphs, of which the most common and abundant is quartz.

Silicate. One of a group of minerals containing silicon and oxygen. This is the most important mineral group, for about 25 percent of the known minerals and 40 percent of the common ones are silicates. It is estimated that about 95 percent of the earth's crust is composed of silicate minerals. See Silicate Structure and Classification.

Silicate Structure and Classification. The basic structural unit of all silicate minerals is the silicon-oxygen tetrahedron. The silicon ion is surrounded by four oxygen ions at the apices of a regular tetrahedron. The silicon-oxygen bond consumes exactly half of the bonding energy of each oxygen ion. The other half is available for bonding to another silicon ion, thus joining tetrahedral groups through the shared oxygen. The great diversity of silicate structures arises from this sharing, which may involve one, two, three or all four of the oxygen ions. The following classification is based on the manner in which the basic tetrahedral groups are united.

Independent tetrahedral groups, nesosilicates. Isolated silicon-oxygen tetrahedra joined together by other ions. The ratio of Si:O = 1:4 as indicated by the formula of olivine, $(Mg, Fe)_2SiO_4$.

Double tetrahedral groups, or sorosilicates. Two silicon-oxygen tetrahedra joined by sharing one oxygen with the ratio Si:O = 2:7. *Hemimorphite, with the formula $Zn_4(Si_2O_7)(OH)_2 H_2O$, is one of these.

Ring structures, cyclosilicates. Three types of ring structures in which three, four or six tetrahedra are joined together to form closed rings; in all types two oxygen ions of each tetrahedron are shared with adjacent tetrahedra, resulting in a ratio of Si:O = 1:3. The six-numbered ring is represented by beryl, $Be_2Al_2Si_6O_{18}$.

Chain structure, inosilicates. Two types of chains of indefinite length formed by linking silicon-oxygen tetrahedra. In one, the single chain, two oxygen ions in each tetrahedron are shared, yielding a ratio of Si:O = 1:3. In the double chain two oxygens are shared by half of the tetrahedra and three oxygens by the other half, with a ratio of Si:O = 4:11. The single-chain structure is found in the pyroxenes, of which enstatite, $Mg_2(Si_2O_6)$, is an example. The double chain characterizes the amphiboles. An example is tremolite, $Ca_2Mg_5(Si_8O_{22})(OH)_2$.

Sheet structures, phyllosilicates. Three oxygens shared by each tetrahedron to form a sheet of indefinite extent yielding a ratio of Si:O = 2:5. This sheet is the controlling structural feature in the micas. There are many sheet silicates, all characterized by a prominent cleavage parallel to the plane of the sheet. Talc, $Mg_3(Si_4O_{10})(OH)_8$, is a typical example.

Three-dimensional frameworks, tectosilicates. Each silicon-oxygen tetrahedron shares all four oxygens with other tetrahedra, and thus the ratio of Si:O = 1:2. This is the structure of quartz and other forms of SiO_2.

However, the three-dimensional framework is the underlying structure for minerals other than those of the silica group. Aluminum may substitute for one or more silicon ions. When this takes place other ions can and must enter the structure to maintain electrical neutrality as in the important feldspar group of minerals. For example, in orthoclase, $K(AlSi_3O_8)$, aluminum substitutes for one-fourth of the silicon. This leaves a negative imbalance compensated for by the presence of a positive potassium ion. (C.S.H.)

Silicification. Organic remains that have been replaced by silica (SiO_2). This replacement may result from the filling of cavities and tissue with groundwater

saturated (or supersaturated) with silica, or from the solution of the original hard parts and the deposition of silica in their place. This type of preservation may occasionally result in an almost perfect specimen. More frequently, silicification obliterates many details of the original structure. Striking examples may be seen in the silicified *Permian fauna from the Glass Mountains of Texas. When subjected to a special type of acid treatment, these limestones yield an amazing variety of perfectly silicified brachiopods, crinoids, bryozoans and mollusks. Due to the high degree of *replacement, even the most delicate features are unchanged.

Another interesting occurrence of silicification has been reported from ancient lake deposits in California's Mojave Desert. Here limestone nodules have yielded perfectly preserved insects, spiders and other arthropods. Many of these silicified fossils are so perfectly preserved that even the body appendages and hairs may be studied in detail. (W.H.M.)

Silicified Wood. Petrified wood in which silica has replaced the wood in a manner that preserves the original form and structure. The silica may be in the form of *chalcedony or *opal.

Sill. A table-like body of rock of molten origin (*magma) that has been injected into crustal rocks. Sills form parallel to any layered structure in the existing rock, such as bedding, and are thin in relation to their extent. They can form at any angle to the earth's surface but are generally most readily recognized in regions where the crust has not been subjected to extensive bending or buckling.

Sills are classified by their origin. A sill formed by a single injection of magma is termed simple. Repeated injections of magma of the same composition produce multiple sills. Repeated intrusions of molten rock of different composition form composite sills. An interesting type of sill, the differentiated, forms when a thick body of magma has moved into the earth's crust and requires a long time to solidify. During the slow cooling process, crystals of a particular mineral able to form at a high temperature begin to crystallize. Since the newly formed crystals are of greater density than the surrounding magma, they settle to the floor of the sill. By the time all the rest of the magma becomes solidified, a layer rich in one particular mineral will be noticeable in the solidified sheet of magma. A sill of this type is exposed in the Palisades on the New Jersey bank of the Hudson River facing New York City. Approximately 6 m above its base can be seen a layer of rock rich in olivine crystals. Sills of much greater extent, such as the Great Whin Sill of northern England, are found throughout the world. (J.W.S.)

Sillimanite. A mineral, an aluminum silicate, one of the three polymorphic forms of Al_2SiO_5; the others are *kyanite and *andalusite. Sillimanite is a metamorphic mineral found in the highest grade of *metamorphic rocks. It is commonly in long, slender crystals, frequently fibrous, and has thus been called fibrolite. See Mineral Properties.

Silt. A clastic sediment consisting mostly of particles having a diameter between 1/16 and 1/256 mm. It is composed of fragments smaller than fine sand yet larger than coarse clay. Silt particles may become cemented to form siltstone, which resembles shale in texture and composition but does not exhibit shale's lamination or its fissility.

Silurian Period. The third oldest period in the *Paleozoic Era; it began about 430 million years ago and lasted some 30 million years. Sir Roderick Murchison named it for the Silures, an ancient Celtic tribe that once lived in the part of Wales where the system was first studied. In addition to Great Britain, Silurian rocks occur in France, Germany, Scandinavia, the Baltic Islands and northwestern Russia. They are also exposed in Asia, Australia, New Zealand and South America. Silurian rocks in North America resemble those of the underlying *Ordovician System and occur mainly in the northern and eastern United States and in eastern Canada, Newfoundland and north and southwest Hudson Bay.

During Early Silurian time thick deposits of conglomerates and sandstones were laid down from western New York to Alabama. These formations suggest that the land along the East Coast was still well above sea level at the beginning of the period. Middle Silurian deposits consist primarily of shales, limestones and dolomites. A good example of Middle Silurian rocks is exposed in the gorge of the Niagara River in New York State and in Ontario, Canada. Niagara Falls owes its existence to the Lockport Dolomite, a resistant Middle Silurian formation that forms the cap rock over which the falls plunge. Considerable Middle Silurian volcanic activity is indicated by lava flows and beds of volcanic ash in Maine and also in New Brunswick, Canada. In Late Silurian time the seas began to withdraw from what is now Ohio, Michigan and New York, leaving behind isolated bodies of salt water. Thick evaporite (salt) deposits were laid down in these landlocked arms of the sea and have been mined in Michigan and New York. Most Silurian rocks occur in the northeastern United States, for western North America appears to have been dry land throughout Silurian time. The Silurian closed quietly in North America; in Europe, however, crustal unrest, which extended some 4000 miles through the British Isles, Scandinavia and westward to Greenland, was produced by the elevation of the Caledonian Mountains.

The presence of reef-building corals and thick, widespread deposits of limestone and dolomite suggest that Silurian climates were generally warm and mild. And the evaporite deposits of Late Silurian age indicate a period of extreme aridity in the areas in which they occur. Marine life of the Silurian time is characterized by expansion of the bryozoans, the brachiopods, reef-building corals, mollusks and graptolites. Trilobites had already reached the peak of their development and were reduced in variety and numbers, but their relatives, the eurypterids, were abundant. These scorpion-like arthropods typically ranged from a few centimeters to 30 cm (1 foot) in length, but some were as much as 2 m (7 feet) long. Life apparently established a foothold on land near the end of the period, for the earliest known land plants have been found in Upper Silurian rocks in Australia. Fossil scorpions and centipedes occur in rocks of Late Silurian age in Wales and may have been the first air-breathing animals on earth. Although primitive fish fossils are rare, they appear to have been the only vertebrates in Silurian seas.

Economic products from Silurian rocks include salt from the evaporite deposits of Ohio, Michigan and New York; iron ore from the Birmingham, Ala., area; and oil and natural gas in several parts of the United States. (W.H.M.)

Silver. A chemical element that occurs as a mineral. It is a *native metal widely distributed in small amounts as a secondary mineral in the oxidized zone of *ore deposits. Elsewhere it is found in larger amounts as a primary mineral deposited from hydrothermal solutions. As a primary mineral native silver is found in three types of deposits: (1) associated with zeolites, calcite, fluorite and quartz, as at the occurrence in Kongsberg, Norway — magnificent specimens of crystallized silver have come from this locality; (2) associated with arsenides and sulfides of cobalt, nickel and silver as at Cobalt, Ontario, and the old German silver-mining area of Saxony; (3) associated with uraninite, as in the deposit at Great Bear Lake, Northwest Territories, Canada. See Mineral Properties. (C.S.H.)

Sima. A word combining the first two letters of "silica" and "magnesia," referring to the abundance of such oxides in the oceanic crust and in the lower portion of the crust underlying continents. The sima consists of rock rich in silica and magnesia. Within the crust of continental landmasses the sima is separated from an upper crustal layer, *sial, by a boundary known as the *Conrad Discontinuity.

Sinkhole (or Sink, Swallow Hole). A depression at the earth's surface, usually resulting from the solution of limestone by groundwater. Sinkholes are also found in areas underlain by gypsum, anhydrite and salt. They may range in depth from 1 to 50 m and in area from a few square meters to a few square kilometers. They may be saucer-shaped or have vertical walls, and they are the commonest feature of *karst topography. They may contain ponds, lakes or bogs, and the flat bottoms and gentle slopes of some are cultivated. The fact that some are dry proves that they may leak, and streams that enter certain sinks (called swallow holes) disappear underground. Sinkhole or karst plains are commonly developed where horizontal or gently dipping beds of limestone crop out at the earth's surface.

There are two kinds of sinkholes: collapse sinks, formed by the collapse of the roof of a cavern; and solution sinks, which grow from the surface down because of the downward movement of water containing carbon dioxide. The latter are commonly located at vertical joints. As seen from the air, the area around Mammoth Cave, Ky., is dimpled with an estimated 70,000 sinkholes. Sinkholes are found on all continents. (R.L.N.)

Skarn. A *metamorphic rock containing minerals composed chiefly of calcium and silicon and lesser amounts of aluminum, iron or magnesium. Skarns are formed from impure *limestones or *dolomites that have undergone *contact metamorphism. The texture of skarn is granular due to its abundance of mineral grains of the same size. The word "skarn" derives from an old Swedish mining term for the valueless minerals associated with ore deposits.

Slate. A fine-grained rock derived by the *metamorphism of rocks such as shale, or from volcanic ash. Slate is characterized by its capacity to split into large, thin, flat sheets. This is due to the alignment of its thin, platelike minerals, known as mica, into parallel planes. Alignment of the mica grains occurs during metamorphism of the original rock. Its capacity to be split into

Silver

1 A common form of native silver, wire silver displays interesting abstract shapes. This specimen comes from Norway.

Sinkhole

2 Two sinkhole ponds 30 m deep (100 feet) are located at the edge of a gypsum escarpment at Bottomless Lakes State Park, N.M.

1

sheets has made it an important quarry stone used for floor tiles, flagging, roofing, the foundation of billiard tables and, in former days, school blackboards. It is also used as electrical insulation. North Wales has long been noted for its large supply of slate, and most Welsh cottages are slate-roofed. In the United States, the most important sources of roofing slate have been in Virginia and Pennsylvania. Other significant production centers are in Washington County, N.Y., and Rutland County, Vt. (J.W.S.)

Slickenside. A smooth, polished or striated rock resulting from the sliding or grinding movement between two bodies of rock separated by a fracture (*fault). Associated with some slickensided surfaces are very small, steplike structures with step faces perpendicular to the striations. It was thought that these steps could be used to determine the direction of movement between the two bodies of rock. Motion was thought to be away from the step faces since it was believed that an opposite motion would grind them away. This assumption has been questioned on the basis of slickensides produced in laboratory studies and in nature. (J.W.S.)

Slipoff Slope. A gentle channelward slope on the inner bend of a meandering stream that is downcutting. It develops under progressive lowering of base level where stream energy is increasing and *meanders are being incised. Thus the outer bank is undercut to form a cliff, and deposition takes place on the inner bank. The ratio of uplift to incisement and deposition is such that the point bar is raised and slopes gently toward the channel, giving an asymmetric terrace on opposite sides of the stream. See Entrenched Meanders.

Slough. A bog or swampy depression that was once part of a drainageway. Sloughs, common on *floodplains and *deltas, represent an abandoned channel now filled with silt and mud.

Slump. The sliding downward and outward of a large body of *regolith or *bedrock along a cliff or a steep slope. Commonly slump occurs along wave-cut cliffs, the banks of rivers and road cuts, where the material is unsupported and yields to the pull of gravity. The upper part of the slumped area is lower than the original surface, and the lower part is pushed above it because the material rotates on a plane that is concave upward. An arc-shaped clifflet is found above the highest part of the slump material, and beyond it the ground may be cracked and fissured. Movement of a block of regolith is facilitated by heavy rains when water is added to the slump material and the friction between grains is reduced, allowing movement along a slip plane. The material at the foot of a slumped area may break up into smaller fragments, and slow flowage—called earth flow or debris flow—may take place when the material is wet. Large slumps occurred during construction of the Panama Canal, delaying completion. (R.L.N.)

Smithsonite. A mineral, zinc carbonate, a member of the *calcite mineral group and similar to calcite in its crystal structure. Crystals are extremely rare and the mineral usually occurs in botryoidal, stalactitic or honeycombed masses called dry bone ore. Smithsonite is a *secondary mineral found in limestone regions, where it formed by the oxidation of the primary zinc sulfide *sphalerite. Where large masses of the mineral are found, it is a valuable ore of zinc. Most frequently, smithsonite is a dirty brown to gray. But in a few places it has a pleasing green or blue color and is cut and polished for ornamental purposes. See Mineral Properties. (C.S.H.)

Snell's Law. See Index of Refraction.

Snow. One of the commonly recognized forms of atmospheric *precipitation. A form of ice, it is composed of small translucent or transparent six-sided ice crystals. It forms from water vapor in the atmosphere when the temperature is below freezing. A snowflake may consist of as many as 50 interlocking ice crystals and measure more than 2.5 cm in diameter. The numerous crystal faces on an ice crystal reflect light and make it appear white. Viewed through a microscope, ice crystals are seen to vary in size, shape and pattern; it has been said that no two snowflakes are identical. After snow has fallen to the ground it may change into *névé and finally into glacial ice. (R.L.N.)

Snowline. The line in any region above which perennial snow occurs. Essentially the snowline is controlled by the regional climate, as shown by the fact that it occurs at lower altitudes on mountains as one goes from the Equator to the polar regions. It is about 4500 m on Mount Kenya in Africa on the Equator, 3300 m in the Alps, 1000 m on Mount St. Elias in Alaska and at or near sea level in the polar regions. There is also an increase in its altitude with increasing distance from the source of moisture; thus it is about 1300 m lower on the Olympic Mountains in Washington near the Pacific than at the same latitude 800 km inland in Glacier National Park. It is also controlled

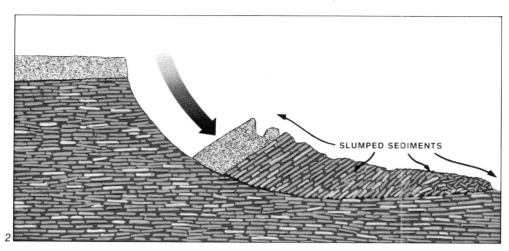

2

3

Slickenside

1 Polished grooves on the base rock of a scarp are caused by abrasion as movement occurs along the fault plane.

Slump

2 Landslides frequently occur when blocks of sediment lubricated by water move downslope, rotating backward as they move.

Snowline

3 On high mountains, such as the French Alps near Argentières, the lower limit of perennial snow is marked by a snowline. Above it, snowfields, glaciers and patches of snow persist throughout the summer season.

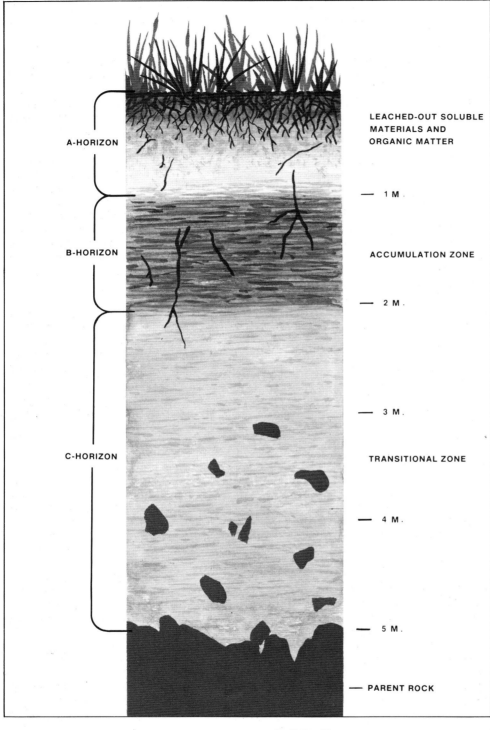

A-HORIZON

LEACHED-OUT SOLUBLE
MATERIALS AND
ORGANIC MATTER

— 1 M.

B-HORIZON

ACCUMULATION ZONE

— 2 M.

— 3 M.

C-HORIZON

TRANSITIONAL ZONE

— 4 M.

— 5 M.

— PARENT ROCK

Soil Profile
*A mature soil has three distinct
horizons. The A-horizon, at the top,
contains decomposed organic matter
and chemically and physically altered
rock. The B-horizon in a humid climate
contains material moved down from the
A-horizon; in a dry climate it contains
material that moved up from below.
The C-horizon contains both altered
and unaltered rock and grades
downward into the unaltered parent
rock.*

by exposure, so that in the Northern Hemisphere it is generally lower on the north side of mountains than on the south, and it tends to be lower on the leeward than on the windward side. It is also influenced by topography, being lower in shaded valleys than on ridges. An area of perennial snow (found only at high altitudes near the Equator but as low as sea level in the polar regions) is called a snowfield. (R.L.N.)

Sodalite. A mineral, sodium aluminum silicate with chlorine, a comparatively rare rock-forming mineral belonging to the *feldspathoid group. Its principal occurrence is in *nepheline syenites associated with nepheline, cancrinite and other feldspathoids. Although sodalite may be white, gray or green, it is usually blue or lavender-blue. Large blue masses are often cut and polished for ornamental purposes. See Mineral Properties.

Soda Niter. A mineral, sodium nitrate, the most abundant of the mineral nitrates, commonly known as Chile saltpeter. Soda niter is soluble in water and thus occurs only in arid and desert regions, principally in northern Chile and adjacent parts of Bolivia, where it extends over large areas interbedded with sand, salt and gypsum. It is recovered from this locality to be used as a source of nitrates in the manufacture of explosives and fertilizer. See Mineral Properties.

Soil. The upper part of the *regolith, which contains humus and can support plant growth. Soil is the product of the physical, chemical and biological alteration of *bedrock or transported regolith. Its characteristics depend upon climate, topography, drainage, vegetation, source material, length of formation time and other factors. Soil is formed slowly, hence the special concern of pedologists (soil scientists) over soil *erosion. Soils are of interest to geologists because (1) they give clues concerning ancient climates, (2) some ore deposits are associated with them, (3) they are the source of the sediments from which some *sedimentary rocks are formed and (4) they may be used to determine the relative ages of the deposits on which they form. (R.L.N.)

Soil Horizon. See Horizons, A-,B-, and C-.

Soil Profile. The succession of layers or horizons in the *soil and the material immediately below it. Each horizon has different physical, chemical and biological characteristics. They are essentially parallel to the topography. Three major horizons — the A-horizon (top), B-horizon (intermediate) and C-horizon (bottom) — all developed from the parent material, are found in soil that is mature and well-developed. The total thickness of the three horizons is less than 1 m in certain places, more than 10 m in others. See Horizons, A-, B-, and C-.

Solar Energy. See Energy Resources.

Solar Radiation. Electromagnetic radiation from the sun, sometimes referred to as insolation (an acronym for *in*coming *sol*ar radi*ation)*. About half the total energy of the solar beam falls within the visible part of the spectrum. Most of the other half of the energy is in the near infrared, while a small portion lies in the ultraviolet part of the spectrum. Meteorologists often refer to the radiation in the visible and near visible wavelengths as shortwave radiation to distinguish it from the longwave or infrared radiation.

The rate at which solar radiation is received on a surface perpendicular to the solar beam, at the earth's average distance from the sun and outside the earth's atmosphere, is called the solar constant. The value of the solar constant is approximately two langleys per minute (a langley is one gram-calorie per square centimeter). Thus the solar constant is equivalent to the power required to raise the temperature of one gram of water from 14°C to 16°C (57°F to 61°F) in one minute, or about 1400 watts per square meter.

The sun, however, shines on only half of the earth's surface at any given moment. Moreover, only the earth's cross-sectional area, equal to one-half the area of a hemisphere, is perpendicular to the solar beam. It follows that each square centimeter of the earth's surface receives .5 x .5 or .25 the energy flux of the solar constant, or 0.5 langley per minute on the average. Since the earth's *albedo (the ratio of the energy reflected to that received) is about 0.3, 30 percent of the solar radiation is not effective in heating the earth or its atmosphere. This reduces the rate of energy absorption by the earth and its atmosphere to 0.35 langley per minute.

The way in which this energy is absorbed and re-emitted by the earth and its atmosphere determines the earth's heat balance, which in turn controls the earth's overall climate. (M.F.H.)

Solar Wind. The flow of plasma (an electrically conductive gas made up of equal numbers of positively and negatively charged particles) produced by the continuous outward expansion of the solar corona. The solar corona is the outermost, high-temperature, low-density envelope of the sun. It may be photographed at solar eclipse or by means of the coronagraph, which employs an occulting disc to block out the image of the body of the sun. The shape of the corona varies during the sunspot cycle, and the velocity of the solar wind also varies with activity on the sun, ranging from 300 to about 850 km per second.

The solar wind was predicted in 1951 by the German physicist Ludwig Biermann, who showed that light pressure from the sun is not sufficient to explain the observed deflection of comet tails away from the sun. He suggested that a plasma of solar electrons and protons could account for this effect. Although Biermann's work led to the discovery of the electron-proton emission from the sun, it is now believed that the deflection of comet tails is an effect of a plasma-magnetic field interaction. The interaction between the solar wind and the earth's magnetic field produces the known asymmetrical shape of the magnetosphere. (M.F.H.)

Solifluction. A slow downslope movement of *regolith saturated with water. It is characteristic of cold climates, and especially of polar regions and high mountains. In such regions the ground is frozen, and when a thin surface layer thaws out during the summer, the *meltwater, augmented perhaps from nearby snowpatches and glaciers, cannot sink downward into the frozen ground. The thawed layer therefore becomes saturated with water and moves downslope over the frozen base, generally at less than 12 cm (5 inches) a year. Solifluction lobes and sheets formed by the process have flat, gently sloping surfaces and terminate in clifflets usually less than 3 m (10 feet) high. Solifluction deposits are poorly sorted and lack good stratification. Solifluction is an important process in polar areas, where it forms long, smooth slopes and denudes the land. (R.L.N.)

Sonar. An abbreviation of "sound navigation and ranging." This technique locates underwater objects by means of sound signals emitted from a source and reflected back to the source. It was perfected during World War II mainly to locate submarines. See Echo Sounding, Acoustics Underwater.

Sorosilicate. One of a group of silicate minerals characterized by the linking of two silicon-oxygen tetrahedra through the sharing of one oxygen atom. The ratio of silicon to oxygen is 2:7. An example is akermanite ($Ca_2MgSi_2O_7$). See Silicate Structure and Classification.

Space Missions. See Apollo; Luna Program; Lunar Geology; Mars, Geology of; Surveyor.

Spatter Cone (or Agglutinate Cone). A volcanic mound or cone built of clots of lava (spatter) that fall to the ground as a viscous liquid. Later clots adhere to earlier ones as they solidify, and a steep-sided cone may be formed. Spatter cones may be 60 or more meters high, but are usually less than 15 m. The small ones (called hornitos) are commonly found on the surfaces of *lava flows. A spatter rampart is formed if the vent is elongated. A spatter-and-cinder cone is formed when the eruption is partly spatter and partly material that was solid when it landed. Large spatter cones are found in the Craters of the Moon National Monument in Idaho. See Agglutinate. (R.L.N.)

Specific Gravity. A number expressing the ratio of the weight of a given volume of a substance to the weight of an equal volume of another substance taken as a standard. Since every mineral has a characteristic specific gravity, the property is important in mineral identification. For solids the standard is water at 4°C. Thus if a mineral weighs 10 grams and an equal volume of water weighs 2 grams, the specific gravity is 5. Although it is easy to weigh a mineral fragment, it is extremely difficult to determine the weight of the same volume of water. It can be done by using the principle that a substance immersed in water is buoyed up and weighs less than in air, and that the weight lost equals the weight of water displaced. Thus, in making a specific gravity determination, the fragment is first weighed in air and then is weighed while suspended in water; the difference between the two readings is the weight of the equal volume of water. The formula is: (Weight in air/Loss of weight in water) = Specific gravity. See Density; Heavy Liquid. (C.S.H.)

Specific Heat. The heat capacity per unit mass. A measure of the amount of heat in calories required to raise a one-gram mass of material one degree Centigrade. Laboratory determination of specific heat values of various earth materials plays an important part in the theoretical evaluation of temperature at different levels within the earth as well as the rate at which heat is being conducted to the surface of the earth.

Specularite. The name given to the mineral *hematite when it occurs in an aggregate of tiny tabular crystals with a brilliant metallic luster.

Speleology. The study of *caves and caverns. It has become a multidiscipline science involving discovery, exploration, surveying, archaeology, zoology, botany, paleontology. meteorology and geology. Much research, especially in discovery, exploration and surveying, is carried out by amateurs and semiprofessionals; several jounals in the United States and Europe are devoted exclusively to the study.

Speleothem. Crystalline deposits in caves, including *stalactites and stalagmites, columns and other features formed by dripstone and flowstone.

Sperrylite. A mineral, platinum arsenide, rare but recovered in a few places as an ore of platinum. Its principal occurrence is at Sudbury, Ontario, Canada, where it is associated with the copper-nickel ores and is a major source of platinum. It is also a platinum-ore mineral in the Bushveld Igneous Complex, South Africa.

Spessartite. A mineral, a manganese-aluminum *garnet with a dark red to brownish red color.

Sphalerite. A mineral, zinc sulfide, the principal ore mineral of zinc. It is common and widely distributed, found in veins and replacement deposits in limestones associated with galena, chalcopyrite, pyrite and calcite. Sphalerite is white or colorless when pure, but this is rarely the case; usually some iron is present, substituting in the structure for zinc. The mineral darkens with increasing iron content, varying from light yellow through brown to black. Black sphalerite was early confused with galena, and it was named "blende" (German meaning blind or deceiving) because although resembling galena it yielded no lead. Sphalerite has perfect dodecahedral cleavage (six directions), galena perfect cubic cleavage (three directions); on this basis a distinction between the two minerals can be made easily. See Mineral Properties; Appendix 3. (C.S.H.)

Sphene (or Titanite). A mineral, calcium titanium silicate, formerly known as sphene. It is a common *accessory mineral in *igneous rocks, particularly *nepheline syenites. It is also present in *gneisses, *schists and crystalline *limestones. On the Kola Peninsula, in the Soviet Union, titanite associated with nepheline and apatite occurs in huge masses and is mined as a source of titanium. Fine crystals are found at Binnental and St. Gothard, Switzerland; Arendal, Norway; and in Ontario and Quebec, Canada. When cut, transparent yellow or green titanite makes a lovely gemstone because of its high luster. However, its low hardness ($5-5\frac{1}{2}$) prevents extensive use as a gem. See Mineral Properties. (C.S.H.)

Spheroidal Weathering. The processes on and beneath the earth's surface that change *bedrock blocks of granite, and other rocks bounded by joints, into spheroidal cores of solid rock. These cores are surrounded by concentric layers of weathered rock. The corestones range from a few centimeters to a few meters in diameter. The shells, generally up to a few inches thick, can usually be separated from the core and from each other by a hammer blow. Spheroidal weathering is actually an instance of small-scaled chemical *exfoliation.

Water and air enter the joints, causing hydrolysis, hydration and oxidation of the silicate minerals and the formation of clay and other minerals. The formation of these new minerals involves an increase in volume, and stresses are set up that split layers and shells from the block. Since corners are attacked more vigorously than edges, and edges more vigorously than faces, the blocks gradually become spheroidal. (R.L.N.)

Spherulite. A small spherical or spheroidal aggregate composed of fibrous minerals radiating from a common center. Spherulites are common in *obsidian and in the groundmass of other siliceous lavas. Some spherulites are composed entirely of feldspar, but more commonly they are intergrowths of feldspar with quartz or tridymite.

Spinel. A mineral, the magnesium aluminum member of the *spinel group of minerals. Its color varies with chemical composition and may be white, red, blue, green, brown or black. When transparent and finely colored, spinel is used as a gem. The red gem material, which is most common, is known as ruby spinel. In fact, when ruby lacks outward crystal form it greatly resembles red spinel, and the two minerals have been confused. The famous stone "Black Prince's Ruby" was determined to be spinel long after it was set in the British crown.

Spinel is a common *metamorphic mineral found in crystalline limestones and gneisses. It is also an accessory mineral in some dark igneous rocks. Because it is chemically inert with high hardness and high specific gravity, it accumulates in *placers. Ruby spinels are found in this way in the gem gravels of Ceylon, Thailand, Burma and Madagascar. See Mineral Properties. (C.S.H.)

Spinel Group. A group of closely related mineral oxides with similar crystal structure, crystal form and analogous chemical compositions. The most important members of the group are: *spinel, $MgAl_2O_4$; *magnetite, $FeFe_2O_4$; *franklinite, $ZnFe_2O_4$; and *chromite, $FeCr_2O_4$. The general formula has been written AB_2O_4, where A can be magnesium, iron, zinc or manganese and B can be aluminum, iron or chromium. There is usually considerable substitution of the A elements for each other and of the B elements for each other. Thus the chemical composition of each mineral as named is subject to considerable variation. For example, in spinel itself iron, zinc and manganese may substitute for magnesium (Mg), and iron and chromium may substitute for aluminum (Al). The octahedron is the dominant crystal form of all members of the group. (C.S.H.)

Spit. A narrow beach consisting of a ridge of sand or gravel projecting into open water from a mainland or an island. Spits range from tens of meters to more than 30 km in length. They are formed, like other beaches, by the activity of waves and currents. Many spits are attached to marine cliffs; they are generally composed of coarse material near the cliffs but progressively finer material farther out. This is because shore currents transport finer material more rapidly than coarse, and because coarse material is reduced by *abrasion during transportation. Spits are frequently found on irregular coastlines marked by promontories, bays and islands. In America they are common along the New England, New York and New Jersey coasts and in England along the Norfolk and Suffolk coasts. (R.L.N.)

Spodumene. A mineral, a lithium aluminum silicate, found in *pegmatites associated with other lithium minerals. It is a major ore mineral of lithium, an element used in ceramics, storage batteries and welding flux. But the major use is as an additive to help grease retain its lubricating properties over a wide range of temperatures. Although spodumene is a comparatively rare mineral, in some pegmatites it is found in very large crystals. The largest reported, measuring over 12 m in length and weighing many tons, came from the Etta Mine in the Black Hills of South Dakota. Spodumene is usually white or gray, but pink, yellow and green transparent crystals have been found. The pink variety, kunzite, and the green variety, hiddenite, are cut into beautiful gemstones. Kunzite, the more common of the gem varieties, is found at Pala, Calif., and at various

Spheroidal Weathering
1 Called Devils Marbles, these boulders
near Tennants Creek, Australia, were
once cubic blocks that were slowly
rounded by solutions penetrating
inward at their corners and edges.

Spit
2 This recurved sand barrier east of
Puerto Penasco, Sonora, Mexico, was
formed by sediments deposited by
longshore currents.

Spring

1 Groundwater emerging at the surface is commonly found in three types of locations: (a) *in fractures in massive rock such as granite*, (b) *in cavernous rock such as limestone*, (c) *at the upper surface of an impermeable layer of rock.*

2 *The Al Khorj Oasis is an important source of fresh water in the arid Saudi Arabian desert.*

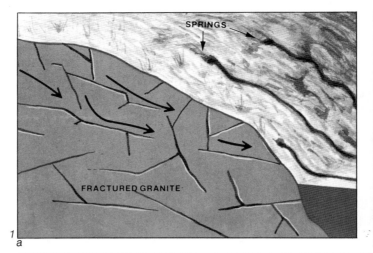

SPRINGS

FRACTURED GRANITE

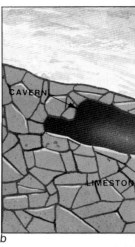

CAVERN

LIMESTONE

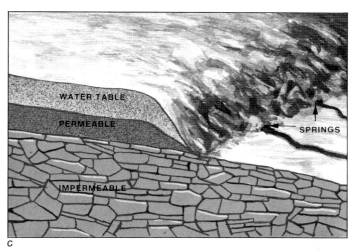

c

localities in Brazil and Madagascar. Hiddenite has been found at Stony Point, N.C. See Mineral Properties. (C.S.H.)

Spring. A natural opening at the earth's surface out of which water issues, ranging in volume from a trickle to millions of liters per day. Springs are usually differentiated from seeps by a more abundant and concentrated flow of water. They result from a variety of subsurface conditions. If a permeable rock is underlain by an impervious rock that crops out lower down on a slope, a spring may occur at the contact. Rain falling on the permeable rock moves downward to the impermeable rock, then moves laterally and is discharged as a spring where the contact intersects the slope.

Large springs occur in *lava flows and *limestone. Lava flows contain tunnels, cracks and other features that make them permeable, and limestone often contains large connected openings that sometimes reach the surface and may be the source of such springs. One of the largest springs in the United States, Silver Springs in Florida, which discharges about 2 billion liters daily, issues from limestone.

Springs may issue from *joints in *bedrock that reach the surface of the ground low down on slopes if they are fed with rainwater higher up. Thus Furnace Creek in Death Valley, Calif., one of the driest places in the United States, has a plentiful supply of water because there is such a joint system in a nearby mountain that receives considerable rainfall.

Ancient temples and cities were sometimes located near springs, and during pioneer days in the United States, homes and forts were often built near them. (R.L.N.)

Spur, Truncated. See Truncated Spur.

Squeeze-up. A bulbous, linear or wedge-shape feature formed when viscous lava is extruded or thrust onto the surface of a *lava flow through cracks in the lava crust. Bulbous squeeze-ups are usually not more than 2 m (5 feet) in diameter and 60 cm (2 feet) in height. Linear squeeze-ups may be as much as 6 m (20 feet) long and 30 cm (1 foot) high. Wedge-shape squeeze-ups are thrust up through the medial cracks of some pressure ridges, widening as they push upward. The largest are approximately 9 m (30 feet) long, 3 m (10 feet) high, 1 m (4 ft) thick at the bottom and about 2 cm wide at the top. (R.L.N.)

Stability. In meteorology, usually the equilibrium state of the atmosphere with respect to vertical motions of the air. The atmosphere is statically stable if a volume

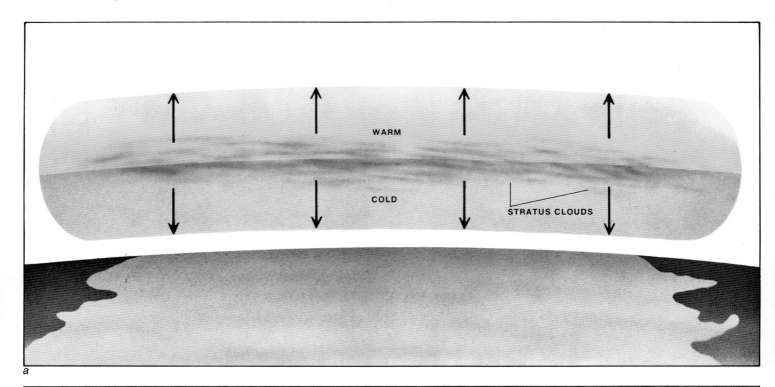

a

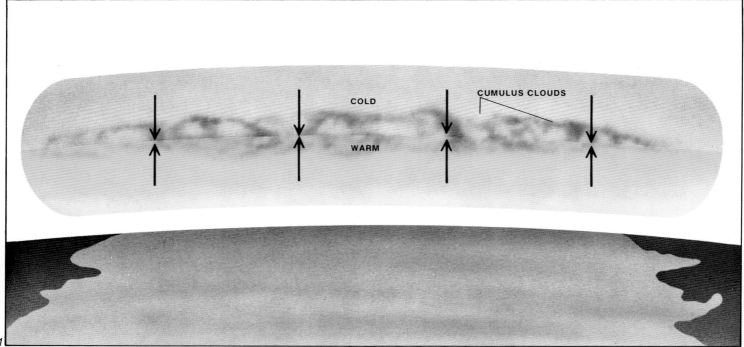

1
b

Stability

1 *An air mass moving over a surface
colder than itself (a) is thermody-
namically stable because the cooled,
heavy lower surface remains at the
bottom. Clouds that form under these
conditions are of the stratus type. When
cold air covers a warmer air mass (b)
the light, warm air rises through the
cold, becoming turbulent and unstable.
Clouds that form are of the cumulus
type.*

Stack

2 *Once connected to the land, this stack
is the wave-eroded remnant of a
collapsed arch off Flowerpot Island,
Tobermory, Ontario, Canada.*

of air, on being displaced upward or downward, is subjected to a buoyant force acting in the opposite direction. For stability, rising air volumes must be colder than the surrounding atmosphere; their greater density causes them to sink again. Sinking air volumes become warmer than their surroundings in a stable atmosphere and are forced to rise. The atmosphere is statically unstable if rising volumes remain warmer and sinking volumes remain colder than their environment. It is in neutral equilibrium when a volume of air displaced vertically undergoes no acceleration from buoyant forces.

The stability of the atmosphere depends upon the environmental temperature lapse rate, that is, upon the rate of decrease of temperature with increasing elevation prevailing at a given time and place. If the air is dry (unsaturated with water vapor) and remains so, the atmosphere is stable when the environmental lapse rate is 9.8°C per km; it is unstable if the lapse rate is greater; and it is in neutral equilibrium if it is exactly 9.8°C per km, the dry-adiabatic lapse rate. The reason is that volumes of unsaturated air displaced upward in the atmosphere are cooled adiabatically at this rate as they expand under decreasing pressure. (An *adiabatic process is one in which no heat or mass is exchanged across the boundaries of the air parcel.) Sinking volumes of unsaturated air are both compressed and warmed at the same rate.

When the air is saturated, rising air volumes cool at a rate less than the dry-adiabatic rate because the latent heat of condensation is added to the air. Similarly, sinking air warms at a lesser rate than the dry-adiabatic rate because the warming due to compression is offset by cooling due to evaporation. For saturated air, the atmosphere is stable if its lapse rate is less than the moist adiabatic lapse rate.

The stability of the atmosphere often determines whether or not cumulus clouds occur on a summer afternoon. When the earth's surface is heated by the sun, the air just above the surface is warmed and begins to rise. If the atmosphere is relatively moist and unstable, the rising air remains warmer than its environment and continues to rise to great heights, perhaps until it reaches the base of the stratosphere. In the stable case, the rising air soon becomes colder than its environment, even though condensation occurs, and is forced down until its temperature is that of the surrounding atmosphere. (M.F.H.)

Stack. A small island of *bedrock a short distance offshore and projecting above a *wave-cut platform. It is an erosional remnant, usually composed of sedimentary

1

Stalactite and Stalagmite

1 *Water laden with calcium carbonate constantly dripping off the ceiling of this cavern eventually deposited travertine "icicles," or stalactites, while water that dripped to the floor built up blunted features called stalagmites. Columns are formed when stalactites and stalagmites join.*

Staurolite

2 *The cruciform intergrowth of two crystals shown here is a distinctive form of this metamorphic mineral.*

3 *Two types of interpenetrating twin crystals are shown here along with a simple crystal.*

rock, detached from a headland undergoing marine erosion. Although surrounded by water at high tide, it may be high and dry at low tide. A stack with a tunnel cut through it is sometimes called a perforated stack. During the marine erosion of a coast, the more resistant parts will generally form headlands. When two caves on opposite sides of a headland unite, a natural arch or bridge results; when the seaward part of the headland remains after collapse, a stack is formed. With the passage of time, marine erosion reduces the stack until all that remains is a wave-cut platform.

Stacks are found off coasts around the world. Along the English coast some stacks are as much as 150 m high. Stacks near John O'Groats, the Old Man of Hoy, Orkney Islands and the Needles off the Isle of Wight are well-known British examples. American examples are found at Seal Rocks and Cannon Beach, Oreg., near San Simeon, Calif., and at Arches State Park, Calif. (R.L.N.)

Stage. The smallest *time-stratigraphic unit in general usage. A stage consists of the strata deposited during an *age of geologic time that typically contain distinctive fossil assemblages and that may be recognized and correlated in widely separated areas. Two or more stages may be combined to form a *series.

Stalactite and Stalagmite. Stalactites are icicle-like forms usually composed of calcium carbonate and found hanging from the ceilings of limestone caves. They range in length from a fraction of a centimeter to many meters. They are chemical precipitates deposited from subsurface water that is supersaturated with calcium bicarbonate and oozes from cracks and joints in the ceilings of caves. The calcium carbonate (precipitated by evaporation and/or loss of dissolved carbon dioxide) is usually calcite but may be aragonite, depending upon the temperature of the cave when deposition occurred.

Stalagmites, usually blunter than stalactites, grow from the floor upward and result from the water dripping to the cave floor. When a stalactite and a stalagmite join, a column is formed. Stalactites can be longer than stalagmites because their length is limited by the tensile strength of calcite, whereas the length of stalagmites is limited by the crushing strength of calcite, which is much greater. It is stalactites and stalagmites — whether white, translucent or, due to impurities, shades of gray, yellow or brown — that make caves spectacularly beautiful.

Conditions favoring the formation of stalactites composed of calcium carbonate are: (1) limestone or marble above a cave, (2) narrow joints and fissures in which subsurface water can only move slowly, (3) water containing dissolved carbon dioxide seeping into the cave, (4) low humidity, facilitating precipitation of calcium carbonate by evaporation in addition to precipitation caused by loss of dissolved carbon dioxide. Rates of growth vary from one part of a cave to another and from season to season. The rates also vary for large stalactites and stalagmites over longer intervals because of climatic changes during their growth. Thus it is not possible to determine accurately the age of a stalactite or a stalagmite by its weight and current rate of growth. The maximum age of a stalagmite can be determined, however, if it has grown on material that can be dated by carbon-14 measurements.

Stalactites may also form in tunnels, subways and under bridges, where water issues from the mortar in the masonry; and in mines and under overhanging cliffs. Stalactites may be composed of pyrite, gypsum, halite, ice, lava and other substances but calcite is probably more common than all the others combined.

Material chemically precipitated in a cavern from dripping water is called dripstone, and material chemically precipitated by water flowing down the walls or along the floor of a cavern is known as flowstone. See Cave; Speleology. (R.L.N.)

Staurolite. A mineral, hydrous iron aluminum silicate, found as a product of *metamorphism in *schists, associated with garnet and kyanite. It is frequently in well-formed single crystals that clearly display *orthorhombic symmetry. Even more common are cruciform *twin crystals composed of two interpenetrating individuals. There are two types of twins; in one the two individuals cross at about 60°; in the other they cross at 90°. The right-angle crosses are sold as amulets under the name fairy stone. This is the only use of staurolite aside from the occasional transparent stone from Brazil that is cut as a gem. See Mineral Properties. (C.S.H.)

Steinkerns. Internal molds. See Casts and Molds.

Steptoe. A hill or mountain surrounded by and projecting above one or more *lava flows. It indicates the kind of rock buried by the lava and tells something about the relief of the pre-lava topography. The name comes from Steptoe Butte, near Colfax, Wash., which is composed of quartzite and projects above the Columbia Plateau lava flows.

2

3

Stibnite

1 Swordlike crystals of this sulfide of antimony up to 45 cm (18 inches) in length have been found in the Japanese antimony veins from which this specimen came.

Stilbite

2 Bundles of parallel crystals reminiscent of sheaves of wheat are a growth form characteristic of this zeolite mineral.

Stratification

3 The layers of this sedimentary rock in Greenland were once unconsolidated deposits of sand, gravel and mud that were laid down in a series of individual beds, one on top of the other.

Stibnite. A mineral, an antimony sulfide, the chief ore of antimony. It most often is bladed, granular or in massive aggregates, but occasionally is found in long, slender, black crystals with a brilliant metallic luster. Stibnite is characterized by perfect cleavage, low hardness (2) and easy fusibility in a match flame. It is found in low-temperature hydrothermal veins and replacement deposits associated with other antimony minerals, cinnabar, barite, realgar, orpiment and gold. See Mineral Properties.

Stilbite. A mineral, hydrous calcium sodium aluminum silicate, one of the *zeolites. It is found most characteristically in sheaf-like aggregates of thin tabular crystals. It is a secondary mineral usually found in cavities in basalts associated with other zeolites and with datolite, calcite, apophyllite and prehnite. Notable localities are in Iceland, India, Nova Scotia, and in the United States in northeastern New Jersey. See Mineral Properties.

Stock. A small igneous intrusion generally covering an area of less than 100 square kilometers (40 square miles). Stocks resemble *batholiths in that they increase in area with depth. Stocks usually cut across layered structures inherent in the intruded rock but it is not always apparent that they do so. In the Henry Mountains of southern Utah, Mount Ellsworth has formed from the eroded remnant of a stock. The Gorongoza mass northeast of Beira, Mozambique, is a granitic stock that rises some 1220 m (4000 feet) above the surrounding territory.

Stoping. See Magmatic Stoping.

Strain. Deformation or movement that occurs in rock subjected to forces greater than the rock's strength. Strain can be measured as the percentage of change in shape or volume of the object undergoing deformation.

Stratification. A layered or bedded arrangement of the constituent particles of a rock. It is the most distinctive feature of *sedimentary rocks, but layering can also be seen in ancient *lava flows and certain *metamorphic rocks. Stratification develops in a series of sedimentary rocks as sediments are deposited in individual strata, one on top of another. The top and bottom of each bed is marked by a more or less distinct surface called a bedding plane, which marks the termination of one deposit and the beginning of another. The layering typically results from variation in the supply of sediment or in the velocities of the currents (wind, water or ice) depositing it.

Stratigraphic Unit. Rock strata grouped together for the purpose of mapping, description, correlation or study. It may be established on the basis of fossil content (or biostratigraphic units), age (or time-stratigraphic units) or some other distinctive property. It should not be confused with a geologic time unit such as an era or period. See Formation, Geologic.

Stratigraphy. The branch of geology that deals with the description and interpretation of rocks as revealed by their age, sequence, structure, physical properties, distribution and mode of origin. It is based mainly on the study of sedimentary rocks, but may also include igneous and metamorphic rocks. It makes possible a stratigraphic classification whereby rocks are systematically organized into *stratigraphic units defined on the basis of their more distinctive physical, chemical and structural characteristics. These include *time units (eras, periods, epochs, ages) and *time-stratigraphic units (erathems, systems, series and stages), *rock-stratigraphic units (groups, formations, members, beds) and *biostratigraphic units (assemblage zone, biozone). In order to maintain a uniform and systematic procedure of stratigraphic classification, specific rules on nomenclature and classification, established by the American Commission on Stratigraphic Nomenclature, are now widely used by stratigraphers. (W.H.M.)

Stratosphere. The layer above the troposphere, extending upward to the base of the mesosphere at an altitude of 50 km (30 miles). The height of the tropopause, the boundary between the troposphere and the stratosphere, varies considerably with latitude, season and weather conditions; it averages about 11 km (7 miles). The temperature is typically constant from the base of the stratosphere to a height of 32 km (20 miles), then it increases to about -2°C (28°F) at 47 km (30 miles). The warmth of the upper stratosphere is the result of strong absorption of ultraviolet radiation near the top of the ozone layer. Since the temperature in the stratosphere remains constant or increases upward, this layer is marked by great stability.

Because airplanes flying in the stratosphere release water vapor and carbon dioxide, plans for a proposed fleet of supersonic aircraft have focused attention on the possibility of contaminating the stratosphere. Gases and small particles remain for considerably longer periods in the stratosphere than in the troposphere. The amount of water vapor naturally present in the stratosphere, although believed to be quite small, is not known with certainty.

Many of the questions remain unanswered because of the paucity of observational data. (M.F.H.)

Stratum. A definite layer of sedimentary rock consisting of approximately the same type of rock material (consolidated or unconsolidated) that has been spread out on the earth's surface. It may be of any thickness and possesses distinct unifying characteristics that distinguish it from adjacent strata. Although certain igneous and metamorphic rocks may exhibit "layering," the term is more properly limited to sedimentary material.

Streak. The color of the powder of a mineral. If the mineral is not too hard (less than 7 on Mohs scale) the streak is obtained by rubbing the mineral on a piece of unglazed porcelain, called a streak plate. Although many minerals have a wide range of color, the streak is usually constant and is thus useful in mineral determination.

Stream (or River). A natural, channelized flow of water. A stream may be classified in several ways. One way is based on how often water flows in it, which depends on the relation of the stream bed to the *water table. If the bed lies below the water table, the stream carries water all the time and is said to be a *perennial stream. A river with its bed lying above the water table carries discharge only during and immediately after rainfall and is termed an *ephemeral stream. A river in which the water table fluctuates so that part of the time it lies above the bed and at other times below it is called an *intermittent stream.

A stream may also be classified by its origin. A stream that flows down the original slope of the land surface is a *consequent stream. Where the original surface has been eroded away and the stream flows down a more recent, lower surface, it is called a *resequent stream. One that flows in a direction opposite to the original slope is an *obsequent stream. A river that follows some weakness in the land such as a *fault, *joint or easily eroded bed is a *subsequent stream. A stream with a direction that has no discernible cause is an *insequent stream.

An *antecedent stream is one that maintains its course even through mountains; such a river cut its path before the mountain appeared. On the other hand, a *superimposed stream is one whose channel is eroded through a geologic structure or rock that was present but was hidden by overlying strata. Such a river established its course on the uppermost layers and maintained the same channel on encountering the hidden structure beneath. It is often hard to distinguish an antecedent river from one that is superimposed.

A stream may also be described according to its channel. One with a sinuous channel is *meandering; one with many divided channelways is a *braided stream. A *beheaded stream is one whose tributaries have been captured by another river. A *misfit stream is one that flows in a valley that is too large or too small for its discharge. If a river seems too small for its valley it is underfit; if it is too large it is overfit. The ideal is a *graded stream—one so well adjusted to its discharge and sediment load that it neither erodes its channel nor deposits in it. (M.M.)

Stream Capture. See Piracy.

Stream Erosion, Cycle of. A theoretical sequence of stream development proposed by W.M. Davis, which states that rivers evolve through successive stages called youth, maturity and old age, each having its own distinctive characteristics. Youthful rivers have narrow, steep-walled, V-shape valleys and steep gradients with waterfalls and rapids. Their ability to erode leads to rapid headward erosion. They have no *floodplains since the high velocities at which they flow give them excess energy to cut down rapidly into their channels. With time, downcutting results in a smooth longitudinal slope with a decreasing gradient. Falls and rapids disappear; the smooth, low gradient provides less energy; so instead of cutting down, the river widens its valley. Since its ability to work has decreased, it deposits material in the widened channel. When it has built up a wide floodplain on which it winds back and forth, so that the width of the valley floor equals the width of the *meander belt, it is said to be mature. The river reaches old age when it is meandering on a very wide floodplain several times the width of the meander belt, with *oxbow lakes, *sloughs and *levees. Theoretically, it then has a very low gradient, which makes it incapable of eroding or transporting material. All tributaries are mature and graded. However, many geologists today believe that this theory does not recognize the dynamic equilibrium between the energy of a river and the resistance of the rock over which it is flowing as factors contributing to the shape of its valley. (M.M.)

Stream Frequency. The number of streams per unit area, or the ratio of the total number of stream channels over the area of the *drainage basin. Stream frequency depends upon climate, vegetation and rock type, and is a measure of the dissection of an area or its drainage texture.

Stream Regimen. The habit of a river in regard to discharge, velocity, channel characteristics and ability to transport a load. Streams that have reached an equilibrium or steady state (*graded stream) are said to be in regimen.

Stream (or Fluvial) Terrace. A flat surface lying above a present *floodplain and separated from it by a low cliff, formed when *rejuvenation enables the stream to cut into its former floodplain. Terraces are generally of *alluvium but may be cut in *bedrock (*strath terrace). Terraces may be paired, that is, may exist at the same level on both sides of the stream, or may be unpaired. Paired terraces result from rapid incisement, whereas single terraces reflect a slower, continuous downcutting while lateral erosion is still taking place. Paired terraces may be formed by alternating climatic changes where dry conditions lead to *aggradation. Increase in rainfall leads to growth of vegetation, reduction of erosion of valley slopes and degradation and entrenchment of the channel. Unpaired terraces imply slow, continuous change and are characterized by terraces with unequal elevations and gradual *slip-off slopes. Cut-and-fill terraces are produced by repeated episodes of deposit and incisement, with lateral erosion. They are caused by sudden changes in climate or runoff regimen. Thus they evidence a change in a stream's climatic, tectonic, eustatic or hydrologic environment.

Stream Tin. The tin oxide *cassiterite, found in the form of rolled pebbles in *placer deposits.

Stress. The force per unit area that acts on or within a body. Stress may be produced by forces within a solid that tend to push individual components closer together (*compression) or pull them apart (*tension). Stress may also result from forces that act as a couple (two equal but oppositely directed forces acting within a shared plane but not along the same line). Stress produced by a couple tends to slide or shear one particle past another within a solid. All structural features found on or in the earth are the result of stress. Such features as *faults or *folds are produced by stress that has exceeded the strength of the rock being acted upon, causing it to break or buckle. See Compressive Stress; Shear Stress; Tensile Stress. (J.W.S.)

Strike. The direction or trend of any layered or planar feature on the surface of the earth, for example, a bedding or fault plane. The strike is represented by a line

Strike
The direction of the outcropping edge perpendicular to the dip of these tilted beds in the Virgin Mountains, Ariz., constitutes its strike.

Submarine Canyon

1 The floors of submarine canyons along the continental shelf are often thick with sediments that have fallen from the shelf above. Here, curved, rodlike sea whips, spherical sea urchins and conical piles made by burrowing worms cover a section of the shelf floor near Cutaline Island, off the coast of California.

2 A section of an undersea cliff, along the east edge of the Mexican Basin in the Gulf of Mexico, was photographed at a depth of 2200 m (7200 feet). The flowerlike form in the foreground is probably a stalked crinoid, or "sea lily."

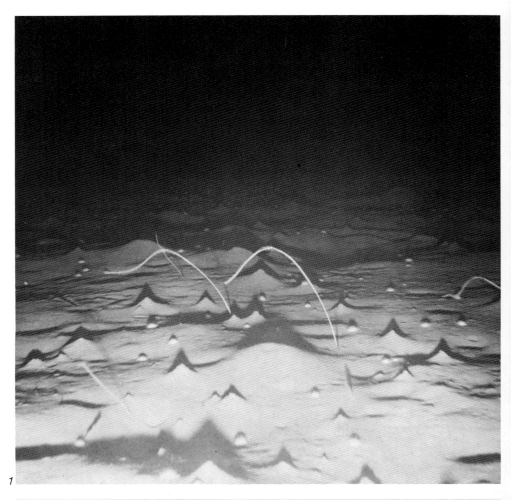

produced by the intersection of a horizontal plane with the surface of the feature.

Strike Joint. A *joint that strikes parallel to a planar feature.

Strontianite. A mineral, strontium carbonate, a member of the aragonite group. It is similar in crystal structure and crystal form to *aragonite, and like the other members of the group will effervesce in cold hydrochloric acid. Strontianite is a low-temperature hydrothermal vein mineral common in limestone associated with barite, celestite and calcite. See Mineral Properties.

Subduction Zone. See Benioff Zone.

Submarine Canyon. A narrow, deep, steep-sided submarine valley. Such canyons occur on all *continental margins. They may cut a few thousand meters into the continental slope and may extend landward into the *continental shelf and seaward across the upper *continental rise. A number of submarine canyons, such as the Hudson and Delaware canyons on the Atlantic coast of the United States, are evidently extensions of terrestrial river systems. The origin of these canyons has been debated for decades. Discarded hypotheses include formation by *tsunamis, *faulting, submarine springs and *submarine slides; most marine geologists now accept the theory that the canyons were formed by a combination of river-cut valleys and *turbidity currents in periods when the sea level was low, particularly during the Ice Ages. At that time most of the continental shelf was exposed and was traversed by rivers carrying great quantities of sediment. This sediment, it is believed, cascaded down the continental slope in the form of turbidity currents, which cut the canyons. (R.A.D.)

Submarine Landslide. An underwater movement of a mass of earth; it is basically the same as a terrestrial landslide. Most often it is generated by an earthquake, and the slide in turn causes a seismic sea wave or *tsunami. In some situations a combination of a terrestrial landslide and a submarine landslide may occur, as along the mountainous coasts of California or on volcanic islands. A large surge wave may be associated with these slides and may damage a coastal area.

Submarine Valley. An elongated, trough-shaped feature of the ocean floor. Among the various types of submarine valleys, submarine canyons are the best known and most common. Valleys that occur on the *continental shelf are called shelf channels. They are extensions of rivers or land valleys that empty into the sea and most of them eventually align themselves with submarine canyons. They are only tens of meters deep and are mainly perpendicular to the adjacent coast. Delta-front troughs and slope gullies are associated with river *deltas. The former are rather broad, U-shaped valleys extending from a delta across the *continental margin. Slope gullies, the smallest of the submarine valleys, are formed by a slumping action that takes place on the unstable slopes of rapidly expanding deltas. Glacial troughs are broad and a few hundred meters deep; they were carved out of the continental shelf by advancing glaciers during the Pleistocene. Fault valleys are narrow, deep and relatively straight; as the name indicates, they are caused by *faulting activity in the earth's crust. (R.A.D.)

Subsequent Stream. A stream whose direction of flow is determined by erosion along some line of weakness such as rock type, *fault, *joint or rock *foliation. Streams in tilted or folded rock tend to be subsequent, eroding along the weaker layers of shale and (in humid regions) limestone, leaving the resistant layers as ridges. Such streams are well adjusted to the underlying rock and structure.

Subsurface Water (or Underground Water, Subterranean Water). Any water beneath the earth's surface. It is held in pores and spaces between grains and fragments in the *regolith, and in cracks, channels, chambers and other openings in *bedrock. It extends downward thousands of meters to where the pressure is so great as to close all openings.

Subsurface water is of great economic importance in many areas of the world. Surveys show that in the United States alone about 230 billion liters (61 billion gallons) a day were drawn from the ground in 1965. This was used for industrial, irrigation, domestic and other purposes, and represented about 20 percent of all the water used in the United States.

Subsurface water is also of great geologic importance. As an erosion agent it dissolves minerals, transports them in solution, deposits some of them in the rocks through which it moves, and carries the rest to rivers, which take them to the sea. In this way the seas have slowly increased in salinity throughout geologic time. In the seas certain dissolved minerals are used by plants and animals in the formation of hard substances, and accumulations of such material eventually form marine rocks after the plants and animals die. Thus subsurface waters help to form younger rocks from older rocks. The weathering of the rocks is partly due to the infiltration of water and dissolved gases that come from the surface. The deposition of mineral matter from subsurface water in spaces in unconsolidated sediments is chiefly responsible for the hardening of sediments into *sedimentary rocks. Limestone caves and their *stalactites and stalagmites, and some ore bodies, also owe their origin to the movement of subsurface water.

Precipitation is overwhelmingly the most important source of subsurface water. The precipitation that falls on the earth's surface evaporates, runs off the land into streams and oceans, or filters downward into the rocks, replenishing the subsurface water. Some of this underground water surfaces in the form of *springs.

There are three kinds of subsurface water: meteoric, connate and juvenile. Water that originates from rain, melted snow and ice is called meteoric (from the Greek word meaning "high in the air"). Water in the spaces between the grains and fragments of sediments at the time of their deposition is called connate. Some of it remains entrapped even after the sediments have been hardened into sedimentary rocks. Although of geologic interest, it is quantitatively far less important than meteoric water. Connate water in marine deposits is too saline for agricultural or domestic use. Salt water brought up in oil wells is generally thought to be connate. The salt licks in Kentucky that were so important to Indians, pioneers and animals in early America are not due to buried salt beds but to the movement to the earth's surface of connate waters with their dissolved salts. Water expelled from the magma as it moves upward and while it solidifies is called juvenile. Some of it is added to the atmosphere and hydrosphere but some remains trapped in the rocks. Quantitatively it is of little importance at the present time. See Water Table. (R.L.N.)

Sulfate. A compound containing the sulfate radical SO_4. There are many sulfate minerals, but only a few are common. These are barite, celestite, anglesite and anhydrite, which are anhydrous; and gypsum, alunite, chalcanthite and epsomite, which are hydrous.

Sulfide. One of a group of minerals in which sulfur is combined with one or more metals. Sulfides are unoxidized — that is, contain no oxygen — in contrast to the *sulfates, which are oxidized compounds. Sulfides are an important mineral group, for among them are many of the ores of the common metals such as copper, lead, zinc, nickel, mercury, molybdenum and cobalt.

Sulfide enrichment. The enrichment of a sulfide ore body by the deposition of secondary sulfides or by the replacement of one sulfide by another that has higher content of a valuable metal. An example is the replacement of pyrite or chalcopyrite by chalcocite. See Secondary Enrichment.

Sulfur. A chemical element that occurs as a mineral in the native state. Sulfur occurs abundantly in the minerals of the *sulfide and *sulfate groups, but native sulfur throughout most of historic time has been the principal commercial source of the element. It has been called brimstone, meaning "the stone that burns," because it burns with a blue flame to sulfur dioxide, a gas with a suffocating odor. This property and its yellow color distinguish it from other minerals.

Sulfur is often found at or near the crater rims of volcanoes, where it forms from gases given off by fumaroles. Such volcanic sulfur is an important commercial source in Mexico, Hawaii, Japan and Argentina. In Chile it is mined from a volcanic peak at an elevation of 5800 m. Native sulfur is also formed from sulfates by the action of sulfur-forming bacteria. This is believed to be the origin of the sulfur found in Tertiary limestones associated with anhydrite and gypsum. The most notable deposits of this type are near Girgenti, Sicily, where mining has been done for centuries, producing the finest specimens of crystallized sulfur. Of similar origin is the sulfur found in the cap rock of salt domes in Texas and Louisiana. Here, associated with gypsum, anhydrite and calcite, the sulfur lies tens of meters below the surface and is extracted by the *Frasch process. These deposits, some of which lie several kilometers offshore in the Gulf of Mexico, have been the world's major source of sulfur since the beginning of the 20th century. (C.S.H.)

Sunstone (or Aventurine Oligoclase). A variety of *oligoclase feldspar containing inclusions of hematite that give a golden shimmer to the mineral.

Supergene. A term applied to ore minerals formed by descending solutions, in contrast to *hypogene which refers to minerals deposited from ascending solutions.

Superimposed (or Superposed) Stream. A river that cuts across geologic structures or resistant rock and that originally flowed on sediments covering the underlying structures. The river established its course

Syenite

1 An intergrowth of abundant light-colored feldspar grains with a smaller proportion of dark elongated amphibole grains gives this igneous rock its characteristic appearance.

Sulfur

2 In this specimen from Sicily, the transparent crystals of sulfur plainly display their orthorhombic symmetry.

Superimposed Stream

3 In (a) a stream flows over sediments that have covered some underlying structure. In (b) the stream has eroded away the top overlying sediments and reached the underlying folds. In (c) the stream is able to keep its channel as it erodes through the folded strata far below its original bed even though they may be resistant, ridge-forming sandstone.

on the overlying rock and then, when it cut through to the rock beneath, kept the same course. The Green River cut through a thick Tertiary sediment cover to superimpose its course through Utah's Uinta Mountains. There are many similar examples in the Rocky Mountains where the superimposed stretch forms picturesque canyons. See Antecedent Stream.

Supratidal Zone. The most landward of marine environments. A narrow area immediately above the *intertidal zone, it is covered by marine water during storms only, or it may receive spray from the intertidal zone. It is inhabited by organisms such as certain blue-green algae that can tolerate long periods of dryness.

Surveyor. A series of unmanned spacecraft that landed on the moon. Each Surveyor carried a television camera and a small electrically operated steam shovel, both of which could be remotely controlled from the earth. The principal scientific finding was that the lunar surface is covered with craters down to the resolution limit of the Surveyor television camera, that is, less than 1 cm. It was concluded that most if not all of these craters were of impact origin. The Alpha-particle-scattering experiment on the last three missions provided the first chemical analyses of lunar material, which basically agreed with subsequent analyses of returned samples. (C.H.S. and J.L.W.)

Suspended Load. Solid debris carried by a stream and supported by the water; it is composed of material finer than *bed load. Particles transported as suspended load have settling velocities that are less than the erosion velocity of the river. The amount of load carried in suspension depends upon the size and weight of the grains, the settling velocity of water, *specific gravity of water and turbulence. The distribution of the suspended load varies with distance above the bed, with greatest concentration near the bottom. Generally there are more sand grains in suspension near the bed, but concentration of silt does not seem to vary vertically. There is a high correlation of suspended load to discharge. At high flows, streams carry a high suspended load. (M.M.)

Syenite. A *plutonic rock composed essentially of microcline or orthoclase *feldspar. Usually hornblende and/or biotite is present, as well as a small amount of plagioclase. Either quartz or nepheline may be present in amounts not exceeding 5 percent. Common accessory minerals are apatite, sphene, zircon and magnetite.

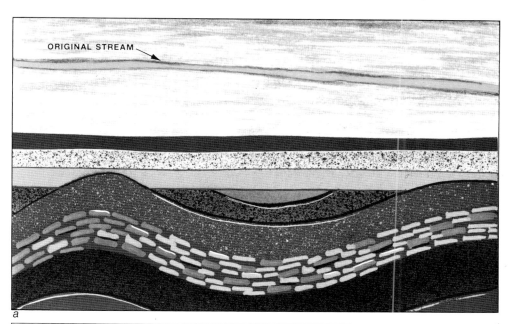

a ORIGINAL STREAM

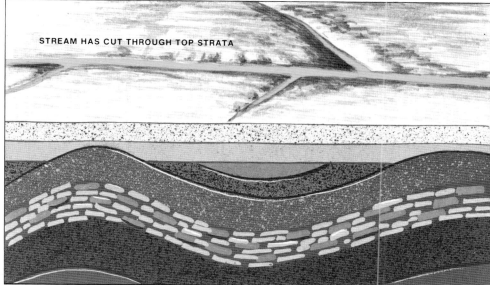

b STREAM HAS CUT THROUGH TOP STRATA

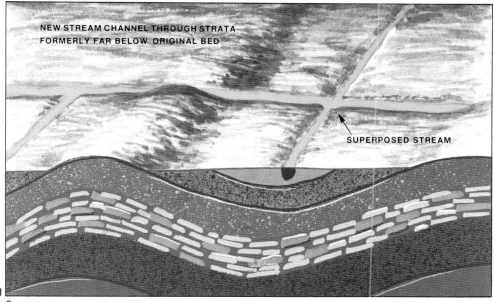

c NEW STREAM CHANNEL THROUGH STRATA FORMERLY FAR BELOW ORIGINAL BED — SUPERPOSED STREAM

Syncline
A downward-plunging fold near Barstow, Calif., is faulted on the left and overlain in the center by Quaternary gravels.

Sylvanite. A mineral, a gold and silver telluride and an ore of these metals. At Kalgoorlie, West Australia, and at Cripple Creek, Colo., sylvanite, associated with the gold telluride, has been an important ore mineral. Aside from these localities it is a rare mineral. See Mineral Properties.

Syncline. A downfolded structure within the outer layer of the earth's *crust. In the simplest synclines the layered rocks forming the sides (limbs) are inclined upward away from the center of curvature of the syncline. In traversing a synclinal structure from one limb to the other on the surface of the ground, the older rocks occur first, with the rocks becoming progressively younger nearer the central portion of the syncline.

Synclinorium. A large downfolded structure at least several kilometers in breadth. The sides or limbs of such a *fold are not smooth like those of a simple fold but are made up of a series of lesser folds. To the west of and parallel with the Appalachian Mountains, the crustal rocks are deformed into this type of fold structure and are referred to as the Allegheny Synclinorium.

System. A *time-stratigraphic unit smaller than an *erathem. It designates the rocks formed during a *period of geologic time. Like the periods, most systems derive their names from the regions where they were first studied and described, although a few are named for some conspicuous feature associated with the rocks in their type of region. Thus the Cretaceous System (from the Latin *creta*, meaning "chalky") was so named because of the chalky nature of the Cretaceous strata along the English Channel between England and France.

Systematic Joint. A *joint that forms essentially parallel to another joint.

T

Taconite. A low-grade iron ore composed essentially of fine-grained quartz, iron silicates and iron oxides. It contains 25 – 30 percent iron and is the primary iron formation from which high-grade iron ore (containing 50 percent or more iron) has been derived. With the near exhaustion of high-grade ore, taconite is becoming increasingly important as a source of iron to feed the blast furnaces of the steel industry. It is being mined on a large scale in the United States and "upgraded" by separating the iron minerals from the waste silica. Over 50 million tons of nearly pure *hematite and *magnetite are thus being "manufactured" each year. Although the process is costly, the concentrate contains over 60 percent iron and is considered a reserve for the indefinite future. (C.S.H.)

Talc. A mineral, hydrated magnesium silicate, one of the softest of minerals, with a greasy or soapy feel. Talc has many uses in powdered form, the most familiar being talcum powder. The finely ground mineral is also used as an ingredient in paint, ceramics, rubber, insecticides and foundry facings. To a lesser extent it is used in its massive form, *soapstone, for laboratory table tops, electrical switchboards and sanitary appliances. It is a secondary mineral formed by the alteration of magnesium silicates such as olivine, pyroxenes and amphiboles. It is characteristic of low-grade *metamorphic rocks and a major constituent of talc schists; as massive soapstone it may make up nearly an entire rock mass. See Mineral Properties; Appendix 3. (C.S.H.)

Talus (or Scree). The accumulation of fallen fragments sloping away from the foot of the cliff or steep slope where it collected. The slope at which talus accumulates is the *angle of repose for the material(25° – 35°). Jointed rocks and *frost action favor this development. The fragments tend to be small when the joint system is closely spaced and large when it is widely spaced. The fragments commonly break when they land on the slope, and the larger generally roll farther down the slope than the smaller. A talus apron is formed when the material collects uniformly at the base of a long cliff, and a talus cone when it accumulates more rapidly at one point than another. Beneath high cliffs — as at Moraine Lake, Canada; Crater Lake, Ore.; and Glacier National Park — talus slopes may be hundreds of meters high, but only a few meters high beneath recently blasted road cuts. No vegetation occurs on a talus slope if the fragments are large or are accumulating rapidly, but it does spring up if the frag-

ments collect slowly, particularly if the talus-forming processes are no longer active. The consolidation of talus results in a talus *breccia. (R.L.N.)

Tantalite. A mineral, an iron, manganese and tantalum oxide. It is a major ore of the element *tantalum and to a lesser extent of niobium. Pure tantalite is rare, for usually some niobium is present and a series extends to *columbite, the niobium endmember. The specific gravity when pure is 7.9 but decreases with increasing amounts of niobium. Tantalite can be recognized by its high specific gravity, black color and one direction of good cleavage. Both tantalite and columbite occur in granites and related pegmatites associated with wolframite, cassiterite and microlite. Tantalum is a metal resistant to acid corrosion and is used in chemical equipment and in surgery for skull plates and for sutures. It is also used in electronic tubes and in tool steels. See Mineral Properties; Appendix 3. (C.S.H.)

Tarn. A small lake in a bedrock depression resulting from differential glacial erosion. Tarns are commonly found on the floors of the glacially formed hollows known as *cirques. A few tarns in a chain, connected by a stream and located in a glaciated valley, are called paternoster lakes because of the resemblance of the chain to rosary beads.

Tar Sand. See Energy Resources.

Tectonics. A branch of earth science dealing with the larger structural features found on or near the earth's surface. As in *structural geology, geoscientists working in the field of tectonics are interested in the classification and, more important, the mode of the origin of earth structures. See Plate Tectonics.

Tectosilicate. One of the group of silicate minerals in which all the oxygen ions of the silicon-oxygen tetrahedra are shared with other tetrahedra. The ration of silicon to oxygen is thus 1:2, as in quartz, SiO_2. The important *feldspar group as well as the *feldspathoid and *zeolite groups belong to the tectosilicates. In them some of the silicon is replaced by aluminum so that although all oxygens are shared by tetrahedra, the ratio of silica to oxygen is not 1:2, but the ratio of silicon plus aluminum to oxygen is 1:2. See Silicate Structure.

Tektite. A small rounded mass composed almost entirely of glass. The exact origin of tektites is unknown, but many authorities consider them meteoric.

Talus

1 Rocks broken off cliffs by frost action form cones at the base of steep ravines at Bow Lake, Alberta, Canada.

Tektites

2 These small, glassy objects, presumed to be of meteoric origin, have been found extensively in Australia, Java, the Philippines and Indochina. Some are more than 700,000 years old.

Telluride. A chemical compound in which tellurium is combined with one or more metals. The principal mineral tellurides are *calaverite and *sylvanite.

Temperature. In general, the degree of hotness or coldness of a substance as measured by one of several types of thermometers. The Kelvin scale (abbreviated "K"), commonly used in scientific work, is identical with the scale used on a gas thermometer utilizing a perfect gas. The zero point on the Kelvin scale, known as absolute zero, is, according to kinetic theory, the temperature at which the molecules of a gas have zero kinetic energy. For convenience, a *Kelvin*, the unit of temperature on the Kelvin scale, has the same size as the degree on the Celsius scale, and the freezing point of water (0°C) is 273.16°K.

Since the earth is spherical in shape, the distribution of incoming *solar radiation (called insolation) depends on the latitude. Thus more heat is received at low latitudes than at high latitudes. The resulting temperature contrast between the equatorial and polar latitudes supplies the energy for the general circulation of the atmosphere. Although the temperature distribution over the earth's surface is basically zonal because of the latitudinal variation of insolation, the zonal pattern is greatly modified by the character of the earth's surface. In winter the isotherms (lines connecting points of equal temperature) bend sharply toward the Equator over the land areas. In summer they bend toward the poles. The oceans tend to maintain an even temperature throughout the year. The land areas, on the other hand, show large temperature changes between seasons, warming up during the summer and becoming cold in the winter.

The seasonal differences in temperature between continents and ocean are explained by the different ways in which land and water respond to insolation. On land, the incoming radiation is absorbed by a very thin layer. The layer therefore heats rapidly, more rapidly than an equivalent layer of water because the specific heat (the amount of heat required to raise the temperature of one gram of a substance by 1°C) of soil is less than that of water. Water has a very high specific heat. The same amount of energy that causes the temperature of a gram of dry sandy soil to increase 4° or 5°C causes an increase of only 1°C in a gram of water. Moreover, the sun's rays penetrate far below the surface of the water, warming it less, but to depths as great as 100 m. Mixing of the water also transfers heat downward; and ocean currents transport heat energy toward the poles. The heat capacity (a measure of a material's ability to absorb heat) of the ocean is much greater than that of the land.

The strong temperature contrast between the continents and the oceans produces the seasonal circulations known as monsoons; these are strongest over southern and eastern Asia, but they also occur on other tropical coasts. (M.F.H.)

Tennantite. A mineral, the arsenic end-member of the *tetrahedrite-tennantite series. See Mineral Properties.

Tensile Stress. See Tension.

Tension. A force acting in a direction such that it would pull a solid body apart if it exceeded the breaking strength of the solid. The tension per unit area is called tensile stress. It is measured in pounds per square inch or in newtons per square centimeter. Rock on or very near the earth's surface is a brittle substance capable of being fractured and offering little resistance to breakage by tensional forces. Should the tensile stress exceed the strength of the solid acted upon, a fracture will develop perpendicular to the applied force. Many fractures in rock outcrops result from such tensional forces. Rock such as sandstone will be pulled apart by a tensile stress of 70 newtons per square centimeter. (J.W.S.)

Tephrite. A volcanic rock composed essentially of calcium-rich plagioclase but containing some nepheline or leucite; a feldspathoidal basalt.

Terrace, Elevated Marine. A terrace formed where a marine cliff and the wave-cut marine *terrace in front of it are raised above sea level. The uplift may be due to crustal movements, to a *eustatic change of sea level or both; but if a terrace changes significantly in elevation from point to point along a coast, the uplift must have been due to crustal movement. Whole flights of such terraces, like the steps of giant stairways, are found along some coasts. A series of thirteen terraces extends about 400 m above sea level on the Palos Verdes Hills in southern California. The uplift has taken place so recently that the marine shells on the platforms are identical with those living in the nearby ocean.

Particularly well-developed uplifted marine terraces, marking former sea levels, are found on the coasts of Oregon, Chile, Peru and New Zealand, at Gibraltar and on the islands of the Pacific. (R.L.N.)

Terrace, Wave-cut (or Wave-cut Platform or Bench). A subaqueous coastal land form resulting from the action of waves and/or longshore currents. Two

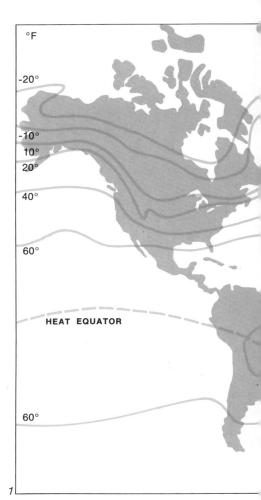

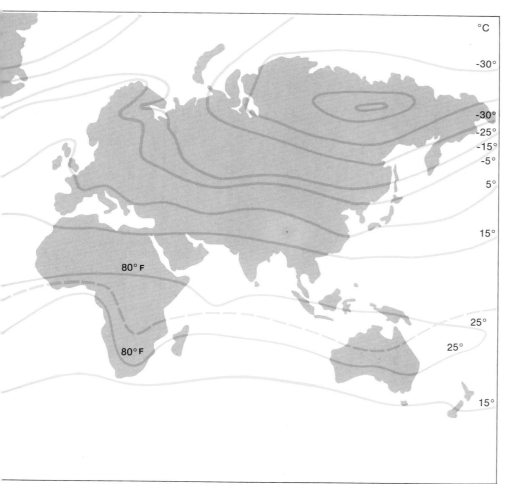

°C

-30°

-30°
-25°
-15°
-5°

5°

15°

80° F

25°

80° F

25°

15°

Temperature

1 The average isotherms for the month of January illustrate the pronounced effect on air temperature of the landmasses of the Northern Hemisphere. Landmasses comprise 39 percent of the Northern Hemisphere but only 19 percent of the Southern Hemisphere.

Terrace, Elevated Marine

2 An elevated marine terrace or elevated wave-cut platform formed when the sea was higher with respect to the land than it is now. The marine cliffs in the foreground have been cut since sea level became established at its present position.

Terrace, Wave-cut
Waves cut through sedimentary rocks to form this platform at La Jolla, Calif., seen here at low tide. The terrace terminates in a marine cliff on the landward side. The more resistant rocks form low ridges in the platform; the less resistant, shallow valleys.

types of marine terrace are wave-cut terraces and wave-built terraces. A wave-cut terrace is cut into bedrock or regolith. The wave-built terrace is composed of unconsolidated sediments which slope off into deeper water from the outer limit of the wave-cut terrace.

The wave-cut terrace results from *marine erosion cutting horizontally into the land. The vertical width of the cut is equal to the distance between the highest and lowest wave action. The bedrock or regolith exposed in the marine cliff above the upper limit of wave action is brought down by subaerial processes into the zone where the waves can attack and remove it. The platform slopes seaward because the seaward portion is older than the landward, has been abraded longer and has therefore been lowered more. Moreover, the larger waves break farther offshore than the smaller and are of course more effective in eroding the platform. The platform may be completely covered by water at high tide but exposed at low tide in a broad strip that may be tens of meters wide. A beach commonly builds up close to the base of the marine cliff. Offshore it may be bare in places, but veneered with marine deposits in other places. Marine *stacks may project above it, small pool-filled depressions may be found on it and a series of miniature *hogbacks with erosional valleys between them will develop where sedimentary rocks of differing resistance are cut.

The width of the wave-cut platform depends on the resistance of the rocks, the strength of the waves and the duration of the action. The platform may be only 1 m wide or as much as many kilometers. Elevated wave-cut platforms 15 and 25 km wide are found on the coast of Peru, and a platform around the Island of Helgoland in the North Sea is 25 to 30 km wide in places. It was formerly thought that in time marine processes could reduce a continent to an erosional plain. But it is now realized that sea level is continuously changing and that marine erosion is a self-limiting process, its effectiveness decreasing as the wave-cut platform is widened. Wave-cut platforms are found wherever the coastline has suffered significant erosion, as on the coasts of California, Oregon and Washington, of Great Britain and around Helgoland. See Terrace, Elevated Marine. (R.L.N.)

Terra Rossa (Italian for "red earth"). The clayey red regolith developed when *limestone at or near the surface dissolves and the insoluble impurities in it accumulate, forming a red residual surface deposit. Flat topography favors its accumulation. In many places it mantles the limestone and fills the *joints and funnel sinks found in it. Because the terra rossa is generally formed on white or gray limestone that apparently contains no red clay, it is thought that the red color is due to oxidation of minerals containing iron after their liberation from the limestone. If so, terra rossa resembles the *laterites of the Tropics. A classic terra rossa area is the Karst region of Yugoslavia, where it mantles white limestone. (R. L. N.)

Terrestrial Radiation. The total infrared radiation emitted from the earth's surface, sometimes called earth radiation. Atmospheric scientists make a distinction between this terrestrial radiation and atmospheric radiation, that is, the infrared radiation emitted by or propagated through the atmosphere. Solar radiation reaching the earth (called insolation), atmospheric radiation and terrestrial radiation determine the heat balance of the earth as a whole.

Terrigenous Sediments. A major category of *deep-sea sediments which consists of material that has been eroded from the continents.

Tertiary Period. The oldest period of the *Cenozoic Era. It began about 65 million years ago and lasted for approximately 63 million years. The name means "third derivation" and is a remnant of an 18th-century classification in which rocks were grouped into four major divisions of time. The period is customarily divided into five epochs of unequal duration; with the oldest first, these are the *Paleocene, *Eocene, *Oligocene, *Miocene and *Pliocene. Their names are based on the proportion of modern species of marine invertebrates that occur among Tertiary fossils. Thus, according to Sir Charles Lyell who proposed the system, the Eocene (meaning "dawn of recent") had fewer modern species than the Pliocene, which translates as "more recent." In another classification system the Tertiary is considered an era of which the first three epochs constitute the *Paleogene Period while the last two are placed in the *Neogene Period.

Tertiary rocks consist largely of loosely consolidated nonmarine and marine sediments, some of which are quite fossiliferous. The major rock divisions were first established in Europe, where they are widely distributed and easily studied and grouped. They are best developed in gentle structural basins such as the London, Paris and Vienna basins. Tertiary deposits are also well known in Italy, southern France, Belgium, Holland and Germany. There are marine rocks in the northern part of Africa and extensive Tertiary deposits in southern South America. A combination of marine, nonmarine and volcanic rocks characterizes the Tertiary of New Zealand and the East Indies, and there is evidence of extensive volcanic activity in Australia.

In North America, Tertiary rocks occur in a narrow belt along the Pacific coast from Baja California to Alaska. Mostly marine in origin, these strata are very thick and have been severely fractured and folded. On the Atlantic Coastal Plain, exposures of Tertiary rocks extend from New Jersey to Florida and continue along the Gulf Coastal Plain into Mexico. Both marine and nonmarine in origin, they are especially thick in the Gulf Coast region of Texas and Louisiana. In places the strata have been pierced by cylindrical plugs of salt that have arched the rocks upward to form *salt domes that may be associated with oil and gas. On the interior of the continent, nonmarine deposits occur from Oklahoma and Texas north to Alberta in the Great Plains region, in the Rocky Mountain area from New Mexico to Saskatchewan, and in the Great Basin province of southern Idaho, eastern Oregon and Nevada. These strata are responsible for the spectacular scenery of the Badlands of South Dakota and Bryce Canyon National Park, Utah. They are also the source of many *fossil mammals and have revealed much about the nature of Tertiary life.

There is evidence of considerable Tertiary volcanic activity in western North America. Lavas, volcanic ash and other types of *igneous rocks cover an area of more than 780,000 square kilometers (about 300,000 square miles) and volcanic peaks such as Mount Shasta, Mount Hood, Lassen Peak and Mount Rainier are also the products of this volcanism. In the Columbia and Snake River region of Oregon, Washington, Idaho and Nevada, huge lava plateaus have been formed of countless lava flows stacked one on the other to cover an area of 520,000 square kilometers (200,000 square miles) to a depth of hundreds of meters (several thousand feet). A widespread series of crustal movements began in Middle Tertiary (Miocene) time and continued with increasing intensity until the end of the Tertiary. This uplift culminated in the Cascadian *orogeny, a mountain-building movement, which elevated the Himalayas in Asia, the Alps in Europe, the Coast Ranges of California and Oregon, and the Cascade Range of Washington and Oregon.

Tertiary climates in North America were in general warmer, more humid and more equable than the climate of today. In Early Tertiary time, tropical and subtropical climates extended as far north as the Canadian border, and arid conditions are indicated for the Great Plains area in the latter part of the period. Near the end of the Tertiary the climate grew steadily colder, heralding the coming of the first *Pleistocene Ice Age.

Because the history of Tertiary life is much like that of the *Quaternary Period, the plant and animal life of both periods are treated under the *Cenozoic.

Much oil is produced from Tertiary rocks in South America, the East Indies, Middle East, the Soviet Union, California and the Gulf Coast region of Louisiana and Texas. Salt is produced from salt domes on the Gulf Coastal Plain, and petroleum occurs in conjunction with some of these. Tertiary coal is mined in Montana, Wyoming, Oregon and Washington, and *lignite occurs in South Dakota and the Gulf Coast region. The Green River Formation of Early Tertiary (Eocene) age is an important source of *oil shale, and deposits of copper, gold and silver from the Tertiary Period have been mined in Mexico, Bolivia, Peru and the Rocky Mountain region of the United States. *Diatomite is produced in Maryland, Virginia and California. (W. H. M.)

Tetragonal System. In crystallography, the crystal system to which belong those crystals having a single fourfold symmetry axis. The crystal forms in the tetragonal system are referred to three mutually perpendicular axes, two of which are of equal length, whereas the third, coincident with the fourfold symmetry axis, is either longer or shorter.

Tetrahedrite. A mineral, copper antimony sulfide, an ore of copper. It is usually found in veins associated with copper, silver, lead and zinc minerals. Tetrahedrite usually contains some arsenic, and with increasing amounts it grades toward *tennantite, the arsenic end-member of a solid-solution series. For both minerals the occurrence, crystal form (characterized by tetrahedrons) and physical properties are so similar it is impossible to distinguish them without chemical tests. Although copper is always the principal metal, iron and zinc are usually present substituting for copper. In the argentiferous variety, freibergite, silver may amount to as much as 18 percent, making the mineral a valuable ore of that metal. See Mineral Properties. (C. S. H.)

Tetrahedron. In crystallography, a crystal form in the *isometric system, composed of four faces, each of which intersects the three crystallographic axes at equal lengths. In geometrically perfect tetrahedrons each face is an equilateral triangle.

1

A

B

2

Thalweg (or Talweg). A line joining the deepest points along a stream channel. As the axis of a valley bottom, it may also be spoken of as the valley thalweg. The thalweg of a stream is usually sinuous. In *meandering streams it moves toward the outside bank of each bend, crossing over near the point at which the stream begins to change its direction of curvature.

Thecodont. An extinct group of reptiles that somewhat resembled the *rhynchocephalians. Many were terrestrial, although some lived in streams and lakes. Thecodonts were limited in both numbers and varieties, and were confined to a single geologic period — the *Triassic. In spite of their short stay on earth, these primitive forms are of considerable evolutionary importance and are believed to have given rise to the crocodiles, dinosaurs and pterosaurs. See Fossil Reptiles.

Therapsid. A mammal-like group of reptiles well developed for a terrestrial existence. Although the remains of these primitive reptiles are not particularly important as *guide fossils, study of the therapsids has provided much valuable information about the origin of the mammals. Members of this group appeared first in the Middle Permian and persisted until the Middle Jurassic. See Fossil Reptiles.

Thermal Conductivity. Measurement of the amount of heat that will traverse a predetermined thickness of material in a specified period of time. For rock materials, thermal conductivity is expressed in millicalories per centimeter per second per degree centigrade. Values have been determined for various rocks, ranging from 3 to 15 millicalories per centimeter per second per degree centigrade.

Thermal Spring (or Hot Spring). A spring discharging water that is heated by natural processes to 5.5°C (10°F) or more above the mean annual temperature of the air. There are over 1000 thermal springs in the western United States and approximately 52 in the eastern United States; many more occur in other parts of the world. Because hot water is a better solvent than cold water, thermal springs generally contain considerable quantities of dissolved mineral matter. Travertine, silica, alum, sulfur and many other substances are precipitated around thermal springs through evaporation, loss of carbon dioxide, lowering of the temperature, the metabolic activity of algae and other organisms, and other processes. The terraces around hot springs are usually formed by deposits of travertine. The most famous and largest terraces in the United States are those at Mammoth Hot Springs in Yellowstone National Park.

It is generally thought that the heat is derived either from (1) hot igneous rocks at depth (as in the hot springs in Yellowstone) or (2) the deep circulation of meteoric waters, heated by the normal increase of temperature with increasing depth, which then move upward so rapidly that they are still warm on reaching the surface (as at Hot Springs, Va., and Warm Springs, Ga.). However, the heat generated by the friction involved in crustal movements, by radioactive disintegration and by the oxidation of sulfides and other chemical reactions is probably also responsible in some localities. When a spring ejects boiling water and steam, it is called a *geyser. (R.L.N.)

Thermocline. A thin zone of great temperature and density change, about 100−200 m below the ocean surface. Because salinity in the open ocean is nearly constant and pressure has only a slight effect on density, temperature is by far the most important factor in seawater density. Surface water is warm and therefore has a low density in comparison to the cold water at lower depths. The rapid temperature and density change in the thermocline makes it a significant boundary for organisms and also for circulation. (R.A.D.)

Thermohaline Current. A deep, slow movement of water generated by differences in density resulting from variations in water temperature and/or salinity from place to place. Below a depth of about 100 m, little circulation is caused by surface winds, but water temperature and salinity vary just enough to result in differences in water density. This creates a density gradient, which in turn causes a current. Such currents move so slowly (on the order of kilometers per year) that they can be detected only by special techniques involving the plotting of temperature, salinity, and oxygen content of water masses at various depths. See Density Current.

Thin Section. In mineralogy, a slice of rock or mineral that has been mounted on a glass slide and then ground exceedingly thin, usually to about 0.03 mm (1/1000 in). In such thin sections most minerals are transparent or translucent, and their optical properties can be studied in transmitted light with a polarizing microscope. Thin sections also offer an important means of studying the textural relationships of minerals in rocks.

Tholoid. A *cumulo-dome extruded in a *crater or *caldera.

Thunderstorm. A local storm always associated with a cumulonimbus cloud and accompanied by lightning and thunder. Gusty winds, heavy rain showers and, less frequently, hail are typical of thunderstorms.

Although usually only a few kilometers in diameter, thunderstorms covering areas 50 km (30 miles) wide have been observed. The larger storms, which extend to a height of 15 km (10 miles) or more, may contain several of the vertical convection currents — each with strong updrafts and downdrafts — that characterize thunderstorms. The large thunderstorms of the intertropical convergence zone, where the trade winds of the two hemispheres meet and the air is forced to rise, provide much of the energy that drives the currents of the atmosphere's general circulation. Some weather stations in the tropics report more than 300 thunderstorms yearly, and it is estimated that more than 2000 storms are in progress over the world at any given moment. It is generally believed, although not yet proved, that thunderstorms maintain the atmospheric electric field.

Thunderstorms are essentially convective phenomena in which the condensation of water vapor plays a vital role by supplying energy, in the form of the latent heat of condensation, for the storm's development. Such storms develop when a deep layer of moist air is given sufficient lift to overcome the normal stability of the atmosphere. If the air is lifted high enough and cooled sufficiently for condensation to occur, latent heat is released; the rising air may thus remain warmer than the surrounding atmosphere and continue to rise freely. The initial lift may be caused by solar heating of the earth's surface, by an advancing cold front, by forced motion of the air over rising terrain, or by the convergence of air currents as, for example, in the zone where the trade winds of the two hemispheres meet. In the central and eastern United States, large and severe thunderstorms, often accompanied by tornadoes, occur in spring and summer in squall lines preceding a cold front. In these situations the atmosphere is potentially unstable because dry air from the west or northwest moves above moist tropical maritime air. See Clouds; Convective Circulation; Lightning; Stability. (M.F.H.)

Tidal Bore. An unusually large single wave that rushes up a river mouth like a wall of water. It occurs in long, narrow *estuaries and rivers as the tide rises. It is caused by the steepening of the wave as it enters the shallow, constricted river area and is opposed by the river's discharge. Although tidal bores are generally less than 1 m high,

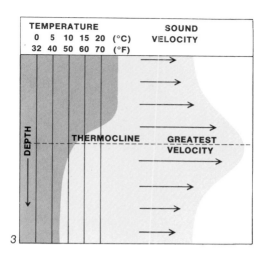

3

4

Thecodont

1 Appearing during the early Triassic and dying out by the end of that period, this order of carnivorous reptiles was an ancestor of the dinosaurs. Saltoposuchus was a small, lightly built thecodont about 1 m in length.

Therapsid

2 The skeleton of extinct reptiles such as Moschops (A) and Dimetridon (B) is in many ways like that of mammals.

Thermocline

3 The speed with which sound travels through water increases to a certain depth and them decreases sharply in part due to decreasing temperature. This occurs at the thermocline, a narrow zone where warm surface water meets the colder deep water and an abrupt change of temperature takes place.

Tidal Bore

4 A wave sometimes reaching a height of 2 m (6 feet) and more than 8 knots (over 9 miles) an hour is created when water rushes, as into a funnel, from the Bay of Fundy into the narrow Petitcodiac River in Nova Scotia.

a

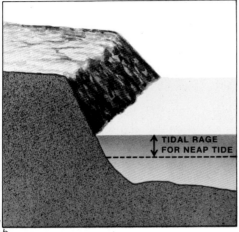

b

Tide

1 During the new and full moon in the spring (a) the combined influence of both sun and moon causes high and low tides. The range (the difference between the level of high and low tide) is very large. At the first and third quarter phases of the moon (b), the sun works against the effect of the moon by raising the moon's low-tide levels and lowering the moon's high-tide levels. This results in neap tides, and the tidal range is quite small.

Till

2 An accumulation of clay, sand, gravel and boulders forming a moraine near Bolzano, Italy, was deposited by a glacier and later eroded into pinnacles by rain. One is topped by a large erratic.

Tombolo

3 These two tombolos connecting two islands to the mainland enclose Duxbury Bay, Mass.

the River Severn in England has a bore of 3 m and the Chien Tang Kuang River in China has bores of 8 m. A famous tidal bore occurs where the St. John River enters the Bay of Fundy. (R.A.D.)

Tidal Power. See Energy Resources.

Tidal Wave. A sea wave of great length caused by submarine earthquakes. See Tsunami.

Tide. The gradual rise and fall of the sea along a coast. A tide may be periodic and predictable, such as astronomical tides, or aperiodic, such as those caused by weather phenomena. The astronomical kind is due to the attractive forces of the earth, moon and sun coupled with the earth's rotation. In accordance with Newton's Law of Universal Gravitation, the attraction between these three celestial bodies causes bulges in bodies of water. These bulges move from place to place as the earth rotates, producing daily changes in water level, the type and amount depending on the location and lunar cycle.

A diurnal tide is one in which there is one high and one low stage each day. A semidiurnal tide has two highs and two lows, whereas a mixed tide is a combination of the two types during one lunar month. Spring tides are tides of maximum range that occur every two weeks, when the earth, moon and sun are aligned and there is consequently a full or new moon. Neap tides are of minimum range and occur when the earth and sun are at right angles with respect to the moon, resulting in a first and third-quarter moon. Although astronomical tides are most evident in the oceans, they also occur in small bodies of water, but their magnitude in the latter is so slight that they go unnoticed. The Great Lakes of North America, for example, have tides of only 2 to 6 cm (about 3/4 inch to 2 inches). When winds blow onshore they tend to cause water to pile up along the coast. Under normal conditions of prevailing winds, this increase is slight and is masked by astronomical tides. During storms, however, and particularly hurricanes or typhoons, these storm tides or wind tides may rise a few meters above normal water level, and they have resulted in devastating coastal flooding. (R.A.D.)

Tiger's Eye. Quartz, pseudomorphous after *crocidolite, used in jewelry and for ornamental objects. In replacing the asbestiform crocidolite, quartz preserves its fibrous nature, giving the mineral an attractive *chatoyancy. The principal source of tiger's eye is the Republic of South Africa.

Till. Material derived from *regolith and *bedrock and deposited directly by glacial ice. Its characteristics depend on the source rocks and other factors. Thus till derived from shale consists largely of clay and silt, with few fragments, whereas till derived from granite is sandy and stony. In general, however, till consists of a mixture of clay, silt, sand and large angular fragments. It is compact, poorly sorted and unstratified. The fragments vary in composition; they may be faceted, polished and striated. Some fragments may weigh hundreds of metric tons; their long axes tend to be parallel to the direction of ice movement. Till may be fossiliferous if the source was regolith containing the remains of organisms. It is scattered unevenly over the millions of square kilometers in North America and Europe that were glaciated during the Pleistocene. Thicknesses more than 300 m have been measured, but the average thickness in northern North America is probably less than 6 m (19 feet). *Moraines, *drumlins and other glacial landforms are composed of till. It can be confused with *mudflows, *colluvium and *pyroclastic breccias, but the presence of faceted and striated fragments and its association with glaciated pavements and with stratified drift assist in its identification. (R.L.N.)

Tillite. A clastic *sedimentary rock formed by the cementation of the unsorted, unstratified glacial deposit called *till. It consists of material deposited directly by and underneath a glacier. The embedded rock fragments are typically angular in shape and may be striated and polished. Tillites are useful in *paleoclimatology, for they indicate glacial climates. The Gowganda Formation, a *Precambrian tillite widespread in parts of Canada, is the earliest known evidence of glaciation. Other tillite deposits are found in Madagascar, South Africa, South America, Australia and Tasmania, indicating drastic climatic changes during the past 200 to 300 million years.

Time-stratigraphic Unit (or Time-rock Unit). A mass of rock defined on the basis of arbitrary time limits rather than on physical characteristics. The time-stratigraphic units represent strata formed during definite portions of geologic time and are the stratigraphic equivalents of *time units as shown below (beginning with the longest):

Time Unit	Stratigraphic Unit
Era	Erathem
Period	System
Epoch	Series
Age	Stage

Time Unit. A subdivision of geologic time. The largest unit of geologic time is an *era, such as the Cenozoic Era. Eras are subdivided into smaller time units called *periods; subdivisions of periods are *epochs. Hence the Eocene Epoch represents a relatively small part of the Tertiary Period, which in turn is a subdivision of the more lengthy Cenozoic Era.

Tin Ore. See Cassiterite.

Tombolo. A beach tying an island to a mainland or to another island. Some islands are attached by a double tombolo; if many islands are joined, the combination of tied islands and tombolos is called a complex tombolo. Tombolos are found along shorelines of submergence, where islands are common. The beach that joins the Rock of Gibraltar to the Spanish mainland, and Chesil Beach, which connects Portland Bill to the British mainland, are tombolos. Many tombolos are found along the Massachusetts coast north of Boston; Nantasket Beach in Boston harbor is a complex tombolo. (R. L. N.)

Tonalite. A synonym for *quartz diorite.

Tongue. A subdivision of a rock formation. Usually of limited geographic extent, tongues typically consist of a section of rock that in one direction wedges out laterally between sediments of different composition and in the opposite direction thickens and becomes a section of a larger body of similar sediments. See Formation, Geologic.

Topaz. A mineral, aluminum silicate, best known as a gemstone. It is usually colorless, but small amounts of impurities may color it red, yellow, green, blue or brown. Although gem-quality topaz may be any of these colors, the wine-yellow variety is most highly prized and is often referred to as "precious topaz." This is meant to distinguish it from less valuable citrine quartz which has a similar color and frequently is sold as topaz. The mineral is characterized by its high hardness (8) and one direction of perfect cleavage.

Topaz is found in pegmatites associated with tourmaline, beryl, cassiterite, fluorite and apatite. It is generally in well-formed crystals, some of which are extremely large. In a search for quartz crystals during World War II, several large, transparent crystals of an unknown mineral were found in Brazil. They were later identified as topaz; the largest weighed 596 pounds and is on exhibit in the American Museum of Natural History, New York City. Brazil also produces the best wine-yellow topaz. The

2

3

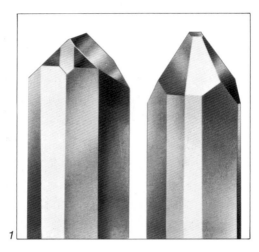

Topaz

1 *A mineral that crystallizes in the orthorhombic system, topaz has crystals that show different combinations of terminal forms.*

Tornado

2 *A roaring funnel of air, often called a "twister," generates winds up to 480 km (300 miles) an hour as it churns past Tracy, Minn.*

stones come from deeply weathered pegmatites and water-worn stream pebbles. Other notable localities are in Russia in the Nerchinsk district and at Mursinsk in the Ural Mountains. The latter locality is famous for its pale blue crystals. See Mineral Properties. (C.S.H.)

Topography. The elevation, relief, size and configuration of the surface features of an area or locality. Topography is usually characterized as rugged or gentle, steep or flat, high or low, of great or small relief and so on. *Endogenic and exogenic processes are responsible for topography. Thus various parts of the earth's surface are characterized by glacial, fluvial, eolian, volcanic and other kinds of topography. Although the earth is between 4 and 5 billion years old and some rocks are more than 3 billion years old, most topography was probably formed in Middle or Late *Cenozoic time and is therefore at most only a few tens of millions of years old. A topographic map shows the elevation, relief, configuration and size of physical features of the landscape by means of *contour lines, shading, color or hachures, and a scale. (R.L.N.)

Tornado. A small, violent storm that rotates cyclonically (in the same direction as the rotating earth) and is recognized by the characteristic funnel cloud pendant from the base of a much larger cumulonimbus cloud which is often termed the parent cloud. The funnel of the rotating vortex is composed of water droplets in its upper portion and of dust and debris extending a short distance upward from the ground. The pressure at the center of a tornado may be 200 millibars lower (about 20 percent) than the pressure of the surrounding atmosphere. The air converging into the vortex thus expands and cools adiabatically (that is, without loss of heat to the surroundings). It is this cooling and the resulting condensation of water vapor that account for the tornado cloud.

In a number of cases, observations show that the larger parent cloud rotates in the same direction as the tornado embedded in it. Since cumulonimbus clouds contain strong updrafts and downdrafts, it is believed that tornadoes develop under conditions of intense, localized convection within a larger mass of air that has cyclonic rotation or vorticity. The initial removal of air, followed by inflow at lower levels, could in this case concentrate the cyclonic spin of the surrounding air and explain the high speeds of the tornadic winds, which may reach 250 or perhaps 300 miles per hour. It has been suggested that severe lightning could contribute the heat needed for the

concentrated convection and removal of air, but tornadoes without lightning have been observed. The sudden release of latent heat of condensation in rising columns of moist air is considered to be a more likely source of energy for tornadoes.

Because of the difficulty of obtaining accurate observations within tornadoes, many questions about their physical characteristics still remain unanswered. Damage and damage paths have supplied indirect estimates of tornado size and the direction of the wind within the funnel near the ground. Tornadoes average only a few hundred meters in diameter, and their average duration on the ground is about 20 minutes. The typical damage track, normally extending from southwest to northeast in the United States, is about 400 m (¼ mile) wide and 15 km (10 miles) long, but large tornadoes have been known to devastate areas more than 1500 m (about 1 mile) wide and 500 km (300 miles) long.

Although tornadoes are occasionally reported in other parts of the world, principally Australia, very severe ones are frequent only in the United States, particularly in the central part of the country, where they are most numerous in April, May and June. It is in this region and at this time of year that the large-scale meteorologic conditions most favorable for tornado development occur. During a typical tornado situation in the midwestern United States, warm moist air from the Gulf of Mexico flows northward in advance of the cold front of a developing cyclone, with very dry air flowing over the tropical air. This is a potentially unstable situation; if the tropical air is pushed upward by the advancing cold front, the rising air columns become warmer than the surrounding atmosphere and remain so, accelerating upward as towering cumulonimbus clouds develop in a squall line ahead of the front. While the most severe tornadoes occur under these conditions, within a polar-front cyclone, tornadoes also occur within tropical cyclones and cumulonimbus clouds not associated with cyclones. (M.F.H.)

Tourmaline. A mineral group of borosilicates with complex and variable chemical composition. It occurs in a variety of colors depending on the composition. Iron tourmaline, the most common and abundant, is black; magnesium tourmaline is brown; the presence of lithium gives rise to lighter colors that may be green, yellow, blue and, more rarely, colorless. The rare lithium-bearing varieties, if transparent and flawless, are used as gemstones and may go under such varietal names as verdelite (green), rubellite (red to pink), indicolite (blue) and achroite (white or

colorless). Tourmaline crystals are usually elongated, and some show a remarkable color zoning from base to top; thus a single crystal may vary from deep green to red to colorless over a length of 5 cm (2 inches). In other crystals the color zoning is arranged in concentric envelopes parallel to the length.

Tourmaline crystallizes in a low symmetry class of the hexagonal system, which permits it to form crystals having a cross section resembling a spherical triangle. Because this crystal habit is so characteristic, it is used to identify the mineral. Another crystallographic feature of tourmaline is that it lacks a center of symmetry and is *piezoelectric; thus if pressure is exerted at the ends of the crystal, one end becomes positively charged and the other end negatively charged. Because of its piezoelectric property, tourmaline is manufactured into gauges to measure transient pressures. A gauge is made from tourmaline plates coated with electrodes connected to a recording device. The voltage recorded is proportional to the pressure. Such gauges have been used to measure the enormous blast pressures from atomic bombs as well as the relatively gentle pressure exerted by waves beating against a seawall.

The black variety of tourmaline is found as an accessory mineral in igneous and metamorphic rocks, but it is best developed and occurs most abundantly in *pegmatites. The rarer, lighter colored varieties are found almost exclusively in pegmatites. They may be firmly embedded in other minerals, but the finest quality gem crystals are found in cavities in the pegmatite known as "pockets." Notable localities for gem tourmaline are in Brazil, Madagascar and the Ural Mountains, and in the United States at Mesa Grande and Pala, Calif., and Paris and Auburn, Maine. See Mineral Properties. (C.S.H.)

Trace Fossil. A structure such as a track, trail or burrow left in sediment by living organisms. Trace fossils most commonly occur in strata composed of sediments deposited in shallow seas. They may reveal life activities of an organism (walking, crawling, burrowing, etc.). The study of trace fossils is called ichnology.

Trachyte. A volcanic rock, the extrusive equivalent of a *syenite. It is composed of alkalic feldspar with small amounts of biotite, hornblende or pyroxene.

Tractive Force. In hydrology, the force that moves loose grains on the bottom of a stream bed. The critical tractive force is that required to initiate movement; it is

measured by the *Duboys equation and depends on the depth and specific weight of the water and the stream gradient.

Trade Winds. A major system of tropical *winds covering roughly the zones between 50° and 30° north and south latitudes. These belts lie just north and south of the *doldrums. In the Northern Hemisphere, the prevailing winds are from the northeast and are termed the northeast trades. In the Southern Hemisphere, the winds come from the southeast and are called the southeast trades. They are known as trade winds because early traders followed them on their sea routes.

Transcurrent Fault. An immense, steeply inclined strike-slip *fault.

Transform Fault. A strike-slip *fault caused by spreading of the sea floor away from the oceanic ridges. See Plate Tectonics.

Transpiration. The loss of moisture to the atmosphere by leaves and other parts of plants. A field of corn during the growing season may transpire an amount of soil moisture almost equal to a depth of 30 cm of water covering the entire field. Trees may transpire twice as much soil moisture. The loss of moisture to the atmosphere from snow, ice and water is called *evaporation. Hydrologists and climatologists use the term evapotranspiration for a soil moisture loss due to both evaporation and transpiration.

Trapezohedron. A crystal form in which every face is a trapezium (a quadrilateral in which no two sides are parallel). It usually refers to the 24-faced forms of the *isometric system.

Traprock. A term applied to any dark, fine-grained *dike or volcanic rock. Most traprocks are of basaltic composition.

Travertine. A chemical *sedimentary rock composed of calcite precipitated from solution by inorganic processes. It occurs in caves as banded deposits of dripstone or flowstone formed by the evaporation of water that contains calcium carbonate in solution. It also forms as calcareous tufa, a light, coarsely crystalline, spongy limestone precipitated by the evaporation of springs, small streams and groundwater.

Tree Mold. A tree mold is a cylindrical opening in a lava flow formed when lava solidifies around a tree. The tree may burn up completely; if not, what remains may rot away. They are commonly found wherever flows have invaded forests.

Trellis Drainage. See Drainage Patterns.

Tremolite. A mineral, a hydrous calcium magnesium silicate, a member of the *amphibole group of minerals. Iron is usually present, substituting for magnesium, and when it amounts to more than 2 percent the mineral is called *actinolite. The mineral is thus considered a member of the tremolite-actinolite series. There is a continuous gradation of color from white, pure tremolite, to dark green, which indicates iron-rich actinolite.

Tremolite is a fibrous mineral, in places asbestiform, and is mined as one of the varieties of *asbestos. Sometimes the fibers are loosely interwoven in the unusual aggregates known as *mountain cork and mountain leather. If the fibers are closely compacted into a hard felted mass, the mineral is *nephrite, one of the varieties of *jade. Tremolite is most frequently found in metamorphosed dolomitic limestones and in talc schists. (C.S.H.)

Triassic Period. The earliest period of the *Mesozoic Era; it began about 225 million years ago and lasted 30 to 35 million years. The name comes from the Latin word *trias,* meaning "three," and was applied by Friederich A. von Alberti in 1834. It refers to the distinct threefold division displayed by Triassic rocks in Germany, where the system was first described. In Europe there are continental deposits in eastern France, western Russia, Spain, England and Germany. Marine sedimentary rocks occur in the Alpine and Mediterranean regions. Triassic rocks are also exposed in parts of Asia, Australia, South Africa and South America. In eastern North America Triassic formations occur in a series of long troughs inland along the Atlantic sea coast, occurring from Nova Scotia to South Carolina. They represent nonmarine sediments, and in places (such as the famous Palisades of the Hudson River) there is evidence that igneous rocks have been intruded into the sedimentary rocks. Western North America was the site of marine deposits in California, Arizona, New Mexico, Idaho, Utah, Montana and Wyoming. Marine formations also extend southward from Alaska and the Yukon Territory into Mexico. In the western interior of the United States, Triassic rocks are mostly of continental origin and occur in western South Dakota, northwestern Texas, New Mexico, western Nebraska and Wyoming. The Chinle Formation of Arizona's Painted Desert is famous for the great fossil logs now preserved in Petrified Forest National Park. The close of the period was marked by moderate uplift that resulted in tilting and fracturing the Triassic (and older)

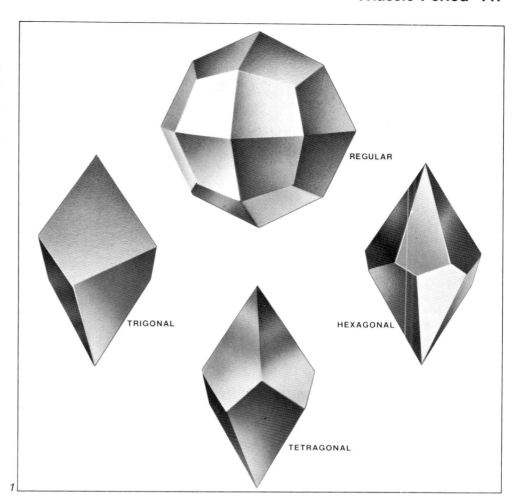

Trapezohedron

1 The crystal form trapezohedron is found in several crystal systems. Shown here are the following trapezohedrons: regular (isometric system), tetragonal (tetragonal system), trigonal and hexagonal (hexagonal system).

Triassic Period

2 Named from the threefold division of its rocks, this period began some 225 million years ago and lasted for about 35 million years. During this time the first dinosaurs appeared. Coelophysis, right, a lightweight theropod 2 m (6½ feet) long, had hollow bones. Its bipedal stance enabled it to move quickly, a feature of primary importance for the capture of prey and to escape enemies. The Triassic phytosaurs, such as Rutiodon, left, were large reptiles, as much as 7—8 m (25 feet) in length, that lived along rivers, preying on fish and other reptiles.

rocks of the Atlantic coastal region of North America; there is no evidence of mountain building in the western part of the continent, however.

Some Triassic rocks (especially dune deposits, salt and gypsum) suggest that arid to semiarid climates existed during much of Triassic time. However, the remains of swamp-dwelling plants, reptiles and amphibians indicate that some areas had a warm, almost subtropical climate. Triassic plants and animals are more advanced than the preceding *Paleozoic forms. However, because of unfavorable conditions of preservation, the record of Triassic life in North America is not very extensive. Plants of the coal-forming type have been found in Virginia and the Carolinas. The cone-bearing trees became abundant, and their remains are found in large numbers in the Petrified Forest of Arizona. Cycads, a group of plants with palmlike leaves, were also abundant at this time. Marine invertebrates included many ammonoids, pelecypods, gastropods and reef-building corals. The torpedo-shaped belemnoids (relatives of the modern squids) were especially numerous. Vertebrates made remarkable gains during this period; sharks were common in the seas and the true bony fishes were increasing in numbers and species. Amphibians, although common, were outnumbered by the reptiles, which were not only numerous but had developed many unusual forms. The phytosaurs, semiaquatic crocodile-like reptiles, were characteristic of this time. The earliest known dinosaurs appeared in the Triassic, but most of them were quite small compared with the giant reptiles of later Mesozoic periods. These early species walked on their hind legs and are known from their skeletons and their three-toed birdlike tracks, which are abundant in certain Triassic shales in the Connecticut Valley. Marine reptiles were represented by two major groups: the ichthyosaurs, streamlined fishlike reptiles, and the plesiosaurs, which had clumsy, flattened, turtlelike bodies and long necks. Both of these forms appear to have descended from land reptiles and represent a secondary adaptation to marine life. Fossil mammals have been reported from the Triassic, but there is disagreement whether these are true mammals or a continuation of the therapsids, mammal-like reptiles that appeared in the *Permian. This confusion arises because of the fragmental nature of the fossil remains.

Triassic economic resources are relatively limited, but salt and gypsum have been mined in Germany, South Dakota, Wyoming, Montana and Nevada. Minor amounts of coal occur in North Carolina and Virginia, and Triassic sandstone ("brownstone") has been quarried for building stone in the eastern United States. (W. H. M.)

Tributary, Barbed. A watercourse that joins a main stream at an angle pointing upstream which is contrary to the usual junction of two streams. This circumstance constitutes *stream capture and a reversal of the flow of the tributary. Barbed tributaries are common in glaciated regions.

Triclinic System. In crystallography, the crystal system in which the forms are referred to three axes of unequal length that make oblique angles with each other. The only symmetry that is possessed by triclinic crystals is a center; some triclinic crystals have no symmetry at all.

Tridymite. A mineral, an oxide of silicon, one of the polymorphs of silica. Tridymite is the stable form between 870° and 1470°C (1598° and 2678°F). *Quartz is the stable form at lower temperatures. Tridymite occurs in small crystals in siliceous volcanic rocks that are impossible to identify without the aid of a microscope. It is an abundant mineral in the lavas of the San Juan district of Colorado.

Trilobite. A member of the class Trilobita, phylum Arthropoda. These extinct marine arthropods range from Early *Cambrian to *Permian in age and are excellent *guide fossils for Cambrian rocks. The name is derived from the three-lobed nature of the body, which was encased in a chitinous exoskeleton or carapace. It is this part of the body that is usually fossilized.

Tripolite. A powdery or earthy siliceous *sedimentary rock produced by the weathering of *chert or siliceous limestone rich with the remains of *diatoms or *radiolarians. Sometimes called diatomaceous earth, tripolite is named for a deposit found in Tripoli, Libya. It is lightweight, porous, crumbles easily and has an abrasive feel. It may be white, gray, red, pink, yellow or buff and occur as a superficial layer on *concretions or in thick strata.

Troposphere. The lowest layer of the atmosphere, defined by vertical temperature distribution. In the troposphere, the temperature characteristically decreases from about 15°C (59°F) at the earth's surface to about -57°C (-71°F) at 11 km, where the tropopause marks the base of the stratosphere. (The height of the tropopause, however, varies markedly with latitude, season and weather conditions.)

The decrease of temperature with height in the troposphere can be broadly explained by the convective mixing of air undergoing both dry and moist adiabatic processes, that is, changes in which no heat or mass is exchanged with the environment. If the air were completely dry, adiabatic mixing would result in a temperature lapse rate of about 10°C (18°F) per km (or 5.5°F per 1000 feet) in the mixed layer. Adiabatic mixing of completely saturated air would produce a smaller lapse rate because latent heat energy would be continuously released during the air's ascent. Actually, the troposphere is neither completely dry nor saturated, and therefore the lapse rate of temperature lies somewhere between the two extremes. An important diabatic (that is, nonadiabatic) process, radiational heating and cooling, may also be a contributing factor to the temperature lapse rate, but in the troposphere it is not the determining factor it is in the stratosphere.

The troposphere is divided into three layers: (1) The surface boundary layer is a thin layer of air adjacent to the earth's surface in which the vertical wind distribution is determined by the vertical temperature lapse rate and by the roughness of the underlying surface. (2) The *Ekman layer is a transition layer between the surface boundary layer and (3) the free atmosphere above. The effects of the earth's surface friction are considered negligible in the free atmosphere. (M. F. H.)

Truncated Spur. A triangular facet with its apex pointing upward, found on the wall of a glaciated valley. Such spurs are formed when a former mainstream valley is occupied by a glacier and the ends of the ridges between tributary stream valleys are eroded backward by glacial erosion.

Tuff, Welded. See Ignimbrite; Welded Tuff.

Tsunami. A Japanese word for tidal waves or sea waves of great length, caused by submarine earthquakes. The waves travel across the open sea at a high speed and can cause heavy damage to coastlines thousands of kilometers from their source. Tsunamis generally occur in the Pacific Ocean because its perimeter is much affected by earthquake activity. The Chilean earthquake of May, 1960, produced tsunamis throughout the Pacific. As far away as Hawaii, waves about 4 m above normal height were recorded. Tsunamis generated by the Kurile Islands earthquakes of 1963 and the Alaska earthquake of 1964 also affected large areas of the Pacific Ocean. At sea, tsunamis may pass undetected. Their wavelengths in deep water may be 100 km or more but their

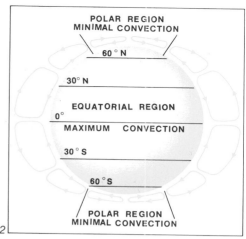

Trilobite

1 These extinct marine arthropods, named for their three-lobed bodies, were found in the Middle Devonian rocks near Ovid, N.Y.

Troposphere

2 The lowest layer of the atmosphere, extending 8 km (5 miles) above the poles and more than 16 km (10 miles) above the Equator, contains air currents that control most of the earth's weather. Humid air rises in great updrafts along the Equator and is transferred by means of circulating convection cells to the far north, where it sinks earthward as a mass of dry, cold air.

Tsunami

3 In 1958 a giant wave, caused by an earthquake-induced landslide, surged 530 m (1740 feet) up the spur of a mountain in Lituya Bay, Alaska, completely denuding the promontory of trees.

1

amplitude (wave height) may be no more than that of normal ocean waves. Because their nature depends on the depth of water through which they travel, tsunamis decrease in wavelength but increase in amplitude (reaching perhaps 100 m in some cases) as they approach shallow coastal waters. This, coupled with the tendency of shallow-water topography to focus the waves on segmented sections of the shoreline, make tsunamis formidable agents of destruction. (J.W.S.)

Tuff. A pyroclastic rock formed of consolidated volcanic fragments generally less than 2 mm (8/100 inch) in diameter. It is essentially consolidated sand and dust resulting from the solidification of liquid droplets erupted from *magma or from the disintegration of older volcanic and nonvolcanic rocks. It varies in color from white to black depending on the chemical composition of the magma. Tuff is classified according to petrographic composition — for example, rhyolitic tuffs, basaltic tuffs. It is also classified by content, thus vitric tuff is rich in glassy material, lithic tuff abounds in crystalline material formed by the explosive disintegration of older rocks, and crystal tuff contains an abundance of crystals that were floating in magma at the time of eruption. Many tuffs are mixtures of these materials.

Volcanic dust may be carried great distances — sometimes even around the world — and tuffs may therefore have wide distribution. A layer of tuff in the Yukon area of Canada covers 520,000 square kilometers (200,000 square miles). Large deposits have also been found in Ordovician rocks of the eastern United States. These widespread deposits are useful in *correlation. The volcanic sand or dust that form tuff may fall on land, oceans or lakes. If it falls on land, it is frequently reworked by running water and mixed with nonvolcanic sediments, thus forming tuffaceous sandstones and shales. Tuff is commonly well stratified because the erupted material varies in color, size or other characteristics. It may also be fossiliferous. A tuffaceous deposit of Miocene age near Florissant, Colo., is one of the best-known fossil insect localities in the world. The insects were apparently swept out of the atmosphere and buried by volcanic dust that fell into a lake. See Breccia; Pyroclastic; Welded Tuff. (R.L.N.)

Tundra. A cold, treeless plain found mainly in the Arctic lowlands of Europe, Asia and North America. Tundra vegetation is characterized by lichens, mosses, grasses, sedges and low shrubs; animal life consists of reindeer, caribou, musk ox, polar fox,

lemming and, near the coast, polar bear. The tundra is underlain by a dark, wet mucky soil and *permafrost. The low temperature is due to the high latitude and the proximity to the frozen Arctic Ocean; the wetness results from low evaporation and poor drainage.

Turbidite. Sediments deposited from *turbidity currents. They are commonly composed of coarser material than would normally be expected in the environment of their accumulation. Virtually all deep-sea sands are attributed to turbidity currents. Turbidites are also characterized by graded bedding (coarse grains on bottom and finer grains on top; organic material and various other types of sedimentary structures.

Turbidity Current. A type of *density current resulting from a density difference between two areas created by suspended sediment in water. The local suspension of sediment causes this more dense sediment-laden water to move downslope in response to gravity, thus displacing adjacent water. Turbidity currents are of great importance in transporting large quantities of sediment from the shallow *continental margin to the *continental rise area. Such currents may be generated by earthquakes, volcanic eruptions, excessive steepening of slopes or any other phenomenon that causes large quantities of sediment to be placed in suspension. (R.A.D.)

Turbulent Flow. The chaotic movement of water masses in all directions, with much mixing and many collisions. Eddies, vortices, waves and "white water" are patterns of turbulence. There are two kinds of turbulence: streaming flow and shooting flow. Streaming flow is ordinary turbulence whereas shooting flow is generated in rapids and falls at very high velocities. Turbulent flow occurs when the combination of density, depth and velocity of the water are much greater than the viscosity. See Reynolds Number.

Turquoise. A mineral, hydrous copper aluminum phosphate. Because of its attractive blue to green color, due to the presence of a small amount of copper, it has long been used as a gemstone. It is a secondary mineral found at or near the earth's surface, and was thus known and used by early peoples. The most ancient mines, on the Sinai Peninsula, are believed to have been worked by the Egyptians in 3400 B.C. In Nishapur, Iran, mines are active today that have been producing turquoise for 2000 years. Other important

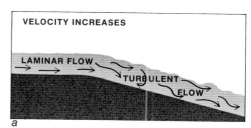

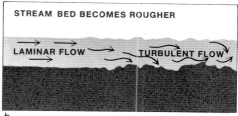

Tundra

1 The tundra on Kotelni Island in western Siberia is a wasteland where only lichens, mosses, grasses and a few dwarf birches and willows grow.

Turbulent Flow

2 As the speed of a stream increases (a), or the channel bed becomes rough (b), water particles are deflected, forming eddies and swirls, and travel in all directions at different velocities, creating turbulence.

localities are found in Tibet, China, and the southwestern United States. American Indians used and treasured turquoise at least as early as A.D. 900, and their art of polishing and setting it in jewelry continues to the present day. See Mineral Properties. (C.S.H.)

Twin Crystal. Two or more crystals of the same substance that have grown together with certain crystallographic directions parallel but others in reverse position. A contact twin is made up of two crystals united along a crystal plane. In a penetration twin, two crystals interpenetrate one another and have an irregular surface of contact. Repeated twins are made up of three or more parts twinned according to the same *twin law. If the successive surfaces on which the individual crystals join are parallel, a polysynthetic twin results; but if the surfaces are nonparallel, a cyclic twin is formed.

The *plagioclase feldspars offer the best example of polysynthetic twinning. In them it is called *albite twinning and is so characteristic that it offers a ready means of identification. Cyclic twinning produces circular forms resulting in a pseudosymmetry for the twinned group not found in the individual crystal. For example, orthorhombic *chrysoberyl and *aragonite and tetragonal *rutile all form cyclic twins having a sixfold axis of pseudosymmetry. (C.S.H.)

Twin Law. In crystallography, a statement defining the relationship of the two individuals in a *twin crystal. This may be (1) as though one individual of a twinned couplet had been derived from the other by reflection over a common plane, the twin plane, or (2) as though one part of the twin had been derived from the other by rotation about a common crystallographic direction, the twin axis. The twin law thus states whether there is a plane or an axis of twinning and gives the crystallographic orientation of the plane or axis.

Type Locality. The geographic location at which a *formation or other stratigraphic unit is typically displayed and from which it derives its name. It is also used to distinguish the place at which a geologic feature (such as a mineral or a fossil species) was originally recognized and described.

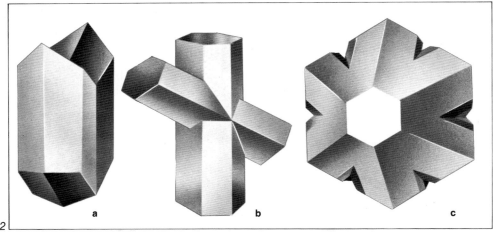

a   b   c

Turquoise

1 This mineral formed by the oxidation of phosphate minerals contains some copper, giving it its characteristic blue color. As a secondary mineral it occurs at or near the surface and was thus found and used by early peoples.

Twin Crystal

2 Three examples of twin laws, among the many under which minerals form twin crystals, are shown here: (a) contact twin in gypsum, (b) penetration twin in staurolite, (c) cyclic twin in chrysoberyl.

Ulexite. A mineral, hydrous sodium calcium borate, occurring in fine fibers that are characteristically aggregated into rounded masses of loose texture known as cotton balls. Ulexite forms in arid regions associated with borax in *evaporite deposits and as a surface efflorescence. A common borate mineral, it is used as a source of boron to produce the sodium borate *borax. See Mineral Properties.

Unconformity. A buried erosion surface separating younger strata from older rocks that were exposed to prolonged erosion before the deposition of the younger. It may be the result of uplift and erosion, a break in sedimentation or nondeposition of sedimentary material. It represents a break or gap in the geologic record, as when rocks of a particular age are absent. The time interval not represented by rocks (or corresponding to rocks missing by comparison with other areas) is the *hiatus of the unconformity. Unconformities are of four basic types, each of which may provide information about past geologic events. If the buried erosion surface lies between two parallel series of strata it is called a disconformity. An angular unconformity exists where the beds below an unconformity are not parallel to those above it. This indicates that the lower rock layers were deformed (by folding or faulting) before being truncated by erosion and covered by younger sediments. An unconformity in which eroded intrusive *igneous rocks or massive *metamorphic rocks are overlain by younger *sedimentary strata is a nonconformity. The more obscure paraconformity resembles a disconformity in that the strata are parallel; however, there is very little physical evidence in it of erosion or a long lapse of deposition. (W. H. M.)

Underfit Stream. See Misfit Stream.

Undertow. A rather slow, uniform seaward movement of water along the bottom just offshore from a beach. An undertow is distinct from the much stronger *rip current and has a different source, although the term is popularly used to describe a rip current. An undertow is caused by backwash activity of waves, or the return flow, due to gravity, of water pushed onto the beach by waves or onshore winds.

Uniformitarianism. The doctrine that holds that the geologic processes of the past operated in much the same manner and at the same rate as they do today. It assumes that today's earth features are the result of present processes acting over long periods of time, and that the history of the

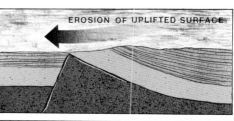

a ORIGINAL UNDISTURBED STRATA

b FAULTING AND UPLIFT

c EROSION OF UPLIFTED SURFACE

d NEW DEPOSITS ARE LAID DOWN — UNCONFORMITY

1

2

Unconformity

1 Movements in the earth's crust can interrupt the continuity of strata of rock. An unconformity occurs when the original layered structure (a) is faulted and uplifted (b). The raised surface is then eroded (c), and eventually covered with sediments (d).

2 Dark Pleistocene basalt flows lie discordantly over Permian marine strata near Panamint Springs, Calif.

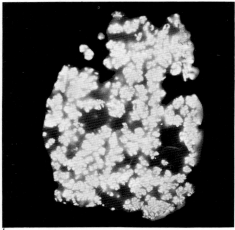

Uraninite

Uraninite crystals (a) are shown in matrix. The distribution of the uranium in the specimen is shown (b) by illumination through an autoradiograph.

earth and its inhabitants is best understood in terms of what is known about the present. Known also as uniformity of process, it has been more simply stated as "the present is the key to the past." The concept was formulated by James Hutton (1726–97), a wealthy Scot who was trained as a physician but became more interested in geology. Hutton's contribution did much to dispel the idea of *catastrophism, for it ruled out the necessity of universal catastrophes to produce biologic and geologic change.

Although basic to much of modern geologic thought, uniformitarianism should not be taken as scientific dogma. Geologic processes of the past may have been similar to those of today, but there is no requirement that they operated at the same rate or scale, and uniformitarianism may not be valid in all geologic interpretation. (W. H. M.)

Unit Cell. The smallest volume or parallel-epiped of a three-dimensional *crystal lattice that contains a complete sample of the crystal.

Upwelling. A nearly vertical movement that carries cold water to the surface of the oceans. It causes the unusually cold water found along some coasts such as that of California and Peru. It is due to coastal winds and the *Coriolis effect resulting from the rotation of the earth. For example, along the west coast of the United States, winds and currents move from the north to the south, but the Coriolis effect deflects the surface water to the right, that is, off shore. As the warm water flows away, it is replaced by cold water from below. This type of circulation causes precipitation of phosphorite nodules and supports abundant organisms by providing much nutrient material. The reverse situation, that is, movement of surface water toward the coast, can cause a sinking phenomenon (downwelling), which carries warm water to much greater depths than normally expected. (R. A. D.)

Uraninite. A mineral, uranium oxide, the principal ore of uranium. It is a primary mineral, and most of the other uranium minerals form by its alteration. Although uraninite is found in crystals, far more common is the massive or botryoidal variety pitchblende.

Uraninite is a sparse constituent of certain igneous rocks and pegmatites. It is found more abundantly in hydrothermal veins associated with nickel, cobalt and copper minerals, as at Great Bear Lake, Canada, and at Joachimsthal, Czechoslovakia. The Shinkalobwe mine in Zaire has been the world's largest producer of uranium. Here in vein deposits uraninite, partly altered to brilliantly colored secondary uranium minerals, is associated with copper and cobalt minerals. Small detrital grains of uraninite are found in the gold-bearing Witwatersrand conglomerate of South Africa. Although the concentration of the mineral is low, the deposits are extensive, making South Africa a major uranium producer.

The composition of uraninite is given as UO_2, but the mineral is always partly oxidized to UO_3. Several other elements are also present, notably lead and helium. The lead is the stable end product of radioactive disintegration of uranium, and the helium (α particles) is emitted during the process. Because the rate of radioactive disintegration is constant, the amount of lead and helium present can be used to determine the time elapsed since crystallization of the mineral and hence the age of the rock in which it is found.

Although uraninite has been known since 1727 it assumed no importance until radium was discovered in it in 1898. For the next 40 years it was mined as the source of this new element. Today uranium has a unique place among the elements, and as a consequence its chief ore, uraninite, is widely sought. Uranium is susceptible to nuclear fission, that is, its atoms may be split apart, with the release of a tremendous amount of energy. This was the energy of the atomic bomb. Today it is being used for generating electric power by nuclear power reactors. With the depletion of fossil fuels such reactors will play an ever-increasing role in the production of energy. And to supply uranium for these reactors, the search for uraninite will continue into the indefinite future. See Mineral Properties; Radioactive Age Determination. (C. S. H.)

Uvarovite. A mineral, the calcium-chromium *garnet with an emerald green color. Uvarovite characteristically occurs in small crystals in veinlets crossing black chromite. It has all the physical characteristics that would make it a lovely gemstone, but crystals are usually too small for cutting.

Valley. A trough or linear depression with higher land on both sides. Generally, a valley has a longitudinal slope seaward and is occupied by a river. A valley may be V- or U-shaped. If it is narrow, deep and has steep, vertical rock walls it is called a canyon or gorge. A deep gorge may also be called a chasm or ravine.

Valley Train. A long, trainlike deposit of sand and/or gravel deposited by meltwater streams in a valley in front of a terminal or a recessional *moraine. It may extend for many miles down valley, the finest sediment being carried the farthest. Valley trains are frequently characterized by *kettle holes, which may be occupied by lakes. Where the kettle holes are large, the deposits were laid down on stagnant ice of variable thickness. Valley trains are being formed today in Alaska, New Zealand, Patagonia, the Alps and elsewhere in valleys occupied by glaciers. (R.L.N.)

Vanadinite. A mineral, lead vanadate, which forms as a secondary mineral in the oxidized portion of lead veins. Its red to orange-red color distinguishes it from similar-appearing *pyromorphite, which forms under the same conditions. See Mineral Properties.

Van Allen Belts. Concentrations of very high-energy protons and electrons trapped within the earth's *magnetic field. The high-energy belts assume the form of crescent-shaped rings surrounding the earth at distances of about 3,000 km and 25,000 km respectively. The charged particles are constrained to spiral back and forth from hemisphere to hemisphere about the lines of magnetic force. The belts are named after J. A. Van Allen, an American physicist and one of the scientists who discovered the belts in 1958 by interpreting radiation measurements from Geiger counters carried on the Explorer 1 satellite. Explorer 1, an early space probe launched by the United States in 1958, was also used extensively to determine atmospheric density at heights near 350 km. (M.F.H.)

Van der Vaals Bond. The weak attractive force that holds together neutral molecules and essentially uncharged units in a crystal lattice. This force, also called the stray-field bond, is the weakest of all the chemical bonds. See Bond, Chemical.

Varves. A pair of thin sedimentary layers, one thick and one thin, deposited within a one-year period. They generally occur in glacial lake sediments and consist of a lower, light band of silt grading upward into a dark band of fine clay. Each varve repre-

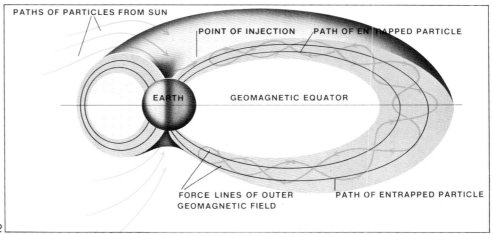

Valley Train
1 Meltwater from a retreating glacier has deposited a broad channel of sand and gravel at Mount McKinley National Park, Alaska.

Van Allen Belt
2 Most of the electrically charged particle radiation from the sun is trapped in zones, called the Van Allen belts, within the earth's magnetosphere. Streams of electrons and protons in the Van Allen belts spiral back and forth along the lines of force.

1

2

sents a record of climatic variation during the year it was deposited. The lighter colored summer layer consists of silt that settled during the warm season; the thinner, darker winter layer of clay was laid down during the colder months. Studies of glacial varves have been used to determine the age of Pleistocene deposits representing the last Ice Age. See Glaciolacustrine Deposits. (W.H.M.)

Vein. Table-like or sheetlike bodies of one or more minerals deposited by groundwater or hydrothermal solutions in the openings of *joints, fissures and *faults. Such veins are commonly banded and may or may not fill the openings completely. They are easily differentiated from the wall rock by their well-defined borders. The arrangement of the minerals in bands parallel to the wall rock, and the open spaces in the middle of many veins, proves that the precipitation of the mineral matter took place from the walls of the openings in toward the center. Quartz, calcite and gypsum are common minerals in the veins formed by ordinary groundwater; gold, silver, copper and other metals occur in the veins formed by hydrothermal solutions. Veins range from a centimeter wide and 30 or so centimeters long to many meters wide and hundreds of meters long. They may be vertical, inclined or horizontal. A vein is younger than the rock in which it is found, and where one vein crosses another, the one that is intersected is older.

Some veins are composed of one or more worthless minerals; others in addition to the worthless *gangue minerals contain valuable ore minerals. The latter type is commonly referred to as a vein deposit or lode. (R.L.N.)

Ventifact. A stone shaped by the abrasive action of windblown sand. Ventifacts are generally polished, pitted and grooved, and feel greasy as a result of their smoothness. They range in size from a centimeter to a meter or so, but their size is limited because windborne sand is generally carried only about 1 m above ground level. Sandblasting cuts a steep facet on the windward side of large fragments. On smaller fragments two or more facets are commonly formed because the wind comes from more than one direction or the fragments have been overturned by *deflation. A ridge or edge is formed where facets intersect. If a ventifact has one edge it is sometimes called an *einkanter* (German for "one edge"); if three edges, a *dreikanter* ("three edges").

The development of ventifacts is favored by strong winds, minimum vegetation, enough sand to cut but not enough to bury the fragments, flat terrain, a limited amount of snow so that the fragments are not snow-covered for long, and time. These conditions are found in hot deserts like the Sahara and Kalahari, in the cold deserts of Antarctica and Greenland, on the leeward side of dry washes in arid regions, and inland from sandy beaches. Ventifacts are also found in the northern part of the United States and Europe. There they were formed in late glacial times before vegetation reestablished itself in areas uncovered by a glacial retreat. The sand generally came from glaciofluvial deposits. These ventifacts usually reveal their antiquity by a *limonite stain and an absence of polish. In areas where ventifacts are common, the *bedrock may also be polished and cut. (R.L.N.)

Vermiculite. Derived from the Latin "to breed worms," the name given to a group of platy hydrous silicates closely related to chlorite and montmorillonite. When rapidly heated to 300° C (570° F) it has the remarkable property of expanding 20 times its size into wormlike forms. The expanded material is used in heat and sound insulation, as a packing medium, a soil conditioner and a potting medium for seeds, and as a filler in paint or paper. Large deposits of vermiculite are found in South Africa, Australia, Brazil and the Soviet Union. In the United States it occurs in the Carolinas and in Montana.

Vesicle (or Gas Cavity). Spheroidal, tubular or roundish holes in glassy or fine-grained *igneous rocks, formed when liquid *lava containing gas bubbles solidifies. Vesicles are generally less than 2 cm in diameter but may be larger, especially where they coalesce. As *magma moves upward in the crust, the pressure decreases and the gases in the magma come out of solution and form bubbles. Many of these bubbles escape into the atmosphere. However, if the magma is very viscous, some bubbles are trapped in the rock when it solidifies. Abundant vesicles are found only in *extrusive igneous rocks, but some do occur in shallow intrusives. Vesicles tend to be spherical if formed in ponded stationary lava and elongated in the direction of flow if formed in moving lava.

Pipe vesicles are hollow, elongated tubes projecting upward from the base of a flow and more or less perpendicular to it. They may be many centimeters long and are usually less than 2 cm in diameter. Most authorities believe they are formed by steam rising into the flow from the wet rocks on which the lava was extruded; some experts contend that they are formed

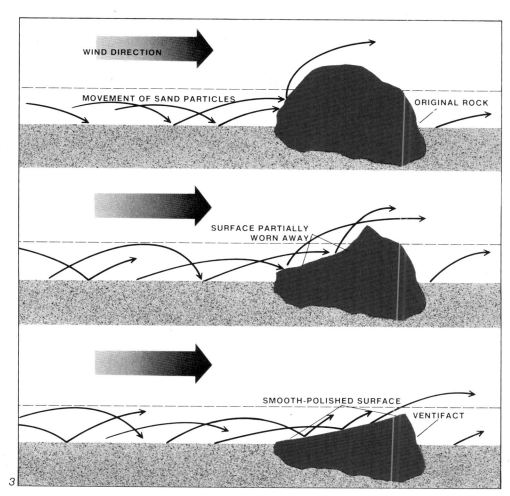

3

4

Varve
1 This name for thin layers of clay and silt seasonally deposited in lakes and often showing dark and light alternating colors is derived from the Swedish for "cycle."

Vein
2 A fracture in gneiss is here filled by light-colored quartz.

Ventifact
3 In deserts and other areas where vegetation is absent, wind-driven sand may cut smooth facets on the surface of rock fragments.

4 These stones on the Saudi Arabian desert have developed sharp edges and surface sheen from the impact of blowing sand grains.

Volcanic Chimney
1 *Steep-walled conduits, such as this volcanic chimney in Mount Mihara, Oshima Island, Japan, may descend several kilometers to a subterranean reservoir of lava.*

Volcanic Eruption
2 *A submarine eruption formed a new island near Fayal Island in the Azores in October, 1957.*

by air trapped beneath the advancing lava. A rock characterized by abundant vesicles, such as *pumice, is said to be vesicular. Vesicular rocks are found not only on the earth but also on the moon. (R.L.N.)

Volcanic Chimney (or Volcanic Conduit). See Volcano.

Volcanic Conglomerate. A consolidated rock consisting of rounded and almost rounded fragments composed of extrusive rocks together with a sandy matrix of *tuff. The rounding results from transportation by running water or, much less commonly, by waves and shore currents. The consolidation is due to cementation. These conglomerates are generally derived from coarse *pyroclastic rocks and debris. (R.L.N.)

Volcanic Eruption. An early classification of volcanic eruptions recognized two types, the quiet and the explosive. This is so general, however, that it is of little use. Today many kinds of eruptions are recognized, of which Icelandic, Hawaiian Strombolian, Vulcanian and Pelean are the most important.

In an Icelandic eruption, large quantities of fluid basaltic lava issue from long fissures and spread over wide areas. After a succession of such eruptions from fissures that change location from time to time, lava plateaus are built up. Examples are the Columbia River lava plateau and India's Deccan Plateau.

In a Hawaiian eruption, large quantities of basaltic lava with only subordinate *pyroclastic material erupt from summit and flank vents. The vents, unlike those in the Icelandic type, do not change location much. Thus *shield volcanoes are constructed, such as Hawaii's Mauna Loa and Kilauea. *Lava fountains and curtains, sometimes more than 450 m high, are commonly associated with this kind of eruption.

In a Strombolian eruption, gases streaming through molten lava in the *crater throw up clots of lava which on impact form *bombs. The eruptions are of minor intensity. They may be nearly continuous or separated by intervals ranging from a few minutes to an hour or more. Periodically more intense eruptions take place, during which *lava flows issue from the crater. Stromboli, on one of the Lipari Islands north of Sicily, exhibits typical Strombolian eruptions, although other types occur there as well.

. The Vulcanian is a very violent eruption characteristic of some composite volcanoes; it usually occurs after a long period of relative quiescence. Dark ash-

Volcanic Neck

1 Shiprock, a steep-sided feature rising 450 m (1300 feet) above the surrounding plateau in New Mexico, is the last remnant of an eroded volcano.

Volcanic Spine

2 The eruption of Mount Pelée, on Martinique, West Indies, in 1902 killed 30,000 people and thrust up a column of stiff lava 300 m (1000 feet) high.

laden clouds of atomized lava rise to great heights. In the final stages large quantities of volcanic gases containing a little ash rise thousands of meters. The abrasive action of these high-velocity, ash-charged gases cores out the chimney of the volcano and deepens and widens the crater. The Pelean is a violent eruption characterized by *nuées ardentes (glowing clouds). These clouds are a product of viscous gas-charged sialic *magma. They are commonly black during daylight but may glow dull red at night. Their movements are controlled by the force of the eruption and by gravity, but they also achieve mobility from the gases emitted from the material of which they are composed. The high temperature and velocity of the clouds make them one of the most destructive of volcanic phenomena. See Volcano, Composite. (R.L.N.)

Volcanic Neck (or Volcanic Plug). An isolated, steep-sided, tower-like, nearly circular hill or mountain composed of the more resistant rocks in the conduit of an extinct volcano. It is an erosional remnant that persists after much or all of the volcano of which it was a part has been destroyed. A neck may be as much as 450 m high and is usually less then 1 km in diameter. It is composed of lava, or of well-consolidated *pyroclastic rocks (often intruded by ramifying dikes) or a combination of both. A large neck, the remnant of a main volcano, is often surrounded by smaller necks from the volcano's parasitic cones. *Dike ridges may radiate kilometers away from necks.

Volcanic necks are found in the United States in New Mexico, Arizona, Utah and Wyoming, and in France, Scotland and elsewhere. A well-known example, Shiprock in New Mexico, is approximately 450 m high; among its dike ridges is one that projects tens of meters above the countryside and stretches away from it for many kilometers. Devil's Tower in Wyoming, generally thought to be a lava neck, towers more than 300 m and is distinguished by the length and excellence of its *columnar jointing. (R.L.N.)

Volcanic Spine. An obelisk-like protrusion of sialic *igneous rocks pushed up through cracks in the carapace of a *cumulo-dome as very viscous lava. It may be hundreds of meters high. The volcanic spine extruded from the dome in the crater of Mount Pelée several months after the disastrous eruption that destroyed St. Pierre, Martinique, reached a record height of approximately 300 m and had a diameter up to 150 m. Its maximum rate of rise was 12 m per day, but continuous collapse rapidly reduced its

Volcano
*Vesuvius, rising 1160 m (3850 feet)
above Italy's Bay of Naples, has had
violent eruptions since 79 A.D.; the
latest occurred in 1944.*

height and in time completely destroyed it. In places its sides were smooth, slicken-sided and grooved by the irregularities on the walls of the vent through which it was extruded, indicating that it was a plastic lava. Smaller spines frequently cover cumulo-domes and block *lava flows. A spine, which is a constructional feature, should not be confused with a piton, a steep volcanic neck that is an erosional landform. (R.L.N.)

Volcano. An opening at the earth's surface from which hot gases, molten liquids and solid materials erupt; also, the structure built up by the accumulation of materials thus ejected from the opening. The word "volcano" comes from Vulcan, the black-smith of the gods in Roman mythology; the ancients thought that Vulcan's forge was beneath Vulcano in the Lipari Islands near Sicily.

Volcanoes range in size from mounds a few meters high to the greatest moun-tains on earth. Hawaii's Mauna Loa, one of the world's largest volcanoes, rises 9000 m (30,000 feet) from the ocean bottom, is about 600 km (360 miles) around at its base and has a volume of about 40,000 cubic kilometers (10,000 cubic miles).

Volcanoes are typically cone- or dome-shaped, depending primarily on the ma-terial ejected. Cone-shaped volcanoes are formed when viscous lavas or fragmen-tal material predominate; dome-shaped volcanoes when fluid lavas predominate. If a volcano is active or has suffered little ero-sion, a bowl- or funnel-shaped depression called a *crater occurs at its summit. Larger depressions called *calderas, due to explosion, collapse or a combination of both, are also found at summits. The vent or opening from which the volcanic material is erupted is generally at the bot-tom of the crater or caldera. A rim separates the crater or caldera walls from the outer slope of the volcano.

A volcano consists of four main units. The volcanic edifice, or superstructure, stands on the topography as it was before the first eruption. The volcanic hearth (magmatic chamber) is located deep in the crust or in the mantle beneath, and from it comes the magma from which the hot gases, molten liquids and solids are derived. Between the superstructure and the hearth is the substructure; extending downward from the vent through the superstructure and substructure to the volcanic hearth is the chimney.

The two principal volcanic gases are steam and carbon dioxide. Other volcanic gases are nitrogen, carbon monoxide, hydrogen, chlorine, hydrochloric acid gas, hydrofluoric acid gas, sulfur dioxide, sulfur trioxide and hydrogen sulfide. The volume of gases emitted during a volcanic eruption may be very great; their emission throughout geologic time has probably been the most important factor in the origin and development of the *atmosphere and the *hydrosphere.

Lava flows issue from vents at the summit and flanks of the volcanic structure. They vary in thickness and in other ways, depending upon their chemical composition. Many kinds of volcanoes are wholly or partly composed of lava. Fragmental material such as volcanic dust, ash, lapilli, blocks and bombs are also ejected from vents, and the consolidation of these materials forms *pyroclastic rocks. The fragmental materials are also important in the formation of many kinds of volcanoes. During eruptions dust may be thrown into the upper atmosphere and deposited over tens of thousands of square kilometers. Darkness may result when sufficient quantities are erupted into the atmosphere.

A volcano is a short-lived geologic feature whose ultimate fate is complete destruction. Running water and, to a lesser extent, glaciers, *mass wasting and waves and shore currents are the erosive agents that destroy a volcano; so do its own explosions. Even during a volcano's greatest activity there are periods of quiescence, when running water will gully its slopes, particularly if unconsolidated fragmental material veneers them; renewed activity will bury these gullies. During the lifetime of the average volcano this growth-erosion-growth cycle will be repeated countless times. But once active growth is over, intensive erosion sets in. The outer part of the volcano is destroyed and the more resistant material filling the chimney forms a neck, which for a time increases in relative height. Later the superstructure, and still later the substructure, may be destroyed, and finally the intrusive igneous rock body, formed from the solidification of the volcanic hearth that fed the volcano, can be exposed at the surface by erosion. The *batholiths and other intrusive rock bodies that crop out at the earth's surface in New England, California and many other places may have given birth to volcanoes when they were molten. In places it can be demonstrated that pyroclastic rocks and lava flows are of the same age and chemical composition as nearby exposed batholiths. It seems likely that in these places we have the solidified magmatic chamber responsible for these volcanic rocks.

Probably every part of the earth's surface has at one time been the site of volcanic activity. The evidence for this is in the distribution of both active and extinct volcanoes, the widespread lava flows on land and under the ocean and, more important, the worldwide distribution of ancient lava flows and pyroclastic rocks often buried by younger rocks. There are about 500 active volcanoes, and some 2500 eruptions have occurred during historic time. In addition, there are several thousand recently extinct volcanoes. They are found on all continents except Australia. The active ones are concentrated chiefly in two elongated belts. The circum-Pacific belt (or the Ring of Fire) of 283 active volcanoes encircles the Pacific Ocean. The Alpine-Himalayan belt of 98 active volcanoes (77 of them in Indonesia) extends from the Canary Islands through the Mediterranean Region to the Bay of Bengal and through Indonesia. There are 84 active volcanoes in Africa and Antarctica and in the Pacific, Atlantic and Indian oceans; and others in the West Indies and Iceland.

A volcano may give warnings when it is about to erupt. If it has a snowfield, the *snowline may get progressively higher. *Fumaroles may suddenly develop or, if already present, may increase in temperature. When there are local earthquakes and small amounts of volcanic gases and dust are emitted, an eruption may be imminent. The rise of magma toward the earth's surface can be detected with magnetic instruments, since magma is nonmagnetic and reduces the local magnetic intensity. At Kilauea on Hawaii a tilting of the ground precedes an eruption, and the top of the volcano may rise about 1 m. Following the eruption, the top of the volcano commonly sinks about 1 m, so it has been said that the volcano "breathes." The thrust of magma from the depths into the uppermost part of the crust is thought to have been responsible for both tilting and uplift.

Among the famous volcanoes of the world are Vesuvius, Etna and Stromboli (called the Lighthouse of the Mediterranean) in the Mediterranean area; Ararat of biblical fame in Turkey; Krakatoa in the East Indies; Kilimanjaro in Africa; Mayon in the Philippine Islands; Fujiyama in Japan; Mauna Loa and Kilauea on the Island of Hawaii; Katmal in Alaska; Paricutín and Popocatepetl in Mexico; Chimborazo and Cotopaxi in the Andes; Mount Pelée in the Caribbean; and Mount Hekla in Iceland. In the continental United States there are Lassen, Shasta, Crater Lake, Hood and Rainier, all in the Cascade Mountains of California, Oregon and Washington. See Cinder Cone; Cumulo-Dome; Shield Volcano; Volcanic Eruptions; Volcano: Active, Dormant, Extinct; Volcano, Composite; Volcano, Compound. See also Color Plates on "Volcanism." (R.L.N.)

Volcano, Active. Depending on their activity, volcanoes are considered active, dormant or extinct. Some authorities define "active" as erupting in the past 50 years, "dormant" as erupting in historic time but not in the past 50 years, and "extinct" as not erupting during historic time. This definition of an extinct volcano is arbitrary, however, since volcanoes that have not erupted during historic time may erupt at any time. There may be little heat in the magmatic chamber of an extinct volcano, but this cannot be determined by superficial observation. If the *geothermal gradient is considerably higher than the average in an area where there is an "extinct" volcano, the classification should perhaps be questioned. A truly extinct volcano will probably have suffered significant erosion. (R.L.N.)

Volcano, Composite (or Strato-volcano). A steep-sided, cone-shaped volcano composed of interbedded lava and pyroclastic rocks. It may be a few thousand meters high and scores of cubic kilometers in volume. Near the summit the slope may be 30° or more, but it is much flatter near the bottom; the sides are therefore concave. The interbedding of lava and pyroclastic rocks produces a rude stratification that explains its name.

Sills and dikes are very often present. These, together with the flows, strengthen the cone and make it more resistant to erosion. A composite volcano is commonly composed of andesitic or more siliceous material. When it is active, the eruption of quietly extruded lava flows is interspersed with more violent eruptions during which pyroclastic detritus is ejected.

A *crater or *caldera is found at the summit if the volcano is active or has suffered only minor erosion. Parasitic *cinder cones and *cumulo-domes are commonly found on the flanks. Composite volcanoes are steeper than *shield volcanoes and contain more pyroclastic material. They are steep-sided like cinder cones but much larger. Even in low latitudes they are commonly covered with snowfields and radiating glaciers because of their great height. Till and other glacial deposits may therefore be interbedded with lava and pyroclastic rocks, as at Crater Lake, Oreg.

Most composite volcanoes are graceful cones of great beauty and grandeur. The following world-famous volcanoes are all composite: Shasta, Hood, St. Helens and Rainier in the Cascade Range; Popocatepetl and Orizaba in Mexico; Fujiyama, Japan; Vesuvius, Italy; Mayon, Philippine Islands; Cotopaxi, Ecuador; Shishaldin, Aleutian Islands; and Kilimanjaro, Tanzania. (R.L.N.)

Volcano
A deposit of fine, dustlike ash, called tephra, blankets Sunset Crater, Ariz., which last erupted in 1064 A.D.

Volcano, Compound. A volcano composed of two or more of the simpler types of volcanic edifices, such as *cinder cones, *spatter cones, *shield volcanoes, composite volcanoes and cumulo-domes. Sicily's Etna, a typical compound volcano, is a huge shield volcano topped by a cinder cone 300 m high. This complex structure indicates a recent change in its eruptive processes. During most of its history *lava flows, erupting quietly, built the shield volcano; more recently its eruptions have been more violent and have formed the cinder cone. Mount Erebus on Ross Island, Antarctica, is an active 4000-m compound volcano with a complex history. The base is a shield volcano with gently sloping sides; a composite volcano has formed on it, and three younger cones are in a *caldera in the composite cone. (R.L.N.)

Volcano, New. The birth of a new volcano is not only a dramatic and awesome spectacle but a rare event. Only about a dozen have been observed in historic time. They have taken place on the ocean bottom as well as on land. (*Parasitic cones formed during historic time, like those on the slopes of Etna in Sicily, are not included in this category since their chimneys are related to the extant main volcano). Three of the new 20th-century volcanoes have been thoroughly studied and documented: Paricutín in Mexico, as well as Surtsey and Heimaey off the coast of Iceland.

Paricutín is located about 300 km west of Mexico City in an area containing large *shield volcanoes and hundreds of basaltic *cinder cones. On February 19, 1943, approximately 300 earthquakes occurred in the vicinity of Paricutín; the next day the volcano was born in a cornfield. It was 36 m (120 feet) high on the second day, 92 m (300 feet) on the fifth day, 222 m (722 feet) by the fifteenth day, and 436 m (1410 feet) a year after its birth. On the first night of its emergence it was a spectacle of breathtaking magnificence. The noise of the explosions could be heard throughout the State of Michoacan and even in Guanajuato, 300 km (186 miles) away. With each explosion literally thousands of red-hot fragments were thrown 600 to 900 m (2000 to 3000 feet) above the crater rim. Lava first erupted on the second day. During the life of the volcano, all but one of the flows issued from the flanks of the cone. During the first year, lava buried the towns of San Juan Parangaricutiro and Paricutín; ash fell frequently in Uruapan about 30 km (20 miles) away, and fine ash fell in Mexico City 300 km (190 miles) away. During its nine years of life it ejected lava and pyroclastics at an average rate of 1.2 million metric tons per day.

Since many oceanic islands are volcanic, and since oceans cover three-fourths of the earth's surface, it is not surprising that new volcanic islands have appeared from time to time in the Mediterranean, Pacific and Atlantic. Usually the new islands reach a height of only tens of meters before being destroyed by wave action. Among the phenomena that accompany submarine eruptions is the boiling of the ocean and the formation of steam columns, dead fish, black columns of tephra erupting from the concentric waves in the ocean, pumice floating on the water, strange odors and abnormal colors in the sea.

The most carefully studied of the new volcanic islands is Surtsey, born during the night of November 14, 1963, about 30 km (20 miles) off the southern coast of Iceland near the Vestmann Islands. For three days preceding the eruption a smell of sulfur was noticed on Heimaey, an island about 20 km (13 miles) from the new volcanic area, and on November 13 a vessel engaged in oceanographic research about 3 km away found that the temperature of the water had suddenly risen from a normal 7°C (44.6°F) to 9.4°C (49°F). On November 14, concentric waves emanated from the new volcanic center, waves began breaking on a new shoal, and the ocean had an unnatural brownish-green color. Later in the day a steam and tephra column rose hundreds of meters in the air. The next night Surtsey was born. It was 40 m (130 feet) high and 600 m (2000 feet) long on November 16, some 60 m (200 feet) high and 600 m long on November 19, and approximately 2100 m (7000 feet) in extent on August 24, 1965. The great quantity of lava erupted makes it certain that Surtsey, regardless of marine erosion, will survive for a long time.

Ten years after the new island of Surtsey was born a new volcano erupted on the island of Heimaey itself.

On January 21, 1973, scientists on the Icelandic mainland began to observe seismic activity of an unknown origin. Two days later, on January 23, a gaping fissure 1800 m (6000 feet) long and 3 m (10 feet) wide silently opened behind the town of Vestmannaeyjar, the largest fishing port in Iceland, and spewed up a molten *lava curtain more than 100 m high. The population of Heimaey was immediately evacuated to the mainland. Three days after the fissure appeared, on January 26, the curtain had subsided to three lava fountains over a length of 300 m (1000 feet) and other volcanic activity intensified, consisting of streams of basaltic lava, volcanic gases and bombs. A cone of ashes formed on the fissure, growing rapidly to a height of 100 m. On January 28 a television camera was installed on a nearby cliff, enabling all of Iceland to watch the progress of the new volcano. By January 30 the summit of the cone was 185 m (600 feet) high; more than 100 houses had disappeared under ashes or had been burned by flying rocks and bombs. On March 25 a lava lake rose in the crater and overflowed, threatening to fill the bay of the port. In order to arrest and divert the flow, an emergency force began to spray it with huge quantities of cooling seawater and built a barrier of ashes in front of it with trucks and steamrollers. After two weeks, their efforts succeeded. The volcanic activity lasted six months, until July 3, 1973, by which time the new volcano, named Eldfell ("mountain of fire"), had reached a height of 225 m (743 feet) and had emitted 250 million cubic meters (875 million cubic feet) of volcanic products. Three hundred houses were destroyed but the port and fishing industry were saved. The cone added 2.5 square kilometers (1.7 square miles) of arable land to the island's area. (R.L.N.)

Volcano, Shield. See Shield Volcano.

Vorticity. A measure of the rotation or spin, usually around the local vertical axis or zenith point, at a given point within a fluid. Within the atmosphere or the ocean, two aspects of this vertical component of the spin have to be considered: (1) the spin relative to the earth beneath, called the relative vorticity; and (2) the spin representing the component of the earth's rotation about the local vertical, sometimes called the earth's vorticity. See Convergence; Divergence. (M.F.H.)

W

Wad. An impure mixture of black hydrous manganese oxides.

Wash Load. In a stream, that part of the sediment load that is finer than the material in the channel bed. It generally consists of very fine-grained particles carried in suspension.

Waterfall (or Cataract). A steep (vertical or almost vertical) fall along a stream course, denoted by a break in a smooth longitudinal profile. It is differentiated from cascade, which is a series of small falls, and from rapids, which are not as steep. Waterfalls are commonly caused by *differential erosion. One type occurs in horizontal sediments where there is a resistant cap rock. The force of the falling water erodes the material beneath it, undercutting the falls and creating a plunge pool, as at Niagara. Intrusive dikes may form vertical obstacles to a stream's erosion, resulting in a waterfall such as the Great Falls on the Yellowstone River in Wyoming.

Falls may also result from tributaries flowing over *hanging valleys to join the main stream. Hanging valleys are found in regions of alpine glaciation where the main valleys were more deeply scoured by ice than the tributary valleys. When the glaciers melt, the tributary valley is left looming high above the main stem, as at Yosemite Falls, in California. Hanging valleys may also be found along coastlines where sea cliffs are retreating faster than small gullies can cut down into their beds.

A third kind of waterfall is tectonic. In the 1959 earthquake at Hegben Lake, Montana, a *fault scarp cut across several streams, creating waterfalls. Such falls can also be created by regional uplift which rejuvenates a stream and causes a wave of downcutting, forming *knickpoints.

All waterfalls are ephemeral features. Because of the tremendous hydraulic force of the water, they retreat upstream until the profile of the stream is smoothed out.

It is difficult to establish the "greatest" waterfall in the world, for falls of the greatest height are rarely those of the greatest volume of water discharge. For example, Angel Falls on the Churun River, Venezuela, at 807 m (2648 feet) the world's highest waterfall, has a volume discharge of less than 560 cubic meters (20,000 cubic feet) per second. In fact, many falls over 300 m (1000 feet) in height have only a modest flow. Sete Quedas, or Guaira' Falls, on the Parana River, by contrast, is only 65 m (213 feet) high but has the greatest volume of water, 13,310 cubic meters (470,000 cubic feet) per second. Considering height and volume together, certain

Wad
Impure black manganese oxides, composed primarily of pyrolusite combine to form wad.

Waterfall

1 Niagara Falls, third greatest in the world in water volume, consists of two falls, both shown here. One is in the United States and the other in Canada.

2 Brazil's Iguazú Falls, 4 km (2.5 miles) wide, is broken up into more than 200 cataracts separated by rocky, forested isles. At peak flooding the volume of water falling into the gorges below averages 1700 cubic meters (60,000 cubic feet) per second.

3 Angel Falls on the Churún River, Venezuela, is the highest waterfall on earth, tumbling 807 m (2648 feet) from the rim of the Guiana plateau to the valley floor below.

falls are clearly the most spectacular—Victoria, Niagara, Paulo Alfonso, and Sete Quedas—and at different times each of these has been proclaimed the world's greatest falls by various observers. In any case, these magnificent natural features are among the most impressive on earth, in addition to being immensely important as sources of hydroelectric power. See Appendix 4; "Most Important Waterfalls."

Water Gap. A *col or pass through a resistant ridge, eroded and occupied by a river. This may result from either antecedence or superposition. In antedecence the river formed its channel before the ridge was uplifted and was able to maintain its course as the ridge rose across its path. In superposition the river formed its channel on rock lying above the structure but kept its course, cutting down into the resistant ridge below. Examples of water gaps may be found in many regions, including the Chalk Downs of England and the Folded Appalachians and Rocky Mountains of the United States. See Wind Gap.

Water Mass. A large body of water in which the particular temperature and/or salinity characteristics produce a unique density. In all oceans there is a tendency for a water mass with specific characteristics to retain its identity over a distance of hundreds or even thousands of kilometers. Such a water mass is typically formed when water flows from one place to another or from one environment into another, as, for example, when freshwater from a large river such as the Mississippi enters the marine environment. One of the classic water masses results when the Mediterrranean Sea moves through the Straits of Gibraltar but still remains fairly well defined in the middle of the Atlantic Ocean. (R.A.D.)

Watershed. See Drainage Basin.

Waterspout. A funnel-shaped uplifting of seawater. A waterspout forms under cumulonimbus clouds and results from air turbulence in continental air masses spreading out over the ocean. Although similar in structure to a tornado, a waterspout is usually smaller and less powerful. Seawater may be lifted only a few meters above the surface; however, the spray is carried much higher. Waterspouts are common in the Gulf of Mexico and in the subtropical waters of the southeastern coast of the United States.

Water Table. The irregular surface in the earth's crust below which the cracks, pores and other openings are filled with water. Above the water table the openings are partly filled with air. In humid climates the water table may approach the ground surface, but in arid areas it may be tens of meters below. It tends to be some distance below the surface of hills but closer to the surface of valleys; it therefore reflects the earth's surface but with less relief. It fluctuates, becoming low during a drought and high during wet periods.

The zone of aeration is that in which interstices and other openings in the mantle and bedrock partly fill with air; it extends from the earth's surface down to the water table. The zone of saturation is that in which interstices and other openings are filled with water; it extends from the water table to depths where the pressure is so great that open spaces cannot exist. Phreatic (from the Greek word for "well") water is the subsurface water below the water table and in the zone of saturation. Groundwater is the subsurface water in the open spaces in *regolith and bedrock. Many geologists restrict the term "groundwater" to the subsurface water below the water table (phreatic water). Suspended water is the water in the zone of aeration found immediately below the earth's surface (belt of soil moisture) which is utilized by plants and is prevented from migrating downward to the water table by capillarity; water found in the interstices immediately above the water table (capillary fringe) which is held then by surface tension and is continuous with the water below the water table; and water in an intermediate belt adhering as films to the boundaries of openings. See Subsurface Water. (R.L.N.)

Water Vapor. Water in gaseous or vapor form; one of the phases of the *hydrologic cycle. Water vapor in the air has its source in *evaporation of water from the earth's surface. Its primary atmospheric sink is a few kilometers above the surface. There, as it changes to liquid through condensation or to ice crystals through sublimation, it becomes cloud. During the evaporation-condensation cycle, water vapor carries latent heat energy, mainly from the oceans, into the atmosphere. In this way a large portion of the solar radiation absorbed at the earth's surface is made available as energy for *atmospheric motions. Water vapor also plays a significant role in the atmosphere's *greenhouse effect by absorbing a portion of the terrestrial radiation and re-emitting some of this energy back to the earth's surface, thus keeping the surface warmer than it would otherwise be.

The term *atmospheric humidity is ordinarily used as the measure of water vapor in the air. Because water vapor enters the atmosphere at the earth's surface and usually condenses within the

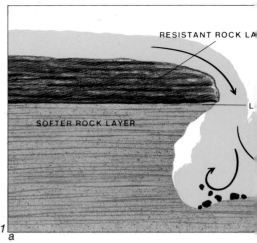

STAGE 1

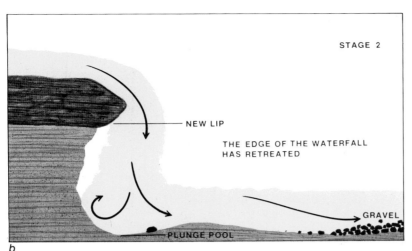

STAGE 2

NEW LIP

THE EDGE OF THE WATERFALL
HAS RETREATED

GRAVEL

PLUNGE POOL

b

Waterfall

1 A fall results when a stream falls over a resistant layer or cap rock (a) that covers a less resistant bed. As the less resistant bed is eroded (b), the undermined cap rock collapses, creating a new lip of the falls farther upstream. Churning water at the base of the falls creates a plunge pool, while coarse gravel is deposited in less turbulent water downstream.

Water Gap

2 The Delaware River flows through a notch it cut in a resistant ridge at the New Jersey–Pennsylvania boundary.

Waterspout

3 Like tornadoes, waterspouts are swirling columns of raindrops that range up to 1500 m (1 mile) in height.

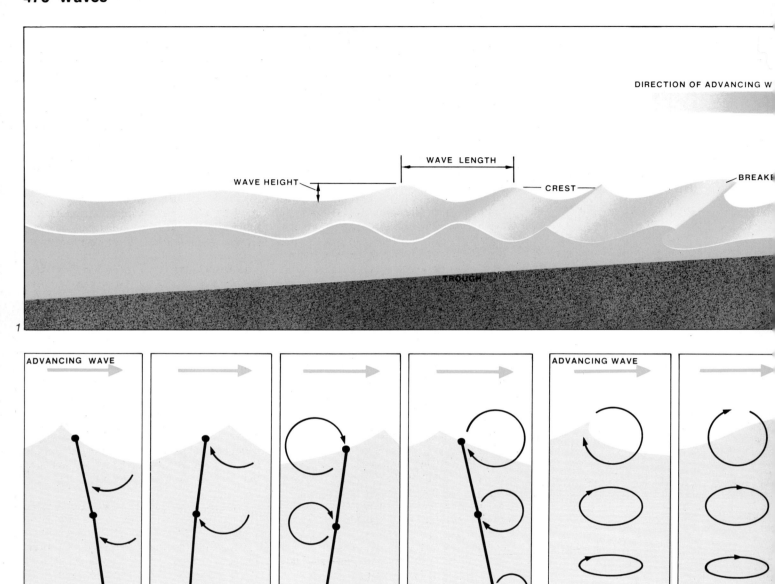

DIRECTION OF ADVANCING W

WAVE LENGTH

WAVE HEIGHT

CREST

BREAKE

TROUGH

1

ADVANCING WAVE

ADVANCING WAVE

DEEP WATER

SHALLOW WATER

SEA FLOOR

2

a

b

Wave

1 The height of a wave is the difference
in elevation between the highest point,
the crest, and the lowest point, the
trough. The wavelength is the distance
from one point on a wave to the
equivalent point on the next (for
example, from crest to crest). As an
advancing wave reaches shallow water,
the lower part of the wave touches
bottom, slowing it down, while the upper
part of the wave moves ahead. As the
crest steepens, it becomes less stable

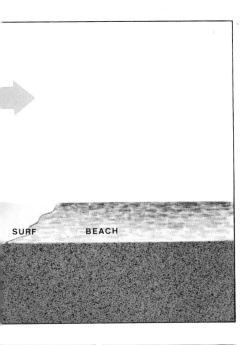

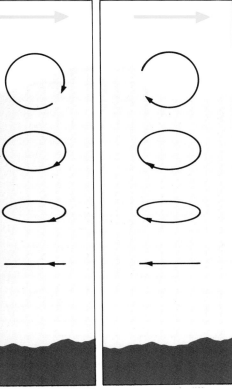

and finally topples over and washes up on the beach as surf.

2 In deep-water waves (a), water particles (shown as black spots) move in circles, their orbits decreasing with depth. In shallow water (b) the wave bottom drags along the ocean floor and the orbits of water particles are flattened into ellipses.

troposphere, only a minute fraction of the total water vapor contained in the atmosphere is found above this layer. About half of all atmospheric water vapor lies below a height of 2 km. The humidity varies with changes of weather and differences in climate around the earth. (M.F.H.)

Wave. A surface disturbance of water. Virtually all waves are formed by changes in air pressure (wind); a few result from an abrupt disturbance to the earth's crust. An ideal wave has a crest and trough that take the smooth shape of a sine curve. It is generally described in terms of wavelength (the distance between two corresponding points on successive waves) and wave height (the vertical distance between crest and trough). A wave may also be characterized by the velocity at which it travels.

The size and velocity of a wind wave is determined by wind velocity, wind duration and the distance of water surface over which the wind blows (the fetch). Two types of deep-water waves may result: a sea wave, which is a wave under the direct influence of the wind, and a swell, which is a wave after the wind has stopped or after a wave has moved beyond the area of active wind. A sea wave has a peaked crest and trough. In shallow water either of these waves may steepen and break, thereby losing its shape; it then becomes what is called surf. A breaking wave is said to spill where the break is slow and the water appears to be spiling over a wall. A plunging breaker is one that steepens and curls over, finally breaking with a crash of water. Surging breakers form from small waves that steepen and rush up on the beach.

Although the wave may move great distances across the water, the water itself will show little or no net movement. Anyone who has watched a floating object knows that it seems to move up and down while waves pass by; actually the motion is basically circular, reflecting the movement of water particles. On wave crests, the water moves in the same direction as the wave, whereas in the trough the water moves in the opposite direction. At the surface, the diameter of the circular motion is equal to the wave height, but the circle becomes smaller with depth and ceases altogether at a depth equal to about half the wavelength.

The steepening and eventual breaking of waves is caused by the interference between the bottom and the motion of water particles as the wave moves into shallow water. Eventually the velocity of the water particles in their circular paths exceeds the velocity of the wave itself and the wave collapses.

A wave is bent or refracted as it moves into shallow water and begins to feel the bottom because it may be unaffected by the bottom at one position but slowed down at another point. It also travels at different speeds at various places along its crest. This results in bending or refracting, and produces waves that tend to parallel bottom contours in shallow water. In coastal engineering, lines called orthogonals are constructed at right angles to wave crests as waves refract in shallow water. These orthogonals show lines of wave energy. They converge at headlands or points and diverge in bays, showing where wave energy is concentrated or dispersed.

In some respects, waves act like light rays or sound waves in that they too may be diffracted or reflected. Wave diffraction is a lateral dispersal of energy as a wave passes an obstruction such as a breakwater or a jetty. The energy of a wave is transmitted laterally in such a way that it causes a certain amount of water motion behind the obstruction as the wave passes. A wave also tends to bounce back from impermeable barriers such as jetties, seawalls or breakwaters; if such barriers are vertical and smooth, and if the wave approaches in an essentially parallel direction, it may undergo little or no loss of energy when it is reflected. (R.A.D.)

Wavellite. A mineral, a hydrous aluminum phosphate, usually found in spherical aggregates formed by radiating crystals. It is a comparatively rare secondary mineral found in lowgrade metamorphic rocks and in limonite and phosphorite deposits. See Mineral Properties.

Weather. The state of the atmosphere, mainly as it affects life and human activities. Weather consists of the short-term changes in the atmosphere, as distinguished from *climate, which has been described as "the synthesis of weather." Although in weather-observation practice the term "weather" is restricted to atmospheric phenomena, in its more general use it includes temperature, humidity, precipitation, cloudiness, visibility and wind, usually with reference to conditions prevailing over periods ranging from a few minutes to a few months. See Clouds; Cyclone; Precipitation; Temperature; Thunderstorm; Tornado; Wind.

Weathering. The alteration of rocks by mechanical and chemical processes at or near the earth's surface. Long-continued action breaks down the solid rocks into fragments, forming unconsolidated and friable material. Essentially weathering in-

volves the interaction between the rocks on the one hand and the atmosphere and groundwater on the other. Weathered rock may extend tens of meters down from the surface; the depth is determined by such factors as rate of erosion, climate, topography, kinds of rocks, permeability and time. When broken up by mechanical processes, a rock is said to have disintegrated; when broken down by chemical processes, it has decomposed.

The processes responsible for the mechanical breakup of rocks are: (1) the expansion that takes place in a confined space when water freezes, (2) the alternate expansion and contraction that results from changes in temperature and (3) the expansion that occurs when deeply buried rock is brought to the surface by erosion.

Chemical weathering results in new minerals. The important chemical reactions are oxidation (which involves oxygen), hydrolysis and hydration (the chemical activity of water), carbonation (the chemical activity of carbon dioxide dissolved in water) and solution (the removal of soluble materials). Chemical weathering is also caused by the biological activity of organisms and the products of their decay. Moreover, in the formation of new minerals by chemical weathering, there is an increase in volume that produces stresses capable of disintegrating rocks. The most important substances formed by chemical weathering are clay minerals; the carbonates of sodium, potassium, calcium and magnesium; silica; and the iron oxides. Several of the new substances are soluble in water; they reach rivers and ultimately oceans, where they are used by marine organisms. The importance of clay minerals to man can hardly be overemphasized, especially in the growth of plants. Chemical weathering takes place more rapidly in warm climates and where water is abundant. It is also more rapid where fragments are small and surface areas consequently greater.

A familiar example of chemical weathering is the rusting of iron. In the presence of moisture and oxygen, iron is changed into rust (limonite), which differs greatly from iron in its physical properties. Another common example of chemical weathering is the green discoloration (malachite) on copper; it results from the action of carbon dioxide and water on the copper.

Weathering is of great importance to man. Without it no soil would be formed on the surface of the earth and the oceans would be devoid of higher forms of life. Many ore minerals owe their origin in part to weathering; examples include bauxite (the most important ore of aluminum), some copper ores as well as placer gold, platinum and diamonds. Weathering changes bedrock into regolith, thereby facilitating erosion by running water and the other *exogenic processes. Evidence of the widespread occurrence of chemical weathering is the yellow and brown stain on and within rocks, the result of the weathering of minerals containing iron, and in the discoloration that takes place in the stones of old buildings. (R.L.N.)

Welded Tuff. A *pyroclastic rock formed by the consolidation of hot pumiceous detritus deposited by the volcanic clouds known as *nuées ardentes. The consolidation results from the welding together of the hot plastic fragments and the activity of contained hot gases. The weight of the deposits creates collapsed bubbles and *pumice and flattened elongated fragments. The flattening produces lamination in the tuff and later flow can cause stretching and distortion. This results in structures easily confused with the flow banding in *rhyolites; many welded tuffs, such as those in the Yellowstone National Park, were once thought to be rhyolitic flows. Because of the contraction resulting from cooling, *columnar joints may be formed. Welded tuffs are found at Crater Lake, Oreg., and many other places. *Ignimbrite plateaus are found in Yellowstone and on the North Island of New Zealand. (R.L.N.)

Wentworth Scale. A scale devised to screen sediments and measure grain size of *sedimentary rocks. The scale ranges from clay particles (diameter less than 1/256 mm) to boulders (diameter greater than 256 mm). Each size is multiplied by a power of 2 to give the next larger size. Terms used, in decreasing order, include boulders, cobbles, pebbles, granules, sand, clay and silt.

Westerly Winds, Prevailing. A belt of winds lying between 35° and 60° N and S latitude. Within the westerly wind belts, winds can blow from any direction, but the westerly components predominate. See Wind.

Widmanstatten Pattern. A system of intersecting bands seen on a cut, polished and acid-etched surface of an iron meteorite. Iron meteorites are composed largely of kamacite and taenite (nickel-iron). Etching will bring out kamacite bands because kamacite is etched more rapidly than the more nickel-rich taenite.

Willemite. A mineral, zinc silicate, found in quantity only at Franklin, N.J. Here associated with *franklinite and *zincite, it has long been mined as an ore of zinc. Most

2

3

Weathering

1 The freezing of water and the expansion of roots have widened this rock crevice.

2 A remarkable, wavelike rock near Hayden, in western Australia, is probably the result of erosion by wind.

Widmanstatten Pattern

3 An iron meteorite, cut, polished and etched with acid, reveals the pattern of its nickel-iron alloys. The ring around the edge of this specimen was caused by heat alteration.

willemite from Franklin fluoresces a vivid yellow-green in ultraviolet light. Elsewhere willemite is found—usually in minor amounts—in crystalline limestones and as a secondary mineral in the oxidized zone of zinc deposits. See Mineral Properties; Appendix 3.

Wind. Air in motion relative to the surface of the earth. Near the earth's surface vertical motions of the air are quite small, hence meteorologists use the term almost exclusively to denote the horizontal motion of the air. Vertical motions are usually identified as such or, in the case of strong vertical currents, as updrafts or downdrafts. Wind is measured by anemometers and wind vanes near the earth's surface; winds aloft are ordinarily found by electronically tracking balloon-borne instruments. Wind direction and speed in the free atmosphere may also be closely estimated from the pressure distribution.

In the free atmosphere the large-scale horizontal motions approximate the *geostrophic current velocity, which represents an exact balance between the *Coriolis acceleration and the pressure force. The geostrophic wind blows parallel to the isobars (lines connecting points of equal pressure), with low pressure to the left in the Northern Hemisphere and to the right in the Southern Hemisphere. Its speed is inversely proportional to the Coriolis parameter; and, being also proportional to the pressure-gradient force, it is inversely related to the spacing between the isobars.

Many characteristic local winds arise from the differential heating and cooling of the air in combination with topographic features: land and sea breezes, mountain and valley winds, fall winds, dust devils, etc. However, wind systems of much larger scale are also identifiable in the atmosphere. In the 17th century the easterly trade winds were identified by the English astronomer Edmund Halley (1656–1742) as a major feature of the general circulation. The trade winds are northeasterly in the Northern Hemisphere and southeasterly in the Southern Hemisphere. Occupying most of the Tropics, or about half the surface area of the globe, the trades constitute the return flow of the Hadley circulation in each hemisphere (see Atmosphere, Planetary circulation of). The Hadley circulations are thermally driven cells with rising motion near the Equator, poleward flow aloft, downward motion near latitude 30 degrees, and return flow toward the Equator, near the earth's surface. Converging, the trades produce the cloud belt that marks the intertropical convergence zone. The trades are the steadiest wind system on earth, showing much less variability than the midlatitude westerlies.

At the earth's surface, the westerlies extend from about 35 to 65 degrees latitude. Their boundary toward the Equator is fairly well defined by the subtropical highs (the high-pressure belts produced by the subsiding air of the Hadley cells), but their poleward boundary is ill-defined and variable.

The polar easterlies constitute shallow, not very well defined wind systems located on the poleward sides of the high-latitude low-pressure belts. The polar easterlies represent the return flow of air that drifts poleward and, cooling, subsides to form the shallow polar *anticyclones.

Above the earth's surface the belts of easterly winds become narrower, and the westerlies extend much farther toward the Equator and the poles. To an observer looking downward on the winter hemisphere, the westerlies would appear as the circulation of an immense vortex, its center at the pole and its outer rim at about latitude 10 degrees. At the base of the stratosphere, near latitude 30 degrees, the core of the westerlies in winter shows an average speed of 40 m per second (90 miles per hour).

Although the westerlies appear as zonal winds when averaged over time and at all longitudes, at any given moment they exhibit wavelike meanderings, called variously planetary waves, long waves in the westerlies and Rossby waves. However, laboratory experiments done with rotating fluids suggest that the waves in the westerlies are not pure Rossby waves (see Atmospheric Waves and Vorticity). The waves in the westerlies develop most strongly after a lengthy period of nearly zonal flow. Since the waves transfer heat from low to high latitudes, it has been inferred that the zonal flow becomes unstable—that is, favorable to the growth of wave disturbances—when the temperature contrast between low and high latitudes reaches a certain critical value. The laboratory experiments referred to above support this view. Experiments in which the fluid is contained in an annulus, or region between two cylinders of different radii, indicate that the convective circulation patterns assumed by a rotating fluid are related both to the meridional temperature contrast and to the rotation rate of the fluid.

Although earth's rotation rate is fixed, the rotation of the horizon plane around a vertical axis, with respect to a fixed reference frame, is variable with latitude: It is greatest at the earth's poles and zero at the Equator. The poleward temperature

Wind
1 In addition to the major wind belts, there are several localized winds in various parts of the world. Here a searing Santa Ana wind, a strong, dry east wind that blows across southern California from the Santa Ana Valley, picks up swirls of dust.

2 A sudden whirlwind, sometimes called a dust devil, lasts just a few minutes but can do much damage. Dust devils usually occur when the overheated earth causes strong upward and downward motions of air in the lower atmosphere; these motions create the whirlwind.

Wolframite

1 These shiny black crystals, typical of this ore mineral of tungsten, come from a high-temperature tin-tungsten vein in Portugal.

Wulfenite

2 Flat tabular crystals of this lead molybdate are found in brilliant hues of red, orange, yellow and brown.

decrease is least in the Tropics and is greatest at middle latitudes. The experiments with rotating fluids show that a slow rotation rate and weak temperature contrast across the annulus favor a circulation similar to that of the Hadley regime of the earth's atmosphere at lower latitudes. A more rapid rotation rate and a stronger temperature contrast favor a "Rossby regime," in which the zonal flow and cellular circulation of the Hadley regime are replaced by wave patterns like those observed in the earth's atmosphere at middle latitudes.

In summer, the westerlies at heights of about 30 km give way to easterly winds that extend from the poles to the Equator. It was once assumed that the easterly winds above the Equator were constant, but observations during the past 15 or 20 years have revealed an interesting periodicity, called the "biennial oscillation," in the winds of the equatorial stratosphere. About every 26 months, the stratospheric flow over the equatorial zone reverses direction, alternately blowing from the east and from the west. It is believed that vertically propagating, internal gravity waves of planetary scale provide the momentum sources for biennial oscillation. (M.F.H.)

Wind Gap. A *col or cut through a resistant ridge, presumably eroded by a stream (*water gap) that is no longer present. Such wind gaps are common in folded regions where *subsequent stream development on softer rock results in the capture of streams trying to erode across resistant layers. Numerous wind gaps can be found in the folded Appalachians in the United States and in England's North and South Downs.

Witherite. A mineral, barium carbonate, a member of the *aragonite group of minerals. It resembles other members of the group in crystal form and crystal structure. Witherite is a comparatively rare mineral, commonly found in mineralized veins associated with *galena. It is a minor source of barium; the more abundant mineral, barite, is the chief source. See Mineral Properties.

Wolframite. A mineral, an iron-manganese tungstate, the principal ore mineral of tungsten. The name is given to a mineral series extending from the pure iron tungstate, *ferberite, to the pure manganese tungstate, *huebnerite. Wolframite, having an intermediate composition, is more common than either endmember. It usually occurs in pegmatite dikes and high-temperature quartz veins associated with granites. The most impor-

tant tungsten-producing countries are China, the Soviet Union, Korea, Bolivia and Australia. See Mineral Properties.

Wollastonite. A mineral, calcium silicate, found usually as a *contact metamorphic mineral in crystalline limestones. Its common mineral associates are diopside, garnet, tremolite, idocrase and epidote. In those places where wollastonite comprises a major portion of the rock mass, it is mined to be used in manufacturing tile. See Mineral Properties; Appendix 3.

Wulfenite. A mineral, lead molybdate, found as a secondary product in the oxidized portion of lead deposits. It is a minor source of molybdenum and is valued by the mineral collector because of its well-formed crystals, high luster and yellow to orange-red color. See Mineral Properties.

XYZ

Xenoblastic. A texture generally found in metamorphic rocks in which the minerals have formed with irregular outlines instead of their characteristic crystal faces. The reasons for this irregular development are still not well understood. See Crystalloblastic Series.

Xenolith. A rock fragment foreign to the *igneous rock in which it is found. During the intrusion of an igneous magma, fragments may be broken from the country rock and completely surrounded by the magma, which later crystallizes. Such fragments are xenoliths.

X-ray Crystallography. The branch of the science of crystallography in which X-rays are used to study the internal structure of crystals. With an X-ray study of single crystals it is possible not only to measure the distance between successive atomic planes and thus determine accurately the dimensions of the *unit cell, but also to determine the positions of the various atoms within the crystal. In the powder method of X-ray investigation of crystals, the X-rays are diffracted from crystalline material that has been ground as finely as possible. The diffracted beams give a pattern, recorded on photographic film, that is characteristic of the material under study. Since every crystalline substance yields a pattern peculiar to itself, the X-ray powder method is a powerful tool in crystal and mineral identification. (C.S.H.)

Yardang. An elongated ridge carved in either *bedrock or *regolith by the abrasive action of windborne sand in a desert region. Yardangs appear as sharp-edged, streamlined ridges separated by steep-sided and round-bottomed troughs. Both ridges and troughs are elongated in the direction of the strongest prevailing wind and are most fully developed in soft rocks. The ridges range from a few centimeters to almost 8 m (25 feet) in height, but in the southern Lut of Iran some are said to stand over 200 m (650 feet) above the troughs. Because sand is generally transported only a few feet above ground level, yardangs are commonly notched at the bottom on the windward side, leaving the upper part to project like a cornice. They are found in Turkestan (where they were also first described); on the Peruvian coast; in Iran; in southern Natal, South Africa; on the Mojave Desert; and elsewhere. (R.L.N.)

Yazoo River. A tributary that flows parallel to a main stream for some distance before joining it. The Yazoo River, the type example, is blocked from joining the Mississippi, which has raised the level of its bed higher than the backswamp area where the Yazoo River flows onto the *floodplain. The Yazoo thus has to continue alongside the main stem, flowing along the edge of the floodplain to the point where the Mississippi swings over to the valley side.

Young's Modulus (or Modulus of Elasticity). The ability of rock or of any other solid to resist deformation. It is a quantitative value that differs for various solids and may be obtained in the laboratory by subjecting a solid cylinder of rock to a compressional load. The solid cylinder of rock of known length and cross-sectional area is compressed until it will no longer deform elastically, that is, return to its original shape after removal of the applied load. The value of the load at this point, measured in pounds or kilograms, is divided by the original cross-sectional area. This determines the stress σ (the Greek letter "sigma") in pounds per square inch or kilograms per square centimeter. Next the strain ε (the Greek letter "epsilon") is calculated by dividing the original length of the cylinder into the amount the cylinder changed in length as a result of the applied load. Young's modulus is then determined as the stress divided by strain, or Y (Young's modulus) $= \sigma / \varepsilon$. Values of Young's modulus are given in dynes per square centimeter or in pounds per square inch. (J.W.S.)

Zeolite. One of a group of minerals with similar chemical composition, association and occurrence. Zeolites are hydrous aluminum silicates of sodium and calcium and, more rarely, of potassium. The list of zeolites covers distinct species, of which only five or six are common. When pure, they all are white, but impurities may color them pink, gray or tan. Their specific gravity is low, ranging from 2.0 to 2.4, and their hardness varies from 3½ to less than 5½. When heated, zeolites fuse with a bubbling action; hence their name, which is derived from the Greek *zeo,* meaning "to boil."

Most zeolites are well crystallized and are characteristically found in cavities and cracks in fine-grained basaltic rocks called *trap rocks. In such an environment they are associated with one another and frequently with prehnite, datolite, apophyllite, pectolite and, more rarely, with amethyst. Notable localities for zeolites are the Deccan traps of India, Northern Ireland, Nova Scotia and, in the United States, the trap-rock ridges extending from Massachusetts through Connecticut into New Jersey. In addition, large deposits of poorly crystallized zeolites have been found in Tanzania and in western United States as an altera-

Xenolith
This isolated block of slate engulfed by granite magma is called a xenolith, Greek for "strange rock."

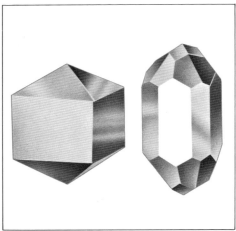

Zircon

1 *Unusually large for zircon, this crystal (from Ontario, Canada) clearly illustrates the shape characteristic of the mineral.*

2 *A typical form of this mineral is a simple tetragonal prism with pyramidal terminations; less common are more complex crystals with several different crystal forms.*

tion of volcanic tuff and volcanic glass.

An interesting and important property of zeolites, resulting from their crystal structure, is their capacity for base exchange. The metal ions, usually sodium and calcium, are held by weak electrical forces to the more rigid aluminosilicate framework. These metal ions can then replace each other without disturbing the basic framework. Thus if a sodium zeolite is immersed in a solution having a high concentration of calcium, the calcium ions will enter the crystal structure while the sodium ions go into the solution. Water softeners act on this principle: Water is hard because it contains many calcium ions. If such water is passed through a tank containing a sodium zeolite, the calcium enters the crystal structure and the sodium goes into the water and makes it "soft." (C.S.H.)

Zinc Blende. A synonym for *sphalerite; used more commonly in England than in America.

Zincite. A mineral, zinc oxide, mined as an ore of zinc at Franklin, N.J. Although zincite has been reported from other localities, its only important occurrence is at Franklin, where it is associated with the other zinc-bearing minerals *franklinite and *willemite. A small amount of manganese present in zincite is probably responsible for its deep red to orange-yellow color, for pure zinc oxide is white. See Mineral Properties.

Zircon. A mineral, zirconium silicate, used as a gemstone and as a source of zirconium. It is a common and widely distributed accessory mineral in igneous rocks and to a lesser extent in metamorphic rocks. Weathering of these rocks liberates the small crystals and grains of zircon; since these are chemically and mechanically stable, they accumulate in stream gravels and beach sands. It is from beach sand concentrations in Australia, Brazil and Florida that zircon is produced commercially. The tiny crystals from beach sands are used as the ore of metallic zirconium and as a source of zirconium oxide, one of the most refractory substances known.

With its high luster and high hardness, zircon makes an attractive gemstone. For centuries it has been recovered from the stream gravels of Ceylon and Burma for this purpose. More recently it has come from similar deposits in Australia and the Soviet Union. The crystals are usually brownish or a red-orange color, but more rarely may be green, violet or colorless. The darker off-color stones can be rendered more attractive by heat-treatment.

When heated under the proper conditions, blue or colorless stones may result. See Mineral Properties; Appendix 3. (C.S.H.)

Zoisite. A mineral, hydrous calcium aluminum silicate, of the same composition and association as the more abundant clinozoisite. The mineral, normally pale green to gray, was found in 1967 in Tanzania, in gem-quality crystals with a beautiful sapphire-blue color. Since then, cut stones of zoisite sold under the name tanzanite have become popular. See Mineral Properties; Appendix 3.

Zone. A *biostratigraphic unit consisting of a subdivision of stratified rock and characterized by one or several distinctive fossils that can be used to identify the zone in which they occur. Zones may bear the name of the fossil (or fossils) they contain.

Zone of Aeration. See Water Table.

Zone of Saturation. See Water Table.

Appendix

Appendix

How to use these tables

To convert units in the vertical column at the left to units in the row across the top, read across the row from the units to be converted to the number found in the column of desired units. Multiply the units to be converted by this number. For example, to convert meters to feet, multiply the number of meters by 3.281. To convert miles to kilometers, multiply the number of miles by 1.609.

Conversion of Lengths

From/to	cm	m	km	in	ft	mile
cm	1	0.01	1×10^{-5}	0.3937	0.03281	6.21×10^{-6}
m	100	1	0.001	39.37	3.281	6.21×10^{-4}
km	1×10^{5}	1000	1	3.94×10^{4}	3281	0.6214
in	2.540	0.02540	2.54×10^{-5}	1	0.08333	1.58×10^{-5}
ft	30.48	0.3048	3.05×10^{-4}	12	1	1.89×10^{-4}
mile	1.61×10^{5}	1609	1.609	6.34×10^{4}	5280	1

Conversion of Areas

From/to	cm^2	m^2	km^2	in^2	ft^2	$mile^2$
cm^2	1	0.0001	1×10^{-10}	0.1550	0.00108	3.86×10^{-11}
m^2	1×10^{4}	1	1×10^{-6}	1550	10.76	3.86×10^{-7}
km^2	1×10^{10}	1×10^{6}	1	1.55×10^{9}	1.08×10^{7}	0.3861
in^2	6.452	6.45×10^{-4}	6.45×10^{-10}	1	0.00694	2.49×10^{-10}
ft^2	929.0	0.09290	9.29×10^{-8}	144	1	3.59×10^{-8}
$mile^2$	2.59×10^{10}	2.59×10^{6}	2.590	4.01×10^{9}	2.79×10^{7}	1

Conversion of Weights

From/to	g	kg	m.ton	grain	oz	lb
g	1	0.001	1×10^{-6}	15.43	0.03527	0.00220
kg	1000	1	0.001	1.54×10^{4}	35.27	2.205
m.ton	1×10^{6}	1000	1	1.54×10^{7}	3.53×10^{4}	2205
grain	0.06480	6.48×10^{-5}	6.48×10^{-8}	1	0.00229	1.43×10^{-4}
oz	28.35	0.02835	2.83×10^{-5}	437.5	1	0.06250
lb	453.6	0.4536	4.54×10^{-4}	7000	16	1

Conversion of Volumes

From/to	cm^3	liter	m^3	in^3	ft^3	yd^3
cm^3	1	0.001	1×10^{-6}	0.06102	3.53×10^{-5}	1.31×10^{-6}
liter	1000	1	0.001	61.02	0.03532	0.00131
m^3	1×10^{6}	1000	1	6.10×10^{4}	35.31	1.308
in^3	16.39	0.01639	1.64×10^{-5}	1	5.79×10^{-4}	2.14×10^{-5}
ft^3	2.83×10^{4}	28.32	0.02832	1728	1	0.03704
yd^3	7.65×10^{5}	764.5	0.7646	4.67×10^{4}	27	1

Temperature Measurement and Scales

Temperature is a measure of the amount of molecular activity in a substance. If molecules are moving slowly, the temperature of the substance containing them is said to be low. If molecular motion is rapid, the temperature is high. In general, when the temperature of a substance is high, the substance tends to expand; when cooled, the substance contracts.

Thermometers are commonly calibrated with the Fahrenheit or Celsius scale. The Fahrenheit scale is used in the United States for public weather information. The Celsius scale is sometimes called centigrade because there are 100 divisions between the freezing and boiling points of water. The Celsius scale is widely used for temperature measurement throughout the world and for practically all scientific work in the United States.

The Kelvin scale is a third scale for temperature measurement. It is called an absolute scale because the zero point is that point at which there is no molecular motion. Kelvin scale divisions are the same size as Celsius divisions; the Kelvin zero is 273 degrees below the Celsius zero.

To convert from Fahrenheit (F) to Celsius (C):

$$°C = \frac{(°F - 32)}{1.8}$$

or

$$°C = 5/9 \ (°F - 32)$$

To convert from Celsius (C) to Fahrenheit (F):

$$°F = (°C \times 1.8) + 32$$

or

$$°F = 9/5 \ C° + 32$$

To convert from Celsius (C) to Kelvin (K):

$$°K = °C + 273$$

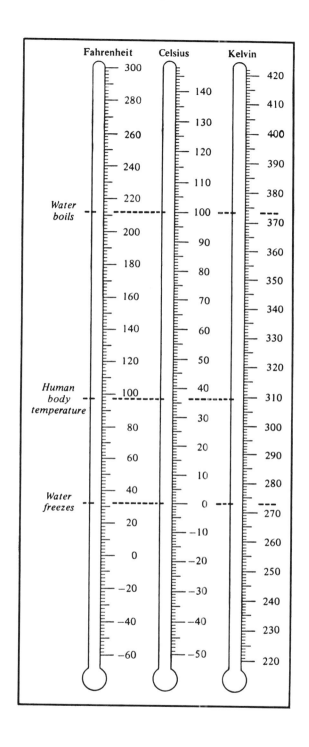

The figure above compares Fahrenheit, Celsius, and Kelvin scales.

Alphabetical List of the Elements

The chemical symbol is the designation of the element used in writing chemical formulas; the atomic number is equal to the number of protons (positive charges) in the nucleus of an atom of the element. The list includes, in addition to the 92 elements found in the rocks of the earth's crust, the transuranian elements that have been made by man in the high-energy particle accelerators known as atom smashers.

Alphabetical List of the Elements

Element	Symbol	Atomic number	Element	Symbol	Atomic number
Actinium	Ac	89	Nitrogen	N	7
Aluminum	Al	13	Nobelium	No	102
Americium	Am	95	Osmium	Os	76
Antimony	Sb	51	Oxygen	O	8
Argon	A	18	Palladium	Pd	46
Arsenic	As	33	Phosphorus	P	15
Astatine	At	85	Platinum	Pt	78
Barium	Ba	56	Plutonium	Pu	94
Berkelium	Bk	97	Polonium	Po	84
Beryllium	Be	4	Potassium	K	19
Bismuth	Bi	83	Praseodymium	Pr	59
Boron	B	5	Promethium	Pm	61
Bromine	Br	35	Protactinium	Pa	91
Cadmium	Cd	48	Radium	Ra	88
Calcium	Ca	20	Radon	Rn	86
Californium	Cf	98	Rhenium	Re	75
Carbon	C	6	Rhodium	Rh	45
Cerium	Ce	58	Rubidium	Rb	37
Cesium	Cs	55	Ruthenium	Ru	44
Chlorine	Cl	17	Samarium	Sm	62
Chromium	Cr	24	Scandium	Sc	21
Cobalt	Co	27	Selenium	Se	34
Columbium	Cb		Silicon	Si	14
(or Niobium)	(Nb)	41	Silver	Ag	47
Copper	Cu	29	Sodium	Na	11
Curium	Cm	96	Strontium	Sr	38
Dysprosium	Dy	66	Sulfur	S	16
Einsteinium	En	99	Tantalum	Ta	73
Erbium	Er	68	Technetium	Tc	43
Europium	Eu	63	Tellurium	Te	52
Fermium	Fm	100	Terbium	Tb	65
Fluorine	F	9	Thallium	Tl	81
Francium	Fr	87	Thorium	Th	90
Gadolinium	Gd	64	Thulium	Tm	69
Gallium	Ga	31	Tin	Sn	50
Germanium	Ge	32	Titanium	Ti	22
Gold	Au	79	Uranium	U	92
Hafnium	Hf	72	Vanadium	V	23
Helium	He	2	Wolframite	W	74
Holmium	Ho	67	Xenon	Xe	54
Hydrogen	H	1	Ytterbium	Yb	70
Indium	In	49	Yttrium	Y	39
Iodine	I	53	Zinc	Zn	30
Iridium	Ir	77	Zirconium	Zr	40
Iron	Fe	26			
Krypton	Kr	36			
Lanthanum	La	57			
Lawrencium	Lw	103			
Lead	Pb	82			
Lithium	Li	3			
Lutetium	Lu	71			
Magnesium	Mg	12			
Manganese	Mn	25			
Mendelevium	Me	101			
Mercury	Hg	80			
Molybdenum	Mo	42			
Neodymium	Nd	60			
Neon	Ne	10			
Neptunium	Np	93			
Nickel	Ni	28			
Niobium	Nb				
(or Columbium)	(Cb)	41			

Abbreviations used:

Crystal Systems: Iso. = isometric;
Tet. = tetragonal;
Hex. = hexagonal;
Rho. = rhombohedral class, hexagonal system;
Orth. = orthorhombic;
Mon. = monoclinic;
Tri. = triclinic.

Mineral Luster: Ada. = adamantine;
Metal. = metallic;
Res. = resinous;
S. Met. = submetallic;
Vit. = vitreous.

Name	Chemical Composition	Crystal System	Mineral Luster	Specific Gravity	Mineral Hardness	Remarks
Actinolite	Ca, Mg, Fe silicate	Mon.	Vit.	3.0-3.2	5-6	An amphibole
Albite	$Na(AlSi_3O_8)$	Tri.	Vit	2.62	6	A feldspar
Alunite	$KAl_3(OH)_6(SO_4)_2$	Rho.	Earth.	2.6–2.8	4	Usually massive
Amblygonite	$LiAlFPO_4$	Tri.	Vit.	3.0–3.1	6	Fusible at 2
Analcime	$Na(AlSi_2O_6)H_2O$	Iso.	Vit.	2.27	5–5½	A feldspathoid
Andalusite	Al_2SiO_5	Orth.	Vit.	3.16–3.20	7½	Infusible
Andradite	$Ca_3Fe_2(SiO_4)_3$	Iso.	Vit.	3.75	7	A garnet
Anglesite	$PbSO_4$	Orth.	Ada.	6.2–6.4	3	3 cleav. direc.
Anhydrite	$CaSo_4$	Orth.	Vit.	2.89–2.98	3–3½	3 cleav. direc.
Ankerite	$Ca(Fe,Mg)(CO_3)_2$	Rho.	Vit.	2.95–3.00	3½	3 cleav. direc.
Anorthite	$CaAl_2Si_2O_8$	Tri.	Vit.	2.76	6	2 cleav.
Apatite	$Ca_5(F,Cl,OH)(PO_4)_3$	Hex.	Vit.	3.15–3.20	5	1 poor cleav.
Apophyllite	$Ca_4K(Si_4O_{10})F \cdot 8H_2O$	Tet.	Vit.	2.3–2.4	4½–5	1 good cleav.
Aragonite	$CaCO_3$	Orth.	Vit.	2.95	3½–4	2 cleav.
Argentite	Ag_2S	Iso.	Metal.	7.3	2–2½	Sectile
Arsenic	As	Rho.	Metal.	5.7	3½	1 good cleav.
Arsenopyrite	FeAsS	Mon.	Metal.	6.07–10.15	5½–6	Silvery
Augite	Ca,Mg,Fe,Al silicate	Mon.	Vit.	3.2–3.4	5–6	Black, 2 cleav.
Autunite	$Ca(UO_2)_2(PO_4)2 \cdot 10–12H_2O$	Tet.	Vit.	3.1–3.2	2–2½	Yellow-green
Azurite	$Cu_3(CO_3)_2(OH)_2$	Mon.	Vit.	3.77	3½–4	Always blue
Barite	$BaSO_4$	Orth.	Vit.	4.5	3–3½	3 cleav. direc.
Beryl	$Be_3Al_2(Si_6O_{18})$	Hex.	Vit.	2.75–2.80	7½–8	Usually green
Biotite	K,Mg,Fe,Al silicate	Mon.	Vit.	2.8–3.2	2½–3	Black mica
Boracite	$Mg_3B_7O_{13}Cl$	Orth.	Vit.	2.9–3.0	7	Pseudo-isometric
Borax	$Na_2B_4O_7 \cdot 10H_2O$	Mon.	Vit.	1.7±	2–2½	1 cleav.
Bornite	Cu_5FeS_4	Iso.	Metal.	5.06–5.08	3	Purple-blue tarnish
Brucite	$Mg(OH)_2$	Rho.	Vit.	2.39	2½	1 good cleav.
Calaverite	$AuTe_2$	Mon.	Metal.	9.35	2½	Easily fusible
Calcite	$CaCO_3$	Rho.	Vit.	2.71	3	Good cleav.
Carnotite	$K_2(UO_2)_2(VO_4)_2 \cdot nH_2O$	Orth.	Vit.	4.1–5.0	Soft	Yellow
Cassiterite	SnO_2	Tet.	Ada.	6.8–7.1	6–7	Colorless to black
Celestite	$SrSO_4$	Orth.	Vit.	3.95–3.97	3–3½	3 cleav. direc.
Cerargyrite	AgCl	Iso.	Waxlike	5.5±	2–3	Perfectly sectile
Cerussite	$PbCO_3$	Orth.	Ada.	6.55	3½	Good cleav.
Chabazite	Ca,Na,Al silicate	Rho.	Vit.	2.05–2.15	4–5	Cubelike crystal
Chalcanthite	$CuSO_4 \cdot 5H_2O$	Tri.	Vit.	2.12–2.30	2½	Soluble in water
Chalcocite	Cu_2S	Orth.	Metal.	5.5–5.8	2½–3	Imperfectly sectile
Chalcopyrite	CuFeS	Tet.	Metal.	4.1–4.3	3½–4	Brittle, yellow
Chlorite	Mg,Fe silicate	Mon.	Vit.	2.6–2.9	2–2½	1 good cleav.
Chromite	$FeCr_2O_4$	Iso.	Metal.	4.6	5½	Luster submetallic
Chrysoberyl	$BeAl_2O_4$	Orth.	Vit.	3.65–3.80	8½	Crystals tabular

Name	Chemical Composition	Crystal System	Mineral Luster	Specific Gravity	Mineral Hardness	Remarks
Chrysocolla	$CuSiO_3 \cdot 2H_2O$	?	Vit.	2.0–2.4	2–4	Bluish-green
Cinnabar	HgS	Rho.	Ada.	8.10	2½	Red
Cobaltite	$CoAsS$	Iso.	Metal.	6.33	5½	In crystals
Colemanite	$Ca_2B_6O_{11} \cdot 5H_2O$	Mon.	Vit.	2.42	4–4½	1 perfect cleav.
Columbite	$(Fe, Mn) Nb_2O_6$	Orth.	S. Met.	5.3–7.3	6	Luster submetallic
Copper	Cu	Iso.	Metal.	8.9	2½–3	Malleable
Cordierite	$(Mg,Fe)_2Al_3 (AlSi_5O_{18})$	Orth.	Vit.	2.60–2.66	7–7½	Blue
Corundum	Al_2O_3	Rho.	Vit.	4.02	9	Various colors
Cristobalite	SiO_2	Tet.	Vit.	2.30	7	In volcanic rocks
Covellite	CuS	Hex.	Metal.	4.60–4.76	1½–2	Blue
Crocoite	$PbCrO_4$	Mon.	Ada.	5.9–6.1	2½–3	Orange-red
Cryolite	Na_3AlF_6	Mon.	Vit.	2.95–3.00	2½	White
Cuprite	Cu_2O	Iso.	Ada.	6.0	3½–4	In red crystals
Datolite	$CaB (SiO_4) (OH)$	Mon.	Vit.	2.8–3.0	5–5½	Usually in crystals
Diamond	C	Iso.	Ada.	3.5	10	In crystals
Diaspore	$AlO (OH)$	Orth.	Vit.	3.35–3.45	6½–7	1 perfect cleav.
Diopside	$CaMg (Si_2O_6)$	Mon.	Vit.	3.2–3.3	5–6	A pyroxene
Dolomite	$CaMg(CO_3)_2$	Rho.	Vit.	2.85	3½–4	Good cleav.
Enargite	Cu_3AsS_4	Orth.	Metal.	4.43–4.45	3	Silvery
Enstatite	$Mg_2 (Si_2O_6)$	Orth.	Vit.	3.2–3.5	5½	A pyroxene
Epidote	Ca,Fe,Al silicate	Mon.	Vit.	3.35–3.45	6–7	Green
Epsomite	$MgSO_4 \cdot 7H_2O$	Orth.	Vit.	1.75	2–2½	Bitter taste
Erythrite	$Co_3(AsO_4)_2 \cdot 8H_2O$	Mon.	Ada.	2.95	1½–2½	Pink, cobalt bloom
Fluorite	CaF_2	Iso.	Vit.	3.18	4	Octahedral cleav.
Franklinite	Fe,Zn,Mn oxide	Iso.	Metal.	5.15	6	Black
Galena	PbS	Iso.	Metal.	7.4–7.6	2½	Cubic cleav.
Garnet	$A_3B_2(SiO_4)_3$	Iso.	Vit.	3.5–4.3	6½–7½	In crystals
Gibbsite	$Al (OH)_3$	Mon.	Vit.	2.3–2.4	2½–3½	Basal cleav.
Glauconite	$K_2 (Mg,Fe)_2Al_6 (Si_4O_{10})_3 (OH)_{12}$	Mon.	Vit.	2.3±	2	In green sands
Glaucophane	$Na_2Mg_3Al_2 (Si_8O_{22}) (OH)_2$	Mon.	Vit.	3.0–3.2	6–6½	An amphibole
Goethite	$HFeO_2$	Orth.	Ada.	4.37	5–5½	1 cleav.
Gold	Au	Iso.	Metal.	15.0–19.3	2½–3	Yellow, soft
Graphite	C	Hex.	Metal.	2.3	1–2	Black, platy
Grossularite	$Ca_3Al_2 (SiO_4)_3$	Iso.	Vit.	3.53	6½	A garnet
Gypsum	$CaSO_4 \cdot _2H_2O$	Mon.	Vit.	2.32	2	White
Halite	$NaCl$	Iso.	Vit.	2.16	2½	Cubic cleav., salty
Hematite	Fe_2O_3	Rho.	Metal	5.26	5½–6½	Red powder
Hemimorphite	$Zn_4 (Si_2O_7)(OH)_2 \cdot H_2O$	Orth.	Vit.	3.4–3.5	4½–5	Cleav.
Heulandite	$Ca (Al_2Si_7O_{18}) \cdot 6H_2O$	Mon.	Vit.	2.18–2.20	3½–4	1 cleav.
Hornblende	Ca,Mg,Fe,Al silicate	Mon.	Vit.	3.2	5–6	Cleav.
Hypersthene	$(Mg,Fe)_2 (Si_2O_6)$	Orth.	Vit.	3.4–3.5	5–6	A pyroxene

Name	Chemical Composition	Crystal System	Mineral Luster	Specific Gravity	Mineral Hardness	Remarks
Idocrase	Ca,Mg,Fe,Al silicate	Tet.	Vit.	3.35−3.45	6½	Prismatic crystals
Ilmenite	$FeTiO_3$	Rho.	Metal.	4.7	5½−6	Slightly magnetic
Iron	Fe	Iso.	Metal.	7.3−7.9	4½	Magnetic
Jadeite	$NaAl(Si_2O_6)$	Mon.	Vit.	3.3−3.5	6½−7	Green, compact
Kaolinite	$Al_4(Si_4O_{10})(OH)_8$	Mon.	Earth	2.60−2.65	2−2½	Earthy
Kernite	$Na_2B_4O_7 \cdot {}_4H_2O$	Mon.	Vit.	1.95	3	2 cleav.
Kyanite	Al_2SiO_5	Tri.	Vit.	3.56−3.66	5−7	Blue-bladed
Lazurite	Na,Ca,Al silicate	Iso.	Vit.	2.40−2.45	5−5½	Pyrite associated
Lepidolite	K,Li,Al silicate	Mon.	Vit.	2.8−3.0	2½−4	A mica
Leucite	$K(AlSi_2O_6)$	?	Vit.	2.45−2.50	5½−6	In crystals
Magnesite	$MgCO_3$	Rho.	Vit.	3.0−3.2	3½−5	Commonly massive
Magnetite	Fe_3O_4	Iso.	Metal.	5.18	6	Strongly magnetic
Malachite	$Cu_2CO_3(OH)_2$	Mon.	Vit.	3.90−4.03	3½−4	Green
Manganite	$MnO(OH)$	Orth.	Metal.	4.3	4	Pneumatic crystals
Marcasite	FeS_2	Orth.	Metal.	4.89	6−6½	In crystals
Margarite	$CaAl_2(Al_2Si_2O_{10})(OH)_2$	Mon.	Vit.	3.0−3.1	3½−5	A brittle mica
Microcline	$K(AlSi_3O_8)$	Tri.	Vit.	2.54−2.57	6	A feldspar
Millerite	NiS	Rho.	Metal.	5.5 ± 0.2	3−3½	Capillary crystals
Mimetite	$Pb_5Cl(AsO_4)_3$	Hex.	Ada.	7.0−7.2	3½	Like pyromorphite
Molybdenite	MoS_2	Hex.	Metal.	4.62−4.73	1−1½	Black, platy
Montmorillonite	$(Al,Mg)_8(Si_4O_{10})(OH)_{10} \cdot 12H_2O$	Mon.	Earth	2.5	1−1½	A clay mineral
Muscovite	$KAl_2(AlSi_3O_{10})(OH)_2$	Mon.	Vit.	2.76−3.10	2−2½	1 perfect cleav.
Natrolite	$Na_2(Al_2Si_3O_{10}) \cdot 2H_2O$	Mon.	Vit.	2.25	5−5½	1 perfect cleav.
Nepheline	$(Na,K)(AlSiO_4)$	Hex.	Vit.	2.55−2.65	5½−6	Greasy appear.
Niccolite	NiAs	Hex.	Metal.	7.78	5−5½	Copper-red
Niter	KNO_3	Orth.	Vit.	2.09−2.14	2	Saltpeter
Olivine	$(Mg,Fe)_2SiO_4$	Orth.	Vit.	3.27−4.37	6½−7	Green rock min.
Opal	$SiO_2 \cdot nH_2O$	Amor.	Vit.	1.9−2.2	5−6	Noncrystal
Orpiment	As_2S_3	Mon.	Vit.	3.49	1½−2	1 cleav., yellow
Orthoclase	$K(AlSi_3O_8)$	Mon.	Vit.	2.57	6	A feldspar
Paragonite	$NaAl_2(AlSi_3O_{10})(OH)_2$	Mon.	Vit.	2.85	2	Like muscovite
Pectolite	$Ca_2NaH(SiO_3)_3$	Tri.	Vit.	2.7−2.8	5	Crystals acicular
Pentlandite	$(Fe,Ni)_9S_8$	Iso.	Metal.	4.6−5.0	3½−4	In pyrrhotite
Phlogopite	$KMg_3(AlSi_3O_{10})(OH)_2$	Mon.	Vit.	2.86	2½−3	Brown mica
Platinum	Pt	Iso.	Metal.	14−19	4−4½	Silvery
Prehnite	$Ca_2Al_2(Si_3O_{10})(OH)_2$	Orth.	Vit.	2.80−2.95	6−6½	Tabular crystals
Proustite	Ag_3AsS_3	Rho.	Ada.	5.55	2−2½	Light ruby silver
Psilomelane	$(Ba,H_2O)_2Mn_5O_{10}$	Orth.	S. Met.	3.7−4.7	5−6	Botryoidal
Pyrargyrite	Ag_3SbS_3	Rho.	Ada.	5.85	2½	Dark ruby silver
Pyrite	FeS_2	Iso.	Metal.	5.02	6−6½	Crystals striated

Name	Chemical Composition	Crystal System	Mineral Luster	Specific Gravity	Mineral Hardness	Remarks
Pyrolusite	MnO_2	Tet.	Metal.	4.75	1−2	Sooty
Pyromorphite	$Pb_5Cl(PO_4)_3$	Hex.	Ada.	6.5−7.1	3½−4	Adamantine luster
Pyrope	$Mg_3Al_2(SiO_4)_3$	Iso.	Vit.	3.51	7	A garnet
Pyrophyllite	$Al_2(Si_4O_{10})(OH)_2$	Mon.	Pearly	2.8−2.9	1−2	Smooth feel
Pyrrhotite	Fe_1-xS	Hex.	Metal.	4.58−4.65	4	Magnetic
Quartz	SiO_2	Rho.	Vit.	2.65	7	No cleav.
Realgar	AsS	Mon.	Res.	3.48	1½−2	1 cleav., red
Rhodochrosite	$MnCO_3$	Rho.	Vit.	3.45−3.60	3½−4½	Pink
Rhodonite	$Mn(SiO_3)$	Tri.	Vit.	3.58−3.70	5½−6	Pink
Rutile	TiO_2	Tet.	Ada.	4.18−4.25	6−6½	Small crystals
Scapolite	Various	Tet.	Vit.	2.65−2.74	5−6	4 cleav. direc.
Scheelite	$CaWO_4$	Tet:	Ada.	5.9−6.1	4½−5	4 cleav. direc.
Serpentine	$Mg_6(Si_4O_{10})(OH)_8$	Mon.	Vit.	2.2	2−5	Green to yellow
Siderite	$FeCO_3$	Rho.	Vit.	3.83−3.88	3½−4	Brown
Sillimanite	Al_2SiO_5	Orth.	Vit.	3.23	6−7	1 cleav.
Silver	Ag	Iso.	Metal.	10.5	2½−3	White, malleable
Smithsonite	$ZnCO_3$	Rho.	Vit.	4.35−4.40	5	Reniform
Sodalite	$Na_4(AlSiO_4)_3Cl$	Iso.	Vit.	2.15−2.30	5½−6	Usually blue
Soda niter	$NaNO_3$	Rho.	Vit.	2.29	1−2	Cooling taste
Sperrylite	$PtAs_2$	Iso.	Metal.	10.50	6−7	See platinum
Spessartite	$Mn_3Al_2(SiO_4)_3$	Iso.	Vit.	4.18	7	A garnet
Sphalerite	ZnS	Iso.	Res.	3.9−4.1	3½−4	6 cleav. direc.
Sphene	$CaTiO(SiO_4)$	Mon.	Ada.	3.40−3.55	5−5½	Wedge-shaped crystals
Spinel	$MgAl_2O_4$	Iso.	Vit.	3.6−4.0	8	In octahedrons
Spodumene	$LiAl(Si_2O_6)$	Mon.	Vit.	3.15−3.20	6½−7	Good cleav.
Staurolite	$Fe_2Al_9O_7(SiO_4)_4(OH)$	Orth.	Vit.	3.65−3.75	7−7½	In cruciform twins
Stibnite	Sb_2S_3	Orth.	Metal.	4.52−4.62	2	Perfect cleav.
Stilbite	$Ca(Al_2Si_7O_{18})\cdot 7H_2O$	Mon.	Vit.	2.1−2.2	3½−4	Sheetlike agg.
Strontianite	$SrCO_3$	Orth.	Vit.	3.7	3½−4	Efferv. in HCl
Sulfur	S	Orth.	Res.	2.05−2.09	1½−2½	Burns
Sylvanite	$(Au,Ag)Te_2$	Mon.	Metal.	8.0−8.2	1½−2	Perfect cleav.
Talc	$Mg_3(Si_4O_{10})(OH)_2$	Mon.	Vit.	2.7−2.8	1	Greasy feel
Tantalite	$(Fe Mn)Ta_2O_6$	Orth.	S. Met.	6.5±	6	See columbite
Tennantite	$(Cu,Fe,Zn,Ag)_{12}As_4S_{13}$	Iso.	Metal.	4.6−5.1	3−4½	In tetrahedrons
Tetrahedrite	$(Cu,Fe,Zn,Ag)_{12}Sb_4S_{13}$	Iso.	Metal.	4.6−5.1	3−4½	In tetrahedrons
Topaz	$Al_2(SiO_4)(F,OH)_2$	Orth.	Vit.	3.4−3.6	8	Perfect cleav.
Tourmaline	Complex porosilicate	Rho.	Vit.	3.00−3.25	7−7½	Trigonal section
Tremolite	$Ca_2Mg_5(Si_8O_{22})(OH)_2$	Mon.	Vit.	3.0−3.3	5−6	Cleav.
Tridymite	SiO_2	Orth.	Vit.	2.26	7	In volcanic rocks
Turquoise	Cu,Al phosphate (hydrous)	Tri.	Vit.	2.6−2.8	6	Blue-green
Ulexite	$NaCaB_5O_9 \cdot 8H_2O$	Tri.	Silky	1.96	1	"Cotton-balls"

Name	Chemical Composition	Crystal System	Mineral Luster	Specific Gravity	Mineral Hardness	Remarks
Uraninite	UO_2	Iso.	Res.	9.0–9.7	5½	Pitchy luster
Uvarovite	$Ca_3Cr_2(SiO_4)_3$	Iso.	Vit.	3.45	7½	Green garnet
Vanadinite	$Pb_5Cl(VO_4)_3$	Hex.	Vit.	6.7–7.1	3	Luster resinous
Wavellite	$Al_3(OH)_3(PO_4)_2 \cdot 5H_2O$	Orth.	Vit.	2.33	3½–4	Radiating aggregates
Willemite	Zn_2SiO_4	Rho.	Vit.	3.9–4.2	5½	From Franklin, N.J.
Witherite	$BaCO_3$	Orth.	Vit.	4.3	3½	Efferv. in HCl
Wolframite	$(Fe,Mn)WO_4$	Mon.	S. Met.	7.0–7.5	5–5½	1 perfect cleav.
Wollastonite	$Ca(SiO_3)$	Tri.	Vit.	2.8–2.9	5–5½	2 cleav.
Wulfenite	$PbMoO_4$	Tet.	Ada.	6.8±	3	Orange-red
Zincite	ZnO	Hex.	Vit.	5.68	4–4½	From Franklin, N.J.
Zircon	$ZiSiO_4$	Tet.	Ada.	4.68	7½	In small crystals
Zoisite	$Ca_2Al_3(SiO_4)_3(OH)$	Orth.	Vit.	3.35	6–6½	Striated crystals

Principal Oceans and Seas	Sea	Area (1000 mi²)	(1000 km²)
	Pacific Ocean		
	with adjacent seas	70,017	181,344
	without adjacent seas	64,186	166,241
	Atlantic Ocean		
	with adjacent seas	36,415	94,314
	without adjacent seas	33,420	86,557
	Indian Ocean		
	with adjacent seas	28,617	74,118
	without adjacent seas	28,350	73,427
	Arctic Ocean		
	with adjacent seas	4,732	12,257
	without adjacent seas	3,662	9,485
	Coral Sea	1,850	4,791
	Arabian Sea	1,492	3,863
	South China (Nan) Sea	1,423	3,685
	Caribbean Sea	1,063	2,754
	Mediterranean Sea	971	2,516
	Bering Sea	890	2,304
	Bay of Bengal	839	2,172
	Sea of Okhotsk	614	1,590
	Gulf of Mexico	596	1,543
	Gulf of Guinea	592	1,533
	Barents Sea	542	1,405
	Norwegian Sea	534	1,383
	Gulf of Alaska	512	1,327
	Hudson Bay	476	1,232
	Greenland Sea	465	1,205
	Arafura Sea	400	1,037
	Philippine Sea	400	1,036
	Sea of Japan	378	978
	East Siberian Sea	348	901
	Kara Sea	341	883
	East China (Tung) Sea	290	752
	Solomon Sea	278	720
	Banda Sea	268	695
	Baffin Bay	266	689
	Laptev Sea	251	650
	Timor Sea	237	615
	Andaman (Burma) Sea	232	602
	North Sea	232	600
	Chukchi (Chuckchee) Sea	225	582
	Great Australian Bight	187	484
	Beaufort Sea	184	476
	Celebes (Sulawesi) Sea	182	472
	Black Sea	178	461
	Red Sea	174	450
	Java (Djawa) Sea	167	433
	Sulu Sea	162	420
	Yellow (Huang, Hwang) Sea	161	417
	Baltic Sea	149	386
	Gulf of Carpentaria	120	311
	Molucca (Maluku) Sea	119	307
	Persian Gulf (Gulf of Iran)	93	241
	Gulf of Siam (Gulf of Thailand)	92	239
	Gulf of Saint Lawrence	92	238
	Gulf of Aden	85	220
	Makassar (Macassar, Makasar) Strait	75	194
	Ceram (Seram) Sea	72	187
	Bay of Biscay	71	184
	Gulf of Oman	70	181
	Aegean Sea	69	179
	Gulf of California	68	177

Sea	Area (1000 mi²)	(1000 km²)
Adriatic Sea	51	132
Flores Sea	47	121
Bali Sea	46	119
Gulf of Bothnia	45	117
Gulf of Tonkin	45	117
Savu Sea	41	105
Irish Sea	40	103
White Sea	35	90
Bass Strait	29	75
English Channel	29	75
Sea of Azov	15	38

Highest Mountain Peaks by Continent

Peak and Massif	Mountain Range or System	Location	Elevation (feet)	(meters)
Africa (over 15000 ft/4570 m)				
Kibo (volc) [Kilimanjaro]	—	Tanganyika, Tanzania	19,340	5890
Mawensi (volc) [Kilimanjaro]	—	Tanganyika, Tanzania	17,100	5210
Batian [Kenya]	—	Kenya	17,050	5200
Nelion [Kenya]	—	Kenya	17,020	5190
Margherita [Stanley]	Ruwenzori	Uganda-Zaire	16,760	5110
Alexandra [Stanley]	Ruwenzori	Uganda-Zaire	16,700	5090
Albert [Stanley]	Ruwenzori	Zaire	16,690	5090
Savoia [Stanley]	Ruwenzori	Uganda	16,330	4980
Elena [Stanley]	Ruwenzori	Uganda	16,300	4970
Elizabeth [Stanley]	Ruwenzori	Uganda	16,170	4930
Philip [Stanley]	Ruwenzori	Uganda	16,140	4920
Moebius [Stanley]	Ruwenzori	Uganda	16,130	4920
Vittorio Emanuele [Speke]	Ruwenzori	Uganda	16,040	4890
Ensonga [Speke]	Ruwenzori	Uganda	15,960	4860
Edward [Baker]	Ruwenzori	Uganda	15,890	4840
Johnston [Speke]	Ruwenzori	Uganda	15,860	4830
Umberto [Emin]	Ruwenzori	Zaire	15,740	4800
Semper [Baker]	Ruwenzori	Uganda	15,730	4790
Kraepelin [Emin]	Ruwenzori	Zaire	15,720	4790
Iolanda [Gessi]	Ruwenzori	Uganda	15,470	4720
Bottego [Gessi]	Ruwenzori	Uganda	15,420	4700
Sella [Luigi di Savoia]	Ruwenzori	Uganda	15,180	4630
Wollaston [Baker]	Ruwenzori	Uganda	15,180	4630
Moore [Baker]	Ruwenzori	Uganda	15,170	4620
Weismann [Luigi di Savoia]	Ruwenzori	Uganda	15,160	4620
Great Tooth [Stanley]	Ruwenzori	Uganda	15,100	4600
Wasuwameso [Mugule]	Ruwenzori	Zaire	15,030	4580
Okusoma [Luigi di Savoia]	Ruwenzori	Uganda	15,020	4580
Trident [Speke]	Ruwenzori	Uganda	15,000	4570
(Emi) Koussi (volc)	Tibesti	Chad	11,470	3490
Thabana Ntlenyana	Drakens(berg)	Lesotho	11,420	3480
Injasuti	Drakens(berg)	Lesotho-South Africa	11,310	3450
Kinyeti	Imatong	Sudan	10,460	3190
Kegueur Terbi	Tibesti	Chad-Libya	10,330	3150
Neiges (volc)	—	Reunion	10,070	3070
Mlanje	Mlanje	Malawi	10,000	3050
Santa Isabel (volc)	—	Equatorial Guinea	9,840	3000
Tahat	Ahaggar (alt Hoggar)*	Algeria	9,570	2920
Maromokotro	Tsaratanana	Malagasy Republic	9,470	2890
Cano (volc)	—	Cape Verde Islands	9,280	2830
Katrinah [Musa]	Sinai	Israel	8,650	2640
Moco	Upanda	Angola	8,600	2620
Konigstein [Brand(berg)]	Kaokoveld	South-West Africa	8,550	2610
Inyangani	Inyanga	Rhodesia	8,520	2600
Binga	Chimanimani	Mozambique-Rhodesia	7,990	2440
Surad Ad	Ogo	Somalia	7,900	2410
Chelia	Aures	Algeria	7,640	2330
Greboun	Air	Niger	7,550	2300
Vogel	Banglang	Nigeria	6,700	2040
Bintimani	Loma	Sierra Leone	6,390	1940
Ruivo de Santana	—	(Madeira Islands), Portugal	6,110	1860
Emlembe	Drakens(berg)	Swaziland	6,100	1860
Nimba	Nimba	Guinea-Ivory Coast-Liberia	5,780	1760
Shanabi	Dorsale	Tunisia	5,070	1540

*alternate name(s)

	Peak and Massif	Mountain Range or System	Location	Elevation (feet)	(meters)
Antarctica (over 1400 ft/4270 m)	—[Vinson]	Sentinel	Antarctica	16,860	5140
	Tyree	Sentinel	Antarctica	16,290	4970
	Shinn	Sentinel	Antarctica	15,750	4800
	Gardner	Sentinel	Antarctica	15,380	4690
	Epperly	Sentinel	Antarctica	15,100	4600
	Kirkpatrick	Queen Alexandra	Antarctica	14,850	4530
	Elizabeth	Queen Alexandra	Antarctica	14,700	4480
	Markham	Queen Elizabeth	Antarctica	14,270	4350
	Bell	Queen Alexandra	Antarctica	14,120	4300
	Mackellar	Queen Alexandra	Antarctica	14,100	4300
Asia (over 25,000 ft/7770 m)	Everest (alt Chumulangma) [Everest]	Nepal Himalaya	China-Nepal	29,030	8850
	K² (alt Chogori, Dapsang, Godwin Austen)	Karakoram	Kashmir-Jammu	28,250	8610
	Kangchenjunga (alt Kanchenjunga): highest peak [Kangchenjunga]	Nepal Himalaya	Nepal-Sikkim	28,170	8590
	Lhotse (Lotzu) [Everest]	Nepal Himalaya	China-Nepal	27,890	8500
	Kangchenjunga: S peak [Kangchenjunga]	Nepal Himalaya	Nepal-Sikkim	27,800	8470
	Makalu I [Makalu]	Nepal Himalaya	China-Nepal	27,790	8470
	Kangchenjunga: W peak [Kangchenjunga]	Nepal Himalaya	Nepal-Sikkim	27,620	8420
	Lhotse Shar (Lhotse: E peak) [Everest]	Nepal Himalaya	China-Nepal	27,500	8380
	Dhaulagiri I	Nepal Himalaya	Nepal	26,810	8170
	Cho Oyu highest peak	Nepal Himalaya	China-Nepal	26,750	8150
	Manaslu highest peak [Manaslu]	Nepal Himalaya	Nepal	26,660	8130
	Nanga Parbat: highest peak	Punjab Himalaya	Kashmir-Jammu	26,660	8130
	Annapurna I	Nepal Himalaya	Nepal	26,500	8080
	Gasherbrum I [Gasherbrum]	Karakoram	Kashmir-Jammu	26,470	8070
	Broad: highest peak [Gasherbrum]	Karakoram	Kashmir-Jammu	26,400	8050
	Gasherbrum II: highest peak [Gasherbrum]	Karakoram	Kashmir-Jammu	26,360	8030
	Gosainthan	Nepal Himalaya	Tibet, China	26,290	8010
	Broad: middle peak [Gasherbrum]	Karakoram	Kashmir-Jammu	26,250	8000
	Gasherbrum III [Gasherbrum]	Karakoram	Kashmir-Jammu	26,090	7950
	Annapurna II	Nepal Himalaya	Nepal	26,040	7940
	Gasherbrum IV [Gasherbrum]	Karakoram	Kashmir-Jammu	26,000	7920
	Gyachung Kang	Nepal Himalaya	China-Nepal	25,990	7920
	Nanga Parbat	Punjab Himalaya	Kashmir-Jammu	25,950	7910
	Kangbachen [Kangchenjunga]	Nepal Himalaya	India-Nepal	25,930	7900
	Manaslu: E pinnacle [Manaslu]	Nepal Himalaya	Nepal	25,900	7900
	Distaghil Sar	Karakoram	Kashmir-Jammu	25,870	7890
	Nuptse (E²) [Everest]	Nepal Himalaya	Nepal	25,850	7880
	Himalchuli: highest peak	Nepal Himalaya	Nepal	25,800	7860
	Khiangyang Kish	Karakoram	Kashmir-Jammu	25,760	7850
	Ngojumba Ri	Nepal Himalaya	China-Nepal	25,720	7840

Peak and Massif	Mountain Range or System	Location	Elevation (feet)	(meters)
Dakura [Manaslu]	Nepal Himalaya	Nepal	25,710	7840
Masherbrum: E peak	Karakoram	Kashmir-Jammu	25,660	7820
Nanda Devi: W peak	Kumaun Himalaya	Uttar Pradesh, India	25,650	7820
Nanga Parbat: N peak	Punjab Himalaya	Kashmir-Jammu	25,650	7820
Chomo Lonzo [Makalu]	Nepal Himalaya	China-Nepal	25,640	7820
Masherbrum: W peak	Karakoram	Kashmir-Jammu	25,610	7810
Rakaposhi	Haramosh (Ridge)	Kashmir-Jammu	25,550	7790
Batura Mustagh I	Karakoram	Kashmir-Jammu	25,540	7790
Gasherbrum II: E peak [Gasherbrum]	Karakoram	Kashmir-Jammu	25,500	7770
Europe (14,800 ft/4510 m)				
Elbrus: W peak (volc)	Caucasus	Russia, USSR	18,480	5630
Elbrus: E peak (volc) [Elbrus]	Caucasus	Russia, USSR	18,360	5590
Shkhara: E peak	Caucasus	Georgia-Russia, USSR	17,060	5200
Dykh(-Tau): W peak	Caucasus	Russia, USSR	17,050	5200
Dykh(-Tau): E peak	Caucasus	Russia, USSR	16,900	5150
Koshtan(-Tau)	Caucasus	Russia, USSR	16,880	5140
Shkhara: W peak	Caucasus	Georgia-Russia, USSR	16,880	5140
Pushkina	Caucasus	Russia, USSR	16,730	5100
Dzhangi(-Tau): NW peak	Caucasus	Georgia, USSR	16,570	5050
Kazbek: E peak	Caucasus	Georgia, USSR	16,560	5050
Dzhangi(-Tau): SE peak	Caucasus	Georgia, USSR	16,520	5030
Katyn(-Tau)	Caucasus	Georgia-Russia, USSR	16,310	4970
Shota Rustaveli	Caucasus	Georgia-Russia, USSR	16,270	4960
Mizhirgi: W peak	Caucasus	Russia, USSR	16,170	4930
Mizhirgi: E peak	Caucasus	Russia, USSR	16,140	4920
Kundyum-Mizhirgi	Caucasus	Russia, USSR	16,010	4880
Gestola	Caucasus	Georgia-Russia, USSR	15,930	4860
Tetnuld	Caucasus	Georgia, USSR	15,920	4850
(Mont-)Blanc	Alps (Alpe, Alpen, Alpes, Alpi)	France-Italy	15,770	4810
Dzhimariy(-Khokh)	Caucasus	Georgia, USSR	15,680	4780
Adish	Caucasus	Georgia, USSR	15,570	4750
Ushba: SW peak	Caucasus	Georgia, USSR	15,450	4710
Ushba: NE peak	Caucasus	Georgia, USSR	15,400	4690
Ullu-Auz(-Bashi)	Caucasus	Russia, USSR	15,360	4680
Panoramnyy	Caucasus	Russia, USSR	15,350	4680
Krumkol	Caucasus	Russia, USSR	15,320	4670
Kazbek: W peak	Caucasus	Georgia, USSR	15,250	4650
Uilpata	Caucasus	Russia, USSR	15,240	4650
Shau(-Khokh)	Caucasus	Georgia, USSR	15,240	4640
Dufourspitze [Rosa]	Alps	Italy-Switzerland	15,200	4630
Grenzgipfel [Rosa]	Alps	Italy-Switzerland	15,190	4630
Kyukyurtlyukol(-Bashi) [Elbrus]	Caucasus	Russia, USSR	15,170	4620
Nordend [Rosa]	Alps	Italy-Switzerland	15,130	4610
Tikhtengen: S peak	Caucasus	Georgia-Russia, USSR	15,130	4610
Tikhtengen: N peak	Caucasus	Georgia-Russia, USSR	15,130	4610
Mayli(-Khokh)	Caucasus	Georgia, USSR	15,100	4600
Dubl: N peak	Caucasus	Russia, USSR	15,030	4580
Zumstein(spitze) [Rosa]	Alps	Italy-Switzerland	15,000	4570
Signal(kuppe) [Rosa]	Alps	Italy-Switzerland	14,960	4560
Dumala(-Tau)	Caucasus	Russia, USSR	14,950	4560
Tyutyun(-Bashi)	Caucasus	Russia, USSR	14,930	4550

Peak and Massif	Mountain Range or System	Location	Elevation (feet)	(meters)
Dom [Mischabel]	Alps	Switzerland	14,910	4540
Aylama	Caucasus	Georgia-Russia, USSR	14,890	4540
Dzhaylyk(-Bashi)	Caucasus	Russia, USSR	14,890	4540
Lyskamm [Rosa]	Alps	Italy-Switzerland	14,890	4540
Tyutyun(-Tau)	Caucasus	Russia, USSR	14,890	4540
Dubl: S peak	Caucasus	Russia, USSR	14,880	4530
Lakutsa	Caucasus	Georgia, USSR	14,830	4520
Karaugom: E peak	Caucasus	Russia, USSR	14,810	4510
Addala Shukhgelmeer	Caucasus	Russia, USSR	14,800	4510
Karaugom: W peak	Caucasus	Russia, USSR	14,800	4510
Spartak	Caucasus	Georgia, USSR	14,800	4510

North America
(14,250 ft/4340 m)

Peak and Massif	Mountain Range or System	Location	Elevation (feet)	(meters)
McKinley: S peak [McKinley]	Alaska	Alaska, USA	20,320	6190
Logan: central peak [Logan]	Saint Elias	Yukon Territory, Canada	19,850	6050
Logan: W peak [Logan]	Saint Elias	Yukon Territory, Canada	19,800	6040
Logan: E peak [Logan]	Saint Elias	Yukon Territory, Canada	19,750	6020
McKinley: N peak [McKinley]	Alaska	Alaska, USA	19,470	5930
Citlatepetl (volc)	Neovolcanica	Puebla-Veracruz, Mexico	18,700	5700
Logan: N peak [Logan]	Saint Elias	Yukon Territory, Canada	18,600	5670
Saint Elias	Saint Elias	Canada-USA	18,010	5490
Popocatepetl (volc)	Neovolcanica	Puebla, Mexico	17,890	5450
Foraker	Alaska	Alaska, USA	17,400	5300
Ixtacihuatl	Neovolcanica	Puebla, Mexico	17,340	5290
Queen [Logan]	Saint Elias	Yukon Territory, Canada	17,300	5070
Lucania	Saint Elias	Yukon Territory, Canada	17,150	5230
King [Logan]	Saint Elias	Yukon Territory, Canada	17,130	5220
Steele	Saint Elias	Yukon Territory, Canada	16,640	5970
Bona	Saint Elias	Alaska, USA	16,500	5030
Blackburn: highest peak	Wrangell	Alaska, USA	16,390	5000
Blackburn: SE peak	Wrangell	Alaska, USA	16,290	4960
Sanford	Wrangell	Alaska, USA	16,240	4950
Wood	Saint Elias	Yukon Territory, Canada	15,880	4840
Vancouver	Saint Elias	Canada-USA	15,700	4790
Churchill	Saint Elias	Alaska, USA	15,640	4770
Slaggard	Saint Elias	Yukon Territory, Canada	15,570	4750
McCauley	Saint Elias	Yukon Territory, Canada	15,470	4720
Fairweather	Saint Elias	Canada-USA	15,300	4660
University	Saint Elias	Alaska, USA	15,030	4580
Hubbard	Saint Elias	Canada-USA	15,010	4580
Bear	Saint Elias	Alaska, USA	14,850	4530
Walsh	Saint Elias	Yukon Territory, Canada	14,780	4500
Malinche	Neovolcanica	Puebla-Tlaxcala, Mexico	14,640	4460
Hunter	Alaska	Alaska, USA	14,570	4440
Alverstone	Saint Elias	Canada-USA	14,530	4430
Browne Tower [McKinley]	Alaska	Alaska, USA	14,530	4430

Peak and Massif	Mountain Range or System	Location	Elevation (feet)	(meters)
Whitney	Sierra Nevada	California, USA	14,490	4420
Aello	Saint Elias	Alaska, USA	14,440	4400
Elbert	Rocky	Colorado, USA	14,430	4400
Harvard	Rocky	Colorado, USA	14,420	4390
Massive	Rocky	Colorado, USA	14,420	4390
Rainier (volc)	Cascade	Washington, USA	14,410	4390
Toluca	Neovolcanica	Mexico, Mexico	14,410	4390
Williamson	Sierra Nevada	California, USA	14,370	4380
La Plata	Rocky	Colorado, USA	14,370	4380
Blanca	Rocky	Colorado, USA	14,320	4360
Uncompahgre	Rocky	Colorado, USA	14,310	4360
Crestone	Rocky	Colorado, USA	14,290	4360
Lincoln	Rocky	Colorado, USA	14,290	4350
Antero	Rocky	Colorado, USA	14,270	4350
Grays	Rocky	Colorado, USA	14,270	4350
Torreys	Rocky	Colorado, USA	14,270	4350
Castle	Rocky	Colorado, USA	14,260	4350
Evans	Rocky	Colorado, USA	14,260	4350
Longs	Rocky	Colorado, USA	14,260	4350
Quandary	Rocky	Colorado, USA	14,260	4350
McArthur[Logan]	Saint Elias	Yukon Territory, Canada	14,250	4340

South America
(over 20,700 ft/6310 m)

Peak and Massif	Mountain Range or System	Location	Elevation (feet)	(meters)
Aconcagua	Andes	Mendoza, Argentina	22,840	6960
Ojos del Salado: SE peak (volc)	Andes	Argentina-Chile	22,560	6870
Bonete	Andes	La Rioja, Argentina	22,550	6870
Pissis	Andes	Catamarca-La Rioja, Argentina	22,240	6780
Huascaran: S peak	Blanca (Andes)	Peru	22,210	6770
Mercedario	Andes	San Juan, Argentina	22,210	6770
Llullaillaco (volc)	Andes	Argentina-Chile	22,100	6730
Libertador [Cachi]	Andes	Salta, Argentina	22,050	6720
Ojos del Salado: NW peak (volc)	Andes	Argentina-Chile	22,050	6720
Tupungato	Andes	Argentina-Chile	21,900	6670
Gonzalez: highest peak	Andes	Argentina-Chile	21,850	6660
Huascaran: N peak	Blanca (Andes)	Peru	21,840	6650
Muerto	Andes	Argentina-Chile	21,820	6650
Yerupaja: N peak	Huayhuash (Andes)	Peru	21,760	6630
Incahuasi	Andes	Argentina-Chile	21,700	6610
Galan	Andes	Catamarca, Argentina	21,650	6600
Tres Cruces: central peak	Andes	Argentina-Chile	21,540	6560
Gonzalez: N peak	Andes	Argentina-Chile	21,490	6550
Sajama	Occidental (Andes)	Bolivia	21,390	6520
Yerupaja: S peak	Huayhuash (Andes)	Peru	21,380	6510
Nacimiento	Andes	Catamarca, Argentina	21,300	6490
Illimani: S peak	Real (Andes)	Bolivia	21,200	6460
Illimani: N peak	Real (Andes)	Bolivia	21,130	6440
Anchohuma [Sorata]	Real (Andes)	Bolivia	21,100	6430
Coropuna: highest peak	Occidental (Andes)	Peru	21,080	6420
Puntiagudo	Andes	Argentina-Chile	21,060	6420
Ramada	Andes	San Juan, Argentina	21,030	6410
Chopilcalqui	Blanca (Andes)	Peru	21,000	6400
Laudo	Andes	Catamarca, Argentina	21,000	6400
Huantsan: S peak	Blanca (Andes)	Peru	20,980	6390
Ausangate: highest peak	Vilcanota (Andes)	Peru	20,950	6380
Toro	Andes	Argentina-Chile	20,930	6380
Illampu [Sorata]	Real (Andes)	Bolivia	20,870	6360
Huandoy: central peak	Blanca (Andes)	Peru	20,850	6360
Huandoy: W peak	Blanca (Andes)	Peru	20,850	6360

Peak and Massif	Mountain Range or System	Location	Elevation (feet)	(meters)
Siula: N peak	Huayhuash (Andes)	Peru	20,850	6360
Tres Cruces: S peak	Andes	Argentina-Chile	20,850	6360
Coropuna: NW peak	Occidental (Andes)	Peru	20,790	6340
Ausangate: E peak	Vilcanota (Andes)	Peru	20,770	6330
Parinacota [Payachata]	Andes	Bolivia-Chile	20,770	6330
Tortolas	Andes	Argentina-Chile	20,760	6330
Solimana	Occidental (Andes)	Peru	20,750	6320
Ampato: highest peak	Occidental (Andes)	Peru	20,700	6310

Principal Deserts

Desert and Location	Estimated Area (sq. mi.)	Estimated Area (sq. km.)
Sahara, Northern Africa	3,000,000	7,800,000
Libyan (part of Sahara Desert), Northeastern Africa	650,000	1,690,000
Australian, West and Central Australia	600,000	1,560,000
Arabian, Arabian Peninsula	500,000	1,300,000
Gobi, Mongolia	400,000	1,040,000
Rub'al Khali (part of Arabian Desert), Southeastern Saudi Arabia	250,000	650,000
Kalahari, Bechuanaland	200,000	520,000
Great Sandy (part of Australian Desert), Northwestern Australia	160,000	416,000
Great Victoria (part of Australian Desert), Southwestern Australia	125,000	325,000
Syrian (part of Arabian Desert), Northern Arabian Peninsula	125,000	325,000
Takla Makan, Southern Sinkiang, China	125,000	325,000
Arunta (part of Australian Desert), Central Australia	120,000	312,000
Karakum, Southern Turkestan, USSR	105,000	273,000
Nubian (part of Sahara Desert), Northeastern Sudan	100,000	260,000
Thar (Great Indian), Northwestern India	100,000	260,000
Kyzylkum, Central Turkestan, USSR	90,000	234,000
Gibson (part of Australian Desert), Western Australia	85,000	221,000
Atacama, Northern Chile	70,000	182,000
An Nafud (part of Arabian Desert), North and Central Saudi Arabia	50,000	130,000
Dasht-i-Lut, Eastern Iran	20,000	52,000
Dasht-i-Kavir, North Central Iran	18,000	46,800
Muyunkum, Eastern Turkestan, USSR	17,000	44,200
Mojave, Southern California, USA	13,500	35,100
Sechura, Northwestern Peru	10,000	26,000
Vizcaino, Baja California, Mexico	6,000	15,600
Painted, Northeastern Arizona, USA	5,000	13,000
Great Salt Lake, Northwestern Utah, USA	4,000	10,400
Colorado, Southeastern California, USA	3,000	7,800

Twenty Longest Rivers

River	Length (mi)	(km)
Nile	4160	6690
Amazon	4080	6570
Mississippi-Missouri	3740	6020
Yangtze	3720	5980
Yenisey	3650	5870
Amur	3590	5780
Ob-Irtysh	3360	5410
Plata-Parana	3030	4880
Yellow	3010	4840
Congo	2880	4630
Lena	2730	4400
Mackenzie	2630	4240
Mekong	2600	4180
Niger	2550	4100
Murray-Darling	2330	3750
Volga	2300	3700
Madeira	1990	3200
Syr (Darya)	1910	3080
St. Lawrence	1900	3060
Rio Grande	1880	3030

Twenty Largest Natural Lakes

Lake	Area (mi²)	(km²)
Caspian (Sea)	143,240	371,000
Superior	32,150	83,270
Aral (Sea)	24,900	64,500
Victoria	24,300	62,990
Huron	23,430	60,700
Michigan	22,400	58,020
Tanganyika	12,350	32,000
Great Bear	12,270	31,790
Baikal	12,160	31,500
Great Slave	10,980	28,440
Erie	9,910	25,680
Winnipeg	9,460	24,510
Nyasa	8,680	22,490
Balkhash	8,490	22,000
Ontario	7,430	19,230
Ladoga	7,100	18,400
Chad	*4,000/10,000	*10,360/25,900
Maracaibo	5,520	14,300
Patos	3,860	10,000
Onega	3,710	9,600

*seasonal variation; minimum and maximum

Greatest Waterfalls (Volume of Water)	Waterfall	River and Location	Average Flow (1000 cfs)	(m³/sec)	Height (ft)	(m)
	Sete Quedas (alt Guaira)	Parana River, Brazil-Paraguay	470	13,310	213	65
	Khone	Mekong(alt. Khong, Lantsang, Tien Giang) River, Cambodia-Laos	410	11,610	70	21
	Niagara	Niagara River, Canada-USA	212	6,000	186	57
	Grande	Uruguay River, Argentina-Uruguay	106	3,000	75	23
	Urubupunga	Parana River, Mato Grosso-São Paulo, Brazil	97	2,750	27	9
	Iguazu(alt Iguacu; for Iguassu)	Iguazu(alt. Iguacu; Iguassu)River, Argentina-Brazil	60	1,700	230	70
	Marimbondo	Grande River, Minas Gerais-São Paulo, Brazil	53	1,500	115	35
	Victoria	Zambezi (alt.Zambesi, Zambeze) River, Rhodesia-Zambia	38	1,090	304	92
	Churchill (for Grand)	Churchill (alt. Hamilton) River, Newfoundland, Canada	35	991	245	75
	Kaveri(alt Cauvery)	Kaveri(alt. Cauvery) River, Madras-Mysore, India	33	934	320	98
	Paulo Afonso	São Francisco River, Alagoas-Bahia, Brazil	25	700	262	80
	Kaieteur	Potaro River, Guyana	23	650	741	226

Highest Waterfalls	Waterfall	Location	Height (ft)	(m)
	Angel	Venezuela	2648	807
	Itatinga	Brazil	2060	628
	Cuquenan	Guyana-Venezuela	2000	610
	Ormeli	Norway	1847	563
	Tysse	Norway	1749	533
	Pilao	Brazil	1719	524
	Ribbon	USA	1612	491
	Vestre Mardola	Norway	1535	468
	Roraima	Guyana	1500	457
	Cleve-Garth	New Zealand	1476	450

A Guide
to Entries
by
Subject

Crystallography
Acute bisectrix
Albite twinning
Anhedral
Anisotrophic
Axial angle
Axial ratio
Becke text
Biaxial crystal
Bonds, chemical
Bragg's law
Covalent bond
Crystal
Crystal axis
Crystal chemistry
Crystal classes
Crystal habit
Crystal lattice
Crystalline
Crystallography
Crystal optics
Crystal symmetry
Crystal system
Crystal twin
Cube
Cubic system
Dichroism
Dimorphous
Dispersion
Dodecahedron
Double refraction
Etch figure
Exsolution
Habit, crystal
Hexagonal system
Homeomorphs
Index of refraction
Inosilicate
Interference color
Interference figure
Ion
Isometric system
Isomorphism
Monoclinic system
Nesosilicate
Nicol prism
Octahedron
Ordinary ray
Orthorhombic system
Paramorph
Parting, mineral
Phyllosilicate
Piezoelectricity
Pleochroism
Polarization of light
Polymorph
Prism
Pseudomorph
Pyramid
Pyritohedron
Pyroxene
Rhombohedron

Silicate structure
Spinel twin
Tectosilicate
Tetragonal system
Trapezohedron
Triclinic system
Twin crystal
Twin law
Unit cell
X-ray crystallography

Economic Geology
Albitization
Asbestos
Assay
Bauxite
Bog iron ore
Bonanza
Cross fiber
Economic geology
Epithermal deposit
Gangue
Gossan
Gouge
Hydraulic mining
Hydrothermal alteration
Hypothermal deposit
Lateral secretion
Mesothermal deposits
Metallogenetic province
Mineral deposits
Mineral properties
Mohs scale of hardness
Molybdenite
Montmorillonite
Ore
Ore body
Ore deposits
Ore dressing
Ore shoot
Oxidized zone
Placer
Placer mining
Porphyry copper
Primary mineral
Protore
Replacement deposit
Secondary enrichment
Sulfide enrichment
Supergene
Taconite
Vein

Environmental Energy and Geology
Anthracite
Bituminous coal
Coal
Energy resources
Environmental geology
Mineral resources
Natural gas
Nuclear breeder reactor

Nuclear energy
Oil shale
Petroleum
Solar energy
Tar sand

Geochemistry
Curie point
Half-life
Isotopes
Radioactive age determination
Specific heat

Geomorphology
Avalanche
Barchan
Barrier beach
Baymouth beach
Beach
Berm
Bolson
Cliff
Coastal plain
Colluvium
Creep
Deflation
Desert
Desert pavement
Diastrophism
Dreikanter
Drowned valley
Dune
Endogenic process
Eolian
Escarpment
Eustatic change of sea level
Exfoliation
Exogenic process
Frost action
Frost heaving
Gorge
Granular disintegration
Hill
Hook
Hydrosphere
Island
Lake
Landslide
Loess
Marine erosion
Mountain
Mudflow
Natural bridge
Plateau
Playa
Playa lake
Quicksand
Rock glacier
Rock slide
Salina
Slump
Solifluction

Spheroidal weathering
Spit
Stack
Talus
Tombolo
Topography
Tundra
Ventifact
Yardang

Geophysics
Aftershock
Asthenosphere
Body wave
Centrosphere
Conrad discontinuity
Density
Earthquake
Earthquake, measurement of
Earthquake prediction
Earth tides
Elastic rebound
Electric log
Electrical conductivity
Epicenter
Equatorial bulge
Equipotential surface
Focus
Geoid
Geomagnetic reversal
Geophone
Geophysics
Geothermal gradient
Gravimeter
Gravitational constant
Gravitational potential
Gravity anomaly
Heat capacity
Heat flow
Heat flow province
Isotherm
Lithosphere
Magnetic anomaly
Magnetic declination
Magnetic field of earth
Magnetic poles of earth
Magnetic susceptibility
Magnetism
Magnetometer
Mantle
Microseism
Mohorovicic discontinuity
Natural remanent magnetism NRM
Nutation, earth's free
Paleomagnetism
Polar wandering
Radiogenic heat
Remote sensing
Seismic belt
Seismicity
Seismic reflection
Seismic refraction

Seismic wave
Seismograph
Seismology
Seismometer
Shadow zone
Shock wave
Sial
Sima
Thermal conductivity
Tsunami

Glacial Geology
Ablation
Arête
Bergschrund
Boulder train
Cirque
Col
Crevasse
Drumlin
Erratic
Esker
Fiord
Glacial drift
Glacial-marine deposit
Glacial striation
Glacial surge
Glacially transported fragment
Glaciated valley
Glaciation
Glacier
Glaciolacustrine deposit
Glaciology
Hanging valley
Horn
Iceberg
Ice-contact slope
Ice-rafted fragment
Ice sheet
Indicator
Kame
Kame moraine
Kame terrace
Kettle hole
Melt water
Moraine
Névé
Nivation
Nunatak
Outwash
Outwash plain
Permafrost
Pluvial lake
Pluvial period
Roche moutonnée
Rock flour
Rock glacier
Snow
Snowline
Tarn
Till
Truncated spur

Valley train
Varve

Historical Geology
Age
Archeozoic era
Cambrian period
Carboniferous period
Catastrophism
Cenozoic era
Cosmogony
Cretaceous period
Devonian period
Diagenesis
Earth, origin of
Eocene epoch
Epoch
Era
Geochronology
Geologic column
Geologic range
Geologic time scale
Holocene epoch
Ice age
Interglacial stage
Jurassic period
Mesozoic era
Miocene epoch
Mississippian period
Neogene period
Neptunism
Oligocene epoch
Ordovician period
Paleocene epoch
Paleogene period
Paleozoic era
Pennsylvanian period
Period
Permian period
Phanerozoic
Pleistocene epoch
Pliocene epoch
Plutonism
Precambrian
Prehistoric time
Proterozoic era
Quaternary period
Silurian period
Tertiary period
Triassic period
Uniformitarianism

Hydrology
Running Water:
Abrasion
Abstraction
Aggradation
Alluvial fan
Alluvium
Annular drainage
Antecedent stream
Arroyo

Attrition
Available relief
Bahada
Base level
Bayou
Bed load
Beheaded stream
Bifurcation
Braided stream
Butte
Canyon
Capacity
Cavitation
Chezy equation
Competency
Consequent stream
Corrasion
Corrosion
Coulee
Cutoff
Degradation
Delta
Dendritic drainage
Differential erosion
Dike ridge
Discharge
Discharge rating curve
Distributary
Drainage basin
Drainage density
Drainage pattern
Dubouys equation
Eddy
Entrenched meander
Ephemeral stream
Erosion
Estuary
Evorsion
Flood
Flood plain
Flow regime
Froude number
Graded stream
Gradient
Gully
Hogback
Hydraulic geometry
Hydraulic radius
Hydraulicking
Hydrograph
Hydrologic cycle
Insequent river
Interfluve
Intermittent stream
Knickpoint
Laminar flow
Levee
Longitudinal profile
Meander
Mesa
Misfit stream
Monadnock

Obsequent stream
Order
Overload flow
Oxbow lake
Parallel drainage
Peneplain
Perennial stream
Piping
Piracy
Playfair's law
Pothole
Profile of equilibrium
Radial drainage
Rectangular drainage
Rejuvenation
Resequent stream
Reynolds number
Rill
River
Runoff
Sediment transport
Sheetwash
Slipoff slope
Slough
Strath
Stream
Stream capture
Stream erosion, cycle of
Stream frequency
Stream regimen
Stream terrace
Subsequent stream
Superimposed stream
Suspended load
Terrace
Thalweg
Tractive force
Trellis drainage
Tributary
Turbulent flow
Underfit stream
Valley
Wash load
Water gap
Waterfall
Watershed
Wind gap
Yazoo

Ground Water:
Aquifer
Artesian well
Cave
Cavern
Cone of depression
Connate water
Darcy's law
Dripstone
Flowstone
Geode
Geyser
Hot-spring deposit

Hydraulic gradient
Infiltration
Juvenile water
Karst topography
Leaching
Meteoric water
Mineral spring
Natural bridge
Perched water table
Permafrost
Permeability
Porosity
Replacement
Sinkhole
Speleology
Speleothem
Spring
Stalactite
Stalagmite
Subsurface water
Thermal spring
Vein
Water table
Zone of aeration
Zone of saturation

Marine Geology
Atoll
Authigenic sediment
Beach
Biogenic sediment
Coccolithophore
Compensation depth
Continental crust
Continental drift
Coral reef
Coralline algae
Coriolis effect
Cosmic sediment
Deep-sea drilling project
Dredges
Fracture zones
Glacial-marine sediment
Glomar Challenger
Guyot
Holocene transgression
Intertidal zone
Island arc
Joides
Littoral zone
Magnetic reversals
Manganese nodules
Mohorovicic discontinuity
Oceanic crust
Oceanic ridge
Ooze
Piston corer
Rift valley
Sand waves
Seamount
Seismic reflection profile
Seismic refraction

Submarine canyon
Submarine landslide
Supratidal zone
Transform fault

Meteorology
Adiabatic process
Air
Air pollution
Albedo
Anticyclone
Atmosphere
Atmospheric density
Atmospheric electricity
Atmospheric humidity
Atmospheric motion
Atmospheric pressure
Atmospheric radiation
Atmospheric tide
Atmospheric wave
Aurora
Carbon dioxide
Chinook
Climate
Cloud
Condensation
Convective circulation
Convergence
Cyclone
Divergence
Doldrums
Ekman layer
Evaporation
Geostrophic wind
Greenhouse effect
Heat balance
Hydrologic cycle
Hydrometeor
Ionosphere
Langley
Lightning
Magnetic storm
Meteorology
Ozone
Polar front
Precipitation
Solar radiation
St. Elmo's fire
Stability
Stratosphere
Temperature
Terrestrial radiation
Thunderstorm
Tornado
Transpiration
Tropical cyclone
Troposphere
Van Allen belts
Vorticity
Water vapor
Weather
Wind

Mineralogy
Actinolite
Adamantine luster
Adularia
Agate
Alabaster
Albite
Alexandrite
Alkali feldspar
Alunite
Amazonite
Amblygonite
Amethyst
Amphibole
Analcime
Andalusite
Andesine
Andradite
Anglesite
Anhydrite
Ankerite
Anorthite
Anorthoclase
Antigorite
Apatite
Apophyllite
Aquamarine
Aragonite
Argentite
Arsenate
Arsenic
Arsenide
Arsenopyrite
Asterism
Augite
Authigenic
Autunite
Aventurine
Azurite
Barite
Bead test
Bentonite
Beryl
Biotite
Blowpipe
Boracite
Borate
Borax
Bornite
Bort
Botryoidal
Brucite
Bytownite
Calaverite
Calcite
Carbonado
Carbonate
Carnotite
Cassitenite
Cat's eye
Celestite
Cerargyrite

Cerussite
Chabazite
Chalcanthite
Chalcedony
Chalcocite
Chalcopyrite
Chatoyancy
Chlorite
Chromate
Chromite
Chrysoberyl
Chrysocolla
Chrysotile
Cinnabar
Cleavage
Cleavelandite
Cobaltite
Colemanite
Collophane
Columbite
Conchoidal fracture
Copper
Cordierite
Corundum
Covellite
Cristobalite
Crocoite
Cryolite
Cuprite
Datolite
Dendrite
Density
Diamond
Diaspore
Differential thermal analysis
Diopside
Dolomite
Druse
Element, native
Electrum
Emerald
Emery
Enargite
Enstatite
Epidote
Epsomite
Erythrite
Fayalite
Feldspar
Feldspathoid
Ferberite
Flame test
Flint
Floss ferri
Fluorescence
Fluorite
Fool's gold
Franklinite
Fulgurite
Galena
Garnet
Garnierite

Gem
Geyserite
Gibbsite
Glaucophane
Goethite
Gold
Graphite
Grossularite
Gypsum
Halide
Halite
Heavy liquid
Hematite
Hemimorphite
Hornblende
Huebnerite
Hyalite
Hypersthene
Iceland spar
Idocrase
Ilmenite
Iridescence
Iron
Jade
Jadeite
Jasper
Jet
Jolly balance
Kaolin
Kaolinite
Kernite
Kyanite
Labradorite
Lapis lazuli
Lazurite
Lepidolite
Leucite
Limonite
Lodestone
Luster, mineral
Magnesite
Magnetic iron ore
Magnetite
Malachite
Manganite
Marcasite
Margarite
Meerschaum
Metal, native
Metamict mineral
Mica
Microcline
Millerite
Mimetite
Mineral
Mineral, accessory
Mineral cleavage
Mineral color
Mineral, dark
Mineral hardness
Mineral, heavy
Mineral, light

Mineral luster
Mineral parting
Mineral, primary
Mineral properties
Moonstone
Mountain cork
Muscovite
Native elements
Native mineral
Natrolite
Nepheline
Nephrite
Niccolite
Niter
Nitrates
Oligoclase
Onyx
Opal
Orthoclase
Oxide
Paragonite
Pectolite
Pentlandite
Peridot
Perthite
Phlogopite
Phosphorescence
Plagioclase
Platinum
Potash feldspar
Prehnite
Proustite
Psilomelane
Pyrargyrite
Pyrolusite
Pyromorphite
Pyrope
Pyrophyllite
Quartz
Realgar
Rhodochrosite
Rhodolite
Rhodonite
Rock crystal
Rock salt
Rose quartz
Rubellite
Ruby
Ruby silver
Rutile
Salt
Sand crystal
Sanidine
Sapphire
Satin spar
Scapolite
Scheelite
Schorl
Selenite
Serpentine
Siderite
Silica

Silicate
Silicified wood
Sillimanite
Silver
Smithsonite
Soda niter
Sodalite
Specific gravity
Specularite
Sperrylite
Sphalerite
Spinel
Spinel group
Spodumene
Staurolite
Stibnite
Stilbite
Streak
Strontianite
Sulfate
Sulfur
Sunstone
Sylvanite
Talc
Tantalite
Telluride
Tennantite
Tetrahedrite
Tiger's eye
Topaz
Tourmaline
Tremolite
Turquoise
Ulexite
Uraninite
Uvarovite
Vanadinite
Vermiculite
Wad
Wavellite
Willemite
Wolframite
Wollastonite
Wulfenite
Zeolite
Zinc blend
Zincite
Zircon
Zoisite

Oceanography
Abyssal hill
Abyssal plain
Abyssal zone
Acoustics, underwater
Amphidromic point
Aphotic zone
Bathyal zone
Bathymetry
Bathyscaph
Bathythermograph
Benioff zone

Benthos
Biozone
Challenger Expedition
Chlorinity
Continental borderland
Continental margin
Continental rise
Continental shelf
Continental slope
Deep-scattering layer
Deep-sea research vehicle
Deep-sea sediment
Deep-submergence rescue vehicle
Density current
Deposit feeder
Desalinization
Diatoms
Drift bottle
Drogues
Echo sounding
Ekman layer
Estuary
Euphotic zone
Eustatic sea level change
Euxinic environment
Filter feeder
Food chain
Hadal zone
Lagoon
Longshore current
Marginal plateau
Marine environment
Mediterranean
Nansen bottle
Nekton
Neritic province
Ocean basin
Oceanic circulation
Oceanography
Oceans
Pelagic environment
Pelagic sediment
Plankton
Relief sediment
Reversing thermometer
Rip current
Salinity
Salt marsh
Seawater density
Sediment coring
Sonar
Suspension feeder
Thermocline
Thermohaline current
Tidal bore
Tides
Tsunami
Turbidity current
Undertow
Upwelling
Water mass
Wave

Paleontology
Amber
Ammonoid
Amphibian
Angiosperm
Archaeopteryx
Arthropod
Belemnoid
Biosphere
Blastoid
Brachiopod
Bryozoa
Calcareous
Carbonization
Casts and molds
Cephalopod
Coelacanth
Coelenterate
Cone-in-cone
Conodont
Coquina
Correlation
Creodont
Crinoid
Crossopterygian
Cycad
Cystoid
Dendrite
Diatoms
Dinosaur
Echinoderm
Echinoid
Eurypterid
Fauna
Foraminifer
Fossil bird
Fossil fish
Fossil fuel
Fossil mammal
Fossil man
Fossil reptile
Fossil, trace
Fossils
Frasch process
Gastrolith
Gastropod
Ginkgo
Graptolite
Guide fossil
Gymnosperm
Ichthyosaur
Macrofossil
Matrix
Microfossil
Micropaleontology
Mollusk
Mosasaur
Nautilus
Ostracoderm
Paleobotany
Paleoclimatology
Paleoecology

Paleogeographic map
Paleogeography
Paleogeology
Paleontology
Paleozoology
Peat
Pelecypod
Pelycosaur
Permineralization
Petrified wood
Phytosaur
Placoderm
Plesiosaur
Protist
Pseudofossil
Pterosaur
Pyritization
Radiolarian
Replacement
Rhynchocephalian
Septaria
Silicification
Thecodont
Therapsid
Trilobite

Petrology
Metamorphic Rocks:
Accessory mineral
ACF diagram
Amphibolite
Augen
Aureole
Cataclasite
Cataclastic metamorphism
Contact metamorphism
Country rock
Eclogite
Gneiss
Hornfels
Idioblastic
Isograd
Kinetic metamorphism
Marble
Metamorphic facies
Metamorphic rock
Metamorphism
Metasomatism
Migmatite
Mylonite
Phyllite
Phyllonite
Poikiloblastic
Porphyroblastic
Quartzite
Regional metamorphism
Retrograde metamorphism
Schist
Schistosity
Skarn
Slate
Xenoblastic

Igneous Rocks:
Accessory mineral
Alkalic igneous rocks
Andesite
Anorthosite
Aphanitic
Aplite
Basalt
Dacite
Dark mineral
Dunite
Euhedral
Extrusive igneous activity
Felsite
Ferromagnesian
Gabbro
Granite
Granitization
Granodiorite
Granophyre
Graphic granite
Greisen
Groundmass
Heavy mineral
Hypabyssal rock
Igneous rock
Intrusive igneous activity
Kimberlite
Lamprophyre
Latite
Light mineral
Magmatic differentiation
Metacryst
Miarolitic cavity
Monzonite
Mylonite
Nepheline syenite
Nephelinite
Norite
Pegmatite
Peridotite
Perlite
Petrogenesis
Petrography
Petrology
Phonolite
Porphyroblast
Porphyry
Pyroxenite
Reaction series
Rhyolite
Rock
Spherulite
Syenite
Tephrite
Thin section
Trachyte
Trap rock
Xenolith

Sedimentology
Chert

Conglomerate
Evaporite
Fuller's earth
Sand
Sandstone
Sediment
Sedimentary rock
Sedimentation
Shale
Silt
Soil
Soil horizon
Tripoli
Varve
Wentworth scale

Stratigraphy
Arkose
Basal conglomerate
Biostratigraphic unit
Breccia
Caliche
Cannel
Cap rock
Chalk
Clastic
Clay
Coal
Concretion
Coprolite
Cross-bedding
Cyclothem
Erathem
Facies
Fireclay
Formation, geologic
Glacial stage
Graywacke
Itacolumite
Laterite
Lignite
Limestone
Lithification
Lithology
Marl
Member
Oolite
Phosphorite
Ripple mark
Rock cycle
Rock-stratigraphic unit
Salt dome
Series
Stage
Stratification
Stratigraphic unit
Stratigraphy
Stratum
System
Tillite
Time unit
Time-stratigraphic unit

Travertine
Unconformity

Space Geology
Achondrite
Albedo
Apollo Program
Chondrite
Luna Program
Lunar geology
Lunar Orbiter
Mare
Mars, geology of
Mesosiderite
Meteor crater
Meteoric dust
Meteoric iron
Meteorite
Rille
Surveyor
Tektite
Widmanstatten pattern

Structural Geology
Anticline
Anticlinorium
Apparent dip
Attitude
Basement complex
Basin
Batholith
Bedding
Bedrock
Boudinage
Breccia
Breccia pipe
Cleavage
Columnar jointing
Compaction
Compression
Concordant
Cone sheet
Contour line
Creep
Culmination
Depression
Dike
Dip
Dip joint
Discordant
Dome
Elastic limit
Fault
Fault scarp
Fold
Foliation
Force
Fracture
Gouge
Graben
Horst
Hydrostatic pressure

Joint
Laccolith
Lineation
Lithostatic pressure
Lopolith
Magmatic stoping
Monocline
Nappe
Outcrop
Overthrust
Phacolith
Plunge
Pluton
Regolith
Rift
Ring dike
Salient
Shear
Sheeting
Slickenside
Stock
Strain
Strata
Stress
Strike
Structural geology
Syncline
Synclinorium
Tensile stress
Tension
Young's modulus

Tectonics
Allochthon
Arch
Autochton
Continental accretion
Continental drift
Convection current
Core
Craton
Crust
Diastrophism
Epeirogeny
Geanticline
Geosyncline
Gondwanaland
Isostasy
Mountain
Orogeny
Pangaea
Plate tectonics
Sea floor spreading
Tectonics

Volcanology
Active volcano
Agglomerate
Agglutinate
Amygdule
Bomb
Breccia, flow

Breccia, pyroclastic
Caldera
Cinder cone
Cinder
Collapse depression
Composite volcano
Compound volcano
Crater
Cumulo-dome
Explosion pit
Fumarole
Geothermal energy
Geyser
Hot spring deposits
Ignimbrite
Kipuka
Lava
Lava curtain
Lava flow
Lava fountain
Lava tree
Lava tunnel
Magma
Magmatic deposit
New volcano
Nuée ardente
Obsidian
Parasitic cone
Pele's tears
Pillow lava
Pitchstone
Plateau basalt
Plug dome
Pumice
Pyroclastic material
Pyroclastic rock
Ropy lava
Scoria
Shield volcano
Spatter cone
Squeeze-up
Strato-volcano
Tholoid
Tuff
Vesicle
Volcanic chimney
Volcanic conglomerate
Volcanic eruption
Volcanic neck
Volcanic plug
Volcanic spine
Volcano
Welded tuff

Page numbers appear at the left of each column; picture numbers are set in italics.

Black-and-white drawings by George Kelvin, except fossil drawings and reconstructions by Matthew Kalmenoff. Four-color illustrations (pp. 19–48) by Howard S. Friedman.

19—Solarfilma
22—*1* After A. G. Smith, "Continental Drift," in I. G. Gass, P. J. Smith, and R. C. L. Wilson (editors), *Understanding the Earth,* The M. I. T. Press, 1971
23—*2* NASA
24— After Frank Armitage in Marion Bickford and others, *Geology Today,* CRM Books, 1973
28—Adapted in part from Calvin Woo in Marion Bickford and others, *Geology Today,* CRM books, 1973
34—*1, 3, 4, 5, 6* Vincent Manson
2 Breck P. Kent
35—*7, 9, 10, 11* Vincent Manson
8 Phil Degginger
12 Breck P. Kent
36—*1* David Muench
37—*2* David Muench
3 George Holton
38—*1* Bruce Coleman
2 David Muench
39—*3* Josef Muench
40—*1* David Muench
2 Woodbridge Williams/Alaska Task Force
41—*3* David Muench
44—*1, 3* Bill Ratcliffe
2 Carlo Bevilacqua/Scala
45—*4, 5, 6, 7* Bill Ratcliffe
54—*1* Jerome Wyckoff
57—Robert C. Frampton
58—*1* Jerome Wyckoff
3 John S. Shelton
60—*1, 3* Robert Adlington
62—*1* Vincent Manson
63—*3* American Museum of Natural History
66—*1, 2* NASA
67—*3, 4* NASA
68—*2* Vincent Manson
69—*5* American Museum of Natural History
70—*1* Jerome Wyckoff
3 Vincent Manson
75—*1* National Oceanographic and Atmospheric Administration
76—Wide World Photos
80—*1* June Bams
81—*3* U. S. Navy
83—*1* Robert C. Frampton
2 Swiss National Tourist Office

87—*1* Vincent Manson
2 John S. Shelton
88—*1* John S. Shelton
89—*3* Sixten Jonsson
90—*1* Lamont-Doherty Geological Observatory
91—*3* John S. Shelton
92—*1* Allen Rokach
93—*2* American Museum of Natural History
94—*1* Vincent Manson
3 Robert Adlington
95—*4* American Museum of Natural History
96—*2* Uno Samuelson/Tiofoto
4 Allen Rokach
99—Lennart Olson/Tiofoto
100—*1, 3* Robert Adlington
103—*1* Jerome Wyckoff
106—*1* Hans W. Silvester/Bavaria-Verlag
108—J. Allan Cash Ltd.
112—*1* C. S. Hurlbut
2 B. M. Shaub
113—*3* Ray-Delvert
114—*2* American Museum of Natural History
115—*3* John S. Shelton and Robert C. Frampton
117—*1* John S. Shelton
118—*1* National Oceanographic and Atmospheric Administration
119—*4, 5* National Oceanographic and Atmospheric Administration
120—*a* National Center for Atmospheric Research
121—*b* Robert C. Frampton
c National Oceanographic and Atmospheric Administration
122—*1* U. S. Bureau of Mines
123—*3* U. S. Bureau of Mines
124—*1* Allen Rokach
2 British Museum
125—*3* Wide World Photos
4 American Museum of Natural History
126—Paul Popper Ltd.
128—*1* American Museum of Natural History
129—*3* National Center for Atmospheric Research
5 American Museum of Natural History
131—*2* Allen Rokach
135—*2a* B. M. Shaub
2b Vincent Manson
136—*1* John S. Shelton
137—*4* Paul Popper Ltd.
138—*2* De Wys Inc.
140—*2* Jerome Wyckoff
141—*3* American Museum of Natural History
143—*2* Ed Cooper
144—*2* Josef Muench

145—*3a* Floyd R. Getsinger
3b American Museum of Natural History
3c B. M. Shaub
148—*2* Walter Dawn/National Audubon Society
150—*2* National Oceanographic and Atmospheric Administration
151—*4* Kurt Severin
155—U. S. Naval Electronics Laboratory
156—*1* After E. P. Shepard, *Submarine Geology*, 2nd edition, Harper & Row, 1963
157—*3* John S. Shelton
158—*1* B. M. Shaub
4 Jerome Wyckoff
160—*1* B. M. Shaub
161—*3* Johns-Manville Photo
4 Jerome Wyckoff
5 Robert C. Frampton and John S. Shelton
164—*1* De Wys Inc.
165—*2* Allen Rokach
166—*1, 3* John S. Shelton
2 Robert C. Frampton and John S. Shelton
167—*4* Canadian Department of National Defense
168—*2* Ray-Delvert
3 Jerome Wyckoff
4 Ed Cooper
173—*2* Robert C. Frampton
177—*5* Allen Rokach
178—After H. U. Sverdrup, *Oceanography for Meteorologists*, Prentice-Hall, 1958
180—*2* Data: U. S. Bureau of Mines
181—*4* Josef Muench
182—*2* Spence Air Photos
3 Josef Muench
183—*4* Tad Nichols
184—*1* David Muench
185—*2* Ed Cooper
3 Geological Survey of Canada
187—*4* Ingmar Holmasen
188—*1* Union Pacific Railroad Photo
189—*2* Ed Cooper
190—*2* John S. Shelton
193—Spence Air Photos
195—*1* Toni Schneiders/Coleman and Hayward
197—*2* C. S. Hurlbut
3 Vincent Manson
198—*2* Tiofoto
3 John S. Shelton
199—*4* Tiofoto
200—Allen Rokach
203—*2* American Museum of Natural History
204—*1,3* American Museum of Natural History
2 Princeton University Museum of Natural History

208—*1, 2* American Museum of Natural History
212—*1, 2, 3* American Museum of Natural History
213—*4, 5* American Museum of Natural History
214—*1* R. L. Nichols
215—*3* George Hunter
216—Helga Fietz/Bavaria-Verlag
219—*1* Durling Studio
221—*1* Allen Rokach
2 American Museum of Natural History
226—*1* Svat Macha/De Wys Inc.
227—*2* John H. Gerard/National Audubon Society
228—*1* Jerome Wyckoff
229—*3* Paul Popper Ltd.
230—U. S. Forest Service
233—*1* Scripps Institution of Oceanography
2 American Museum of Natural History
234—*1* Carl Frank
235—*3* John S. Shelton
237—*1* Robert Clemenz
2 Emil Javorsky
238—*2* Allen Rokach
239—*3* John S. Shelton
240—Harry Groom
243—*2* Emil Javorsky
244—*3* American Museum of Natural History
246—*1* John S. Shelton
2 Bradford Washburn
247—*3* Jerome Wyckoff
248—*1* American Museum of Natural History
249—*2* United Press International
3 Paul Popper Ltd.
253—*1* Sven Gillsäter/Tiofoto
261—Rare Art, Inc.
262—*1* Allen Rokach
2 Josef Muench
267—*1* Jerome Wyckoff
268—*2* Louis Maher/De Wys Inc.
269—*3* Geological Survey of Canada
273—*1* Canadian Government Travel Bureau
274—*1* After J. L. Hough, *Geology of the Great Lakes*, University of Illinois Press, 1958
275—*2* United Press International
277—Camera Hawaii
278—*1* Camera Hawaii
279—*2* Werner Stoy/Camera Hawaii
280—Jerome Wyckoff
282—*1, 2* Jerome Wyckoff
283—*4* John S. Shelton
284—*1* Lick Observatory
285—*3* NASA
286—*2, 3* NASA
287—*4* NASA

288—*1, 2* NASA
289—*3* NASA
291—*1, 2* NASA
297—*3* American Museum of Natural History
298—*1* After C. O. Hines, *Science,* 1963, and B. J. O'Brien, *Science,* 1965
2 American Museum of Natural History
3 Allen Rokach
299—*4* C. S. Hurlbut
301—*3* B. M. Shaub
4 American Museum of Natural History
302—*3* After Elizabeth Burckmyer in Von Engeln, *Geomorphology*, Macmillan, 1942
304—*2* NASA
305—*3* NASA
4 Jerome Wyckoff
307—*1* John S. Shelton
2, 3 Toni Schneiders
308—*1* Jerome Wyckoff
2 Allen Rokach
310—*1* John S. Shelton
2, 3 American Museum of Natural History
311—*4* Lamont-Doherty Geological Observatory
312—Allen Rokach
316—*2* Allen Rokach
317—*4* Jerome Wyckoff
319—*3* Bradford Washburn
320—*2* American Museum of Natural History
321—*4, 5* Ed Cooper
322—Tad Nichols
325—American Museum of Natural History
326—Josef Muench
329—*1* René Catala
2 Ron Church/Photography Unlimited
330—*1* American Museum of Natural History
2 U. S. Navy
333—*1* Joel E. Arem
2 B. M. Shaub
334—*2* After M. J. Keen, *Introduction to Marine Geology*, Pergamon Press, 1968
335—*3* Ron Church/Photography Unlimited
4 United Press International
338—John S. Shelton
341—Ansel Adams
342—*1* Alfred Ehrhardt
343—*3* Jerome Wyckoff
349—Vincent Manson
350—*1* Allen Rokach
351—*2* Douglas P. Wilson
3 Lamont-Doherty Geological Observatory

352—2 After A. N. Strahler, *Physical Geography*, John Wiley, 1960
354—1 Josef Muench
355—3 Allen Rokach
4 Robert C. Frampton
357—1 American Museum of Natural History
2 Jerome Wyckoff
358—2 John S. Shelton
359—3, 4 Lamont-Doherty Geological Observatory
5 After E. P. Shepard, *Submarine Geology*, 2nd edition, Harper & Row, 1967
361—3 Elmer B. Rowley/Durling Studio
4 National Science Foundation
362—2 Josef Muench
365—2 Stuttgart Museum
3 After R. F. Flint, *Glacial and Pleistocene Geology*, John Wiley, 1957
366—3 After J. W. Northrop, III and A. A. Meyerhof, "Validity of Polar and Continental Movement Hypotheses" *American Association of Petroleum Geologists Bulletin*, vol. 47, 1963
368—2 Allen Rokach
372—2, 4 Vincent Manson
377—2 American Museum of Natural History
378—Vincent Manson
382—1 Lamont-Doherty Geological Observatory
383—2 NASA
384—1 top Photo Klopfenstein
1 bottom B. M. Shaub
387—1, 2 Joel E. Arem
388—1 J. Allan Cash Ltd.
389—2 NASA
390—2 John S. Shelton
3 Jerome Wyckoff
391—5 Jerome Wyckoff
393—1 John S. Shelton
2 De Wys Inc.
394—1 Ed Cooper
2 Vincent Manson
397—Ray-Delvert
398—Andreas Feininger
400—1 American Museum of Natural History
2 John S. Shelton
403—1, 2, 3 After J. B. Hersey, "Continuous Reflection Profiling," in M. N. Hill (editor), *The Sea*, John Wiley, 1963
404—4 Lamont-Doherty Geological Observatory
405—5 American Museum of Natural History
407—2 Jerome Wyckoff
3 Allen Rokach
408—1 After Iris Vanders and Paul Kerr, *Mineral Recognition*, John Wiley, 1967

2 American Museum of Natural History
3 Geological Survey of Canada
411—1 B. M. Shaub
2 Jerome Wyckoff
412—1 Jerome Wyckoff
413—3 Jerome Wyckoff
414—After C. E. Kellogg, "Great Soil Groups," U. S. Department of Agriculture, 1936
417—1 Robin Smith Photography Ltd.
2 John S. Shelton
418—1 After C. R. Longwell and R. F. Flint, *Introduction to Physical Geology*, John Wiley, 1962
2 J. Allan Cash Ltd.
421—2 De Wys Inc.
422—1 U. S. Department of the Interior
423—2 Vincent Manson
424—1, 2 Vincent Manson
3 Tore Johnson/Tiofoto
426—Robert C. Frampton
428—1 K. O. Emery
2 D. M. Owen/Woods Hole Oceanographic Institution
430—1, 2 American Museum of Natural History
432—John S. Shelton
435—1 Jerome Wyckoff
2 American Museum of Natural History
437—1 Data: U. S. Weather Bureau
2 John S. Shelton
438—Robert C. Frampton
441—4 New Brunswick Travel Bureau
443—2 Donald McLeish Collection/Paul Popper Ltd.
3 John S. Shelton
445—2 Wide World Photos
449—1 American Museum of Natural History
2 After A. N. Strahler and A. H. Strahler, *Environmental Geoscience*, John Wiley, 1973
3 U.S. Department of Commerce/Coast and Geodetic Survey
450—1 Novosti Press Agency
452—1 American Museum of Natural History
455—2 John S. Shelton
456—American Museum of Natural History
459—1 Bradford Washburn
2 After R. Jastrow, *Journal of Geophysical Research*, vol. 64, 1959
460—1, 2 Allen Rokach
461—4 Jerome Wyckoff
462—1 L. B. Foster/Naval Photographic Center
463—2 Willard Parsons
464—1 Spence Air Photos
465—2 American Museum of Natural History

466—Erich Fischer
469—David Muench
473—American Museum of Natural History
474—1 Louis Goldman
2 Carl Frank
475—Karl Weidmann
476—2 John S. Shelton
3 National Oceanographic and Atmospheric Administration
478—1, 2 After M. F. Harris *et al*, *Investigating the Earth*, Houghton Mifflin, 1973
480—1 Allen Rokach
481—2 Australian News and Information Bureau
3 Floyd R. Getsinger
482—1 John S. Shelton
483—2 De Wys Inc.
484—1, 2 Vincent Manson
487—Allen Rokach
488—1 Vincent Manson
504-511, 513-14—Tables based on data in Victor Showers, *The World in Figures*, John Wiley, 1973

This book was prepared and produced by Chanticleer Press, Inc.

Publisher: Paul Steiner
Editor-in-Chief: Milton Rugoff
Project Editor: Lynne Williams
Copy Editor: Carol O'Neill
Art Director: Gudrun Buettner
Production: Emma Staffelbach
Design: Diane Kosowski, John Lynch
Illustrators: George Kelvin, Howard S. Friedman, Matthew Kalmenoff

Printed in Israel / Bound in Holland